W9-AFF-647

ANNUAL REVIEW OF
NEUROSCIENCE

ANNUAL REVIEW OF NEUROSCIENCE

W. MAXWELL COWAN, *Editor*
Washington University School of Medicine

ZACH W. HALL, *Associate Editor*
University of California School of Medicine

ERIC R. KANDEL, *Associate Editor*
College of Physicians and Surgeons of Columbia University

VOLUME 1

1978

ANNUAL REVIEWS INC. 4139 EL CAMINO WAY PALO ALTO, CALIFORNIA 94306

ANNUAL REVIEWS INC.
Palo Alto, California, USA

REPRINTS The conspicuous number aligned in the margin with the title of each article in this volume is a key for use in ordering reprints. Available reprints are priced at the uniform rate of $1.00 each postpaid. The minimum acceptable reprint order is 10 reprints and/or $10.00 prepaid. A quantity discount is available.

International Standard Serial Number: 0147-006X
International Standard Book Number: 0-8243-2401-3

PRINTED AND BOUND IN THE UNITED STATES OF AMERICA

PREFACE

One of the more remarkable developments in Biology in the last two decades has been the unprecedented growth of the traditional fields of Neurophysiology, Neuroanatomy, Neurochemistry and Physiological Psychology, and concurrently, the gradual emergence of a new, interdisciplinary approach to the study of the nervous system which has come to be known as Neuroscience. Future historians of science will no doubt identify several elements that collectively contributed to this development, but even at this distance four factors can be recognized.

The first, and perhaps the most significant, contributory factor has been the growing appreciation by both scientists and those who support their work, that few things are more important than understanding how the nervous system controls behavior. And with this, there has been a growing sense that the study of the nervous system represents one of the great intellectual challenges of our time. By common consent Neuroscience has acquired much of the intellectual prestige and excitement that Molecular Biology had in the 1950s and 1960s, and it promises at least as much in terms of solid scientific achievement.

The second factor has been the entry into the field of many scientists who had made their mark in Molecular and/or Cellular Biology, and of others who a few years ago would have been attracted into these disciplines. This itself has had a leavening influence upon the traditional neurological sciences, but equally importantly, it has also brought to these areas a novel and diverse set of technical approaches which have already contributed much, and promise even more for the future. The analysis of gene expression in neurons, the study of the long-term consequences of synaptic activity, and the elucidation of the molecular structure of neuronal membranes and their associated ion channels and receptors, are but a few examples of this contribution.

The third factor has been the vigorous application of these and other technical developments to long-standing problems in Neuroscience. Here we might cite the striking advances in neuroanatomy that have followed the introduction of techniques for identifying transmitter-specific neuronal populations, of quantitative methods for studying the morphology of individual neurons, and highly sensitive techniques for tracing neuronal connections based on the phenomenon of axonal transport. Comparable technical developments in almost every other area have opened up new fields for investigation, and have thrown new light on subjects that have hitherto been refractory to experimental analysis.

Lastly, there has been the increasing awareness that further advances in the field are most likely to be made by those who can address problems without the prejudices and limitations that derive from a technique-bound perspective. This realization, more than any other factor, has given birth to "Neuroscience" as an interdisciplinary approach. It is difficult to pin-point exactly when the jurisdictional boundaries between the various neurological disciplines began to be systematically transgressed. However, as even the most casual perusal of the relevant literature will reveal, such transgressions have now become the rule rather than the exception. For example, the behavioral biologist is concerned to know which neurons are involved in the behavior he is studying, how they are interconnected, and how they function during different phases of the behavioral paradigm. The neurophysiologist is as concerned about the morphology of the neurons he is investigating and their connectivity, as the neuroanatomist, and today both are interested in knowing what synaptic transmitters are involved, and what types of enduring changes may occur following their release. The neurochemist is no longer simply interested in the regional distribution of various enzymes or substrates, but in the chemical characterization of individual neurons, or parts of neurons, and in the changes that can be detected in them during development or after periods of activity.

Considerations of this kind led, during the late 1960s and early 1970s, to the formation of several national and international societies for Neuroscience, most of which have as their expressed goals the bringing together of workers in the various traditional neurological disciplines, and the promotion of interdisciplinary work in the field. The rapid growth of these societies is one of the most tangible measures of the growing interest in Neuroscience. The experience of the Society for Neuroscience in North America may be mentioned as being fairly typical. When the Society was first formed some eight or nine years ago, it had a membership of just over 200. By the time of its annual meeting in 1977, the membership had grown to more than 5100, and no fewer than 1900 papers and poster presentations were listed in the program, in addition to the usual symposia and special lectures. A comparable measure of the growth of the field is the exuberant expansion of the neuroscience literature and the quite phenomenal proliferation of new journals and books covering the area. In one year alone, twelve new "neuroscience journals" appeared, and three of the leading publications more than doubled their annual output of papers.

The present volume, and the Annual Review of Neuroscience series which it initiates, are further reflections of this interest. Recognizing the importance of Neuroscience and its newfound independence, the Directors of Annual Reviews, Inc. asked the Editor-in-Chief, Mr. William Kaufmann, to explore the possibility of producing such a series. Encouragement was forthcoming from leading neuroscientists in the United States and Canada about the desirability of such a series, and two meetings were planned to coincide with the annual meeting of the Society for Neuroscience in New York City, in November 1975. At both meetings there was general agreement that such a series was desirable and would serve a broad cross-section of the neuroscience community. Indeed, the only serious reservation expressed was that the creation of this new series might have important repercussions on some of the established series that had traditionally covered certain aspects of Neuroscience. However, it was felt that the degree of overlap between the projected and the existing series was not likely to be greater than that between other related series (such as the Annual Reviews of Biochemistry, Biophysics and Bioengineering, Genetics, and Microbiology), and also that much of the potential overlap could be obviated if, when the time came to plan the initial volume of the Annual Review of Neuroscience, representatives of the Editorial Committees of some of the related series were to be present. In addition to giving the projected series their general support, those attending the New York meetings suggested a number of topics for consideration, and the names of several individuals from among whom the Editors and Members of the Editorial Committee could later be selected. Encouraged by these responses, the Directors gave final authorization in the Spring of 1976 to publishing such a series and duly appointed an Editorial Committee. This committee met in Palo Alto in April 1976, together with representatives of the Annual Reviews of Pharmacology, Physiology, Medicine, and Psychology. At this meeting topics and authors for the present volume were selected, and an overall plan for the series was drawn up. According to this "master plan" each of the major areas in the field will be systematically reviewed over a period of every four or five years, and at the same time each volume will contain a number of reviews of exciting new areas, and of subjects that may be of more focused interest.

In defining the field of Neuroscience, the Editorial Committee has taken a fairly liberal position. Neuroscience is taken to include all aspects of neural structure and function, from its genetic determination to the highest expression of its activity in human behavior. The new series will cover both the vertebrate and invertebrate nervous systems, and it will deal with certain aspects of the related clinical disciplines of neurology, neurological surgery and psychiatry, as well as with the so-called basic neurosciences. The breadth of intended coverage is clearly indicated by the Table of Contents of the present volume. It includes some largely technical papers and some that are mainly theoretical; some reviews deal with topics that are of clinical interest, while others are concerned with some of the most basic aspects of the field. The majority cover the selected topic rather broadly, while at the same time presenting a particular viewpoint, but some follow the traditional pattern in other Annual Reviews and cover only the more recent literature. Regrettably, it has not been possible to publish all the intended reviews, partly because of limitations of space and partly because of the late arrival of certain manuscripts. Moreover, the publication schedule requires the articles be printed in the order in which they are finally received. Thus, some of the natural relationships between the various topics may be obscured by their appearance in different parts of the volume. However, the titles are sufficiently self-explanatory and the interested reader should have no difficulty in identifying related subjects. A list of chapters planned for next year's volume, as well as a list of chapters from other current Annual Review volumes that are of potential interest to neuroscientists, appears on the pages following the Contents page.

It is the hope of the Editorial Committee that this first volume and, indeed, the entire series, will make as significant a contribution to Neuroscience as other Annual Reviews have made to their respective fields.

W. M. COWAN, EDITOR

ANNUAL REVIEWS INC. is a nonprofit corporation established to promote the advancement of the sciences. Beginning in 1932 with the *Annual Review of Biochemistry,* the Company has pursued as its principal function the publication of high quality, reasonably priced Annual Review volumes. The volumes are organized by Editors and Editorial Committees who invite qualified authors to contribute critical articles reviewing significant developments within each major discipline.

Annual Reviews Inc. is administered by a Board of Directors whose members serve without compensation.

Annual Reviews are published in the following sciences: Anthropology, Astronomy and Astrophysics, Biochemistry, Biophysics and Bioengineering, Earth and Planetary Sciences, Ecology and Systematics, Energy, Entomology, Fluid Mechanics, Genetics, Materials Science, Medicine, Microbiology, Neuroscience, Nuclear Science, Pharmacology and Toxicology, Physical Chemistry, Physiology, Phytopathology, Plant Physiology, Psychology, and Sociology. In addition, two special volumes have been published by Annual Reviews Inc.: *History of Entomology* (1973) and *The Excitement and Fascination of Science* (1965).

Annual Review of Neuroscience
Volume 1, 1978

CONTENTS

CHAPTERS PLANNED FOR THE NEXT *ANNUAL REVIEW OF NEUROSCIENCE*

Volume 2 (1979)

Mechanisms of Visual Transduction, *Wayne L. Hubbell and Deric Bownds*

The Physiology of the Retina, *Akimichi Kaneko*

Visual Pathways, *R. W. Rodieck*

The Functional Organization of the Superior Colliculus, *Robert H. Wurtz and Joanne Albano*

Visual Cortical Areas, *David C. Van Essen*

Motor Control, *Emilio Bizzi*

Vestibular Mechanisms, *W. Precht*

The Biochemistry of Nerve Terminals and Transmitter Release, *Regis B. Kelly*

Synaptic Modulatory Function, *Irving Kupfermann*

Ion Channels in Development, *Nicholas C. Spitzer*

Sodium Channels in Myelinated Axons, *J. M. Ritchie*

Excitatory Properties of Membranes: Basic Mechanisms in Artificial Lipid Bilayers, *M. Eisenberg*

Gene Regulation in Neurons: The Use of Axonally Transported Materials as Indicators of Gene Action, *James Schwartz*

The Brain as a Target for Hormone Action, *Bruce S. McEwen and Donald W. Pfaff*

Opiate Receptors and Endorphins, *Solomon H. Snyder*

Central Catecholamine Neuron Systems: Anatomy and Physiology of the Norepinephrine and Epinephrine Systems, *R. Y. Moore and F. E. Bloom*

The Biology of Affective Disorders, *Edward J. Sachar*

The Development of Behavior in Human Infants, Premature and Newborn, *Peter H. Wolff*

Memory, *Brenda A. Milner*

Neuronal Modeling, *Donald H. Perkel*

Invertebrate Developmental Neurobiology, *John S. Edwards and John Palka*

SOME RELATED ARTICLES IN OTHER *ANNUAL REVIEWS*

From the *Annual Review of Biochemistry*, Volume 47 (1978)

Nerve Growth Factor, R. A. Bradshaw

Structural and Functional Properties of the Acetylcholine Receptor Protein in Its Purified and Membrane-Bound States, T. Heidmann and J. P. Changeux

Hypothalamic Regulatory Hormones, A. V. Schally, D. H. Coy, and C. A. Meyers

From the *Annual Review of Biophysics and Bioengineering*, Volume 7 (1978)

Excitation and Contraction Processes in Muscle, C. Caputo

Excitation and Interactions in the Retina, L. Cervetto and M. G. F. Fuortes

Calcium-Dependent Potassium Activation in Nervous Tissues, R. W. Meech

From the *Annual Review of Medicine*, Volume 29 (1978)

Recent Membrane Research and Its Implications for Clinical Medicine, V. T. Marchesi and A. N. Brady

From the *Annual Review of Pharmacology and Toxicology*, Volume 18 (1978)

The Generation and Conduction of Activity in Smooth Muscle, E. E. Daniel and S. Sarna

Small Intensely Fluorescent (SIF) Cells and Nervous Transmission in Sympathetic Ganglia, Olavi Eränkö

Neuropharmacology of Amino Acid Inhibitory Transmitters, G. A. R. Johnston

Retrograde Transport of Macromolecules in Axons, K. Kristensson

The Opiate Receptors, E. J. Simon and J. M. Hiller

Endogenous Peptides and Analgesia, L. Terenius

Hypothalamic Hormones: Subcellular Distribution and Mechanisms of Release, L. C. Terry and J. B. Martin

From the *Annual Review of Physiology*, Volume 40 (1978)

Brainstem Control of Spinal Pain-Transmission Neurons, H. L. Fields and A. I. Basbaum

Physiology of the Hippocampus and Related Structures, F. H. Lopes da Silva and D. E. A. T. Arnolds

The Physiology of Circadian Pacemakers, M. Menaker, J. S. Takahashi, and A. Eskin

Brainstem Mechanisms for Rapid and Slow Eye Movements, T. Raphan and B. Cohen

Localization and Release of Neurophysins, S. M. Seif and A. G. Robinson

From *Annual Review of Psychology,* Volume 29 (1978)

Behavioral Genetics, J. C. DeFries and R. Plomin

Visual Perception, R. N. Haber

Visual Sensitivity, D. I. A. MacLeod

Biofeedback and Visceral Learning, N. E. Miller

Ann. Rev. Neurosci. 1978. 1:1–17

ENVIRONMENTAL DETERMINATION OF AUTONOMIC NEUROTRANSMITTER FUNCTIONS[1]

❖11500

Paul H. Patterson
Department of Neurobiology, Harvard Medical School, Boston, Massachusetts 02115

Many of the decisions neurons make during development depend on cues provided by the environment. One of the primary choices a developing neuron makes is which neurotransmitter to produce, thereby determining the effect its synapses will have on target cells. It has proved possible to influence experimentally the choice of transmitters and the type of synapses made by some neural crest derivatives through manipulation of their cellular or fluid environment.

THE NEURAL CREST

The vertebrate neural crest is a transient embryonic structure lying along the dorsal margin of the neural tube. This seemingly homogenous population of cells migrates primarily in two directions: either dorsolaterally, into the epithelial area, or ventrally, into the mesenchyme. Many of the dorsal cells disperse as melanocytes, whereas the ventral stream of cells follows definite pathways characteristic of each axial level and gives rise to such diverse cell types as autonomic neurons, sensory neurons, and glia (Weston 1970), neuroendocrine cells such as the adrenal medulla and calcitonin-producing cells (LeDouarin & LeLievre 1970), and cells that form skeletal and connective tissues of the head (Johnston, Bhakdinaronk & Reid 1974). Furthermore, even within some of these categories there is a rich diversity of cell types. For instance, mammalian autonomic ganglia were divided into 3 classes by Langley (1921): the sympathetic, parasympathetic, and enteric ganglia.

[1]Abbreviations used: ACh, acetylcholine; CA, catecholamine; CAT, choline acetyltransferase; CM, conditioned medium; FIF, formaldehyde-induced fluorescence; NE, norepinephrine; NGF, nerve growth factor; RNA, ribonucleic acid; TH, tyrosine hydroxylase.

0147-006X/78/0325-0001$01.00

1. Most neurons in sympathetic ganglia produce norepinephrine (NE), but minority populations use dopamine (Björklund et al 1970, Fuxe et al 1971), acetylcholine (ACh) (Uvnäs 1954, Sjöqvist 1963, Aiken & Reit 1969), or perhaps epinephrine (Ciaranello & Axelrod 1975, Phillipson & Moore 1975, Elfvin, Hökfelt & Goldstein 1975).
2. Most neurons in parasympathetic ganglia produce ACh, but adrenergic cells are also occasionally present (Jacobowitz 1967, Ehinger & Falck 1970).
3. The complexity of enteric ganglia may explain the number of putative transmitters now being localized to neurons intrinsic to the gut: ACh (Dale 1937), NE (Costa, Furness & Gabella 1971), serotonin (Dreyfus, Sherman & Gershon 1977, Dreyfus, Bornstein & Gershon 1977, Gershon et al 1977), substance P (Pearse & Polak 1975), somatostatin (Hökfelt et al 1975), vasoactive intestinal polypeptide (Said & Rosenberg 1976), and ATP (Burnstock 1975).

Thus, even though cells that form the neural crest have already been greatly restricted in their differentiative capabilities as a consequence of primary embryonic induction (Weston 1970), the population still has a number of possible fates. This review will consider primarily the environmental influences that determine adrenergic and cholinergic properties. This neuronal decision has crucial functional consequences because these two transmitters have antagonistic effects on many autonomic target tissues such as the heart and blood vessels.

INFLUENCES ON CREST CELLS DURING THEIR MIGRATION

Can the environment through which neural crest cells migrate influence their eventual expression of transmitter production? The characteristic migratory routes and final sites of localization for the ventrally moving crest cells have been defined at each axial level. The most recent and useful technique (LeDouarin 1969, 1974, see also Triplett 1958) is transplantation of quail crest cells into the comparable site of a chick host (or vice versa). Quail donor cells can be distinguished from the chick host cells by the differential staining properties of the interphase nuclei.

Transplantation experiments suggest that the cellular environment and/or extracellular matrix at each axial level can guide the ventral stream of crest cells along specific routes to its final sites. When labeled crest cells from one axial level are grafted into a host at a different (heterotopic) level, many cells follow the routes expected for the host (and differentiate accordingly) rather than the pathways formerly expected of the donor cells (Weston 1963, 1970, Johnston, Bhakdinaronk & Reid 1974, Noden 1975, LeDouarin et al 1975). This will be discussed later in more detail. However, a second important point is that some donor cells do find their way to their original target areas in the host. Such heterogeneity within the crest population has been documented for melanocyte migration and cephalic mesenchyme formation (Twitty 1945, LeDouarin & Teillet 1974, LeDouarin, Teillet & LeLièvre 1977). Therefore, although the migratory route itself can influence crest cell movements, some of the cells may already be predetermined to somehow find their way to their original target.

Shortly after reaching the sites of formation of the primary sympathetic ganglionic chain, some cells of the ventral stream can be shown by the formaldehyde-induced fluorescence (FIF) technique to contain significant levels of catecholamines (Enemar, Falck & Hakanson 1965). In an attempt to determine if the expression of adrenergic properties was induced during migration or by the final destination, Cohen (1972) used the FIF technique to assay catecholamines in crest cells prevented from reaching their destination. He explanted chick crest cells, which had not yet migrated, to the chorioallantoic membrane of the egg, either alone or with other embryonic tissues. Some of these were tissues that crest cells pass on the ventral migratory route (ventral neural tube and somitic mesenchyme); others were not part of normal migratory routes (mesoderm from embryonic heart, wing, or tail buds). Only the former combination produced large numbers of cells exhibiting catecholamine fluorescence and neuronlike processes. Thus, crest cells do not have to reach their normal destination to develop properties of adrenergic neurons, and tissues normally found along the migratory route can provide an environment favorable for the expression of such properties.

Were the heterotypic (neural crest and noncrest) tissue grafts simply permissive for the expression of previously determined function, or did the heterotypic tissues actively participate in the determination? If there were active interactions, did the somitic mesenchyme and neural tube each act directly on the crest cells or with each other first? An interesting result obtained by Cohen was that the somitic mesenchyme was not effective in promoting FIF in crest cells unless the ventral neural tube was also present. In order to study such interactions further, Norr (1973) cultured heterotypic tissues on millipore filters in vitro. If a dorsal neural tube-containing crest was cultured with somites previously "conditioned" by 36-hr transfilter contact with ventral neural tube, 43 percent of the explants contained fluorescent cells with neuronlike processes. Without the ventral tube conditioning however, the somites did not stimulate FIF significantly above dorsal tube plus crest controls (11 percent vs 6–18 percent, respectively). Norr also confirmed that the source of the mesoderm was important for stimulation of FIF. Thus, these in vitro results support Cohen's findings that tissues along the migratory route can provide an environment favorable for expression of FIF by crest cells. They further suggest that the ventral neural tube and somitic mesenchyme may interact to produce such an environment.

Does the notochord play a role in these tissue interactions as it does in other embryonic inductions (cf. Lash 1967)? LeDouarin and colleagues (LeDouarin 1977) grafted neural crest plus neural tube in hind-gut mesoderm (nonsomitic) and grew the grafts on the chorioallantoic membrane. FIF was lacking if the notochord was not included. However, if crest plus neural tube and notochord were implanted, fluorescent neurons were present in 3 of 7 explants. Perhaps the notochord promotes FIF in combinations involving such nonsomitic mesenchyme. However, Cohen (1972) found that the notochord was not important for FIF using somitic mesenchyme. Thus, the importance of the notochord in the development of neural crest FIF is not completely clear at present. It is worth noting that the notochord itself shows specific catecholamine fluorescence at these stages (Kirby & Gilmore 1972, Lawrence & Burden 1973). The presence of neurotransmitters in embryonic structures before the onset of synaptic activity, or even before neurons have stopped dividing (Cohen 1974, Rothman, Gershon & Holtzer 1977), has provoked much speculation as to its developmental significance (Baker & Quay 1969, Filogamo & Marchisio 1971, Vernadakis & Gibson 1974, McMahon 1974).

Given that nerve growth factor (NGF) is so important at later stages of sympathetic and sensory neuron development (see Black 1978, Varon & Bunge 1978), it was of interest to determine if there is a role for NGF at these early embryonic stages. Addition of NGF to the complete system of crest plus ventral neural tube and somitic mesenchyme had no effect, but if ventral tube was removed too soon, some stimulatory effects of NGF on FIF were observed (Norr 1973). Bjerre (1973) and Bjerre & Björklund (1973) reported that development of FIF was increased by NGF and decreased by its antiserum in explants of cranial neural tissue from early embryos (stages 5–9); the results were statistically significant but not pronounced. However, since the origin and normal development fate of the cranial fluorescent cells are in doubt, the significance of the NGF effect is not clear. Given the difficulty in assessing these experiments, it might be helpful to see what effects injections of NGF antiserum have on early development in vivo.

Another question raised by the Norr study, as well as by those of Bjerre (1973) and Newgreen & Jones (1975), is why FIF was seen in a significant number of controls (neural tube plus crest). That is, did some crest cells not require heterotypic interactions to express FIF? Reconsideration of this question is most strongly suggested by the recent results of Cohen (1977). He explanted neural tube plus crest into culture, and after 24–48 hr removed the neural tube, leaving behind the cells that had migrated onto the dish. After 4 days, weak FIF was observed in some cells, and by 2 weeks intense fluorescence was present in large numbers of cells. Some of these resemble chromaffin or SIF cells, and others have long varicose processes characteristic of adrenergic neurons. Thus some neural crest cells can differentiate adrenergically without the continued presence of ventral neural tube or somitic mesenchyme. It is possible that the collagen substrate, the culture medium, or the initial presence of the neural tube (or cells derived from it) provided a favorable environment for the expression of adrenergic properties that was not provided by the chorioallantoic membrane in vivo or the millipore filter in vitro.

However, these recent studies do question the significance of heterotypic tissue interactions in adrenergic development. Part of the apparent discrepancy among the crest transplantation studies may be due to a heterogeneity in the crest cell population that was previously alluded to in the discussion of migration. Further study of dispersed primary crest cell cultures, perhaps with cloning (Cohen & Konigsberg 1975), is definitely warranted. It will also be of interest to assess quantitatively the effect of somite, ventral neural tube, etc, on the expression of various neuronal phenotypes under such controlled conditions.

DEVELOPMENT OF TRANSMITTERS AFTER GANGLION FORMATION

As previously discussed, the technique of making chick-quail chimeras has provided evidence that the migratory environment can affect crest cell movements. This technique has also provided evidence that the site of ganglion formation can exert a major influence on transmitter development. When quail "trunk" neural tube plus crest (which would normally give rise to adrenergic sympathetic neurons and adrenergic adrenal medullary cells) was transplanted into the "vagal" neural tube level of the chick embryo, quail cells populated the enteric ganglia of the chick gut (LeDouarin & Teillet 1974, LeDouarin et al 1975). The quail cells that migrated to this abnormal site did not fluoresce as they normally would have. Thus the presumptive adrenergic fate of this population was suppressed. In addition, electrical stimulation of the vagus nerve to the duodenum in the heterotopically grafted chimeras evoked normal, cholinergically driven, gut-muscle contraction. This result is consistent with the interpretation that at least some of the presumptive adrenergic population had become cholinergic neurons. An alternative explanation is that some cells are predetermined to be cholinergic (and some adrenergic), and the environment selects the appropriate cells. This is discussed again later.

Similar conclusions were drawn from the results of the reverse experiment: Presumptive cholinergic quail crest (mesencephalon and anterior rhombencephalon) was transplanted to the presumptive adrenergic adrenal medullary region of the chick (LeDouarin & Teillet 1974). Twenty-four of 27 host adrenal glands contained numerous fluorescing quail cells. Electron microscopy of such glands showed many quail cells with the secretory granules characteristic of adrenal medullary cells. Thus, a neural crest population that does not normally give rise to adrenergic cells does so when allowed to migrate from the appropriate axial level.

In these experiments it is not clear whether the influence on transmitter expression occurs during or after cell migration. To study this question, LeDouarin and colleagues avoided migration completely by transplanting vagal or trunk crest directly into aneural hind-gut mesoderm and culturing the combination on the chorioallantoic membrane. Thus presumptive cholinergic or adrenergic crest populations were placed directly in a target tissue where cholinergic enteric ganglia normally develop without fluorescent neurons. Under these circumstances neither the trunk (LeDouarin & Teillet 1974, Smith, Cochard & LeDouarin 1977) nor the vagal (Smith, Cochard & LeDouarin 1977) neural crest gave rise to fluorescent cells in the gut. However, both donors formed enteric ganglia that contained silver-stained neurons (LeDouarin & Teillet 1974), heavy acetylcholinesterase staining, and significant choline acetyltransferase (CAT) activity (Smith, Cochard & LeDouarin 1977). Aneural gut transplanted without crest showed none of these features. Thus normal adrenergic expression was suppressed and cholinergic function promoted without contact with any of the structures normally encountered during migration except the embryonic gut tissue itself. These results also raise the possibility of the presence of cholinergic neurons in the nonfluorescent grafts of Cohen and Norr.

Two series of experiments demonstrate that even after crest cells have formed a ganglion, where one transmitter normally predominates, they may be induced to produce predominantly another transmitter instead. This adds to the hypothesis presented earlier the possibility that while some crest cells may be truly uncommitted before migration, others may have reversible predilections. That is, cells may be predisposed to one transmitter and under permissive conditions express that phenotype, but upon exposure to other conditions reverse or change that decision. Again, the problem concerns if and when cells become irreversibly committed to one transmitter choice. Two examples of "reversibility" follow. The first is an in vivo transplantation experiment, and the second involves culture of sympathetic ganglion cells.

The ganglion of Remak in the chick lies in the dorsal mesentery of the gut and some sections of it contain cholinergic but no adrenergic neurons. If the cholinergic sections are taken out of the embryo just as they form (4–5 days incubation) and are cultured on the chorioallantoic membrane, they develop into full ganglia with no FIF. However, if similar sections of a 4-day quail ganglion of Remak are "back-transplanted" into the trunk neural tube region of a 2.5-day chick, the cells of the graft do not remain together but migrate out to the following sites: sympathetic ganglia, aortic plexus, and adrenal medulla. Quail cells at each of these sites show FIF (LeDouarin, Teillet & LeLièvre 1977). Similar results have been obtained by "back-transplanting" cholinergic ciliary ganglia just after they have formed (N. LeDouarin, personal communication). Thus the transmitter produced by this crest population can be influenced even after the cells have localized and stopped migrating.

Had any of the cells that fluoresced after the transplantion previously begun their differentiation as cholinergic neurons in the ganglion of Remak? This leads the level of inquiry to the decisions made by single cells. As previously mentioned, all of the transplantation studies can be explained by supposing either (*a*) that environmental factors induced a new or different transmitter in pluripotent or reversibly committed cells, or (*b*) that the environment selected among heterogeneous dividing or postmitotic cells that were already irreversibly committed to one transmitter. This selection could be accomplished by death or turn-off of cells not favored. The induction hypothesis implies that the choice of transmitter is environmentally imposed on individual neurons; the selection hypothesis supposes that the environment simply permits the expression of certain prior decisions but not others. In either case, the transplantation studies have clearly demonstrated the profound influence that the developmental environment can have on the expression of transmitter functions in vivo.

The second series of experiments involves cultures of dissociated rat sympathetic neurons; similar questions to those just considered are being investigated on single cells and on populations. The type of transmitter (ACh or NE) and the type of synapses formed by these cells can be experimentally controlled. Because these neurons do not appear to divide in culture (Mains & Patterson 1973a), such changes are postmitotic. It is possible to grow the neurons either in the near absence of other cell types or in the presence of a variety of nonneuronal cells of known origin (Bray 1970, Mains & Patterson 1973a). When grown in the virtual absence of other cell types these neurons can develop many of the properties expected of adrenergic neurons. They synthesize and accumulate NE from tyrosine (Mains & Patterson 1973a) and develop this ability with a time course that qualitatively parallels that seen in vivo. This development of NE metabolism represents a neuronal differentiation, since it differs in magnitude and time course from overall neuronal growth, which is measured by the synthesis and accumulation of protein, lipid, and RNA from radioactive precursors (Mains & Patterson 1973b). These neurons take up, store, and release NE as do adrenergic neurons in vivo (Claude 1973, Rees & Bunge 1974, O'Lague et al 1974, Burton & Bunge 1975, Patterson, Reichardt & Chun 1975); they also conduct action potentials and are sensitive to ACh applied by iontophoresis (O'Lague et al 1975). Rees & Bunge (1974) have shown that these cells form morphological synapses with each other that appear to be adrenergic: They exhibit membrane thickenings following aldehyde fixation and small granular vesicles following permanganate fixation to localize vesicular stores of NE. Finally, as expected, these cultures do not synthesize and accumulate detectable levels of other putative transmitters such as γ-aminobutyric acid, serotonin, or histamine from their labeled precursors. However, in older cultures synthesis of small amounts of ACh from radioactive choline can be detected (Mains & Patterson 1973a).

In contrast, when the neurons are cocultured with appropriate (see below) nonneuronal cells, the mixed cultures produce as much as 1000-fold more ACh than cultures that only contain neurons (Patterson & Chun 1974). Furthermore, the ACh is secreted at functional cholinergic synapses made by the neurons on each other (O'Lague et al 1974, 1975, Johnson et al 1976, Ko et al 1976), on skeletal myotubes (Nurse & O'Lague 1975), and on cardiac myocytes (Furshpan et al 1976). The neurons do not have to be in contact with the nonneuronal cells for these cholinergic changes to occur; medium that has been conditioned by cultures of appropriate nonneuronal cells (CM) also causes dramatic increases in (*a*) CAT activity as measured in neuronal extracts, (*b*) ACh synthesis and accumulation from ^{3}H-choline by living cells and (*c*) cholinergic synapse formation between the neurons (Patterson, Reichardt & Chun 1975, MacLeish 1976, Landis et al 1976, Patterson & Chun 1977a). At the same time, however, growth in CM causes a marked reduction in adrenergic properties: NE synthesis and the proportion of synaptic vesicles containing small dense cores following permanganate fixation are considerably lower than controls (Landis et al 1976, Patterson & Chun 1977a). Similarly, coculture with appropriate nonneuronal cells reduces the proportion of small granular vesicles observed after loading with exogenous NE (Johnson et al 1976). In addition, the CM effects are graded. A higher proportion of CM (Landis et al 1976, Patterson & Chun 1977a) or a greater number of nonneuronal cells (MacLeish 1976) gives a higher ratio of ACh to NE synthesis, a higher incidence of cholinergic transmission, and a higher proportion of synapses that lack small granular vesicles and appear cholinergic. Thus the presence of certain nonneuronal cells, or a medium conditioned by them, has a profound effect on the type of transmitter chosen by the sympathetic neurons. It is also worthy of note that other components of the culture medium such as buffers and sera can influence ACh synthesis (Ross & Bunge 1976, Patterson & Chun 1977a). One possibility is that embryo extracts and sera contain molecules that can be modified by nonneuronal cells in vitro (or have been modified in vivo), and that these in turn affect transmitter production.

It is often difficult to interpret effects of coculture or CM because of the possibility of nonspecific "feeder" effects on cell growth or survival (cf Eagle & Piez 1962). However, in the sympathetic neuron cultures, the effects of CM are rather specific. Doses of CM that increase the ratio of ACh to NE production 100 to 1000-fold do not appreciably affect neuronal survival or growth (as monitored by number of somas and total neuronal protein and lipid) (Patterson & Chun 1977a). This constancy in survival and growth is not because only a few neurons in the population are involved in these transmitter changes: Single cell experiments have shown that most if not all of the neurons are influenced in this manner by the culture environment. It is possible to grow single neurons in microcultures containing various concentrations of CM or nonneuronal cells (Reichardt, Patterson & Chun 1976); as many as 80–90% of individual neurons grown on heart or skeletal muscle can be cholinergic, while as few as 0% are cholinergic under "control" conditions (Reichardt & Patterson 1977, Nurse 1977). Furthermore, by growing cultures in various concentrations of CM it is possible to obtain populations of single cells in which almost every cell synthesizes NE or in which a high proportion of the cells synthesize ACh (Reichardt & Patterson 1977). Finally, there is no evidence for a significant population of "silent" neurons that make no transmitter under these conditions (Reichardt & Patterson 1977). These observations, and the constancy in neuronal number and growth in the mass cultures, suggest that CM affects the choice of transmitter and type of synapse formed by individual sympathetic neurons and does not select certain cell types for survival. This appears to be the first clear example of determination (as opposed to differentiation) of a cell phenotype occurring postmitotically (without further cell division).

Under certain conditions, mass neuronal cultures produce both ACh and NE, and the question arises whether individual neurons can display both transmitter functions simultaneously. Electrophysiological (Furshpan et al 1976) and electron microscopic (Landis 1976) observations on young single neurons grown in microcultures on rat cardiac myocytes for about 2 weeks indicate that this is possible. Three categories of neurons were identified: (*a*) neurons that formed hexamethonium-sensitive cholinergic synapses (with no small granular vesicles) on their own somas and elicited atropine-sensitive cholinergic responses in the heart cells; (*b*) neurons that elicited propranolol-sensitive adrenergic responses in the heart cells and had neuronal varicosities containing high proportions of small granular vesicles; and (*c*) neurons that were "dual-functional"; they elicited first a cholinergic response and then an adrenergic response in beating myocytes, and their varicosities contained only occasional small granular vesicles. The incidence of these dual-function neurons has not yet been determined satisfactorily, but they are not exceedingly rare (Furshpan et al 1976). On the other hand, biochemical assays of transmitter production on older (3–5 weeks) single neurons have provided few if any examples of dual-function cells (Reichardt & Patterson 1977); the vast majority of neurons made detectable amounts of either NE or ACh, depending on the culture conditions. The difference between the electrophysiological and biochemical results with respect to dual-function neurons may reflect an inability of the biochemical methods to detect a low level of one of the two transmitters in cells that were actually bifunctional. Alternatively, the young dual-function neurons detected electrophysiologically may have been passing through a transient stage that leads to one of the two differentiated states observed in the older cells with biochemical methods. It will clearly be of interest to study a single neuron over time and see if it can be shifted from one transmitter function to another by addition or withdrawal of CM at various intervals.

The fact that most mature neurons did not synthesize detectable amounts of more than one transmitter suggests that regulatory mechanisms force sympathetic cells to respond in a "flip-flop" fashion to the cholinergic "factor" present in CM (Reichardt & Patterson 1977). Data from clonal cell lines of crest origin may be of interest in this context. Amano, Richelson & Nirenberg (1972) found that clones of the mouse C1300 line contain high levels of tyrosine hydroxylase (TH) or CAT activity, but not both. On the other hand another clonal line, rat pheochromocytoma PC12 (Greene & Tischler 1976), produces high levels of both NE and ACh and responds to CM by an increase in CAT-specific activity (Schubert, Heinemann & Kidokoro 1977, Greene & Rein 1977). This line has been subcloned to analyze transmitter synthesis further, and several types of clones have been obtained (Greene & Rein 1977). Analysis of single cells may well be necessary to determine if cholinergic and adrenergic properties are simultaneously expressed. Other neuronal clonal lines (Prasad et al 1973, Schubert et al 1974) and glial-neuronal hybrid lines (Hamprecht, Traber & Lamprecht 1974) can express more than one transmitter-synthetic function simultaneously. Since the stages in normal development are not yet firmly established, it is not possible to state that these cells are in some early, normally transient state. It would be instructive to see if these cells could be made to produce only one transmitter through alteration of their culture environment.

In order to define the possible developmental stages that sympathetic neurons go through, it is important to consider the potential for inducibility and reversibility of cholinergic function at various ages. Present information about normal development in the rat superior cervical ganglion suggests that most of the neurons express adrenergic functions at a relatively low level before birth, that is, before the neurons are put into culture (Champlain et al 1970, Eränkö 1972a,b). Moreover, the cultured neurons synthesize NE during the first week in vitro (Mains & Patterson 1973b), and all their varicosities contain high proportions of small granule vesicles at one week (Johnson et al 1976), even when grown in CM from day one (S. Landis, in preparation). These observations suggest that most, if not all, of the neurons are adrenergic when initially placed in culture and, as previously discussed, they can either continue to develop adrenergically or, in the presence of CM, become cholinergic. Since most, if not all, of the neurons can become cholinergic (see above), many can reverse their differentiative fate after beginning to express another phenotype. This early expression of adrenergic function may represent the "protodifferentiation" stage described by Rutter et al (1968).

The next stage involves a major (postnatal) differentiation, either adrenergic (Mains & Patterson 1973b) or cholinergic (O'Lague et al 1975, Johnson et al 1976, MacLeish 1976, Patterson & Chun 1977b), and occurs in the following two weeks. Do the neurons have the capacity to reverse their phenotype at this point, after they have fully differentiated to become either cholinergic or adrenergic? Since ganglionic explants develop CAT activity and form cholinergic synapses in vitro (Crain & Peterson 1974, Purves et al 1974, Hill et al 1976), it is possible to assay the development of CAT and adrenergic enzymes in explants from rats of various ages. Recent findings show that CAT develops in vitro only in explants taken from very young rats; explants from older animals develop only adrenergic enzymes (Hill & Hendry 1977, Ross, Johnson & Bunge 1977). Thus, after full adrenergic maturation the neurons may not respond to cholinergic cues. On the other hand, it is also possible to argue that (*a*) the neurons that responded in the newborn ganglia were selectively killed by explantation at older ages, (*b*) the responding neurons died in vivo after birth, or (*c*) the cholinergic cues within the ganglia were no longer present in older explants. These alternate interpretations seem unlikely, however, since similar results have been obtained with the dissociated cell cultures where neuronal number and cholinergic influence (CM) remain constant with length of time in culture. These experiments demonstrate that (*a*) the neurons become more refractory to induction of ACh synthesis by CM as they mature adrenergically and (*b*) they become more refractory to induction of adrenergic function by CM withdrawal as they mature cholinergically (Patterson, Chun & Reichardt 1977, Patterson & Chun 1977b). Therefore, the cholinergic-adrenergic decision appears to be a function of the developmental stage. Young neurons can respond to available cues and can differentiate to become either adrenergic or cholinergic (or sometimes both). They can alter their decision even after beginning to express such functions, but they become less plastic as they mature. This may represent a third and final stage in which the decision to produce a particular transmitter is not easily reversible but where its production can be quantitatively modulated by various stimuli (Thoenen 1975).

To understand normal development it is also important to know what types of nonneuronal cells exert this influence on the neurons. It is known that a glial cell line and nonneuronal cells from sympathetic ganglia (Patterson & Chun 1974), as well as cells cultured from a variety of tissues taken from the newborn rat (Ross & Bunge 1976, Patterson & Chun 1977a), induce neuronal CAT activity. Large quantitative differences have been observed in the ability of cells from the various nonneuronal sources to induce cholinergic functions, and some of these differences are intriguing. Targets that receive cholinergic or mixed innervation, such as skeletal and heart muscle, are excellent inducers, whereas targets that receive only adrenergic innervation, such as brown fat (Cottle 1970) and liver (Forssmann & Ito 1977), are relatively poor inducers. There is also the interesting possibility that cues from targets may play a role in the development of other parts of the nervous system. For instance, spinal cord cells cultured with skeletal muscle cells, or CM from them, show increased CAT activity (Giller et al 1973, 1977, Nelson 1975), whereas some spinal cord cells cultured with liver display FIF (Bird & James 1975). Although the concept that a target organ can modify the transmitter used by its innervation is not a new one, it is still much too early to evaluate its merits. A somewhat puzzling observation is that primary embryonic (Patterson & Chun 1977a) and periosteal (Ross & Bunge 1976) fibroblasts induce ACh synthesis in sympathetic neurons (just as they produce NGF; see Young et al 1975). Ultimately it will be necessary to isolate the molecules involved in these phenomena, compare their properties, and determine their distribution in the developing animal. Such purifications are underway in several laboratories.

As described in more detail elsewhere (Black 1978, Varon & Bunge 1978), NGF is essential for the survival, growth, and differentiation of adrenergic sympathetic neurons. What, then, is the role of NGF in the development of the sympathetic cholinergic neurons? Recent work has compared the role of NGF in the development of sympathetic neurons grown in the absence of nonneuronal cells under conditions where the vast majority of the neurons are adrenergic (0% CM) or cholinergic (62% CM). Immature neurons ($<$ 2 weeks old) in both types of cultures have an absolute requirement for NGF; in its absence all the neurons die, and NGF causes a dose-dependent increase in survival. Both types of culture become less dependent on NGF with time (Chun & Patterson 1977a–c; see also Lazarus et al 1976). NGF also increases, in a dose-dependent fashion, the ability of the neurons in adrenergic cultures to produce catecholamines (CA). This increase is not simply a consequence of the effect of NGF on neuronal survival or growth, because it is seen even when the CA production is expressed per neuron, or per total neuronal protein or lipid phosphate (Chun & Patterson 1977a). Therefore, as in earlier in vivo experiments (Levi-Montalcini & Angeletti 1968, Thoenen et al 1971), NGF stimulates not only survival and growth, but enhances adrenergic differentiation in a dose-dependent manner. Perhaps unexpectedly, NGF also stimulates the differentiation of cholinergic cultures; ACh production per neuron increases as the NGF concentration is raised (Chun & Patterson 1977c, see also Hill & Hendry 1977). In fact, NGF stimulates the production of both transmitters to the same extent. That is, the ratio of ACh to CA, which is a reflection of the relative rates of synthesis of these transmitters, remains constant over a wide range of NGF concentrations for both adrenergic (ACh/CA = 0.02) and cholinergic (ACh/CA = 200) cultures (Chun & Patterson 1977c). Thus, with respect to transmitter production, NGF is permissive rather than instructive in that it is necessary for survival and stimulates growth and differentiation but does not tell the neurons which transmitter to produce. That NGF does not determine what transmitter a cell will make is further suggested by the fact that dorsal root ganglion cells that are sensitive to NGF for certain periods of embryonic life do not appear to produce significant levels of NE or ACh.

In summary, NGF appears to enhance the differentiation of immature sympathetic neurons along either adrenergic or cholinergic paths. In fact, at least at young ages, some cultured neurons appear to secrete both transmitters simultaneously. The particular phenotype chosen depends on other developmental cues. CM produced by certain types of nonneuronal cells can direct the neurons in the cholinergic direction. CM is a qualitatively different developmental signal from NGF because it is not necessary for neuronal survival and does not appreciably stimulate growth. By analogy there would be a similar cue for adrenergic function, which is not NGF. This factor may be produced by some combination of neural tube, somitic mesenchyme, and perhaps notochord. However, it is not yet clear whether these tissues actively signal the crest cells or merely provide a permissive environment for the expression of a previously determined phenotype. However, even if cells have been previously instructed as to neurotransmitter synthesis, this order can be countermanded in young neurons in culture (and may be in vivo). This reversal can occur after ganglion formation in vivo and initial differentiation in vitro has begun. If there is an early adrenergic influence, then why does a cholinergic minority population develop in some sympathetic ganglia? One can speculate that either only a small number of the neurons receive the countermanding cholinergic signal because it is selectively localized (perhaps in certain targets or in certain ganglionic nonneuronal cells), or all the neurons receive the cholinergic signal, but most are prevented from responding to it by factors that are not present in the culture experiments. Preganglionic innervation is not present in culture, and neuronal activity is known to be important in adrenergic sympathetic development (see Black 1978). Such activity could conceivably play a role in determination as well as differentiation.[1] Another factor not controlled in the culture experiments is the hormonal content of the sera used. There are clear examples of hormonal regulation of transmitter expression in developing sympathetic ganglia. For instance, glucocorticoids can increase the number of SIF cells (Eränkö & Eränkö 1972, Eränkö et al 1972, Hervonen & Eränkö 1975) and the amount of epinephrine in sympathetic ganglia both in vivo and in vitro (Ciaranello et al 1975). They can also modulate the induction of TH and dopamine-β-hydroxylase activities mediated by NGF or preganglionic neurons (Otten & Thoenen 1976a,b). In sum, clear evidence has been obtained for the presence of environmental influences on the determination of autonomic neurotransmitter functions. It seems likely that future investigations of such factors will be pursued at the molecular level through the synergistic combination of in vivo embryological experiments and culture techniques.

ACKNOWLEDGMENTS

I thank Eleanor Livingston and Doreen McDowell for excellent help with the manuscript, numerous friends for their advice, and several authors for sending unpublished papers. I also thank the American and Massachusetts Heart Associations and the NINCDS for support.

[1]Recent experiments have in fact demonstrated that the level of the membrane potential can affect the adrenergic-cholinergic decision (Walicke & Patterson 1977).

Literature Cited

Aiken, J. W., Reit, E. 1969. A comparison of the sensitivity to chemical stimuli of adrenergic and cholinergic neurons in the cat stellate ganglion. *J. Pharmacol. Exp. Ther.* 169:211–23

Amano, T., Richelson, E., Nirenberg, M. 1972. Neurotransmitter synthesis by neuroblastoma clones. *Proc. Natl. Acad. Sci. USA* 69:258–63

Baker, P. C., Quay, W. B. 1969. 5-hydroxytryptamine metabolism in early embryogenesis, and the development of brain and retinal tissues. A review. *Brain Res.* 12:273–95

Bird, M., James, D. 1975. The culture of previously dissociated embryonic chick spinal cord cells on feeder layers of liver and kidney, and the development of paraformaldehyde-induced fluorescence upon the former. *J. Neurocytol.* 4:633–46

Bjerre, B. 1973. The production of catecholamine-containing cells *in vitro* by young chick embryos studied by the histochemical fluorescence method. *J. Anat.* 115:119–31

Bjerre, B., Björklund, A. 1973. The production of catecholamine-containing cells in vitro by young chick embryos: effects of nerve growth factor and its antiserum. *Neurobiology* 3:140–61

Björklund, A., Cegrell, L., Falck, B., Ritzen, M., Rosengren, E. 1970. Dopamine-containing cells in sympathetic ganglia. *Acta Physiol. Scand.* 78:334–38

Black, I. B. 1978. Regulation of autonomic development. *Ann. Rev. Neurosci.* 1: 183–214

Bray, D. 1970. Surface movements during the growth of single explanted neurons. *Proc. Natl. Acad. Sci. USA* 65:905–10

Burnstock, G. 1975. Purinergic transmission. In *Handbook of Psychopharmacology,* 5:131–94, ed. L. Iversen, S. D. Iversen, S. H. Snyder. New York: Plenum

Burton, H., Bunge, R. P. 1975. A comparison of the uptake and release of [^{3}H] norepinephrine in rat autonomic and sensory ganglia in tissue culture. *Brain Res.* 97:157–62

Champlain, J. de, Malmfors, T., Olson, L., Sachs, C. 1970. Ontogenesis of peripheral adrenergic neurons in the rat: pre- and postnatal observations. *Acta Physiol. Scand.* 80:276–88

Chun, L. L. Y., Patterson, P. H. 1977a. The role of nerve growth factor in the development of rat sympathetic neurons in vitro. I. Survival, growth and differentiation of catecholamine production. *J. Cell Biol.* In press

Chun, L. L. Y., Patterson, P. H. 1977b. The role of nerve growth factor in the development of rat sympathetic neurons in vitro. II. Developmental studies. *J. Cell Biol.* In press

Chun, L. L. Y., Patterson, P. H. 1977c. The role of nerve growth factor in the development of rat sympathetic neurons in vitro. III. Effect on acetylcholine production. *J. Cell Biol.* In press

Ciaranello, R. D., Axelrod, J. 1975. Effects of dexamethasone on neurotransmitter enzymes in chromaffin tissue of the newborn rat. *J. Neurochem.* 24:775–78

Claude, P. 1973. Electron microscopy of dissociated rat sympathetic neurons in vitro. *J. Cell Biol.* 59:57a

Cohen, A. M. 1972. Factors directing the expression of sympathetic nerve traits in cells of neural crest origin. *J. Exp. Zool.* 179:167–82

Cohen, A. M. 1974. DNA synthesis and cell division in differentiating avian adrenergic neuroblasts. In *Dynamics of Degeneration and Growth in Neurons,* pp. 359–70, ed. K. Fuxe, L. Olson, Y. Zotterman. New York: Pergamon

Cohen, A. M. 1977. Independent expression of the adrenergic phenotype by neural crest cells in vitro. *Proc. Natl. Acad. Sci. USA* 74:2899–2903

Cohen, A. M., Konigsberg, I. R. 1975. A clonal approach to the problem of neural crest determination. *Dev. Biol.* 46:262–80

Costa, M., Furness, J. B., Gabella, G. 1971. Catecholamine-containing nerve cells in the mammalian myenteric plexus. *Histochemie* 25:103–6

Cottle, W. H. 1970. The innervation of brown adipose tissue. In *Brown Adipose Tissue,* ed. O. Lindberg, pp. 155–78. New York: Elsevier.

Crain, S. M., Peterson, E. R. 1974. Development of neural connections in culture. *Ann. NY Acad. Sci.* 228:6–33

Dale, H. 1937. Acetylcholine as a chemical transmitter of the effects of nerve impulses. I. History of ideas and evidence. Peripheral autonomic actions. Functional nomenclature of nerve fibers. *Mt. Sinai J. Med. NY* 4:401–15

Dreyfus, C. F., Bornstein, M. B., Gershon, M. D. 1977. Synthesis of serotonin by neurons of the myenteric plexus in situ and in organotypic tissue culture. *Brain Res.* 128:125–39

Dreyfus, C. F., Sherman, D. L., Gershon, M. D. 1977. Uptake of serotonin by intrinsic neurons of the myenteric plexus grown in organotypic tissue culture. *Brain Res.* 128:109–23

Eagle, H., Piez, L. 1962. The population-dependent requirement by cultured mammalian cells for metabolites which they can synthesize. *J. Exp. Med.* 116:29–43

Ehinger, B., Falck, B. 1970. Uptake of some catecholamines and their precursors into neurons of the rat ciliary ganglion. *Acta Physiol. Scand.* 78:132–41

Elfvin, L. G., Hökfelt, T., Goldstein, M. 1975. Fluorescence microscopical, immunohistochemical and ultrastructural studies on sympathetic ganglia of the guinea pig, with special reference to the SIF cells and their catecholamine content. *J. Ultrastruct. Res.* 51:377–96

Enemar, A., Falck, B., Hakanson, R. 1965. Observations on the appearance of norepinephrine in the sympathetic nervous system of the chick embryo. *Dev. Biol.* 11:268–83

Eränkö, L. 1972a. Ultrastructure of the developing sympathetic nerve cell and the storage of catecholamines. *Brain Res.* 46:159–75

Eränkö, L. 1972b. Postnatal development of histochemically demonstrable catecholamines in the superior cervical ganglion of the rat. *Histochem. J.* 4:225–36

Eränkö, L., Eränkö, O. 1972. Effect of hydrocortisone on histochemically demonstrable catecholamines in the sympathetic ganglia and extra-adrenal chromaffin tissue of the rat. *Acta Physiol. Scand.* 84:125–33

Eränkö, O., Eränkö, L., Hill, C. E., Burnstock, G. 1972. Hydrocortisone-induced increase in the number of small intensely fluorescent cells and their histochemically demonstrable catecholamine content in cultures of sympathetic ganglia of the newborn rat. *Histochem. J.* 4:49–58

Filogamo, G., Marchisio, P. C. 1971. Acetylcholine system and neural development. *Neurosci. Res.* 4:29–64

Forssmann, W. G., Ito, S. 1977. Morphology of efferent liver innervation in primates. *J. Cell Biol.* 74:299–314

Furshpan, E. J., MacLeish, P. R., O'Lague, P. H., Potter, D. D. 1976. Chemical transmission between rat sympathetic neurons and cardiac myocytes developing in microcultures: evidence for cholinergic, adrenergic, and dual-function neurons. *Proc. Natl. Acad. Sci. USA* 73:4225–29

Fuxe, K., Goldstein, M., Hökfelt, T., Joh, T. H. 1971. Cellular localization of dopamine-β-hydroxylase and phenylethanolamine-N-methyltransferase as revealed by immunohistochemistry. *Prog. Brain Res.* 34:127–38

Gershon, M. D., Dreyfus, C. F., Pickel, V. M., Joh, T. H., Reis, D. J. 1977. Serotonergic neurons in the peripheral nervous system: identification in gut by immunohistochemical localization of tryptophan hydroxylase. *Proc. Natl. Acad. Sci. USA* 74:3086–89

Giller, E. L., Neale, J. H., Bullock, P. N., Schrier, B. K., Nelson, P. G. 1977. Choline acetyltransferase activity of spinal cord cell cultures increased by co-culture with muscle and by muscle-conditioned medium. *J. Cell Biol.* 74:16–29

Giller, E. L., Schrier, B. K., Shainberg, A., Fisk, H. R., Nelson, P. G. 1973. Choline acetyltransferase activity is increased in combined cultures of spinal cord and muscle cells from mice. *Science* 182:588–89

Greene, L. A., Rein, G. 1977. Synthesis storage and release of acetylcholine by a nerve-growth-factor-responsive line of rat pheochromocytoma cells. *Nature* 268:349–51

Greene, L. A., Tischler, A. 1976. Establishment of a noradrenergic clonal line of rat adrenal pheochromocytoma cells which respond to nerve growth factor. *Proc. Natl. Acad. Sci. USA* 73:2424–28

Hamprecht, B., Traber, J., Lamprecht, F. 1974. Dopamine-β-hydroxylase activity in cholinergic neuroblastoma x glioma hybrid cells; increase of activity by N^6, O^2-dibutyryl adenosine 3':5'-cyclic monophosphate. *FEBS Lett.* 42:221–26

Hervonen, H., Eränkö, O. 1975. Fluorescence histochemical and electron microscopical observations on sympathetic ganglia of the chick embryo cultured with and without hydrocortisone. *Cell Tissue Res.* 156:145–66

Hill, C. E., Hendry, I. A. 1977. Development of neurons synthesizing noradrenaline and acetylcholine in the superior cervical ganglion of the rat in vivo and in vitro. *Neuroscience* In press

Hill, C. E., Purves, R. D., Watanabe, H., Burnstock, G. 1976. Specificity of innervation of iris musculature by sympathetic nerve fibers in tissue culture. *Pfluegers Arch.* 361:127–34

Hökfelt, T., Johansson, O., Efendic, S., Luft, R., Arimura, A. 1975. Are there somatostatin-containing nerves in the rat gut? Immunohistochemical evidence for a new type of peripheral nerve. *Experientia* 31:852–54

Jacobowitz, D. 1967. Histochemical studies of the relationship of chromaffin cells and adrenergic nerve fibers to the cardiac ganglia of several species. *J. Pharmacol. Exp. Ther.* 158:227–40

Johnson, M., Ross, D., Meyers, M., Rees, R., Bunge, R., Wakshull, E., Burton, H. 1976. Synaptic vesicle cytochemistry changes when cultured sympathetic neurons develop cholinergic interactions. *Nature* 262:308–10

Johnston, M. C., Bhakdinaronk, A., Reid, Y. C. 1974. An expanded role of the neural crest in oral and pharyngeal development. In *4th Symp. Oral Sensation Percept.,* ed. J. F. Bosma, pp. 37–52. Fogarty Int. Cent. Proc. No. 21. Washington, DC: GPO

Kirby, M. L., Gilmore, S. A. 1972. A fluorescence study on the ability of the notochord to synthesize and store catecholamines in early chick embryos. *Anat. Rec.* 173:469–78

Ko, C.-P., Burton, H., Johnson, M. I., Bunge, R. P. 1976. Synaptic transmission between rat superior cervical ganglion neurons in dissociated cell cultures. *Brain Res.* 117:461–85

Landis, S. C. 1976. Rat sympathetic neurons and cardiac myocytes developing in microcultures: correlation of the fine structure of endings with neurotransmitter function in single neurons. *Proc. Natl. Acad. Sci. USA* 73:4220–24

Landis, S. C., MacLeish, P. R., Potter, D. D., Furshpan, E. J., Patterson, P. H. 1976. Synapses formed between dissociated neurons: the influence of conditioned medium. *Sixth Ann. Soc. Neurosci.,* Abstr. 280, p. 197

Langley, J. N. 1921. *The Autonomic Nervous System.* Cambridge, England: Heffer. 80 pp.

Lash, J. W. 1967. Differential behavior of anterior and posterior embryonic somites in vitro. *J. Exp. Zool.* 165:47–56

Lawrence, I. E., Burden, H. W. 1973. Catecholamines and morphogenesis of the chick neural tube and notochord. *Am. J. Anat.* 137:199–208

Lazarus, K. J., Bradshaw, R. A., West, N. R., Bunge, R. P. 1976. Adaptive survival of rat sympathetic neurons cultured without supporting cells or exogenous nerve growth factor. *Brain Res.* 113:159–64

LeDouarin, N. 1969. Particularités du noyau interphasique chez la Caille japonaise (*Coturnix coturnix* japonica). Utilisation de ces particularités comme "marquage biologique" dans des recherches sur les interactions tissulaires et les migrations cellulaires au cours de l'ontogènese. *Bull. Biol. Fr. Belg.* 103:435–52

LeDouarin, N. M. 1974. Cell recognition based on natural morphological nuclear markers. *Med. Biol.* 52:281–319

LeDouarin, N. M. 1977. The differentiation of the ganglioblasts of the autonomic nervous system studied in chimeric avian embryos. Sigrid Jusélius Symposium, Helsinki, Aug. 23–25, 1976. In *Cell Interactions in Differentiation,* ed. M. Karkinen-Jääskeläinen, pp. 171–90. London: Academic

LeDouarin, N., LeLièvre, C. 1970. Démonstration de l'origin neurale des cellules à calcitonine du corps ultimobranchial chez l'embryon de Poulet. *C. R. Acad. Sci. D* 270:2857–60

LeDouarin, N. M., Renaud, D., Teillet, M. A., LeDouarin, G. H. 1975. Cholinergic differentiation of presumptive adrenergic neuroblasts in interspecific chimeras after heterotopic transplantations. *Proc. Natl. Acad. Sci. USA* 72:728–32

LeDouarin, N. M., Teillet, M.-A. M. 1974. Experimental analysis of the migration and differentiation of neuroblasts of the autonomic nervous system and of neuroectodermal mesenchymal derivatives, using a biological cell marking technique. *Dev. Biol.* 41:162–84

LeDouarin, N. M., Teillet, M.-A. M., LeLièvre, C. 1977. Influence of the tissue environment on the differentiation of neural crest cells. 30th Annu. Meet. Soc. Gen. Physiol. Woods Hole, Sept. 13–16, 1976. In *Cell and Tissue Interactions,* ed. J. Lash, M. Burger. New York: Raven. In press

Levi-Montalcini, R., Angeletti, P. U. 1968. Nerve growth factor. *Physiol. Rev.* 48:534–69

MacLeish, P. R. 1976. Synapse formation in cultures of dissociated rat sympathetic neurons grown on dissociated rat heart cells. PhD thesis. Harvard Univ., Cambridge. 142 pp.

Mains, R. E., Patterson, P. H. 1973a. Primary cultures of dissociated sympathetic neurons. I. Establishment of long-term growth in culture and studies of differentiated properties. *J. Cell Biol.* 59:329–45

Mains, R. E., Patterson, P. H. 1973b. Primary cultures of dissociated sympathetic neurons. III. Changes in metabolism with age in culture. *J. Cell Biol.* 59:361–66

McMahon, D. 1974. Chemical messengers in development: a hypothesis. *Science* 185:1012–21

Nelson, P. G. 1975. Central nervous system synapses in cell culture. *Cold Spring Harbor Symp. Quant. Biol.* 40:359–71

Newgreen, D. F., Jones, R. O. 1975. Differentiation in vitro of sympathetic cells from chick embryo sensory ganglia. *J. Embryol. Exp. Morphol.* 33:43–56

Noden, D. 1975. An analysis of the migratory behavior of avian cephalic neural crest cells. *Dev. Biol.* 42:106–30

Norr, S. 1973. In vitro analysis of sympathetic neuron differentiation from chick neural crest cells. *Dev. Biol.* 34:16–38

Nurse, C. A. 1977. The formation of cholinergic synapses between dissociated rat sympathetic neurons and skeletal myotubes in cell culture. PhD thesis. Harvard Univ. Cambridge, 124 pp.

Nurse, C. A., O'Lague, P. H. 1975. Formation of cholinergic synapses between dissociated sympathetic neurons and skeletal myotubes of the rat in cell culture. *Proc. Natl. Acad. Sci. USA* 72:1955–59

O'Lague, P. H., MacLeish, P. R., Nurse, C. A., Claude, P., Furshpan, E. J., Potter, D. D. 1975. Physiological and morphological studies on developing sympathetic neurons in dissociated cell culture. *Cold Spring Harbor Symp. Quant. Biol.* 40:399–407

O'Lague, P. H., Obata, K., Claude, P., Furshpan, E. J., Potter, D. D. 1974. Evidence for cholinergic synapses between dissociated rat sympathetic neurons in cell culture. *Proc. Natl. Acad. Sci. USA* 71:3602–6

Otten, U., Thoenen, H. 1976a. Selective induction of tyrosine hydroxylase and dopamine-β-hydroxylase in sympathetic ganglia in organ culture: role of glucocorticoids as modulators. *Mol. Pharmacol.* 12:353–62

Otten, U., Thoenen, H. 1976b. Modulatory role of glucocorticoids on NGF-mediated enzyme induction in organ cultures of sympathetic ganglia. *Brain Res.* 111:438–41

Patterson, P. H., Chun, L. L. Y. 1974. The influence of nonneuronal cells on catecholamine and acetylcholine synthesis and accumulation in cultures of dissociated sympathetic neurons. *Proc. Natl. Acad. Sci. USA* 71:3607–10

Patterson, P. H., Chun, L. L. Y. 1977a. The induction of acetylcholine synthesis in primary cultures of dissociated rat sympathetic neurons. I. Effects of conditioned medium. *Dev. Biol.* 56:263–80

Patterson, P. H., Chun, L. L. Y. 1977b. The induction of acetylcholine synthesis in primary cultures of dissociated rat sympathetic neurons. II. Developmental aspects. *Dev. Biol.* In press

Patterson, P. H., Chun, L. L. Y., Reichardt, L. F. 1977. The role of nonneuronal cells in the development of sympathetically-derived neurons. In *Cellular Neurobiology,* ed. Z. Hall, R. Kelly, C. F. Fox, pp. 95–104. New York: Liss

Patterson, P. H., Reichardt, L. F., Chun, L. L. Y. 1975. Biochemical studies on the development of primary sympathetic neurons in cell culture. *Cold Spring Harbor Symp. Quant. Biol.* 40:389–97

Pearse, A. G. E., Polak, J. M. 1975. Immunocytochemical localization of substance P in mammalian intestine. *Histochemistry* 41:373–75

Phillipson, O. T., Moore, K. E. 1975. Effects of dexamethasone and nerve growth factor on phenylethanolamine N-methyltransferase and adrenalin in organ cultures of newborn rat superior cervical ganglion. *J. Neurochem.* 25: 295–98

Prasad, K. N., Mandal, B., Waymire, J. C., Lees, G. J., Vernadakis, A., Weiner, N. 1973. Basal level of neurotransmitter-synthesizing enzymes and effect of cyclic AMP agents on the morphological differentiation of isolated neuroblastoma clones. *Nature New Biol.* 241: 117–19

Purves, R. D., Hill, C. E., Chamley, J. H., Mark, G. E., Fry, D. M., Burnstock, G. 1974. Functional autonomic neuromuscular junctions in tissue culture. *Pfluegers Arch.* 350:1–7

Rees, R., Bunge, R. P. 1974. Morphological and cytochemical studies of synapses formed in culture between isolated rat superior cervical ganglion neurons. *J. Comp. Neurol.* 157:1–11

Reichardt, L. F., Patterson, P. H. 1977. Neurotransmitter synthesis and uptake by individual rat sympathetic neurons developing in microcultures. *Nature* In press

Reichardt, L. F., Patterson, P. H., Chun, L. L. Y. 1976. Norepinephrine and acetylcholine synthesis by individual sympathetic neurons under various culture conditions. *Sixth Ann. Soc. Neurosci. Abstr.* 327, p. 197

Ross, D., Bunge, R. P. 1976. Choline acetyltransferase in cultures of rat superior cervical ganglion. *Sixth Ann. Soc. Neurosci.* Abstr. 1094, p. 769

Ross, D., Johnson, M., Bunge, R. 1977. Evidence that development of cholinergic characteristics in adrenergic neurons is age dependent. *Nature* 267:536–39

Rothman, T. P., Gershon, M. D., Holtzer, H. 1977. Cell division in the development of adrenergic neurons. *7th Ann. Soc. Neurosci.* In press

Rutter, W. J., Clark, W. R., Kemp, J. D., Bradshaw, W. S., Sanders, T. G., Ball, W. D. 1968. In *Epithelial-Mesenchymal Interactions: 18th Hahnemann Symp.* pp. 114–31

Said, S. I., Rosenberg, R. N. 1976. Vasoactive intestinal polypeptide: abundant immunoactivity in neural cell lines and normal nervous tissue. *Science* 192: 907–8

Schubert, D., Heinemann, S., Carlisle, W., Tarikas, H., Kimes, B., Patrick, J., Steinbach, J. H., Culp, W., Brandt, B. L. 1974. Clonal cell lines from the rat central nervous system. *Nature* 249: 224–27

Schubert, D., Heinemann, S., Kidokoro, Y. 1977. Cholinergic metabolism and synapse formation by a rat pheochromocytoma cell line. *Proc. Natl. Acad. Sci. USA* 74:2579–83

Sjöqvist, F. 1963. The correlation between the occurrence and localization of acetylcholinesterase-rich cell bodies in the stellate ganglion and the outflow of cholinergic sweat secretory fibres to the fore paw of the cat. *Acta Physiol. Scand.* 57:339–51

Smith, J., Cochard, P., LeDouarin, N. M. 1977. Development of choline acetyltransferase and cholinesterase activities in enteric ganglia derived from presumptive adrenergic and cholinergic levels of the neural crest. *Cell Differ.* In press

Thoenen, H. 1975. Transynaptic regulation of neuronal enzyme synthesis. See Burnstock 1975, pp. 443–75

Thoenen, H., Angeletti, P. U., Levi-Montalcini, R., Kettler, R. 1971. Selective induction by nerve growth factor of tyrosine hydroxylase and dopamine-β-hydroxylase in the rat superior cervical ganglion. *Proc. Natl. Acad. Sci. USA* 68:1598–1602

Triplett, E. L. 1958. The development of the sympathetic ganglia, sheath cells, and meninges in amphibians. *J. Exp. Zool.* 138:283–311

Twitty, V. C. 1945. The developmental analysis of specific pigment patterns. *J. Exp. Zool.* 100:141–78

Uvnäs, B. 1954. Sympathetic vasodilator outflow. *Physiol. Rev.* 34:608–18

Varon, S. S., Bunge, R. P. 1978. Trophic mechanisms in the peripheral nervous system. *Ann. Rev. Neurosci.* 1:327–61

Vernadakis, A., Gibson, D. A. 1974. Role of neurotransmitter substances in neural growth. In *Perinatal Pharmacology: Problems and Priorities,* ed. J. Dancis, J. C. Hwang, pp. 65–76. New York: Raven

Walicke, P., Patterson, P. H. 1977. *Proc. Natl. Acad. Sci. USA* In press

Weston, J. A. 1963. A radioautographic analysis of the migration and localization of trunk neural crest cells in the chick. *Dev. Biol.* 6:279–310

Weston, J. A. 1970. The migration and differentiation of neural crest cells. *Adv. Morphog.* 8:41–114

Young, M., Oger, J., Blanchard, M. H., Asdourian, H., Amos, H., Arnason, B. G. W. 1975. Secretion of a nerve growth factor by primary chick fibroblast cultures. *Science* 187:361–62

Ann. Rev. Neurosci. 1978. 1:19–34

CIRCADIAN PACEMAKERS IN THE NERVOUS SYSTEM

❖11501

Gene D. Block and Terry L. Page
Stanford University, Hopkins Marine Station, Pacific Grove, California 93950

INTRODUCTION

Although circadian phenomena have long been observed, certainly as far back as de Mairan's eighteenth-century observations of rhythmic movements of plants in aperiodic environments, the concept of endogenous oscillators functioning as clocks has only been developed in the last 30 years. During this time, a number of empirical generalizations have been extensively documented that lay the groundwork for a substantial formalism. This formalism has led to an appreciation of the functional utility of circadian pacemakers in that they not only ensure the organism of an appropriately phased temporal organization but also measure time necessary for seasonal reproduction and for celestial navigation and orientation.

Paralleling the development of analytical models, physiological investigations have begun to describe the anatomical substrate of circadian organization. In the metazoa the central nervous and neuroendocrine systems are the main source of circadian oscillations.

Perhaps the most thoroughly studied problem in the physiology of circadian rhythms has been the search for the cells responsible for generating the circadian behaviors. The early demonstration that single eukaryotic cells (Bruce & Pittendrigh 1956, Hastings & Sweeney 1958) and isolated tissues (Andrews & Folk 1964, Andrews 1971) could exhibit rhythmicity suggested that a number of sites within the organisms could provide the timing cue for the circadian rhythms. Thus, it is still a major question to what extent the rhythmicity observed in any particular function is exclusively under the control of a particular group of cells. Many early investigations were directed toward localizing the "biological clock." Although it was indeed often possible to localize circadian oscillators that contributed substantially to the rhythmicity of some function, the results also frequently reflected the complexities in circadian organization that were becoming apparent in formal analysis. The results of both physiological and formal approaches suggested that the circadian systems of multicellular organisms are composed of many circadian pacemakers, and circadian organization is derived from the integrative relationships within this multioscillator system (Pittendrigh 1960, 1967).

0147-006X/78/0325-0019$01.00

Probably the single most intriguing problem we face in understanding the organization of multioscillator systems is how the internal phase relationships between pacemakers are maintained. Are the phase relationships between the several oscillators maintained by the entrainment of each oscillator to external stimuli, or is control exerted hierarchically by means of a "master clock"? Or does the temporal order arise (Pittendrigh 1974, Menaker 1974) from a combination of mutual and hierarchical entrainment among a population of oscillators within the animal?

While no single circadian system has been completely characterized, several general features of circadian organization have emerged from physiological studies, upon which we focus our attention. We first review evidence suggesting that circadian oscillators that contribute substantially to the control of rhythmic processes can be localized in the nervous systems of a number of phylogenetically diverse organisms. Second we consider in some detail the evidence that there are multiple pacemakers within the circadian system and that it is the interaction between these pacemakers that ultimately determines the phase and frequency of rhythmic processes in the animal. This paper is not intended to be an exhaustive review of the literature, but is directed toward identifying the key issues that have emerged from the study of circadian organization in the metazoa.

LOCALIZATION OF CIRCADIAN PACEMAKERS

We face two basic issues in discussing the localization of circadian pacemakers in multicellular animals. First, can restricted portions of the nervous system function as self-sustaining oscillators? Experimental results indicate that, at best, oscillators have been localized to particular nuclei, ganglia, or organs. In each case at least hundreds, and usually thousands, of cells are involved. It is not known whether a single neuron can sustain a circadian oscillation, whether several cells are needed, or whether the periodicity is an emergent property of a large number of interacting cells.

The second problem concerns the integration of "localized oscillators" into the circadian system. Here the complexity of the issue makes the questions somewhat less well-defined. There are many possible schemes for circadian organization, involving many possible functional roles for circadian oscillators. The evidence discussed in this section focuses on the localization of the primary driving oscillator for the circadian system. In the following section, we address in some detail the evidence that this primary oscillator may itself be composed of 2 or more oscillators, and that many metazoan circadian systems may also utilize other second-order oscillators, entrained by the driving oscillator, to control various rhythmic processes.

Restricted Portions of the Nervous System as Self-Sustaining Oscillators

In organisms as diverse as molluscs, insects, and birds restricted portions of the nervous system can function as circadian oscillators. This conclusion is probably best exemplified by studies demonstrating that nervous tissue isolated in culture can maintain a circadian rhythm. For example, Jacklet (1969, 1974) has demonstrated the presence of a beautifully precise circadian rhythm in the isolated eye of the marine gastropod *Aplysia.* If the eye and attached optic nerve are removed from the animal and placed in sterile seawater (or culture medium) optic nerve activity, which occurs spontaneously as compound action potentials (CAPs), exhibits a circadian rhythm both in frequency and amplitude. Other circadian rhythms of neural activity have been demonstrated in the eye of *Navanax* (Eskin & Harcombe 1977), in the spike activity of the genital nerve of the isolated abdominal ganglion of *Aplysia* (Strumwasser, 1974), and in spike frequency in the second root of the sixth ganglion of the isolated crayfish abdominal nerve cord (Block 1976).

These results do not demonstrate, however, that any single neuron is capable of generating a circadian oscillation, nor have recent efforts to approach this issue provided any resolution. The eye of *Aplysia,* for example, is composed of several thousand cells (Jacklet et al 1972), including primary receptors, secondary interneurons, and support cells. Jacklet & Geronimo (1971) have suggested that a large number of these cells are involved in producing the circadian period that is postulated to emerge from the mutual interactions of a population of coupled oscillators each of which, independently, have periods much shorter than 24 hours. This claim is based on experiments in which systematic surgical reduction of the distal retina gave some evidence of a progressive decrease in the period of the rhythm. On the other hand, Sener (see references in Strumwasser 1974) has failed to observe this effect. Although a complete account of the experiments has not yet been published, he reports that eyes reduced to their extreme base (about 100 cells remaining) can still sustain circadian periodicities.

Although there is evidence that single identified neurons can express a circadian rhythm (Strumwasser 1965, Aréchiga & Wiersma 1969), there is only a single case in which evidence suggests that a single isolated neuron can sustain a circadian rhythm. Strumwasser (1971) has reported that the parabolic burster (cell R-15) in the *Aplysia* abdominal ganglion exhibits a circadian rhythm of subthreshold membrane potentials when chemically isolated from the rest of the ganglion by use of low Ca^{2+} and tetrodotoxin. However, whether or not R-15 is a circadian oscillator —or whether it even exhibits a circadian rhythm in the isolated ganglion—has recently been questioned (Beiswanger & Jacklet 1975, Lickey et al 1976, Smith 1976) on the basis of its failure to exhibit rhythmic behavior during long-term recording in culture medium, even though the cell remains active and apparently healthy for several weeks.

The Role of Localized Oscillators in the Circadian System

A large proportion of the physiological research on circadian rhythms has been directed towards localizing cells acting as circadian pacemakers that exert primary control over some particular rhythmic process. Most investigations have employed lesions, followed by an assay for the presence or absence of rhythmicity in a specific behavioral or physiological function. These experiments, however, often suffer from a major interpretive difficulty. The rhythmic expression of the behavior may involve pathways and processes quite distinct from the pacemaker itself, and therefore arrhythmicity in a previously rhythmic behavior is not necessarily an exclusive result of pacemaker lesions. As pointed out by Pittendrigh (1976), the only parameters of the overt behavior that can be considered to truly reflect the state of the pacemaker are the phase and freerunning period of the system *in steady state.*

This interpretive obstacle has been overcome in some systems. In two cases where the oscillator exerts its effect hormonally it has been possible by transplantation of the putative pacemaker to demonstrate the location of the oscillator. The first convincing demonstration came from the now classic work of Truman & Riddiford (1970, Truman 1972) on silkmoths. They worked with two species in which the emergence of the pharate adult from its pupal case is gated by a circadian oscillator. When placed into constant conditions, populations of both species exhibit a daily peak of eclosion. The time of emergence for each species is different however. When raised on a 24-hr light cycle consisting of 17 hr of light followed by 7 hr of darkness, one species, *Hyalophora cecropia,* emerges just after dawn while the other, *Anthereae pernyi,* emerges near dusk. Removal of the brains of these animals did not prevent eclosion, "brainless" moths emerged at random times throughout the day. If the brain was removed from the head but reimplanted in the abdomen, normal rhythmicity was restored. While these results strongly implicated the brain as the site of a clock that controlled the time of emergence, the conclusive evidence was provided by experiments in which brains were removed from the head of one species and transplanted into the abdomen of the other. Animals that had received these "switched brain" transplants exhibited normal eclosion behavior for the host species; however, the phase of the eclosion rhythm was characteristic of the donor and not the host. *A. pernyi* that had received *H. cecropia* brains emerged just after dawn, while *H. cecropia* that received *A. pernyi* brains emerged near dusk (Figure 1). The

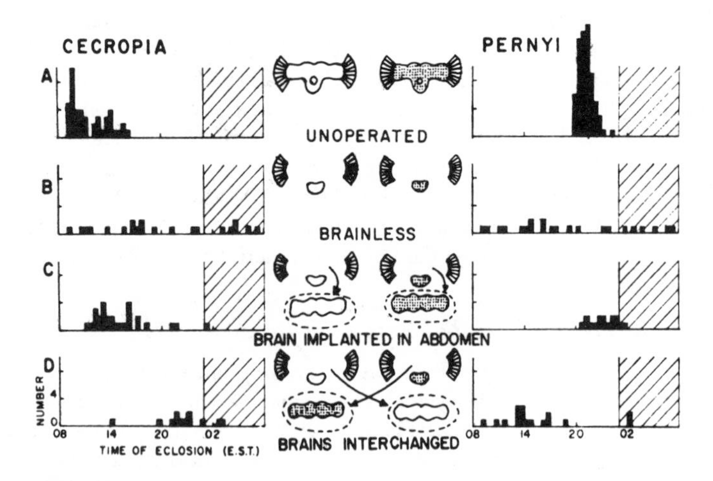

Figure 1 The eclosion of *Hyalophora cecropia* and *Antheraea pernyi* moths in a 17 hr light: 7 hr dark regimen showing the effects of brain removal, transplantation of the brain to the abdomen, and interchange of brains between the two species. After brain exchange the host emerges at the eclosion time characteristic of the donor species (from Truman 1971).

transplantation of phase as well as the restoration of rhythmicity in the host moths leaves virtually no doubt that the circadian oscillator that controls the time of emergence is located in the brain.

In a similar series of experiments on the house sparrow, *Passer domesticus,* it has been demonstrated that the pineal gland is most likely the locus of the circadian oscillator that controls the rhythmicity of perch-hopping behavior. Complete removal of the pineal in a bird freerunning in constant darkness immediately abolishes this rhythm (Gaston & Menaker 1968). However, when all known neutral connections with the pineal were destroyed by severing the pineal "stalk," or when the pineal's innervation by the sympathetic nervous system (via the superior cervical ganglia) was chemically blocked with massive doses of 6-hydroxydopamine, rhythmicity was unaffected (Zimmerman & Menaker 1975). Thus, whatever the role of the pineal in sustaining rhythmicity, its effect must be exerted hormonally. Powerful evidence that the pineal is, in fact, the source of the oscillation controlling rhythmic perch-hopping was provided by transplanting pineals from two groups of donor birds maintained on different light cycles into host birds previously made arrhythmic by pinealectomy (Zimmerman 1976). Rhythmicity was immediately restored in recipient birds by transplantation of the donor's pineal into the anterior chamber of the host's eye. Critically, the phase of the host's rhythm following the implant corresponded quite closely with the phase of the donor's rhythm at the time of surgery. The transplantation of phase, as well as the restoration of rhythmicity—as in the moth case—provides a convincing demonstration that the pineal is the driving oscillator controlling perch-hopping behavior.

A somewhat different situation is found in the cockroach. Efforts to localize the pacemaker that controls the rhythm of locomotor activity (reviewed by Brady 1969, 1971) have resulted in two findings that seem clear: (*a*) ablation of both optic lobes or section of the optic tracts invariably abolishes rhythmicity (Nishiitsutsuji-Uwo & Pittendrigh 1968, Roberts 1974, Sokolove 1975), and (*b*) section of the circumesophogeal connectives likewise abolishes the locomotor rhythm (Roberts et al 1971). These results have led to the suggestion that the optic lobes are the site of origin of the oscillation, and their effects upon thoracic motor centers are electrically mediated via axons in the circumesophogeal connectives. However, as Nishiitsutsuji-Uwo & Pittendrigh (1968) pointed out, these experiments do not provide conclusive evidence that the optic lobes act as a circadian pacemaker, only that they are necessary for expression of the overt rhythm. Nor is the residual uncertainty alleviated by the finer localization of the putative pacemaker site to the region of the lobula (Roberts 1974, Sokolove 1975). A recent study, however, has provided evidence that strongly supports the notion that the optic lobes are involved in pacemaker activity. Page et al (1977) have shown that although either the right or the left lobe alone is sufficient to maintain rhythmicity, ablation of either one of the lobes has the consistent effect of lengthening the freerunning period, while sham operations, such as optic nerve section, have no effect on period. The fact that the optic lobes are not only necessary to sustain the rhythm but are also instrumental in the determination of its frequency strongly suggests that the optic lobes contain at least part of the circadian pacemaker controlling locomotor activity.

It is clear in the three cases discussed above that circadian oscillators that play a major role in setting the phase and period of overt rhythms can be localized in the nervous system. However, we have not addressed the question of whether these localized oscillators are involved in the control of many of the various processes in the animals that exhibit circadian rhythms—that is, can localized oscillators exert primary control not only over a particular rhythmic function but also have general control over the circadian system? This problem has been examined most extensively in mammals. There is growing evidence, primarily from rodents, that the suprachiasmatic nuclei of the hypothalamus contain a circadian oscillator that has a profound influence on mammalian circadian organization. Destruction of the suprachiasmatic nuclei can abolish the circadian rhythms of locomotor and drinking activity (Stephan & Zucker 1972, Stetson & Watson-Whitmyre 1976), adrenal corticosteroid levels (Moore & Eichler 1972), and N-acetyltransferase levels in the pineal (Moore 1974). Furthermore, two processes that utilize the circadian system for time measurement—photoperiodically controlled gonadal recrudescence (Elliott et al 1972) and estrous cyclicity (Alleva et al 1971, Fitzgerald & Zucker 1976)—are disrupted by suprachiasmatic nuclei lesions (Stetson & Watson-Whitmyre 1976, Rusak & Morin 1976). Apparently, destruction of the suprachiasmatic nuclei disrupts control of the circadian system over the normally rhythmic processes of the organism. Aside from the fact that the suprachiasmatic nuclei are involved in the circadian control of so many dissimilar functions, there are two other lines of evidence that suggest these nuclei contain the primary oscillator in the circadian system. First, direct visual input to the suprachiasmatic nuclei via the recently discovered retinohypothalamic tract (Moore & Lenn 1972, Moore 1973) has been shown to be a sufficient pathway for entrainment by light (Moore 1974). Second, destruction of known inputs to the suprachiasmatic nuclei via the retinohypothalamic tract (Moore 1974) and the raphé nuclei (Block & Zucker 1976), does not abolish rhythmicity. The simplest interpretation of these results is that the suprachiasmatic nuclei function as the exclusive pacemaker in the circadian system and, via distributed outputs, maintain circadian organization in mammals. Nevertheless, recent evidence from Rusak (1977) indicates a more complex role for the suprachiasmatic nuclei in controlling rhythmic processes. When hamsters were maintained on running-wheels for several months after suprachiasmatic nuclei lesions, a wide range of effects were observed, including: (*a*) clear 8 or 12 hr periodicities, (*b*) unstable activity patterns that were occasionally rhythmic, or (*c*) complete arrhythmicity in a few cases. Rusak interprets these data as suggesting a multioscillator organization where circadian oscillators in the suprachiasmatic nuclei play an integrative role in coordinating the activity of independent pacemakers that directly control rhythmic processes. This hypothesis is consistent with other evidence for multiple pacemakers in mammals. For example, tissue culture studies show circadian rhythms of oxygen consumption and steroid output persist in isolated adrenal glands (Andrews & Folk 1964, Andrews 1971), and, as we discuss below, Aschoff (1969, Aschoff et al 1967) has shown that rhythms in body temperature activity in humans are controlled by separate pacemakers (Figure 2).

There are other examples of circadian systems in which localized oscillators control several dissimilar functions. In the sparrow, for example, the pineal is not only necessary to sustain the freerunning perch-hopping rhythm but also the rhythms of body temperature (Binkley et al 1971) and uric acid excretion (Menaker 1974). In the cricket, ablation of the optic lobes abolishes circadian rhythms in locomotion (Sokolove & Loher 1975), stridulation (Loher 1972, Sokolove & Loher 1975), and spermatophore production (Loher 1974). In the crayfish the supraesophageal ganglion has been implicated in the locomotor rhythm (Page & Larimer 1975a) and retinal pigment migration rhythms (Page & Larimer 1975b, Aréchiga et al 1973).

MULTIPLE OSCILLATORS

In the previous section we considered a phylogenetically diverse group of animals in which restricted regions of the nervous system are necessary not only for the expression of rhythmicity but for determining its phase and frequency. Clearly, then, circadian oscillators are localizable. In many organisms these localized oscillators often exert primary control over a number of dissimilar functions. These results could reflect the existence of a "master clock" within each individual which, by virtue of distributed outputs, can impose rhythmicity on a variety of physiological and biochemical processes. Nevertheless, the simple scheme of a single oscillator sequentially triggering, in appropriate order, diverse processes is inconsistent with much of the evidence (described below). There is a growing conviction that many, if not all, circadian systems in multicellular organisms are composed of many circadian oscillators that provide for the temporal organization of the organism through mutual and hierarchical coupling (Pittendrigh 1974, Menaker 1974). Furthermore, it has been suggested that there are potentially significant functional consequences (Pittendrigh & Daan 1976a,b), as well as selective advantages (Pittendrigh 1976), of a multioscillator organization.

For convenience, the evidence for multiple oscillators within the nervous system can be separated into two categories: (*a*) evidence for bilateral redundant pacemakers in bilaterally symmetric organisms, and (*b*) evidence for multiple "nonredundant" oscillators.

Bilaterally Paired Circadian Pacemakers

The localization of the circadian pacemaker to paired structures in bilaterally symmetric animals raises the possibility that an autonomous pacemaker may reside in each half of the organism. This is without question the case in *Aplysia,* where it is possible to record from each of the two eyes from a single animal and demonstrate that each contains a competent circadian oscillator (Jacklet 1969). The evidence is also reasonably strong for a cockroach (Nishiitsutsuji-Uwo & Pittendrigh 1968, Page et al 1977), where it has been shown that destruction of either the right or left side of the putative oscillator site (optic lobes) leaves the rhythm intact, but bilateral destruction results in aperiodicity.

The demonstration of bilateral autonomous pacemakers raises two questions. First, are the two pacemakers, left and right, functionally equivalent, and, second, are they in some way coupled to form a compound pacemaker that formally behaves as though it were a single oscillator?

In *Aplysia,* where it has been possible to record from both eyes simultaneously, it appears that the rhythm of CAPs in both eyes is virtually identical in phase, period, and wave form (for example see Figure 1 in Jacklet 1974, Rothman & Strumwasser 1976). In the cockroach it has not been possible to measure the frequencies of both the right and left pacemakers in the same individual. Nevertheless, systematic measurements of the freerunning periods of a number of individuals with only one optic lobe intact have shown that, at least as measured by their *average* frequency, left and right pacemakers are not significantly different—both have periods of about 23.95 hr at 24.5°C. Firm conclusions are premature, but the data from these studies suggest that anatomically redundant pacemakers are, to a first approximation, also functionally redundant.

A priori it would seem reasonable to expect bilaterally paired oscillators to be mutually coupled—any frequency instability or differences in freerunning period would thus be shared between the two pacemakers. However, the question of mutual synchronization is clearly not trivial, as recent evidence from *Aplysia* demonstrates. Both Jacklet (1971) and Lickey et al (1976) have presented data that indicate that any interaction between the two ocular pacemakers, if present, must be very weak. The experiment that Jacklet performed involved capping one eye to exclude ambient light, then subjecting the animal to a phase shift in the light/dark (LD) cycle to which the *Aplysia* had been previously entrained. After one day in the new monocularly applied light cycle the eyes were removed from the animal, and the phase of the rhythms in both eyes assayed in constant darkness. Jacklet found that only the eye exposed to the new LD cycle had shifted, while the capped eye had remained in phase with the original, unshifted light cycle. Lickey et al (1976) have evidence that supports Jacklet's conclusion that the eyes are not strongly coupled. In their experiment, one eye of *Aplysia* maintained in constant darkness was locally illuminated by means of a small light sewn over the eye. The illuminated eye was exposed to 3–5 LD cycles phase-shifted with respect to previous binocularly applied light cycles. The results obtained from subsequent electrophysiological recording revealed that in four of the five *Aplysia,* the illuminated eye had entrained to the monocularly applied LD cycle, while in every case the nonilluminated eye gave no evidence of entrainment. The apparent lack of interaction between the eyes suggests that *Aplysia* may predominantly rely on the submission of each ocular pacemaker to entrainment by the external LD cycle to maintain the two oscillators in an appropriate phase relationship. More recently, Hudson (personal communication) has obtained evidence for weak coupling between the eyes. If *Aplysia* are allowed to freerun in constant darkness for several weeks, subsequent recording from the eyes indicates that the pacemakers have stabilized 180° out of phase. This suggests some interaction between the eyes.

In contrast, in the cockroach, where it is believed a competent circadian pacemaker resides in each of the optic lobes, the evidence suggests the bilaterally paired oscillators are mutually coupled (Page et al 1977). The hypothesis of mutual entrainment has been proposed to account for three separate facts:

1. The lack of any clear evidence that the activity rhythm dissociates into two components with different freerunning periods during extended freeruns or during spontaneous changes in period. This would be expected if each oscillator were independently free to vary its frequency.
2. Either of the compound eyes is competent to entrain *both* the ipsilateral and contralateral pacemakers.
3. Removal of one of the optic lobes consistently results in an increase in the freerunning period of the activity rhythm, whereas similar surgical disturbances that fail to interfere with the pacemaker in the lobe (e.g. optic nerve section) have no effect on period.

On the basis of these results, it has been suggested that the bilaterally paired pacemakers are mutually coupled and accelerate each other to form a compound pacemaker whose freerunning period is shorter than either constituent oscillator (Page et al 1977). No one of these observations demands as a unique explanation the hypothesis of mutual entrainment; however, this hypothesis is the simplest one that accounts for all the observations.

Multiple "Nonredundant" Pacemakers

Perhaps more intriguing than evidence for bilaterally paired pacemakers are results, now available from several species, that suggest that circadian systems may be composed of multiple and "nonredundant" oscillators, i.e. oscillators with distinct properties and functions. The presence of "driver" and "slave" oscillators has long been suggested on the basis of formal properties of rhythms. Pittendrigh (1960), for example, has provided convincing evidence that the eclosion rhythm of *Drosophila pseudoobscura* is composed of a light-sensitive driving oscillator that gates emergence via hierarchical entrainment of a second oscillator. In the same paper Pittendrigh also presented evidence (although not as compelling as that for *Drosophila*) that the circadian activity rhythm in the cockroach may be controlled by a similar system.

Most of the evidence from formal analysis that suggests the existence of multiple oscillators involves the observation of multiple frequencies in the rhythmic expression of one or more behaviors in a single individual. One striking example is the report by Aschoff (Aschoff 1969, Aschoff et al 1967) that in a significant number of human subjects the activity (sleep-wakefulness) cycle can freerun with a substantially different period than the rhythm in body temperature (Figure 2). It is difficult to escape the conclusion that two different oscillators are controlling these two rhythmic processes.

The clearest and most reproducible examples of multiple frequencies are those that have been referred to as "splitting" (Pittendrigh 1960, 1967), in which a single behavior (running wheel activity in every case thus far described) breaks up into two components that freerun with different frequencies. With appropriate lighting conditions, the behavior has been found to be reproducible in hamsters (Pittendrigh

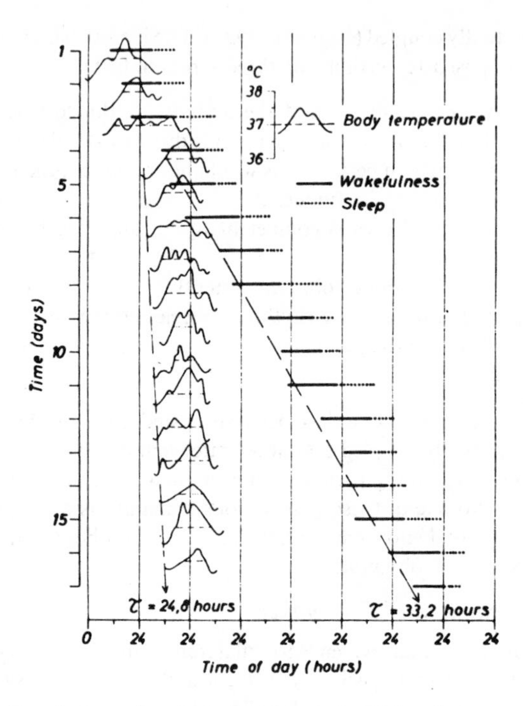

Figure 2 Circadian rhythms of rectal temperature and activity of a human subject enclosed in isolation without time cues. Consecutive periods are drawn below each other. These data provide unequivocal evidence for multiple oscillators within the human circadian system (from Aschoff 1969).

1960, 1967, 1974, Pittendrigh & Daan 1976a) and in the primate *Tupaia* (Hoffman 1971); the behavior has been seen occasionally in a number of other rodents (see Pittendrigh 1974). Generally, the split components freerun with different periods until they are approximately 180° out of phase, whereupon the rhythm stabilizes with a freerunning period different from that of the unsplit system. This condition may persist for several weeks. The activity may also undergo spontaneous refusion in which the two activity components reestablish a normal phase relationship. Splitting is generally interpreted as evidence that the activity is controlled by two self-sustaining oscillators that are mutually coupled and as a consequence tend to maintain one of two possible stable phase relationships (either in-phase or in anti-phase). Thus, the activity rhythms of these organisms are controlled by two oscillators with different characteristics; the analytical properties of the behavioral rhythm reflect not only the properties of the individual pacemakers but the nature of their interaction as well (Pittendrigh & Daan 1976b, Daan & Berde 1977).

Work on the pineal-parietal eye complex of lizards has yielded similar evidence for multiple oscillators, although the results are complex. Underwood (1977) has found that removal of the pineal and parietal eye of the iguanid lizard *Sceloporus olivaceus* freerunning in constant light can lead to one of three results: (*a*) apparent aperiodicity, (*b*) a large increase in freerunning period, or, (*c*) in one instance, splitting. The last case is particularly interesting. The two activity bands of the split rhythm had quite different freerunning periods and "crossed" several times over the course of about 100 cycles. The results clearly demonstrate that multiple oscillators are involved in control of the activity rhythm, although not enough data are yet available to assess the role of the pineal-parietal eye complex in maintaining the oscillators in a stable phase relationship.

There are a number of other physiological studies that support the notion that metazoan circadian systems are composed of multiple oscillators. Results obtained from the sparrow, *Passer domesticus,* present a particularly clear and interesting case. As we noted in the previous section, removal of the pineal of a bird freerunning in continual darkness immediately abolishes rhythmicity. However, when placed in an LD cycle, pinealectomized birds will exhibit a rhythm in perch-hopping behavior. Three facts strongly suggest that this rhythmicity results from the entrainment of one or more oscillators. First, the onset of activity exhibits a large positive-phase angle with respect to the onset of light (activity anticipates dawn); second, animals returned to constant darkness from an LD cycle only gradually become arrhythmic, often exhibiting as many as 10 days (cycles) of rhythmicity; and, third, transient rhythmicity is induced in pinealectomized birds given a single light pulse (Gaston & Menaker 1968, Gaston 1971). These observations—in conjunction with the results described above, which demonstrate that the pineal acts as the driving oscillator for the system—suggest that the pineal functions as a master clock that controls activity via entrainment of a secondary oscillator.

Recent work by Rence & Loher (1975) suggests that the organization of the cricket circadian system may be somewhat similar to the sparrow. Crickets with their optic lobes removed (which, like cockroaches, are arrhythmic in constant conditions) can exhibit a 24-hr stridulation rhythm when placed in a 24-hr temperature cycle of 10°C amplitude. Loher suggests, on the basis of this observation and the fact that "lobeless" crickets are aperiodic on a 30-hr temperature cycle, that a residual temperature-sensitive oscillator with a circadian period remains following optic lobe ablation. Unfortunately, evidence has not been presented that demonstrates that the rhythm persists for even a single cycle following return of these animals to constant conditions; thus, it may be premature to conclude that an oscillator is present.

Finally, there is evidence for multiple "nonredundant" pacemakers in *Aplysia.* Block & Lickey (1973, Lickey et al 1977) found that although removal of the ocular oscillators tends to disrupt the locomotor rhythm, some, but not all, eyeless *Aplysia* were capable of exhibiting freerunning activity. Although there appears to be a pacemaker outside of the eyes, it remains to be demonstrated whether the pacemakers in the eyes are in any way coupled to a residual pacemaker system that remains following eye removal.

We have just begun to gain an appreciation of the complexities of multioscillator circadian organization. By far the most striking observation at this time is the apparent variability in this organization among different organisms. In the case of bilaterally paired pacemakers, virtual extremes in coupling arrangements have already been observed. In the cockroach there is evidence for strong mutual coupling of pacemakers, whereas in *Aplysia* interocular interactions are weak, with phase relationships maintained primarily by access to a common *Zeitgeber.* Although any explanation of these differences may be premature, it is perhaps worth noting that *Aplysia californica* are diurnally active gastropods, occupying shallow coastal waters where it might be impossible to avoid daily light cycles. In contrast, cockroaches, which are nocturnal and often obscured from daylight, may go many days without receiving a light pulse, and thus require synchronization via internal mechanisms.

The fact that circadian organization involves multiple and "nonredundant" oscillators is now clear in several vertebrate classes, as well as in some invertebrates. Yet the evidence for this general conclusion is derived from a diversity of seemingly unrelated experimental designs and observations; for that reason it is very difficult to extract any general features of multioscillator organization from these data. Does the variability between species suggest fundamental differences in their circadian organization, or does it simply reflect the incompleteness of the localization and characterization of the components within the circadian systems of these organisms? It is clear that there are species differences in the anatomical organization of circadian systems. What is not clear is whether or not there are comparable differences in the functional properties of individual components within each system and in the coupling relationships between them.

Where internal synchronization between pacemakers has been demonstrated, questions still remain about the exact nature of these interactions. In the sparrow, the loss of sustained freerunning capacity following pinealectomy leaves open the question of whether residual pacemakers are incapable of sustained freerunning activity or whether the pineal oscillator maintains synchronization among a population of extrapineal pacemakers. It is also uncertain whether the pineal entrains the residual pacemaker(s) hierarchically or whether feedback on the pineal is present. In cockroaches where mutual entrainment of optic lobe pacemakers has been demonstrated, it is still unclear whether each pacemaker contributes equally in determining the coupled freerunning period or whether some dominance exists. Understanding the interactions among pacemakers in a multiple oscillator system provides one of the most challenging problems to the physiological study of circadian rhythms.

PHOTORECEPTORS IN THE CIRCADIAN SYSTEM

Although the analytical properties of the entrainment process are reasonably well understood for a few organisms (e.g. see Pittendrigh 1960, Pittendrigh & Minis 1964, Pittendrigh & Daan 1976a), very little is known about the mechanistic basis of entrainment (although see Eskin 1972, 1977), or about the photoreceptors that mediate it.

A number of attempts have been made to identify photoreceptors for entrainment, and this work has recently been extensively reviewed (*Photochemistry Photobiology,*

1976, Vol. 23). The results of these studies are complex. In a majority of organisms, extraretinal photoreceptors are sufficient for entrainment, although in mammals, cockroaches, and crickets the eyes are required. In a few cases (sparrows, lizards, and *Aplysia*), both retinal and extraretinal elements have been shown to contribute to entrainment.

The fact that most organisms utilize extraretinal photoreceptors in entrainment is a remarkable result whose functional significance remains obscure, although there has been some speculation. It has been suggested, for instance, that in holometabolous insects extraretinal pathways may be necessary to provide photic information to the pacemaker before and during metamorphosis, a time when organized photoreceptive structures are absent or not yet functional (Truman 1976, Pittendrigh 1976). In cases where both retinal and extraretinal elements are involved, it often appears that the photoreceptors perform nonredundant functions; multiple photoreceptors may serve to expand spectral sensitivity (Block et al 1974), lower the threshold for entrainment (Menaker 1968, Underwood 1973), or detect different stimulus parameters of the light cycle (Underwood & Menaker 1976). It has also been suggested that the ubiquity of extraretinal photoreceptors may reflect an association between the photoreceptor pigment and the pacemaker (Pittendrigh 1976). Unfortunately, with one possible exception (Adler 1971), extraretinal photoreceptors have not yet been identified, and as a consequence we know very little about their sensory physiology. The identification and physiological analysis of extraretinal photoreceptors critical for entrainment is likely to provide insights into this problem.

ACKNOWLEDGMENTS

We thank Dr. C. S. Pittendrigh for his critical reading of an early draft of the manuscript. One of us (T. L. P.) was supported by a stipend from a grant from the National Institute of Aging (AG00490-02) to C. S. Pittendrigh. G. D. B. was supported by a National Institutes of Health Fellowship (IF32NS5035-02).

Literature Cited

Adler, K. 1971. Pineal End Organ: role in exthaoptic entrainment of circadian locomotor rhythm in frogs. In *Biochronometry,* ed. M. Menaker, pp. 342–50. Nat. Acad. Sci., Wash., DC

Alleva, J. J., Waleski, M. V., Alleva, F. R. 1971. A biological clock controlling the estrous cycle of the hamster. *Endocrinology* 88:1368–79

Andrews, R. V. 1971. Circadian rhythms in adrenal organ cultures. *Gegenbaurs morphol. Jahrb., Leipzig* 117:89–98

Andrews, R. V., Folk, G. E. 1964. Circadian metabolic patterns in cultured hamster adrenal glands. *Comp. Biochem. Physiol.* 11:393–409

Aréchiga, H., Fuentes, B., Barrera, B. 1973. Circadian rhythm of responsiveness in the visual system of the crayfish. In *Neurobiology of Invertebrates,* ed. J. Solánki, pp. 403–21. Budapest: Akad. Kiadó

Aréchiga, H., Wiersma, C. A. G. 1969. Circadian rhythm of responsiveness in crayfish visual units. *J. Neurobiol.* 1:71–85

Aschoff, J. 1969. Desynchronization and resynchronization of human circadian rhythms. *Aerosp. Med.* 40:844–49

Aschoff, J., Gerecke, U., Wever, R. 1967. Desynchronization of human circadian rhythms. *Jpn. J. Physiol.* 17:450–57

Beiswanger, C. M., Jacklet, J. W. 1975. In vitro tests for a circadian rhythm in the electrical activity of a single neuron in *Aplysia californica. J. Comp. Physiol.* 103:19–37

Binkley, S., Kluth, E., Menaker, M. 1971. Pineal function in sparrows: circadian rhythms and body temperature. *Science* 174:311–14

Block, G. D. 1976. Evidence for an entrainable circadian oscillator in the abdominal ganglia of crayfish. *Neurosci. Abstr.* 2:315

Block, G. D., Hudson, D. J., Lickey, M. E. 1974. Extraocular photoreceptors can entrain the circadian oscillator in the eye of *Aplysia. J. Comp. Physiol.* 89:237–50

Block, G. D., Lickey, M. E. 1973. Extraocular photoreceptors and oscillators can control the circadian rhythm of behavioral activity in *Aplysia. J. Comp. Physiol.* 84:367–74

Block, M., Zucker, I. 1976. Circadian rhythms of rat locomotor activity after lesions of the midbrain raphé nuclei. *J. Comp. Physiol.* 109:235–47

Brady, J. 1969. How are insect circadian rhythms controlled? *Nature* 223: 781–84

Brady, J. 1971. The search for the insect clock. See Adler 1971, pp. 517–24

Bruce, V. G., Pittendrigh, C. S. 1956. Temperature independence in a unicellular "clock". *Proc. Natl. Acad. Sci. USA* 42:676–82

Daan, S., Berde, C. 1977. Two coupled oscillators: simulations of the circadian pacemaker in mammalian activity rhythms. *J. Theor. Biol.* In press

Elliott, J. A., Stetson, M. H., Menaker, M. 1972. Regulation of testis function in golden hamsters: A circadian clock measures photoperiodic time. *Science* 178:771–73

Eskin, A. 1972. Phase shifting a circadian rhythm in the eye of *Aplysia* by high potassium pulses. *J. Comp. Physiol.* 80:353–78

Eskin, A. 1977. Neurophysiological mechanisms involved in photoentrainment of the circadian rhythm from the *Aplysia* eye. *J. Neurobiol.* In press

Eskin, A., Harcombe, E. 1977. Eye of *Navanax:* optic activity, circadian rhythm, and morphology. *Comp. Biochem. Physiol.* In press

Fitzgerald, K. M., Zucker, I. 1976. Circadian organization of the estrous cycle of the golden hamster. *Proc. Natl. Acad. Sci. USA* 73:2923–27

Gaston, S. 1971. The influence of the pineal organ on the circadian activity rhythm in birds. See Adler 1971, pp. 541–48

Gaston, S., Menaker, M. 1968. Pineal function: the biological clock in the sparrow? *Science* 160:1125–27

Hastings, J. W., Sweeney, B. M. 1958. A persistent diurnal rhythm of luminescence in *Gonyaulax polyedra. Biol. Bull. Woods Hole* 115:440–58

Hoffman, K. 1971. Splitting of the circadian rhythm as a function of light intensity. See Adler 1971, pp. 134–46

Jacklet, J. W. 1969. Circadian rhythm of optic nerve impulses recorded in darkness from isolated eye of *Aplysia. Science* 164:562–63.

Jacklet, J. W. 1971. A circadian rhythm in optic nerve impulses from an isolated eye in darkness. See Adler 1971, pp. 35–62

Jacklet, J. W. 1974. The effects of constant light and light pulses on the circadian rhythm in the eye of *Aplysia. J. Comp. Physiol.* 90:33–45

Jacklet, J. W., Alvarez, R., Bernstein, B. 1972. Ultrastructure of the eye of *Aplysia. J. Ultrastruct. Res.* 38:246–61

Jacklet, J. W., Geronimo, J. 1971. Circadian rhythm: population of interacting neurons. *Science* 174:299–302

Lickey, M. E., Block, G. D., Hudson, D. J., Smith, J. T. 1976. Circadian oscillators and photoreceptors in the gastropod, *Aplysia. Photochem. Photobiol.* 23: 253–73

Lickey, M. E., Wozniak, J., Block, G. D., Hudson, D. J., Augter, G. 1977. The consequences of eye removal for the circadian rhythm of behavioral activity in *Aplysia. J. Comp. Physiol.* In press

Loher, W. 1972. Circadian control of stridulation in the cricket, *Teleogryllus commodus* Walker. *J. Comp. Physiol.* 79:173–90

Loher, W. 1974. Circadian control of spermatophore formation in the cricket *Teleogryllus commodus* Walker. *J. Insect Physiol.* 20:1155–72

Menaker, M. 1968. Light perception by extraretinal receptors in the brain of the sparrow. *Proc. 76th Annu. Conv. Am. Psychol. Assoc.* 3:299–300

Menaker, M. 1974. Aspects of the physiology of circadian rhythmicity in the vertebrate central nervous system. In *The Neurosciences: Third Study Program,* ed. F. O. Schmitt, F. G. Worden, pp. 479–89. Cambridge: MIT

Moore, R. Y. 1973. Retinohypothalamic projection in mammals: a comparative study. *Brain Res.* 49:403–9

Moore, R. Y. 1974. Visual pathways and the central neural control of diurnal rhythms. See Menaker 1974, pp. 537–42

Moore, R. Y., Eichler, V. B. 1972. Loss of a circadian adrenal corticosterone rhythm following suprachiasmatic lesions in the rat. *Brain Res.* 42:201–6

Moore, R. Y., Lenn, N. J. 1972. A retinohypothalamic projection in the rat. *J. Comp. Neurol.* 146:1–14

Nishiitsutsuji-Uwo, J., Pittendrigh, C. S. 1968. Central nervous control of circadian rhythmicity in the cockroach. III. The optic lobes, locus of the driving oscillation? *Zeit. Vel. Physiol.* 58:1–46

Page, T. L., Caldarola, P. C., Pittendrigh, C. S. 1977. Mutual entrainment of bilaterally distributed circadian pacemakers. *Proc. Natl. Acad. Sci. USA* 74:1277–81

Page, T. L., Larimer, J. L. 1975a. Neural control of circadian rhythmicity in the crayfish I. The locomotor activity rhythm. *J. Comp. Physiol.* 97:59–80

Page, T. L., Larimer, J. L. 1975b. Neural control of circadian rhythmicity in the crayfish II. The ERG amplitude rhythm. *J. Comp. Physiol.* 97:81–96

Pittendrigh, C. S. 1960. Circadian rhythms and the circadian organization of living systems. *Cold Spring Harbor Symp. Quant. Biol.* 25:159–84

Pittendrigh, C. S. 1967. Circadian rhythms, space research, and manned space flight. *Life Sci. Space Res.* 5:122–34

Pittendrigh, C. S. 1974. Circadian oscillations in cells and the circadian organization of muticellular systems. See Menaker 1974, pp. 437–58

Pittendrigh, C. S. 1976. Circadian clocks: What are they? In *The Molecular Basis of Circadian Rhythms,* ed. J. W. Hastings, H. Schweiger Berlin: Dahlem Konferenzen pp. 11–48

Pittendrigh, C. S., Daan, S. 1976a. A functional analysis of circadian pacemakers in nocturnal rodents. IV. Entrainment: pacemaker as clock. *J. Comp. Physiol.* 106:291–331

Pittendrigh, C. S., Daan, S. 1976b. A functional analysis of circadian pacemakers in nocturnal rodents. V. Pacemaker structure: a clock for all seasons. *J. Comp. Physiol.* 106:333–55

Pittendrigh, C. S., Minis, D. H. 1964. The entrainment of circadian oscillations by light and their role as photoperiodic clocks. *Am. Nat.* 98:26–94

Rence, B., Loher, W. 1975. Arrhythmically singing crickets: thermoperiodic reentrainment after bilobectomy. *Science* 190:385–87

Roberts, S. K. 1974. Circadian rhythms in cockroaches. Effects of optic lobe lesions. *J. Comp. Physiol.* 88:2-30

Roberts, S. K., Skopik, S. D., Driskill, R. J. 1971. Circadian rhythms in cockroaches: does brain hormone mediate the locomotor cycle? See Adler 1971, pp. 505–15

Rothman, B. S., Strumwasser, F. 1976. Phase shifting the circadian rhythm of neuronal activity in the isolated *Aplysia* eye with puromycin and cycloheximide. *J. Gen. Physiol.* 68:359–84

Rusak, B. 1977. The role of the suprachiasmatic nuclei in the generation of circadian rhythms in the golden hamster, *Mesocricetus Auratus. J. Comp. Physiol.* In press

Rusak, B., Morin, L. P. 1976. Testicular responses to photoperiod are blocked by lesions of the suprachiasmatic nuclei in golden hamsters. *Biol. Reprod.* 15: 366–74

Selverston, A., Mulloney, B. 1974. Synaptic and structural analysis of a small neural system. See Menaker 1974, pp. 389–95

Smith, J. T. 1976. *Long term in vitro recording from the optic nerve and neuron R15 of Aplysia.* PhD dissertation. Univ. Oregon, Eugene. 181 pp.

Sokolove, P. G. 1975. Localization of the cockroach optic lobe circadian pacemaker with microlesions. *Brain Res.* 87:13–21

Sokolove, P. G., Loher, W. 1975. Role of eyes, optic lobes, and pars intercerebralis in locomotory and stridulatory circadian rhythms of *Teleogryllus commodus. J. Insect Physiol.* 21:785–99

Stephan, F. K., Zucker, I. 1972. Circadian rhythms in drinking behavior and locomotor activity of rats are eliminated by hypothalamic lesions. *Proc. Natl. Acad. Sci. USA* 69:1583–86

Stetson, M. H., Watson-Whitmyre, M. 1976. Nucleus suprachiasmaticus: the biological clock in the hamster. *Science* 191:197–99

Strumwasser, F. 1965. The demonstration and manipulation of a circadian rhythm in a single neuron. In *Circadian Clocks,* ed. J. Aschoff, pp. 442–62. Amsterdam: North-Holland

Strumwasser, F. 1971. The cellular basis of behavior in *Aplysia. J. Psychiatr. Res.* 8:237–57

Strumwasser, F. 1974. Neuronal principles organizing periodic behaviors. See Menaker 1974, pp. 459–78

Truman, J. W. 1971. Circadian rhythms and physiology with special reference to neuroendocrine processes in insects. In *Proc. Int. Symp. Circadian Rhythmicity.*

Wageningen, Netherlands: Pudoc Press. pp. 111–35

Truman, J. W. 1972. Physiology of insect rhythms. II. The silk moth brain as the location of the biological clock controlling eclosion. *J. Comp. Physiol.* 81:99–114

Truman, J. W. 1976. Extraretinal photoreception in insects. *Photochem. Photobiol.* 23:215–25

Truman, J. W., Riddiford, L. M. 1970. Neuroendocrine control of ecdysis in silk moths. *Science* 167:1624–26

Underwood, H. 1973. Retinal and extraretinal photoreceptors mediate entrainment of the circadian locomotor rhythm in lizards. *J. Comp. Physiol.* 83:187–222

Underwood, H. 1977. Circadian organization in lizards: the role of the pineal organ. *Science* 195:587–89

Underwood, H., Menaker, M. 1976. Extraretinal photoreception in lizards. *Photochem. Photobiol.* 23:227–43

Zimmerman, N. H. 1976. *Organization within the circadian system of the house sparrow: hormonal coupling and the location of a circadian oscillator.* PhD dissertation. Univ. Texas, Austin. 156 pp.

Zimmerman, N. H., Menaker, M. 1975. Neural connections of sparrow pineal: role in circadian control of activity. *Science* 190:477–79

Ann. Rev. Neurosci. 1978. 1:35–59

NEURAL CONTROL OF BEHAVIOR

❖11502

David Bentley
Department of Zoology, University of California, Berkeley, California 94720

Masakazu Konishi
Division of Biology, California Institute of Technology, Pasadena, California 91125

A major goal of neurobiology is to provide an understanding of animal behavior in terms of the operation of the nervous system. Much of neurobiological research is therefore concerned at some level with the control of behavior. We have restricted our comments to attempts to relate a cellular level of neural analysis, involving neurons whose physiology, morphology, and connectivity are relatively well known, to fairly complete behavioral acts and to the overall organization of behavior. Consequently, examples will be drawn almost entirely from work on invertebrate nervous systems. Rather than review any single research area in detail, we hope to focus on the nature of behavior as seen from a cellular level, and to identify emerging principles and problems provided by this approach.

A neural explanation of behavior can be offered at several different levels. Behavior is expressed as a series of movements and postures. An initial step in neural analysis is to identify the sensory, neural, and effector elements involved, to describe their patterns of connection and activity, and to determine the properties and interactions that create those patterns. These are the problems that most research projects on "small systems" have addressed in the last few years, and very substantial progress has been made in understanding the neural basis of patterns of movements. Recent reviews can be found in Fentress (1976), Wiersma (1974), Kandel (1976), Usherwood & Newth (1975) and Hoyle (1977).

A higher level and more difficult question is that of how the complete behavioral repertoire available to an animal is represented in the nervous system, and how access to the various effector systems from this representation is mediated. Although behavior is observed as a continuous stream of effector events, many lines of evidence—including invariant repetitions of patterns, specific activation of appropriate response systems, and genetic fractionation of motor patterns—indicate that it is

0147-006X/78/0325-0035$01.00

discontinuously encoded in the nervous system. An approach to this problem involves investigation of how neural networks generating motor patterns are themselves interrelated and controlled. Is there a minimal unit of centrally initiated motor output? Are there separate or multiple control lines for motor pattern activation? Limited (but promising) information on these questions is becoming available.

Beyond the analysis of how movement patterns are generated and controlled is the further question of what determines the sequence of behavior patterns. This can be approached from the point of view of predicting oncoming behavior or from analysis of what determines when any particular behavior pattern will be activated. This area has primarily been the domain of psychologists and ethologists, but it is beginning to be invaded by neurobiologists. Some progress has been made in revealing which behavior patterns are potentially available to an animal at any particular time, which patterns are capable of immediate release, how sensory input is analyzed, and what precipitates the activation of a motor pattern.

NEURAL BASES OF PATTERNED EFFECTOR ACTIVATION

Rhythmical Movements

Most behavior patterns are composed of postures, rhythmically repeated movements, one-time or episodic movements, and nonrepetitive sequences of movements. Rhythmical movements have been favored experimental subjects for neurobiologists because they are readily related to normal behavior, are relatively easily elicited, and provide for many repetitions of experimental paradigms. Prior to 1960, it was generally thought that the pattern of such movements was determined by the interaction between sensory input and central elements, particularly motor neurons. The work of Wilson (1961, 1968a) established that the basic pattern of coordination can be generated instead by central neurons. By selective lesion and stimulation experiments, he showed in the locust flight system that phasic sensory input was not necessary for production of rhythmical, properly coordinated firing in flight motoneurons. This type of organization has now been observed in numerous behaviors of several phyla including insect walking (Pearson 1972), ventilation (Burrows 1974), and singing (Bentley 1969, Kutsch & Huber 1970); crustacean swimmeret beating (Wiersma & Ikeda 1964) and stomatogastric movements (Maynard 1972, Mulloney & Selverston 1974, Selverston & Mulloney 1974); arachnid ventilation (Wyse 1972); annelid swimming (Figure 1; Kristan, Stent & Ort 1974); and molluscan swimming (Willows 1967), buccal movements (Kater & Rowell 1973; Siegler, Mpitsos & Davis 1974), and gill ventilation (Kupfermann et al 1971). A growing variety of vertebrate movement patterns, including walking in tetrapods; swimming in fish, amphibians (Grillner 1975), and reptiles (Stein 1978); singing in frogs (Schmidt 1974); and patterned discharge of fish electric organs (Bennet 1970) can also be maintained without patterned sensory input. Central pattern generation appears to be a general feature of many, perhaps all rhythmical behaviors.

Following the discovery of central generators, attention was focused on the cellular mechanisms and interactions underlying such patterns. The first system described in detail was the stomatogastric ganglion of the lobster, a small ganglion of about 30 neurons that operates the stomach (Maynard 1972, Mulloney & Selverston 1974). It proved possible to identify each neuron, determine interactions by pair-wise intracellular recordings, and test hypotheses about the roles of cells by selective excitation or inhibition with intracellular current injection. The ganglion generates two motor patterns: the pyloric pattern is based upon a small set of

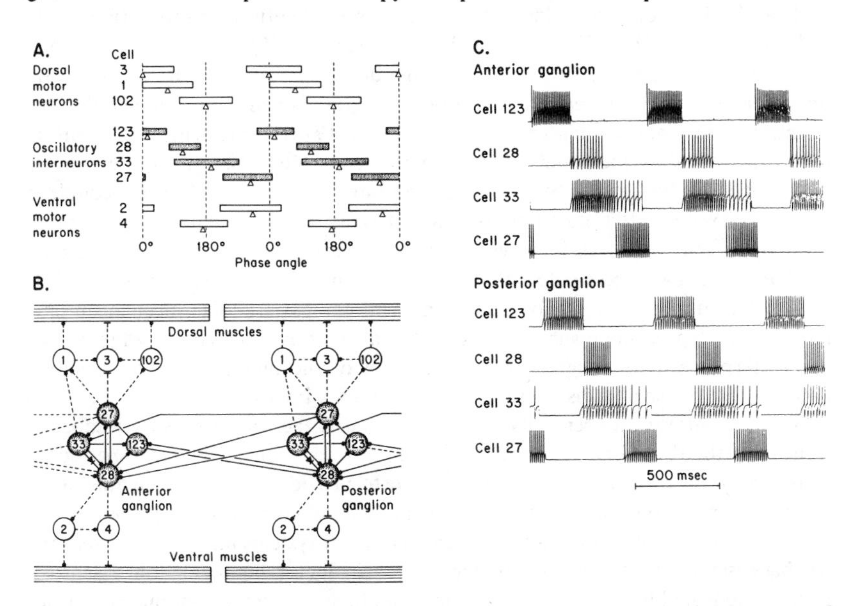

Figure 1 Neural network generating the swimming rhythm of the leech. Panel *A:* Phase diagram of the swimming activity cycles of motor neurons and interneurons. Each bar indicates the duration of the impulse burst of the cell; the triangle under the bar points to the burst midpoint, or middle spike. The burst midpoint of cell 3 has been arbitrarily assigned to phase angle 0°. Panel *B:* Summary circuit diagram of identified synaptic connections between interneurons, motor neurons, and longitudinal muscles responsible for the swimming rhythm. The dashed lines represent connections that were not included in the electronic analog circuit. Meaning of symbols: *T* joint = excitatory synapse; filled circle = inhibitory synapse; diode = rectifying electrical junction. Panel *C:* Impulse bursts generated by an electronic analog model of the circuit formed by the connections shown in solid lines in panel *B.* Each trace is the output of the "neuromime" element corresponding to one of the eight oscillatory interneurons. The model envisaged an anterior and a posterior ganglion separated by four segments, and hence a total interganglionic impulse conduction time of 80 msec. Lack of a sufficient number of neuromime elements prevented modeling of the interganglionic inhibitory inputs shown in panel *B* to reach the interneurons of the posterior ganglion from the rear. However, these inputs were simulated in the analog model by introducing into the posterior ganglion additional intraganglionic inhibitory connections from cell 27 to cells 33 and 28, from cell 33 to cell 28, and from cell 28 to cell 123. These additional connections incorporated a transmission delay of 160 msec, in order to mimic both a phase lag of 80 msec and a conduction time of 80 msec that would be appropriate for inputs reaching the posterior ganglion from a ganglion four segments to its rear (from Friesen, Poon & Stent 1976).

endogenously rhythmical cells; the discharge of other cells is keyed from this basic rhythm. In the gastric pattern, on the other hand, the rhythm is an emergent property of the network of cells; no single cell is inherently rhythmical, and the pattern arises primarily through reciprocally inhibitory connections. A second motor pattern generator, that underlying leech swimming, has recently been characterized at the level of connections and properties of identified neurons (Figure 1, Kristan, Stent & Ort 1974, Friesen, Poon & Stent 1976). It, too, is a network oscillator based upon inhibitory interactions, but for normal operation it depends upon extensive intersegmental connections. Progress is being made in cellular description of several other pattern generators (Stein 1978). Some depend upon unexpected mechanisms including nonspiking neurons (Mendelson 1971, Pearson & Fourtner 1975, Burrows & Siegler 1976) and burst-terminating electrical coupling (Getting & Willows 1973). Perhaps the most striking general conclusion to be drawn from these analyses of various systems is their lack of dependence on a unitary mechanism. It has become clear that there are numerous methods, some of which are fundamentally different, for generating a motor pattern. The mechanism employed may depend upon such factors as the range of output frequencies required, the pattern speed, the degree of specialization of the motoneurons for a single pattern, and the spatial distribution of the elements involved.

If motor patterns are centrally generated, what is the role of sensory feedback? Following the discovery of central coordination, reafference suffered a period of disregard during which it was often assumed to provide a smoothed maintenance of central excitation, especially in movements where wide variation in load is not encountered. Recent work in a variety of rhythmical systems indicates that sensory feedback sculptures the central pattern. Even in locust flight, identified sensory neurons activated by elevation of the wings modulate the motor output on a cycle-by-cycle basis (Wendler 1974) by inhibiting identified wing elevator motoneurons while monosynaptically exciting depressor motoneurons (Burrows 1975). In movements such as cockroach stepping (Pearson 1972, Pearson, Wong & Fourtner 1976) or snail buccal scraping (Kater & Rowell 1973), where varying resistances occur, sensory input delivers strong, intracycle positive feedback during the effector power stroke. It may also delay the return stroke until the load is relieved and therefore modulate the rhythm of the pattern as well. A very similar arrangement is found in mammalian walking (Grillner 1975). A more subtle use of sensory input is in correcting "errors" in centrally generated motor programs that inevitably arise through genetic or developmental misinformation or noise. Motor "errors" are unmasked when normal sensory input is unavailable, as in the well-known circular walking paths of persons lacking visual cues, or in roll-plane instability of locusts flying in the dark (Wilson 1968b).

An implication of the use of sensory feedback for correcting errors is that there is an expected feedback that the actual feedback should match. The central representation of this expected feedback can be said to constitute a sensory template (Konishi 1965) or sensory tape (Hoyle 1970). This is another possible mechanism for determining effector activation patterns since a rather unspecified motor output could be modulated until feedback matching the template was produced. Bird song is organized on this principle. All the song birds studied so far produce quite structureless songs when deafened before the onset of singing. When raised intact in sound isolation, the same species can produce either normal or much more structured songs than deaf birds (Konishi 1965, Mulligan 1966, Kroodsma 1977). These results indicate that the vocal motor system has little or no intrinsic capability to develop patterned output without auditory feedback and that the bird uses auditory feedback in order to compare its vocal output with an internal reference. As our understanding of the neural basis of acquired effector patterns improves, it seems likely that this type of organization will play an increasingly important role.

Episodic Movements

Some nonrhythmical types of movements can be termed episodic. These are single, continuous movements that can occur alone or can be coupled with other patterns. A neural analysis at the cellular level has now been obtained for several movements of this type, including *Aplysia* gill-withdrawal (Figure 2; Kupfermann, Carew & Kandel 1974) and inking (Carew & Kandel 1977), crayfish tail flips (Wiersma 1938, Zucker, Kennedy & Selverston 1971, Wine & Krasne 1972), and locust jumping (Heitler & Burrows 1977, Burrows & Rowell 1973). In general, the form of these motor patterns also seems to be determined by fixed central connections. A particularly clear example of this "hard-wired" representation of movements is seen in crayfish tail flips; stimulation of the anterior half of the animal can activate an interneuron (MG) whose discharge results in a backward movement, whereas stimulation of the abdomen excites a different interneuron (LG) that initiates a forward somersault with a half twist (Figure 7; Wine & Krasne 1972). Mittenthal & Wine (1973) showed that the different movements are caused by a different geometry assumed by the tail during the flip, and that this in turn is caused by differential, monosynaptic connectivity of the two interneurons with identified motoneurons in various tail segments. A feature of considerable interest in this superficially simple system is the discovery of extensive ancillary circuits that are activated concurrently with those generating the movement. These circuits act, primarily through selective inhibition, to control habituation and to prevent premature reactivation of the movement (Krasne, Wine & Kramer 1977, Wine & Mistick 1977).

A significant departure from this form of central representation of the movement is seen in the locust jump system. Here, sensory feedback plays an essential role in building up tension in the jumping muscles to the point where the behavior can occur (Heitler & Burrows 1977). When less "ballistic" behaviors are analyzed at a cellular level, it again seems likely that the importance of reafference in determining the form of movements will increase.

Episodic movement circuits that have been analyzed also differ from those underlying rhythmical movements in being remarkably plastic. They have provided the best-understood examples of use-dependent alteration of transmission across identified synapses (Kandel 1967, 1976, Rowell 1974, Krasne 1976, Matsumoto & Murphey 1977).

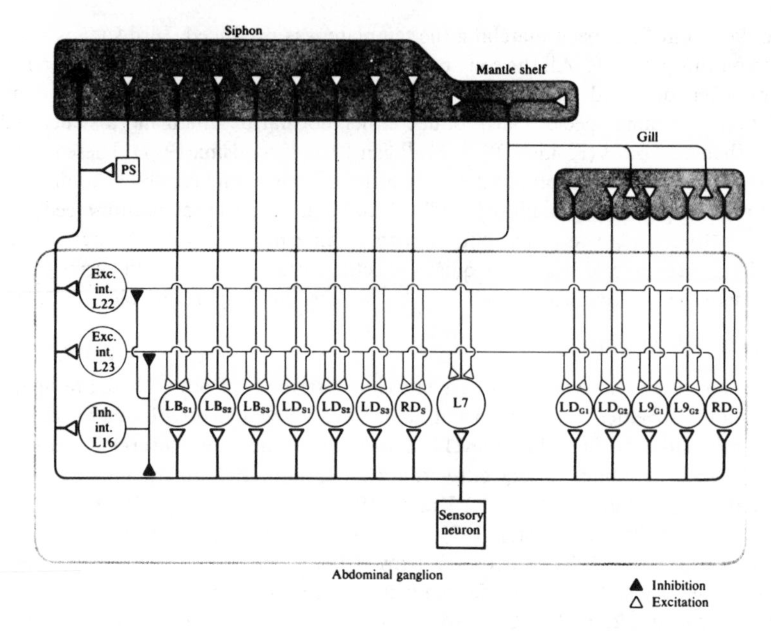

Figure 2 Schematic neural circuit of the defensive-withdrawal reflex of *Aplysia* that indicates the sensory, interneuronal, and motorneuronal components of the total reflex. The population of sensory neurons innervating the siphon skin consists of about 24 cells. These neurons have direct connections to the motor neurons as well as indirect connections via two excitatory and one inhibitory interneuron. The three interneurons and the 13 central motor cells in the abdominal ganglion are all unique, identified cells. The peripheral siphon motor cells (PS) are not uniquely identified. The same population of sensory neurons activates both the siphon and gill motor neurons. This accounts for the fact that the two acts of the reflex pattern occur simultaneously (from Kandel 1976).

Posture

Another element of most behaviors is posture. Postures are stable body geometries that usually reflect active contraction of muscles and a continuous pattern of neural discharge. Many postures have particular behavioral significance and are specifically assumed. Postures arise from a central command for joint or body position that is maintained by resistance reflexes. The receptor systems mediating these reflexes are among the best studied in invertebrate (Kuffler & Eyzaquirre 1955, Alexandrowicz 1967, Kennedy, Evoy, & Fields 1966) and vertebrate neurophysiology. Most act as length detectors in load-compensating servo-mechanisms, or as tension detectors. Recent work in this area with identified sensory and motoneurons has emphasized selective connections of receptors with specific motoneurons, intersegmental reflex gradients, uncoupling of resistance reflexes during active movement, and premotor control (Sokolove & Tatton 1975, Tatton & Sokolove 1975, Barnes, Spirito & Evoy 1972, Page & Sokolove 1972). The contribution of visual systems to maintenance of posture and orientation is also a heavily investigated and relatively well-understood area in both invertebrate (Horridge 1975, Poggio & Reichardt 1973, Kien 1977, Collett & Land 1975, Sandeman, Kien & Erber 1975) and vertebrate preparations, although the complete sequence of neurons and operations from receptor to motoneuron has not yet been worked out in any system. The origin of the central command underlying specific postures is discussed below.

Nonrepetitive Movements

More sophisticated behavior patterns consist of an ordered sequence of nonrepeated acts that can be rhythmical movements, episodic movements, or postures. Such sequences are seen, for example, in grooming, nest building, courtship, foraging, and prey capture. In these behaviors, the mechanism for establishing the order of subunits, and the interactions between central and peripheral elements for controlling subunit duration and switching are of primary interest. Neurobiological work on nonrepetitive sequences is just beginning. In grasshopper courtship songs composed of several different phrases and lasting many seconds, a long time on the cellular neurobiological time scale, Elsner (1974, 1975) has identified and characterized the output of the key motoneurons and has shown the basic patterns to be centrally generated. Moth eclosion from the pupa involves a sequence of several motor patterns generally lasting over an hour (Figure 4). Truman and his co-workers (Truman & Riddiford 1970, Truman & Sokolove 1972, Truman 1976) have begun a description of the motor patterns, have determined when the neural networks subserving them are constructed, and, most importantly, have shown that the entire block of patterns is activated by a hormone (discussed below). The most complex sequential behavior described at the level of identified neural output is that underlying cricket ecdysis (molting). This behavior lasts over four hours and involves practically all of the body musculature in a very complicated series of movement patterns (Carlson & Bentley 1977, Carlson 1977). The performance is generated by about 48 relatively independent motor programs that are responsible for the mechanical tasks. The motor programs are in turn controlled by at least two hierarchically organized layers of coordinating elements. There is a continual interplay of sensory input and central activity in effecting a smooth ecdysis, and it also seems likely that neuroendocrine factors are involved. Analysis of the neural mechanisms underlying such sequences will be an important step in understanding the organization of behavior. These pilot studies indicate that analysis on the cellular level will be possible.

HIGHER NEURAL CONTROL OF MOTOR PROGRAMS

The impression emerging from the study of simpler behavior patterns is that they arise from specific connections among particular ensembles of neurons. How are such ensembles controlled? Does the animal have, in any sense, a discontinuous and restricted set of motor programs that represent its complete behavioral repertoire? If so, how are such programs turned on? At what level in the nervous system does separate representation of different programs occur?

The most promising neural data on these questions comes from the study of "command" interneurons in crayfish. Command interneurons can be defined as interneurons whose activation alone suffices to elicit a recognizable fragment of behavior through excitation and/or inhibition of a constellation of motoneurons. Such cells were first proposed by Wiersma (1938) and Wiersma & Ikeda (1964) on the basis of strong, indirect evidence and were directly demonstrated by Kennedy, Evoy & Hanawalt (1966). Although most of the information on command interneurons is derived from crayfish, neurons with similar effects have been found in mollusks (Getting 1976, 1977, Willows 1968, Kupfermann, Carew & Kandel 1974, Kandel 1976), annelids (Weeks & Kristan, personal communication), and insects (Bentley 1977). Electrical stimulation of specific brain areas also releases various normal movements in vertebrates, e.g. species-specific behaviors in the domestic chicken (von Holst & von St. Paul 1960), vocal signals in the Japanese quail (Potash 1970) and the squirrel monkey (Jürgens & Ploog 1970), sonar cries in bats (Suga et al 1973), swimming (Grillner, Perret & Zangger 1976), and walking (Grillner 1975). The vertebrate results are hard to interpret without knowing which part of the neural circuit subserving the behavior is stimulated. The dissimilarity between the stimulus pattern and the response pattern, as well as other evidence, often suggests that systems that trigger and maintain the motor pattern are being driven. The nature of these systems is still unknown. For cat walking, a midbrain "command" pathway controlling spinal rhythm generators has been proposed (Grillner 1975, see also Stein 1978).

In crayfish, the important features of command interneuron organization are:

1. They control all types of motor acts including postures (Wiersma 1952, Atwood & Wiersma 1967, Kennedy, Evoy & Hanawalt 1966, Bowerman & Larimer 1974a), episodic movements (Wiersma 1938, Zucker, Kennedy & Selverston 1971, Wine & Krasne 1972), and rhythmic movements (Wiersma & Ikeda 1964, Davis & Kennedy 1972, Larimer & Kennedy 1969, Bowerman & Larimer, 1974b).
2. They control all levels of movement in terms of the amount of the body musculature involved. Some fibers control the entire body configuration (Wiersma 1952), whereas others influence only one or two segments (Larimer & Kennedy 1969). Moreover, single cells encode for complex movements that could also be encoded by the combined activation of other known command neurons (Larimer & Kennedy 1969).
3. There are related clusters of command neurons controlling the same general motor output. For example, at least 18 neurons specify abdominal geometry (Figure 3; Bowerman & Larimer 1974a). Where such clusters occur, detailed investigation has shown that the fibers encode for somewhat different outputs. There are five interneurons that release rhythmical swimmeret beating but each specifies a different range of frequencies (Davis & Kennedy 1972).
4. A single interneuron, when driven at increasing frequencies, can elicit a sequence of related behaviors. For example, a neuron that produced a defensive posture at low frequencies of firing initiated backward walking when driven at a higher frequency (Bowerman & Larimer 1974a).

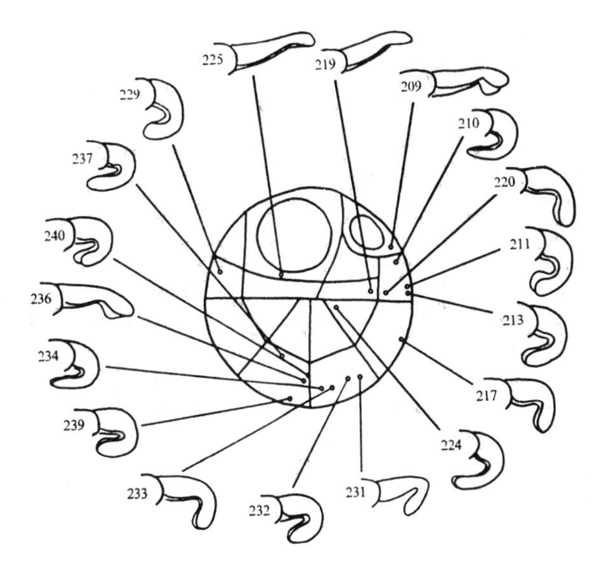

Figure 3 The number and range of abdominal geometries that were evoked by stimulation of tonic command fibres isolated from a single crayfish. Both the outline of the evoked abdominal geometry and the location of the responsible element within the connective are presented for each of the 18 tonic command fibres (from Bowerman & Larimer 1974a).

5. Compared to the range of movements known to be available to crayfish, the catalogue of command fibers is becoming quite extensive. A partial listing includes interneurons for escape tail flips, swimmeret beating, swimming, forward walking, backward walking, turning, feeding, grooming, heartbeat, stomach movements (Dando & Selverston 1972), and numerous specific cheliped, leg, abdominal, and uropod geometries.

Many important questions about command fibers are unanswered. Little information is available on how they are actually employed by the animal (Schrameck 1970, Larimer & Eggleston 1971). Is the animal unable to initiate a movement for which it does not have a command fiber, so that the full set of command fibers represents a discrete "alphabet" of possible movements? Are they used singly, or in combination? What determines the combinations that can be concurrently active? How general is this type of organization? Despite many uncertainties, it appears that in the crayfish there is a premotor level of extensive encoding of possible movements in the connections of specific interneurons. Analysis of this level of organization is the most promising opportunity so far of determining how the motor repertoire of an animal is represented in the nervous system.

DETERMINATION OF THE SEQUENCE OF BEHAVIORS

Each animal has a certain behavioral repertoire. At any one time, one behavior is selected from that repertoire for expression. Over the course of time, this series of selections specifies a behavior sequence. Ethologists, psychologists, and more recently, neurobiologists have tried to determine how this selection is made. The selection can be viewed as the outcome of a series of constraints limiting the expression of behavior. In order of decreasing time scale, these constraints are:

1. The receptor/neural/effector machinery subserving a particular behavior is often only present during a restricted portion of an animal's life.
2. If this machinery is present, it may be blocked so that it cannot be activated even by the appropriate stimulus; it is not "releasable."
3. If a behavior is releasable, it may be further gated by a circadian clock.
4. If the previous gates are "open," the presence of specific external, or internal stimuli may be necessary for triggering the behavior.
5. Only if all these factors are "go," will different behaviors compete for selection. In this situation, additional factors must come into play. These may include a priority hierarchy or stochastic choice.

Some information on each of the above constraints is now available from cellular analysis.

Network Assembly

Locust eggs are deposited under the soil surface, and the newly hatched vermiform larvae wriggle toward the surface. If, during this behavior, a container with the larvae is inverted, they will change course 180° and continue toward the new surface (Bernays 1971). However, once the larvae are allowed to reach the surface, this behavior pattern disappears within a very short time and cannot be reactivated by reburying. This is an extreme example of a general phenomenon. Many behavior patterns occur only during well-defined periods of an animal's life. Division of the life periods into vegetative, dispersal, and reproductive phases is usually clear, and many further subdivisions are common. Each of these has its own complement of behaviors. In longer-lived animals, seasonal changes in behaviors can be equally pronounced.

A typical example of this can be seen in orthopteran flight behavior. Crickets, for example, undergo a gradual development, produce wings at the last ecdysis, have a migratory flight period for about two weeks in some species, and then lose the flight response. Bentley & Hoy (1970) recorded from identified flight motor neurons at different life-stages while the animals were exposed in a wind tunnel to a flight-inducing stimulus. They found that early in post-embryonic life, no flight motor pattern could be elicited but that the first signs of flight coordination appeared well before adulthood. The number of units active and their burst rhythmicity, duration, and relative timing increasingly resembled the adult pattern until fully coordinated patterns could be produced by nymphs just before the molt to adulthood. A similar sequence of events has been described in the development of the locust flight pattern (Kutsch 1974, Altman & Tyrer 1974, Altman 1975), but it is shifted to a later stage in the life-cycle. Remarkably, the change from uncoordinated to normal relative timing of antagonistic units appears to occur within less than two days (Altman 1975). These studies indicate that although the neurons that will generate adult behavior patterns may be present at hatching, the connectivity and/or adult physiology necessary to produce a functioning network does not arise until much later in the life-cycle at a time just before the behavior is required.

Cricket song patterns are produced by neural assemblies in the thoracic ganglia (Bentley 1969, Huber 1975, Bentley 1977). Unlike the flight pattern, song pattern is not released in nymphs when they are exposed to an appropriate stimulus. However, when lesions are made in a particular part of the brain, a perfectly coordinated song pattern can be generated by mature nymphs (Bentley & Hoy 1970). This shows that the song neural network, like that underlying flight, is fully assembled by the end of nymphal life. However, it is somehow blocked or suppressed so that it is not activated by normal stimuli. This mechanism may be used to prevent premature activation of the pattern.

Similarly interesting developments are seen in silk moths (Figure 4). These animals undergo larval development and then pupate. In the pupa, development to the adult form occurs, and at the end of this period emergence (eclosion) to adulthood takes place. The act of eclosion is a relatively complex and long-lasting performance that is mediated by a centrally generated motor program (Truman & Sokolove 1972). Some elements of the program can be elicited early in the period of adult development in response to peeling the pupa (Truman 1976). Like the flight pattern and song pattern, the neural network underlying this behavior appears to be assembled shortly before it is actually employed in the behavioral repertoire. As the time of eclosion approaches, it becomes increasingly difficult to elicit the eclosion program by peeling the pupa, which suggests that this program is also being blocked or suppressed. This interpretation is supported by the observation that some adult behavior patterns can be released at this stage by the application of picrotoxin, which is thought to block inhibitory synapses in insects (Truman 1976).

What factors mediate the activation of assembled neural circuits? Additional insights have been provided by the analysis of silk-moth eclosion. At the normal time of eclosion, the motor program is released by a hormone (Truman & Riddiford 1970). Even in an isolated nervous system, application of the hormone will activate a coordinated eclosion program. The hormone must interact with neurons to change

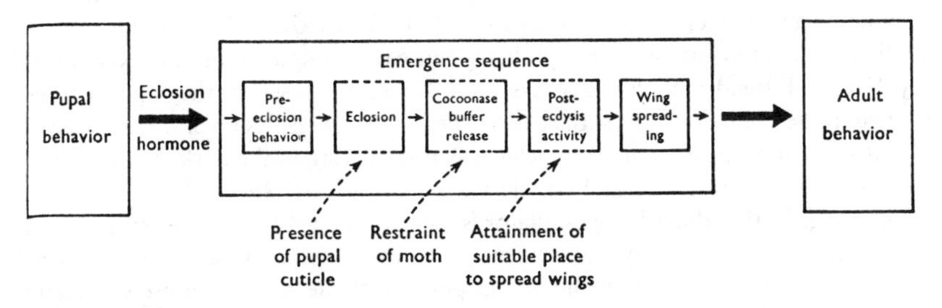

Figure 4 A schematic representation of the control of the behavior involved in the shedding of the pupal cuticle of the silkmoth and the escape from the cocoon. The emergence sequence comprises a centrally programmed series of five elements of behavior. The progression from one part to the next is effected by this central program, but the duration of certain parts can be influenced by environmental factors. The activation of this program is triggered by the eclosion hormone (from Truman 1971).

their properties, and a probable site of this interaction is suppression of the neurons that are blocking release of the program. Thus the program seems to be triggered under normal circumstances by disinhibition. Recently, it has been shown that cAMP mimics the action of the eclosion hormone and that the effect is enhanced by a substance (theophylline) that prevents the destruction of the cyclic nucleotides (Truman, Fallon & Wyatt 1976). Progress is also being made in identifying and characterizing neurons involved in the transformation from larval to adult behavior (Truman & Reiss 1976, Taylor & Truman 1974, Kammer & Rheuben 1976, Casaday & Camhi 1976). During this transformation, some new neurons arise, and after eclosion many larval neurons degenerate. Some identified larval neurons are modified into adult neurons by substantial reorganization of their arborizations. Hormones are well known to have powerful effects on the expression of vertebrate and invertebrate behavior. In the silk moth and certain opisthobranch mollusks (Davis, Mpitsos & Pinneo 1974b, Kupfermann 1967) it is now becoming possible to examine the link between hormone release, the activity of identified neurons, and the expression of behavior.

The conclusion from this work is that many behaviors are only transiently available to an animal and that availability is constrained on several levels including (*a*) construction of the appropriate neural machinery, (*b*) blockage of access to that machinery by inhibition until the appropriate time for release, (*c*) disinhibition of the system, which may involve direct release of the behavior or may render it triggerable by a suitable stimulus.

Circadian Control

On a finer time scale, many behaviors that are potentially releasable are organized by an internal clock. The clock mediates a circadian distribution of the times of expression of different behaviors. The complete sequence of events from clock mechanism to behavior is not known in any system (Block & Page 1978). Among invertebrates, clocks in mollusks, crustaceans, and insects are generally associated with the optical processing part of the nervous system. Only in *Aplysia* have the neurons and connections responsible for generating a circadian rhythm in the eye been determined (Jacklet 1969, Beiswanger & Jacklet 1975). The most intensively studied clocks of invertebrates are probably those of orthopteran insects (Nishiitsutsuji-Uwo & Pittendrigh 1968, Roberts 1974, Loher 1972) where they appear to be located in the brain optic lobes. In cockroaches and crickets, lesion of the outer layer of this structure causes animals to behave as if they were in the dark. More central lesions, however, result in arrhythmic behavior (but see Rence & Loher 1975), which suggests that the driving oscillator is located in one of the more central areas of the optic lobes. Recently, experiments with microlesioning techniques suggest that cell bodies of neurons in the lobula portion of the optic lobe are critical in generation of the rhythm (Roberts 1974, Sokolove 1975).

In orthoptera, the bulk of evidence indicates that the output of the clock is neural, not hormonal. This is in contrast to some vertebrates' clocks. In house sparrows, for example, Menaker & Zimmermann (1976) found that removal of the pineal body abolishes the free-running rhythm of locomotor activity, but that severance of the input and output fibers to and from the pineal does not affect the rhythm. By far the most remarkable finding is that the rhythm of one bird could be transferred to another by transplanting the pineal (M. Menaker, personal communication). Therefore, the output of this system appears to be chemically mediated.

In both vertebrates and invertebtates, multiple behavioral rhythms seem to be established by a single clock. In rats, lesion of the retinohypothalamic projection to the suprachiasmatic nuclei results in the loss of circadian locomotor, drinking, and sleep-wakefulness cycles (Stephan & Zucker 1972, Block & Zucker 1976, Moore & Klein 1974, Ibuka & Kawamura 1975). The same lesion also causes persistent estrous in the female rat (Barraclough et al 1964), which indicates that the same clock might provide a time base for the 4–5 day sexual cycle. In crickets locomotion, singing (calling song), and spermatophore production all exhibit a strong circadian rhythm (Sokolove & Loher 1975). All three behaviors exhibit the same free-running rhythm, and all lose rhythmicity following separation of the optic lobes from the brain suggesting, along with other evidence, that all three are driven off the same clock at different phase points (Figure 5).

Links between the clock output and motor-pattern-generating neurons have been difficult to trace. There is some evidence that another part of the insect brain, the

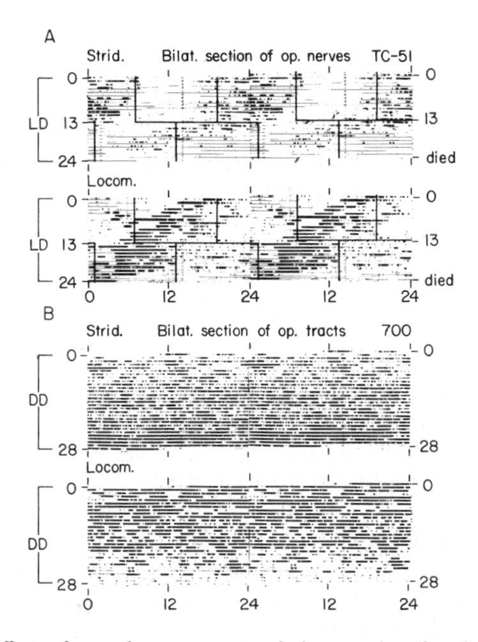

Figure 5 Similar effects of central nervous system lesions on circadian rhythms for locomotion and stridulation (calling song) in crickets. *A.* Bilateral section of nerves from the ommatidia to optic lobes results in a free-running rhythm of less than 24 hr for both behaviors. *B.* Bilateral section of the tract between the optic lobes and the protocerebrum results in aperiodic expression of both locomotion and stridulation. These results support the hypothesis that both behaviors are controlled by the same (bilaterally represented) driving oscillator, and that it is located in the optic lobes (from Sokolove & Loher 1975).

pars intercerebralis, acts as a secondary center connecting the clock output either neurally or hormonally to motor centers (Sokolove & Loher 1975). Lesions in this brain region can release steady singing (Huber 1960); fibers have been found that connect the head ganglia to the thoracic ganglia and that initiate normal calling song when stimulated (Bentley 1977), so that it may be possible to follow such neurons back to the clock. Explanation of the circadian system at a cellular level still seems far from realization but is clearly a critical step in understanding how the sequencing of behavior is organized.

Sensory Triggering

Of behavior patterns that are "releasable" at any particular time, many are triggered by a specific external and/or internal stimulus configuration. Ethologists have placed great emphasis on the role of "sign stimuli" in triggering behavior and on the neural "innate releasing mechanisms" that respond to the stimuli. In mammals, dramatic progress has been made in recent years in the analysis of sensory processing, especially in the visual system. In fact, this analysis is yielding considerable information about the basic organization of the brain. Despite this, it has been practically impossible to obtain direct, physiological tests of connection between cells and even more difficult to determine the motor consequences of sensory processing.

In invertebrates it has now been possible in several systems to follow sensory information analysis at a cellular level from the receptors to the initiation of motor acts. These systems vary considerably in the complexity of the stimuli that activate them. In *Aplysia,* for example, a mechanical stimulus activates sensory neurons that directly drive gill withdrawal motor neurons, and interneurons (Figure 2; Kupfermann, Carew & Kandel 1974). There is no pattern recognition and no threshold for a unitary motor response. A more powerful stimulus induces a more vigorous withdrawal. In crayfish, activation of the LG interneuron system induces a tail-flip escape response. This is a more complex situation because the response is an all-or-nothing act that engages the whole organism. The sensory system must evaluate the input and come to a clear decision about whether it is sufficiently "threatening" to warrant escape. The optimal stimulus features are a novel, mechanical disturbance of a certain minimum magnitude occurring in a short time span. The stimulus is filtered through a convergent hierarchy of interneurons whose connectivity and synaptic properties have been well characterized (Figure 7; Zucker, Kennedy, & Selverston 1971, Krasne, Wine & Kramer 1977). Temporal stimulus requirements are imposed by a rapidly antifacilitating synapse, and by the time course of summating postsynaptic potentials. Magnitude constraints arise from the extent and intensity of interneuron discharge necessary to bring a single cell, LG, to threshold. Firing of this cell signals the presence of the stimulus that the system is set up to detect.

The most complex sensory analyzer that is linked to a motor output, and whose neurobiological basis is relatively well understood, is the locust movement detector system (Rowell 1971, O'Shea, Rowell & Williams 1974, O'Shea & Williams 1974, Burrows & Rowell 1973, Heitler & Burrows 1977, Rowell & O'Shea 1976, Rowell, O'Shea & Williams 1977). This system can initiate jumping and responds optimally to novel, moving objects of a relatively small size anywhere in the visual field. It is more complex than the magnitude-sensitive systems previously mentioned because it requires a pattern of activity in an array of similar receptors; that is, some receptors must be active and some receptors must be not active. Moreover the analyzer must work for stimuli located anywhere in the visual field.

The sensory analyzing system is located in the optic lobe of the brain (Figure 6). Detection of the appropriate stimulus configuration is expressed as the discharge of a single, identified interneuron (LGMD). This neuron sends its axon into the brain, where it drives another identified neuron (DCMD) through a powerful electrical synapse. DCMD sends its axon to the ganglion of the thoracic segment that bears the jumping legs, and monosynaptically excites and inhibits a network of identified motoneurons that generate the jump motor program. In the optic lobe, the stimulus filtering is achieved by a complex neural network involving (*a*) a rapidly decrementing (anti-facilitating) synapse, (*b*) lateral inhibition of sustained wide-field responses, (*c*) feed-forward inhibition of transient wide-field responses, and (*d*) summation of "on" and "off" responses. Although the most peripheral neural elements in this system are still unidentified, the properties and connections responsible for the specific feature detection appear to be characterized.

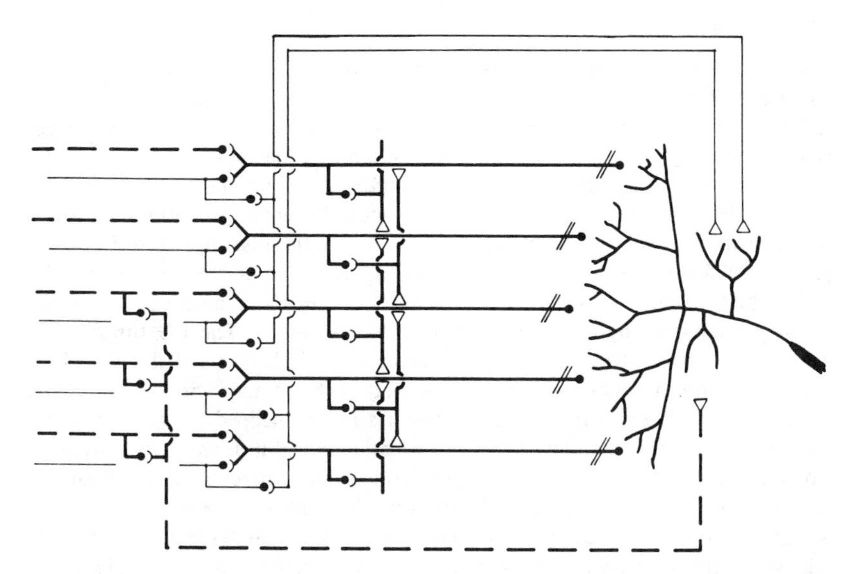

Figure 6 The sensory feature analyzing neural arrangement in the optic lobe of the locust that controls the Lobula Giant Movement Detector (LGMD) interneuron. Separate "on" and "off" neural arrays that scan the visual field converge on "on/off" neurons in the medulla. The "on/off" neurons converge on the dendritic fan of LGMD where they have a rapidly habituating synapse. A lateral inhibition network between the "on/off" neurons selects for small-field stimuli. Further, the "on" and "off" fibers also stimulate separate sets of wide-field medullar interneurons that, if activated, inhibit the LGMD. Consequently only novel, small-field stimuli can optimally drive LGMD. LGMD drives DCMD, a brain interneuron that excites the Fast Extensor Tibia motoneuron (from Rowell, O'Shea, & Williams 1977).

These two sensory analyzers, and others that are known in less detail, have several basic features in common. They are hierarchically organized. The neurons are arranged in a series of arrays of cells with similar connections and properties. The information transfer from array to array is based upon combination logic (*and, or, & and not*) that results in magnitude, generalization, or specificity requirements for discharge of the postsynaptic cells. This basic arrangement is elaborated by lateral inhibition, feed-forward, feed-back, and by synapses with special properties, particularly rapid antifacilitation. In both these feature-detecting systems, the information ultimately converges on a single cell so that the firing of this cell signals that the feature is present. Therefore, these sensory analysers can be viewed as the converse of a motor pattern generator. In the latter, a single, tonically firing command cell activates a neural assembly to elaborate a complex pattern of discharge in many output lines. In the former, a complex pattern of discharge in many input lines is filtered through a neural assembly to ultimately activate a single output neuron. It is clear that the early stages of this organization have much in common with information processing in the better-understood mammalian sensory systems; a question of considerable import is whether recognition in these systems is also based upon convergence on single cells.

Decision and Choice

Once feature-detecting neurons have been activated, how is the decision to mobilize a particular behavior pattern actually taken? The simplest case is in the crayfish LG tail flip. Here, the feature-detecting cell upon which the sensory analysis converges is the command neuron from which motor information diverges. This may be because the feature in question is so important for the animal that its presence alone generally warrants activation of the motor program. Recent investigations, however, reveal a variety of decision-weighing subtleties even in this system (Figure 7; Krasne & Wine 1975, Krasne et al 1977). Depending upon the circumstances in which the animal finds itself, the command neuron threshold to mechanical stimuli can be strongly modulated by other parts of the nervous system. The LG threshold is considerably elevated if the animal is out of water, where a tail flip would be relatively ineffective in producing locomotion, or if the animal is firmly held by the cephalothorax. Alternatively, if the animal has lost a cheliped (claw), it is much more likely to flip its tail and escape in a situation where it might previously have stood its ground. The neural elements mediating these changes are not well known but they do involve descending inhibition from the brain and the use of alternative nongiant interneuron pathways for effecting the response. Therefore, even this relatively elementary decision-making pathway is tempered by a variety of extrinsic factors.

What of less urgent response systems where the feature-detecting neuron is a different cell from the command neuron? In the locust, activation of LGMD alone often does not result in a jump. Since the presence of other stimuli, such as acoustical or substrate vibration, increases the probability of a jump, several different feature-indicating neurons may cooperate to elicit a response. In the thorax, mobilization of the motor program is accomplished in three stages; there is no single neuron whose activation alone constitutes a decision to jump (Heitler & Burrows 1977). A

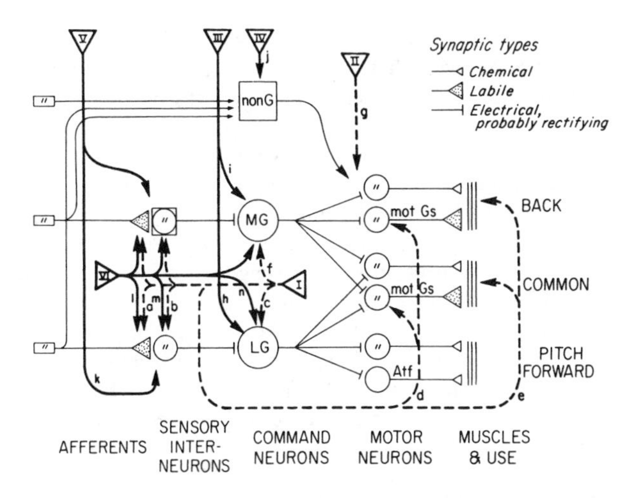

Figure 7 Inhibitory pathways (bold, solid lines) that influence the escape decision process in crayfish. Two behavioral outputs (backward or forward movements) are controlled by three command systems (LG, Lateral Giant; MG, Medial Giant; non G, nongiant interneuron) that operate via a variety of identified sensory, interneuron, and motor elements. Several extrinsic pathways have been detected that modulate these systems. For example, III originates in the brain and most rostral thoracic ganglia and acts directly upon LG. These pathways are discussed in detail in Krasne, Wine & Kramer (1977), from which this figure is taken.

variety of tactile stimuli can activate the first stage, flexion of the tibiae about the femora. This flexion may be followed by the second stage, a co-contraction of extensor and flexor tibial muscles, mediated by reafference pathways. Finally, this co-contraction either may die away or there may be a sudden relaxation of the flexors resulting in a jump. The second stage, co-contraction, is initiated by the first extensor spike, and the third stage, sudden flexor relaxation, is produced by an unidentified inhibitory input. Whether a stimulus is successful in causing a jump seems to depend on other stimuli present, on the central state of arousal, and perhaps on competing motor programs. Thus this system is more a decision network with several "gates"; failure to proceed at any of several points will abort the motor program. To what degree this system is mutually exclusive with single decision neurons, and which arrangement is the more prevalent, will be important points for future research.

Animals rarely attempt to execute two mutually exclusive behavior patterns simultaneously. Therefore, there must be mechanisms in the nervous system for choosing between such acts when both could potentially occur. Choice has been a major topic in behavior, and the implications of two models, hierarchies and stochastic selection between equivalent modes, have been discussed at length (see below). Little information is available from cellular analysis, although there are some promising leads (Camhi 1977). Rowell (1964, 1969) described an insect grooming reflex and showed that expression of this motor pattern is inhibited by a variety of sources in the central nervous system. By cutting a single tract (the meso-metathoracic connective), for example, the responsiveness of the pattern could be raised from activation by 13% of stimuli to 84%. This suggests that grooming may be a low priority behavior that is inhibited by practically any other ongoing behavior, and that a network of such inhibitory links could establish behavior priorities.

The most detailed analysis of priorities between different behaviors in an animal amenable to single-cell analysis has been carried out in a predatory marine mollusk, *Pleurobranchaea.* Escape swimming, feeding, egg laying, righting, mating, and local withdrawal have been examined (Davis & Mpitsos 1971, Davis, Mpitsos & Pinneo 1974a,b). Certain of these behaviors are strongly hierarchically arranged, and others have essential equivalence. The mechanism of interaction has been shown in one case to be hormonal suppression (Davis, Mpitsos & Pinneo 1974b). Considerable progress has been made in identifying and characterizing neurons involved in these motor programs (Siegler, Mpitsos & Davis 1974, W. J. Davis, personal communication) so that it may be possible to directly examine neuronal pathways that mediate behavioral choice.

FUTURE DIRECTIONS IN CELLULAR APPROACH

In the last two decades such major concepts as that of central coordination guided the progress of integrative neurophysiology. What are the prevailing ideas leading the study of the cellular basis of behavior today? The current trends in the cellular analysis of behavior strongly indicate that more effort will go into the identification of neuronal circuits subserving natural behavior and that attempts will be made to find cellular explanations for more complex and dynamic aspects of behavior. The ultimate goal of this endeavor is to explain the behavior of a whole animal in terms of neuronal interaction.

To the student of small systems a chart showing the connections among identifiable neurons, the time course of their discharge or inhibition during behavior, and how synaptic interactions among them result in the generation of the final motor pattern constitutes a satisfactory explanation of behavior. It should be obvious that this approach will become increasingly difficult as it tackles behaviors involving many neurons and more complex connectivity. To prepare a chart of the above sort for, say, a mere 1000 neurons with complex connectivity is not only difficult but also quite useless in understanding behavior, unless the data are reduced to manageable and meaningful forms by some rules. What are the rules? How can arbitrary rules be avoided? They must have direct or indirect relationships to the operating rules of the nervous system.

Behavior is the product of neural processes that must obey the rules of the nervous system. Therefore, the analysis of behavior should help discover and define them. Viewed from another perspective, the nervous system must obey the rules of behavior because it is made to serve the animal. For example, the diagonal sequence of stepping in tetrapod walking is due to the need to keep balance on three legs. The neural system subserving walking must incorporate the rules of stepping in its design. Walking involves many neurons, but the modes of their interaction can be organized into a simple scheme based on the rules of coordination within each limb and between different limbs. In general, behavioral acts that are organized on simple rules such as fixed dominant-subordinate relationships or sequences allow the establishment of a direct link between behavioral and neural rules. Promising starts have now been made in this type of analysis.

Ethologists have long sought for the rules governing the occurrence of identifiable movements during complex behavior such as courtship, grooming, and singing. The behavior of a whole animal in its natural environment is perhaps governed by complex rules of priority and equivalence. Some of them are invariant and others change according to various internal and external factors. The concept of hierar-

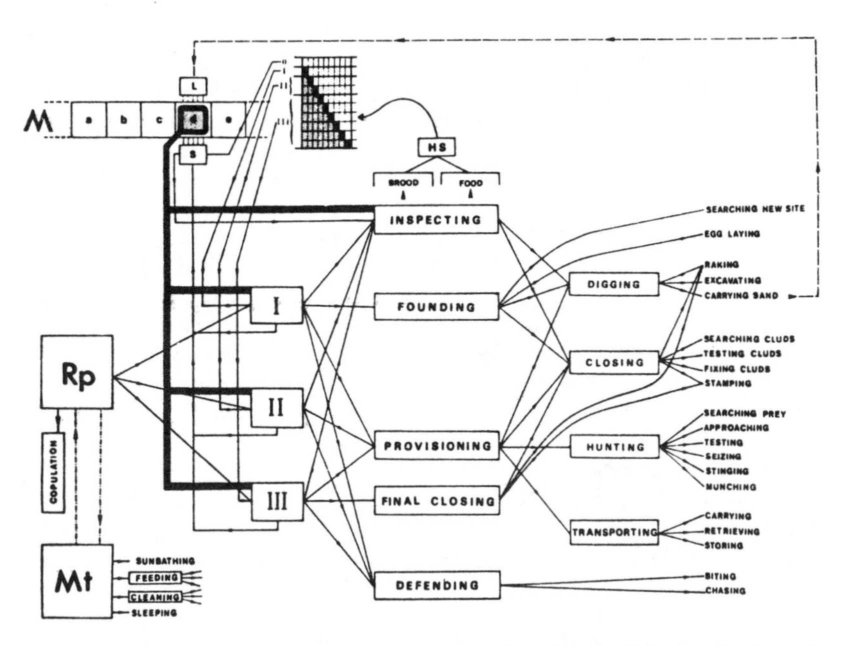

Figure 8 Diagram of the hierarchical organization of provisioning behaviour in the wasp, *Ammophila adriaansei.* The fixed action patterns used for the various aspects of provisioning behavior (right column) are controlled by four sub-systems that are themselves under the control of sub-systems of higher order (central column). Of these sub-systems "founding" is exclusively controlled by the system for phase I, "final closing" by that of phase III; the other systems can be activated in each of the three phases. The phases themselves are controlled by a superimposed "reproductive system" (Rp), which itself competes with a "maintenance system" (Mt). Each phase begins with nest inspection; if during that visit a larva is found, its size and the amount of food still available determine via a heterogeneous summation unit (HS) which phase will occur, i.e. the amount of caterpillars to be brought in; the black blocks in the inserted grid represent the caterpillars stored during the three phases. If no larva is present (as well as when a phase is completed) a scanning mechanism (S) searches the memory (M) and focuses the attention of the wasp on the oldest yet uncompleted nest, which will then be inspected. If no such nest is available, a new one is founded (phase I). In the memory (M), *a* to *e* represent the nests already existing; the situation of a new nest is learned (unit L) during digging. During a phase, attention is focused on one nest only (heavy bars) (from Baerends 1976).

chical organization of behavior has held a strong appeal for ethologists (Figure 8). As Dawkins (1976) points out, the concept did not enjoy continual support among ethologists mainly because it got associated with the idea of unitary drives, which became unacceptable.

The recent view in this field has been to regard behavioral events as stochastic processes. Many studies have measured the probability of transition from one behavior to another and tried to discover trends in the contingency table so produced. Just as tabulation of the transition probability of discharge of 1000 neurons would be confusing and unwieldy, the same would apply for a behavioral contingency table. Dawkins (1976) discusses how hierarchical rules help organize a large table into a simpler and comprehensive form; he also introduces an interesting way of looking at behavioral hierarchy based on the idea of decision. If behavioral acts occur in a fixed sequence, the decision to perform the last of the series can be said to have been made when the first act was selected. Using this definition, a hierarchy of decisions can be envisaged: Animals initially make general decisions such as whether to move and then make progressively narrower decisions such as which part of the body to involve. Behavioral events that involve some elements of stochastic processes will be one of the greatest challenges for future integrative neurophysiology.

There are reasons to believe that nervous systems are designed on modular and hierarchical principles (Szentagothai & Arbib 1975). This must be reflected in the hierarchical organization of behavior. Tinbergen (1951) made an early attempt to correlate his model of behavioral hierarchy with Weiss' scheme of hierarchical organization in the nervous system (Weiss 1941). The study of small systems would seem to predict that the modular and hierarchical organization of behavior has a structural and functional representation in the nervous system.

There is little doubt that the cellular analysis of behavior will make some of our dreams come true. Many of the design features of the nervous system can be uncovered only by recording from identifiable neurons. However, there are also features that may be revealed only by behavioral analysis. An excellent case in point is the demonstration of binocular global depth perception in man with random-dot stereograms (Julesz 1971). It is difficult to imagine how this could have been found with either neurophysiological or anatomical methods.

Literature Cited

Alexandrowicz, J. S. 1967. Receptor organs in the thoracic and abdominal muscles of crustacea. *Biol. Rev. Cambridge Philos. Soc.* 42:288–326

Altman, J. S. 1975. Changes in flight motor pattern during development of Australian plague locust, *Chortoicetes terminifera. J. Comp. Physiol.* 97:127–42

Altman, J. S., Tyrer, N. M. 1974. Insect flight as a system for the study of the development of neuronal connections. In *Experimental Analysis of Insect Behaviour*, ed. L. B. Browne, pp. 159–79. New York: Springer. 366 pp.

Atwood, H. L., Wiersma, C. A. G. 1967. Command interneurons in the crayfish central nervous system. *J. Exp. Biol.* 46:249–61

Baerends, G. P. 1976. The functional organization of behaviour. *Anim. Behav.* 24:726–38

Barnes, W. J. P., Spirito, C. P., Evoy, W. H. 1972. Nervous control of walking in the crab, *Cardisoma guanhumi.* II. Role of resistance reflexes in walking. *Z. Vgl. Physiol.* 76:16–31

Barraclough, C. A., Yrarranzaval, S., Hatton, R. 1964. A possible hypothalamic

site of action of progesterone in the facilitation of ovulation in the rat. *Endocrinology* 75:838–45
Beiswanger, C. M., Jacklet, J. W. 1975. In vitro tests for a circadian rhythm in the electrical activity of a single neuron in *Aplysia californica. J. Comp. Physiol.* 103:21–37.
Bennett, M. V. L. 1970. Comparative physiology: electric organs. *Ann. Rev. Physiol.* 32:471–528
Bentley, D. 1969. Intracellular activity in cricket neurons during generation of song patterns. *Z. Vgl. Physiol.* 62: 267–83
Bentley, D. 1977. Control of cricket song patterns by descending interneurons. *J. Comp. Physiol.* 116:19–38
Bentley, D., Hoy, R. R. 1970. Post-embryonic development of adult motor patterns in crickets: a neural analysis. *Science* 170:1409–11
Bernays, E. A. 1971. The vermiform larva of *Schistocerca gregaria* (Forskal): form and activity (Insecta: orthoptera). *Z. Morphol. Tiere* 70:183–200
Block, G. D., Page, T. L. 1978. Circadian pacemakers in the nervous system. *Ann. Rev. Neurosci.* 1:19–34
Block, M., Zucker, I. 1976. Circadian rhythms of rat locomotor activity after lesions of the midbrain raphe nuclei. *J. Comp. Physiol.* 109:235–47
Bowerman, R. F., Larimer, J. L. 1974a. Command fibers in the circumesophageal connectives of crayfish. I. Tonic fibers. *J. Exp. Biol.* 60:95–117
Bowerman, R. F., Larimer, J. L. 1974b. Command fibers in the circumesophageal connectives of crayfish. II. Tonic fibers *J. Exp. Biol.* 60:119–34
Burrows, M. 1974. Modes of activation of motoneurons controlling ventilatory movements of the locust abdomen. *Philos. Trans. R. Soc. London Ser. B* 269:29–48
Burrows, M. 1975. Monosynaptic connections between wing stretch receptors and flight motoneurons of the locust. *J. Exp. Biol.* 62:189–200
Burrows, M., Rowell, C. H. F. 1973. Connections between descending visual interneurons and metathoracic motoneurons in the locust. *J. Comp. Physiol.* 85: 221–34
Burrows, M., Siegler, M. V. S. 1976. Transmission without spikes between locust interneurones and motoneurones. *Nature* 262:222–24
Camhi, J. 1977. Behavioral switching in cockroaches—transformations of tactile reflexes during righting behavior. *J. Comp. Physiol.* 113:283–302
Carew, T. J., Kandel, E. R. 1977. Inking in *Aplysia californica.* I. Neural circuit of an all-or-none behavioral response. *J. Neurophysiol.* 40:692–707
Carlson, J. R. 1977. The imaginal ecdysis of the cricket (*Teleogryllus oceanicus*). II. The roles of identified motor units. *J. Comp. Physiol.* 115:319–36
Carlson, J. R., Bentley, D. 1977. Ecdysis: neural orchestration of a complex behavioral performance. *Science* 195: 1006–8
Casaday, G. B., Camhi, J. M. 1976. Metamorphosis of flight motor neurons in moth Manduca-sexta. *J. Comp. Physiol.* 112:143–58
Collett, T. S., Land, M. F. 1975. Visual spatial memory in a hoverfly. *J. Comp. Physiol.* 100:59–84
Dando, M. R., Selverston, A. I. 1972. Command fibres from the supra-oesophageal ganglion to the stomatogastric ganglion in *Panulirus argus. J. Comp. Physiol.* 78:138–75
Davis, W. J., Kennedy, D. 1972. Command interneurons controlling swimmeret movements in the lobster. I. Types of effects on motoneurons. *J. Neurophysiol.* 35:1–12
Davis, W. J., Mpitsos, G. J. 1971. Behavioral choice and habituation in the marine mollusk *Pleurobranchaea californica. Z. Vgl. Physiol.* 75:207–32
Davis, W. J., Mpitsos, G. J., Pinneo, J. M. 1974a. The behavioral hierarchy of the mollusk *Pleurobranchaea.* I. The dominant position of the feeding behavior. *J. Comp. Physiol.* 90:207–24
Davis, W. J., Mpitsos, G. J., Pinneo, J. M. 1974b. The behavioral hierarchy of the mollusk *Pleurobranchaea.* II. Hormonal suppression of feeding associated with egg-laying. *J. Comp. Physiol.* 90: 225–43
Dawkins, R. 1976. Hierarchical organization: a candidate principle for ethology. In *Growing Points in Ethology,* ed. P. P. G. Bateson, R. H. Hinde, pp. 7–54. London: Cambridge
Elsner, N. 1974. Neuroethology of sound production in Gomphocerine grasshoppers (Orthoptera: Acrididae). I. Song patterns and stridulatory movements. *J. Comp. Physiol.* 88:67–102
Elsner, N. 1975. Neuroethology of sound production in gomphocerine grasshoppers (orthoptera–acrididae). I. Neuromuscular activity underlying stridulation. *J. Comp. Physiol.* 97:291–322

Fentress, J. C. 1976. *Simpler Networks and Behavior.* Sunderland, Mass: Sinauer. 403 pp.

Friesen, W. O., Poon, M., Stent, G. S. 1976. Oscillatory neuronal circuit generating a locomotory rhythm. *Proc. Natl. Acad. Sci. USA* 73:3734–38

Getting, P. A. 1976. Afferent neurons mediating escape swimming of the marine mollusc, *Tritonia. J. Comp. Physiol.* 110:271–86

Getting, P. A. 1977. Neuronal organization of escape swimming in *Tritonia. J. Comp. Physiol.* In press

Getting, P. A., Willows, A. O. D. 1973. Burst formation in electrically coupled neurons. *Brain Res.* 63:424–29

Grillner, S. 1975. Locomotion in vertebrates: central mechanisms and reflex interaction. *Physiol. Rev.* 55:247–306

Grillner, S., Perret, C., Zangger, P. 1976. Central generation of locomotion in the spinal dogfish. *Brain Res.* 109:255–69

Heitler, W. J., Burrows, M. 1977. The locust jump. *J. Exp. Biol.* 66:203–41

Horridge, G. A. 1975. *The Compound Eye and Vision of Insects.* Oxford: Clarendon Press. 595 pp.

Hoyle, G. 1970. Cellular mechanisms underlying behavior. *Adv. Insect Physiol.* 7:349–444

Hoyle, G., ed. 1978. *Identified Neurons and Behavior of Arthropods.* New York: Plenum. In press

Huber, F. 1960. Untersuchungen über die Funktion des Zentralnervensystems und insbesondere des Gehirns bei der Fortbewegung und der Lauterzeugung der Grillen. *Z. Vgl. Physiol.* 44:60–132

Huber, F. 1975. Principles of motor co-ordination in cyclically recurring behaviour in insects. In *"Simple" Nervous Systems,* ed. P. N. R. Usherwood, D. R. Newth, pp. 381–414. London: Arnold. 490 pp.

Ibuka, N., Kawamura, H. 1975. Loss of circadian rhythm in sleep-wakefulness cycle in the rat by suprachiasmatic nucleus lesions. *Brain Res.* 96:76–81

Jacklet, J. W. 1969. Circadian rhythm of optic nerve impulses recorded in darkness from isolated eye of *Aplysia. Science* 164:562–63

Julesz, B. 1971. *Foundations of Cyclopean Perception.* Chicago: Chicago Univ. Press. 406 pp.

Jürgens, U., Ploog, D. 1970. Cerebral representation of vocalization on the squirrel monkey. *Exp. Brain Res.* 10:532–54

Kammer, A. E., Rheuben, M. B. 1976. Adult motor patterns produced by moth pupae during development. *J. Exp. Biol.* 65:65–84

Kandel, E. R. 1967. Cellular studies of learning. In *The Neurosciences: A Study Program,* ed. G. C. Quarton, T. Melnechuk, F. O. Schmitt, pp. 666–89. New York: Rockefeller Univ. Press. 962 pp.

Kandel, E. R. 1976. *Cellular Basis of Behavior.* San Francisco: Freeman. 727 pp.

Kater, S. B., Rowell, C. H. F. 1973. Integration of sensory and centrally programmed components in generation of cyclical feeding activity of *Helisoma trivolvis. J. Neurophysiol.* 36:142–55

Kennedy, D., Evoy, W. H., Fields, H. L. 1966. The unit basis of some crustacean reflexes. *Symp. Soc. Exp. Biol.* 20:75–109

Kennedy, D., Evoy, W. H., Hanawalt, J. T. 1966. Release of coordinated behavior in crayfish by single central neurones. *Science* 143:917–19

Kien, J. 1977. Comparison of sensory input with motor output in locust optomotor system. *J. Comp. Physiol.* 113:161–80

Konishi, M. 1965. The role of auditory feedback in the control of vocalization in the white-crowned sparrow. *Z. Tierpsychol.* 22:770–83

Krasne, F. B. 1976. Invertebrate Systems as a means of gaining insight into the nature of learning and memory. In *Neural Mechanisms of Learning and Memory,* ed. M. R. Rosenzweig, E. L. Bennett, pp. 401–29. Cambridge, Mass: MIT. 637 pp.

Krasne, F. B., Wine, J. J. 1975. Extrinsic modulation of crayfish escape behaviour. *J. Exp. Biol.* 63:433–50

Krasne, F. B., Wine, J. J., Kramer, A. 1978. The control of crayfish escape behavior. See Hoyle 1978, in press

Kristan, W. B., Stent, G. S., Ort, C. A. 1974. Neuronal control of swimming in the medicinal leech. III. Impulse patterns of the motor neurons. *J. Comp. Physiol.* 94:155–76

Kroodsma, D. E. 1977. A re-evaluation of song development in the song sparrow. *Anim. Behav.* 25:390–99

Kuffler, S. W., Eyzaquirre, C. 1955. Synaptic inhibition in an isolated nerve cell. *J. Gen. Physiol.* 39:155–84

Kupfermann, I. 1967. Stimulation of egg-laying: possible neuroendocrine function of bag cells of abdominal ganglion of *Aplysia californica. Nature* 216: 814–15

Kupfermann, I., Carew, T. J., Kandel, E. R. 1974. Local, reflex, and central commands controlling gill and siphon

movements in *Aplysia. J. Neurophysiol.* 37:996–1019
Kupfermann, I., Pinsker, H., Castellucci, V., Kandel, E. R. 1971. Central and peripheral control of gill movements in *Aplysia. Science* 174:1252–56
Kutsch, W. 1974. The development of the flight pattern in locusts. See Altman & Tyrer 1974, pp. 149–58
Kutsch, W., Huber, F. 1970. Central vs. peripheral control in cricket stridulation. *Z. Vgl. Physiol.* 67:140–59
Larimer, J. L., Eggleston, A. C. 1971. Motor programs for abdominal positioning in crayfish. *Z. Vgl. Physiol.* 74:388–402
Larimer, J. L., Kennedy, D. 1969. The central nervous control of complex movements in the uropods of crayfish. *J. Exp. Biol.* 51:135–50
Loher, W. 1972. Circadian control of stridulation in the cricket *Teleogryllus commodus* Walker. *J. Comp. Physiol.* 79: 173–90
Matsumoto, S. G., Murphey, R. K. 1977. Sensory deprivation during development decreases the responsiveness of cricket giant interneurones. *J. Physiol.* 268:533–48
Maynard, D. M. 1972. Simpler networks. *Ann. N. Y. Acad. Sci.* 193:59–72
Menaker, M., Zimmerman, N. 1976. Role of the pineal in the circadian system of birds. *Am. Zool.* 16:45–55
Mendelson, M. 1971. Oscillator neurons in crustacean ganglia. *Science* 171: 1170–73
Mittenthal, J. E., Wine, J. J. 1973. Connectivity patterns of crayfish giant interneurons: visualization of synaptic regions with cobalt dye. *Science* 177: 182–84
Moore, R. Y., Klein, D. B. 1974. Visual pathways and the central neural control of a circadian rhythm in pineal serotonin N-acetyltransferase activity. *Brain Res.* 71:17–33
Mulligan, J. A. 1966. Singing behavior and its development in the song sparrow *Melospiza melodia. Univ. Calif. Berkeley Publ. Zool.* 81:1–73
Mulloney, B., Selverston, A. I. 1974. Organization of the stomatogastric ganglion of the spiny lobster. III. Coordination of the two subsets of the gastric system. *J. Comp. Physiol.* 91:53–78
Nishiitsutsuji-Uwo, J., Pittendrigh, C. S. 1968. Central nervous control of circadian rhythmicity in the cockroach. III. The optic lobes, locus of the driving oscillator? *Z. Vgl. Physiol.* 58:14–46
O'Shea, M., Rowell, C. H. F., Williams, J. L. D. 1974. The anatomy of a locust visual interneurone: the descending contralateral movement detector. *J. Exp. Biol.* 60:1–12
O'Shea, M., Williams, J. L. D. 1974. Anatomy and output connections of the lobular giant movement detector neuron (LGMD) of the locust. *J. Comp. Physiol.* 41:257–66
Page, C. H., Sokolove, P. G. 1972. Crayfish muscle receptor organ: role in regulation of postural flexion. *Science* 175: 647–50
Pearson, K. G. 1972. Central programming and reflex control of walking in the cockroach. *J. Exp. Biol.* 56:173–93
Pearson, K. G., Fourtner, C. R. 1975. Nonspiking interneurons in walking system of the cockroach. *J. Neurophysiol.* 38:33–52
Pearson, K. G., Wong, R. K. S., Fourtner, C. R. 1976. Connexions between hair-plate afferents and motoneurons in the cockroach leg. *J. Exp. Biol.* 64:251–66
Poggio, T., Reichardt, W. 1973. Considerations on models of movement detection. *Kybernetik* 13:223–27
Potash, L. M. 1970. Vocalization elicited by electrical brain stimulation in *Corturnix japonica. Behaviour* 36:149–67
Rence, B., Loher, W. 1975. Arhythmically singing crickets: thermoperiodic reentrainment after bilobectomy. *Science* 190:385–87
Roberts, S. K. 1974. Circadian rhythms in cockroaches. Effects of optic lobe lesions. *J. Comp. Physiol.* 88:21–30
Rowell, C. H. F. 1964. Central control of insect segmental reflex. I. Inhibition by different parts of the central nervous system. *J. Exp. Biol.* 41:559–72
Rowell, C. H. F. 1969. Central control of an insect segmental reflex. II. Analysis of the inhibitory input from the metathoracic ganglion. *J. Exp. Biol.* 50:191–201
Rowell, C. H. F. 1971. The orthopteran descending movement detector (DMD) neurones: a characterisation and review. *Z. Vgl. Physiol.* 73:167–94
Rowell, C. H. F. 1974. Boredom and attention in a cell in the locust visual system. See Altman & Tyrer 1974, pp. 87–99
Rowell, C. H. F., O'Shea, M. 1976. The neuronal basis of a sensory analyser, the acridid movement detector system. I. Effects of simple incremental and decremental stimuli in light and dark adapted animals. *J. Exp. Biol.* 65:273–88

Rowell, C. H. F., O'Shea, M., Williams, J. L. D. 1977. The neuronal basis of a sensory analyser, the acridid movement detector system. IV. The preference for small field stimuli. *J. Exp. Biol.* 68:157–86

Sandeman, D. C., Kien, J., Erber, J. 1975. Optokinetic eye movements in the crab, *Carcinus maenas.* II. Responses of optokinetic interneurons. *J. Comp. Physiol.* 101:259–74

Schmidt, R. D. 1974. Neural correlates of frog calling. Independence from peripheral feedback. *J. Comp. Physiol.* 88: 321–33

Schrameck, J. E. 1970. Crayfish swimming: alternating motor output and giant fiber activity. *Science* 169:698–700

Selverston, A. I., Mulloney, B. 1974. Organization of the stomatogastric ganglion of the spiny lobster. II. Neurons driving the medial tooth. *J. Comp. Physiol.* 91:33–51

Siegler, M. V. S., Mpitsos, G. J., Davis, W. J. 1974. Motor organization and generation of rhythmic feeding output in buccal ganglion of *Pleurobranchaea. J. Neurophysiol.* 37:1173–96

Sokolove, P. G. 1975. Localization of the cockroach optic lobe circadian pacemaker with microlesions. *Brain Res.* 87:13–21

Sokolove, P. G., Loher, W. 1975. Role of eyes, optic lobes, and pars intercerebralis in locomotory and stridulatory circadian rhythms of *Teleogryllus commodus. J. Insect Physiol.* 21:785–800

Sokolove, P. G., Tatton, W. G. 1975. Analysis of postural motoneuron activity in crayfish abdomen. I. Coordination by premotor connections. *J. Neurophysiol.* 38:313–32

Stein, P. S. G. 1978. Motor systems, with specific reference to the control of locomotion. *Ann. Rev. Neurosci.* 1:61–81

Stephan, F. K., Zucker, I. 1972. Circadian rhythms in drinking behavior and locomotor activity of rats are eliminated by hypothalamic lesions. *Proc. Natl. Acad. Sci. USA* 69:1583–86

Suga, N., Schlegel, P., Shimozawa, T., Simmons, J. A. 1973. Orientation sounds evoked from echolocating bats by electrical stimulation of the brain. *J. Acoust. Soc. Am.* 54:793–97

Szentagothai, J., Arbib, M. A. 1975. *Conceptual Models of Neural Organization.* Cambridge, Mass: MIT. 205 pp.

Tatton, W. G., Sokolove, P. G. 1975. Intersegmental gradient of motoneuron activity in an invertebrate postural system. *Brain Res.* 85:86–91

Taylor, H. M., Truman, J. W. 1974. Metamorphosis of the abdominal ganglion of the tobacco horn worm, *Manduca sexta.* Changes in populations of identified motor neurons. *J. Comp. Physiol.* 90:367–88

Tinbergen, N. 1951. *The Study of Instinct.* London: Oxford Univ. Press. 228 pp.

Truman, J. W. 1971. Physiology of insect ecdysis. I. The eclosion behaviour of saturniid moths and its hormonal release. *J. Exp. Biol.* 54:805–14

Truman, J. W. 1976. Development and hormonal release of adult behavior patterns in silkmoths. *J. Comp. Physiol.* 107: 39–48

Truman, J. W., Fallon, A. M., Wyatt, G. R. 1976. Hormonal release of programmed behavior in silk moths; probable mediation by cyclic AMP. *Science* 194: 1432–34

Truman, J. W., Reiss, S. E. 1976. Dendritic reorganization of an identified motoneuron during metamorphosis of the tobacco hornworm moth. *Science* 192: 477–78

Truman, J. W., Riddiford, L. M. 1970. Neuroendocrine control of ecdysis in silkmoths. *Science* 167:1624–26

Truman, J. W., Sokolove, P. G. 1972. Silk moth eclosion: hormonal triggering of a centrally programmed pattern of behavior. *Science* 175:805–14

Usherwood, P. N. R., Newth, D. R. 1975. *"Simple" Nervous Systems.* London: Arnold. 490 pp.

von Holst, E., von St. Paul, U. 1960. Vom Wirkungsgefüge der Triebe. *Naturwissenschaften* 18:409–22

Weiss, P. 1941. Self-differentiation of the basic patterns of coordination. *Comp. Psychol. Monogr.* 17:1–96

Wendler, G. 1974. The influence of proprioceptive feedback on locust flight coordination. *J. Comp. Physiol.* 88:173–200

Wiersma, C. A. G. 1938. Function of the giant fibres of the central nervous system of the crayfish. *Proc. Soc. Exp. Biol. Med.* 38:661–62

Wiersma, C. A. G. 1952. Neurons of arthropods. *Cold Spring Harbor Symp. Quant. Biol.* 17:155–63

Wiersma, C. A. G., ed. 1974. Invertebrate neurons and behavior. In *The Neurosciences, Third Study Program,* ed. F. O. Schmitt, F. G. Worden, pp. 341–431. Cambridge, Mass: MIT. 1107 pp.

Wiersma, C. A. G., Ikeda, K. 1964. Interneurons commanding swimmeret movements in the crayfish Procambarus clarkii (Girard). *Comp. Biochem. Physiol.* 12:509–25

Willows, A. O. D. 1967. Behavioral acts elicited by stimulation of single, identifiable brain cells. *Science* 157:570–74

Willows, A. O. D. 1968. Behavioral acts elicited by stimulation of single identifiable nerve cells. In *Physiological and Biochemical Aspects of Nervous Integration,* ed. F. D. Carlson, pp. 217–44. F 'glewood Cliffs, N. J: Prentice-Hall. 391 pp.

Wilson, D. M. 1961. The central nervous control of flight in a locust. *J. Exp. Biol.* 38:471–90

Wilson, D. M. 1968a. The nervous control of insect flight and related behavior. *Adv. Insect Physiol.* 5:289–338

Wilson, D. M. 1968b. Inherent asymmetry and reflex modulation of the locust flight motor pattern. *J. Exp. Biol.* 48: 631–41

Wine, J. J., Krasne, F. B. 1972. The organization of escape behaviour in the crayfish. *J. Exp. Biol.* 56:1–18

Wine, J. J., Mistick, D. C. 1977. The temporal organization of crayfish escape behavior: delayed recruitment of peripheral inhibition. *J. Neurophysiol.* 40: 904–45

Wyse, G. A. 1972. Intracellular and extracellular motor neuron activity underlying rhythmic respiration in *Limulus. J. Comp. Physiol.* 81:259–76

Zucker, R. S., Kennedy, D., Selverston, A. I. 1971. Neuronal circuit mediating escape responses in crayfish. *Science* 173:645–49

Ann. Rev. Neurosci. 1978. 1:61–81

MOTOR SYSTEMS, WITH SPECIFIC REFERENCE TO THE CONTROL OF LOCOMOTION

❖11503

Paul S. G. Stein
Department of Biology, Washington University, St. Louis, Missouri 63130

INTRODUCTION

The central nervous system (CNS) must produce specific patterns of motor neuron impulses during a coordinated movement. The desire to understand how these patterns are produced and controlled has led investigators to examine the properties of interneurons within the CNS. A set of theoretical principles has emerged from these studies that are applicable to both invertebrates and vertebrates (Wilson 1967, Evarts et al 1971, Gurfinkel & Shik 1973, R. B. Stein et al 1973, Grillner 1975, Herman et al 1976, Shik & Orlovsky 1976, Wetzel & Stuart 1976, P. S. G. Stein 1977). This review will discuss how these principles apply to locomotion and will demonstrate that they may also serve as useful working hypotheses in the study of other motor systems.

The central hypothesis is that there is a neural pattern generator residing within the CNS that serves to produce the basic motor program (Wilson 1972). Information derived from sensory input may modify the output of the pattern generator so that the motor output is adapted to the particular mechanical properties of the organism and its environment. The principles are the following:

1. There are command neurons that activate the pattern generator.
2. The pattern generator is composed of a set of local control centers.
3. There are neurons within these centers that coordinate muscular synergies and generate timing signals.
4. Coordination among the centers can be produced by centrally derived signals, sensory-derived signals, or both.
5. Sensory signals can measure motor performance and be utilized to improve performance.
6. The pattern generator and/or the command signals can adjust the properties of sensory pathways during the movement.

0147-006X/78/0325-0061$01.00

It is important to emphasize that these are working hypotheses, not dogma, and, when applied to any specific motor system, they may have to be modified to provide an accurate theory concerning the function of that specific motor system.

ACTIVATION OF MOTOR PATHWAYS

The activation of a single interneuron in some invertebrates and fishes is sufficient to excite an entire coordinated behavior. Such an interneuron, termed a command neuron, was initially described in the crayfish by Wiersma and his co-workers (Wiersma 1947, 1952, Hughes & Wiersma 1960, Wiersma & Ikeda 1964, Atwood & Wiersma 1967). The concept of the command neuron was successfully applied to a number of motor systems in the crayfish, most notably in the laboratories of Kennedy (Kennedy, Evoy & Hanawalt 1966, Evoy & Kennedy 1967, Kennedy et al 1967, Davis & Kennedy 1972) and Larimer (Bowerman & Larimer 1974a,b, 1976, Larimer 1976, Larimer & Gordon 1977). Much of this experimental work has centered upon the nature of the motor output driven by single command interneurons. It is apparent that the synaptic *output* of such cells is extremely complex and highly organized. Much less information is available concerning the synaptic *input* to command neurons. The available data indicates that a command neuron for a behavior will be active during the natural expression of that behavior (Kennedy, Evoy & Hanawalt 1966, Davis & Kennedy 1972, Wine & Krasne 1972, Davis 1977, Glantz 1977, Gillette & Davis 1977, Gillette, Kovac & Davis 1977, Larimer & Gordon 1977, Zottoli 1977).

Although a single interneuron is sufficient to activate a coordinated behavior, and while that single interneuron may also be active during the expression of the behavior, it is unlikely that only one command interneuron is activated during the natural occurrence of that behavior. In fact, the available evidence suggests that a small collection of command interneurons may be activated during a natural behavior (Davis & Kennedy 1972, Sokolove 1973, Davis 1976, 1977, Gillette, Kovac & Davis 1977).

There are at least two classes of command neurons, namely the trigger command neurons and the gate command neurons. A trigger command neuron elicits a motor sequence that lasts longer than the stimulus itself. The impulses in the trigger cell have a distinct timing relationship with motor neuron events. In contrast, a gate command neuron will elicit a motor behavior only when the activity in the command cell is maintained. If the gate command cell is no longer active, then the behavior is no longer observed.

Trigger Command Cells

The term *trigger cells* was initially utilized by Willows & Hoyle (1969) to describe a population of cells that, when stimulated with extracellular electrodes, evoked a swimming behavior in *Tritonia* that lasted longer than the stimulus. The term *trigger command cell* can apply equally well to many of the giant interneurons found in invertebrates and lower vertebrates. In particular, the giant axons of the crayfish have been well studied (Wiersma 1947, Zucker, Kennedy & Selverston 1971, Wine

& Krasne 1972, Zucker 1972, Mittenthal & Wine 1973, Krasne & Wine 1977). It is known that a single impulse in a crayfish giant axon can elicit a complex sequence of motor neuron and interneuron activity that will outlast the impulse discharge of the giant axon. There is considerable specificity in both the input and the output connections of the giant axons. Sensory input to the head region can activate the medial giant axon, whereas sensory input to the abdominal region can activate the lateral giant axon. An impulse in the crayfish medial giant axon can cause a different movement of the abdomen than an impulse in the lateral giant axon (Larimer et al 1971, Wine & Krasne 1972). Thus, when stimulation is applied to the head region, a medial giant axon-mediated tail flip can be initiated that will propel the crayfish backwards and away from the stimulus. When stimulation is applied to the abdominal region, a lateral giant axon-mediated tail flip can be initiated that will propel the crayfish in a backward somersault and away from the stimulus (Wine & Krasne 1972). This system is a premier example of how an understanding of a motor control system can be reached via a detailed study of the properties of specific interneurons in the CNS (see Krasne & Wine 1977 for a review of this system).

The difficulties that can be encountered in the study of command neurons are exemplified in the work on swimming in the marine mollusc, *Tritonia.* There is now evidence that the group of cells in the *Tritonia* brain that were initially identified as trigger command cells (Willows & Hoyle 1969, Willows 1976) may not be involved in the triggering of swimming behavior (Getting 1975, 1976). Getting has demonstrated that the Willows result may be due to the spread of extracellular stimulating current to a group of sensory cells that lie underneath the so-called trigger cell group. Getting's work suggests that the command cells for *Tritonia* may be postsynaptic to the sensory cell population and that the former have not yet been identified.

Gate Command Cells

The term *gate command cell* applies to an interneuron that must be continually activated in order to obtain the continued expression of the evoked behavior. Most of the published literature on *command neurons* has, in fact, dealt with gate command neurons. Gate command neurons can activate tonic motor systems involved in posture (Kennedy, Evoy & Hanawalt 1966, Bowerman & Larimer 1974a), and phasic motor systems involved in locomotion and other movements (Hughes & Wiersma 1960, Davis & Kennedy 1972, Bowerman & Larimer 1974b). In both situations, a change in the frequency of command neuron stimulation can lead to a change in the intensity of the response. In the case of a postural system, the magnitude of a postural adjustment can be altered by changing the frequency of command stimulation (Kennedy et al 1967). In the case of a locomotor system, the frequency of the movement cycle can be altered by a change in the frequency of command stimulation (Davis & Kennedy 1972). It is interesting to note that Davis & Kennedy (1972) also found command neurons that produced a constant movement frequency independent of command frequency, but as the command neuron frequency was altered there was a change in the impulse frequency of motor neurons. For most gate command neurons, it is assumed that there is no specific timing information carried by the command, e.g. a random stimulation regime of the command cell can produce the normal activation of the motor program (Wilson & Wyman 1965). It is not known how generalizable this result may be since there have been no other studies utilizing random stimulation regimes and few studies that have recorded from gate command neurons during the natural activation of behavior. Davis and his colleagues (1976, 1977, Gillette & Davis 1977, Gillette, Kovac & Davis 1977) have presented evidence that one type of command cell may get phasic motor feedback from the program generator and may therefore carry timing information. The theoretical importance of this result is immense, and it will be important to look for similar effects in other motor systems. Another example of the importance of examining both the inputs and the outputs of a command neuron is provided in the work of Glantz (1974a,b, 1977). He has studied in detail the properties of the command neuron for the defense reflex of the crayfish (Wiersma 1952). This reflex is elicited by a moving visual stimulus and displays habituation when the stimulus is repeatedly applied. Glantz has shown that the command neuron displays many of the response characteristics of the whole animal's behavior. From this and other work it is implied that the decision to perform a motor act is made at the level of the command neuron or at a level presynaptic to the command neuron.

Extrapolation of the Command Concept to the Vertebrates

In fishes there are single reticulospinal neurons that are command neurons in the sense utilized in the invertebrate literature. These include the Muller and Mauthner cells of the lamprey (Rovainen 1967) and the Mauthner cells of teleost fishes (Diamond 1971, Kimmel & Eaton 1976, Eaton, Bombardieri & Meyer 1977, Eaton et al 1977, Zottoli 1977). In particular, stimulation of the Mauthner cell can elicit a tail flip that causes a rapid movement of the fish. Zottoli (1977) has recorded the Mauthner cell action potential during a tail flip induced by auditory stimulation in a freely moving fish. It therefore appears that the axon of the Mauthner cell is functionally similar to the giant axons of the invertebrates.

In the tetrapod vertebrates, there is no clear evidence that the neural output of a single cell is sufficient to produce a coordinated motor output. It appears, however, that a group of cells can act as if it has a command function. For example, it has been known that stimulation of the white matter of the cat spinal cord can elicit stepping (Sherrington 1910, Grillner & Zangger 1974, Grillner 1976) and scratching (Deliagina et al 1975) movements. Such stimulation, when applied to the dorsolateral funiculus of the turtle spinal cord, can elicit swimming movements (Lennard & Stein 1977). And when applied to specific regions of the brainstem, the electrical pulses can elicit treadmill locomotion in the cat (Shik, Severin & Orlovsky 1966, Mori, Shik & Yagodnitsyn 1977) and swimming in fishes (Kashin, Feldman & Orlovsky 1974). Whether or not such central structures are actually utilized by the organism in the expression of the behavior is not known, although it is likely that at least some reticulospinal pathways act as command tracts. Corroborative data for this point of view can be obtained by ablation studies of specific spinal tracts, thus, the ventrolateral funiculus appears to be critical for the expression of lordosis in the rat (Kow, Montgomery & Pfaff 1977). Similar studies on bird song have suggested that certain regions of the bird brain may serve a command function (Nottebohm, Stokes & Leonard 1976).

There are also excellent data on oculomotor control by the primate superior colliculus that suggests that cells in the deeper layers of the colliculus may act as command cells for saccades. These data have been obtained from recording (Schiller & Stryker 1972, Sparks, Holland & Guthrie 1976) and stimulation (Robinson 1972, Schiller & Stryker 1972) studies. The former demonstrate that cells in the deeper layers of the superior colliculus have a "motor field" for a saccade of a given amplitude and direction (i.e. they fire with highest frequency for the saccade of appropriate amplitude and direction, they fire with a lower frequency for saccades with amplitude and direction near the preferred region, and they do not fire when other saccades are performed). Moreover, this correlation can be found for some cells that are activated by either a visual or a vestibular stimulus (Schiller & Stryker 1972). The stimulation studies demonstrate that with low current stimulation a saccade is elicited that has a direction and amplitude that matches the motor field of cells in that region of colliculus. Both the motor field maps and the stimulation maps are independent of the position of the eye in the orbit. In addition, there is a correspondence of these maps with the visual receptor map obtained from cells in more superficial layers of the colliculus (Cynader & Berman 1972, Schiller & Stryker 1972). These data taken together suggest that information coming into the superficial layers might be processed and then lead to a motor decision in the deeper layers. Such a decision would allow the animal to fix the fovea on the object detected in visual space. This view of collicular functioning has been challenged by the work of Wurtz and his co-workers (Wurtz & Goldberg 1972, Mohler & Wurtz 1976). They suggest that the *intermediate* layers may take information from both the deep and superficial layers and make higher-order decisions.

Higher Order Command Functions

It is to be expected that as the task, and the organism, display more complexity the simple concept of a command neuron must be modified and expanded. Significant progress toward this end has been made by Mountcastle and his co-workers (Mountcastle et al 1975, Mountcastle 1976, Lynch et al 1977). Single unit recordings were obtained from the parietal cortex of alert monkeys during a series of complex eye and hand-movement tasks. Units were found that were related to specific classes of movements if the animal was motivated to make that movement while attending to a particular task. If the animal made the same movement while not attending to the task, the unit would not be active. Lynch et al (1977) argue that it is unlikely that these cells are corollary discharge neurons and suggest that the parietal lobe has the appropriate anatomical connections to act as a motor command area in situations where visual attention is required. These arguments are strengthened by clinical data from humans with parietal lobe lesions. These patients display oculomotor apraxia, i.e. while they are able to move their eyes spontaneously and display no paralysis of the eyes, they are not able to fixate on a particular object in visual space upon verbal command.

A similar approach to command function has been taken by Evarts and his co-workers. It was Evarts' pioneering effort in the late 1960s that first demonstrated the power of unit recording during voluntary hand movements in primates (Evarts 1966, 1968). Recently Evarts & Tanji (1976, Tanji & Evarts 1976) have extended this work by demonstrating that units in the motor cortex display activity that is contingent upon the intent of the next movement. This work is a prime example of how combining unit recording with a behavioral task can yield information about the working of a motor system that could not be obtained in an inert, nonbehaving animal.

Future Directions of Command Neuron Experiments

There is a strong body of evidence to support the notion that interneurons in the CNS serve to activate pattern generators. A major direction of future research will be devoted to understanding both the behavioral context in which these command neurons are activated and the synaptic input to the command neurons. It is also clear that further hypotheses need to be developed concerning command function. It is known that there are certain behavioral hierarchies that exist in an organism such that when one behavior is expressed, others will be inhibited (Davis et al 1974, Davis et al 1977). Kovac & Davis (1977) have established the mechanism responsible for one such hierarchy by demonstrating that a corollary discharge cell from the "feeding pattern generator" can inhibit the expression of withdrawal behavior. One possible circuit which could explain this effect is that the corollary discharge neuron from the feeding network could directly inhibit the withdrawal command neuron. W. J. Davis & R. Gillette (1977) have also shown that when the motivation to feed is altered, the synaptic input to feeding command cells is also altered.

It is also important to examine hormonal effects on command function. Since hormones can either directly stimulate or facilitate a behavior (Truman 1976, Pfaff et al 1974), it is possible that the use of labeled hormones to identify command neurons may become a powerful technique. Pfaff and his co-workers have utilized this approach for both the lordosis reflex (Pfaff et al 1974) and bird song (Arnold, Nottebohm & Pfaff 1976).

ORGANIZATION OF THE PATTERN GENERATOR

The CNS can produce organized motor programs without the assistance of sensory feedback. However, sensory information can influence central programs so that the output of the CNS is adaptive to the particular needs of the task. The central program generator for locomotion is distributed along the neuraxis and in some organisms can be subdivided into local control centers. The role of each center is to organize the details of muscular synergies for those motor units innervated by the center. Each center may also send centrally derived neural information to other centers. The concept of a local control center is also useful in the analysis of other forms of motor output (cf Doty 1976).

Identification of the Local Control Center

One type of experiment that has been used to establish that a given piece of nervous tissue contains all or part of a local control center involves removing the tissue from the CNS and placing it in a saline-filled dish. If it is possible to record a motor output from the isolated preparation that replicates that observed in the intact organism, it is clear that the local control center resides in that piece of nervous tissue. Note that this procedure does not guarantee that the entire local control center has been removed but only a sufficiently large piece of the center so that it still produces a normal motor output (see D. F. Russell 1976 for an illustration of this warning). The isolation experiments have been successful in a number of preparations, namely for a locomotor system (Ikeda & Wiersma 1964), a feeding system (Selverston et al 1976), for postural control (Sokolove & Tatton 1975), and vocalization (Schmidt 1976). This list is by no means exhaustive, but it indicates the diversity of motor systems in which the isolated preparation has been successful.

It is important to recognize that for some motor systems it may be difficult to apply the concept of the local control center. In particular, the neural circuitry underlying the tail flip in the crayfish involves a monosynaptic connection between the command neuron and the motor neuron (Krasne & Wine 1977). In this case it could be argued that the command neuron also acts as the control center, but one could also choose not to apply the term *control center* to this particular motor system. It is also clear that for most motor systems the control center is a collection of interneurons that synaptically activate motor neurons. It is important to recognize, however, that some local control centers may lack a large interneuronal population, may consist mainly of motor neurons, and the critical synaptic interactions may take place among the motor neuron population (Selverston et al 1976).

Properties of Cells within the Local Control Center

The simplest control center has one synergic group of interneurons that drives the motor behavior of one synergic group of motor neurons. Such a center has been found to control cardiac musculature in the crustaceans (Hartline 1967, Watanabe, Obara & Akiyama 1967, Mayeri 1973, Friesen 1975). In the best studied example, the cardiac ganglion in the lobster, there are four interneurons that act synergistically to excite five motor neurons. For this system it is known that (*a*) the rhythm is generated mainly among the interneurons, (*b*) slow potentials among the interneurons may be critical for the rhythm formation, and (*c*) there are extensive electrical and chemical synaptic connections in the ganglion. Recently, Friesen (1975) has demonstrated that there is a hierarchical organization among the interneurons with some of the interneurons having a role mainly in the generation of timing, while one of the interneurons has a major role in the excitation of motor neuron impulses.

A control center responsible for feeding that displays slightly greater complexity is found in the snail buccal ganglion (Kater 1974). In this system, there is a synergic set of interneurons that are electrically coupled to each other and are responsible for the generation of the feeding rhythm. Some members of this interneuronal pool chemically excite retractor motor neurons, while other members of the pool chemically inhibit protractor motor neurons. In contrast, in *Aplysia,* the same interneuron can chemically excite one population of motor neurons and chemically inhibit another population of motor neurons (Koester et al 1974).

An even more complex control center is found in the stomatogastric ganglion of the lobster (Maynard 1972, Selverston et al 1976). In this control center, the organization of motor neuron synergies is mainly the result of connections among the motor neurons themselves. There are actually two distinct rhythms that are controlled by the ganglion, a faster rhythm that controls the pyloric movements, and a slower rhythm that controls the movements of the internal teeth of the stomach. There are three main synergies that are generated by the motor neurons of the pyloric system. The cells of one of these synergic groups appear to have endogenous bursting capacities. The cells of the other two groups are inhibited by this first group and mutually inhibit each other. In the cells controlling the teeth, the gastric mill cells, the interactions are even more complex. The gastric system is broken down into subsets of strongly interacting cells, and within these subsets there are many inhibitory chemical synapses and excitatory electrical synapses. There does not appear to be any particular cell type that has endogenous bursting capacities. The rhythm in this portion of the system appears to result from mutual inhibition and from rings of inhibitory connections (for theoretical approach see Kling & Szekely 1968, Wilson & Waldron 1968, Perkel & Mulloney 1974). Neural modeling of the circuitry of the ganglion has revealed that the known connections and properties of the cells can account for the production of the coordinated rhythm (Perkel & Mulloney 1974, Warshaw & Hartline 1976). In particular, Warshaw & Hartline (1976) have demonstrated that the synaptic connections among the cells in the pyloric region are sufficient to produce the rhythmicity even in the absence of any assumptions about the endogenous bursting capacities of any synergic group of cells. From this result it can be inferred that in this system there are redundant mechanisms for the generation of timing and that perhaps in the intact system they reinforce each other to produce a more reliable rhythm. Readers should consult Selverston et al (1976) for a complete description of the system.

A still more complex system exists for the control of swimming in the leech (Kristan, Stent & Ort 1974, Kristan & Calabrese 1976, Friesen, Poon & Stent 1976). In this organism, swimming consists of alternate movements between the dorsal and ventral musculature. There are approximately 21 segments that participate in the swimming rhythm, and during swimming the wavelength of the movement is approximately one body length. An isolated chain of eight or more ganglia can respond to stimulation of a peripheral nerve with a proper motor neuron rhythm (Kristan & Calabrese 1976). This rhythm not only consists of alternation of dorsal motor neurons with ventral motor neurons but also involves the peripheral inhibitors to both the dorsal and ventral musculature. The excitatory motor neurons are not strictly coactive with the inhibitory motor neurons to the antagonist muscles. Therefore each segmental ganglion will produce motor output with four distinct phases. Corresponding to these four phases are four interneurons per hemiganglion that are connected to each other via a ring of inhibitory synapses (Friesen, Poon & Stent 1976). In addition, there are some inhibitory connections across the ring. Many of the connections seen within the ganglion are also observed between ganglia. In fact, modeling of the system indicates that the interganglionic connections are necessary to produce a rhythm with a period in the appropriate physiological range. It is important to note that the leech system involves the control of axial musculature, and there is a major portion of the interneuronal circuitry that travels intersegmentally. It will be important to contrast this system with one for appendicular control in which the intersegmental connections may not be as necessary for the generation of an appropriately timed motor output from a single ganglion.

An equally complex control system exists for the limbs of the cockroach (Pearson 1972, 1976a,b, 1977, Pearson & Iles 1973, Pearson & Fourtner 1975, Pearson & Duysens 1976). A deafferented CNS can produce a proper walking motor output (Pearson 1972). There are intrasegmental interneurons that do not produce action potentials and have a strong synaptic influence on the motor neurons (Pearson & Fourtner 1975, Fourtner 1976, Pearson 1976a). The occurrence of local interneurons without action potentials is a feature of many local centers (Rakic 1975, Pearson 1976a). There is a suggestion that these nonspiking interneurons may also serve as distribution nodes for reflex input to the local control center (Pearson, Wong & Fourtner 1976). Evidence for nonspiking interneurons controlling motor neuron discharges also exists for the control of gill bailers in the crustacea (Mendelson 1971).

One of the classic systems for the study of motor control is the flight system of the locust (Wilson 1961, 1967, 1968a,b, 1972). In this pioneering work, Wilson was able to demonstrate that the deafferented CNS could produce a proper flight rhythm but with a low flight frequency. A proper flight frequency could be induced by stimulation of wing sensory cells even with a random stimulation regime (Wilson & Gettrup 1963, Wilson & Wyman 1965). Burrows (1975, 1976, 1977) has extended the work of Wilson with the aid of intracellular recording techniques, and has demonstrated that some of the interneurons that control flight may also be involved in the control of other motor systems such as respiration.

While there has been much progress in the understanding of local control centers in the invertebrates, there has also been much progress with interneuronal recordings from local control centers in the vertebrates, in particular in the cat spinal cord. The cat spinal cord has been one of the major preparations of twentieth century neurophysiology commencing with the early work of Sherrington (1910) and Graham Brown (1911) that established that "spinal cats" could produce a coordinated motor output. In particular, Brown (1911) demonstrated that the deafferented hindlimb could produce the type of alternation of flexion and extension that is seen in the intact cat during walking. More recently, Grillner and his co-workers have dealt with the local control properties of the cat spinal cord (Grillner & Zangger 1974, Edgerton et al 1976, Sjostrom & Zangger 1976). Chemical stimulation of the spinal cord can produce rhythmic alternation between extensor activity and flexor activity in paralyzed cats (Jankowska et al 1967, Grillner 1976) even when there is no rhythmic sensory input (Grillner & Zangger 1974, Edgerton et al 1976, Sjostrom & Zangger 1976). Intracellular recordings from spinal motoneurons reveal that there are rhythmic bursts of EPSPs, which lead to action potentials followed in some motor neurons by IPSPs (Edgerton et al 1976). At least one source of the IPSPs are action potentials in the group Ia inhibitory interneurons (Hultborn 1972, Jankowska & Roberts 1972, Feldman & Orlovsky 1975). In addition, Sjostrom & Zangger (1976) have demonstrated that the deafferented "low spinal" cat treated with L-dopa can display coactivation of alpha and gamma motor neuron activity. Similarly, Feldman & Orlovsky (1975) have shown that a de-efferented lumbosacral spinal cord will display coactivation of the Ia inhibitory interneurons with motor neurons that receive similar Ia input. This demonstrates that the Ia inhibitory interneurons are directly driven by the local control center and may in fact be considered a part of the center. Further evidence for this viewpoint has been obtained by Hultborn, Illert & Santini (1976), who have shown that there is direct reciprocal inhibition between Ia inhibitory interneurons that synapse upon antagonist motoneurons. All this indicates that the spinal cord can produce a complex locomotor output when properly activated, and the spinal cat continues to be an excellent preparation for the study of motor control.

Coordination Among Several Control Centers

There are many motor systems in which different portions of body musculature innervated by different regions of the neuraxis are activated during distinct phases of a behavior. As discussed in the previous section, many local regions of the neuraxis contain a local control center that is responsible for the generation of the synergies of the local portion of the motor apparatus. The CNS must therefore coordinate the activities of the separate local centers so that the different body parts move at the appropriate time.

One possible mechanism utilizes interneurons that are driven by one local control center and therefore carry a central representation of efferent activity. If these interneurons can synaptically influence other local centers, then coordination among the centers is possible in the isolated CNS. Such interneurons have been isolated in the swimmeret system of the crayfish and have been termed *coordinating neurons* (Stein 1971, 1974, 1976). It is likely that interneurons of this type exist in other locomotory systems, e.g. the cockroach walking system (Pearson & Iles 1973) and cat locomotion (Halbertsma et al 1976).

A second possible mechanism utilizes mechanical linkages among different muscular systems to coordinate the activity of separate control centers. Each center can then utilize sensory information derived locally to measure the performance of other centers. In particular, the forces upon a given limb during locomotion depend upon which other limbs are also in contact with the ground. If the force detected by sensory structures in the limb is small, then the organism will be able to lift the leg off the substrate without impairing its equilibrium (Gray 1968). Pearson and his co-workers have demonstrated that force detectors in the limb of both a cockroach and a cat can act to inhibit the lift-off of a limb if they detect a load on the limb that is too large (Pearson & Iles 1973, Pearson & Duysens 1976). Thus mechanical linkages within the body can combine with reflex mechanisms to assist in the coordination of motor output derived from several local centers.

Another example of mechanical linkages assisting in the coordination of two muscular systems is found in the work of Bizzi and his co-workers on eye-head coordination in the monkey (Bizzi, Kalil & Tagliasco 1971, Bizzi, Polit, & Morasso 1976). In this work, monkeys were trained to shift their gaze from one point in space to another. If the head is restrained, the eyes will gaze directly at the new point in space by turning and remaining deviated in the orbit. If the head is free to move, the head will face the new point, and the eyes will become centered in the orbit. In the latter case, the neural signal to move drives a simultaneous coactivation of both eye and neck muscles. Because of the mechanical properties of the system, the eyes move first. As the head begins to move, there is a compensatory counter-rotation of the eyes in the orbit that summates with the saccadic movement. The major sensor for the counter-rotation is the vestibular apparatus that drives the open-loop vestibulo-ocular reflex. Thus the vestibular apparatus measures the head velocity and reflexly drives the eyes with an equal and opposite velocity. If the vestibular apparatus is removed, other mechanisms can be utilized to produce eye-head coordination (Dichgans et al 1973).

SENSORY MODIFICATION OF MOTOR BEHAVIOR

Most motor programs are generated within the CNS, and the role of peripheral feedback mechanisms is to make corrective adjustments in the central program so that the program is adapted to the particular demands of the animal's current status (Wilson 1972). If the animal is performing a postural task, then it may be necessary to produce a change in the amplitude of the force produced by one set of postural muscles (Fields 1966, Fields, Evoy & Kennedy 1967, Kennedy 1969). If the animal is performing a rhythmic task such as locomotion, then the organism can alter the amplitude of a movement without affecting the rhythm (amplitude modulation, or AM) or the organism can alter the timing of the movement (frequency modulation, or FM). In each of these cases, the modifications will result in a more adaptive behavior.

Amplitude Modulation

Animals with certain classes of sensory deprivation cannot locomote in a straight line. Wilson (1968b, 1972) has attributed this result to basic asymmetries in morphology that can arise as an asymmetry of either CNS output or the peripheral motor structure. These imbalances can be caused by either genetic information, developmental error, or injury. During locust flight, an imbalance of motor output around the roll axis can be corrected by visual input (Wilson 1968b). A locust that is free to move around its longitudinal axis will systematically roll in one direction in the dark. If a light is turned on, then the locust will stop rolling and will orient the dorsal surface of its body toward the light. A similar system exists for the control of yaw movements during locust flight (Camhi 1976). A change in statocyst information induced by a roll of the lobster (Davis 1968) can induce a change in the direction of swimmeret beating without a change in the rhythm.

Similarly, stimulation of certain bulbospinal pathways in the "mesencephalic cat" can elicit an amplitude modulation of the hindlimb electromyogram (EMG) during treadmill locomotion (Orlovsky 1972). Some of these pathways monosynaptically activate motoneurons, while others act via a few synapses. Another form of amplitude modulation is possible in the cat with the reflexes driven by Ia afferents, such as the monosynaptic reflex of the Ia afferent onto synergic alpha motoneurons and the di-synaptic reflex of the Ia afferents onto Ia inhibitory interneurons that, in turn, inhibit the antagonist alpha motoneurons. In both of these reflexes, it is possible that the amplitude of muscle activation could be altered by a change in Ia input. Such a change will act to compensate for load (Kennedy 1969, Marsden, Merton & Morton 1977) if there is alpha-gamma coactivation. There is evidence that there is alpha-gamma coactivation during treadmill locomotion (Severin, Orlovsky & Shik 1967) and during a L-dopa-induced locomotor rhythm (Sjostrum & Zangger 1976). In contrast, direct recordings from primary afferents during locomotion in the intact, freely moving cat reveal that there is Ia activation some of the time during the contraction of its synergistic muscle (Prochasza, Westerman & Ziccone 1976). At other times, the Ia afferent is activated when the synergistic muscle is stretched as a result of shortening of the antagonist. At still other times, the Ia afferent will be active at two phases of the locomotor cycle. The Ia population therefore cannot be regarded as a homogenous population. It will be important to test all inferences gathered from dissected preparations with data on freely moving animals (Wetzel & Stuart 1976).

Similar evidence for coactivation of the receptors and the working muscles has been obtained in the slow extensor musculature of the crayfish. Such coactivation is the result of activation of slow extensor motor neuron #4, which innervates both types of muscles (Fields, Evoy & Kennedy 1967). Recording from the muscle receptor organ (MRO) afferent during a reflex-induced extension of the abdomen shows that the afferent fires during the movement but is quiet at the end of the movement (Fields 1966). Moreover, if the movement is stopped at a midway position, the firing of the MRO afferent is proportional to the distance yet to be covered, i.e. it acts as an error detector (Sokolove 1973). The MRO afferent excites slow extensor motor neuron #2, which only innervates the working muscles. Thus this information can act to compensate for changes in load such as would occur when the crayfish abdomen is in air as opposed to its being in water, where the organism is buoyant. Similar compensations for load are also seen during feeding behavior (Kater & Rowell 1973, Macmillan, Wales & Laverack 1976).

Phase Modulation

During locomotion, each limb is active at a particular phase of the locomotory cycle. If the animal encounters an unexpected obstacle, then it may be maladaptive to maintain that particular phase. Pearson and co-workers (Pearson & Iles 1973, Pearson & Duysens 1976) have developed an interesting hypothesis that can explain how sensory input can modify the walking program when an animal is moving over uneven terrain or is suffering from weakness in, or injury to, one or more limbs. They postulate that there are load-detectors in the limb that detect the amount of force on the limb during the stance portion of the walking cycle. In a healthy animal walking over an even substrate, the force on any given limb should be small towards the end of the period when the animal is standing on it. This is so because the other limbs by this time will have touched down and taken up a major share of the load. There will, therefore, be little activity in the limb load-detectors. If the situation is modified so that at the end of the stance there is still a significant load on that limb, and if the animal lifts the leg, the balance of the organism would be disrupted, and the animal might fall over. If the load detectors inhibit the onset of the lift-off, then an inappropriate lift-off will be prevented. Such a circuit exists in both the cockroach and the cat (Pearson & Iles 1973, Pearson & Duysens 1976).

Frequency Modulation

Frequency modulation has been attributed to the stretch receptors of the locust (Wilson & Gettrup 1963). Other sensory cells may also be involved (Burrows 1975). A change in flight frequency can also occur when the wing is mechanically forced to move at a different frequency (Wendler 1974). A change in the frequency of electrical pulses emitted by electric fish can be elicited by the presentation of an electrical signal a few Hz away from the fish's natural signal (Bullock, Hamstra & Scheich 1972). The fish will respond by altering the frequency of its own signal so that it is further away in frequency from the imposed signal; this behavior has been termed the Jamming Avoidance Response. Scheich (1977) has recorded from many of the interneurons involved in this complex reflex and has presented a model of how the sensory processing in the temporal domain leads to a signal that, when presented to the pacemaker center, produces the appropriate frequency shift.

Parameter Adjustment

Another important role for sensory input in motor behavior is the adjustment of parameters, especially those involved in open-loop reflexes. The vestibulo-ocular reflex (VOR) is one such reflex in which the head velocity signal measured by the vestibular system can cause an eye movement of equal and opposite velocity. Such a reflex is critical in eye-head coordination (Bizzi, Kalil & Tagliasco 1971) and in the stabilization of the eyes in visual space during head movements. The gain of the VOR can be modified by the application of telescopic lenses to the monkey (Miles & Fuller 1974). The sensory signal for gain modification is the velocity of slippage of the visual image on the retina. It has been postulated that the cerebellum might be involved in the recalibration of the VOR (Ito 1972). A cat without certain portions of the cerebellum does not show an adaptive change in the VOR but will still demonstrate the reflex (Robinson 1976). Recently Ito, Nisumaru & Yamamoto (1977) have shown that output from the cerebellum can affect reflexes driven by specific canals to specific muscles but will not affect other reflexes.

CENTRAL CONTROL OF SENSORY PATHWAYS

The sensory pathways of an organism are stimulated during a movement. In some motor systems it may be necessary to adjust the gain of these pathways so that adaptive sensorimotor interactions can be produced during the active behavior. It has been possible to demonstrate in a few motor systems that such a gain adjustment is the direct result of command neuron stimulation (Evoy 1976).

Attenuation of Excitation in Sensory Pathways During Active Movement

Activity of tactile receptors in the crayfish abdomen can cause an impulse in the lateral giant command neuron that will in turn elicit a tail flip (Zucker, Kennedy & Selverston 1971). In an intact crayfish, the movement caused by the tail flip will cause reexcitation of the tactile receptors. Reflex activation of sensory interneurons resulting from tactile receptor action potentials is reduced during the tail flip (Krasne & Bryan 1973, Kennedy, Calabrese & Wine 1974, Krasne & Wine 1977). This reduction will occur even in an isolated nerve cord and is the result of a depolarizing IPSP, termed the depolarizing afferent potential (DAP), that is produced in the axon terminals of the tactile receptor. The DAP is polysynaptically driven by action potentials in the giant axon. A major role of the DAP is attenuation of excitation in sensory pathways during a movement. A second role of the DAP is to protect the crayfish from antifacilitation of the chemical EPSP produced by the tactile afferent input to sensory interneurons (Zucker 1972, Krasne & Bryan 1973, Kennedy, Calabrese & Wine 1974). This antifacilitation can occur when the tactile receptor is repeatedly stimulated, and it is a neural correlate of the behavioral habituation observed when the animal is repeatedly stimulated. When the animal makes an active movement, there is a DAP in the axon terminals of the primary afferents that then reduces transmitter release and, in turn, prevents antifacilitation. Thus this central control of afferent pathways "protects" the animal from habituating in response to its own movement (Krasne & Bryan 1973, Kennedy, Calabrese & Wine 1974, Krasne & Wine 1977).

Sokolove & Tatton (1975) have also shown in the crayfish that the slow flexor motor neurons and the accessory inhibitor neuron are coactivated. This latter cell directly inhibits the MRO stretch receptor neuron (Fields, Evoy & Kennedy 1967). Therefore the stretch receptor excitation of extensor motor neuron #2 will be attenuated during flexor motor activity.

A similar phenomenon has been demonstrated in the lateral line system of fish (I. J. Russell 1976). In this system there is peripheral inhibition of the tactile afferents via inhibitory efferents that travel in the lateral line (Flock & Russell 1976). Moreover, there is a reduction of transmission at the synapse between the lateral line afferents and sensory interneurons (I. J. Russell 1976). These inhibitory actions can be elicited by the stimulation of a number of motor pathways including the Mauthner cell.

Attenuation along sensory pathways is also observed in the echolocating system of the bat. When the bat vocalizes, it produces an intense sound. The signal of importance to the bat is the weak echo that is reflected from an object in the environment (Simmons, Howell & Suga 1975). The bat attenuates the sensory signal resulting from its own vocalization by at least two mechanisms. First, the bat contracts its middle ear muscles prior to vocalization (Suga & Jen 1975); this makes the ear less sensitive to the vocalization. In addition, there is a neural attenuation in the brainstem auditory pathways during a vocalization (Suga & Schlegel 1972). All these factors assist the bat in detecting the weak returning echo.

Phase-Dependent Changes in Gain in Reflex Pathways

During the rhythmic motor program of locomotion, it is possible that a reflex that is adaptive during one phase of the movement may not be adaptive during another phase of the movement. In such a case, the central program for the movement should adjust the gain of reflex pathways during the behavior so that the most adaptive behavior can be produced at each phase of the movement.

One example in which the gain of a reflex pathway is altered during locomotion occurs in the tri-neuronal reflex pathway involved in the Ia inhibition of the antagonist motoneurons. The intermediate neuron in this pathway is the Ia inhibitory interneuron (Hultborn 1972, Jankowska & Roberts 1972). This cell receives a powerful input from Ia afferents and can follow Ia stimulation up to very high frequencies. But this cell is also strongly driven by the locomotor pattern generator, since the Ia inhibitory interneuron will discharge in phase with agonist motoneurons even in a de-efferented preparation ("fictive locomotion," Feldman & Orlovsky 1975). This implies that the gain of the di-synaptic pathway will be radically changed during the different phases of the step cycle.

A second example in which the gain of a reflex pathway can be altered is in the phase-dependent reflex reversal observed with cutaneous stimulation to the cat hindlimb (Duysens & Pearson 1976, Forssberg et al 1976). In these preparations, cutaneous stimulation is applied to a particular region of the cat limb. During one phase of the cycle, the reflex effect may be to excite flexors, whereas during another phase of the cycle the reflex effect may be to excite extensors. Since these effects can be demonstrated in a "low spinal" kitten (Forssberg et al 1976), it is presumed that these effects are mediated by the local control center. Similar effects are also seen in fishes (Forssberg et al 1976) and in insects (Bassler 1976).

CONCLUSIONS

This review has stressed the viewpoint that a set of principles utilized in the study of locomotion may also serve as useful working hypotheses for the study of other motor systems (Stein 1977). Support for this view comes not only from the "simpler" invertebrate preparations (Kennedy 1976, Kennedy & Davis 1977), but also from the "more complex" vertebrate preparations (Grillner 1975, Bullock 1976, Wetzel & Stuart 1976). These principles are not to be regarded as dogma that must apply to all systems; instead they are to be regarded as testable hypotheses that must be refined to meet the needs of each experimental system. With that point of view in mind, it is my hope that these principles will continue to be useful in the field of motor control in the years to come.

ACKNOWLEDGMENTS

The author was supported by NSF grant BNS-75-18040 during the preparation of this review.

Literature Cited

Arnold, A. P., Nottebohm, F., Pfaff, D. W. 1976. Hormone concentrating cells in vocal control and other areas of the brain of the zebra finch (*Peophila guttata*). *J. Comp. Neurol.* 165:487–511

Atwood, H. L., Wiersma, C. A. G. 1967. Command interneurons in the crayfish central nervous system. *J. Exp. Biol.* 46:249–61

Bassler, U. 1976. Reversal of a reflex to a single motoneuron in the stick insect, *Carausius morosus. Biol. Cybern.* 24: 47–49

Bizzi, E., Kalil, R. E., Tagliasco, V. 1971. Eye-head coordination in monkeys: evidence for centrally patterned organization. *Science* 173:452–54

Bizzi, E., Polit, A., Morasso, P. 1976. Mechanisms underlying achievement of final head position. *J. Neurophysiol.* 39: 435–44

Bowerman, R. F., Larimer, J. L. 1974a. Command fibers in the circumoesophageal connectives of crayfish. I. Tonic fibres. *J. Exp. Biol.* 60:95–117

Bowerman, R. F., Larimer, J. L. 1974b. Command fibers in the circumoesphageal connectives of crayfish. II. Phasic fibres. *J. Exp. Biol.* 60:119–34

Bowerman, R. F., Larimer, J. L. 1976. Command neurons in crustaceans. *Comp. Biochem. Physiol. A.* 54:1–5

Brown, T. G. 1911. The intrinsic factors in the act of progression in the mammal. *Proc. Roy. Soc. London Ser. B* 84: 308–19

Bullock, T. H. 1976. In search of principles in neural integration. In *Simpler Networks and Behavior,* ed. J. C. Fentress, pp. 52–60. Sunderland, Mass.: Sinauer Assoc. 403 pp.

Bullock, T. H., Hamstra, R. H. Jr., Scheich, H. 1972. The jamming avoidance response of high frequency electric fish. II. Quantitative aspects. *J. Comp. Physiol.* 77:23–48

Burrows, M. 1975. Monosynaptic connexions between wing stretch receptors and flight motoneurons of the locust. *J. Exp. Biol.* 62:189–219

Burrows, M. 1976. Neural control of flight in the locust. In *Neural Control of Locomotion,* ed. R. M. Herman, S. Grillner, P. S. G. Stein, D. G. Stuart, pp. 419–38. New York: Plenum. 822 pp.

Burrows, M. 1977. Flight mechanisms in the locust. In *Identified Neurons and Behavior of Arthropods,* ed. G. Hoyle, pp. 339–56. New York: Plenum

Camhi, J. M. 1976. Non-rhythmic sensory inputs: influence on locomotory outputs in arthropods. See Burrows 1976, pp. 561–86

Cynader, M., Berman, N. 1972. Receptive-field organization of the monkey superior colliculus. *J. Neurophysiol.* 35:187–201

Davis, W. J. 1968. Lobster righting responses and their neural control. *Proc. Roy. Soc. London Ser. B* 170:435–56

Davis, W. J. 1976. Organizational concepts in the central motor networks of invertebrates. See Burrows 1976, pp. 265–92

Davis, W. J. 1977. The "command" neuron. See Burrows 1977, pp. 293–305

Davis, W. J., Gillette, R. 1977. Neural correlate of behavioral plasticity in command neurons. *Science.* In press

Davis, W. J., Kennedy, D. 1972. Command interneurons controlling swimmeret movements in the lobster. I. Types of effects on motoneurons. *J. Neurophysiol.* 35:1–12

Davis, W. J., Mpitsos, G. J., Pinneo, J. M., Ram, J. L. 1977. Modification of the behavioral hierarchy of *Pleurobranchaea:* I. Satiation and feeding motivation. *J. Comp. Physiol.* 117:99–125

Davis, W. J., Mpitsos, G. J., Siegler, M. V. S., Pinneo, J. M., Davis, K. B. 1974. Neuronal substrates of behavioral hierarchies and associative learning in the mollusk *Pleurobranchaea. Am. Zool.* 14:1037–50

Deliagina, T. G., Feldman, A. G., Gelfand, I. M., Orlovsky, G. N. 1975. On the role of central program and afferent inflow in the control of scratching movements in the cat. *Brain Res.* 100:297–313

Diamond, J. 1971. The Mauthner cell. In *Fish Physiology,* ed. W. S. Hoar, D. J. Randall, 5:265–346. New York: Academic. 600 pp.

Dichgans, J., Bizzi, E., Morasso, P., Tagliasco, V. 1973. Mechanisms underlying recovery of eye-head coordination in monkeys. *Exp. Brain Res.* 18:548–62

Doty, R. W. Sr. 1976. The concept of neural centers. See Bullock 1976, pp. 251–65

Duysens, J., Pearson, K. G. 1976. The role of cutaneous afferents from the distal hindlimb in the regulation of the step cycle of thalamic cats. *Exp. Brain Res.* 24:245–55

Eaton, R. C., Bombardieri, R. A., Meyer, D. L. 1977. The Mauthner-initiated startle response in teleost fish. *J. Exp. Biol.* 66:65–81

Eaton, R. C., Farley, R. D., Kimmel, C. B., Schabtach, E. 1977. Functional development in the Mauthner cell system of embryos and larvae of the zebra fish. *J. Neurobiol.* 8:151–72

Edgerton, V. R., Grillner, S., Sjöström, A., Zangger, P. 1976. Central generation of locomotion in vertebrates. See Burrows 1976, pp. 439–64

Evarts, E. V. 1966. Pyramidal tract activity associated with a conditioned hand movement in the monkey. *J. Neurophysiol.* 29:1011–27

Evarts, E. V. 1968. Relation of pyramidal tract activity to force exerted during voluntary movement. *J. Neurophysiol.* 31:14–27

Evarts, E. V., Bizzi, E., Burke, R. E., DeLong, M., Thach, W. T. Jr. 1971. Central control of movement. *Neurosci. Res. Program Bull.* 9:1–170

Evarts, E. V., Tanji, J. 1976. Reflex and intended responses in motor cortex pyramidal tract neurons of monkey. *J. Neurophysiol.* 39:1069–80

Evoy, W. H. 1976. Modulation of proprioceptive information in crustacea. See Burrows 1976, pp. 617–45

Evoy, W. H., Kennedy, D. 1967. The central nervous organization underlying control of antagonistic muscles in the crayfish. I. Types of command fibers. *J. Exp. Zool.* 165:223–38

Feldman, A. G., Orlovsky, G. N. 1975. Activity of interneurons mediating reciprocal Ia inhibition during locomotion. *Brain Res.* 84:181–94

Fields, H. L. 1966. Proprioceptive control of posture in the crayfish abdomen. *J. Exp. Biol.* 44:455–68

Fields, H. L., Evoy, W. H., Kennedy, D. 1967. Reflex role played by efferent control of an invertebrate stretch receptor. *J. Neurophysiol.* 30:859–74

Flock, A., Russell, I. 1976. Inhibition by efferent nerve fibers: action on hair cells and afferent synaptic transmission on the lateral line organ of the turbot *Lota lota. J. Physiol. London* 257:45–62

Forssberg, H., Grillner, S., Rossignol, S., Wallen, P. 1976. Phasic control of reflexes during locomotion in vertebrates. See Burrows 1976, pp. 647–74

Fourtner, C. R. 1976. Central nervous control of cockroach walking. See Burrows 1976, pp. 401–18

Friesen, W. O. 1975. Synaptic interactions in the cardiac ganglion of the spiny lobster *Panulirus interruptus. J. Comp. Physiol.* 101:191–205

Friesen, W. O., Poon, M., Stent, G. S. 1976. An oscillatory neuronal circuit generating a locomotory rhythm. *Proc. Natl. Acad. Sci. USA* 73:3734–38

Getting, P. A. 1975. *Tritonia* swimming: triggering of a fixed action pattern. *Brain Res.* 96:128–33

Getting, P. A. 1976. Afferent neurons mediating escape swimming of the marine mollusc, *Tritonia. J. Comp. Physiol.* 110:271–86

Gillette, R., Davis, W. J. 1977. The role of the metacerebral giant neuron in the feeding behavior of *Pleurobranchaea. J. Comp. Physiol.* 116:129–59

Gillette, R., Kovac, M. P., Davis, W. J. 1978. Command neurons receive synaptic feedback from the motor network they excite. *Science.* In press

Glantz, R. M. 1974a. Habituation of the motion detectors of the crayfish optic nerve: their relationship to the visually evoked defense reflex. *J. Neurobiol.* 5:489–510

Glantz, R. M. 1974b. Defense reflex and motion detector responsiveness to approaching targets: the motion detector trigger to the defense reflex pathway. *J. Comp. Physiol.* 95:297–314

Glantz, R. M. 1977. Visual input and motor output of command interneurons of the defense reflex pathway in the crayfish. See Burrows 1977, pp. 259–74

Gray, J. 1968. *Animal Locomotion.* London: Weidenfeld & Nicolson. 479 pp.

Grillner, S. 1975. Locomotion in vertebrates —central mechanisms and reflex interaction. *Physiol. Rev.* 55:247–304

Grillner, S. 1976. Some aspects on the descending control of the spinal circuits generating locomotor movements. See Burrows 1976, pp. 351–75

Grillner, S., Zangger, P. 1974. Locomotor movements generated by the deafferented spinal cord. *Acta Physiol. Scand.* 91:38A–39A

Gurfinkel, V. S., Shik, M. L. 1973. The control of posture and locomotion. In *Motor Control,* ed. A. A. Gydikov, N. T. Tankov, D. S. Kosarov, pp. 217–34. New York: Plenum. 259 pp.

Halbertsma, J., Miller, S., van der Meche, F. G. A. 1976. Basic programs for the phasing of flexion and extension movements of the limbs during locomotion. See Burrows 1976, pp. 489–517

Hartline, D. K. 1967. Impulse identification and axon mapping of the nine neurons in the cardiac ganglion of the lobster *Homarus americanus. J. Exp. Biol.* 47:327–40

Herman, R. M., Grillner, S., Stein, P. S. G., Stuart, D. G., eds. 1976. *Neural Control of Locomotion.* New York: Plenum. 822 pp.

Hughes, G. M., Wiersma, C. A. G. 1960. The coordination of swimmeret movements in the crayfish *Procambarus clarkii* (Girard). *J. Exp. Biol.* 37:657–70

Hultborn, H. 1972. Convergence on interneurons in the reciprocal Ia inhibitory pathway to motoneurons. *Acta Physiol. Scand. Suppl.* 375:1–42

Hultborn, H., Illert, M., Santini, M. 1976. Convergence on interneurones mediating the reciprocal Ia inhibition of motoneurones. I. Disynaptic Ia inhibition of Ia inhibitory interneurones. *Acta Physiol. Scand.* 96:193–201

Ikeda, K., Wiersma, C. A. G. 1964. Autogenic activity in the abdominal ganglia of the crayfish: the control of swimmeret movements. *Comp. Biochem. Physiol.* 12:107–15

Ito, M. 1972. Neural design of the cerebellar motor system. *Brain Res.* 40:81–84

Ito, M., Nisimaru, N., Yamamoto, M. 1977. Specific patterns of neuronal connexions involved in the control of the rabbit's vestibulo-ocular reflexes by the cerebellar flocculus. *J. Physiol. London* 265:833–54

Jankowska, E., Jukes, M. G. M., Lund, S., Lundberg, A. 1967. The effect of DOPA on the spinal cord. 5. Reciprocal organization of pathways transmitting excitatory action to alpha motoneurones of flexors and extensors. *Acta Physiol. Scand.* 70:369–88

Jankowska, E., Roberts, W. J. 1972. Synaptic actions of single interneurons mediating reciprocal Ia inhibition of motoneurones. *J. Physiol. London* 222:623–42

Kashin, S. M., Feldman, A. G., Orlovsky, G. N. 1974. Locomotion of fish evoked by electrical stimulation of the brain. *Brain Res.* 82:41–47

Kater, S. B. 1974. Feeding in *Helisoma trivolvis:* the morphological and physiological bases of a fixed action pattern. *Am. Zool.* 14:1017–36

Kater, S. B., Rowell, C. H. F. 1973. Integration of sensory and centrally programmed components in the generation of cyclical feeding activity of *Helisoma trivolvis. J. Neurophysiol.* 36:142–55

Kennedy, D. 1969. The control of output by central neurons. In *The Interneuron,* ed. M. A. B. Brazier, pp. 21–36. Berkeley: Univ. Calif. 552 pp.

Kennedy, D. 1976. Neural elements in relation to network function. See Bullock 1976, pp. 65–81

Kennedy, D., Calabrese, R. L., Wine, J. J. 1974. Presynaptic inhibition: primary afferent depolarization in crayfish neurons. *Science* 186:451–54

Kennedy, D., Davis, W. J. 1977. Organization of invertebrate motor systems. In *Cellular Biology of Neurons, Handbook of Physiology, The Nervous System, Sect. 1, Vol. 1, Pt. 2,* ed. E. R. Kandel, pp. 1023–87. Bethesda: Am. Physiol. Soc.

Kennedy, D., Evoy, W. H., Dane, B., Hanawalt, J. T. 1967. The central nervous organization underlying control of antagonistic muscles in the crayfish. II. Coding of position by command fibers. *J. Exp. Zool.* 165:239–48

Kennedy, D., Evoy, W. H., Hanawalt, J. T. 1966. Release of coordinated behavior in crayfish by single central neurons. *Science* 154:917–19

Kimmel, C. B., Eaton, R. C. 1976. Development of the Mauthner cell. See Bullock 1976, pp. 186–202

Kling, U., Szekely, G. 1968. Simulation of rhythmic nervous activities. I. Function of networks with cyclic inhibitions. *Kybernetik* 5:89–103

Koester, J., Mayeri, E., Liebeswar, G., Kandel, E. R. 1974. Neural control of circulation in *Aplysia.* II. Interneurons. *J. Neurophysiol.* 37:476–96

Kovac, M. P., Davis, W. J. 1978. Behavioral choice: neural mechanism in *Pleurobranchaea. Science.* In press

Kow, L. M., Montgomery, M. O., Pfaff, D. W. 1977. Effects of spinal cord transections on lordosis reflex in female rats. *Brain Res.* 123:75–88

Krasne, F. B., Bryan, J. S. 1973. Habituation: regulation via presynaptic inhibition. *Science* 182:590–92

Krasne, F. B., Wine, J. J. 1977. The control of crayfish escape behavior. See Burrows 1977, pp. 275–92

Kristan, W. B. Jr., Calabrese, R. L. 1976. Rhythmic swimming activity in neurones of the isolated nerve cord of the leech. *J. Exp. Biol.* 65:643–68

Kristan, W. B. Jr., Stent, G. S., Ort, C. A. 1974. Neuronal control of swimming in the medicinal leech. III. Impulse patterns of motor neurons. *J. Comp. Physiol.* 94:155–76

Larimer, J. L. 1976. Command interneurons and locomotor behavior in crustaceans. See Burrows 1976, pp. 293–326

Larimer, J. L., Eggleston, A. C., Masukawa, L. M., Kennedy, D. 1971. The different connections and motor outputs of lateral and medial giant fibers in the crayfish. *J. Exp. Biol.* 54:391–402

Larimer, J. L., Gordon, W. H. III. 1977. Circumesophageal interneurons and behavior in crayfish. See Burrows 1977, pp. 243–58

Lennard, P. R., Stein, P. S. G. 1977. Swimming movements elicited by electrical stimulation of turtle spinal cord: low-spinal and intact preparations. *J. Neurophysiol.* 40:768–78

Lynch, J. C., Mountcastle, V. B., Talbot, W. H., Yin, T. C. T. 1977. Parietal lobe mechanisms for directed visual attention. *J. Neurophysiol.* 40:362–89

Macmillan, D. L., Wales, W., Laverack, M. S. 1976. Mandibular movements and their control in *Homarus gammarus.* III. The effect of load. *J. Comp. Physiol.* 106:207–21

Marsden, C. D., Merton, P. A., Morton, H. B. 1977. The sensory mechanism of servo action in human muscles. *J. Physiol. London* 265:521–35

Mayeri, E. 1973. Functional organization of the cardiac ganglion of the lobster, *Homarus americanus. J. Gen. Physiol.* 62:448–72

Maynard, D. M. 1972. Simpler networks. *Ann. NY Acad. Sci.* 193:59–72

Mendelson, M. 1971. Oscillator neurons in crustacean ganglia. *Science* 171: 1170–73

Miles, F. A., Fuller, J. H. 1974. Adaptive plasticity in the vestibuloocular responses of the rhesus monkey. *Brain Res.* 80:512–16

Mittenthal, J. E., Wine, J. J. 1973. Connectivity patterns of crayfish giant interneurons: visualization of synaptic regions with cobalt dye. *Science* 179: 182–84

Mohler, C. W., Wurtz, R. H. 1976. Organization of monkey superior colliculus: intermediate layer cells discharging before eye movements. *J. Neurophysiol.* 39:722–44

Mori, S., Shik, M. L., Yagodnitsyn, A. S. 1977. Role of pontine tegmentum for locomotor control in mesencephalic cats. *J. Neurophysiol.* 40:284–95

Mountcastle, V. B. 1976. The world around us: neural command functions for selective attention. *Neurosci. Res. Program Bull. Suppl.* 14:1–47

Mountcastle, V. B., Lynch, J. C., Georgopoulos, A., Sakata, H., Acuna, C. 1975. Posterior parietal association cortex of the monkey: command functions for operations within extrapersonal space. *J. Neurophysiol.* 38:871–908

Nottebohm, F., Stokes, T. M., Leonard, C. M. 1976. Central control of song in the canary, *Serinus canarius. J. Comp. Neurol.* 165:457–86

Orlovsky, G. N. 1972. The effect of different descending systems on flexor and extensor activity during locomotion. *Brain Res.* 40:359–71

Pearson, K. G. 1972. Central programming and reflex control of walking in the cockroach. *J. Exp. Biol.* 56:173–93

Pearson, K. G. 1976a. Nerve cells without action potentials. See Bullock 1976, pp. 99–110

Pearson, K. G. 1976b. The control of walking. *Sci. Am.* 235(6):72–87

Pearson, K. G. 1977. Interneurons in the ventral nerve cord of insects. See Burrows 1977, pp. 329–37

Pearson, K. G., Duysens, J. 1976. Function of segmental reflexes in the control of stepping in cockroaches and cats. See Burrows 1976, pp. 519–37

Pearson, K. G., Fourtner, C. R. 1975. Nonspiking interneurons in the walking system of the cockroach. *J. Neurophysiol.* 38:33–52

Pearson, K. G., Iles, J. F. 1973. Nervous mechanisms underlying intersegmental co-ordination of leg movements during walking in the cockroach. *J. Exp. Biol.* 58:725–44

Pearson, K. G., Wong, R. K. S., Fourtner, C. R. 1976. Connexions between hair-plate afferents and motoneurons in the cockroach leg. *J. Exp. Biol.* 64:251–66

Perkel, D. H., Mulloney, B. 1974. Motor pattern production in reciprocally inhibitory neurons exhibiting postinhibitory rebound. *Science* 185:181–83

Pfaff, D. W., Diakow, C., Zigmond, R. E., Kow, L. 1974. Neural and hormonal determinants of female mating behavior in rats. In *The Neurosciences: Third Study Program,* ed. F. O. Schmidt, F. G. Worden, pp. 621–46. Cambridge, Mass: MIT. 1107 pp.

Prochazka, A., Westerman, R. A., Ziccone, S. P. 1976. Discharges of single hindlimb afferents in the freely moving cat. *J. Neurophysiol.* 39:1090–1104

Rakic, P. 1975. Local circuit neurons. *Neurosci. Res. Program Bull.* 13:291–446

Robinson, D. A. 1972. Eye movements evoked by collicular stimulation in the alert monkey. *Vision Res.* 12:1795–1808

Robinson, D. A. 1976. Adaptive gain control of vestibuloocular reflex by the cerebellum. *J. Neurophysiol.* 39:954–69

Rovainen, C. M. 1967. Physiological and anatomical studies on large neurons of the central nervous system of the sea lamprey (*Petromyzon marinus*). I. Muller and Mauthner cells. *J. Neurophysiol.* 30:1000–23

Russell, D. F. 1976. Rhythmic excitatory inputs to the lobster stomatogastric ganglion. *Brain Res.* 101:582–88

Russell, I. J. 1976. Central inhibition of lateral line input in the medulla of goldfish by neurones which control active body movements. *J. Comp. Physiol.* 111: 335–58

Scheich, H. 1977. Neural basis of communication in the high frequency electric fish, *Eigenmannia virescens* (jamming avoidance response). III. Central integration in the sensory pathway and control of the pacemaker. *J. Comp. Physiol.* 113:229–55

Schiller, P. H., Stryker, M. 1972. Single-unit recording and stimulation in superior colliculus of the alert rhesus monkey. *J. Neurophysiol.* 35:915–24

Schmidt, R. S. 1976. Neural correlates of frog calling: isolated brainstem. *J. Comp. Physiol.* 108:99–113

Selverston, A. I., Russell, D. F., Miller, J. P., King, D. G. 1976. The stomatogastric nervous system: structure and function of a small neural network. *Prog. Neurobiol.* 7:215–90

Severin, F. V., Orlovsky, G. N., Shik, M. L. 1967. Work of the muscle receptors during controlled locomotion. *Biophysics USSR* 12:575–86

Sherrington, C. S. 1910. Flexion-reflex of the limb, crossed extension reflex, and reflex stepping and standing. *J. Physiol. London* 40:28–121

Shik, M. L., Orlovsky, G. N. 1976. Neurophysiology of locomotor automatism. *Physiol. Rev.* 56:465–501

Shik, M. L., Severin, F. V., Orlovsky, G. N. 1966. Control of walking and running by means of electrical stimulation of the midbrain. *Biophysics USSR* 11:756–65

Simmons, J. A., Howell, D. J., Suga, N. 1975. Information content of bat sonar echoes. *Am. Sci.* 63:204–15

Sjöström, A., Zangger, P. 1976. Muscle spindle control during locomotor movements generated by the deafferented spinal cord. *Acta Physiol. Scand.* 97: 281–91

Sokolove, P. G. 1973. Crayfish stretch receptor and motor unit behavior during abdominal extensions. *J. Comp. Physiol.* 84:251–66

Sokolove, P. G., Tatton, W. G. 1975. Analysis of postural motoneuron activity in crayfish abdomen. I. Coordination by premotoneuron connections. *J. Neurophysiol.* 38:313–31

Sparks, D. L., Holland, R., Guthrie, B. L. 1976. Size and distribution of movement fields in the monkey superior colliculus. *Brain Res.* 113:21–34

Stein, P. S. G. 1971. Intersegmental coordination of swimmeret motoneuron activity in crayfish. *J. Neurophysiol.* 34: 310–18

Stein, P. S. G. 1974. The neural control of interappendage phase during locomotion. *Am. Zool.* 14:1003–16

Stein, P. S. G. 1976. Mechanisms of interlimb phase control. See Burrows 1976, pp. 465–87

Stein, P. S. G. 1977. A comparative approach to the neural control of locomotion. See Burrows 1977, pp. 227–39

Stein, R. B., Pearson, K. G., Smith, R. S., Redford, J. B., eds. 1973. *Control of Posture and Locomotion.* New York: Plenum. 635 pp.

Suga, N., Jen, P. H. S. 1975. Peripheral control of acoustic signals in the auditory system of echolocating bats. *J. Exp. Biol.* 62:277–311

Suga, N., Schlegel, P. 1972. Neural attenuation of responses to emitted sounds in echolocating bats. *Science* 177:82–84

Tanji, J., Evarts, E. V. 1976. Anticipatory activity of motor cortex neurons in relation to direction of intended movements. *J. Neurophysiol.* 39:1062–68

Truman, J. W. 1976. Hormonal release of differentiated behavior patterns. See Bullock 1976, pp. 111–20

Warshaw, H. S., Hartline, D. K. 1976. Simulation of network activity in stomatogastric ganglion of the spiny lobster, *Panulirus. Brain Res.* 110:259–72

Watanabe, A., Obara, S., Akiyama, T. 1967. Pacemaker potentials for the periodic burst discharge in the heart ganglion of a stomatopod, *Squilla oratoria. J. Gen. Physiol.* 50:839–62

Wendler, G. 1974. The influence of proprioceptive feedback on locust flight coordination. *J. Comp. Physiol.* 88:173–200

Wetzel, M. C., Stuart, D. G. 1976. Ensemble characteristics of cat locomotion and its neural control. *Prog. Neurobiol.* 7:1–98

Wiersma, C. A. G. 1947. Giant nerve fiber system of the crayfish. A contribution to comparative physiology of synapse. *J. Neurophysiol.* 10:23–38

Wiersma, C. A. G. 1952. Neurons of arthropods. *Cold Spring Harbor Symp. Quant. Biol.* 17:155–63

Wiersma, C. A. G., Ikeda, K. 1964. Interneurons commanding swimmeret movements in the crayfish, *Procambarus clarkii* (Girard). *Comp. Biochem. Physiol.* 12:509–25

Willows, A. O. D. 1976. Trigger neurons in the mollusk *Tritonia.* See Burrows 1976, pp. 327–49

Willows, A. O. D., Hoyle, G. 1969. Neuronal network triggering a fixed action pattern. *Science* 166:1549–51

Wilson, D. M. 1961. The central nervous control of flight in a locust. *J. Exp. Biol.* 38:471–90

Wilson, D. M. 1967. An approach to the problem of control of rhythmic behavior. In *Invertebrate Nervous Systems,* ed. C. A. G. Wiersma, pp. 219–29. Chicago: Univ. Chicago Press. 370 pp.

Wilson, D. M. 1968a. The flight control system of a locust. *Sci. Am.* 218(5):83–90

Wilson, D. M. 1968b. Inherent asymmetry and reflex modulation of the locust flight pattern. *J. Exp. Biol.* 48:631–41

Wilson, D. M. 1972. Genetic and sensory mechanisms for locomotion and orientation in animals. *Am. Sci.* 60:358–65

Wilson, D. M., Gettrup, E. 1963. A stretch reflex controlling wingbeat frequency in grasshoppers. *J. Exp. Biol.* 40:171–85

Wilson, D. M., Waldron, I. 1968. Models for the generation of the motor output pattern in flying locusts. *Proc. IEEE* 56:1058–64

Wilson, D. M., Wyman, R. J. 1965. Motor output patterns during random and rhythmic stimulation of locust thoracic ganglia. *Biophys. J.* 5:121–43

Wine, J. J., Krasne, F. B. 1972. The organization of escape behavior in the crayfish. *J. Exp. Biol.* 56:1–18

Wurtz, R. H., Goldberg, M. E. 1972. Activity of superior colliculus in behaving monkey. III. Cells discharging before eye movements. *J. Neurophysiol.* 35:575–86

Zottoli, S. J. 1977. Correlation of the startle reflex and Mauthner cell auditory responses in unrestrained goldfish. *J. Exp. Biol.* 66:243–54

Zucker, R. S. 1972. Crayfish escape behavior and central synapses. II. Physiological mechanisms underlying behavioral habituation. *J. Neurophysiol.* 35:621–37

Zucker, R. S., Kennedy, D., Selverston, A. I. 1971. Neuronal circuit mediating escape responses in crayfish. *Science* 173:645–50

Ann. Rev. Neurosci. 1978. 1:83–102

PAIN

❖11504

Frederick W. L. Kerr and Peter R. Wilson
Department of Neurologic Surgery, Mayo Clinic and Foundation,
Rochester, Minnesota 55901

INTRODUCTION

Over the past decade, advances in our understanding of pain mechanisms have taken place at an unprecedented rate and over a wide front. Particularly noteworthy contributions include stimulus-produced analgesia, identification of the opiate receptor, and elucidation of the mechanisms and sites of action of morphine analgesia. Finally, the discovery and characterization of an entirely new endogenous analgesic system, unknown until 1975, yet fitting smoothly into the pattern of these previous findings, has literally transformed thinking in this field in a manner that could not have been anticipated ten years ago.

This review attempts to summarize and assess these and other areas in which important progress is occurring.

NEUROANATOMY

Primary Afferents (PA)

It is well known that at least two populations of neurons—small, dark and large, paler elements—are present in dorsal root ganglia. Neurochemical support for this has now been provided by Hökfelt et al (1975a), who have demonstrated immunohistochemically in the rat that approximately 10 percent of these neurons, mostly those of small size, contain somatostatin, a polypeptide with potent depressant activity, identified and characterized (H-Ala-Gly-Cys-Lys-Asn-Phe-Phe-Trp-Lys-Thr-Phe-Thr-Ser-Cys-OH) by Burgus et al (1973) in extracts of the hypothalamus. Hökfelt et al (1975a), have subsequently shown that there is a dense plexus of somatostatin-positive fibers in the substantia gelatinosa of the cord and suggested that it may act as an inhibitory transmitter for the sensory system.

Substance P (SP) was first isolated by von Euler & Gaddum (1931) from the brain. It was shown by Chang & Leeman (1970) to be an undecapeptide (H-Arg-Pro-Lys-Pro-Gln-Gln-Phe-Phe-Gly-Leu-Met-NH_2) and was synthesized by Tregear et al (1971). It has striking depolarizing effects on motoneuron membranes (Otsuka et al 1972, Konishi & Otsuka 1974) and has been demonstrated in dorsal roots by

0147-006X/78/0325-0083$01.00

Otsuka et al (1972) and Takahashi et al (1974); it was demonstrated in the dorsal horn by Takahashi & Otsuka (1975), by use of bioassay techniques. Hökfelt et al (1975b), by employing immunohistochemical methodology, have localized SP in the skin and in a most elegant manner in laminae I, II, and III of the spinal cord and in the tract of Lissauer, where it appears to be concentrated in the endings of PA. Following dorsal rhizotomy there was a marked decrease in the intensity of the fluorescence in the s. gelatinosa, while compression of dorsal roots (DR) resulted in the accumulation of SP in small cell bodies of the ganglia and in some of the DR fibers. Since SP is a more potent excitatory transmitter in the spinal cord than glutamate, its role in sensory transmission may be particularly significant, as noted by Hökfelt et al (1975b). Hökfelt et al (1976) have subsequently shown that two separate populations of primary afferent neurons are present, one containing somatostatin and the other SP.

In recent years the classical concept that small primary afferents enter the cord in a lateral position has been challenged; however, evidence both in man (Sindou et al 1974) and in monkey (Kerr 1975a, Snyder 1977) supports the older view; some of the controversy appears to be due to species differences, since according to Snyder there is no lateral accumulation of fine fibers in cat dorsal roots.

It has been known for many years that unmyelinated fibers are present in ventral roots as noted by Applebaum et al (1976), though their source and function remained uncertain. Coggeshall et al (1973) and Applebaum et al (1976) have demonstrated that approximately 50% of unmyelinated fibers in ventral roots are sensory (the other 50% are preganglionic efferents); one third of the sensory fibers are associated with high-threshold somatic receptors, while two thirds have visceral receptive fields (Clifton et al 1976). The suggestion that they may be responsible for the failure of dorsal rhizotomy to relieve pain in humans (Coggeshall et al 1975), however, is not supported by either ablation or stimulation studies in man, as discussed in detail by White & Sweet (1955).

Dorsal Horn

In the marginal zone, Narotzky & Kerr (1977) reported that PA input to large marginal neurons was predominantly to the intermediate and more distal dendrites, whereas lateral Lissauer tractotomy, which presumably transects axons of gelatinosa neurons, results in predominant degeneration of axo-somatic contacts; this suggests that the excitatory effects of PA on marginal neurons are modulated by gelatinosa cells whose function as inhibitory interneurons has been proposed by Denny-Brown et al (1973). A considerable degree of convergence from separate dorsal roots was indicated, in agreement with the electrophysiological observations of Christensen & Perl (1970). Marginal neurons do not form a homogenous population as inferred from their axonal projections; some of their axons enter the lateral funiculus (Cajal 1909), others ascend in the anterolateral quadrant of the cord (Kuru 1949, Kumazawa et al 1975) to the diencephalon (Trevino et al 1973, Albe-Fessard et al 1974), while still others project to more caudal levels of the cord (Burton & Loewy 1976). The possibility that bifurcated axons might result in both an ascending and a descending projection should be considered.

Detailed three-dimensional reconstructions of neurons in the s. gelatinosa by Sugiura (1975) have shown apparent differences between the axonal distribution of cells in lamina II and those in lamina III. Most axons of lamina II cells ran to lamina I but were lost when they entered the tract of Lissauer; some followed a ventral course into "various laminae of the dorsal horn," and a few of these, after passing through the full thickness of the dorsal gray, entered the lateral funiculus and bifurcated into ascending and descending branches. Similar reconstruction of the axonal and dendritic distribution of three neurons in lamina III of a kitten by Mannen & Sugiura (1976) showed a significantly different picture; in each instance the axon and its branches were restricted to a distance of less than 1 mm from the cell body; they thus correspond to Golgi type II or local circuit elements. Cajal (1909), on the other hand, strongly emphasized that short-axon neurons were extremely rare in the spinal gray and were limited to occasional elements in the s. gelatinosa. Detailed studies of the synaptic organization of the glomeruli of the s. gelatinosa have been reported by Gobel (1974).

INPUT TO THE DORSAL HORN The classical, generally accepted manner of distribution of PA in this area dates from Cajal's studies (1909, reviewed by Réthelyi & Szentagothai 1973; Kerr 1975a); according to this view, the large PA's enter along the medial aspect of the dorsal gray and reverse direction to approach the s. gelatinosa from its ventral surface. Only after entering this area do they divide into numerous terminal brushes oriented radially; gelatinosa neurons, together with apical dendrites of deeper-lying neurons (laminae IV, V, and possibly VI), are arranged in alternating radial sheets between the former. Small PA's enter the s. gelatinosa from its dorsal aspect, usually after bifurcating and ascending or descending for one to two segments. Réthelyi (1977) has proposed an entirely different type of organization, based on studies of Golgi-stained material in cats. He concluded that fine PA's, probably of the unmyelinated type, are the only ones ending in lamina II (which he regards as the entire s. gelatinosa), and he correlated these conclusions with the finding of Kumazawa & Perl (1977) that cutaneous receptors with C-afferents are excitatory to gelatinosa neurons. He suggests that the long conduction times observed would be hard to relate to myelinated PA's. However, longer delays might occur in the branches of large PA's, especially in the subterminal candelabrum. As Réthelyi noted, much of the supporting evidence is indirect and circumstantial at this time.

However, using different techniques (namely, suppressive silver and axonal transport), LaMotte (1977) has reported that small fibers end in laminae I, II, and III, whereas large fibers are said to end in laminae IV, V, and VI. Again, the evidence is indirect and further information is required before Cajal's 1909 classical concept, based on direct evidence from Golgi-stained material and repeatedly confirmed (Szentagothai 1964, Sprague & Ha 1964, Scheibel & Scheibel 1968), can be replaced.

OUTPUT FROM THE DORSAL HORN Neurons responsible for projections to higher levels have been identified in the cat and monkey by Trevino & Carstens (1975), by use of the horseradish peroxidase (HRP) transport method. Following relatively large injections into the thalamus of monkeys, the majority of labeled cells were found in laminae IV and V (laterally) on the contralateral side, with smaller numbers in laminae I, VII, and VIII; the close correspondence between these observations and electrophysiologic localization by antidromic firing was noted.

Several aspects of the composition and location of nociceptive tracts in the spinal cord have been investigated. Lippman & Kerr (1972), using light- and electron-microscopic methods, looked for unmyelinated fibers in the spinothalamic tract, and, finding none, concluded that the relay to higher levels of C-fiber nociceptive input of dorsal roots appears to be over thin myelinated fibers in the tract.

Recurrence of pain following anterolateral cordotomy may be due to conduction in other areas of the cord. The existence of a ventral spinothalamic tract with similar connections to the lateral spinothalamic tract was described by Kerr (1975b). Nijensohn & Kerr (1975), in an axonal degeneration study of ascending projections in the dorsolateral funiculus in the primate, found no evidence for a significant supplementary pathway for nociception. However, Moffie (1975) has reported that good control of pain was achieved in two instances of percutaneous cordotomy in which the lesion was unintentionally made in the dorsolateral quadrant of the cord, which suggests that nociceptive relay may in fact ascend in this area. Conduction via multiple relays in the spinal gray may also account for late recurrence of pain. Basbaum (1973) has reported the presence of such relays in the spinal gray of the rat; however, because of marked species differences, extrapolation to primates and man must necessarily be cautious.

Descending Systems to the Spinal Cord

In addition to the well-known cortically originating systems that modulate transmission through the dorsal horn, Basbaum et al (1977) have reported evidence for a hitherto undescribed descending system in the dorsolateral funiculus of the spinal cord of the rat. It was believed to originate in the n. raphé magnus and to be, at least in part, serotoninergic. Transection of the dorsolateral funiculus decreased or abolished stimulus-produced analgesia (SPA) and had a comparable effect on the analgesia produced by systemic (intraperitoneal) administration of morphine. From the similarity of effects they inferred that both SPA and morphine analgesia may share the same neural substrate. They also inferred that morphine analgesia was not mediated directly via the n. raphé magnus, but that this was a relay nucleus for a descending system originating in the periaqueductal gray (see below). However, as indicated at a later point, there is good evidence that morphine can also act directly at the segmental level (Yaksh & Rudy 1976).

Brainstem and Thalamus

The organization of the periaqueductal gray and the n. raphé magnus are of special interest because of their role in SPA. In a detailed study using anterograde transport and autoradiography Ruda (1975) demonstrated widespread projections from this area; of these the following are of particular interest in the present context: (*a*) descending connections to the medullary reticular nuclei and to the n. raphé magnus, both of which project to the spinal cord, and (*b*) ascending projections to the intralaminar nuclei of the thalamus in a pattern very similar to that known for the spinothalamic tract.

At the thalamic level Burton & Jones (1976), using the axonal transport technique, have shown that the medial component of the posterior nuclear complex (PO area) projects to the retroinsular field, and they have reviewed the evidence for the participation of this system in pain mechanisms. This observation is especially significant in view of the long-standing controversy regarding cortical representation for pain and correlates well with recent physiological observations (Vycklicky et al 1972, Chatrian et al 1975; see below).

PHYSIOLOGY

Receptors

The structural and physicochemical mechanisms involved in transduction by cutaneous receptors remain unknown (Iggo 1976). Rather typically, low threshold receptors (mechanical and thermal) are inactivated by noxious stimuli, whereas nociceptors are much more resistant to damage (Burgess 1974).

Perl (1976) has shown that nociceptors supplied by fine (A-δ or C) fibers have two distinguishing features: (*a*) their threshold to natural stimuli is high in comparison to other receptors in the same tissue, and (*b*) they undergo a form of "sensitization", so that they respond at a lower stimulus intensity after requiring strong-intensity stimulation for initial activation. Thus, threshold may not be reached for a thermal nociceptor until 50°C or higher, but on repeated stimulation the threshold will be found to have dropped several degrees. A tenfold increase in discharge frequency in response to the same stimulus resulted from sensitization. Preliminary studies using perfused rabbit-ear preparations suggested that an unidentified agent was released by the injury. The possible relationship of this phenomenon to the hyperalgesia that occurs following thermal and other injuries to the skin was noted.

Peripheral Nerves

Activity of single A-δ and C-fiber activity in human peripheral nerves has been recorded by several groups using tungsten microelectrodes. Torebjörk (1974) and Torebjörk & Hallin (1974) were able to correlate electrically elicited pricking pain with A-delta fiber activity, whereas similarly induced discharges in C fibers were perceived as a prolonged, chronic, dull, sometimes burning pain. As these authors, Van Hees & Gybels (1973) and Van Hees (1976) have observed, C fiber activity can be elicited by electrical stimulation of the skin without pain perception; however, the great majority of unmyelinated fibers in man are concerned with nociception as opposed to only some 50% in the cat (Torebjörk 1974). In accord with this are the observations of Dubner & Beitel (1976), who demonstrated that activity in both A-δ heat nociceptors and C-polymodal nociceptors is necessary for escape behavior to occur in the monkey.

With regard to the balance between activity in large and small fibers in peripheral nerves, which has been postulated to be a major factor in the afferent modulation of pain, Dyck et al (1976) concluded from a study of human peripheral neuropathies that painfulness correlated better with the pathological appearance and rate of breakdown of myelinated fibers than with the ratio of surviving large to small fibers. They concluded, furthermore, that this neither supported nor refuted the gate control theory.

Spinal Cord

Important new data have been reported by Kirk & Denny-Brown (1970) and by Denny-Brown et al (1973) with regard to the functional properties of primary afferents, dermatomes, and substantia gelatinosa. They have shown in the monkey that adjacent dorsal root ganglia have a facilitatory influence on the function of an isolated test root, that individual dorsal roots have an area of distribution in the skin twice as large as generally believed, and that any locus in the ventral cutaneous distribution is supplied by at least five separate dorsal roots. They also concluded, on the basis of discrete lesions and testing of cutaneous sensation, that the medial aspect of the tract of Lissauer exerts a facilitatory effect, whereas the lateral part is inhibitory. In view of the origin of the axons, the latter effects were ascribed to gelatinosa neurons, and since they were reversed by a subconvulsive systemic dose of strychnine, it was suggested that glycine may be the transmitter.

The demonstration by Christensen & Perl (1970) that a significant proportion of the marginal neurons (lamina I) in the cat, is related to cutaneous nociceptors (thermal and mechanical), indicates that specificity for nociceptive input is maintained after arrival in the CNS. The observations have been confirmed and extended by Willis et al (1974) in the monkey and Giesler et al (1976) in the rat, while Mosso & Kruger (1973) have shown that marginal neurons in the spinal nucleus of the trigeminal nerve have similar functions. They have been activated antidromically from the thalamic level (Trevino et al 1973), and their role in nociceptive mechanisms is now well established.

Recording of activity in the s. gelatinosa has resisted most efforts (Wall 1973), but recently Perl (1976) has shown that some gelatinosa cells (in lamina II) have response characteristics similar to those of C-polymodal nociceptors, responding to noxious thermal and mechanical stimuli, but not to lower intensity stimulation. These findings are in good accord with the report by Nashold et al (1976) that discrete focal coagulations of the s. gelatinosa gave relief of otherwise intractable pain in patients with brachial plexus avulsions.

Most reports indicate that nociceptive neurons are also located in lamina V, though it should be noted that units projecting to rostral levels are widely distributed throughout the spinal gray, including laminae VI and VIII (Trevino et al 1973, Trevino & Carstens 1975).

Mayer et al (1975) concluded, on the basis of stimulus thresholds, refractory periods, and the report of pain in studies on primates and humans in which the anterolateral quadrant (ALQ) of the cord was stimulated to activate spinothalamic tract neurons, that neurons in lamina V are, of themselves, sufficient to mediate nociception in man. Based on the conduction velocities they recorded, the nociceptive axons in the ALQ are in the myelinated fiber caliber range. This correlates well with the report by Lippman & Kerr (1972) referred to earlier with regard to the absence of unmyelinated fibers in the spinothalamic tract.

A significant proportion of the units in lamina V responsible for nociceptive relay is referred to by Price & Mayer (1974, 1975) as wide dynamic-range neurons, since they respond to both non-noxious and noxious stimuli, their firing rate increasing in parallel with stimulus intensity.

Cervero et al (1976) have divided dorsal horn neurons into three classes: Class I, which are excited by cutaneous mechanoreceptors, Class II, which are excited by both mechano- and nociceptors, and Class 3, which respond only to noxious input. Class 3 is further divided into 3a, which receive A-δ input, and 3b, excited by both A-δ and C fibers. The Class 2 cells would appear to correspond to the wide dynamic-range neurons of Price & Mayer (1975), but Iggo (1974) and Cervero et al (1976) note their low threshold and maintained activity in response to natural stimulation of the skin militates against their having a significant role in pain mechanisms. They also consider that Class 3 neurons, which are located predominantly in lamina I and occasionally in lamina II are, on neurophysiological grounds, more likely to be involved in nociceptive processing than Class 2 cells.

To summarize, it appears at this time that there are two populations of nociceptive neurons in the dorsal horn; the marginal neurons that are commonly but not always activated by high-threshold polymodal nociceptors, and lamina V neurons, which also have polymodal input but with a relatively wide range of thresholds. What their individual roles are in the nociceptive process remains to be determined, but it would appear that both populations integrate and relay nociceptive signals to higher levels.

Kumazawa & Perl (1977) have investigated the relationship between A-δ and C-fiber peripheral input and the resultant activity of interneurons in the dorsal horn. They reported that A-δ input is distributed to cells in the marginal zone, whereas neurons responding to C-fiber input are located in the substantia gelatinosa.

Good evidence is now available for convergence of visceral and somatic activity on single neurons in deeper layers of the dorsal horn. (Pomeranz et al 1968, Hancock et al 1975).

Trevino et al (1973) have established that neurons projecting to the thalamus are present in layers I, IV, V, and VI of the dorsal horn and also in lamina VII. This widespread representation of the source of the spinothalamic system is supported by the retrograde axoplasmic transport studies of Trevino & Carstens (1975) referred to earlier.

A clinical observation of particular interest with regard to the function of the lateral spinothalamic tract was made by Noordenbos & Wall (1976), in a patient in whom all but the anterolateral quadrant (ALQ) of the cord on one side was severed. One month after the injury the patient was able to identify temperature and pinprick on the side opposite the intact ALQ as would be expected, but she also reported an unpleasant sensation to pinprick and a stinging sensation on electrical stimulation of the skin on the side of the preserved ALQ. However, there was no description of thermal, crush, or deep pressure-pain sensation. Touch, pressure, and passive movement could also be identified relatively well on the side ipsilateral to the intact

ALQ, thus indicating a considerably greater role for the ALQ than generally accepted.

EFFERENT CONTROL OF SPINAL SENSORY MECHANISMS Modulation of activity of spinal sensory input by more rostral systems is a well-recognized phenomenon. Marked specificity of effects appears to exist, as shown by single unit analyses reported by several groups. Sensorimotor cortex stimulation effects on identified spinothalamic neurons were described by Coulter et al (1974); low-threshold units (lamina IV) were inhibited, but high-threshold neurons in lamina V were unaffected. Handwerker et al (1975) described tonic inhibitory effects on nociceptive units that were abolished by cold block of the cord rostral to the recording site.

The recently identified areas (periventricular gray and raphé nuclei) from which stimulus-produced analgesia is obtained, have precisely the opposite effects on spinal sensory neurons. Thus, Oliveras et al [1974 (midbrain), 1975] and Beall et al [1976 (medulla)] have shown that inhibitory effects from these sites are exerted specifically on nociceptive neurons in the dorsal horn, whereas low-threshold units are usually unaffected.

Medulla

Neurons activated specifically by high-threshold cutaneous receptors have been identified in the medullary reticular formation by Benjamin (1970), Casey (1971), and Casey and co-workers (1974), who indicate that these units are concerned with avoidance behavior. It is important to recognize that

1. Single unit studies indicate only that a cell is activated by a stimulus—in the present context, a peripheral one of high intensity. However, the unit may be either in a reflex pathway (e.g. vasopressor or pupillomotor) or in a pathway involved in the perception of pain.
2. No description of activity clearly indicative of pain from stimulation of the lower reticular formation has been found in our review of the literature. For further details the excellent review of the reticular formation and nociception by Bowsher (1976) should be consulted.

Midbrain

At the mesencephalic level, Niquist & Greenhoot (1974) and Young & Gottschaldt (1976) have described units in the reticular formation that respond to high-threshold receptors. In a study of the PO group in the cat, Curry (1972) found only 3% of units with nociceptive characteristics and concluded that it was not a "pain center." However, deeper levels of anesthesia than used by previous investigators may well have biased the results.

Thalamus

Ishijima & Sano (1971) have found a good correlation between activation of C-fibers and activity in nonspecific thalamic nuclei; similarly, myelinated fiber activity was associated with discharges in the ventroposterolateral nucleus (VPL). Furthermore, they have reported strong inhibitory effects of the latter system on the former.

Cortex

Evidence for nociceptive input to the cerebral cortex is now available from several different sources. Berkley & Parmer (1974) have reported that ablation of the second somatosensory cortex in the cat significantly increased the escape threshold to noxious stimulation; the effect was enhanced by excision of adjacent gyri. Studies employing tooth pulp stimulation in the cat (Vycklicky et al 1972) and in the human (Chatrian et al 1975) provide further evidence that the cortex is involved in nociception.

EXPERIMENTAL PHYSIOPATHOLOGY

The great majority of experimental studies on pain mechanisms have been carried out in acute experimental preparations, either under controlled levels of analgesia or in the decerebrate state; the inevitable shortcomings of both situations are evident. However, many forms of pain encountered in clinical experience are subacute or chronic, and for these usually no experimental model is either available or acceptable on ethical grounds. Thus, our concepts of mechanisms for most human pain problems are based on extrapolations from studies on acute nociception that may bear only a remote relationship to chronic pain. Recent efforts to correct this situation have been reported. Wall & Gutnick (1974) and Devor & Wall (1976) described the electrophysiological characteristics of the output from experimental peripheral neuromas in rats, showing that a significant delay occurs in the arrival of impulses at dorsal roots following stimulation of the neuroma, and that the delay is not due to slow conduction in sprouts in the lesion but apparently because only fine fibers (A-δ) sprout to form neuromas.

Dickhaus et al (1977) reported that after crushing the plantar nerves in cats, unmyelinated fibers associated with thermal cutaneous receptors regenerated to reach numbers identical to those found in the normal nerves, but their receptive field thresholds were significantly lower than normal. They suggested that this increased sensitivity may be the basis for the hyperpathia observed clinically following nerve regeneration.

NEUROPHARMACOLOGY

Monoamines and Pain

Nociceptive mechanisms are markedly affected by changes in the levels of cerebral monoamines. Akil & Liebeskind (1975), using the tail-flick test in the rat, have shown that stimulus-produced analgesia (SPA) induced by an electrode in the region of the periaqueductal gray is markedly depressed by tetrabenazine, which depletes the three major cerebral monoamines. This effect was reversed by administration of serotonin or L-dopa. The role of individual monoamines was also studied. Blockade of dopamine receptors with pimozide impaired SPA for periods of seven hours, while stimulation of dopamine receptors by apomorphine produced a marked increase in SPA. Depression of serotonin by para-chloro-phenyl-alanine decreased

SPA; the effect was reversed by administration of the precursor (5-hydroxytryptophan).

Depletion of noradrenaline, by blocking its synthesis from dopamine with disulfiram, resulted in a significant increase in SPA at 6 hr. To evaluate the relative roles of dopamine and noradrenaline, haloperidol was used to produce a general blockade of catecholamine receptors. Since this resulted in a significant decrease in SPA, it was concluded that dopamine effects were more potent than those due to noradrenaline. Dopamine also appears to play an important role in human pain, as evidenced by the increases in threshold and tolerance to chronic pain during L-dopa administration reported by Battista & Wolff (1973); however, Hodge & King (1976) have reported that L-dopa administration resulted in an increase in pain and contraction of areas that had previously been rendered analgesic by rhizotomies. The possibility that central degenerative phenomena secondary to the operative ablation may alter aminergic modulating mechanisms would seem to be worth considering. There is also long-standing evidence that morphine analgesia is modulated by monoamine levels. Radouco-Thomas et al (1957) and Schumann (1958) showed that reserpine antagonized experimental morphine analgesia, and these observations have been confirmed in numerous subsequent reports (e.g. Wei & Shen 1971).

Other Neurotransmitters

Identification of neurotransmitters in the somatosensory system has met with little success until recently. The two main candidates at this time are substance P (SP) and glutamate.

As mentioned above, SP is fairly widely distributed in the CNS. In addition to its presence in dorsal roots and s. gelatinosa, it is found in considerable concentration in the hypothalamus and the substantia nigra. Its release is triggered by K^+ and is Ca^{2+}-dependent (Iversen et al 1976). Henry (1976) has reported that approximately 50% of the neurons in laminae V and VI are excited by microiontophoretic injection of SP and that most of the units with nociceptive thermal cutaneous fields were activated, while there was a negative correlation for tactile elements. He suggested that failure of some nociceptive units to respond to SP might be explained if it is the transmitter for primary afferents; the nonresponsive units might then be second- or higher-order neurons.

Some problems exist, however, in accepting SP as a neurotransmitter at this time. The response of most neurons to SP is characterized by a gradual increase in firing frequency that continues for a considerable period after iontophoresis is discontinued. This had led Krnjevic & Morris (1974) to suggest that rather than being a transmitter (since two critical characteristics are speed and briefness of action), SP may act as a long-term modulator of sensory processes. But it should be noted that the studies of Hökfelt et al (1975b) show SP to be predominantly localized in laminae I, II, and III, with little evidence for its occurrence in laminae V and VI, where Henry's (1976) recordings were obtained. There are also some indications that if it is a transmitter for primary afferents, only some of these would employ SP (Hökfelt et al 1975b). In a recent study, Randic & Miletic (1977) report that all neurons with high-threshold peripheral fields that were exposed to SP by microiontophoretic injection responded intensely, whereas low-threshold units responded either weakly or not at all. However, long latencies ranging up to 16 sec were recorded, which again raises doubts about the significance of SP as a neurotransmitter and suggests the possibility that it may be a precursor.

The evidence for considering glutamate as a sensory neurotransmitter is not persuasive at this time and is based mainly on its higher concentration in dorsal roots and ganglia than in ventral roots (Duggan & Johnston 1970). Finally, as noted earlier, Hökfelt et al (1975a) have suggested that somatostatin might be an inhibitory transmitter for primary afferents.

STIMULUS–PRODUCED ANALGESIA (SPA)

Electrical stimulation of either peripheral nerves or specific loci in the CNS has been shown to result in marked degrees of analgesia.

Peripheral Nerve Stimulation

Peripheral nerve stimulation, introduced by Wall & Sweet (1967), was reported to elevate pain thresholds significantly during the period of stimulation, though the precise mechanism by which this is achieved is still controversial. In a recent summary of its status as a therapeutic tool, Long & Hagfors (1975) concluded that it can be an effective treatment in selected patients; the best results are obtained with pain secondary to peripheral nerve injury. The results are less predictable when the source of pain is more remote. The proportion of patients responding to this treatment on a long-term basis (i.e. for a year or more) was reported to be small but significant.

Central Stimulation

Centrally induced analgesia was first described by Reynolds (1969), who found that electrical stimulation in the periaqueductal gray induced analgesia of sufficient magnitude to permit abdominal surgery in conscious rats. His observation was confirmed and extended by Mayer et al (1971), Liebeskind (1976), and Liebeskind et al (1977). Mayer & Liebeskind (1974) reported that the antinociceptive effect can be equivalent to between 10 and 50 mg kg^{-1} of morphine and occurs in all animals tested to date, including man (Adams 1976, Richardson & Akil 1977 a, b, Hosobuchi et al 1977).

The optimal site for eliciting SPA appears to be the periaqueductal gray (PAG) in or near the dorsal raphé nucleus and in the gray matter that continues rostrally into the diencephalic periventricular area. Not only somatic, but also visceral pain, can be modulated from this area (Giesler & Liebeskind 1976).

As noted earlier, the n. raphé magnus of the medulla is also an important site for SPA (Proudfitt & Anderson 1975, Oliveras et al 1975, Fields et al 1976, 1977), with its descending, mainly serotoninergic pathway to the cord. Beall et al (1976) have shown that stimulation of the n. raphé magnus was more effective in reducing the effects of the A-δ component of an afferent volley than those of the A-β component.

Striking evidence of the intimate relationships between SPA and morphine analgesia was provided by the demonstration that both tolerance and cross-tolerance develop (Mayer & Hayes 1975), and that naloxone blocks SPA in animals (Akil et al 1976) and man (Adams 1976, Hosobuchi et al 1977).

In summary, much evidence now indicates that there is an intrinsic system in the brainstem that is centered in midline structures extending from the gray matter surrounding the third ventricle rostrally, through the PAG, to the raphé of the medulla. This system is mutually interconnected, and its activation results in powerful inhibitory effects on nociceptive transmission, effects that occur without apparent changes in the level of consciousness or in other sensory modalities.

These are not the only sites from which analgesia can be obtained, though at this time they appear to exert the most potent effects. Gol (1967) showed that stimulation of the septal area in one patient gave striking relief from pain due to neoplastic invasion; the effect outlasted the stimulation by many hours. However, the results were inconsistent in other patients. Stimulation in the area of the medial forebrain bundle has also been reported to relieve pain (Balagura & Ralph 1973), but the mechanisms involved are probably different to those of the PAG.

MORPHINE AND ENDOGENOUS MORPHINE-LIKE SUBSTANCES (MLS)

Remarkable advances have occurred in our understanding the analgesic action of opiates since Tsou & Jang (1964) showed that μg quantities of morphine injected into the periventricular gray had analgesic effects equivalent to a systemic dose 500 times greater. Injections into other brain areas were ineffective, while analgesia resulting from systemically administered morphine was blocked by μg injections of nalorphine into the periventricular gray. Herz & Teschemacher (1971), Yaksh et al (1976), and Jacquet & Lajtha (1976) have obtained similar results but find, in addition, that injections of morphine into the PAG are equally effective. It is also of interest that, while Tsou & Jang (1964) did not detect an analgesic effect from injections of morphine into the spinal subarachnoid space, Yaksh & Rudy (1976) reported pronounced, long-lasting analgesia by this route; this may correlate with the high concentrations of opiate receptors demonstrated in the s. gelatinosa by LaMotte et al (1976) and in both this and the marginal zone by Atweh & Kuhar (1977).

The marked stereospecificity of opiate analgesics had long been regarded as an indication that these drugs had an affinity for a receptor with correspondingly specific characteristics; hence there developed the concept of an opiate receptor (Terenius 1973, Simon et al 1973, Snyder 1974), which was fully confirmed by Pert et al (1974), who localized it to membranes (but not vesicles) in the synaptosomal fraction of brain homogenates.

However, it seemed evident that such a receptor in the CNS must be present to interact with a naturally occurring ligand, and that the morphine affinity was fortuitous as well as fortunate. Consequently, a search for an endogenous ligand in the brain was undertaken by several groups and successfully concluded by Hughes

(1975), Terenius & Wahlström (1975), and Pasternak et al (1975). The substance, named enkephalin by Hughes, is a pentapeptide with the amino acid sequence: NH_2-Tyr-Gly-Gly-Phe-Met-COOH. This is the more active form; a weaker pentapeptide with the same sequence but containing leucine instead of methionine is also present. It produces profound analgesia when injected into the cerebral ventricles of the rat (Belluzzi et al 1976), though its effect is relatively brief (some 10 min) due, it is believed, to degradation by polypeptidases; the analgesia is blocked by the morphine antagonist, naloxone. The distribution of enkephalin (Hughes 1975, Hökfelt et al 1977) closely parallels that of opiate receptors (Kuhar et al 1973, Pert & Yaksh 1974) and, in turn, several of the sites in which high concentrations of both are present correspond to areas from which SPA is best obtained.

In addition to methionine and leucine enkephalin, a number of other MLS have been described. Among these, several polypeptides have been separated from the pituitary hormone β-lipotropin by Cox et al (1976) and have been named α-, β-, and γ-endorphins, each consisting of sequences of 15 or more peptides. Cox and co-workers (1975) believe that the endorphins are the active agents and that the enkephalins are breakdown products. Loh et al (1976), report that β-endorphin is 18–33 times as potent an analgesic as morphine, as determined in studies on the rat; methionine enkephalin was said to be only weakly analgesic in the same experiments. They also note that chronic administration of β-endorphin gives rise to signs comparable to those of morphine dependence.

The relationship of enkephalins and related MLS to morphine, SPA, and other varieties of modulation of nociception is of great interest, both in terms of possible mechanisms and for their potential therapeutic applications. Some of the known correlations can be expressed as follows:

1. Stimulation of certain areas (periaqueductal and periventricular gray, and n. raphé magnus) results in marked elevation in pain threshold.
2. These areas are connected by identified fiber systems (Ruda 1975).
3. Injection of morphine in μg quantities into some of them produces marked elevation in pain threshold.
4. Opiate-receptor binding is high in the areas tested so far (Kuhar et al 1973) where SPA is most marked.
5. Enkephalin concentrations are also high in these areas (Hughes 1975).
6. Morphine-, SPA-, and enkephalin-induced analgesia are reversed by the morphine antagonist naloxone, though not all reports agree on this point.

There is mounting evidence that these various phenomena (SPA, morphine and enkephalin analgesia) share a common physiological substrate. While iontophoretic administration of morphine to various brain loci (Zieglgänsberger & Bayerl 1976) and of Met-enkephalin (Bradley et al 1976) has produced predominantly inhibitory effects on neurons, it would be anticipated that at least those neurons giving rise to the descending inhibitory pathways to the spinal cord should be excited. It is therefore of particular interest that neurons in the periaqueductal gray (Urca et al

1977) and in the raphé nuclei of the medulla (Anderson et al 1977, Oleson et al 1977) have been shown to increase their firing rate very significantly in response to activation of the above descending modulatory systems.

The potential role of these MLS for the control of pain is of much interest; some possible problems that may be encountered and responses that are already appearing may be summarized here. The brief duration of the effects (5–10 min) would appear to be a serious obstacle. This objection seems to be met by at least two recent reports: Walker et al (1977) administered a D-alanine[2] analog of metenkephalin intraventricularly in the rat and obtained profound naloxone-reversible dose-dependent analgesia that lasted for approximately one hr, but this was accompanied by stupor and severe catatonic effects.

Tseng et al (1976) injected β-endorphin intravenously in mice and reported analgesic effects three to four times more potent than molar equivalent doses of morphine; the effects lasted for 30–60 min, were dose dependent, and were blocked by pretreatment with naloxone. This observation appears to indicate that the blood-brain barrier will not be an obstacle should therapeutic application of at least some of these peptides become feasible.

Whether, and to what extent, addiction to the enkephalins/endorphins may develop is as yet unknown, but Wei & Loh (1976) have already noted that intraventricular injections of enkephalins in the rat result in signs of both tolerance and dependence. The possibility that enkephalins may have some other drawbacks must be carefully weighed. Thus, Urca et al (1977) have evidence that they have epileptogenic effects on the rat cerebral cortex, and pronounced catatonic posturing usually occurs after intraventricular injections in the rat.

CONCLUDING REMARKS

Concepts in the field of pain are changing rapidly and in many areas. The long-debated issue of specificity vs pattern theory appears to be inclining progressively toward a specific system in which nociceptors with corresponding afferent fibers activate nociceptive neurons in the CNS. Some of the later are pure high-threshold units—i.e. responding only to noxious stimuli—but a considerable degree of convergence between noxious and non-noxious input on certain neurons tends to obscure some of the specificity. However, the issue appears to be resolved for such units by their increase in firing rate in response to noxious stimulation.

Further evidence for specificity comes from studies in which morphine (Kitahata et al 1974, LeBars et al 1975) or SPA (Oliveras et al 1974, Mayer & Liebeskind 1974) inhibit nociceptive neurons while leaving mechanoreceptors unaffected. A number of the features that support the concept that pain is a specific sensation have been dealt with in depth by Perl (1971), and most subsequent developments have added weight to his arguments.

That interactions do exist between pain and other somesthetic and visceral modalities is not in doubt, and it is in part these interactions and the ubiquity of pain that make identification and separation of this modality at once difficult and, to an extent, artificial.

A complex modulatory system of descending control that appears to use aminergic pathways operates against this background of nociceptor input. This system, in turn, is controlled by newly identified endogenous transmitters or modulators.

Although these new contributions are impressive and hold much promise, both in terms of understanding and, thus, of controlling pain, pain control has always been elusive. The best of analgesics, the narcotics, exact their toll in addiction and progressively fail as tolerance develops. Surgical ablations may be highly effective in selected cases, but there still remain large numbers of patients for whom no form of treatment offers relief. It is for these unfortunate individuals that rapid progress, such as is currently occurring in this field, offers hope.

Literature Cited

Adams, J. E. 1976. Naloxone reversal of analgesia produced by brain stimulation in the human. *Pain* 2:161–66

Akil, H., Liebeskind, J. C. 1975. Monoaminergic mechanism of stimulation-produced analgesia. *Brain Res.* 94:279–96

Akil, H., Mayer, D. J., Liebeskind, J. C. 1976. Antagonism of stimulation-produced analgesia by naloxone, a narcotic antagonist. *Science* 191:961–62

Albe-Fessard, D., Levante, A., Lamour, Y. 1974. Origin of spino-thalamic tract in monkeys. *Brain Res.* 65:503–9

Anderson, S. D., Basbaum, A. I., Fields, H. L. 1977. Responses of medullary raphé neurons to peripheral stimulation and to systemic opiates. *Brain Res.* 123: 363–68

Applebaum, M. L., Clifton, G. L., Coggeshall, R. E., Coulter, J. D., Vance, W. H., Willis, W. D. 1976. Unmyelinated fibres in the sacral-3 and caudal-1 ventral roots of the cat. *J. Physiol. London* 256:557–72

Atweh, S., Kuhar, M. J. 1977. Autoradiographic localization of opiate receptors in rat brain. I. Spinal cord and lower medulla. *Brain Res.* 124:53–67

Balagura, S., Ralph, T. 1973. The analgesic effect of electrical stimulation of the diencephalon and mesencephalon. *Brain Res.* 60:369–79

Basbaum, A. I. 1973. Conduction of the effects of noxious stimulation by short-fiber multisynaptic systems of the spinal cord in the rat. *Exp. Neurol.* 40:699–716

Basbaum, A. I., Marley, N. J. E., O'Keefe, J., Clanton, C. H. 1977. Reversal of morphine and stimulus-produced analgesia by subtotal spinal cord lesions. *Pain* 3:43–56

Battista, A. F., Wolff, B. B. 1973. Levodopa and induced-pain response. A study of patients with Parkinsonian and pain syndromes. *Arch. Intern. Med.* 132: 70–74

Beall, J. E., Martin, R. F., Applebaum, A. E., Willis, W. D. 1976. Inhibition of primate spinothalamic tract neurons by stimulation in the region of the nucleus raphé magnus. *Brain Res.* 114:328–33

Belluzzi, J. D., Grant, N., Garsky, V., Sarantakis, D., Wise, C. D., Stein, L. 1976. Analgesia induced *in vivo* by central administration of enkephalin in the rat. *Nature* 260:625–26

Benjamin, R. M. 1970. Single neurons in the rat medulla responsive to nociceptive stimulation. *Brain Res.* 24:525–29

Berkley, K. J., Parmer, R. 1974. Somatosensory cortical involvement in responses to noxious stimulation in the cat. *Exp. Brain Res.* 20:363–74

Bowsher, D. 1976. Role of the reticular formation in responses to noxious stimulation. *Pain* 2:361–78

Bradley, P. B., Briggs, I., Gayton, R. J., Lambert, L. A. 1976. Effects of microiontophoretically applied methionine-enkephalin on single neurones in rat brainstem. *Nature* 261:425–26

Burgess, P. R. 1974. Patterns of discharge evoked in cutaneous nerves and their significance for sensation. *Adv. Neurol.* 4:11–18

Burgess, P. R., Perl, E. R. 1973. Cutaneous mechanoreceptors and nociceptors. In *Handbook of Sensory Physiology, Vol. 2, Somatosensory System,* ed. A. Iggo, pp. 39–78. Berlin: Springer. 851 pp.

Burgus, R., Ling, N., Butcher, M., Guillemin, R. 1973. Primary structure of somatostatin, a hypothalamic peptide that inhibits the secretion of pituitary growth hormone. *Proc. Natl. Acad. Sci. USA* 70:684–88

Burton, H., Jones, E. G. 1976. The posterior thalamic region and its cortical projection in New World and Old World monkeys. *J. Comp. Neurol.* 168:249–302

Burton, H., Loewy, A. D. 1976. Descending projections from the marginal cell layer and other regions of the monkey spinal cord. *Brain Res.* 116:485–91

Cajal, S. R. 1909. *Histologie du Système Nerveux de l'Homme et des Vertébrés.* Vol. 1. Paris: Maloine. 986 pp.

Casey, K. L. 1971. Responses of bulboreticular units to somatic stimuli eliciting escape behavior in the cat. *Int. J. Neurosci.* 2:29–34

Casey, K. L., Keene, J. J., Morrow, T. 1974. Bulboreticular and medial thalamic unit activity in relation to aversive behavior and pain. *Adv. Neurol.* 4:197–205

Cervero, F., Iggo, A., Ogawa, H. 1976. Nociceptor-driven dorsal horn neurones in the lumbar spinal cord of the cat. *Pain* 2:5–24

Chang, M. M., Leeman, S. E. 1970. Isolation of a sialogogic peptide from bovine hypothalamic tissue and its characterization as substance P. *J. Biol. Chem.* 245:4784–90

Chatrian, G. E., Canfield, R. C., Knauss, T. A., Lettich, E. 1975. Cerebral responses to electrical tooth pulp stimulation in man. An objective correlate of acute experimental pain. *Neurology* 25:745–57

Christensen, B. N., Perl, E. R. 1970. Spinal neurons specifically excited by noxious or thermal stimuli: marginal zone of the dorsal horn. *J. Neurophysiol.* 33:293–307

Clifton, G. L., Coggeshall, R. E., Vance, W. H., Willis, W. D. 1976. Receptive fields of unmyelinated ventral root afferent fibres in the cat. *J. Physiol. London* 256:573–600

Coggeshall, R. E., Applebaum, M. L., Fazen, M., Stubbs, T. B. III, Sykes, M. T. 1975. Unmyelinated axons in human ventral roots, a possible explanation for the failure of dorsal rhizotomy to relieve pain. *Brain* 98:157–66

Coggeshall, R. E., Coulter, J. D., Willis, W. D. Jr. 1973. Unmyelinated fibers in the ventral root. *Brain Res.* 57:229–33

Coulter, J. D., Maunz, R. A., Willis, W. D. 1974. Effects of stimulation of sensorimotor cortex on primate spinothalamic neurons. *Brain Res.* 65:351–56

Cox, B. M., Goldstein, A., Li, C. H. 1976. Opioid activity of a peptide β-lipotropin. *Proc. Natl. Acad. Sci. USA* 73:1821–23

Cox, B. M., Opheim, K. E., Teschemacher, H., Goldstein, A. 1975. A peptidelike substance from pituitary that acts like morphine. 2. Purification and properties. *Life Sci.* 16:1777–82

Curry, M. J. 1972. The exteroceptive properties of neurones in the somatic part of the posterior group (PO). *Brain Res.* 44:439–62

Denny-Brown, D., Kirk, E. J., Yanagisawa, N. 1973. The tract of Lissauer in relation to sensory transmission in the dorsal horn of spinal cord in the Macaque monkey. *J. Comp. Neurol.* 151:175–200

Devor, M., Wall, P. D. 1976. Type of sensory nerve fibre sprouting to form a neuroma. *Nature* 262:705–8

Dickhaus, H., Zimmermann, M., Zotterman, Y. 1977. The development in regenerating cutaneous nerves of C-fibre receptors responding to noxious heating of the skin. In *Sensory Functions of the Skin,* ed. Y. Zotterman, pp. 415–23. Oxford: Pergamon. 576 pp.

Dubner, R., Beitel, R. E. 1976. Peripheral neural correlates of escape behavior in rhesus monkey to noxious heat applied to the face. In *Advances in Pain Research and Therapy,* ed. J. J. Bonica, D. Albe-Fessard, 1:155–60. New York: Raven. 1012 pp.

Duggan, A. W., Johnston, G. A. R. 1970. Glutamate and related amino acids in cat spinal roots, dorsal root ganglia and peripheral nerves. *J. Neurochem.* 17:1205–8

Dyck, P. J., Lambert, E. H., O'Brien, P. C. 1976. Pain in peripheral neuropathy related to rate and kind of fiber degeneration. *Neurology* 26:466–71

Euler, U. S. von, Gaddum, J. H. 1931. An unidentified depressor substance in certain tissue extracts. *J. Physiol. London* 72:74–87

Fields, H. L., Anderson, S. D., Clanton, C. H., Basbaum, A. I. 1976. Nucleus raphé magnus: a common mediator of opiate and SPA. *Trans. Am. Neurol. Assoc.* 101:208–9

Fields, H. L., Basbaum, A. E., Clanton, C. H., Anderson, S. D. 1977. Nucleus raphe magnus inhibition of spinal cord dorsal horn neurons. *Brain Res.* 126:441–53

Giesler, G. J. Jr., Liebeskind, J. C. 1976. Inhibition of visceral pain by electrical stimulation of the periaqueductal gray matter. *Pain* 2:43–48

Giesler, G. J. Jr., Menetrey, D., Guilbaud, G., Besson, J.-M. 1976. Lumbar cord neurons at the origin of the spinothalamic tract in the rat. *Brain Res.* 118:320–24

Gobel, S. 1974. Synaptic organization of the substantia gelatinosa glomeruli in the spinal trigeminal nucleus of the adult cat. *J. Neurocytol.* 3:219–43

Gol, A. 1967. Relief of pain by electrical stimulation of the septal area. *J. Neurol. Sci.* 5:115–20

Hancock, M. B., Foreman, R. D., Willis, W. D. 1975. Convergence of visceral and cutaneous input onto spinothalamic tract cells in the thoracic spinal cord of the cat. *Exp. Neurol.* 47:240–48

Handwerker, H. O., Iggo, A., Zimmerman, M. 1975. Segmental and supraspinal actions on dorsal horn neurons responding to noxious and non-noxious skin stimuli. *Pain* 1:147–65

Henry, J. L. 1976. Effects of substance P on functionally identified units in cat spinal cord. *Brain Res.* 114:439–51

Herz, A., Teschemacher, H.-J. 1971. Activities and sites of antinociceptive action of morphine-like analgesics and kinetics of distribution following intravenous, intracerebral and intraventricular application. *Adv. Drug Res.* 6:79–119

Hodge, C. J., King, R. B. 1976. Medical modification of sensation. *J. Neurosurg.* 44:21–28

Hökfelt, T., Elde, R., Johansson, O., Luft, R., Arimura, A. 1975a. Immunohistochemical evidence for the presence of Somatostatin, a powerful inhibitory peptide in some primary sensory neurons. *Neurosci. Lett.* 1:231–35

Hökfelt, T., Elde, R., Johansson, O., Luft, R., Nilson, G., Arimura, A. 1976. Immunohistochemical evidence for separate populations of somatostatin-containing and substance P-containing primary afferent neurons in the rat. *Neuroscience* 1:131–36

Hökfelt, T., Kellerth, J.-O., Nilsson, G., Pernow, B. 1975b. Experimental immunohistochemical studies on the localization and distribution of substance P in cat primary sensory neurons. *Brain Res.* 100:235–52

Hökfelt, T., Ljungdahl, A., Elde, R., Nilsson, G., Terenius, L. 1977. Immunohistochemical analysis of peptide pathways possibly related to pain and analgesia: Enkephalin and substance P. *Proc. Natl. Acad. Sci. USA* 74:3081–85

Hosobuchi, Y., Adams, J. E., Linchitz, R. 1977. Pain relief by central gray stimulation and its reversal by naloxone. *Science* 197:183–86

Hughes, J. 1976. Isolation of an endogenous compound from the brain with pharmacological properties similar to morphine. *Brain Res.* 88:295–308

Hughes, J., Smith, T. W., Kosterlitz, H. W., Fothergill, L. A., Morgan, B. A., Morris, H. R. 1975. Identification of two related pentapeptides from the brain with potent opiate agonist activity. *Nature* 258:577–79

Iggo, A. 1974. Activation of cutaneous nociceptors and their actions on dorsal horn neurons. *Adv. Neurol.* 4:1–9

Iggo, A. 1976. Is the physiology of cutaneous receptors determined by morphology? *Prog. Brain Res.* 43:15–31

Ishijima, B., Sano, K. 1971. Response of specific and nonspecific thalamic nuclei to the selective A-δ and C-fiber stimulation, and their interactions in cat's brain. *Neurol. Med. Chir.* 11:84–100

Iversen, L. L., Jessell, T., Kanazawa, I. 1976. Release and metabolism of substance P in rat hypothalamus. *Nature* 264:81–83

Jacquet, Y. F., Lajtha, A. 1976. The periaqueductal gray: site of morphine analgesia and tolerance as shown by 2-way cross tolerance between systemic and intracerebral injections. *Brain Res.* 103:501–13

Kerr, F. W. L. 1975a. Neuroanatomical substrates of nociception in the spinal cord. *Pain* 1:325–56

Kerr, F. W. L. 1975b. The ventral spinothalamic tract and other ascending systems of the ventral funiculus of the spinal cord. *J. Comp. Neurol.* 159:335–56

Kirk, E. J., Denny-Brown, D. 1970. Functional variation in dermatomes in the macaque monkey following dorsal root lesions. *J. Comp. Neurol.* 139:307–20

Kitahata, L. M., Kosaka, Y., Taub, A., Bonikos, K., Hoffert, M. 1974. Lamina-specific suppression of dorsal-horn unit activity by morphine sulphate. *Anesthesiology* 41:39–48

Konishi, S., Otsuka, M. 1974. The effects of substance P and other peptides on spinal neurons of the frog. *Brain Res.* 65:397–410

Krnjeviç, K., Morris, M. E. 1974. An excitatory action of substance P on cuneate neurones. *Can. J. Physiol. Pharmacol.* 52:736–44

Kuhar, M. J., Pert, C. B., Snyder, S. H. 1973. Regional distribution of opiate receptor binding in monkey and human brain. *Nature* 245:447–50

Kumazawa, T., Perl, E. R. 1977. Differential excitation of dorsal horn and substantia gelatinosa marginal neurons by primary afferent units with fine (Aδ and C) fibers. See Dickhaus et al 1977, pp. 67–88

Kumazawa, T., Perl, E. R., Burgess, P. R., Whitehorn, D. 1975. Ascending projections from marginal zone (lamina I) neurons of the spinal dorsal horn. *J. Comp. Neurol.* 162:1–12

Kuru, M. 1949. *Sensory pathways in the spinal cord and brainstem of man.* Tokyo: Sogensya. 39 pp.

LaMotte, C. 1977. Distribution of the tract of Lissauer and the dorsal root fibers in the primate spinal cord. *J. Comp. Neurol.* 172:529–62

LaMotte, C., Pert, C. B., Snyder, S. H. 1976. Opiate receptor binding in primate spinal cord: distribution and changes after dorsal root section *Brain Res.* 112: 407–12

Le Bars, D., Menétrey, D., Conseiller, C., Besson, J. M. 1975. Depressive effects of morphine upon lamina V cell activities in the dorsal horn of the spinal cat. *Brain Res.* 98:261–77

Liebeskind, J. C. 1976. Pain modulation by central nervous system stimulation. See Dubner & Beitel 1976, pp. 445–53

Liebeskind, J. C., Giesler, G. J., Urca, G. 1977. Evidence pertaining to an endogenous mechanism of pain inhibition in the central nervous system. See Dickhaus et al 1977, pp. 561–73

Lippman, H. H., Kerr, F. W. L. 1972. Light and electron microscopic study of crossed ascending pathways in the anterolateral funiculus in the monkey. *Brain Res.* 40:496–99

Loh, H. H., Tseng, L. F., Wei, E., Li, C. H. 1976. β-Endorphin is a potent analgesic agent. *Proc. Natl. Acad. Sci. USA* 73:2895–98

Long, D. M., Hagfors, N. 1975. Electrical stimulation in the nervous system: the current status of electrical stimulation of the nervous system for relief of pain. *Pain* 1:109–23

Mannen, H., Sugiura, Y. 1976. Reconstruction of neurons of dorsal horn proper using Golgi-stained serial sections. *J. Comp. Neurol.* 168:303–12

Mayer, D. J., Hayes, R. L. 1975. Stimulation-produced analgesia: development of tolerance and cross-tolerance to morphine. *Science* 188:941–43

Mayer, D. J., Liebeskind, J. C. 1974. Pain reduction by focal electrical stimulation of the brain: an anatomical and behavioral analysis. *Brain Res.* 68:73–93

Mayer, D. J., Price, D. D., Becker, D. P. 1975. Neurophysiological characterization of the anterolateral spinal cord neurons' contribution to pain perception in man. *Pain* 1:51–58

Mayer, D. J., Wolfle, T. L., Akil, H., Carder, B., Liebeskind, J. C. 1971. Analgesia from electrical stimulation in the brainstem of the rat. *Science* 174:1351–54

Moffie, D. 1975. Spinothalamic fibres, pain conduction and cordotomy. *Clin. Neurol. Neurosurg.* 78(4):261–68

Mosso, J. A., Kruger, L. 1973. Receptor categories represented in spinal trigeminal nucleus caudalis. *J. Neurophysiol.* 36:472–88

Narotzky, R. E., Kerr, F. W. L. 1977. Marginal neurons of the spinal cord: types, afferent synaptology and functional considerations. *Brain Res.* In press

Nashold, B. S., Urban, B., Zorub, D. S. 1976. Phantom pain relief by focal destruction of the substantia gelatinosa of Rolando. See Dubner & Beitel 1976, pp. 959–63

Nijensohn, D. E., Kerr, F. W. L. 1975. The ascending projections of the dorsolateral funiculus of the spinal cord in the primate. *J. Comp. Neurol.* 161:459–70

Noordenbos, W., Wall, P. D. 1976. Diverse sensory functions with an almost totally divided spinal cord. A case of spinal cord transection with preservation of part of one anterolateral quadrant. *Pain* 2:185–95

Nyquist, J. K., Greenhoot, J. H. 1974. A single unit analysis of mesencephalic reticular formation responses to high intensity cutaneous input in cat. *Brain Res.* 70:157–64

Oleson, T. D., Twombly, D. A., Liebeskind, J. C. 1977. Effects of pain-attenuating brain stimulation and morphine on electrical activity in the raphe nuclei of the awake rat. *Pain.* In press

Oliveras, J. L., Besson, J. M., Guilbaud, G., Liebeskind, J. C. 1974. Behavioral and electrophysiological evidence of pain inhibition from midbrain stimulation in the cat. *Exp. Brain Res.* 20:32–44

Oliveras, J. L., Redjemi, F., Guilbaud, G., Besson, J. M. 1975. Analgesia induced by electrical stimulation of the inferior centralis nucleus of the raphé in the cat. *Pain* 1:139–45

Otsuka, M., Konishi, S., Takahashi, T. 1972. The presence of a motoneuron-depolarizing peptide in bovine dorsal roots of spinal nerves. *Proc. Jpn. Acad.* 48: 342–46

Pasternak, G. W., Goodman, R., Snyder, S. H. 1975. An endogenous morphine-like factor in mammalian brain. *Life Sci.* 16:1765–69

Perl, E. R. 1971. Is pain a specific sensation: *J. Psychiatr. Res.* 8:273–87

Perl, E. R. 1976. Sensitization of nociceptors and its relation to sensation. See Dubner & Beitel 1976, pp. 17–28

Pert, C. B., Snowman, A. M., Snyder, S. H. 1974. Localization of opiate receptor binding in synaptic membranes of rat brain. *Brain Res.* 70:184–88

Pert, A., Yaksh, T. 1974. Sites of morphine induced analgesia in the primate brain: relation to pain pathways. *Brain Res.* 80:135–40

Pomeranz, B., Wall, P. D., Weber, W. V. 1968. Cord cells responding to fine myelinated afferents from viscera, muscle and skin. *J. Physiol. London* 199:511–32

Price, D. D., Mayer, D. J. 1974. Physiological laminar organization of the dorsal horn of *M. mulatta*. *Brain Res.* 79:321–25

Price, D. D., Mayer, D. J. 1975. Neurophysiological characterization of the anterolateral quadrant neurons subserving pain in *M. mulatta*. *Pain* 1:59–72

Proudfitt, H. K., Anderson, E. G. 1975. Morphine analgesia: blockade by raphé magnus lesions. *Brain Res.* 98:612–18

Radouco-Thomas, S., Radouco-Thomas, C., LeBreton, E. 1957. Action de la noradrenaline et de la reserpine sur l'analgesis experimentale. *Arch. Exp. Pathol. Pharmakol.* 232:279–81

Randic, M., Miletic, V. 1977. Effect of substance P in cat dorsal horn neurons activated by noxious stimuli. *Brain Res.* 128:164–69

Réthelyi, M. 1977. Preterminal and terminal axon arborizations in the substantia gelatinosa of cat's spinal cord. *J. Comp. Neurol.* 172:511–28

Réthelyi, M., Szentagothai, J. 1973. Distribution and connections of afferent fibers in the spinal cord. See Burgess & Perl 1973, pp. 207–52

Reynolds, D. V. 1969. Surgery in the rat during electrical analgesia induced by focal brain stimulation. *Science* 164:444–45

Richardson, D. E., Akil, H. 1977a. Pain reduction by electrical brain stimulation in man. Pt. 1: Acute administration in periaqueductal and periventricular sites. *J. Neurosurg.* 47:178–83

Richardson, D. E., Akil, H. 1977b. Pain reduction by electrical brain stimulation in man. Pt. 2: Chronic self-administration in the periventricular gray matter. *J. Neurosurg.* 47:184–94

Ruda, M. 1975. *Autoradiographic study of the efferent projections of the midbrain central gray in the cat.* PhD thesis. Univ. Penn. 73 pp.

Scheibel, M. E., Scheibel, A. B. 1968. Terminal axonal patterns in cat spinal cord. 2. The dorsal horn. *Brain Res.* 9:32–58

Schumann, W. 1958. Beeinflussung der analgetischen Wirkung des Morphins durch Reserpin. *Arch. Exp. Pathol. Pharmakol.* 235:1–9

Simon, E. J., Hiller, J. M., Edelman, I. 1973. Stereospecific binding of the potent narcotic analgesic [^{3}H] etorphine to rat brain homogenate. *Proc. Natl. Acad. Sci. USA* 70:1947–49

Sindou, M., Quoex, C., Baleydier, C. 1974. Fiber organization at the posterior spinal cord-rootlet junction in man. *J. Comp. Neurol.* 153:15–26

Snyder, R. 1977. The organization of the dorsal root entry zone in cats and monkeys. *J. Comp. Neurol.* 174:47–70

Snyder, S. H. 1974. The opiate receptor. *Neurosci. Res. Prog. Bull.* 13:1–27

Sprague, J. M., Ha, H. 1964. The terminal fields of dorsal root fibers in the lumbosacral spinal cord of the cat and the dendritic organization of the motor nuclei. *Prog. Brain Res.* 11:120–54

Sugiura, Y. 1975. Three dimensional analysis of neurons in the substantia gelatinosa Rolandi. *Proc. Jpn. Acad.* 51:336–41

Szentagothai, J. 1964. Neuronal and synaptic arrangements in the substantia gelatinosa Rolandi. *J. Comp. Neurol.* 122:219–39

Takahashi, T., Konishi, S., Powell, D., Leeman, S. E., Otsuka, M. 1974. Identification of the motoneuron-depolarizing peptide in bovine dorsal root as hypothalamic substance P. *Brain Res.* 73:59-69

Takahashi, T., Otsuka, M. 1975. Regional distribution of substance P in the spinal cord and nerve roots of the cat and the effect of dorsal root section. *Brain Res.* 87:1–11

Terenius, L. 1973. Characteristics of the "receptor" for narcotic analgesics in synaptic plasma membrane fraction from rat brain. *Acta Pharmacol. Toxicol.* 33:377–84

Terenius, L., Wahlström, A. 1975. Morphine-like ligand for opiate receptors in human CSF. *Life Sci.* 16:1759–64

Torebjörk, H. E. 1974. Afferent C units responding to mechanical, thermal and chemical stimuli in human non-glabrous skin. *Acta Physiol. Scand.* 92: 374–90

Torebjörk, H. E., Hallin, R. G. 1974. Identification of afferent C units in intact human skin nerves. *Brain Res.* 67:387–403

Tregear, G. W., Niall, H. D., Potts, J. T. Jr., Leeman, S. E., Chang, M. M. 1971. Synthesis of substance P. *Nature New Biol.* 232:87–89

Trevino, D. L., Carstens, E. 1975. Confirmation of the location of spinothalamic neurons in the cat and monkey by the retrograde transport of horseradish peroxidase. *Brain Res.* 98:177–82

Trevino, D. L., Coulter, J. D., Willis, W. D. 1973. Location of cells of origin of spinothalamic tract in lumbar enlargement of the monkey. *J. Neurophysiol.* 36:750–61

Tseng, L.-F., Loh, H. H., Li, C. H. 1976. β-Endorphin as a potent analgesic by intravenous injection. *Nature* 263: 239–40

Tsou, K., Jang, C. S. 1964. Studies on the site of analgesic action of morphine by intracerebral micro-injection. *Sci. Sin.* 13:1099–1109

Urca, G., Frenk, H., Liebeskind, J. C., Taylor, A. 1977. Morphine and enkephalin: analgesic and epileptic properties. To be published

Van Hees, J. 1976. Human C-fiber input during painful and nonpainful skin stimulation with radiant heat. See Dubner & Beitel 1976, pp. 35–40

Van Hees, J., Gybels, J. 1973. L'activité unitaire des fibers C enregistreé dans un nerf cutané chez l'homme et sa relation avec la douleur. (The unit activity of C fibers recorded in the human cutaneous nerve and its relationship to pain.) *Acta Neurol. Belg.* 73:39–43

Vyklicky, L., Keller, O., Brožek, G., Butkhuzi, S. M. 1972. Cortical potentials evoked by stimulation of tooth pulp afferents in the cat. *Brain Res.* 41:211–13

Walker, J. M., Berntson, G. G., Sandman, C. A., Coy, D. H., Schally, A. V., Kastin, A. J. 1977. An analog of enkephalin having prolonged opiate-like effects in vivo. *Science* 196:85–87

Wall, P. D. 1973. Dorsal horn electrophysiology. See Burgess & Perl 1973, pp. 253–70

Wall, P. D., Gutnick, M. 1974. Ongoing activity in peripheral nerves: the physiology and pharmacology of impulses originating from a neuroma. *Exp. Neurol.* 43:580–93

Wall, P. D., Sweet, W. H. 1967. Temporary abolition of pain in man. *Science* 155:108–9

Wei, E., Loh, H. 1976. Physical dependence on opiate-like peptides. *Science* 193: 1262–63

Wei, E. L., Shen, F.-H. 1971. Catecholamines and 5-hydroxytryptamine. In *Narcotic Drugs: Biochemical Pharmacology.* ed. D. H. Clouet. pp. 229–53. New York: Plenum. 506 pp.

White, J. C., Sweet, W. H. 1955. *Pain. Its Mechanisms and Neurosurgical Control.* pp. 31–36. Springfield, Ill: Thomas. 736 pp. 1st ed.

Willis, W. D., Trevino, D. L., Coulter, J. D., Maunz, R. A. 1974. Responses of primate spinothalamic tract neurons to natural stimulation of hindlimb. *J. Neurophysiol.* 37:358–72

Yaksh, T. L., Rudy, T. A. 1976. Analgesia mediated by a direct spinal action of narcotics. *Science* 192:1357–58

Yaksh, T. L., Yeung, J. C., Rudy, T. A. 1976. Systematic examination in the rat of brain sites sensitive to the direct application of morphine: observation of differential effects within the periaqueductal gray. *Brain Res.* 114:83–103

Young, D. W., Gottschaldt, K.-M. 1976. Neurons in rostral mesencephalic reticular formation of the cat responding specifically to noxious mechanical stimulation. *Exp. Neurol.* 51:628–36

Zieglgänsberger, W., Bayerl, H. 1976. The mechanism of inhibition of neuronal activity by opiates in the spinal cord of the cat. *Brain Res.* 115:111–28

Ann. Rev. Neurosci. 1978. 1:103–27

SENSORY EVOKED POTENTIALS IN CLINICAL DISORDERS OF THE NERVOUS SYSTEM

❖11505

A. Starr

Department of Neurology, University of California Irvine Medical Center, Orange, California 92668

INTRODUCTION

The diagnostic tools presently used by clinicians in the field of neuroscience provide useful information about the structure of the nervous system, but fall short with respect to providing insights into function. For this, the clinician must still rely on history, physical examination, and the patient's own description of the experience of his disorder. The search for relatively noninvasive, simple tests of brain function has stimulated the recent development of sensory evoked potentials recorded from the scalp as low-risk, clinically applicable procedures capable of providing new and objective information about a variety of nervous system functions. In this review I summarize the various applications of auditory, visual, and somatosensory evoked potentials to the study of clinical neurological disorders in man, beginning with a brief description of the stimuli employed, normal patterns of evoked potentials, stimulus and subject variables that influence these potentials, and their probable neural generators.

Clinical applications are analyzed with reference to each sensory system rather than according to disease states. Table 1 contains a listing of the various types of auditory, visual, and somatosensory evoked potentials based on their presumed sites of origin from along the sensory pathway. The evoked potential wave forms are included in the figures and the technical details of their recording are in Table 2.

Sensory evoked potentials were among the earliest measures used to study the functions of the brain (Lindsley 1969, Davis 1976, Bergamini & Bergamasco 1967). While techniques for quantifying intracellular and single neuronal activity now predominate in the study of other animals, interest in evoked potentials persists particularly for the analysis of human brain functions. Moreover, the development

0147-006X/78/0325-0103$01.00

Table 1 Sensory evoked potentials in man: analysis of evoked potential tests and their probable generators in relation to the afferent pathway

Sensory pathway components	Auditory		Visual		Somatosensory	
	Test	Generator	Test	Generator	Test	Generator
Receptor	Electrocochleogram auditory far field potentials	Hair cells	Electroretinogram	Mostly rod function	Not available	—
Primary afferent neuron	Electrocochleogram	VIII nerve	Not available	—	Whole nerve action action potentials	Peripheral nerve–large myelinated fibers
	Auditory nerve and brainstem potentials					
Ascending path	Auditory brainstem potentials	VIII nerve through inferior colliculus	Not available	—	Spinal evoked potentials	Dorsal column
					Far-field potentials	Medial lemniscus Diencephalon Cerebellum
Primary sensory cortex	Middle latency potentials	Unknown	Visual evoked potentials	Primary visual cortex	Somatosensory evoked potentials	Specific sensory-motor cortex
Nonspecific cortex	Long-latency auditory potentials Sustained potentials	Unknown	Visual evoked potentials	Unknown	Long-latency somatosensory potential	Unknown

of "averaging" techniques for extracting low-level stimulus-related signals from the background electroencephalogram and other biological potentials now provides a reliable basis for the quantitative study of evoked potentials. Some of the clinical reasons for investigating sensory evoked potentials in humans are:

1. Evoked potentials can provide quantitative and objective measures of sensory function (Sokol 1976, Picton et al 1977). The establishment of reliable correlations in normal subjects between attributes of the physical stimulus (i.e. intensity, frequency, wavelength), sensory perception, (i.e. loudness, pitch, hue, etc), and the latency or amplitude of the various components of evoked brain potentials permits the application of these measures to individuals who are unable to accurately describe their sensory experiences, such as infants, retarded individuals, some patients with neurological disorders, or even normal but anxious subjects. The information obtained about such an individual's sensory function may provide important clinical information ("Does the child hear?" "Is there a problem of visual acuity?"), public health data (the sensory effect of exposure to toxins), or it may be of medicolegal value when an objective definition of disordered function is essential.

2. Sensory evoked potentials are relevant in clinical neurology as an objective test of brain function (Regan 1972). Changes in these potentials may localize the lesion to a particular site along the afferent pathway from receptor to cortex. The analysis depends on the presence of a precise relationship between particular anatomical structures and components of the evoked potential wave form. This is an expanding area of research interest in which information from both clinical-pathological correlations in humans and experimental studies in animals provides the framework for accurate clinical application of evoked potential measures. Knowledge of the generators of the various sensory evoked potentials would be of immense help in defining the locus of lesions producing sensory deficits (i.e. numbness), coma (Greenberg et al 1977), or dementia (Visser et al 1976).

3. Sensory evoked potentials can provide insight into normal physiological processes related to maturation (Hecox 1975) and aging (Dustman & Beck 1969). The finding that psychological factors such as "attention," "habituation," and "significance" can influence sensory evoked potentials (Picton et al 1976), has been utilized by several investigators to gain insight into affective and thought disorders in man (Shagass 1972, Callaway 1977) and, more recently, as an objective measure of general brain functions (John et al 1977).

TECHNICAL CONSIDERATIONS

There are many sources of interference that can obscure the detection of low-amplitude sensory evoked potentials. These include electrical impulses from other monitoring devices on the patient or patient-generated events such as muscle potentials, electrocardiogram, or eye movements. Special computer circuits that reject samples containing the artifact from the averaging process or filter circuits that attenuate unwanted potentials have been utilized to reduce this problem.

Once an average evoked potential is obtained, several strategies are employed to assess both its reliability and relation to the sensory stimulus. These include (*a*)

Table 2 Sensory evoked potentials in man: recording parameters

	Electrode placement[a]	Band pass (kHz) (3 dB down points)	Stimulus	Time base (msec)	Trials (*n*)	Amplitude (μV)	Test name
Auditory							
Receptor	I Mastoid-reference	Variable	Tones	Up to 10	2,000	<1.0	Cochlear microphonic
Afferent	I Mastoid-reference	0.1–10.0	Clicks	3	2,000	<1.0	Auditory nerve and brainstem potentials
Ascending	Vertex-I mastoid	0.1–3.0	Click	10	2,000	<1.0	Auditory brainstem potentials
	Vertex-C mastoid	0.1–3.0	Tone	20	2,000	<1.0	Frequency-following potentials
Specific cortex	Vertex-mastoid	0.01–0.3	Tone/Click	60	100	<2.0	Middle latency potentials
Nonspecific cortex	Vertex-mastoid	0.001–0.1	Tone	500	50	<10.0	Long latency potentials
Visual							
Receptor	Cornea-reference	0.001–1.0	Light flash	100	1	<500.0	Electroretinogram
	or lower canthus–reference	0.001–1.0	Light flash	100	25	<50.0	Electroretinogram
Afferent	—	—	Not available	—	—	—	—

Table 2 *(Continued)*

Ascending	—	—	Not available	—	—	—	—
Specific cortex	O_z–vertex	0.001–0.1	Light flash or pattern reversal	200	50	<20.0	Visual evoked potentials
Nonspecific cortex	O_z–vertex	0.001–0.1	Light flash or pattern reversal	500	50	<20.0	Visual evoked potentials
Somatosensory							
Receptor	—	—	Not available	—	—	—	—
Afferent	Over peripheral nerve	0.1–1.0	Shock	20	100	<5.0	Nerve action potential
Ascending	Over spinal column	0.1–3.0	Shock	20	8,000	<1.0	Spinal cord potentials
	Vertex-reference	0.1–3.0	Shock	20	2,000	<1.0	Far-field potentials
Specific cortex	Sensory cortex contralateral to stimulus (C3 or C4)-reference	0.01–1.0	Shock	75	100	<10.0	Somatosensory evoked potentials
Nonspecific cortex	Vertex-reference	0.001–0.1	Shock	500	100	<10.0	Somatosensory evoked potential

[a] I = Ipsilateral to acoustic stimulus; C = contralateral to acoustic stimulus.

reproducibility of the potential wave forms in duplicate averages, (*b*) determining the absence of the potential wave forms by alternating the sign of the averaging process between addition and subtraction, and (*c*) the loss of evoked potential wave form when averages are performed in the absence of the sensory stimulus.

The sensory signals must be carefully calibrated, and the patient's level of arousal and clinical neurological deficit defined. Finally, if sensory evoked potentials are to be useful in a clinical environment, the procedures should be (*a*) rapid, (*b*) simple to perform, and (*c*) relatively inexpensive.

Somatosensory Evoked Potentials

STIMULUS Somatosensory evoked potentials are typically elicited by electrical stimulation of peripheral nerve trunks. This technique provides a precise onset for averaging purposes, but the number and types of nerve fibers activated are difficult to quantify. A major drawback of electrical stimulation of peripheral nerves is that it can be extremely uncomfortable. There is need for the development of precisely controlled natural forms of stimulation for the clinical evaluation of somatosensory functions.

NORMAL EVOKED POTENTIALS

Afferent: peripheral nerve The potentials ascending in the peripheral nerve can be recorded from skin electrodes overlying the nerve or from needle electrodes inserted close to the nerve trunk (Figure 1, *primary afferent, neuron*). Their latency can be used to calculate the conduction velocity of the ascending somatosensory impulses. The presence of abnormally slow conduction velocities is evidence of a peripheral nerve disorder and can by itself be associated with alterations in somatosensory evoked potentials from central structures (Desmedt & Noel 1973). The measurement of peripheral-nerve conduction velocity is a prerequisite for evaluating abnormalities of somatosensory evoked potentials in the clinical setting.

Ascending pathway: spinal cord and brainstem The shortest latency somatosensory evoked potentials originating in the central nervous system can be detected from the skin surface overlying the spinal cord (Cracco 1973). Their amplitudes are less than 1 μV, so that as many as 8000 stimulus trials are needed to insure a satisfactory average. The ascent of the evoked potential up the spinal cord can be monitored at several points to provide a measure of spinal cord conduction velocity. An indirect method of estimating spinal cord conduction has been suggested by Dorfman (1977), using the difference in latency of scalp-derived potentials from stimulating peripheral nerves in the arm and leg. More recently, far-field recordings of activity in somatosensory pathways of the spinal cord and brainstem have been made with electrodes located on both the scalp and a distant reference site such as the hand or knee (Cracco & Cracco 1976). The potentials recorded in this manner from stimulation of the median nerve at the wrist consist of a sequence of components of less than 1.0 μV in amplitude with peak latencies of 9, 11, 14, and 19 msec (Figure 1, *ascending path*). The neural origins of the components are still uncertain, but there is evidence that they derive from activity in peripheral nerve fibers and ascending brainstem, diencephalic, and cerebellar somatosensory pathways, respectively (Wiederholt & Iragui-Madoz 1977).

Thalamus and specific cortex The somatosensory evoked potentials recorded from the scalp that occur between 15 and 65 msec after stimulation of a peripheral nerve in the upper extremity appear to derive from activation of specific sensory areas within the cerebral hemispheres (Figure 1, *sensory cortex*). Their amplitudes are maximal from scalp regions overlying the primary sensory-motor cortex *contralateral* to the limb stimulated (Goff et al 1977). Moreover, the potentials evoked by stimulating the peripheral nerves of the legs are maximal medial to those sites where potentials evoked by upper-limb stimulation occur (Desmedt 1971). These somatosensory evoked potentials can be up to 10 μV in amplitude and require only 60–120 stimulus repetitions to elicit clear averages. The potentials consist of a sequence of positive and negative components that have been variously designated. A nomencla-

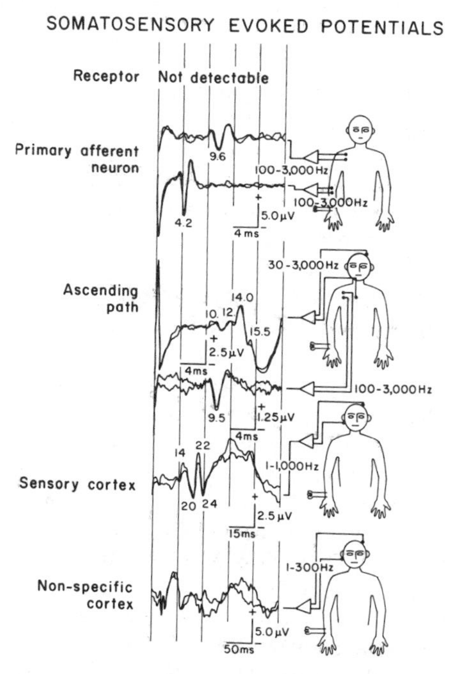

Figure 1 Somatosensory potentials evoked by electrical stimulation of the median nerve at the wrist. The intensity of the stimulus was adjusted to be just below that necessary to elicit a contraction of the thenar muscles. The evoked potentials are designated by their presumed generators listed on the left, i.e. "receptor, primary afferent neuron, ascending path, sensory cortex, and nonspecific cortex." The recording sites for these potentials are depicted on the figures to the right with the particular amplifier bandpass settings in Hz. The numbers inscribed above the component peaks of the evoked potential wave forms refer to their latency in msec. Note that the time base and amplitude calibrations vary.

ture has recently been proposed that signifies a component's polarity (positive, P, and negative, N) and latency in msec (i.e. P15, N100, etc). Stimulation of the median nerve at the wrist elicits a P15, N20, P28, or a dual positive complex P25 and P30, N35 and P45 (Giblin 1964, Liberson 1966, Desmedt 1971). Stimulation of the posterior tibial nerve at the ankle will elicit a similar set of potentials, though they are somewhat delayed because of the additional length of the ascending pathway. The initial positive component is often difficult to detect with stimulation of the lower extremity (Dorfman 1977). The new nomenclature may prove inadequate, since latency is affected by the location of the stimulating electrodes along the length of a peripheral nerve.

Nonspecific cortex Somatosensory evoked potentials occurring after 60 msec are distributed over both hemispheres (Figure 1, *nonspecific cortex*). The components from 65 to 100 msec appear to be a mixture of both muscular and neural activities, whereas the components from 140 to 500 msec appear to be of neural origin (Goff et al 1977).

EVOKED POTENTIALS, STIMULUS AND SUBJECT VARIABLES Stimulus strength has minimal influence on the latency of somatosensory evoked potentials. Moreover, maximal response amplitudes are achieved with current strengths only slightly above threshold (Uttal & Cook 1964). Thus, the precise quantification of suprathreshold electrical stimulation of peripheral nerves may not be a significant issue in clinical testing. Recovery functions of somatosensory potentials from stimulation of the median nerve require 100–200 msec for full return of the amplitude of the specific cortical components (latency < 65 msec), whereas the longer-latency components may require up to 3000 msec for recovery (Namerow 1970, Allison 1962).

Prolonged stimulation may produce a decrement in amplitude of the N40 component (Giblin 1964) that could be pertinent in situations requiring prolonged testing, such as evoked-potential monitoring during surgical procedures.

Aspects of the subject's waking state can also influence the evoked potential. Thus sleep can affect the amplitude of the long-latency components (> 100 msec), whereas the shortest-latency N14 response is little influenced by sleep or even deep anesthesia (Goff et al 1966, Desmedt & Manil 1970). Muscle potentials from the scalp contribute to some of the components, particularly between 65 and 100 msec following the stimulus. Their amplitude is largest over the forehead, neck, and temporalis muscles (Goff et al 1977). Perceptual factors such as attention to the stimulus can produce enhancement of the N120, P190, and P300 components (Desmedt 1971). The latency of the initial negative component to median nerve stimulation is slightly less in the newborn than in adults. However, considering the reduced length of the somatosensory pathways in infants, conduction velocity during childhood is actually quite slow, and adult values are achieved only at about 8 years of age (Desmedt, et al 1976). Senescence has little effect on the latency of the potentials (Luders 1970). Among other variables, movement of the stimulated limb has been reported to be associated with a decrease in evoked-potential amplitudes. A similar attenuation of evoked potentials can occur during the simultaneous application of natural cutaneous stimulation to the skin surface innervated by the stimulated nerve. The mechanism of attenuation probably involves the "masking" of one sensory stimulus by another (Giblin 1964).

NEURAL GENERATORS OF THE EVOKED-POTENTIAL COMPONENTS The fiber pathways that are essential for detecting somatosensory potentials from the scalp are in the dorsal column and medial lemniscus. Individuals with anterolateral spinal-cord tract lesions that produce isolated loss of pain and temperature functions have normal somatosensory evoked potentials, whereas individuals with dorsal-column spinal-cord lesions that produce loss of vibration and position sense have altered somatosensory evoked potentials (Halliday & Wakefield 1963, Namerow 1969, Giblin 1964). There are some clear correlations between somatosensory evoked potentials recorded from the scalp and the cortical surface. Broughton (1969) and, more recently, Allison et al (1977) showed that the P20 and N30 components recorded anterior to the central sulcus become of opposite polarity posterior to the sulcus. These results are compatible with a cortical dipole source in the precentral sulcus oriented in a rostral-caudal direction. The P25 component that can be recorded from the scalp in some subjects appears to be generated at the central sulcus by a dipole oriented in an orthogonal direction, which may account for the variability of detecting this component in different individuals. The earlier positive component at 15 msec (P15), detected in scalp recording, is not present in recording from the cortical surface, which suggests its origin in subcortical structures (Broughton 1969).

CLINICAL UTILITY OF SOMATOSENSORY EVOKED POTENTIALS

Peripheral nerve function Measurement of the change in latency of the initial scalp-derived negative component from stimulation at various points along a peripheral nerve can provide a measure of the nerve's conduction velocity (Desmedt 1971). This technique is particularly applicable to individuals with advanced peripheral neuropathies in whom compound nerve-action potentials may be difficult to detect.

Spinal cord function The techniques of recording ascending activity in the spinal cord from surface electrodes located over the spinal column has been used by Cracco (1975) to localize the level of spinal cord pathology in infants. The technique may also be used to define somatosensory functions in infants.

The presence of potentials that can be recorded from the scalp following stimulation of the nerves in the legs has been utilized by Perot (1972) as a rapid and objective clinical measure of spinal cord function in individuals rendered unconscious or uncooperative from trauma. Normal somatosensory evoked potentials indicate integrity of dorsal column function, whereas their absence, prolonged latency, or diminished amplitude alerts the clinician to the presence of a spinal cord lesion. This technique has also been utilized to monitor spinal cord function in the operating room in individuals undergoing laminectomy for removal of spinal cord tumors or in individuals undergoing correction of spinal column curvature (Allen & Starr 1977). An awareness of changes in spinal cord function may assist the surgeon in preventing some of the undesirable side effects of operative manipulation of the spinal cord. The effects of anesthesia, fluctuations in blood pressure, and manipulation of the spinal cord, dura, and roots on the evoked potentials need to be defined in greater detail before the utility of this technique can be fully realized.

Slowed conduction in the somatosensory pathways of the spinal cord and the brainstem Partial lesions of the somatosensory pathway, such as those occurring in demyelinating disease, can be associated with a prolongation of latency and decrease in amplitude of the potentials recorded at the scalp. The evoked potentials recorded from patients with the clinical diagnosis of multiple sclerosis may even be delayed when sensation is normal (Namerow 1968, Desmedt & Noel 1973). Thus, as is the case for visual and auditory brainstem evoked potentials (see the appropriate sections in this article), somatosensory evoked potentials can be used to define clinically inapparent lesions of the somatosensory pathways in individuals suspected of having multiple sclerosis. Variations in the methods of somatosensory testing, including recovery to paired stimuli or the effects of differing rates of stimulation on the amplitude of the potentials, may enhance the detection of lesions of the ascending somatosensory pathways (Sclabassi et al 1974). Knowledge of the state of peripheral nerve function must be known, since prolonged latency of evoked potentials also occurs with peripheral nerve disorders. Furthermore, a prolonged latency of evoked potential in the absence of peripheral nerve lesions cannot be equated with a specific disease such as multiple sclerosis, since other pathological processes such as vascular lesions of the brainstem (Noel & Desmedt 1975) and infiltrating tumors of the ascending pathway will also have the same effect.

Disorders in the somatosensory pathways of the cerebral hemispheres The effects of cerebral lesions on evoked potentials depend on (*a*) the extent and type of sensory loss, (*b*) the time interval between the lesion and evoked potential testing, and (*c*) the locus of the lesion (Giblin 1964, Halliday 1967b). Lesions of the cerebral hemisphere that result in a loss of sensation (touch, pin, position sense) are associated with a loss of evoked potentials from both the affected and normal hemispheres if the stimulus is applied to the limbs with decreased sensibility (Williamson et al 1970, Liberson 1966, Green & Hamilton 1976). In contrast, stimulation of the unaffected limbs results in the bilateral appearance of normal evoked potentials (Tsumoto et al 1973). Thus, the primary somatosensory pathway within the cerebral hemisphere must be intact for the bilateral representation of evoked potentials. Giblin (1964) described a group of patients in whom sensory loss was particularly evident during simultaneous bilateral sensory testing (a phenomenon called "extinction"). These patients also failed to detect light touch during unilateral stimulation of the affected limb, though somatosensory evoked potentials were normal. Finally, there are patients with normal sensation in whom the amplitude of the evoked potentials is significantly altered. They may be increased (Halliday 1967a, Giblin 1964, Tsumoto et al 1973), decreased (Giblin 1964), or even have additional components not usually encountered (Giblin 1964).

Disorders of evoked potentials in epilepsy The amplitude of somatosensory evoked potentials may be up to tenfold larger in myoclonic epilepsy. Even the earliest latency components are thus enhanced (N20, P30), whereas their latencies are unaffected (Dawson 1947, Halliday 1967a). Halliday noted that the evoked potentials were particularly enhanced if the patients were actively experiencing myoclonic jerks at the time of the tests.

Functional sensory loss Normal somatosensory evoked potentials have been recorded in patients with a sensory defect due to hypnosis (Halliday & Mason 1964) and in patients suspected of having hysterical hemianesthesia.

Visual Evoked Potentials

STIMULUS A wide variety of stimuli have been employed to study human visual evoked potentials, including diffuse, patterned, and colored lights. Factors such as the rate of presentation, the portion of the visual field stimulated, and monocular vs binocular presentation are significant variables (Regan 1975). The various types of stimuli and methods of analysis have particular applications. For instance, the onset of an infrequently presented signal evokes a "transient" set of potentials in which one can measure the latency and amplitudes of the various components as indices of visual function. In contrast, "steady-state" evoked potentials can be detected by analyzing only those components that have spectral energies at the fundamental or a harmonic of the stimulus rate. If Fourier analysis is used, these potentials can be detected at extremely low levels ($<1.0\ \mu V$) and both the phase and amplitude of the components can be precisely specified. A significant advantage of "steady-state" potentials for clinical testing is that they can be defined after a brief period of stimulation.

NORMAL EVOKED POTENTIALS

Receptor: the retina Retinal potentials or the electroretinogram (ERG) evoked by diffuse light flash can be detected, without averaging, by electrodes placed directly on the cornea or sclera. These same retinal potentials can also be recorded by electrodes placed on the skin surface close to the eye, using a computer to average the low-amplitude potentials from background activity (Figure 2, *receptor*). The reader is referred to standard texts for details of the ERG. The assessment of retinal function by the ERG has been suggested as a necessary prerequisite for evaluating abnormalities of visual evoked potentials in clinical disorders. However, the relationship between the two types of potentials is complex since the ERG primarily reflects functions of the rods and their retinal connections, whereas the scalp-derived visual evoked potentials seem to reflect primarily the activity of cones and their central connections.

Ascending pathway, optic nerve, and lateral geniculate Efforts at recording the activity of optic nerve and lateral geniculate nucleus by far-field recording techniques analogous to those used in recording activity of ascending pathways in the somatosensory and auditory systems have not been successful. The high amplitude and prolonged time course of the retinal potentials are probably the major factors contributing to this failure.

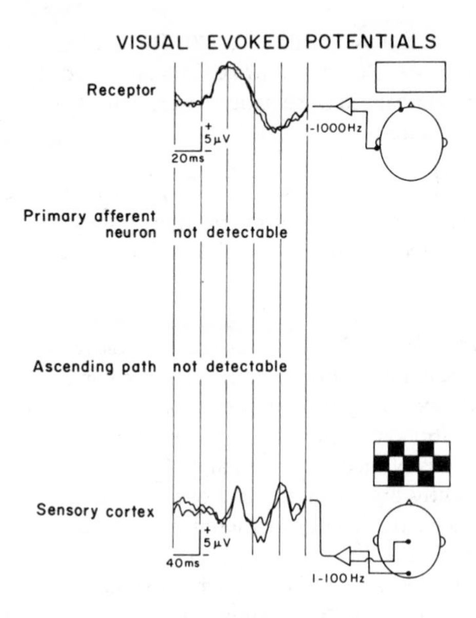

Figure 2 Visual potentials evoked by light flash (upper trace) or pattern reversal stimulation (lower trace). The format for this figure is the same as in Figure 1.

Specific cortex Scalp-derived visual evoked potentials (Allison et al 1977) to brief diffuse light flashes that can be distinguished from retinal potentials occur between 40 and 145 msec following stimulus onset and are best recorded over the occipital area. The latency, amplitude, and even occurrence of the components have not been consistently described in the literature. Factors such as stimulus luminance (DeVoe et al 1968), level of arousal (Oosterhuis et al 1969), and electrode location (Biersdorf & Nakamora 1971) influence the form and latency of the flash evoked potentials and are probably responsible for the variations reported in the literature.

In contrast, pattern evoked potentials are reliable both in form and latency between subjects (Harter & White 1968). A checkerboard stimulus in which the black and white squares reverse position at a rate of 2 sec^{-1} evokes a prominent positive potential over the occipital region with a latency of about 100 msec. The location of the recording electrode and stimulus luminance (over two log units) have only minimal effects on the latency of the potential evoked by the full-field reversing checkerboard signal (Halliday et al 1970). However, changes in the location of the pattern within the visual field do affect the distribution of potentials (Shagass et al 1976, Halliday & Michael 1970, Jeffreys & Axford 1972a,b, Michael & Halliday 1971). Stimulation of the right visual field with patterned or even diffuse light will elicit a positive potential between 80 and 100 msec of maximal amplitude over the left occiput that reverses polarity near the midline. The opposite distribution of potentials occurs upon stimulation of the left visual field. Furthermore, stimulation of the upper and lower halves of the visual fields by patterns evokes differing amplitude and polarity distributions of potentials over the occipital region. The prominent positive component at 100 msec seen with full field is also present on activating just the lower half of the visual field, whereas stimulation of only the upper half visual field evokes a small negative component at this latency.

Steady-state visual evoked potentials to repetitive diffuse or patterned stimulation have three distinct amplitude maxima (10 Hz, 16–18 Hz, and 45–60 Hz) that Regan (1972) suggests correspond to separate neural channels of visual processing.

Nonspecific cortex Transient visual evoked potentials to diffuse or patterned light flash that occur between 90 and 500 msec after stimulus onset are distributed widely over the scalp (Allison et al 1977). One of these components (P130) appears to be a mixture of neural and myogenic components. A positive component at 300 msec (the P300) appears in stimulus situations in which the subject must "attend" to the signal.

EVOKED POTENTIALS AND STIMULUS VARIABLES Luminance has significant influence on both the latency and amplitude of diffuse flash-evoked potentials but has little effect on potentials evoked by patterned stimuli. Factors such as the visual angle of the components of the pattern and their location within the visual field affect the amplitude of the potentials (Harter 1971). Visual spacings of 10–20 min are the most effective in evoking potentials if the stimulus falls on the central 3° of the visual field, whereas spacings of 50–60 min are most effective in the parafoveal regions. The clarity of focus of the patterned stimuli on the retina affects the amplitude but not the latency of evoked potentials. Insertion of lens to distort the patterns leads to decreased amplitudes, whereas a lens that corrects refractive errors will increase the amplitude of pattern evoked potentials (Harter & White 1968).

EVOKED POTENTIALS AND SUBJECT VARIABLES The effect of sleep on visual evoked potentials has not been as thoroughly investigated as other sensory evoked potentials, but changes in amplitude and form do occur (Kooi et al 1964). Muscle potentials from the scalp contribute to visual evoked potentials and have similar latency and distribution to the muscle potentials generated by auditory and somatosensory stimulation (Allison et al 1977). Perceptual factors, such as attention, can produce enhancement of certain components of flash- or pattern evoked potentials. In particular, the positive components occurring at a latency of 300–400 msec distributed over the central parietal or frontal regions are enhanced during perceptual tasks (Courchesne et al 1975). Both flash- and pattern evoked potentials change with maturation and senility. Diffuse light flash evokes a simple evoked potential form in young infants that is delayed in latency and of lower amplitude than the evoked potentials recorded from adults (Ellingson 1966). Evoked potentials can define the changes in visual acuity that occur during maturation (Marg et al 1976). With senescence the latency of pattern reversal evoked potential lengthens without any change in amplitude (Celesia & Daly 1977).

NEURAL GENERATORS OF THE EVOKED POTENTIAL There are excellent studies reviewed by Creutzfield & Kuhnt (1973) on the neuronal basis for visual evoked cortical responses in experimental animals. In contrast, details of cortical and subcortical visual evoked potentials in humans and their relation to the scalp-derived components are lacking.

CLINICAL UTILITY OF VISUAL EVOKED POTENTIALS

Ocular function The excellent correlation between amplitude of pattern evoked visual potentials and the clarity of focus of the image on the retina provides a precise means for the objective definition of visual acuity. This technique may be of value in selecting corrective lenses for young children with refractive errors. Astigmatic errors can also be detected by using evoked potential to patterns of various orientation (Regan 1977). The correction of such defects may prevent "meridional amblyopia," that is, amblyopia restricted to a particular plane of orientation (Freeman & Thibos 1975). Evoked potential studies in "amblyopia ex anopsia" have, in general, been unrewarding in defining the underlying mechanisms of this disorder. Diffuse flash-evoked potentials are normal, whereas pattern evoked potentials are larger than normal if the stimulus acts on the parafoveal region (Spekreijse et al 1972). This finding raises the possibility that visual pathways from parafoveal regions are more extensively developed in amblyopic than in normal eyes. Finally, the definition of color blindness can be objectively specified by use of evoked potentials to pattern reversal stimuli of appropriate spectral composition. Regan & Spekreijse (1974) have shown that red-green color-blind individuals do not generate evoked potentials when checkerboard patterns reverse between these two hues but have quite normal potentials when presented with only the red or green checkerboard.

Optic nerve function Abnormalities of evoked potentials to diffuse light flash were found in patients with visual loss due to optic nerve involvement from tumors or acute demyelination (Richey et al 1971, Vaughan & Katzman 1964). Moreover, these potentials were also abnormal in some individuals with multiple sclerosis, even when their vision was not impaired, which raises the possibility of using visual evoked potentials as a diagnostic test for multiple sclerosis. Several groups of investigators found that between 50 and 100% of patients with multiple sclerosis without visual loss had abnormalities of the flash-evoked potentials (Richey et al 1971, Feinsod et al 1973, Feinsod & Hoyt 1975, Namerow & Enns 1972). However, the marked variability of diffuse flash-evoked potentials in normal subjects may have accounted for its lack of acceptance as a clinical test for defining optic nerve disorders in multiple sclerosis. Recently, the consistency and reliability of the pattern-reversal evoked potentials led Halliday and his associates (1972, 1973a,b) to reassess this technique in patients with multiple sclerosis. They first noted that pattern-reversal evoked potentials were delayed in latency in >90% of patients with an acute retrobulbar neuritis and that the delay persisted even after the patients' visual acuity and fields returned to normal (Halliday et al 1973a). Moreover, in patients suspected of having multiple sclerosis but without visual complaints, pattern-reversal evoked potentials have been reported to be delayed in latency in between 59 and 96% by three separate groups of investigators (Asselman et al 1975, Halliday et al 1973b, Celesia & Daly 1977). The sensitivity of the technique is enhanced if the criteria of abnormality for unilateral optic nerve involvement are expanded to include differences in latency between the two eyes (Celesia & Daly 1977). Steady-state evoked potentials to repetitive diffuse or patterned light stimuli are also abnormal in patients with multiple sclerosis independent of clinical evidence of optic nerve involvement (Regan et al 1976, Milner et al 1974). The abnormality was restricted to the steady-state potentials evoked by repetitions at 18–20 Hz, whereas the potentials to more rapid repetitions at 45–60 Hz were normal. There are no studies in patients with multiple sclerosis to compare the ability of transient pattern reversal and steady-state stimulation to detect optic nerve disorders. Steady-state potentials are apparently remarkably sensitive since they can define small quadrantic defects of glaucomatous eyes (Cappin & Nissim 1975).

The enthusiasm for using pattern evoked potentials as a diagnostic aid in multiple sclerosis must be tempered by the awareness that *any* lesion of the optic nerve can alter these potentials (Halliday et al 1976, Feinsod et al 1976). Halliday's suggestion that the type of abnormality of the evoked potentials (i.e. latency, form, or amplitude) may distinguish between demyelination and compression of the optic nerve needs further study.

Disorders of the central visual pathway Vaughan and his collaborators (1963) explored the use of visual evoked potentials as an objective measure of central lesions of the visual pathway. They utilized an amplitude difference between the two occipital poles of >50% to full-field diffuse light stimulation as an indicator of homonymous visual pathway alteration. Visual evoked potentials have also been described as normal in individuals with hemianopsia (Asselman et al 1975). The characterization that there are definite hemispheric asymmetries in normal subjects of potentials evoked by stimulation of the half visual fields will allow a reassessment of central lesions on visual evoked potentials and visual field defects (Shagass et al 1976). Certainly an abnormal hemispheric distribution of steady-state potentials has been clearly demonstrated in patients with central lesions producing visual field defects (Regan & Heron 1969, Wildberger et al 1976, Bodis-Wollner 1977).

Epilepsy Visual evoked potentials have been recorded in individuals with photosensitive epilepsy (Hishikawa et al 1970, Harden & Pampiglione 1971). In neuronal storage disease the amplitude of diffuse flash-evoked potentials was markedly enhanced even though the ERG was either absent or depressed in amplitude. In other forms of photosensitive epilepsy the amplitude of flash-evoked potentials will vary, depending, in part, on whether the stimulus can precipitate an epileptic discharge (Hishikawa et al 1970).

Operating room Flash-evoked visual potentials have been used to monitor optic nerve function during operation on individuals with orbital (Wright et al 1973) or chiasmatic lesions. Feinsod and his associates (1976) showed that visual evoked potentials increase in amplitude following removal of tumors compressing the optic nerve. Allen & Starr (1977) describe the sudden appearance of visual evoked potentials in the course of surgery in individuals with pituitary tumors who had been without vision or evoked potentials prior to the operation. The effects of anesthesia and blood pressure on these potentials need to be defined in detail before their full utility can be appreciated.

Miscellaneous Visual evoked potentials have been used to define possible alterations of central connections of the visual pathways in albinos (Creel et al 1974). Moreover, these procedures have been utilized to help explore such diverse neurological problems as dyslexia (Symann-Louett et al 1977) and classic migraine (Regan 1972).

Auditory Evoked Potentials

STIMULUS Auditory evoked potentials are elicited by clicks or brief tone bursts that can be varied in intensity, frequency, repetition rate, "rise" and "fall" times of the tones, duration, or monaural and binaural presentations.

NORMAL EVOKED POTENTIALS

Receptor and afferent input Both cochlear microphonic and VIII nerve activity can be detected from an electrode located in the middle ear on the bony promontory of the cochlea in a procedure called electrocochleography (Eggermont et al 1974). Requirements for accurate and safe placement of the electrodes demand the skills of an otorhinolaryngologist. Recently it has become possible to detect both cochlear microphonic and VIII nerve activity from scalp electrodes located in the ear canal or on the mastoid of the stimulated ear (Sohmer & Pratt 1976) by averaging responses to a great number of acoustic signals (Figure 3, *receptor* and *primary afferent neuron*).

Ascending activity The electrical events generated in the central auditory pathway can be recorded in the far field from scalp electrodes if many stimulus presentations are averaged (Jewett 1970). The resultant potentials have been variously designated as auditory brain responses (ABR) or brainstem evoked responses (BER). Click signals evoke seven low-amplitude ($<1.0\ \mu V$) potentials in the initial 10 msec following stimulus presentation with the positive components at the vertex designated in sequence by Roman numerals (Figure 3, *ascending path*). Approximately 2000 clicks are required to obtain a reliable average, but since the clicks are presented at rates between 5 and 30 sec the recording time is not excessive. The concept of "active" and "reference" electrodes in the far-field detection of electrical events is not useful since all electrode sites, even those remote from the scalp, are "active." The designation of the electrodes as "active" and "less active" would be more accurate. While a variety of electrode locations have been employed, the practice in our laboratory is to record between the vertex and mastoid ipsilateral to the stimulus site.

Low-frequency tone bursts evoke another form of activity from the ascending auditory pathway that is termed the frequency following response or FFR (Moushegian et al 1973). These potentials occur at the same frequency as the stimulus tone in the range below 1 kHz (Figure 3, *frequency following potential*). Care must be taken to insure that the recorded potentials are not contaminated by the stimulus voltages applied to the earphones.

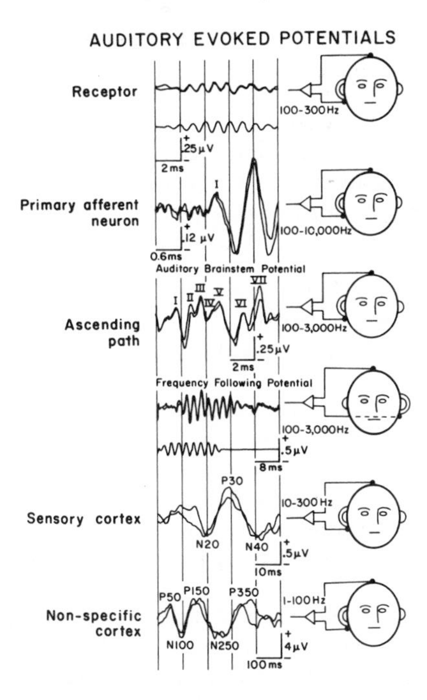

Figure 3 Auditory potentials evoked by clicks and tones. The format is the same as Figure 1. The voltage wave form applied to the earphone for tone stimulation is depicted just below the tracings of the "receptor" potentials (cochlear microphonic) and "ascending path" potentials (frequency following potential). Tone bursts were also used to elicit sensory cortex and nonspecific cortex potentials. Clicks were used to evoke primary afferent neuron and auditory brainstem potentials. The components of these latter two potentials are designated by Roman numerals. The letters and numbers above the potential wave forms in the lower two traces refer to the polarity (positive, P, or negative, N) and latency in msec of the components.

Thalamus and specific cortex A set of potentials occurring between 10 and 50 msec after stimulus presentation can be detected from scalp electrodes and have been termed "middle-latency components" (Davis 1976). These components are candidates for thalamic and primary auditory cortical activity (Figure 3, *sensory cortex*). They are best elicited by filtered clicks or brief tones, and their detection is enhanced by appropriate filters (10–100 Hz). The study of these middle-latency components is complicated by the existence of scalp-derived sound-evoked muscle potentials [i.e. "the inion response" or the "micro reflex" (Bickford 1972)] at this same latency. However, the two types of potentials can be distinguished by the differential effects of signal intensity and level of arousal.

Nonspecific cortex Long-latency sound-evoked potentials can be detected by an electrode located at the vertex and consist of P50, N100, P150, and N200 components (Figure 3, *nonspecific cortex*). These potentials are of maximal amplitude at the vertex and are best elicited by tone-burst stimuli that are at least 30 msec in duration (David 1976, Picton et al 1977).

Sustained potentials A steady potential shift to sustained tone signals can be detected from scalp electrodes at the vertex (Keidel 1971). Their detection requires either DC recording or relatively long time constants. The potential can be distinguished from the contingent negative variation (CNV) of Walter by the differential effects of sleep and perceptual task (Picton et al 1977).

EVOKED POTENTIALS AND STIMULUS VARIABLES An increase in signal intensity is associated with both a decrease in latency and a growth in amplitude of cochlear microphonic and VIII nerve responses, auditory brainstem potentials, middle latency, and certain components of the long latency potentials. The orderly relation between signal intensity and evoked potential latency provides a means for the objective definition of auditory function.

Stimulus frequency may have significant influence on the evoked potentials. Auditory brainstem responses are most easily elicited by click signals containing energy above 2 kHz. In contrast, the FFR can only be elicited by tones below 1 kHz (Marsh et al 1975). The middle- and long-latency potentials are relatively independent of stimulus frequency.

Stimulus repetition rate will affect all of the evoked potentials. The amplitude of the early components of the brainstem potentials (Waves I–III) are significantly attenuated at click repetitions greater than 20 sec^{-1}, whereas Wave V is little affected (Don et al 1977). Long-latency cortical evoked potentials are attenuated in amplitude if stimulus presentation is more rapid than 1 per 10 sec (Davis et al 1966).

EVOKED POTENTIALS AND SUBJECT VARIABLES Sleep affects the amplitude of long-latency evoked potentials but has no effect on cochlear microphonic, VIII nerve, brainstem or middle-latency responses (Mendel et al 1975, Amadeo & Shagass 1973). Muscle potentials from the scalp can be a major contaminant of both the middle- and long-latency evoked potentials but do not influence brainstem potentials other than increasing the background recording "noise" (Goff et al 1977). Behavioral tasks requiring "attention" to the stimulus will enhance the amplitude of the P300 component of long-latency cortical responses but is without effect on middle-latency, brainstem, VIII nerve, or cochlear microphonic potentials (Picton & Hillyard 1974). There is a systematic decrease in latency of the auditory brainstem potentials with maturation (Starr et al 1977, Schulman-Galambos & Galambos 1975), and adult values are achieved between 1 and 2 years of age (Hecox & Galambos 1974, Salamy & McKean 1976). Slow cortical evoked potentials change in a more complex manner both in form and latency during this same developmental period (Barnet et al 1975, Davis & Onishi 1969).

NEURAL GENERATORS OF AUDITORY EVOKED POTENTIALS There is evidence that the potentials recorded by electrocochleography represent activity of the hair cells and VIII nerve since they correspond in many respects to these same potentials recorded from the cochlea in animal experiments. Short-latency auditory evoked potentials (>10 msec) appear to originate from the brainstem portions of the auditory pathway. Studies in animals show that activity recorded from particular brainstem nuclear regions occurs at the same times as do the components of the far-field potentials (Jewett 1970, Lev & Sohmer 1972). Moreover, the effects of focal brainstem lesions in both animals and man (Starr & Hamilton 1976, Buchwald & Huang 1975) suggest that Wave I originates from the VIII nerve, Wave II from the region of the cochlear nucleus, Wave III from the region of the superior olive and trapezoid body, and Waves IV and V from the midbrain. The generators of Waves VI and VII are unknown. The neural generators for middle-latency (10–50 msec) and long-latency (50–500 msec) auditory evoked potentials are uncertain. Celesia & Puletti (1971) recorded sound-evoked potentials from the exposed auditory cortex in man and described components occurring between 12 and 22 msec following stimulation restricted to primary auditory cortex and long-latency potentials from more widespread cortical regions. The correlation between evoked potentials recorded from the scalp and those recorded from exposed cortex has been poor (Celesia 1968).

CLINICAL APPLICATION AND AUDITORY EVOKED POTENTIALS

Cochlear function Auditory evoked potentials can serve as an objective measure of hearing (Davis 1976, Picton et al 1977). Electrocochleography, while providing reliable and accurate measures of cochlear function, requires that the patient be sedated or even anesthetized. In contrast, auditory brainstem potentials are relatively simple to record and provide information as to both threshold and type of hearing loss (sensorineural, conductive). There is no need to employ any special sedation, and the potentials are present independent of level of arousal. However, auditory brainstem potentials seem to reflect the functions of the basal end of the cochlea and thus do not provide an accurate reflection of low-frequency hearing. The use of the FFR (frequency following response) may correct this deficiency. Middle-latency evoked potentials are also suitable for defining auditory sensitivity to a wide range of tonal frequencies, but the detection of the potentials may be contaminated by muscle activity. Finally, long-latency cortical evoked potentials are affected by the subject's level of arousal, which may interfere with the determination of hearing threshold (Zerlin & Davis 1967).

The availability of a wide variety of evoked potential methods for reliable and objective hearing evaluation is a major clinical advance. The problem as to which is the most suitable test has not yet been resolved.

Central auditory pathway disorders Evidence that the various components of auditory brainstem potentials depend upon the functional integrity of particular portions of the auditory pathway in its course from the cochlea to the cortex has obvious application for the localization of brainstem disorders in clinical situations. The latency separation between component peaks is relatively independent of signal intensity or cochlear function (Starr 1977). This measure of "central conduction time" in the auditory pathway has been used to localize and define abnormalities of the brainstem (Stockard & Rossiter 1977). Furthermore, changes in the amplitude of the various components have also been associated with lesions of appropriate portions of the auditory pathway (Starr & Achor 1975). The characterization of absolute amplitude in far-field recordings is complicated by the poor signal size relative to background noise (Thornton 1975) and relative amplitudes between the various peaks have been used instead to detect abnormalities in brainstem potentials. Measurement of auditory brainstem potentials in clinical neurological disorders is relatively new, but there is evidence of clinical relevance in the definition of acoustic neuromas (Thornton & Hawkes 1976, Brackman & Selters 1977, Terkildsen et al 1977), brainstem tumors, infarcts, and demyelinating diseases (Starr & Achor 1975, Stockard & Rossiter 1977, Sohmer et al 1974, Robinson & Rudge 1975). Auditory brainstem potentials can help assess brainstem function in patients in coma and in the evaluation of "brain death" (Greenberg et al 1977, Starr 1976). The low amplitude of the far-field potentials requires careful recording techniques, since the potentials can be obscured by a wide variety of artifacts. An advantage of auditory brainstem potentials is that they can be measured rapidly at the bedside to provide quantitative information about brainstem function for the clinician.

Thalamic and cortical disorders Middle-latency and slow cortical evoked potentials have had little application to neurological disorders (Rapin & Graziani 1967) other than as an objective method of hearing function. The paucity of study may reflect the uncertainties as to the neural generators of these potentials.

PERSPECTIVES

The relative simplicity of recording sensory evoked potentials and the present optimism as to their clinical utility will certainly lead to increasing clinical use. Several areas are particularly well suited for investigation.

First, the *establishment* of a reliable relation between the site of neurological lesion and alterations in evoked potentials will provide important clues as to the generators of the evoked-potential components.

Second, assessing the development of sensory and neurological functions in the infant, and their subsequent change with senescence, is likely to replace behavioral testing which has serious limitations.

Third, the measurement of long-latency evoked-potential components related to "attention" (P300) or "expectancy" (CNV) will be used to analyze dementia and memory impairment.

Fourth, while event-related potentials that precede motor behavior were not discussed in this review, their investigation could provide a means of quantifying the wide variety of movement disorders that occur in clinical neurology.

There is a need, however, for improvements in technology. The computers should become small, easy to use, and provide automatic stimulus control and analysis of the evoked-potential wave forms.

It is difficult to predict whether the present enthusiasm for clinical application of evoked-potential measures in man will persist. There is general agreement, however, that evoked potentials are one of the best techniques for objective and noninvasive study of the human brain.

ACKNOWLEDGMENTS

I would like to express my appreciation to Dr. K. Squires, who provided a critical review of the manuscript and the compilation of the set of evoked potentials. I would also like to thank Drs. R. Galambos and R. B. Livingston, who encouraged me to delve into the clinical study of evoked potentials in neurology.

Research was supported by Grant #PHS NS11876–03.

Literature Cited

SOMATOSENSORY-EVOKED POTENTIALS

Allen, A., Starr, A. 1977. Sensory evoked potentials in the operating room. *Neurology* 27:358

Allison, T. 1962. Recovery functions of somatosensory evoked responses in man. *Electroencephalogr. Clin. Neurophysiol.* 14:331–43

Allison, T., Goff, W. R., Williamson, P. D., Van Gilder, J. C. 1977. On the neural origin of early components of the human somatosensory evoked potential. In *Progress in Clinical Neurophysiology*, ed. J. E. Desmedt, Vol. 7. Basel: Karger

Bergamini, L., Bergamasco, B. 1967. *Cortical Evoked Potentials in Man.* Springfield, Ill: Thomas. 116 pp.

Broughton, R. J. 1969. Discussion after paper by H. G. Vaughan, Jr. In *Average Evoked Potentials: Methods, Results, and Evaluation,* ed. E. Donchin, D. B. Lindsley, pp. 79–84. Wash. DC: US GPO

Cracco, J. B., Cracco, R. Q., Graziani, L. J. 1975. The spinal evoked response in infants and children. *Neurology* 25:31–36

Cracco, R. Q. 1973. Spinal evoked response: peripheral nerve stimulation in man: far field potentials. *Electroencephalogr. Clin. Neurophysiol.* 35:379–86

Cracco, R. Q., Cracco, J. B. 1976. Somatosensory evoked potential in man: far field potentials. *Electroencephalogr. Clin. Neurophysiol.* 41:460–66

Dawson, G. D. 1947. Investigation on a patient subject to myoclonic seizures after sensory stimulation. *J. Neurol. Neurosurg. Psychiatry* 10:134–40

Desmedt, J. E. 1971. Somatosensory cerebral evoked potentials in man. In *Handb. Electroencephalogr. Clin. Neurophysiol.*, ed. A. Remond, 9:55–82. Amsterdam: Elsevier

Desmedt, J. E., Brunko, E., Debecker, J. 1976. Maturation of the somatosensory evoked potentials in normal infants and children, with special reference to the early N_1 component. *Electrocephalogr. Clin. Neurophysiol.* 40:43–58

Desmedt, J. E., Manil, J. 1970. Somatosensory evoked potentials of the normal human neonate in REM sleep, in slow wave sleep and in waking. *Electroencephalogr. Clin. Neurophysiol.* 29:113–26

Desmedt, J. E., Noel, P. 1973. Average cerebral evoked potentials in the evaluation of lesions of the sensory nerves and of the central somatosensory pathway. *New Dev. Electromyography Clin. Neurophysiol.* 2:352–71

Dorfman, L. J. 1977. Indirect estimation of spinal cord conduction velocity in man. *Electroencephalogr. Clin. Neurophysiol.* 43:26–34

Giblin, D. R. 1964. Somatosensory evoked potentials in healthy subjects and in patients with lesions of the nervous system. *Ann. NY Acad. Sci.* 112:93–142

Goff, G. D., Matsumiya, Y., Allison, T., Goff, W. R. 1977. The scalp topography of human somatosensory and auditory evoked potentials. *Electroencephalogr. Clin. Neurophysiol.* 42:57–76

Goff, W. R., Allison, T., Shapiro, A., Rosner, B. S. 1966. Cerebral somatosensory responses evoked during sleep in man. *Electroencephalogr. Clin. Neurophysiol.* 21:1–9

Green, J. B., Hamilton, W. J. 1976. Anosognosia for hemiplegia: Somatosensory evoked potential studies. *Neurology* 26:1141–44

Greenberg, R. P., Becker, D. P., Miller, J. D., Mayer, D. J. 1977. Evaluation of brain function in severe head trauma with multimodality evoked potentials. Pt. 2. Localization of brain dysfunction and correlation with posttraumatic neurological conditions. *J. Neurosurg.* 47:163–77

Halliday, A. M. 1967a. The electrophysiological study of myoclonus in man. *Brain* 90:241–84

Halliday, A. M. 1967b. Changes in the form of cerebral evoked responses in man associated with various lesions of the nervous system. *Electroencephalogr. Clin. Neurophysiol. Suppl.* 25:178–92

Halliday, A. M., Mason, A. A. 1964. The effect of hypnotic anesthesia on cortical responses. *J. Neurol. Neurosurg. Psychiatry* 27:300–12

Halliday, A. M., Wakefield, G. S. 1963. Cerebral evoked potentials in patients with dissociated sensory loss. *J. Neurol. Neurosurg. Psychiatry* 26:211–19

Liberson, W. T. 1966. Study of evoked potential in aphasics. *Am. J. Phys. Med.* 45:135–42

Lindsley, D. B. 1969. Average evoked potentials—achievement, failures and prospects. See Broughton 1969, pp. 1–44

Luders, H. 1970. The effects of aging on the wave form of the somatosensory cortical evoked potential. *Electroencephalogr. Clin. Neurophysiol.* 29:450–60

Namerow, N. S. 1968. Somatosensory evoked responses in multiple sclerosis patients with varying sensory loss. *Neurology* 18:1197–1204

Namerow, N. S. 1969. Somatosensory evoked responses following cervical cordotomy. *Bull. Los Angeles Neurol. Soc.* 34:184–88

Namerow, N. S. 1970. Somatosensory recovery functions in multiple sclerosis patients. *Neurology* 20:813–17

Noel, P., Desmedt, J. E. 1975. Somatosensory cerebral evoked potentials after vascular lesions of the brain-stem and diencephalon. *Brain* 98:113–28

Perot, P. L. Jr. 1972. The clinical use of somatosensory evoked potentials in spinal cord injury. *Clin. Neurosurg.* 20: 367–82

Sclabassi, R. J., Namerow, N. S., Enns, N. F. 1974. Somatosensory response to stimulus trains in patient with multiple sclerosis. *Electroencephalogr. Clin. Neurophysiol.* 37:23–33

Shagass, L. 1972. *Evoked Brain Potentials in Psychiatry.* New York: Plenum. 274 pp.

Tsumoto, T., Hirose, N., Nonaka, S., Takahashi, M. 1973. Cerebrovascular disease: changes in somatosensory evoked potentials associated with unilateral lesions. *Electroencephalogr. Clin. Neurophysiol.* 35:463–73

Uttal, R., Cook, R. 1964. Systematics of the evoked somatosensory cortical potential: a psychophysical-electrophysiological comparison. *Ann. NY Acad. Sci.* 112:60–80

Wiederholt, W. G., Iragui-Madoz, V. J. 1977. Far-field somatosensory potentials in the rat. *Electroencephalogr. Clin. Neurophysiol.* 42:456–65

Williamson, P. D., Goff, W. R., Allison, T. 1970. Somatosensory evoked responses in patients with unilateral cerebral lesions. *Electroencephalogr. Clin. Neurophysiol.* 28:566–75

VISUAL-EVOKED POTENTIALS

Allison, T., Matsumiya, Y., Goff, G. D., Goff, W. R. 1977. The scalp topography of human visual evoked potentials. *Electroencephalogr. Clin. Neurophysiol.* 42:185–97

Asselman, P., Chadwick, D. W., Marsden, C. D. 1975. Visual evoked responses in the diagnoses and management of patients with multiple sclerosis. *Brain* 98: 261–82

Biersdorf, W. R., Nakamura, Z. 1971. Electroencephalogram potentials evoked by hemi-retinal stimulation. *Experientia* 27:402–3

Bodis-Wollner, I. 1977. Recovery from cerebral blindness: evoked potential and psychophysical measurements. *Electroencephalogr. Clin. Neurophysiol.* 43: 178–84

Callaway, E. 1977. *Brain Electrical Potentials and Individual Psychological Differences.* New York: Grune & Stratton. 214 pp.

Cappin, J. M., Nissim, S. 1975. Visual evoked responses in the assessment of field defects in glaucoma. *Arch. Ophthalmol.* 93:9–18

Celesia, C. G., Daly, R. F. 1977. Effects of aging on visual responses. *Arch. Neurol. Chicago* 34:403–7

Courchesne, E., Hillyard, S. A., Galambos, R. 1975. Stimulus novelty, task relevance and the visual evoked potential in man. *Electroencephalogr. Clin. Neurophysiol.* 30:131–43

Creel, D., Witkop, C. J., King, R. A. 1974. Asymmetric visual evoked potentials in human albinos: evidence for visual system abnormalities. *Invest. Ophthalmol.* 13:430–40

Creutzfeld, O. D., Kuhnt, U. 1973. Electrophysiology and topographical distribution of visual evoked potentials in animals. In *Handbook of Sensory Physiology,* ed. R. Jung, 7:595–646. Berlin: Springer

DeVoe, R. C., Ripps, H., Vaughan, H. G. Jr. 1968. Cortical responses to stimulation of the human foea. *Vision Res.* 8:135–47

Dustman, R. E., Beck, E. C. 1969. The effects of maturation and aging on the waveform of visually evoked potentials. *Electroencephalogr. Clin. Neurophysiol.* 26: 2–11

Ellingson, R. J. 1966. Development of visual evoked responses in human infant recorded by a response averager. *Electroencephalogr. Clin. Neurophysiol.* 21: 403–4

Feinsod, M., Abramsky, O., Auerbach, E. 1973. Electrophysiological examination of the visual system in multiple sclerosis. *J. Neurol. Sci.* 20:161–75

Feinsod, M., Hoyt, W. F. 1975. Subclinical optic neuropathy in multiple sclerosis. *J. Neurol. Neurosurg. Psychiatry.* 38: 1109–14

Feinsod, M., Selhorst, J. B., Hoyt, W. F., Wilson, C. B. 1976. Monitoring optic nerve function during craniotomy. *J. Neurosurg.* 44:29–31

Freeman, R. D., Thibos, L. N. 1975. Visual evoked response in humans with abnormal visual experience. *J. Physiol. London* 247:711–24

Halliday, A. M., Halliday, E., Kriss, A., McDonald, W. I., Mushin, J. 1976. The pattern-evoked potential in compression of the anterior visual pathways. *Brain* 99:357–74

Halliday, A. M., McDonald, W. I., Mushin, J. 1972. Delayed visual evoked response in optic neuritis. *Lancet* 1:982–85

Halliday, A. M., McDonald, W. I., Mushin, J. 1973a. Delayed pattern-evoked responses in optic neuritis in relation to visual acuity. *Trans. Ophthalmol. Soc. UK* 93:315–24

Halliday, A. M., McDonald, W. I., Mushin, J. 1973b. Visual evoked response in diagnosis of multiple sclerosis. *Br. Med. J.* 4:661–64

Halliday, A. M., Michael, W. F. 1970. Changes in pattern-evoked responses in man associated with the vertical and horizontal meridians of the visual field. *J. Physiol. London* 208:499–513

Harden, A., Pampiglione, G. 1971. ERG, VER and EEG in twelve children with late infantile neuronal lipidosis. *Adv. Exp. Med. Biol.* 24:287–93

Harter, M. R. 1971. Evoked cortical responses to checkerboard patterns: effects of check size as a function of retinal eccentricity. *Vision Res.* 10: 1365–76

Harter, M. R., White, C. T. 1968. Effects of contour sharpness and check size on visually evoked cortical potentials. *Vision Res.* 8:701–11

Harter, M. R., White, C. L. 1970. Evoked cortical responses to checkerboard patterns: effect of check size as a function of visual acuity. *Electroencephalogr. Clin. Neurophysiol.* 28:48–54

Heron, J. R., Regan, D., Milner, B. A. 1974. Delay in visual perception in unilateral optic atrophy after retrobulbar neuritis. *Brain* 97:69–78

Hishikawa, Y., Yamamoto, J., Furuya, E., Yamada, Y., Miyazaki, K., Kaneko, Z. 1967. Photosensitive epilepsy. Relationships between the visual evoked responses and the epileptiform discharges induced by intermittent photic stimulation. *Electroencephalogr. Clin. Neurophysiol.* 23:320–34

Jeffreys, D. A., Axford, J. G. 1972. Source locations of pattern-specific components of human visual evoked potentials. II. Component of extrastriate cortical origin. *Exp. Brain Res.* 16: 22–40

Jeffreys, D. A., Axford, J. G. 1972. Source locations of pattern-specific components of human visual evoked potentials. II. Component of extrastriate cortical origin. *Exp. Brain Res.* 16:22–40

John, E. F., Karmel, B. Z., Corning, W. C., Easton, P., Brown, D., Ahn, H., John, M., Harmony, T., Prichep, L., Toro, A., Gerson, I., Bartlett, F., Thatcher, R., Kaye, H., Valdes, P., Schwartz, E. 1977. "Neurometrics": The use of numerical taxonomy to evaluate brain functions. *Science* 196:1393–1410

Kooi, K. A., Baghi, B. K., Jordan, R. N. 1964. Observations on photically evoked occipital and vertex waves during sleep in man. *Ann. NY Acad. Sci.* 112:270–80

Marg, E., Freeman, D. N., Peltzman, P., Goldstein, P. J. 1976. Visual acuity development in human infants: evoked potentials measurements. *Invest. Ophthalmol.* 15:150–53

Michael, W. F., Halliday, A. M. 1971. Differences between the occipital distribution of upper and lower field pattern-evoked responses in man. *Brain Res.* 32:311–24

Milner, B. A., Regan, D., Heron, J. R. 1974. Differential diagnosis of multiple sclerosis by visual evoked potential recording. *Brain* 97:755–72

Namerow, N., Enns, N. 1972. Visual evoked responses in patients with multiple sclerosis. *J. Neurol. Neurosurg. Psychiatry* 35:829–33

Oosterhuis, H. J. G. H., Ponsen, L., Jonkman, E. J., Magnus, O. 1969. The average visual response in patients with cerebrovascular disease. *Electroencephalogr. Clin. Neurophysiol.* 27:23–34

Regan, D. 1972. *Evoked Potentials in Psychology, Sensory Physiology and Clinical Medicine.* New York: Wiley-Interscience. 328 pp.

Regan, D. 1975. Recent advances in electrical recording from the human brain. *Nature* 253:401–7

Regan, D. 1977. Clinical applications of steady state evoked potentials: Speedy methods of refracting the eye and assessing visual acuity in amblyopia. In *Cerebral Evoked Potentials in Man,* ed. J. E. Desmedt. London: Oxford Univ. Press. In press

Regan, D., Heron, J. R. 1969. Clinical investigation of lesions of the visual pathway: a new objective technique. *J. Neurol. Neurosurg. Psychiatry* 32: 479–83

Regan, D., Milner, B. A., Heron, J. R. 1976. Delayed visual perception and delayed visual evoked potentials in the spinal form of multiple sclerosis and in retrobulbar neuritis. *Brain* 99:43–66

Regan, D., Spekreijse, H. 1974. Evoked potential indications of color blindness. *Vision Res.* 14:89–95

Richey, E. T., Kooi, K. A., Tourtellotte, W. W. 1971. Visually evoked responses in multiple sclerosis. *J. Neurol. Neurosurg. Psychiatry* 34:275–80

Shagass, C., Amadeo, M., Roemer, R. A. 1976. Spatial distribution of potentials evoked by half-field pattern-reversal and pattern-onset stimuli. *Electroencephalogr. Clin. Neurophysiol.* 41: 609–22

Sokol, S. 1976. Visually evoked potentials: theory, techniques and clinical applications. *Surv. Ophthalmol.* 21:18–44

Spekreijse, H., Khoe, L. H., Van der Twell, L. H. 1972. A case of amblyopia; electrophysiology and psychophysics of luminance and contrast. *Adv. Exp. Med. Biol.* 24:141–56

Symann-Louett, N., Gascon, G., Matsumiya, Y., Lombroso, C. 1977. Wave form difference in visual evoked responses between normal and reading disabled children. *Neurology* 27:156–59

Vaughan, H. G. Jr., Katzman, R. 1964. Evoked responses in visual disorders. *Ann. N.Y Acad. Sci.* 112:305–19

Vaughan, H. G. Jr., Katzman, R., Taylor, J. 1963. Alterations of visual evoked response in the presence of homonymous visual defects. *Electroencephalogr. Clin. Neurophysiol.* 15:737–46

Visser, S. L., Stam, F. C., Van Tilburg, W., Op Den Velde, W., Blom, J. L., De-Rijke, W. 1976. Visual evoked response in senile and presenile dementia. *Electroencephalogr. Clin. Neurophysiol.* 40: 385–92

Wildberger, H. G. H., Van Lith, G. H. M., Wijngaarde, R., Mak, G. T. M. 1976. Visually evoked cortical potentials in the evaluation of homonymous and bitemporal visual field defects. *Br. J. Ophthalmol.* 60:273–78

Wright, J. E., Arden, G., Jones, B. R. 1973. Continuous monitoring of the visually evoked response during intra-orbital surgery. *Trans. Ophthalmol. Soc. UK* 93:311–14

AUDITORY-EVOKED POTENTIALS

Amadeo, M., Shagass, C. 1973. Brief latency click-evoked potentials during waking and sleep in man. *Psychophysiology* 10:244–50

Barnet, A. B., Ohlrich, E. S., Weiss, I. P., Shanks, B. 1975. Auditory evoked potentials during sleep in children from ten days to three years of age. *Electroencephalogr. Clin. Neurophysiol.* 39:29–41

Bickford, R. G. 1972. Physiological and clinical studies of microreflexes. *Electroencephalogr. Clin. Neurophysiol. Suppl.* 31:93–108

Brackmann, D., Selters, W. A. 1977. Acoustic tumor detection with brainstem electric response audiometry. *Arch. Otolaryngol.* 103:181–87

Buchwald, J. S., Huang, C. M. 1975. Far-field acoustic response: origins in the cat. *Science* 189:382–84

Celesia, C. G. 1968. Auditory evoked responses. *Arch. Neurol.* Chicago 19: 430–37

Celesia, C. G., Puletti, F. 1971. Auditory input to the human cortex during states of drowsiness and surgical anesthesia. *Electroencephalogr. Clin. Neurophysiol.* 31:603–9

Davis, H. 1976. Principles of electric response audiometry. *Ann. Otol. Rhinol. Laryngol.* Suppl. 28, Vol. 85, No. 3, Pt. 3

Davis, H., Mast, T., Yoshie, N., Zerlin, S. 1966. The slow response of the human cortex to auditory stimuli: recovery process. *Electroencephalogr. Clin. Neurophysiol.* 21:105–13

Davis, H., Onishi, S. 1969. Maturation of auditory evoked potentials. *Int. Audiol.* 8:24–33

Davis, H., Zerlin, S. 1966. Acoustic relations of the human vertex potentials. *J. Acoust. Soc. Am.* 39:109–16

Don, M., Allen, A., Starr, A. 1977. The effect of click rate on the latency of auditory brainstem responses in humans. *Ann. Otol. Rhinol. Laryngol.* 88:186

Eggermont, J. J., Odenthal, D. W., Schmidt, P. H., Spoor, A. 1974. Clinical electrocochleography. *Acta Oto-Laryngol. Suppl.* 316:62–74

Hecox, K. 1975. Electrophysiological correlates of human auditory development. In *Infant Perception: From Sensation to*

Cognition, Vol. II, ed. L. B. Cohen, P. Salapatek, pp. 141–91. New York: Academic

Hecox, K., Galambos, R. 1974. Brain stem auditory evoked responses in human infants and adults. *Arch. Otolaryngol.* 99:30–33

Jewett, D. L. 1970. Volume-conducted potentials in response to auditory stimuli as detected by averaging in the cat. *Electroencephalogr. Clin. Neurophysiol.* 28:609–18

Keidel, W. D. 1971. D. C. potentials in the auditory evoked responses in man. *Acta. Oto-Laryngol.* 71:242–48

Lev, A., Sohmer, H. 1972. Sources of averaged neural responses recorded in animal and human subjects during cochler audiometry (Electrocochleogram). *Arch. Klin. Exp. Ohren Nasen Kehlkopfheilkd.* 201:79–90

Marsh, J. T., Brown, W. S., Smith, J. C. 1975. Far-field recorded frequency-following responses: correlates of low pitch auditory perception in humans. *Electroencephalogr. Clin. Neurophysiol.* 38: 113–19

Mendel, M. I., Hosick, E. C., Windman, T. R., Davis, H., Hirsh, S. K., Dinges, D. F. 1975. Audiometric comparison of the middle and late components of the adult auditory evoked potentials awake and asleep. *Electroencephalgr. Clin. Neurophysiol.* 38:27–33

Moushegian, G., Rupert, A. L., Stillman, R. D. 1973. Scalp recorded early responses in man to frequencies in the speech range. *Electroencephalogr. Clin. Neurophysiol.* 36:665–67

Picton, T. W., Hillyard, S. A. 1974. Human auditory evoked potentials. II. Effects of attention. *Electroencephalogr. Clin. Neurophysiol.* 36:191–99

Picton, T. W., Hillyard, S. A., Galambos, R. 1976. Habituation and attention in the auditory system. In *Handbook of Sensory Physiology.* ed. W. D. Keidel, W. D. Neff, 3:343–89. New York: Springer

Picton, T. W., Woods, D. L., Baribeau-Braun, J., Healey, T. M. G. 1977. Evoked potential audiometry. *J. Otolaryngol.* 6:90–119

Rapin, I., Graziani, L. J. 1967. Auditory-evoked responses in normal, brain-damaged, and deaf infants. *Neurology* 17:881–94

Robinson, K., Rudge, P. 1975. Auditory evoked responses in multiple sclerosis. *Lancet* 1:1164–66

Salamy, A., McKean, C. M. 1976. Postnatal development of human brainstem potentials during the first year of life. *Electroencephalogr. Clin. Neurophysiol.* 40:418–26

Schulman-Galambos, C., Galambos, R. 1975. Brain stem auditory-evoked responses in premature infants. *J. Speech Hear. Res.* 18:456–65

Sohmer, H., Feinmesser, M., Szabo, G. 1974. Sources of electrocochleographic responses as studied in patients with brain damage. *Electroencephalogr. Clin. Neurophysiol.* 37:663–69

Sohmer, H., Pratt, H. 1976. Recording of cochlear microphonic potential with surface electrodes. *Electroencephalogr. Clin. Neurophysiol.* 40:253–60

Starr, A. 1977. Clinical relevance of brain stem auditory evoked potentials in brain stem disorders in man. In *Progress in Clinical Neurophysiology,* ed. J. E. Desmedt, 2:45–57. Basel: Karger

Starr, A. 1976. Auditory brainstem responses in brain death. *Brain* 99:543-54

Starr, A., Achor, J. 1975. Auditory brainstem responses in neurological disease. *Arch. Neurol. Chicago* 32:761–68

Starr, A., Amlie, R. N., Martin, W. H., Sanders, S. 1977. Development of auditory function in newborn infants revealed by auditory brainstem potentials. *Pediatrics.* In press

Starr, A., Hamilton, A. 1976. Correlation between confirmed sites of neurological lesions of far-field auditory brainstem responses. *Electroencephalogr. Clin. Neurophysiol.* 41:595–608

Stockard, J. J., Rossiter, V. S. 1977. Clinical and pathological correlates of brain stem auditory response abnormalities. *Neurology* 27:316–25

Terkildsen, K., Huis in't Veld, F., Osterhammel, P. 1977. Auditory brain stem responses in the diagnosis of cerebellopontine angle tumours. *Scand. Audiol.* 6:43–47

Thornton, A. R. D. 1975. Statistical properties of surface-recorded electrocochleographic responses. *Scand. Audiol.* 4:91–102

Thornton, A. R. D., Hawkes, C. H. 1976. Neurological applications of surface recorded electrocochleography. *J. Neurol. Neurosurg. Psychiatry* 39: 586–92

Zerlin, S., Davis, H. 1967. The variability of single evoked vertex potentials in man. *Electroencephalogr. Clin. Neurophysiol.* 23:468–72

Ann. Rev. Neurosci. 1978. 1:129–69

CENTRAL CATECHOLAMINE NEURON SYSTEMS: ANATOMY AND PHYSIOLOGY OF THE DOPAMINE SYSTEMS

❖11506

R. Y. Moore
Department of Neurosciences, University of California, San Diego,
La Jolla, California 92093

F. E. Bloom
Arthur V. Davis Center for Behavioral Neurobiology, The Salk Institute,
La Jolla, California 92037

INTRODUCTION

The central nervous system contains three types of catecholamines: dopamine (DA), norepinephrine (NE), and epinephrine (E). Over the past 25 years, progressive analysis of the synthesis, storage, release, conservation, and eventual catabolism of these three catecholamines has pioneered the conceptualization of synaptic transmitter metabolism and the modes of action of most psychotropic drugs. Nevertheless, despite the elegance and detail of these neurochemical and pharmacological data, our understanding of the stucture and function of the catecholamine-containing cells of the central nervous system has emerged much more slowly. This relatively slower rate of progress appears to have resulted in part from the fact that methods for the analysis of the catecholamine pathways required techniques not previously available, and in part from the fact that once these techniques were available and in widespread use they produced a body of cytological and physiological data that indicated that catecholamine neuron systems have properties that differ dramatically from those of "classical" central systems.

In this chapter, we seek to review the present status of the central dopamine neuron systems with particular reference to their structure and function. We also place in perspective those structural and functional properties of these neuron systems that set them at variance with certain other well-defined neuronal circuits.

0147-006X/78/0325-0129$01.00

In a subsequent review we shall deal with the other central catecholamine-containing neuron systems, but it may be helpful here to examine those general properties of the catecholaminergic systems that mark them as unique.

With few exceptions, we do not consider the powers and limitations of the techniques employed to study catecholaminergic neurons of the central nervous system. Recent reviews of the methods of fluorescence histochemistry (Björklund & Moore 1977), orthograde and retrograde nerve circuit tracing (LaVail & LaVail 1972, Graybiel 1975), and iontophoresis (Bloom 1974, 1975a) should serve as an introduction to that body of technical information. Instead, we focus here on a combined structural and functional analysis of the dopaminergic system as defined by these and other methods. The literature search from which these observations were derived was closed in March 1977. Although we have attempted to be inclusive, an inescapable facet of these attractive fields is that more material is available than can be selected for detailed analysis. Literature citations have been selected to guide readers to the most inclusive material. We particularly regret that we have been unable to cite many important papers that have contributed to the material presented in this review. Finally, for similar reasons, no attempt was made to cover studies on invertebrates and lower vertebrates, on the ontogeny of the dopaminergic system, on its plasticity and response to injury, or on its role in behavior. Undoubtedly these important topics will be covered in future reviews in this series. Other recent reviews of central catecholamine structure (Lindvall & Björklund 1977, Björklund & Moore 1978) and function (Bloom 1975a, b, 1977, Hoffer & Bloom 1976) may yield broader sources of reference.

However, one important issue central to the cytologic and functional tests that we cover is that of the identity of the cell to be evaluated as a catecholamine target. That is, the function of a catecholamine neuron is established by input that determines the activity of the system and by the location and type of interaction produced on effector elements.

For example, a primary concern in evaluating reports of iontophoretic tests in any brain region is the identity of the cells tested. Such identifications can rely on the discharge pattern or the response of test cells to stimulation of specific antidromic or orthodromic projections; alternatively, cells tested can be identified post hoc by marking the recording sites with any of several methods and examining the recording sites cytologically after the experiment.

An example of the importance of test cells identification can be seen in the clear differences that occur in responsiveness to NE between tests on "all-cells-in-a-region" and tests on specific identifiable cell types within a region (Bloom 1974, Hoffer & Bloom 1976). In general, the response of identified cells to NE is inhibitory with the exception of border cells in the ventromedial nucleus of the hypothalamus (Hori & Nakayama 1973, Krebs & Bindra 1971), the cells of the paramedian reticular nucleus (Boakes et al 1971), the lateral vestibular nucleus (Kirsten & Sharma 1976), and some cells in the pontine raphe nucleus (Couch 1970), all of which respond to NE in an excitatory manner. In no cases do identified cells exhibit significant instances of mixed responses (i.e. some cells firing faster, some cells slower) as seen when "all-cells-in-a-region" are artificially lumped together.

Secondly, identification of tested cells is even more important when drug responses are to be compared to a specific synaptic input to a test cell, or in attempts to determine the molecular basis of the synaptic response or the effects of drugs. In this case, the test cells must be identified in order to establish cytologically that the pathway under examination does indeed contact the cells to be tested. As we shall describe in a later review in this series, several cytologically defined noradrenergic synaptic projections (that is, from known NE-source neurons to specific postsynaptic cells) have now been studied electrophysiologically. Thus, it is now possible to confirm directly the synaptic inference of iontophoretic responses.

Thirdly, identification of the test cells can define which of the iontophoretic responses observed may rarely, if ever, be utilized by normal synaptic connections [e.g. the excitatory receptors on cells where evidence for catecholaminergic synapses is minimal (Kirsten & Sharma 1976, Yamamoto 1967)]. Finally, identification of test cells permits data to be accumulated on homogenous cell populations in order to evaluate the antagonists or potentiators of the test synapses or test substances. These important rules apply to the evaluation of all catecholamine cells, but as will be obvious, such experiments are generally restricted to the nigrostriatal DA system and to the cellular targets of the locus coeruleus NE system.

Our initial intent was to cover the anatomy and physiology of all central catecholaminergic systems in a single review. However, even with the restrictions of content noted above, it soon became evident that the available data could not be contained within a reasonable space. For that reason, the dopamine neuron systems are reviewed here, and the norepinephrine and epinephrine systems will be covered in a companion review to appear in the 1979 volume of the *Annual Review of Neuroscience.*

THE DOPAMINE SYSTEMS

Anatomy: Overview

The dopamine (DA) neuron systems are more complex in their anatomy, more diverse in localization and apparent function, and more numerous, both in terms of definable systems and in numbers of neurons, than the other catecholamine systems. DA neuron systems are principally located in the upper mesencephalon and diencephalon. They appear to vary anatomically from systems of neurons without axons (retina, olfactory bulb) and with very restricted projections, to systems with extensive axonal arborizations. The principal morphologic difference, however, between the DA and NE systems is that the DA systems appear to be "local" systems with highly specified, topographically organized projections. Until recently (Ungerstedt 1971), the morphology of the DA systems appeared to be quite simple. The major projection was that of the substantia nigra, pars compacta, upon the neostriatum. In addition, ventral tegmental area projections upon the olfactory tubercle and adjacent structures were known (Ungerstedt 1971), and the tuberohypophysial system, first described by Fuxe (1964), had been intensively analyzed (Fuxe & Hökfelt 1969, Björklund et al 1970, 1973a, b, Jonsson et al 1972). With the introduction of the glyoxylic acid fluorescence histochemical method (Lindvall

& Björklund 1974a), the increasing application of ultrastructural analysis (e.g. Ajika & Hökfelt 1973, Bloom, 1973), and the advent of new, powerful neuroanatomical methods (see above), there has been a very rapid, marked increase in our understanding of the extent and organization of DA neuron systems. This is reviewed in the sections to follow. A summary of the organization of the major DA systems is given in Table 1.

The Meso-Telencephalic DA System

INTRODUCTION The localization of DA neurons in the upper mesencephalon is shown diagrammatically in Figure 1. These cells form a continuous group extending rostrally from the level of the mammillary complex to the level of the decussation of the brachium conjunctivum. The cells are continuous from medial to lateral and do not appear to form distinct groupings or nuclei. For that reason we do not use the nuclear designations A8, A9, and A10 of Dahlström & Fuxe (1964). Instead, the term "meso-telencephalic" more accurately designates that the entire mesencephalic DA system projects largely, if not exclusively, upon the telencephalon. It is

Table 1 Dopamine neuron systems in the mammalian brain

System	Nucleus of origin	Site(s) of termination
Meso-telencephalic[a]		
Nigrostriatal	Substantia nigra, pars compacta; ventral tegmental area	Neostriatum (caudate-putamen), globus pallidus
Mesocortical	Ventral tegmental area; substantia nigra, pars compacta	Isocortex (mesial frontal, anterior cingulate, entorhinal, perirhinal) Allocortex (olfactory bulb, anterior olfactory nucleus, olfactory tubercle, piriform cortex, septal area, nucleus accumbens, amygdaloid complex)
Tubero-hypophysial	Arcuate and periventricular hypothalamic nuclei	Neuro-intermediate lobe of pituitary, median eminence
Retinal	Interplexiform cells, of retina	Inner and outer plexiform layers of retina
Incerto-hypothalamic	Zona incerta, posterior hypothalamus	Dorsal hypothalamic area, septum
Periventricular	Medulla in area of dorsal motor vagus, nucleus tractus solitarius, periaqueductal and periventricular gray	Periventricular and peri-aqueductal gray, tegmentum, tectum, thalamus, hypothalamus
Olfactory bulb	Periglomerular cells	Glomeruli (mitral cells)

[a] The term meso-telencephalic refers to the projections of the midbrain dopamine neurons upon the telencephalon. See text for description.

possible, however, to discriminate two distinct components to the system. The first, arising largely in the substantia nigra and terminating in the neostriatum, is referred to as the nigrostriatal system. The second, arising in both the substantia nigra and the ventral tegmental area and projecting widely to both isocortical and allocortical (Bailey & Von Bonin 1951) regions of the telencephalon, is termed the mesocortical DA system. Further support for this more unified view of the DA neuron systems of the upper midbrain is obtained from ontogenetic studies (Olson & Seiger 1972, Seiger & Olson 1973) that show the origin of these neuronal groups as a single unit. This description of mesencephalic DA neurons refers only to those contained in the substantia nigra and ventral tegmental area and does not include the periventricular and periaqueductal DA neurons that are described separately below.

CELL BODIES The cell bodies are first noted in fluorescence histochemical material at the mesencephalic-diencephalic junction, dorsal to the cerebral peduncle and ventral to the medial lemniscus (Figure 1*A*). More caudally, the substantia nigra is well developed and the DA cells form a compact, thin group extending from above the caudal mammillary complex to the lateral edge of the medial lemniscus (Figure 1*B*). At this level and at all subsequent levels where the pars reticulata of the substantia nigra is present, a number of cells are located within the pars reticulata. In addition, some cells are located quite laterally in the region termed "pars later-

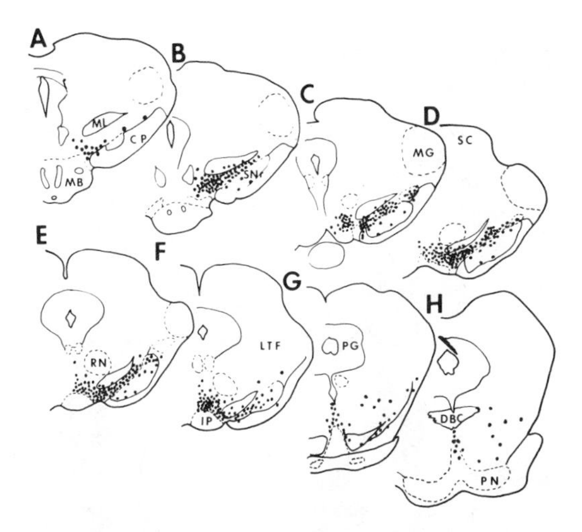

Figure 1 Diagrams from a series of coronal sections through the rat midbrain prepared by the GA fluorescence histochemical method. The black dots represent the location and relative density of DA neurons. *A–H* represent rostral to caudal levels. Abbreviations: CP, cerebral peduncle; DBC, decussation of the brachium conjunctivum; IP, interpeduncular nucleus; LTF, lateral tegmental field; MB, mammillary body; MG, medial geniculate; ML, medial lemniscus; PG, periaqueductal gray; PN, pontine nuclei; RN, red nucleus; SC, superior colliculus; SNr, substantia nigra, pars reticulata. The pars compacta lies just above the pars reticulata. (From J. H. Fallon & R. Y. Moore, unpublished observations.)

alis" by Hanaway et al (1970) in their study of the cytoarchitecture of the rat substantia nigra. As the mammillary complex disappears (Figure 1*C,D*), large numbers of DA cells are found across the ventral tegmental area. There is no distinct border either in fluorescent histochemical material or in Nissl-stained material (Huber et al 1943, Hanaway et al 1970) between the substantia nigra, pars compacta, and the ventral tegmental area. The appearance of the DA cell groups is similar caudally, but, at the level of the interpeduncular nucleus, the DA neurons in the ventral tegmental area are more numerous than those in the substantia nigra. At these levels (Figure 1*E,H*), scattered DA neurons are present in the lateral reticular formation, but these appear to be continuous with, and a component of, the substantia nigra DA neurons.

In fluorescence microscopy the DA cells appear medium-sized and intensely fluorescent (Figure 2). They are multipolar, and those in the ventral portion of the nucleus have long dendrites extending into the pars reticulata, where they course in bundles and have a pronounced varicose appearance, suggesting the possibility that a DA dendrite is presynaptic to some other element (Björklund & Lindvall 1975). The cells of the ventral tegmental area are generally more elongated, with horizontal dendrites. In Golgi material two types of neuron are noted in the pars compacta (Juraska et al 1977). One is a medium-sized neuron conforming in morphology and dendritic pattern to DA cells described above. The second is a small

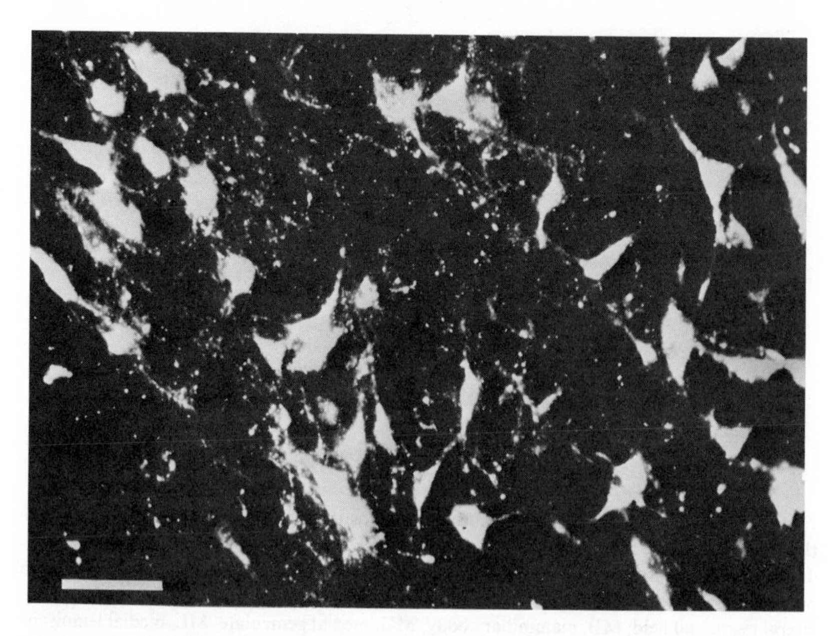

Figure 2 DA neurons of the substantia nigra, pars compacta. Vibratomeformaldehyde varient of the Falck-Hillarp method. Marker bar = 50 μm (From J. H. Fallon & R. Y. Moore, unpublished.)

neuron with nondirectional dendritic fields that is believed to be an interneuron. Both types are described in ultrastructural studies (Rinvik & Grofova 1970, Gulley & Wood 1971, Agid et al 1973, Hökfelt & Ungerstedt 1973). The large neurons clearly are the DA neurons as they are affected selectively by 6-hydroxydopamine (6-OHDA) and show preferential uptake of labeled CA's (Sotelo 1971, Sotelo et al 1973, Hökfelt & Ungerstedt 1973). Nigral DA neurons exhibit no special ultrastructural features. They contain the usual organelles, including a vesicular component of the Golgi apparatus, 300–400 Å in diameter, but coated vesicles and large, dense-core vesicles are also present. In addition, occasional neurons exhibit large concentric whorls of cisternae of endoplasmic reticulum (Gulley & Wood 1971). Nissl substance is abundant. We are not aware of any ultrastructural studies of the ventral tegmental area.

AFFERENT CONNECTIONS The afferent connections of the substantia nigra have been studied recently by Bunney & Aghajanian (1977). A prominent afferent input is present from the neostriatum to the pars reticulata. This projection has been recognized for a number of years (cf Voneida 1960, Grofova 1975, Grofova & Rinvik 1970) and can probably be viewed as an input onto DA neurons because of their dendritic extensions into the reticulata. In addition, there is an indirect neostriatal input via the globus pallidus (Hattori et al 1975, Bunney & Aghajanian 1977), along with inputs from the midbrain raphe, central nucleus of the amygdala, prefrontal cortex, and lateral habenula (Bunney & Aghajanian 1977). An input from locus coeruleus has also been described (Jones & Moore 1977). The afferent connections of the ventral tegmental area have been studied extensively and include input from central gray, isthmic and midbrain reticular formation, lateral hypothalamus, and the basal forebrain (cf Nauta 1958, Nauta & Haymaker 1969 for reviews).

The Nigrostriatal Projection

The existence of a direct projection from the substantia nigra to the neostriatum was not fully confirmed until its existence was strongly indicated by Hornykiewicz and his collaborators (cf Hornykiewicz 1966 for review), who noted the marked reduction in DA content of the neostriatum in parkinsonian patients in association with the well-known pathology of the pars compacta of the substantia nigra in such patients. With the introduction of the Falck-Hillarp method (Falck 1962, Falck et al 1962), however, this projection was rapidly established (Andén et al 1964, 1965, 1966a), and it has subsequently been shown that the projection can be demonstrated by a variety of techniques (Hökfelt & Ungerstedt 1969, Moore et al 1971a, Ungerstedt 1971, Fallon & Moore 1976b, Usunoff et al 1976). The following sections describe the pathway and the organization of the terminal field.

THE PATHWAY Axons arising from the pars compacta of the substantia nigra and the medial ventral tegmental area run dorsally into the overlying tegmentum to form a broad bundle beneath and ventromedial to the red nucleus. The axons arise predominantly from cells in the ventral portion of the substantia nigra and in the ventral tegmental area. The bundle ascends through the medial tegmentum and enters the diencephalon in the prerubral field and the dorsal part of the lateral hypothalamic area (Ungerstedt 1971, Moore et al 1971a, Shimizu & Ohnishi 1973). As the nigrostriatal bundle courses rostrally in the diencephalon, it occupies a position partly in the lateral hypothalamic area and partly in the medial portion of the internal capsule. Rostrally, the fibers run dorsally in the internal capsule and distribute throughout the caudate and putamen. In animals such as the cat, where caudate and putamen are separated, axons leave the main bundle to run laterally through the globus pallidus (and entopeduncular nucleus when present) to reach the caudal putamen. More rostrally, the nigrostriatal fibers enter the putamen directly from the internal capsule. The pathway can be shown by the glyoxylic acid (GA) technique (Lindvall & Björklund 1974b, Figure 3) where it is characterized by very thin, nonvaricose, faintly fluorescent fibers. It can also be demonstrated by silver impregnation techniques (Moore et al 1971a, Ibata et al 1973, Shimizu & Ohnishi 1973, Maler et al 1973) and by autoradiography (Fallon & Moore 1976b).

TERMINAL FIELDS The nigrostriatal projection as it is described here has two terminal fields. No description will be made of the non-DA, pars reticulata neuron projections such as those to thalamus and tectum (Shimizu & Ohnishi 1973). The terminal fields of the DA neurons are the striatal nuclei. There is a sparse projection to entopeduncular nucleus and globus pallidus that consists of scattered fine DA terminals. The major projection is to the neostriatal nuclei, the caudate and putamen. This is so dense that in fluorescence histochemical material the neostriatum appears to be virtually covered with fine DA varicosities (Hökfelt & Ungerstedt 1969, Lindvall & Björklund 1974b, Figure 4). As the DA axons of the nigrostriatal projection enter the neostriatum they undergo a massive collateralization in which each branch contains numerous small varicosities. Andén et al (1966b) have calculated that in the rat each nigral DA neuron innervating the neostriatum has an axon between 55 and 77 cm in length that contains approximately 500,000 varicosities. In ultrastructural studies, DA axon terminals have been identified predominantly on the small dendritic branches of small neurons of the neostriatum (Hökfelt & Ungerstedt 1969, Ibata et al 1973, Hattori et al 1973). The terminals are small, about 0.5–0.7 μm in diameter, containing 400–500 Å pleomorphic synaptic vesicles that do not contain a dense core with the usual fixation. Occasional large dense-core vesicles are evident. The very careful electron-microscopic autoradiographic analysis of Hattori et al (1973) and McGeer et al (1975) demonstrated, first, that labeled terminals make synaptic contacts and, second, that these contacts are virtually all of the asymmetric (Gray type I) variety on dendritic spines.

The other striking feature of the nigrostriatal projection is its highly organized topography. The initial observations suggesting this topography came from the studies of Moore et al (1971a) and Carpenter & Peter (1972). Recently, an extensive study using the HRP retrograde transport method (Fallon & Moore 1976b) has shown many of the details of this topography (Figure 5). The most medial and rostral portion of the neostriatum is innervated by DA neurons of the lateral ventral tegmental area. This overlaps the area innervating the nucleus accumbens (Figure 5). More lateral areas are innervated by progressively more lateral nigral DA neurons with the most caudal and ventral portion of the neostriatum innervated by

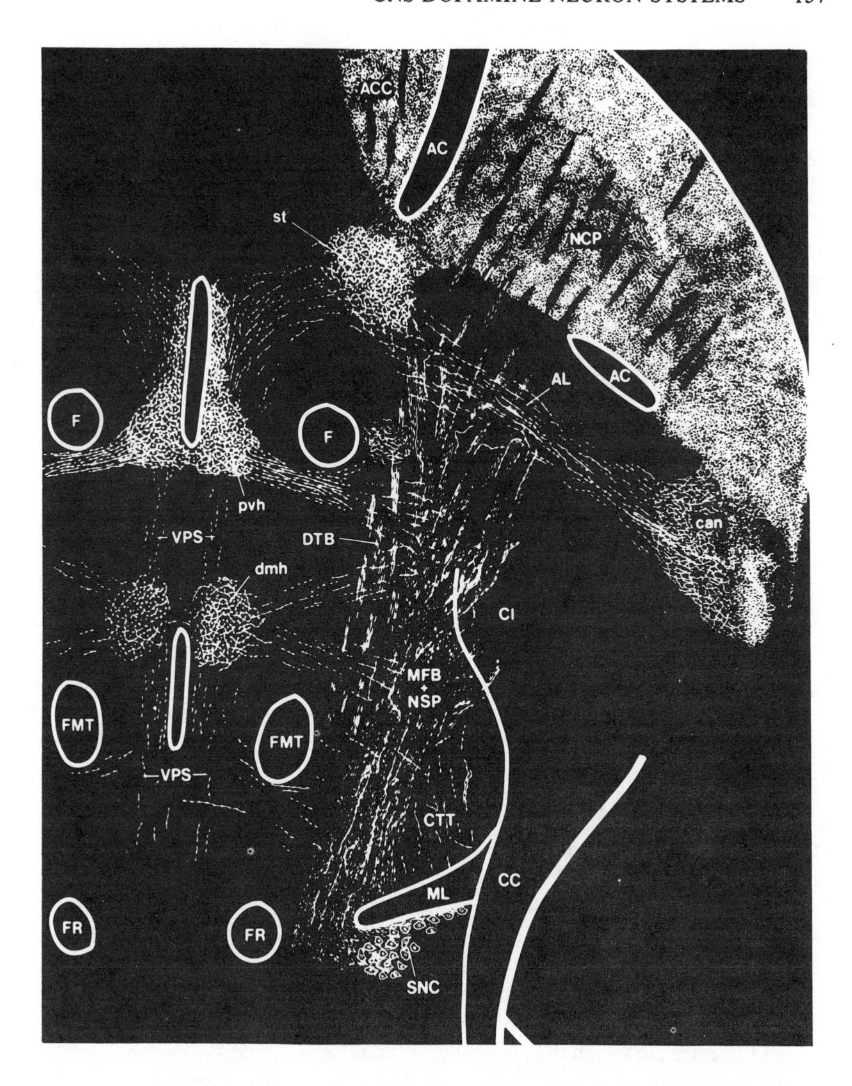

Figure 3 Diagrammatic representation of the nigrostriatal projection in horizontal section. The very fine fibers that arise from the substantia nigra and traverse the medial forebrain bundle to the caudate-putamen are the DA axons of the projection. Abbreviations: AC, anterior commissure; ACC, nucleus accumbens; AL, ansa lenticularis; can, central amygdaloid nucleus; CC, crus cerebri; CI, internal capsule; CTT, central tegmental tract; dmh, dorsomedial hypothalamic nucleus; DTB, dorsal tegmental CA bundle; F, fornix; FMT, mammillothalamic tract; FR, fasciculus retroflexus; MFB, medial forebrain bundle; ML, medial lemniscus; NCP, caudate-putamen (neostriatum); NSP, nigrostriatal projection; pvh, periventricular hypothalamic nucleus; SNC, substantia nigra, pars compacta; St, bed nucleus of the stria terminalis; VPS, ventral periventricular system. (From Lindvall & Björklund 1974b.)

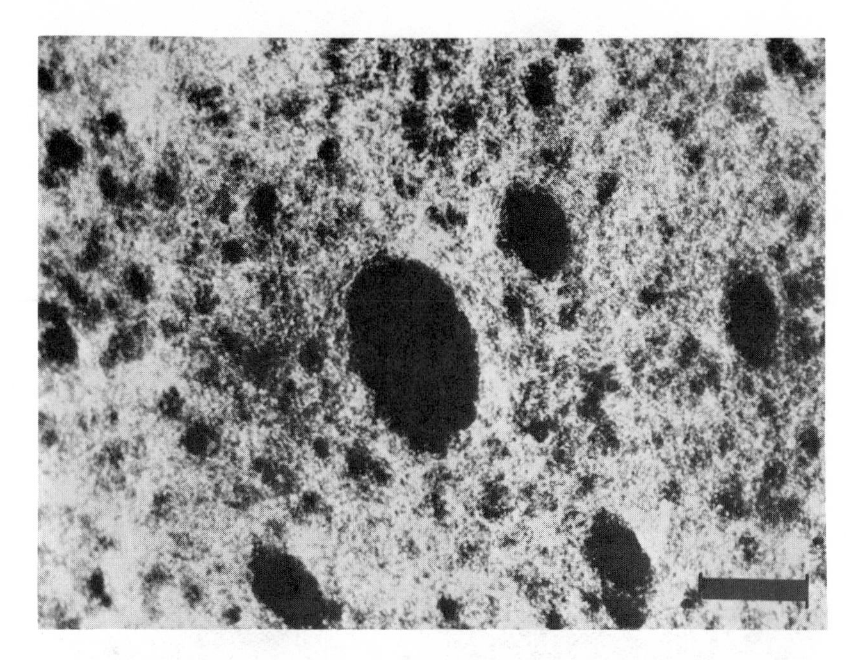

Figure 4 Neostriatum of a normal rat prepared by the GA method showing a dense DA innervation of fine axons and small varicosities. The dark areas represent bundles of axons of the internal capsule traversing the neostriatum. Marker bar = 75 μm. (From J. H. Fallon & R. Y. Moore, unpublished.)

the most caudal meso-telencephalic DA neurons. The rostro-caudal topography of the projection is represented largely by a rostro-caudal orientation of DA neurons in the ventral tegmental area and substantia nigra. The dorso-ventral topography of the projection is represented by a reverse orientation of DA neurons within the ventral tegmental area and substantia nigra. That is, dorsally placed cells in the mesencephalic DA cell group project ventrally on the neostriatum, whereas ventrally placed cells project dorsally (Fallon & Moore 1976b). It is not known at present the extent to which the massively collateralized axonal system from one DA neuron overlaps in its terminal field with that of other DA neurons.

PHYSIOLOGY To examine the cellular function of DA released from nigral and ventro-tegmental neurons on their target cells in the striatum and forebrain, three general approaches can be employed: first, release of DA from axons can be simulated by iontophoretic applications and the effects can be evaluated by extracellular or intracellular electrophysiologic indices (York 1975, Connor 1968, 1970, Siggins 1977). Second, the effects of the DA-containing tracts can also be evaluated—at least in theory—by observation of the electrophysiologic consequences on striatum of nigral or ventro-tegmental stimulation. Thirdly, the function of these fibers might

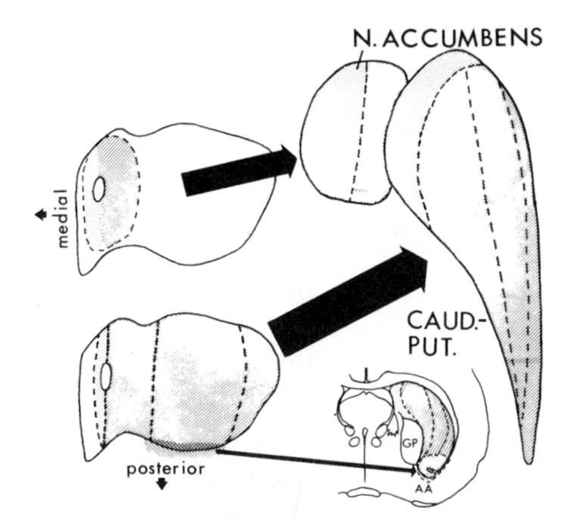

Figure 5 Diagrammatic representation of the projection of the mesencephalic DA neurons on the nucleus accumbens and caudate-putamen (CAUD.–PUT.). As indicated, the projection is organized in a medio-lateral, rostro-caudal axis in both the nucleus of origin and the area of termination, except that the most posterior DA cells project ventrally and caudally to the neostriatum and adjacent anterior amygdaloid area (AA). GP = globus pallidus. (From J. H. Fallon & R. Y. Moore, unpublished observations.)

be inferred from the dysfunctions observed after electrolytic or chemical lesioning of the DA neurons. All three methods have been employed extensively and, with certain important exceptions, there is strong evidence that the dopaminergic neurons of the nigra inhibit their target cells in the caudate.

Iontophoretic studies Of the 20 reports on the effects of DA iontophoresis, in 17 the predominant response recorded extracellularly was inhibitory (Table 2). Excluding the three negative studies, the proportion of cells that are inhibited by DA, regardless of species, anesthetic, or iontophoretic technique, vary from a low of 78% (Bradshaw et al 1973) to a high of 100% (Feltz & De Champlain 1972a). The inhibition typically appears some 2–15 sec after the onset of iontophoresis and may persist for minutes after the termination of the ejecting current. DA applied by iontophoresis inhibited spontaneous activity (Bloom et al 1965, Bunney & Aghajanian 1973, Connor 1970, Herz & Zieglgänsberger 1968, McLennan & York 1967, Siggins et al 1974, Spehlmann 1975, Stone 1976, Stone & Bailey, 1975) and also inhibited activity produced by the iontophoresis of excitatory amino acids (Feltz & De Champlain 1972a, Gonzalez-Vegas 1974, Herz & Zieglgänsberger 1968, Siggins et al 1974, Spehlmann 1975, Spencer & Havlicek 1974, Stone 1976, Stone & Bailey 1975, York 1975). DA iontophoresis also inhibits activity induced by stimulation of orthodromic or antidromic axon projections (Feltz & De Champlain 1972a, Gonzalez-Vegas 1974, Herz & Zieglgänsberger 1968, McLennan & York 1967). Inhibition also occurs when caudate neurons are excited by stimulating the substantia nigra in the cat (Feltz & De Champlain 1972a, McLennan & York 1967)

Table 2 Effects of iontophoretically applied dopamine on caudate neurons

Percentage of responding cells depressed	Number of cells tested	Animal	Preparation	Reference
96	50	Cat	Several anesthetics	Bloom et al 1965
78	55	Cat	Decerebrate	Bloom et al 1965
87	130	Cat	Decerebrate	McLennan & York 1967
84	81	Cat	Decerebrate	Connor 1970
70	87	Rabbit	Gallamine paralyzed	Herz & Zieglgänsberger 1968
100	(243)	Cat	Dial + urethane, 6-OHDA	Feltz & De Champlain 1972a
43	136	Cat	Dial + urethane	Feltz & De Champlain 1972a
86	(89)	Rat	Urethane	Gonzalez-Vegas 1974
94	(91)	Rat	Halothane	Siggins et al 1974
52	64	Cat	Cerveau isole	Spehlmann 1975
28*	40	Rat	Chloral hydrate	Yarbrough 1975
38*	51	Rat	Chloral hydrate, haloperidol	Yarbrough 1975
92	103	Rat	Urethane	Stone 1976, Stone & Bailey 1975
90	40	Rat	Dial	Spencer & Havlicek 1974
42	64	Rat	Penthrane	Spencer & Havlicek 1974
34*	62	Rat	Halothane	Bevan et al 1975
0	26	Cat	Pentobarbitone	Kitai et al 1976

and rat (Gonzalez-Vegas 1974). If the majority of neurons in the substantia nigra contain DA, and if these neurons project to the caudate nucleus among other targets, how can substantia nigra stimulation excite caudate cells if the predominant effect of DA iontophoresis is inhibitory? To understand the background of this apparent paradox requires a somewhat more thorough analysis of the experiments.

If iontophoresis experiments are viewed as simulations of the neural release of DA, the cells that should be tested are obviously those on which the DA pathway terminates. However, as reviewed above, the histological analysis of the caudate nucleus does not yet provide clearcut answers as to which of the several neuronal categories found in the target region should be considered the primary targets of the DA neurons, nor even whether there is any more-or-less exclusive synaptic target to be found there. In such a state of structural uncertainty, the cells tested by iontophoresis may or may not be the correct ones. Indeed, the criticism may be made that iontophoretic testing with extracellular recording may show effects that are not generated directly on the cell being recorded but rather on a covertly placed, functionally interconnected neighboring cell. However, the fact that 70–100% of all cells tested are inhibited tends to deflate this argument. Nevertheless, direct observation of iontophoretic results on transmembrane potentials can be achieved with intracellular recording, and this combination of methods permits comparison of iontophoretic DA with nigral actions both on changes in membrane potential as well as changes in membrane conductance.

In the sole examination of intracellularly recorded actions of iontophoretic DA, Kitai et al (1976) report that DA depolarizes cells within 10 msec of the onset of iontophoretic current and that similar effects were observed when the nigra was stimulated electrically (see below). Two possible explanations for the discrepancies between these results with intracellular recording and those with extracellular recording come immediately to mind:

1. The intracellular and extracellular electrodes sample different cell populations, with the intracellular excitatory actions arising from a more limited cell pool, which may or may not represent the targets of the DA pathway. Several factors make this explanation untenable: Intracellular marking studies (Kitai et al 1976) indicate that the cells responding by depolarization are medium-sized neurons, similar in size to those that constitute ~ 95% of the caudate. Therefore, the populations tested are not easily separated. Furthermore, the distance of the drug delivery pipette to the intracellular recording electrode tip has the same relative separation (30–50 μm) as the distance between the extracellular-iontophoretic pipettes and the cell being recorded. Therefore, the possible existence of covert neighboring interneurons remains to complicate both sets of data. Finally, it is difficult to reconcile the fact that inhibitory responses predominate in the extracellular studies and are not seen at all in the intracellular study; if selection of similarly sized cells by the electrode were random, an occasional depressant effect should have been seen even with the finer tips of the intracellular electrode.

2. The depolarizing actions observed in this intracellular study may have been produced as a by-product of the novel method of iontophoresis employed by Kitai et al (1976), namely very short pulses of relatively high current. In studies attempting to define the dose and time relationships between current applied to the iontophoretic pipette and the release of DA or NE, latencies of response on the order of 2–5 sec are common with ejecting currents in the range normally employed (0.2×10^{-6}A or less). This latency is due to the influence of the retaining currents employed in iontophoretic studies to avoid the inadvertent leakage of test substances from the pipette between trials (see Bradshaw et al 1973, Clarke et al 1973), and brief lags occur with even the most potent amino acids (Siggins 1977). In the case of DA, the differences in latencies observed between effects of DA and those of amino acid excitants or depressants may reflect the added effects of slightly higher biological thresholds (see Ben-Ari & Kelly 1976) and the moderately lower efficiency with which applied currents eject catecholamines from iontophoretic pipettes into the brain (Hoffer et al 1971a). With these physical and biological attributes of DA delivery in mind, it is difficult to envision how significant DA can be released from extracellular drug pipettes and produce an effect on transmembrane potentials within msec. Moreover, the use of high drug delivery currents may lead to direct

electrical effects on the cell being tested that are independent of the contents of the drug pipette. Although current control tests were employed in some cases by Kitai et al (1976), their experimental design (one drug pipette with no balance barrel, or two drug pipettes with no balance barrel) is not compatible with consistent application of this control. Current artifacts may be evaluated by testing the cell for the effects of currents passed through a NaCl barrel that are equal to those passed through the DA barrel. However, in contrast to the reduction of artifacts by continuous current neutralization (Salmoiraghi & Weight 1967), the use of post-hoc polarization tests is somewhat less conclusive since the effects of current without drug may well differ from those of current imbalance in the presence of drug (W. Zieglgänsberger & R. Champagnat, personal communication). However, when continuous current neutralization (see Salmoiraghi & Weight 1967) is used routinely as an index for comparison of the iontophoretic results with DA, only the study by Spencer & Havlicek (1974)—in which less than half of the responsive cells were inhibited—is found clearly not to have employed neutralization. Moreover, in the study by Bevan et al (1975), current neutralization and otherwise routine methods of drug release were employed, and yet the results reported were still mainly excitations.

Therefore, while the most common result attributed to the effects of iontophoretic DA in studies using extracellular recordings is inhibition, the reasons underlying the apparent depolarizing actions of DA in the one study in which intracellular recording was used, remain to be determined. A detailed analysis of the amount of DA delivered by short duration–high current tests vs the longer duration–low current tests more commonly used could elucidate this problem.

Activation of the nigro-striatal dopaminergic pathway The most direct approach to the function of the dopaminergic circuits in the striatum would be to determine the cellular effects of stimulating the dopamine pathway. However, despite the apparent simplicity of this idea, putting it to the test has remained problematic. As in the case of the iontophoretic tests, a variety of results have been reported with different methodologies.

With intracellular recording methods, single stimuli in the substantia nigra have been found to have depolarizing actions (EPSPs) or EPSPs followed by a hyperpolarization (IPSPs), but rarely produce IPSPs alone (Frigyesi & Purpura 1967, Hull et al 1974, Kitai et al 1975). Conversely, trains of stimuli delivered to the nigra (Connor 1968, 1970, Gonzalez-Vegas 1974, Hull et al 1974) generally lead to inhibitory responses (IPSPs or long hyperpolarizations); excitatory responses, however, have also been seen with trains of nigral stimuli (Kitai et al 1975, 1976). These inhibitory and excitatory striatal effects of nigral stimulation generally show latencies of 5–25 msec. The question that we must now consider is which—if any—of these effects reveals the function of the dopaminergic projection.

One method for approaching the dopaminergic effects of nigral stimulation employs specific dopamine antagonists. Thus, Feltz (1971) observed that the excitatory responses to nigral stimulation could be blocked by superfusion of the caudate with high concentrations of haloperidol. However, Feltz (1971) considered the effect "unspecific" because the same drug treatment also blocked excitations produced by iontophoresis of glutamate and by electrical stimulation of non-DA pathways. Moreover, large parenteral doses of haloperidol (10 mg kg^{-1}) had no effect on the nigral-evoked excitations (also see Ben-Ari & Kelly 1976). Kitai et al (1976) also reported blockade of nigral-evoked excitations after iontophoresis of chlorpromazine. However, without tests of the action of chlorpromazine on responses to other transmitters or pathways, it is difficult to rule out antinigral excitations resulting from the well-known local anesthetic action of chlorpromazine (cf Hoffer et al 1971b). Moreover, since very short (100 msec) iontophoretic pulses of chlorpromazine were employed, the previous issues of adequate drug release and current side effects must also be contemplated.

In contrast, Connor (1970) reported that the inhibitions produced by iontophoretic DA and by stimulation of the SN with trains are both blocked by alpha-methyl dopamine, a presumed DA antagonist. This result would have provided direct evidence that the SN-evoked inhibitions were dopaminergic had Connor demonstrated the specificity of the antagonism. An objective appraisal of these blocking experiments at this time would lead one to conclude that a specific pharmacologic blockade of either the inhibitions or excitations has yet to be accomplished. In fact, Feltz & De Champlain (1972b) reported that neither the excitations or the inhibitions produced in caudate with multiple stimuli to nigra are abolished when the DA projection is removed with 6-OHDA. They therefore concluded that neither of these SN-evoked effects could be dopaminergic. Subsequently, Feltz et al (1976) have reported that the nigral-produced inhibition can be antagonized by picrotoxin and bicuculline; they interpet this effect to suggest that GABA-containing axon collaterals of a striato-nigral pathway may have been activated by stimulation of the nigra. Regardless of the eventual explanation of the apparent antinigral effects of GABA-related drugs, it is clear that numerous fast-conducting tracts closely adjoin the nigra, such as the medial lemniscus, the cerebral peduncles, the dentato-rubro-thalamic tract, and ascending components of the central tegmental tract. Presumably their sensitivity to electrical stimulation would be greater than that of nigra cell bodies (see Ranck 1975). Therefore, selective stimulation of the nigra must overcome or avoid the other circuits that lie between the nigra and striatum.

Functional inferences of the dopaminergic nigrostriatal circuit from lesion studies Neither iontophoresis nor electrical stimulation of the nigra provides definitive information on the function of the dopamine pathway. A third approach is the study of deficits after transection of the nigrostriatal DA circuit. Hull et al (1974) reported that lesions of the nigra, or of the DA projection, produce no statistically significant increases in the firing rates of neurons ipsilateral to the lesions, but did produce significant decreases in firing in the caudate on the side contralateral to the lesion. This effect was not correlated with the depletion of striatal DA. While these effects might not support an inhibitory role for DA fibers in the striatum, the result is not straightforward. In addition to the spatial problems attributable to stimulation studies on the nigra, there is also the as yet poorly characterized functional interdependence between the ipsilateral and contralateral substantia nigra (Nieoullon et al 1977), and there is the even larger question as to whether a simple analysis of neuronal discharge rate per se can reflect the degree to which underlying DA function has been impaired. As a result, functional changes that follow for the loss of DA fibers in the striatum may be masked or compensated for by the loss of non-DA fibers either to the striatum or to other structures projecting to the striatum.

A far more selective tool for producing lesions in the DA system is the use of the chemical neurotoxin, 6-OHDA (see Bloom 1975c). Firing rates of striatal neurons increase ipsilateral to lesions produced by the injection of 6-OHDA into the "DA bundle" or the SN in both rat (Arbuthnott 1974, Siggins et al 1976) and cat (Feltz & De Champlain 1972a,b). However, this result cannot be regarded as directly documenting the inhibitory nature of the nigrostriatal DA system; the effect could arise through the loss of a tonic DA inhibitory projection to more remote brain regions that in turn project to the striatum.

Such objections cannot be leveled at iontophoretic studies of DA sensitivity following 6-OHDA lesions. Here, receptor sensitivity provides a cellular index to the degree of denervation supersensitivity that results from the lesion (see Ungerstedt 1971). Under these conditions striatal neurons show enhanced sensitivity to the actions of DA and the DA-agonist apomorphine (Feltz & De Champlain 1972a, Siggins et al 1974). In neither study was there any increased incidence of DA excitations. Long-term treatment with the presumptive DA antagonist haloperidol (Yarbrough 1975) also produced an apparent "supersensitivity" that was again selective for depressant responses to DA or apomorphine (but not GABA). Here again the inference that DA is inhibitory for caudate neurons is sustained.

Anatomical constraints on the physiological analysis of the nigrostriatal system Two anatomical aspects of the nigrostriatal system would appear capable of accounting for a large number of the problems encountered so far: the morphology and topology of the nigral cells projecting to the striatum, and the diameter (and hence the conduction velocity) of these fibers.

The pars compacta of the nigra is a long, thin, ribbon-shaped structure lying obliquely in an anterior-posterior plane. In the horizontal dimension, the pars compacta may be as thin as 0.5 mm even in the cat, a distance that is less than the tip dimensions of most bipolar stimulating electrodes. Even if the adjacent myelinated systems were neglected, selective stimulation of the nigra would seem an ambitious goal. Selective stimulation of the DA pathway is further complicated by the possibility of parallel nondopaminergic pathways coursing between nigra and striatum such as GABA– (Dray et al 1976) or substance P–containing fibers (Hökfelt et al 1977).

If we add to this the further constraints imposed by topographic orientation of the nigral and ventral tegmental striatal afferents (Fallon & Moore 1976b), any random probe of the striatum should reveal that most units will not respond to any given stimulated area in the nigra. Moreover, nigral topographic projections to the caudate may also be organized in the vertical plane (Domesick et al 1976), which presents even further anatomical complexities.

The physiologic results of nigral stimulation also bear close scrutiny with regard to reported conduction latencies. In most studies, latencies for either orthodromic excitation or inhibition fall between 5 and 25 msec (see review by Siggins 1977). Since the nigrostriatal distance in the cat is 15 mm, the corresponding conduction velocities would be 0.6–3 M sec^{-1}, without allowing for synaptic delay.

However, as elaborated above, the great majority of the striatal DA fibers are very fine unmyelinated C-fibers with diameters of less than 0.4–0.5 μ (Fuxe et al 1964, Hökfelt 1968). Their conduction and response times can only be further increased by the extensive branching that occurs after the DA fibers penetrate the neostriatum.

With such fine branching axons, the nigrostriatal DA pathway could not conduct faster than 1 M sec^{-1}, according to the properties of such fine fibers in the peripheral nervous system (Nishi et al 1965, see also Siggins et al 1971). These slower conducton velocities would then give response latencies between nigra and striatum of 30–100 msec, which greatly exceeds the delays reported for either excitations or inhibitions in most studies (see Albe-Fessard et al 1967, Gonzalez-Vegas 1974). If the postsynaptic response to dopamine in striatum is mediated by cAMP (see Bloom 1975b, Greengard 1976, Siggins et al 1974, 1976, 1977), additional synaptic delay time would be needed for the activation of the adenylate cyclase and, subsequently, a protein kinase (see Greengard 1976). For example, in sympathetic ganglion the synaptic delay of the slow IPSP, which may be mediated by DA and cAMP (Greengard & Kebabian 1974) is about 35 msec (see Libet 1967).

If a similar synaptic delay is required to mediate the effects of the fine, divergent DA projection from nigra to striatum, the resultant response latencies would be at least 65 msec (30 msec of conduction plus 35 msec of synaptic delay) or more. Similar long latencies have been reported for the inhibition of cerebellar Purkinje cells and hippocampal pyramidal cells (see below) following activation of their noradrenergic afferents from locus coeruleus, effects that are also considered to be cAMP-mediated.

The Mesocortical System

INTRODUCTION The designation *mesocortical system* is used to imply the projections from the mesencephalic DA cell group of the substantia nigra and ventral tegmental area upon telencephalic areas other than the basal ganglia. All of the areas so innervated can be considered cortical by the criteria of Bailey & Von Bonin (1951) with one exception, the nucleus accumbens. This nucleus arises embryologically with the neostriatum, has neurons similar to the neostriatal neurons and receives a DA innervation that arises from the ventral tegmental DA neurons (Figure 5) immediately adjacent to those innervating the rostral, medial neostriatum. The DA innervation of the nucleus accumbens has not been considered part of the nigrostriatal projection (Ungerstedt 1971). We retain this distinction here in conformity to this and to other anatomical evidence that the nucleus accumbens has connections that differ from those of the neostriatum. We do not distinguish, however, between a mesocortical and a mesolimbic system as others have done (Ungerstedt 1971, Lindvall & Björklund 1974, Lindvall et al 1974a, Berger et al 1974, 1976). This appears to be an artificial distinction with little to recommend it. The distinction has not been made in any uniform way in the literature largely because the term "limbic system" is itself vague and ill defined. Consequently, this projection system of the mesencephalic DA neurons would appear most parsimoniously covered under the single designation, *mesocortical.*

CELL BODIES The cell bodies of the mesocortical DA system are located in the ventral tegmental area and substantia nigra as described above.

PATHWAYS Axons arising from the DA cell bodies traverse the mesencephalic tegmentum to ascend with the nigrostriatal neurons. There are four major branches from the main pathway (cf Lindvall & Björklund 1974b, 1977 for reviews). The first runs laterally along the ventral amygdaloid bundle–ansa peduncularis system to enter the amygdala and the external capsule. External capsule axons descend to the ventral entorhinal area and ascend to the perirhinal and piriform cortex. The main bundle continues rostrally, largely within the medial forebrain bundle complex and gives off a second major group of fibers into the septal nuclei and the interstitial nucleus of the stria terminalis. A third group turns dorsally from the medial forebrain bundle system into the mesial frontal and anterior cingulate cortex. The fourth group is made up of fibers that leave the medial forebrain bundle system to innervate the olfactory tubercle and continue in the medial olfactory stria to the anterior olfactory nucleus and olfactory bulb. The mesocortical system is represented diagrammatically in Figure 6.

TERMINAL FIELDS The mesocortical DA axons provide a terminal field innervation that, in most instances, differs from that of the nigrostriatal system. In a few areas—nucleus accumbens, olfactory tubercle, interstitial nucleus of the stria terminalis, lateral septal nucleus, central amygdaloid nucleus, intercalated amygdaloid cell groups—the usual pattern of extensive axonal collateralization with densely

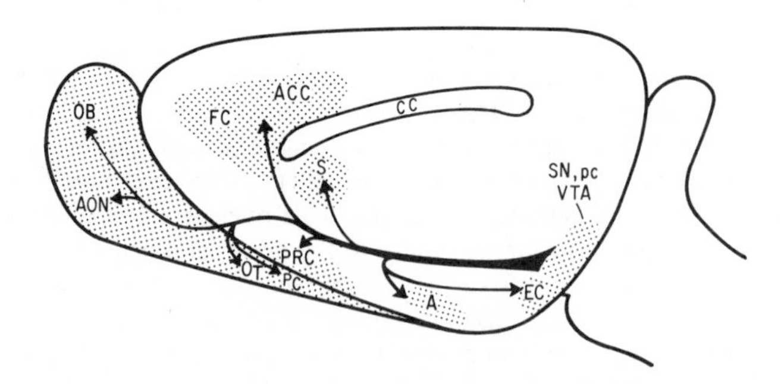

Figure 6 Diagram representing the mesocortical DA system as it would be seen projected in a sagittal plane. See text for description. Abbreviations: A, amygdaloid complex; ACC, anterior cingulate cortex; AON, anterior olfactory nucleus; CC, corpus callosum; EC, entorhinal cortex; FC, frontal cortex; OB, olfactory bulb; OT, olfactory tubercle; PC, piriform cortex; PRC, perirhinal cortex; S, septal area; SN,pc, substantia nigra, pars compacta; VTA, ventral tegmental area. (From observations of Hökfelt et al 1974a, Lindvall et al 1974a, Berger et al 1976, Fallon & Moore 1976b)

packed, numerous, fine varicosities is noted. Within the other areas, however, the pattern of innervation varies markedly. The preterminal axon is thin and without varicosities but appears to become larger within the terminal field. Two major patterns of terminal innervation are encountered. The most striking is that of a dense pericellular array of axons that appear to ensheath the cell body and proximal dendrites. The axons are intensely fluorescent and give rise to fairly large, 1–2 μm, intensely fluorescent varicosities (Figure 7). These are usually not numerous. This pattern is particularly evident in the lateral septal nucleus (Lindvall 1975, Moore 1975) and in the anterior amygdaloid area (Fallon & Moore 1976a). The other pattern of innervation is with single axons traversing the neuropil that branch and give rise to relatively infrequent varicosities similar in appearance to those in the pericellular arrays (Figure 7). This pattern is evident in the lateral septal nucleus, the amygdaloid complex, the anterior olfactory nucleus, the olfactory bulb, and the frontal, cingulate, piriform, perirhinal, and entorhinal cortices.

The density and distribution of the mesocortical DA innervation varies markedly from area to area. The densest innervation is uniformly present in the nucleus accumbens, in the olfactory tubercle that receives a dense innervation to all three

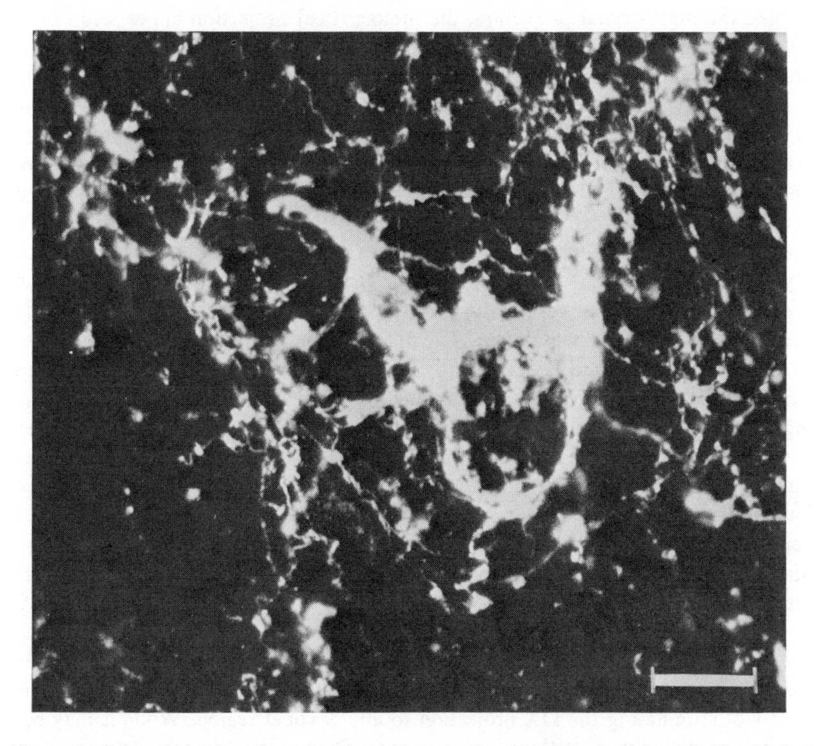

Figure 7 Lateral septal nucleus showing numerous fine DA fibers in the neuropil with 1–2 μm varicosities and a dense pericellular array of fine DA fibers and varicosities in the center. GA method. Marker bar = 25 μm. (From R. Y. Moore, unpublished.)

layers, in the medial part of the lateral septal nucleus, and in the central amygdaloid nucleus (Ungerstedt 1971, Lindvall & Björklund 1974b, Lindvall 1975, Fallon & Moore 1976a). In each case the innervation is of the neostriatal type. The lateral septal nucleus also receives a dense innervation of the pericellular type with some innervation by DA fibers scattered in the neuropil (Lindvall 1975, Moore 1975). The innervation of the piriform cortex is predominantly to the outer layers and is of the scattered neuropil type. Similar innervation is present in the perirhinal cortex, in the basolateral amygdaloid nuclei, the anterior olfactory nucleus, and the olfactory bulb (Fallon & Moore 1976a,b). There is also similar innervation in frontal cortex, predominantly in deep layers, and this is continuous with the innervation of cingulate cortex, which is largely in the second and third layers (Thierry et al 1973a,b, Lindvall et al 1974a, Berger et al 1974, 1976, Lindvall & Björklund 1977; Figure 6). The DA innervation of the frontal cortex has an additional interesting feature in that it appears to correspond to the distribution of thalamo-cortical projections of the dorsomedial thalamic nucleus in all species studied thus far (Lindvall & Björklund 1977). The innervation of the entorhinal cortex is restricted to its ventral portion where the DA axons terminate predominantly in large, basketlike arrays in the second and third layers (Hökfelt et al 1974, Fallon & Moore 1976b).

Like the nigrostriatal projection, the mesocortical projection appears to have a distinct topography. The projection appears to arise from both the ventral tegmental area and the substantia nigra; in the latter the most dorsally situated cells of the pars compacta appear to give rise to the projection (Fallon & Moore 1976b). As with the nigrostriatal projection, dorsal cells project ventrally and ventral cells project dorsally; the medio-lateral topography of the projection is represented by a mediolateral distribution of cells, and the rostro-caudal topography of the projection by a rostro-caudal distribution of cells. Thus, ventral areas such as the olfactory tubercle receive the mesencephalic DA projection from dorsal cells, whereas dorsal areas such as septum or cortex are innervated by ventral cells. Further, in the cortical innervation, the most rostral, medial cortex, the mesial frontal, is innervated by rostral medial cells of the ventral tegmental area. The anterior cingulate area is innervated by a slightly more caudal and lateral group extending into the medial substantia nigra. The piriform-suprarhinal area is innervated by more lateral cells of the substantia nigra, and with a rostro-caudal topography, whereas the entorhinal area is innervated by caudal, medial cells of the ventral tegmental area. These serve as examples, but the precision of the projection holds throughout (Fallon & Moore 1976b). These data indicate that, while the ventral tegmental area–substantia nigra DA neurons can be considered as a single cell group, the cell group has two distinct populations that project separately but topographically upon different telencephalic regions.

PHYSIOLOGY OF THE MESOCORTICAL DA SYSTEM At the present time, there have been no reported instances of attempts to evaluate electrophysiologically the effects of stimulating the DA projection to any cortical region. While it may certainly be anticipated that the cellular analysis of DA fibers in the cortical targets of these projections will be fraught with the same anatomical and physiologic complexities that have plagued the analysis of the nigrostriatal system, the reports published at this time indicate a far more homogenous set of functional results. In neurons of the cat amygdaloid complex, Ben-Ari & Kelly (1976) have reported potent inhibitory actions of iontophoretically applied DA; these actions were partially antagonized by either intravenous or iontophoretically administered alpha-flupenthixol, a potent antipsychotic drug (Ben-Ari & Kelly 1976). However, large intravenous doses of either pimozide or haloperidol, two other potent antipsychotics for which neurochemical evidence of DA antagonism had previously been reported (Iversen 1975), failed to block the inhibitory effects of iontophoretic DA.

In the frontal and cingulate cortex, in which a substantial DA innervation has been noted, the actions of iontophoretic DA are also reported to be inhibitory (Bunney & Aghajanian 1977), but the pharmacological sensitivity of the inhibitory effect of DA here can be distinguished pharmacologically from the inhibitory β-noradrenergic receptors that characterize responsiveness to NE in the more superficial layers of the cingulate cortex.

The Tubero-Hypophysial System

CELL BODIES The cell bodies of the tubero-hypophysial system in the rat are located in the arcuate nucleus of the hypothalamus and the immediately adjacent periventricular nucleus. The neurons are medium-sized and scattered among the other, more numerous, neurons that comprise these nuclei. They correspond to the A12 group of Dahlström & Fuxe (1964) and have been described on numerous occasions (cf Hökfelt & Fuxe 1972, Björklund et al 1973a for reviews). The neurons can be labeled by intraventricular injection of tritiated DA (Scott et al 1976); they differ in their ultrastructure from other elements in the arcuate nucleus because of the labeling with tritiated DA and the absence of neurosecretory activity that is present in the unlabeled neurons of the nucleus.

PATHWAYS The projection from the arcuate-periventricular nucleus neurons is as follows. Axons project ventrally through the arcuate nucleus and then turn medially into the palisade zone (zona externa) of the median eminence (Björklund et al 1970, 1973a,b, Rethelyi & Halasz 1970, Jonsson et al 1972, Ajika & Hökfelt 1973, 1975). There is an exact topography of the projection (Figure 8) such that the anterior portion of the arcuate nucleus projects upon the pars intermedia of the pituitary. In this instance, as with the innervation of the neural lobe, the axons continue beyond the median eminence and traverse the pituitary stalk to reach the neuro-intermediate lobe. Lesions that destroy the most rostral part of the arcuate nucleus result in DA axon denervation of the pars intermedia and some decrease in innervation of the immediately adjacent median eminence, but the neural lobe and remaining median eminence DA axons are intact. A lesion just caudal to the one denervating the pars intermedia has the same effect upon the neural lobe and also partially denervates the median eminence immediately adjacent. A large set of such lesions showed similar discrete effects (Björklund et al 1973b), thus leading to the view of the topography of the projection shown in Figure 8. There is a question as

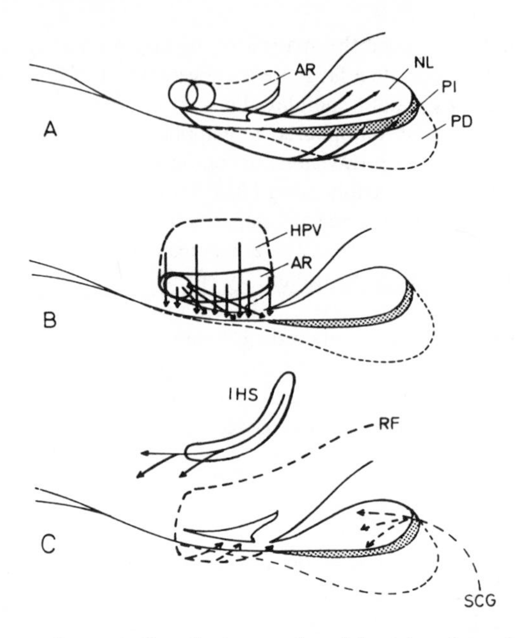

Figure 8 Diagrams demonstrating the topography of the tubero-hypophysial DA neuron projection, as seen in the sagittal plane. See text for description. Abbreviations: AR, arcuate nucleus; HPV, periventricular hypothalamic nucleus; IHS, incerto-hypothalamic system; NL, neural lobe of the pituitary; PD, pars distalis; PI, pars intermedia; RF, reticular formation; SCG, superior cervical ganglion. (From Björklund et al 1973.)

to whether the arcuate-periventricular DA neurons are the only source of DA innervation to the median eminence. As Björklund et al (1970, 1973a) have emphasized, only 20% of hypothalamic DA can be attributed to the arcuate-periventricular projection. Part is contributed by the incerto-hypothalamic system (Björklund et al 1975), but some is undoubtedly contributed by components of the mesotelencephalic DA system. In a recent study Kizer et al (1976) demonstrated significant decreases in DA content in the suprachiasmatic nucleus, the ventromedial nucleus, and the median eminence (40% below control values) after large lesions in the DA cells of the ventral tegmental area. A decrease in median eminence DA would be in accord with the findings of Björklund et al (1970), who found some alteration in fluorescence after complete hypothalamic deafferentation, but it is not in accord with the findings of Jonsson et al (1972), who found no change in median eminence DA innervation after hypothalamic deafferentation. Neither is it in accord with the study of Weiner et al (1972), who produced small islands of mediobasal hypothalamus by deafferentation and found no alteration in DA content of the islands. This is obviously an important issue that warrants further investigation since the data available to date are conflicting.

TERMINAL AREAS As is evident in Figure 9, the tubero-hypophysial system has three terminal areas. The first of these is the neural lobe of the pituitary (Björklund 1968, Björklund & Falck 1969, Björklund et al 1970, 1973b). Within the neural lobe there is a rich plexus of delicate, varicose CA fibers scattered throughout the parenchyma. In addition, there are large, coarse, CA fibers, primarily in association with blood vessels. These are found in all parts of the pituitary and disappear following sympathectomy. The delicate, varicose fibers are unaffected by sympathectomy but are lost, as noted above, after arcuate nucleus or pituitary stalk section. One characteristic of the central CA fiber innervation of neural lobe in virtually all species that have been studied is the presence of large, varicose, highly fluorescent structures ("droplets"; Björklund 1968). These reach sizes in the range of 10–20 μm. Their significance is uncertain, but ultrastructurally (Baumgarten et al 1972) they contain both the usual organelles of a terminal, vesicles and mitochondria, as well as strongly osmiophilic lamellated membrane complexes that resemble myelin bodies and multivesiculated bodies encircling disintegrated vesicles. These observations led Baumgarten et al (1972) to suggest that this central CA innervation to the neuro-intermediate lobe is undergoing a continuous reorganization through degeneration and regeneration cycles. Similar arguments have been put forth for other areas of the nervous system (Sotelo & Palay 1971). The central CA innervation of the neuro-intermediate lobe was identified as dopaminergic by both chemical analysis and microspectrofluorometry (Björklund et al 1970, 1973b). The DA axons innervating the neuro-intermediate lobe were identified ultrastructurally by their uptake of 5-hydroxydopamine to form small, dense core vesicles and by their selective destruction by 6-hydroxydopamine (Baumgarten et al 1972). The DA axons varied in diameter from 0.1 μm in nonvaricose portions to 2 μm in varicosities. Larger varicosities contained the indications of neuronal degeneration noted above. The normal varicosities contained predominantly small granular vesicles (350–500 Å) and medium-sized to large granular vesicles (500–1200 Å). In the neural lobe they appeared randomly distributed and sparse in number in comparison to the neurosecretory endings. Single fibers run in close apposition (80–120 Å) to

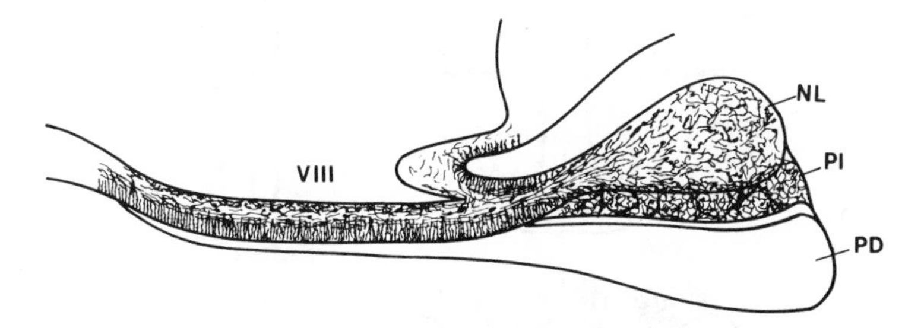

Figure 9 Diagram in the sagittal plane showing the appearance of the NE and DA innervation of the median eminence and pituitary. Abbreviations: NL, neural lobe; PD, pars distalis; PI, pars intermedia; VIII, 3rd ventricle. (From Björklund et al 1973.)

pituicyte processes or neurosecretory endings, but no membrane specializations are evident.

The second innervation, that of the intermediate lobe, is similar in appearance but more dense (Björklund et al 1970, 1973b). The ultrastructure of the DA fibers is identical, but the distribution of their terminals was to the secretory cells of the pars intermedia. DA axon varicosities come into close apposition to the cells (70–120 Å) and accumulations of small vesicles are frequently found at the points of close apposition. No membrane thickenings are evident. In some cases, vesicle-filled varicosities of DA axons are evident invaginated into the secretory cell cytoplasm (Baumgarten et al 1972).

The third innervation is of the median eminence and pituitary stalk. As noted above, the DA fibers from the arcuate and ventral periventricular nuclei run ventrally and then laterally to turn into the median eminence (Björklund et al 1970, 1973, Jonsson et al 1972, Ajika & Hökfelt 1973). In the median eminence and pituitary stalk they make an exceedingly dense plexus of very fine varicosities that is predominantly in the zona externa but is present in the internal and subependymal zones as well (Fuxe 1964, Fuxe & Hökfelt 1969, Björklund et al 1970, 1973, Jonsson et al 1972, Ajika & Hökfelt 1973, 1975). In addition to the DA fibers, there is also an NE projection that arises in the brainstem lateral tegmental NE cell groups and terminates in the internal and subependymal zones of the median eminence and pituitary stalk (Björklund et al 1970, 1973a,b; Figure 8). The DA axon varicosities have been studied ultrastructurally (cf Ajika & Hökfelt 1973, 1975 for reviews). They show sites of close approximation to nonaminergic terminals, ependymal cells, and the pericapillary space of the portal system. In the lateral part of the zona externa they make up approximately 33% of all terminals present (Figure 10). This is of particular interest in that this is the zone of highest concentration of luteinizing hormone-releasing hormone (LH-RH) terminals (Setalo et al 1975, Baker et al 1975). Goldsmith & Ganong (1976) have shown, however, that the two types are separate.

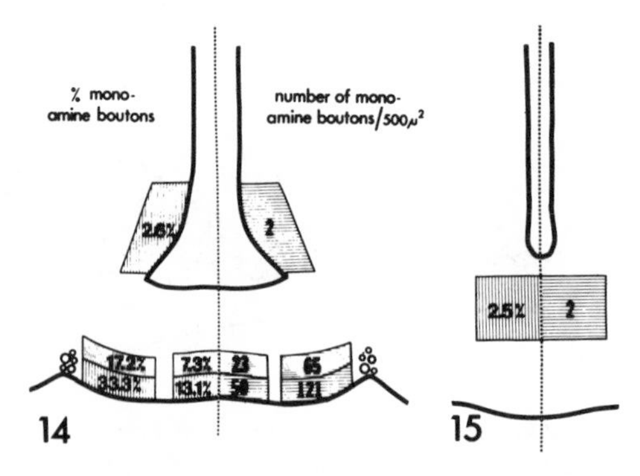

Figure 10 Diagram showing quantitative data on the tubero-hypophysial DA innervation of the median eminence region. (From Ajika & Hökfelt 1973.)

The anatomical evidence now available, then, makes it evident that there is a rich DA neuron innervation of the neuro-intermediate lobe of the pituitary and the median eminence-pituitary stalk. This is highly topographically organized in origin and distribution. In the neural lobe the DA terminals are in approximation to the neurosecretory elements and the pericapillary spaces. In the intermediate lobe the DA axons make close contacts with secretory cells, whereas in the median eminence-pituitary stalk they are in proximity to the pericapillary space of the portal plexus, to ependymal cells, and to other neuron terminals. The tubero-hypophysial system, therefore, should probably be considered as a component of Szentagothai's (1964) parvicellular neurosecretory system. The anatomy of the system seems quite similar in other mammals (cf Cheung & Sladek 1975, Hoffman et al 1976).

PHYSIOLOGY There have been no electrophysiological studies of which we are aware of the tubero-hypophysial DA system. It has been the subject, however, of extensive investigation by workers studying neuroendocrine regulation, and, since this conforms to our stated objective of analyzing the interaction of central catecholamine neurons with target neurons, we briefly review it here. The role of the DA neuron systems appears most clearly defined in respect to the regulation of prolactin secretion. Tubero-hypophysial DA neurons act to inhibit pituitary prolactin secretion (Kamberi et al 1971, Olson et al 1972, Quijada et al 1973). At present, however, it is not clear whether DA acts via influences on neurons secreting a prolactin inhibitory factor (PIF) or whether DA itself is the PIF. The observations of Kamberi et al (1971) suggest an indirect action, whereas MacLeod & Lehmeyer (1974), Shaar & Clemens (1974), and Takahara et al (1974) have obtained evidence for a direct inhibitory effect of DA on pituitary prolactin-producing cells. There is further evidence that indicates the participation of tubero-hypophysial neurons in prolactin secretion. Blocking CA synthesis results in increased plasma prolactin levels (Donoso et al 1973), whereas DA receptor agonists have the opposite effect (Fuxe et al 1975). In addition, Hökfelt & Fuxe (1972) have shown that increased plasma levels are associated with increased activity in the tubero-hypophysial DA neurons, as this is assessed by DA turnover.

The association of a large number of DA terminals with the major site of LH-RH neuron terminals in the median eminence strongly suggests that DA neurons participate in the regulation of gonadotropin production. This has been studied extensively and Fuxe et al (1969a,b), Hökfelt & Fuxe (1972), Lichtensteiger (1969, 1970), and Lichtensteiger & Keller (1974) have provided evidence for an inhibitory effect of tubero-hypophysial neurons on LH-RH secretion. The evidence for this is as follows. The activity of the DA neurons (as estimated by turnover) changes during the estrous cycle of the rat. It is highest during periods of least LH-RH secretion (diestrus) and lowest during periods of greatest LH-RH secretion, particularly proestrus (Ahren et al 1971, Hökfelt & Fuxe 1972, Löfstrom 1976). These cyclic changes are not present in the male or in castrated females (Hökfelt & Fuxe 1972). Further, gonadal steroids produce an increase in DA turnover that is greatest in castrated rats, whereas other hormones such as adrenal steroids do not have this effect (Fuxe et al 1969a). Hökfelt & Fuxe (1972) have proposed that a gonadal steroid–induced activation of the tubero-hypophysial DA system is an important component of an inhibitory feedback action of such steroids from the brain. The model would involve either direct or indirect activation of the DA neurons from gonadal steroid receptor neurons in the hypothalamus (Pfaff & Keiner 1973). This would lead to the release of DA that, presumably, would act to inhibit the release of LH-RH from adjacent terminals in the median eminence. This concept is in accord with the observations of Fuxe et al (1969a,b) of high DA turnover in states of low LH secretion, as with castration. In addition, it is in accord with the studies reviewed above that show appropriate changes in DA turnover with the components of the estrous cycle.

It should be noted, however, that this view has not been universally accepted (cf McCann 1975 for review) and the tubero-hypophysial neurons have also been proposed to be excitatory to gonadotropin production (Schneider & McCann 1969). The analysis of the role of tubero-hypophysial DA neurons in gonadotropin regulation is complicated by the presence of NE neuron terminals in the arcuate nucleus and median eminence as well. There is now quite good accord on the view that the NE neuron innervation has an excitatory effect on LH-RH secretion, and this can explain some of the discrepancies in the literature. The role of the NE neurons will be described in detail in a subsequent review.

No functional significance of the DA neuron innervation of the neural lobe has emerged as yet. The innervation of the intermediate lobe has recently been said to have an inhibitory effect on melanocyte-stimulating hormone secretion (Tilders & Mulder 1975, Tilders et al 1975), but this remains to be firmly established.

The Incerto-Hypothalamic System

CELL BODIES Since the introduction of the fluorescence histochemical methods, a number of catecholamine cell groups have been identified in the diencephalon (Dahlström & Fuxe 1964, Fuxe, Hökfelt & Ungerstedt 1969, Ungerstedt 1971, Jacobowitz & Palkovits 1974, Björklund & Nobin 1973). Microspectrofluorometric analysis has shown that these are nearly all DA neurons (Björklund & Nobin 1973), but their projections had not been identified until recently, save for the neurons of the tubero-hypophysial system (Fuxe & Hökfelt 1966, Björklund et al 1970, 1973a,b, Jonsson et al 1972, Ajika & Hökfelt 1975). With the introduction of the GA method, however, a more precise analysis of the projections of these cell groups has been possible. The cells are found in the caudal hypothalamus and zona incerta, the groups A11 and A13 of Fuxe, Hökfelt & Ungerstedt (1969) and Björklund & Nobin (1973), and in the rostral periventricular hypothalamus (group A14 of Björklund & Nobin 1973). Their distribution is shown in Figure 11. The most caudal cells are located in the posterior hypothalamus and ventral thalamus, principally scattered around the mammillothalamic tract. This group is small in number and is continuous with a fairly dense accumulation of cells in the caudal, medial zona incerta. Consequently, this can be viewed as a single cell group and is the major component of what Björklund et al (1975) have termed the "incerto-hypothalamic dopamine neuron system." The remaining cell bodies they include in this system are located within the periventricular nucleus of the hypothalamus extending from rostral tuberal levels to the level of the anterior commissure (Figure 11).

PATHWAYS AND TERMINAL AREAS The incerto-hypothalamic system is viewed as essentially an intradiencephalic system. The axons of incerto-hypothalamic neurons are very delicate and weakly fluorescent. They exhibit uniform, small varicosities ranging in shape from spherical to fusiform, with the intervaricose segments only partly visible. The caudal portion of the system is present as a loosely arranged bundle among the caudal cells. This can be followed through the posterior hypothalamic area and the zona incerta. At caudal levels the fibers appear to be running more or less parallel to one another, but as the bundle becomes much more dense at the zona incerta this arrangement is lost, and the appearance of the projection is an irregular plexus. The densest areas of apparent terminal innervation are in the medial zona incerta and the dorsal and anterior hypothalamic areas (Figure 11). Some fibers extend laterally into the zona incerta and ventrally into the dorsal component of the dorsomedial hypothalamic nucleus.

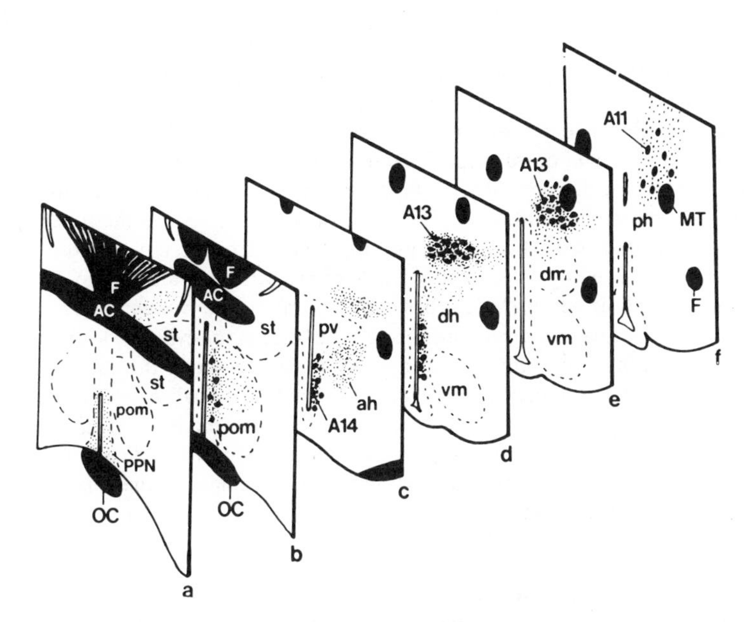

Figure 11 A series of coronal sections through the diencephalon to demonstrate the incerto-hypothalamic system. *a–f*, progressively rostral to caudal levels. See text for description. Abbreviations: AC, anterior commissure; ah, anterior hypothalamic area; A11, A13, A14, diencephalic DA cell groups; dh, dorsal hypothalamic area; dm, dorso-medial nucleus; F, fornix; MT, mammillothalamic tract; OC, optic chiasm; ph, posterior hypothalamus; pom, medial preoptic area; ppn, preoptic periventricular nucleus; pv, paraventricular nucleus; st, bed nucleus of the stria terminalis. (From Lindvall et al 1974a.)

The rostral part, originating from the anterior periventricular cells, has short projections rostrally and caudally into the periventricular nucleus and fibers that extend laterally into the adjacent medial preoptic area (Figure 11). In addition, one component of this projection appears to project dorsally into the most caudal and ventral portion of the lateral septal nucleus.

The separation of this system from other components of the catecholamine neuron systems was made on the basis of two types of evidence. First, the morphology of the axons in GA material is unlike that of any other central group except the tubero-hypophysial fibers innervating the median eminence and pituitary. The fibers are conclusively DA-containing. They have characteristic excitation and emission spectra on microspectrofluorometric analysis, are depleted of transmitter by reserpine, and take up DA by mechanisms not blocked by desipramine in doses that block NE uptake by NE axons (Björklund et al 1975). Secondly, destruction of the locus coeruleus bilaterally, of all ascending NE neurons bilaterally in the midbrain, and unilateral destruction of the ventral tegmental area and substantia nigra has no effect upon the appearance of the projection system (Björklund et al 1975). The lesions were effective, however, in eliminating appropriate components of the ascending catecholamine neuron systems.

Thus, the incerto-hypothalamic system can be viewed as an intradiencephalic system with two components. One originates caudally, predominantly from the zona incerta, and gives rise to short, diffuse projections into its nucleus of origin and the adjacent anterior and dorsal hypothalamic regions. The second component constitutes a periventricular-preoptic system, again with short, diffuse projections into the areas adjacent to the cell bodies. The two components resemble the tubero-hypophysial system in that they are made up of short-axoned DA neurons with a regionally restricted and well-defined mode of projection. In all likelihood, further functional analysis will provide a basis for separating the two components of the system, but this does not appear justified on the basis of the evidence available now.

The incerto-hypothalamic system appears to be present in the cat (Cheung & Sladek 1975), but Hoffman et al (1976) were unable to visualize any large number of CA cell bodies in the hypothalamus of the macaque monkey. Their study was carried out using the conventional Falck-Hillarp method, which does not show the system in the rat, so that this may well represent a technical rather than a species difference. It is likely that the incerto-hypothalamic system, like other CA neuron systems, is a stable feature of the mammalian brain.

PHYSIOLOGY There have been no electrophysiological studies of the incerto-hypothalamic system of which we are aware. As noted by Björklund et al (1975), however, the anatomical similarity of the system to the tubero-hypophysial system and its distribution in the hypothalamus strongly suggests a function in neuroendocrine regulation, but this remains to be demonstrated.

The Retinal System

ANATOMY The retinal DA neuron system was first described by Malmfors (1963), who used the Falck-Hillarp fluorescence histochemical technique. It has now been studied extensively in a large variety of species including teleosts (Ehinger et al 1969, Dowling & Ehinger 1975), amphibians and reptiles (Scheie & Laties 1971), birds (Ehinger 1967), and mammals (Ehinger 1966a,b,c, 1973, Laties & Jacobowitz 1966, Ehinger & Falck 1969a,b, see Ehinger 1976 for review). The original demonstration that the CA present in retinal neurons is DA was by pharmacological manipulation (Ehinger & Falck 1969b), but this has since been confirmed by microspectrofluorometry (Ehinger 1976).

CELL BODIES The cells of the retinal DA system are present throughout the retina in animals without a fovea, but in mammals with a fovea the cells decrease in number centrally and are absent from the macula (Ehinger & Falck 1969b). And, although there are variations in location of cell bodies and processes among species, these are relatively minor. The majority of DA neuron cell bodies are found in the inner part of the inner nuclear layer of the retina, where they constitute approximately 5–10% of the cells in the region (Ehinger & Falck 1969a). The cells are similar in shape in all species examined; typically appearing round or oval in sections cut across the cell layers and radiate in sections through the cell layers (Ehinger & Falck 1969a,b). These cells were first termed outer adrenergic cells by Ehinger (1966c) but later were divided into junctional cells and pleomorphic cells, depending upon their location in the inner nuclear layer of the cebus monkey (Ehinger & Falck 1969a). The pleomorphic cells are much less common than the junctional cells, are scattered through the inner nuclear layer, and are heterogeneous in shape. These cells do not correspond in either form, location of cell bodies, or distribution of processes to any of the cell types described by Poljak (1940) in his classical work on the retina. Recently, however, a new class of retinal neuron has been identified in the cat by Gallego (1971), who has termed them interplexiform cells because their processes extend into both the inner and outer plexiform layers. This terminology has been adopted by Dowling & Ehinger (1975) and Boycott et al (1975) for the DA cells of the retina of the goldfish and cebus monkey and is preferable to the previous terminology.

Two additional DA cell types have been described by Ehinger (1966c). One group, termed eremite cells, is few in number in any species, present in the inner plexiform layer, and made up of cells similar to the interplexiform cells. Their processes are distributed exactly as the interplexiform cells (see below). The second group, termed alloganglionic cells by Ehinger (1966a,b,c), has been found in nearly every species studied. They make up approximately 10% of the interplexiform cell population, are present in the ganglion cell layer, and resemble normal ganglion cells except that they send their processes into the inner plexiform layer (Ehinger 1966b, Ehinger & Falck 1969b). These cells should all be considered, however, as variants of the interplexiform cell.

Terminal distribution The major retinal DA cell type, the interplexiform cell, sends its processes into the inner plexiform layer, where three fiber sublayers are formed: an outer layer at the border with the inner nuclear layer, an intermediate layer, and an inner layer at the border with the ganglion cell layer (Ehinger 1966a).

There is some species variation in this, although the pattern is fairly standard. In addition, some species have scattered fibers in the inner nuclear layer and fibers in the outer plexiform layer immediately adjacent to the inner nuclear layer. Dowling & Ehinger (1975) have recently studied the synaptic organization of DA-containing interplexiform cells of the goldfish and cebus monkey retina with injections of 5,6-dihydroxytryptamine into the vitreous humor. With this technique, the processes of retinal DA neurons can be demonstrated ultrastructurally, due to the degeneration induced by the neurotoxic amine. A diagram summarizing the observations of Dowling & Ehinger (1975) and Dowling et al (1976) is shown in Figure 12. In both species, the amine-containing neurons of the inner nuclear layer send out fine processes into both the inner plexiform layer and the outer plexiform layer. These processes are connected by fine, radially oriented fibers, about 0.5 μm in diameter, that traverse the inner nuclear layer. In the inner plexiform layer the DA interplexiform cell processes make both pre- and postsynaptic contacts with amacrine cell processes. Consequently, these are all dendro-dendritic synapses. In the outer plexiform layer the same arrangement is present; the DA cell processes make pre- and postsynaptic contacts with dendrites of horizontal cells. On the basis of their morphological studies, Dowling & Ehinger (1975) have suggested that the DA interplexiform neurons represent an intraretinal centrifugal neuron that receives input from amacrine cells and terminates upon horizontal cells.

The Periventricular System

The periventricular CA neuron system has been described by Lindvall et al (1974b) and Lindvall & Björklund (1974b, 1977). In the context in which the term is used by these workers, the system contains both DA and NE axons as well as DA cell bodies. The major source of NE axons to the system is from the locus coeruleus; this is not discussed further here, since it will be described in detail in our subsequent review of the NE neuron systems. The periventricular system is distributed along the periventricular and periaqueductal gray from the medulla to the rostral third ventricle.

CELL BODIES The cell bodies that give rise to the periventricular system are located principally in the periventricular and periaqueductal gray from the medulla to the rostral third ventricle. These cell bodies were not evident for the most part with the conventional Falck-Hillarp method but are readily demonstrated with the GA method. They are scattered and no more than a few are noted at any given level, except in the diencephalon where they appear more numerous.

PATHWAYS AND TERMINAL AREAS The periventricular system has been divided into two components, a dorsal periventricular system corresponding in large part to at least one component of the dorsal longitudinal fasciculus and a ventral periventricular system extending along the periventricular region of the hypothalamus (Figure 13). The arrangement of cell bodies along the dorsal periventricular system suggests that it may be composed of neurons with relatively short axons that project predominantly into the periventricular and periaqueductal gray to

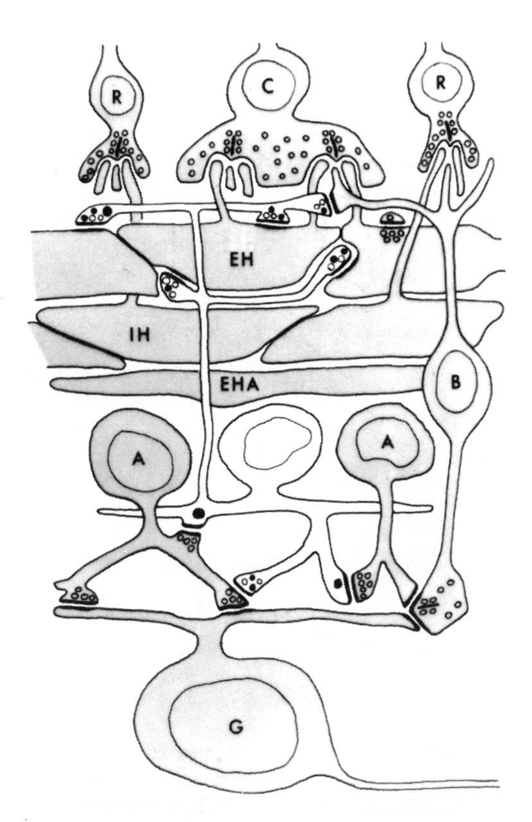

Figure 12 Diagram of the synaptic relations of the retinal interplexiform DA neuron (white neuron) with other retinal elements. Abbreviations: A, amacrine cells; B, bipolar cell; C, cone; EH, external horizontal cell; EHA, external horizontal cell axon; G, ganglion cell; IH, internal horizontal cell; R, rods. (From Dowling et al 1976.)

terminate close to the cell of origin. Lindvall et al (1974b) found no alteration in the number of fibers in the mesencephalic and thalamic dorsal periventricular system after complete interruption of the system at the level of the dorsal raphe nucleus, which suggests that few, long, ascending fibers exist in the system. The dorsal periventricular system appears to innervate areas outside the periventricular region. Lindvall et al (1974b) and Lindvall & Björklund (1974b) describe projections to the tectum, pretectal area, thalamus, and, possibly, the septal area (Figure 13). The ventral periventricular bundle is formed in the supramammillary region and runs rostrally through the posterior hypothalamic area into the dorsomedial hypothalamic nucleus. At this level, fibers from the dorsal periventricular bundle join it, and the ventral periventricular bundle ascends in the lateral part of the periventricular hypothalamic nucleus, innervating that nucleus, some of the paraventricular nucleus, and the bed nucleus of the stria terminalis (Figure 13).

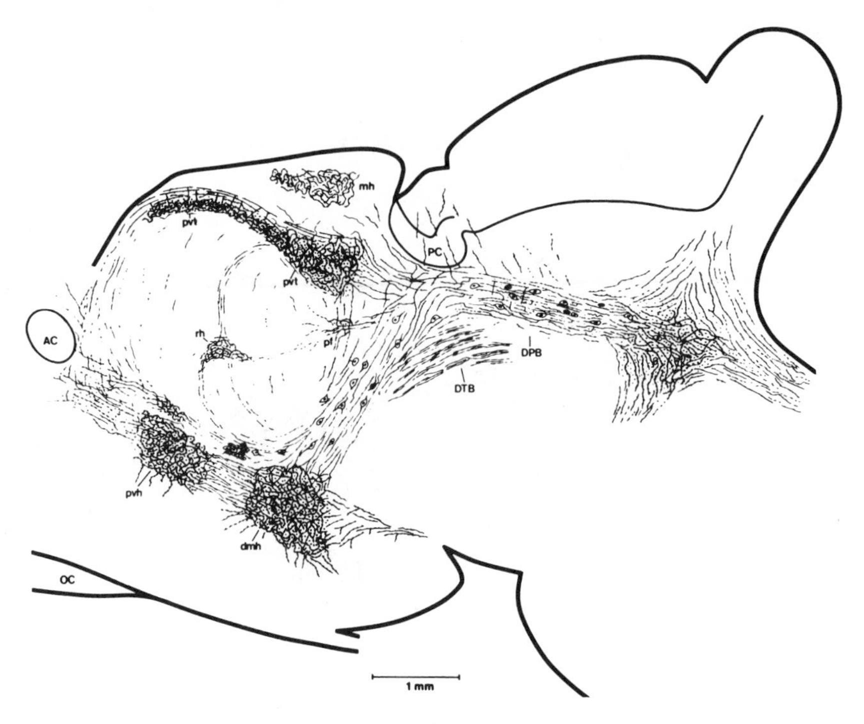

Figure 13 Diagram of a sagittal representation of the periventricular CA neuron system. See text for description. Abbreviations: AC, anterior commissure; dmh, dorsomedial hypothalamic nucleus; DPB, dorsal periventricular CA bundle; DTB, dorsal tegmental CA bundle; mh, medial habenular nucleus; OC, optic chiasm; PC, posterior commissure; pf, parafascicular nucleus; pvt, paraventricular thalamic nucleus; pvh, periventricular hypothalamic nucleus; rh, rhomboid thalamic nucleus. (From Lindvall et al 1974b.)

Other DA Neuron Systems

Two additional DA neuron systems have been described by Hökfelt (1975) on the basis of immunohistochemical analysis. The first of these is a component of the periglomerular cell system of the olfactory bulb. A population of periglomerular cells shows a positive immunohistochemical reaction for tyrosine hydroxylase but not for dopamine-β-hydroxylase or phenylethanolamine-N-methyltransferase (Hökfelt et al 1975). This pattern of reactivity suggests that the cells are DA-producing. The morphology of the neurons is typical of periglomerular cells in general. The second cell group detected by the same immunohistochemical procedure is a component of the dorsal medullary cells in the region of the nucleus tractus solitarius and the dorsal motor nucleus of the vagus. The projections of these cells are unknown, and, although their identification as DA neurons appears reasonably secure from the specificity of the immunohistochemistry, this remains to be confirmed by other methods.

PHYSIOLOGY There are no electrophysiological studies in the literature concerning the tubero-infundibular, incerto-hypothalamic, periventricular, olfactory bulb, or caudal medullary DA neurons. In one study, Dowling et al (1976) have shown that DA neurons may function to enhance the "center-surround" effect of ganglion cell activation by visual stimuli.

CONCLUDING COMMENTS

The purpose of this review has been to summarize our current knowledge of the anatomy and physiology of central neuron systems that use DA as a neurotransmitter. At present, progress on the anatomical side is significantly ahead of the physiology, but this undoubtedly reflects the natural development of the field. Among the known central CA systems, DA neuron systems are most numerous both in terms of the number of systems that have been described and the total number of neurons comprising the systems. And, although they are diverse in location and in some aspects of morphology, DA systems uniformly have a precise, discrete topography and a restricted terminal distribution area. Consequently, on morphological grounds, it would appear appropriate to view the DA systems as components of a set of broader systems with distinct, separate functions. The contrasts among these broader systems are evident from the observations recorded in the body of this review. For example, the mesencephalic DA projection upon the neostriatum is a component of a system that participates in the regulation of motor control. This is an extremely dense, topographically organized projection to the neostriatum which makes up a significant proportion of the terminals making synapses upon neostriatal neurons. Functionally the projection appears to be inhibitory, but this is not completely established. In contrast, the tubero-hypophysial DA system projects in an equally topographic manner upon the median eminence, the infundibular stem, the intermediate lobe and the posterior pituitary. The density of the innervation is greater than that to the neostriatum in some areas innervated, but no synaptic contacts are made. Rather, the DA neuron terminals are in proximity to those of the magnocellular neurosecretory system, cells of the intermediate lobe, and terminals of releasing hormone-producing neurons in the median eminence and infundibular stem. The function of this system appears to be to regulate neurosecretion and the production and release of hypothalamic and pituitary hormones. In addition, the topography of the system in relation to other neural elements in the areas innervated suggests that it is probably a series of subsystems, each with a discrete function. As a last example, to demonstrate the diversity of DA neuron systems, the interplexiform retinal DA cell represents a neuron that lacks an axon but, through its dendritic interactions with other retinal neurons, has demonstrable effects upon visual sensory input. Thus, these DA systems have a distinct localization of their neuronal perikarya, a topographic distribution of processes, and a distinct function within another well-known and discrete functional system. There is much yet to be learned about the anatomy and, particularly, the physiology of the DA neuron systems, but there is nothing to indicate that further information will alter the general conclusions presented here.

As will be presented in the companion to this review, to be published in the next *Annual Review of Neuroscience,* the anatomy and physiology of the norepinephrine and epinephrine neuron systems differ significantly from the DA systems. With the presentation of this material it will be possible, then, to make comparisons between the CA systems and arrive at a series of general statements concerning the organization and function of CA neuron systems in the mammalian brain.

ACKNOWLEDGMENTS

The preparation of this review and some of the work presented in it has been supported in part by USPHS Grant NS-12080 and NSF Grant BNS76-09318. We are grateful to Dr. George Siggins for the opportunity to examine his unpublished review on the electrophysiology of the dopamine innervation of the neostriatum and to Drs. Olle Lindvall and Anders Björklund for providing us a copy of their review of the organization of catecholamine neuron systems to be published in the *Handbook of Psychopharmacology.*

Literature Cited

Agid, Y., Javoy, F., Glowinski, J., Bouvet, D., Sotelo, C. 1973. Injection of 6-hydroxydopamine into the substantia nigra of the rat. II. Diffusion and specificity. *Brain Res.* 58:291–301

Ahren, K., Fuxe, K., Hamberger, L., Hökfelt, T. 1971. Turnover changes in the tubero-infundibular dopamine neurons during the ovarian cycle of the rat. *Endocrinology* 88:1415–24

Ajika, K., Hökfelt, T. 1973. Ultrastructural identification of catecholamine neurones in the hypothalamus periventricular-arcuate nucleus-median eminence complex with special reference to quantitative aspects. *Brain Res.* 57:97–117

Ajika, K., Hökfelt, T. 1975. Projections to the median eminence and the arcuate nucleus with special reference to monoamine systems: effects of lesions. *Cell Tissue Res.* 158:15–35

Albe-Fessard, D., Raieva, S., Santiago, W. 1967. Sur les relations entre substance noire et noyau caudé. *J. Physiol. Paris* 59:324–25

Andén, N.-E., Carlsson, A., Dahlström, A., Fuxe, K., Hillarp, N.-A., Larsson, K. 1964. Demonstration and mapping out of nigro-neostriatal dopamine neurons. *Life Sci.* 3:523–30

Andén, N.-E., Dahlström, A., Fuxe, K., Larsson, K. 1965. Further evidence for the presence of nigro-neostriatal dopamine neurons in the rat. *Am. J. Anat.* 116:329–34

Andén, N.-E., Dahlström, A., Fuxe, K., Larsson, K., Olson, L., Ungerstedt, U. 1966a. Ascending monoamine neurons to the telencephalon and diencephalon. *Acta Physiol. Scand.* 67:313–26

Andén, N.-E., Fuxe, K., Hamberger, B., Hökfelt, T. 1966b. A quantitative study of the nigro-neostriatal dopamine neurons system in the rat. *Acta Physiol. Scand.* 67:306–12

Arbuthnott, G. W. 1974. Spontaneous activity of single units in the striatum after unilateral destruction of the dopamine input. *J. Physiol.* 239:121–22

Bailey, P., Von Bonin, G. 1951. *The Isocortex of Man.* Urbana: Univ. Illinois Press

Baker, B. L., Dermody, W. C., Reel, J. Jr. 1975. Distribution of gonadotropin-releasing hormone in the rat brain as observed with immunocytochemistry. *Endocrinology* 97:125–35.

Baumgarten, H. G., Björklund, A., Holstein, A. F., Nobin, A. 1972. Organization and ultrastructural identification of the catecholamine nerve terminals in the neural lobe and pars intermedia of the rat pituitary. *Z. Zellforsch. Mikrosk. Anat.* 126:483–517

Ben-Ari, Y., Kelly, J. S. 1976. Dopamine-evoked inhibition of single cells of the feline putamen and basolateral amygdala. *J. Physiol. London* 256:1–21

Berger, B., Tassin, J. P., Blanc, G., Moyne, M. A., Thierry, A. M. 1974. Histochemical confirmation for dopaminergic innervation of the rat cerebral cortex after destruction of the noradrenergic ascending pathways. *Brain Res.* 81: 332–37

Berger, B., Thierry, A. M., Tassin, J. P., Moyne, M. A. 1976. Dopaminergic innervation of the rat prefrontal cortex: a fluorescence histochemical study. *Brain Res.* 106:133–45

Bevan, P., Bradshaw, C. M., Szabadi, E. 1975. Effects of desipramine on neuronal responses to dopamine, noradrenaline, 5-hydroxytryptamine and acetylcholine in the caudate nucleus of the rat. *Br. J. Pharmacol.* 54:285–93

Björklund, A. 1968. Monoamine-containing fibers in the neurointermediate lobe of the pig and rat. *Z. Zellforsch. Mikrosk. Anat.* 89:573–89

Björklund, A., Falck, B. 1969. Pituitary monoamines of the cat with special reference to the presence of an unidentified monoamine-like substance in the adenohypophysis. *Z. Zellforsch. Mikrosk. Anat.* 93:254–64

Björklund, A., Falck, B., Hromek, F., Owman, C., West, K. A. 1970. Identification and terminal distribution of the tubero-hypophyseal monoamine systems in the rat by means of stereotaxic and microspectrofluorimetric techniques. *Brain Res.* 17:1–23

Björklund, A., Falck, B., Nobin, A., Stenevi, U. 1973a. Organization of the dopamine and noradrenaline innervations of the median eminence-pituitary region in the rat. In *Neurosecretion—The Final Neuroendocrine Pathway,* ed. F. Knowles, L. Vollrath, pp. 209–22. New York: Springer

Björklund, A., Lindvall, O. 1975. Dopamine in dendrites of substantia nigra neurons: suggestions for a role in dendritic terminals. *Brain Res.* 83:531–37

Björklund, A., Lindvall, O., Nobin, A. 1975. Evidence of an incerto-hypothalamic dopamine neurone system in the rat. *Brain Res.* 89:29–42

Björklund, A., Moore, R. Y. 1978. *The Central Adrenergic Neuron.* New York: Raven Press. In press

Björklund, A., Moore, R. Y., Nobin, A., Stenevi, U. 1973b. The organization of tubero-hypophyseal and reticulo-infundibular catecholamine neuron systems in the rat brain. *Brain Res.* 51:171–91

Björklund, A., Nobin, A. 1973. Fluorescence histochemical and microspectrofluorometric mapping of dopamine, and noradrenaline cell groups in the rat diencephalon. *Brain Res.* 51:193–205

Bloom, F. E. 1973. Ultrastructural identification of catecholamine-containing central synaptic terminals. *J. Histochem. Cytochem.* 21:333–48

Bloom, F. E. 1974. To spritz or not to spritz: the doubtful value of aimless iontophoresis. *Life Sci.* 14:1819–34

Bloom, F. E. 1975a. Amine receptors in CNS: I. Norepinephrine. In *Handbook of Psychopharmacology,* ed. L. L. Iversen, S. D. Iversen, S. H. Snyder, 6:1–22. New York: Raven Press

Bloom, F. E. 1975b. The role of cyclic nucleotides in central synaptic function. *Rev. Phys. B* 74:1–103

Bloom, F. E. 1975c. Monoaminergic neurotoxins: are they selective? *J. Neural Transm.* 37:183–87

Bloom, F. E. 1977. Central noradrenergic systems: physiology and pharmacology. In *Psychopharmacology–A 20 Year Progress Report,* ed. M. E. Lipton, K. C. Killam, A. DiMascio, pp. 131–41. New York: Raven Press

Bloom, F. E., Costa, E., Salmoiraghi, G. C. 1965. Anesthesia and the responsiveness of individual neurons of the cat's caudate nucleus to acetylcholine, norepinephrine, and dopamine administered by microelectrophoresis. *J. Pharmacol. Exp. Ther.* 150:244–52

Boakes, R. J., Bradley, P. B., Brookes, N., Candy, J. M., Wolstencroft, J. H. 1971. Actions of noradrenaline, other sympathomimetic amines and antagonists on neurones in the brain stem of the cat. *Br. J. Pharmacol.* 41:262–71

Boycott, B. B., Dowling, J. E., Fisher, S. K., Kolb, H., Laties, A. M. 1975. Interplexiform cells of the mammalian retina and their comparison with catecholamine-containing retinal cells. *Proc. R. Soc. London Ser. B* 191:353–68

Bradshaw, C. M., Szabadi, E., Roberts, M. H. T. 1973. Kinetics of the release of noradrenaline from micropipettes: interaction between ejecting and retaining currents. *Br. J. Pharmacol.* 49:667–77

Bunney, B. S., Aghajanian, G. K. 1973. Electrophysiological effects of amphetamine on dopaminergic neurons. In *Frontiers in Catecholamine Research,* ed. E. Usdin, S. Snyder, pp. 957–62. New York: Raven Press

Bunney, B. S., Aghajanian, G. K. 1977. Studies on cerebral cortex neurons. In *Pharmacology of Non-striatal Dopaminergic Neurons,* ed. E. Costa, M. Trabucchi, G. L. Gessa, pp. 65–70. New York: Raven Press

Carpenter, M. B., Peter, P. 1972. Nigrostriatal and nigrothalamic fibers in the rhesus monkey. *J. Comp. Neurol.* 144:93–115

Cheung, Y., Sladek, J. R. Jr. 1975. Catecholamine distribution in the feline hypothalamus. *J. Comp. Neurol.* 164: 339–60

Clarke, G., Hill, R. G., Simmonds, M. A. 1973. Microiontophoretic release of drugs from micropipettes: use of 24Na as a model. *Br. J. Pharmacol.* 48:156–61

Connor, J. D. 1968. Caudate unit responses to nigral stimuli: evidence for a possible nigro-neostriatal pathway. *Science* 160: 899–900.

Connor, J. D. 1970. Caudate nucleus neurones; correlation of the effects of substantia nigra stimulation with iontophoretic dopamine. *J. Physiol. London* 208:691–703

Couch, J. R. 1970. Responses of neurons in the raphe nuclei to serotonin, norepinephrine, and acetylcholine and their correlation with an excitatory input. *Brain Res.* 19:137–50

Dahlström, A., Fuxe, K. 1964. Evidence for the existence of monoamine-containing neurons in the central nervous system. I. Demonstration of monoamines in the cell bodies of brain stem neurons. *Acta Physiol. Scand.* Suppl. 62(232):1–55

Domesick, V. B., Beckstead, R. M., Nauta, W. J. H. 1976. Some ascending and descending projections of the substantia nigra and ventral tegmental area in the rat. *Neurosci. Abstr.* 2:61

Donoso, A. O., Bishop, W., McCann, S. M. 1973. The effects of drugs which modify catecholamine synthesis on serum prolactin in rats with median eminence lesions. *Proc. Soc. Exp. Biol. Med.* 143: 360–63

Dowling, J. E., Ehinger, B. 1975. Synaptic organization of the amine-containing interplexiform cells of the goldfish and cebus monkey retinas. *Science* 188: 270–73

Dowling, J. E., Ehinger, B., Hedden, W. L., 1976. The interplexiform cell: a new type of retinal neuron. *Invest. Ophthalmol.* 15:916–26

Dray, A., Gonye, T. J., Oakley, N. R. 1976. Caudate stimulation and substantia nigra activity in the rat. *J. Physiol. London* 259:825–49

Ehinger, B. 1966a. Distribution of adrenergic nerves in the eye and some related structures in the cat. *Acta Physiol. Scand.* 66:123–28

Ehinger, B. 1966b. Adrenergic nerves to the eyes and to related structures in man and in the cynomolgus monkey (Macaca irus). *Invest. Ophthalmol.* 5:42–52

Ehinger, B. 1966c. Adrenergic retinal neurons. *Z. Zellforsch. Mikrosk. Anat.* 71:146–52

Ehinger, B. 1967. Adrenergic nerves in the ovian eye and ciliary ganglion. *Z. Zellforsch. Mikrosk. Anat.* 82:577–88

Ehinger, B. 1973. Ocular adrenergic neurons of the flying fox *Pteropus giganteus Brünn* (Megachiroptera). *Z. Zellforsch. Mikrosk. Anat.* 139:171–78

Ehinger, B. 1976. Biogenic amines as transmitters in the retina. In *Transmitters in the Visual Process,* ed. S. L. Bonting, pp. 145–63 Oxford: Pergamon Press

Ehinger, B., Falck, B. 1969a. Adrenergic retinal neurons of some new world monkeys. *Z. Zellforsch. Mikrosk. Anat.* 100:364–75

Ehinger, B., Falck, B. 1969b. Morphological and pharmohistochemical characteristics of retinal neurons of some mammals. *Albrecht Von Graefes Arch. Klin. Exp. Ophthalmol.* 178:295–300

Ehinger, B., Falck, B., Laties, A. M. 1969. Adrenergic neurons in teleost retina. *Z. Zellforsch. Mikrosk. Anat.* 97:285–97

Falck, B. 1962. Observations on the possibilities of the cellular localization of monoamines by a fluorescence method. *Acta Physiol. Scand. Suppl.* 197(56): 1–25

Falck, B., Hillarp, N.-A. Thieme, G., Torp, A. 1962. Fluorescence of catecholamines and related compounds condensed with formaldehyde. *J. Histochem. Cytochem.* 10:348–54

Fallon, J. H., Moore, R. Y. 1976a. Catecholamine neuron innervation of the rat amygdala. *Anat. Rec.* 184:399

Fallon, J. H., Moore, R. Y. 1976b. Dopamine innervation of some basal forebrain areas in the rat. *Neurosci. Abstr.* 2:486

Feltz, P. 1971. Sensitivity to haloperidol of caudate neurones excited by nigral stimulation. *Eur. J. Pharmacol.* 14: 360–64

Feltz, P., De Champlain, J. 1972a. Enhanced sensitivity of caudate neurons to microiontophoretic injections of dopamine in 6-hydroxydopamine treated cats. *Brain Res.* 43:601–5

Feltz, P., De Champlain, J., 1972b. Persistence of caudate unitary responses to nigral stimulation after destruction and functional impairment of the striatal dopaminergic terminals. *Brain Res.* 43:595–600.

Feltz, P., De Champlain, J., Dessama, J.-M. 1976. The question of nigro caudate inhibition which persists after inactivation of dopamine-releasing synapses: a possible inhibitory process related to gamma-amino butyric acid. *Neuropsychopharmacol. Excerpta Med. Int. Congr. Ser.* No. 359, pp. 453–58

Frigyesi, T. L., Purpura, D. P. 1967. Electrophysiological analysis of reciprocal caudate-nigral relations. *Brain Res.* 6: 440–56

Fuxe, K. 1964. Cellular localization of monoamines in the median eminence and infundibular stem of some mammals. *Z. Zellforsch. Mikrosk. Anat.* 61:710–24

Fuxe, K., Agnati, L. F., Corrodi, H., Jonsson, G., Hökfelt, T. 1975. Action of dopamine receptor agonists in forebrain and hypothalamus. *Adv. Neurol.* 9: 223–42

Fuxe, K., Hökfelt, T. 1966. Further evidence for the existence of tubero-infundibular dopamine neurons. *Acta Physiol. Scand.* 66:243–44

Fuxe, K. Hökfelt, T. 1969. Catecholamines in the hypothalamus and pituitary gland. In *Frontiers in Neuroendocrinology,* ed. L. Martini, W. F. Ganong, pp. 47–96. New York: Oxford Press

Fuxe, K., Hökfelt, T., Nilsson, O. 1964. Observations on the localization of dopamine in the caudate nucleus of the rat. *Z. Zellforsch. Mikrosk. Anat.* 63:701–6

Fuxe, K., Hökfelt, T., Nilsson, O. 1969a. Castration, sex hormones and tubero-infundibular dopamine neurons. *Neuroendocrinology* 5:107–20

Fuxe, K., Hökfelt, T., Nilsson, O. 1969b. Factors involved in the control of the activity of the tubero-infundibular neurons during pregnancy and lactation. *Neuroendocrinology* 5:257–70

Fuxe, K., Hökfelt, T., Ungerstedt, U. 1969. Distribution of monoamines in the mammalian central nervous system by histochemical studies. In *Metabolism of Amines in the Brain,* ed. G. Hooper, pp. 10–22. London: McMillan

Gallego, A. 1971. Horizontal and amacrine cells in the mammal's retina. *Vision Res.* 3:33–50

Goldsmith, P. C., Ganong, W. F. 1976. Ultrastructural localization of luteinizing hormone releasing or hormone in the median eminence of the rat. *Brain Res.* 97:181–93

Gonzalez-Vegas, J. A. 1974. Antagonism of dopamine-mediated inhibition in the nigro-striatal pathway: a modes of action of some catatonia-inducing drugs. *Brain Res.* 80:219–28

Graybiel, A. M. 1975. Wallerian degeneration and anterograde tracer methods. In *The Use of Axonal Transport for Studies of Neuronal Connectivity,* ed. W. M. Cowan, M. Cuenod, pp. 174–216. Amsterdam: Elsevier

Greengard, P. 1976. Possible role for cyclic nucleotides and phosphorylated membrane proteins in postsynaptic actions of neurotransmitters. *Nature* 260:101–8

Greengard, P., Kebabian, J. W. 1974. Role of cyclic AMP in synaptic transmission in the mammalian peripheral nervous system. *Fed. Proc.* 33:1059–7

Grofova, I. 1975. The identification of striatal and pallidal neurons projecting to substantia nigra: an experimental study by means of horseradish peroxidase. *Brain Res.* 91:286–91

Grofova, I., Rinvik, E. 1970. An experimental electron microscopic study on the striatonigral projection in the cat. *Brain Res.* 11:249–62

Gulley, R. L., Wood, R. L. 1971. The fine structure of neurons in the rat substantia nigra. *Tissue Cell* 3:675–90

Hanaway, J., McConnell, J. A., Detsky, M. G. 1970. Cytoarchitecture of the substantia nigra in the rat. *Am. J. Anat.* 129:417–38

Hattori, T., Fibiger, H. C., McGeer, P. L. 1975. Demonstration of a pallido nigral projection innervating dopaminergic neurons. *J. Comp. Neurol.* 162:487–504

Hattori, T., Fibiger, H. C., McGeer, P. L., Maler, L. 1973. Analysis of the fine structure of the dopaminergic nigrostriatal projection by electron microscopic autoradiography. *Exp. Neurol.* 41:599–611

Herz, A., Zieglgänsberger, W. 1968. The influence of microelectrophoretically applied biogenic amines, cholinomimetics and procaine on synaptic excitation in the corpus striatum. *Int. J. Neuropharmacol.* 7:221–30

Hoffer, B. J., Bloom, F. E. 1976. Norepinephrine and central neurons. In *Chemical Transmission in the Mammalian Central Nervous System,* ed. C. H. Hockman, D. Bieger, pp. 327–48. Baltimore: Univ. Park Press

Hoffer, B. J., Neff, N. H., Siggins, G. R. 1971a. Microiontophoretic release of norepinephrine from micropipettes. *Neuropharmacology* 10:175–80

Hoffer, B. J., Siggins, G. R., Bloom, F. E. 1971b. Studies on norepinephrine-containing afferents to Purkinje cells of rat cerebellum: II. Sensitivity of Purkinje cells to norepinephrine and related substances administered by microiontophoresis. *Brain Res.* 25:523–34

Hoffman, G. E., Sladek, J. R. Jr., Felten, D. L. 1976. Monoamine distribution in primate brain. III. Catecholamine-containing varicosities in the hypothalamus of Macaca Mulatta. *Am. J. Anat.* 147:501–14

Hökfelt, T. 1968. In vitro sutdies on central and peripheral monoamine neurons at the ultrastructural level. *Z. Zellforsch. Mikrosk. Anat.* 91:1–74

Hökfelt, T. 1975. *Immunohistochemistry of central CA neurons.* Presented at Symp. "Organization and Function of Central Catecholamine Neurons," Univ. Lund, Sweden, December 1975

Hökfelt, T., Elde, R., Johansson, O., Ljungdahl, A., Schultzberg, M., Fuxe, K. 1977. The distribution of peptide-containing neurons in the nervous system. In *Psychopharmacology—A Generation of Progress,* ed. K. Killam, A. DiMascio, M. Lipton. New York: Raven Press. In press

Hökfelt, T., Fuxe, K. 1972. Brain endocrine interactions: on the morphology and the neuro-endocrine role of hypothalamus catecholamine neurons. In *Median Eminence, Structure and Function,* ed. K. M. Knigge, E. E. Scott, A. Weindle, pp. 181–223. Basel: Karger

Hökfelt, T., Fuxe, K., Goldstein, M., Johansson, O. 1974. Immunohistochemical evidence for the existence of adrenaline neurons in the rat brain. *Brain Res.* 66:235–51

Hökfelt, T., Halasz, N., Ljungdahl, A., Johansson, O., Goldstein, M., Park, D. 1975. Histochemical support for a dopaminergic mechanism in the dendrites of certain periglomerular cells in the rat olfactory bulb. *Neurosci. Lett.* 1:85–90

Hökfelt, T., Ungerstedt, U. 1969. Electron and fluorescence microscopical studies on the nucleus caudatus putamen of the rat after unilateral lesions of nigro-neostriatal dopamine neurons. *Acta Physiol. Scand.* 76:415–26

Hökfelt, T., Ungerstedt, U. 1973. Specificity of 6-hydroxydopamine induced degeneration of central monoamine neurons: an electron and fluorescence microscopic study with special reference to intracerebral injection of the nigrostriatal dopamine system. *Brain Res.* 60:269–97

Hori, T., Nakayama, S. 1973. Effects of biogenic amines on thermosensitivity in the rabbit. *J. Physiol. London* 232:71–85

Hornykiewicz, O. 1966. Dopamine and brain function. *Pharmacol. Rev.* 18:925–64

Huber, G. C., Crosby, E., Woodburne, R., Gillilan, L., Brown, J., Tauthai, B. 1943. The mammalian midbrain and isthmus regions. *J. Comp. Neurol.* 78:129–556

Hull, C. D., Levine, M. S., Buchwald, N. A., Heller, A., Browning, R. A. 1974. The spontaneous firing pattern of forebrain neurons. I. The effects of dopamine and non-dopamine depleting lesions on caudate unit firing patterns. *Brain Res.* 73:241–62

Ibata, Y., Nojyo, Y., Matsuura, T., Sano, Y. 1973. Nigro-neostriatal projection. *Z. Zellforsch. Mikrosk. Anat.* 138:333–44

Iversen, L. L. 1975. Dopamine receptors in the brain. *Science* 188:1084–89

Jacobowitz, D. M., Palkovits, M. 1974. Topographic atlas of catecholamine and acetylcholinesterase-containing neurons in the rat brain. I. Forebrain (telencephalon, diencephalon). *J. Comp. Neurol.* 157:13–28

Jones, B. E., Moore, R. Y. 1977. Ascending projections of the locus coeruleus in the rat. II. Autoradiographic study. *Brain Res* 127:23–53

Jonsson, G., Fuxe, K., Hökfelt, T. 1972. On the catecholamine innervation of the hypothalamus, with special reference to the median eminence. *Brain Res.* 40: 271–81

Juraska, J. M., Wilson, C. J., Groves, P. M. 1977. The substantia nigra in the rat: a golgi study. *J. Comp. Neurol.* 172:585–600

Kamberi, I., Schneider, H. P. G., McCann, S. M. 1971. Action of dopamine to release FSH-releasing factor (FRF) from hypothalamic tissue in vitro. *Endocrinology* 73:345–48

Kirsten, E. G., Sharma, J. N. 1976. Characteristics and response differences to iontophoretically applied norepinephrine, d-amphetamine, and acetylcholine on neurons in the medial and lateral vestibular nuclei of the cat. *Brain Res.* 112:77–90

Kitai, S. T., Sugimori, M., Kocsis, J. C. 1976. Excitatory nature of dopamine in the nigro-caudate pathway. *Exp. Brain. Res.* 24:351–63

Kitai, S. T., Wagner, A., Precht, W., Ohno, T. 1975. Nigro-caudate and caudatonigral relationship: an electrophysiological study. *Brain Res.* 85:44–48

Kizer, J. S., Palkovits, M., Brownstein, M. J. 1976. The projections of the A8, A9 and A10 dopaminergic cell bodies: evidence for a nigro-hypothalamic-median eminence dopaminergic pathway. *Brain Res.* 108:363–70

Krebs, H., Bindra, D. 1971. Noradrenaline and chemical coding of hypothalamic neurons. *Nature* 229:178–80

Laties, A. M., Jacobowitz, D. 1966. A comparative study of the autonomic innervation of the eye in monkey, cat, and rabbit. *Anat. Rec.* 156:383–96

LaVail, J. H., LaVail, M. M. 1972. Retrograde axonal transport in the central nervous system. *Science* 176:1416–17

Libet, B. 1967. Long latent periods and further analysis of slow synaptic responses in sympathetic ganglia. *J. Neurophysiol.* 30:494–514

Lichtensteiger, W. 1969. Cyclic variations of catecholamine content in hypothalamic nerve cells during estrus cycle in the rat, with a concomitant study of the substantia nigra. *J. Pharmacol. Exp. Ther.* 165:204–15

Lichtensteiger, W. 1970. Catecholamine-containing neurones in neuroendocrine regulation: principles and changes with microfluorometry. *Prog. Histochem. Cytochem.* 1:185–276

Lichtensteiger, W., Keller, P. J. 1974. Tubero-infundibular dopamine neurons and the secretion of luteinizing hormones and prolactin: extra-hypothalamic influences interaction with cholinergic systems and the effects of urethane anesthesia. *Brain Res.* 74:279–303

Lindvall, O. 1975. Mesencephalic dopaminergic afferents to the lateral septal nucleus of the rat. *Brain Res.* 87:89–95

Lindvall, O., Björklund, A. 1974a. The glyoxylic acid fluorescence histochemical method: a detailed account of the methodology for the visualization of central catecholamine neurons. *Histochemistry* 39:97–127

Lindvall, O., Björklund, A. 1974b. The organization of the ascending catecholamine neuron systems in the rat brain as revealed by the glyoxylic acid fluorescence method. *Acta Physiol. Scand.* 412:1–48

Lindvall, O., Björklund, A. 1977. Organization of catecholamine neurons in the rat central nervous system. In *Handbook of Psychopharmacology,* ed. L. Iversen, S. Iversen, S. H. Snyder. New York: Plenum Press. In press

Lindvall, O., Björklund, A., Moore, R. Y., Stenevi, U. 1974a. Mesencephalic dopamine neurons projecting to neocortex. *Brain Res.* 81:325–31

Lindvall, O., Björklund, A., Nobin, A., Stenevi, U. 1974b. The adrenergic innervation of the rat thalamus as revealed by the glyoxylic acid fluorescence method. *J. Comp. Neurol.* 154:317–48

Löfstrom, A. 1976. Catecholamine turnover alterations in discrete areas of the median eminence of the 4- and 5-day cyclic rat. *Brain Res.* 120:113–31

MacLeod, R. M., Lehmeyer, J. E. 1974. Studies on the mechanism of the dopamine-mediated inhibition of prolactin secretion. *Endocrinology* 94:1077–85

Maler, L., Fibiger, H. C., McGeer, P. L. 1973. Demonstration of the nigrostriatal projection by silver staining after nigral injections of 6-hydroxydopamine. *Exp. Neurol.* 40:505–15

Malmfors, T. 1963. Evidence of adrenergic neurons with synaptic terminals in the retina of rats demonstrated with fluorescence and electron microscopy. *Acta Physiol. Scand.* 58:99–100

McCann, S. M. 1975. Neurohumoral correlates of ovulation. *Fed. Proc.* 29: 1888–94

McGeer, E. G., Hattori, T., McGeer, P. L. 1975. Electron microscopic localization of labeled norepinephrine transported in nigrostriatal neurons. *Brain Res.* 86:478–82

McLennan, H., York, D. H. 1967. The action of dopamine on neurons of the caudate nucleus. *J. Physiol. London* 189:393–402

Moore, R. Y. 1975. Monoamine neurons innervating the hippocampal formation and septum: organization and response to injury. In *The Hippocampus,* ed. R. L. Issacson, K. H. Pribram. 5:215–37. New York: Plenum

Moore, R. Y., Bhatnagar, R. K., Heller, A. 1971a. Anatomical and chemical studies of a nigro-neostriatal projection in the cat. *Brain Res.* 30:119–35

Moore, R. Y., Bjöklund, A., Stenevi, U. 1971b. Plastic changes in the adrenergic innervation of the rat septal area in response to denervation. *Brain Res.* 33: 13–35

Nauta, W. J. H. 1958. Hippocampal projections and related neural pathways to the midbrain in the cat. *Brain* 81:319–40

Nauta, W. J. H., Haymaker, W. 1969. Hypothalamic nuclei and fiber connections. In *The Hypothalamus,* ed. W. Haymaker, E. Andersson, W. J. H. Nauta, pp. 136–209. Springfield: Thomas

Nieoullon, A., Cheramy, A., Glowinski, J. 1977. Release of dopamine from termi-

nals and dendrites of the two nigrostriatal dopaminergic pathways in response to unilateral sensory stimuli in the cat. *Nature.* 269:340–41
Nishi, S., Soeda, H., Koketsu, K. 1965. Studies in sympathetic B and C neurons and patterns of preganglionic innervation. *J. Cell Comp. Physiol.* 66:19–32
Olson, L., Seiger, A. 1972. Early prenatal ontogeny of central monoamine neurons in the rat. Fluorescence histochemical observations. *Z. Anat. Entwicklungsgesch.* 137:301–46
Olson, L., Seiger, A., Fuxe, K. 1972. Heterogeneity of striatal and limbic dopamine innervation: highly fluorescent islands in developing and adult rats. *Brain Res.* 44:283–88
Pfaff, D., Keiner, M. 1973. Atlas of estradiol-concentrating cells in the central nervous system of the female rat. *J. Comp. Neurol.* 151:121–57
Poljak, S. 1940. *The Retina.* Chicago: Univ. Chicago Press 607 pp.
Quijada, M., Illner, P., Kmilick, L., McCann, S. M. 1973. The effect of catecholamines on hormone release from anterior pituitaries and ventral hypothalamic incubated in vitro. *Neuroendocrinology* 13:151–63
Ranck, J. B. Jr. 1975. What do stimulating electrodes in the brain stimulate? *Brain Res.* 98:417–40
Rethelyi, M., Halasz, B. 1970. Origin of the nerve endings in the surface zone of the median eminence in the rat hypothalamus. *Exp. Brain Res.* 11:145–58
Rinvik, F., Grofova, I. 1970. Observations on the fine structure of the substantia nigra in the cat. *Exp. Brain Res.* 11:229–48
Salmoiraghi, G. C., Weight, F. 1967. Micromethods in neuropharmacology: an approach to the study of anesthetics. *Anesthesiology* 28:54–64
Scheie, E., Laties, A. M. 1971. Catecholamine-containing cells in the retinas of Gecko gecko and Rana pipiens. *Herpetologica* 27:77–80
Schneider, H. P. G., McCann, S. M. 1969. Possible role of dopamine as transmitter to promote discharge of LH-releasing factor. *Endocrinology* 85:121–32
Scott, D. E., Saldek, J. R. Jr., Knigge, K. M., Krubisch-Dudley, G., Kent, D. L., Sladek, E. D. 1976. Localization of dopamine in the endocrine hypothalamus of the rat. *Cell Tiss. Res.* 166:401–73
Seiger, A., Olson, L. 1973. Late prenatal ontogeny of central monoamine neurons in the rat; fluorescence histochemical observations. *Z. Anat. Entwicklungsgesch.* 140:281–318
Setalo, G., Vigh, S., Schally, A. V., Arimura, A., Flerko, B. 1975. LHRH-containing neural elements in the rat hypothalamus. *Endocrinology* 96:135–42
Shaar, C. J., Clemens, J. A. 1974. The role of catecholamines in the release of anterior pituitary prolactin in vitro. *Endocrinology* 95:1202–12
Shimizu, N., Ohnishi, S. 1973. Demonstration of nigro-neostriatal tract by degeneration silver method. *Exp. Brain Res.* 17:133–38
Siggins, G. R. 1977. The electrophysiological role of dopamine in striatum: excitatory or inhibitory? See Bloom 1977, pp. 143–57
Siggins, G. R., Hoffer, B. J., Bloom, F. E. 1971. Prostaglandin-norepinephrine interactions in brain: Microelectrophoretic and histochemical correlates. *Ann NY Acad. Sci.* 180:302–23
Siggins, G. R., Hoffer, B. J., Bloom, F. E., Ungerstedt, U. 1976. In *The Basal Ganglia,* ed. M. D. Yahr, pp. 227–48. New York: Raven Press
Siggins, G. R., Hoffer, B. J., Ungerstedt, U. 1974. Electrophysiological evidence for involvement of cyclic adenosine monophosphate in dopamine responses of caudate neurons. *Life Sci.* 16:779–92
Sotelo, C. 1971. The fine structural localization of norepinephrine-3H in the substantia nigra and area postrema of the rat: an autoradiographic study. *J. Ultrastruct. Res.* 36:824–41
Sotelo, C., Javoy, F., Agid, Y., Glowinski, J. 1973. Injection of b-hydroxydopamine in the substantia nigra of the rat. I. Morphological study. *Brain Res.* 58: 269–90
Sotelo, C., Palay, S. 1971. Altered axons and axon terminals in the lateral vestibular nucleus of the rat: possible example of axonal remodeling. *Lab. Invest.* 25: 653–72
Spehlmann, R. 1975. The effects of acetylcholine and dopamine on the caudate nucleus depleted of biogenic amines. *Brain* 98:219–30
Spencer, H. J., Havlicek, V. 1974. Alterations by anesthetic agents of the responses of rat striatal neurons to iontophoretically applied amphetamine, acetylcholine, noradrenaline and dopamine. *Can. J. Physiol. Pharmacol.* 52:808–13
Stone, T. W. 1976. Responses of neurones in the cerebral cortex and caudate nucleus to amantadine, amphetamine and dopamine. *Br. J. Pharmacol.* 56:101–10

Stone, T. W., Bailey, E. V. 1975. Responses of central neurones to amantadine: comparison with dopamine and amphetamine. *Brain Res.* 85:126–29

Takahara, J., Arimura, A., Schally, A. V. 1974. Effect of catecholamines on the TRH-stimulated release of prolactin and growth hormone from sheep pituitaries in vitro. *Endocrinology* 95:490–94

Thierry, A. M., Blane, G., Sobel, A., Stinus, L., Glowinski, J. 1973a. Dopaminergic terminals in the rat cortex. *Science* 182:499–501

Thierry, A. M., Stinus, L., Blane, G., Glowinski, J. 1973b. Some evidence for the existence of dopaminergic neurons in the rat cortex. *Brain Res.* 50:230–34

Tilders, F. J. H., Mulder, A. H. 1975. In vitro demonstration of melanocyte stimulating hormone release inhibiting action of dopaminergic nerve fibers. *J. Endocrinol.* 64:63p–64p

Tilders, F. J. H., Mulder, A. H., Smelik, P. G. 1975. On the presence of an MSH-releasing inhibitory system in the rat neurointermediate lobe. *Neuroendocrinology* 18:125–30

Ungerstedt, U. 1971. Stereotaxic mapping of the monoamine pathways in the rat brain. *Acta Physiol. Scand.* 367:1–48

Usunoff, K. G., Hassler, R., Romansky, K., Usunova, P., Wagner, A. 1976. The nigrostriatal projection in the cat. I. Silver impregnation study. *J. Neurol. Sci.* 28:265–88

Voneida, T. J. 1960. An experimental study of the course and destination of fibers arising in the head of the caudate nucleus in the cat and monkey. *J. Comp. Neurol.* 115:75–87

Weiner, R. I., Shryne, J. E., Gorski, R. A., Sawyer, C. H. 1972. Changes in catecholamine content of rat hypothalamus following differentiation. *Endocrinology* 90:867–73

Yamamoto, C. 1967. Pharmacologic studies of norepinephrine, acetylcholine, and related compounds on neurons in Dieter's nucleus and the cerebellum. *J. Pharmacol. Exp. Ther.* 156:39–47

Yarbrough, G. G. 1975. Supersensitivity of caudate neurons after repeated administration of haloperidol. *Eur. J. Pharmacol.* 31:367–69

York, D. H. 1975. Amine receptors in CNS:II. Dopamine. See Bloom 1975a, pp. 23–61

Ann. Rev. Neurosci. 1978. 1:171–82

OPTICAL METHODS FOR MONITORING NEURON ACTIVITY

❖11507

L. B. Cohen, B. M. Salzberg, and A. Grinvald
Department of Physiology, Yale University School of Medicine, New Haven, Connecticut 06510 and Department of Physiology and Pharmacology, University of Pennsylvania, School of Dental Medicine, Philadelphia, Pennsylvania 19174

INTRODUCTION

A method for monitoring neuron activity that does not require electrodes might be useful in situations where the use of electrodes is difficult because of neuron size, geometry, or number. A neuron will interact with electromagnetic or ultrasonic radiation in many different ways. If any one of these interactions changes during neuron activity, then its measurement could be used as a noninvasive method for monitoring that activity.

The use of an optical method was suggested by the discovery of changes in light scattering, birefringence, and fluorescence that accompany the action potential (Cohen, Keynes & Hille 1968, Tasaki et al 1968). However, because the signal-to-noise ratios of these signals were too small to be used to monitor activity in individual neurons, a search for larger signals was initiated (Davila, Cohen & Waggoner 1972, Cohen, 1973). At present, the largest signals are changes in absorption of light by axons stained with the merocyanine dyes (Ross et al 1977, Salzberg et al 1977), whose structures are shown in Figure 1. These absorption signals are about 100 times larger than the largest signal available in 1971.

The first part of this article presents evidence that the absorption changes of these dyes somehow result from the changes in membrane potential rather than from the ionic currents or the increases in membrane permeability that occur during an action potential. This is followed by a review of our efforts to find larger signals. The use of a dye may lead to pharmacologic effects or to photodynamic damage, and both possible problems are discussed briefly.

Because the light from many different parts of a preparation can be measured simultaneously, it should, in principle, be possible to monitor activity in many neurons at once. It is this application that has inspired our interest in optical

0147-006X/0325-0171$01.00

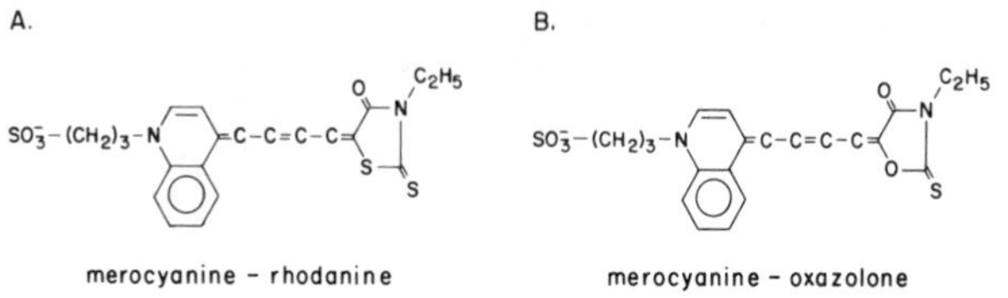

A. merocyanine - rhodanine

B. merocyanine - oxazolone

Figure 1 The structures of the two dyes used in the experiments illustrated in this article. The merocyanine-rhodanine (*A*) is dye XVII of Ross et al (1977). This dye was used in the experiments illustrated in Figures 2 and 3. The merocyanine-oxazolone (*B*) was used in the experiments illustrated in Figures 4, 5, and 6.

recording of membrane potential. Our attempts to develop a technique that can be used to monitor activity of many neurons in an invertebrate central nervous system are discussed. We think that such an apparatus might be a useful tool for solving certain problems in neurobiology.

This article is a summary of experiments carried out by William Ross, Vicencio Davila, and the authors. We are not aware of publications by other investigators on this topic. However, optical measurements of membrane potential have been used in several other areas of biology, e.g. in attempts to demonstrate potential changes in the internal membrane systems of muscle. For reviews see Waggoner (1976) or Cohen & Salzberg (1978).

EVIDENCE THAT AN OPTICAL RECORDING IS RELATED TO AN ELECTRODE MEASUREMENT OF POTENTIAL

The evidence that an optical measurement of potential is related to an electrode measurement comes from a direct comparison of the two measurements using giant axons from the squid, *Loligo pealii.* The large diameter of the axon allows the use of low resistance intracellular electrodes to control the potential uniformly over the relatively large membrane area (7.5 mm^2) that is illuminated. The axons or ganglia were incubated for 10–15 min in a saline that contains the dye; the dye solution was then washed out and the experiment carried out with the dye that was bound to the preparation. Using axons and ganglia, we found the largest signals at wavelengths between 720 and 750 nm, while W. N. Ross and L. Reichardt (personal communication) found that the peak was near 675 nm in tissue cultured cells from the rat superior cervical ganglion. A simultaneous measurement of absorption and membrane potential during an action potential in a giant axon is shown in Figure 2. The dots are the light measurement; the thin line is the electrode measurement. Clearly, the absorption change has a time course very similar to that of the potential change. Because of this similarity it was likely that the change in absorption was signalling changes in membrane potential rather than ionic current or membrane permeability. The results of experiments that used voltage clamp steps were in agreement with this conclusion. The absorption change during both hyperpolarizing and depolarizing steps had the same shape as the potential (Figure 3 of Ross et al 1977). Although

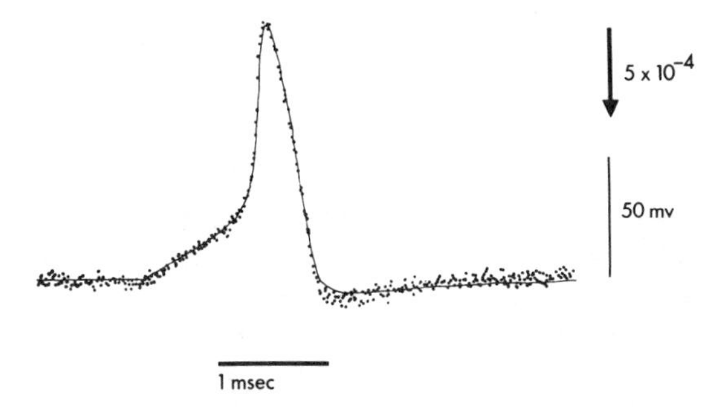

Figure 2 Changes in absorption (dots) of a stained giant axon during a membrane action potential (smooth trace). The change in absorption and the action potential had the same time course. (However, there seems to be a small but significant discrepancy between optical and electrode measurements at the end of the sweep. This may reflect a small, slow component in the dye XVII signal that is often seen in voltage-clamp experiments when relatively high dye concentrations are used.) The direction of the vertical arrow adjacent to the optical trace indicates the direction of an increase in absorption; the size of the arrow represents the stated value of a change in absorption, ΔA, in a single sweep divided by the resting absorption due to the dye, A_r. The response time constant of the light measuring system was 5 μsec, 32 sweeps were averaged. (From Ross et al 1977. The apparatus used in the axon experiments is described in detail in this paper.)

there were large ionic currents and large increases in permeability during the depolarization, they did not seem to affect the absorption signal. In experiments with four potential steps during a sweep, when the size of the absorption change was plotted against the size of the potential step the result shown in Figure 3 was obtained. Absorption changes and potential were linearly related over the range of potentials tested. While the relation between potential and absorption was linear when 1 msec steps were used, it is not known whether the same relation would be obtained for longer steps. A linear relation was also found for most other dyes, although for some dyes a more complicated result was obtained (see Figures 4 and 5 of Cohen et al 1974, Figure 4 of Ross et al 1977). While it had been suggested that some probes, especially anilinonapthalenesulfonates, exhibit fluorescence changes during action potentials that are not potential dependent (Tasaki et al 1969, Conti et al 1971, Tasaki, Watanabe & Hallett 1972), further experiments demonstrated that this suggestion was incorrect (Cohen et al 1970, Conti & Tasaki 1970, Patrick et al 1971, Davila et al 1974, Cohen et al 1974, Conti, 1975). Establishing potential dependence is of some interest in the context of monitoring neuron activity because a signal that is potential dependent should monitor synaptic and electrotonic potentials in the same way that it monitors action potentials.

SIGNAL SIZE

The absorption change illustrated in Figures 2 and 4 is small. The length of the arrow to the right of the optical traces represents the stated value of the change in absorption divided by the resting absorption in a single trial. Even for one of the best dyes, the fractional change in absorption during an action potential is between one part in 10^4 and one part in 10^3. In order to measure a synaptic potential the noise would have to be less than 10^{-5} of the resting intensity. Thus, signal size is at present a major factor limiting the usefulness of optical measurements of neuron activity. The increase in signal size since 1971 was important in enabling us to do the experiments illustrated in Figures 4, 5, and 6, but even more sensitive dyes would facilitate these and permit experiments now prohibitively difficult.

There are now a large number of intrinsic and extrinsic optical signals that might be used to monitor membrane potential. They have been classified into three types (Cohen & Salzberg 1978): 1. fast signals that depend on extrinsic dyes (e.g. Figure 2), 2. redistribution signals that result from the reequilibration of charged permeant dyes according to changes in membrane potential, and 3. intrinsic signals dependent upon changes in the native optical properties of neurons. Since the intrinsic signals are more than two orders of magnitude smaller than the largest dye signals (Cohen & Salzberg 1978), it would be difficult to use intrinsic signals for monitoring activity with the experimental apparatus in present use. While redistribution signals have been relatively large ($\sim$1.0% mV^{-1}) when compared to the fast signals ($\sim$0.01% mV^{-1}), the redistribution signals are much slower (Cohen & Salzberg 1978). If we use the signal size during a 3 msec step in a squid axon to rank probes, then the two shown in Figure 1 are about 5 times better than the best redistribution dye. With the dyes in Figure 1 it is possible to monitor single action potentials in a giant axon with a signal-to-noise ratio in the optical signal of better than 50:1 (Ross et al 1977).

The search for larger signals that led to the two dyes shown in Figure 1 has not been systematic. As soon as it was shown (Tasaki, Carnay & Watanabe 1969, Cohen

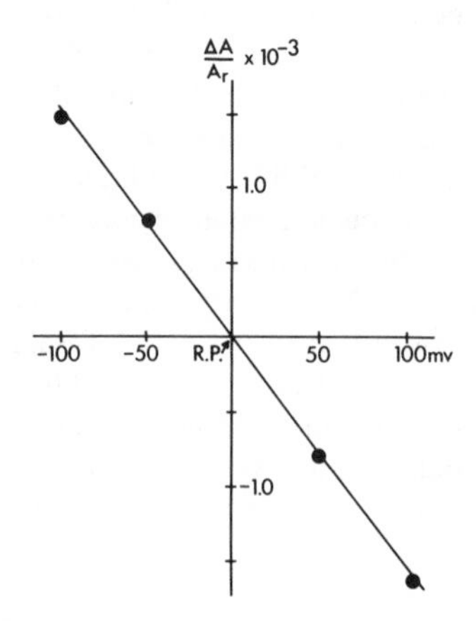

Figure 3 Change in absorption as a function of membrane potential in a squid axon. The absorption change was linearly related to the potential over the range tested. (From Ross et al 1977).

et al 1970) that some extrinsic signals were larger than the known intrinsic signals, we began trying commercially available dyes. These were mainly fluorescent dyes used in histology. Then we were given four photosensitizing dyes by Eastman Kodak and one of these, a merocyanine (now marketed as Merocyanine 540) had a relatively large signal (Davila et al 1973). Since we were not able to obtain additional dyes from Eastman Kodak, we became dependent upon our invaluable collaboration with Alan Waggoner and Jeff Wang, of Amherst College, for the synthesis of analogues. We soon found that two other classes of photosensitizing dyes, cyanines and oxonols, also gave relatively large signals (Cohen et al 1974). Later, Britton Chance pointed out that Nippon Kankoh-Shikiso Kenkyusho Co. Ltd. (NK) sold several thousand photosensitizing dyes. At about the same time Sims et al (1974) and Chance & Baltscheffsky (1975) found that there were changes in absorption that were concomitant with the changes in fluorescence. When we used the NK catalog and analogues synthesized by Wang and measured absorption as well as fluorescence, we found relatively large optical signals with the merocyanines shown in Figure 1. These merocyanines have absorption signals about 10 times larger than the signals obtained with Merocyanine 540 and cause much less photodynamic damage to the membrane (see below). We have now tried (Ross et al 1977) most of the dyes in the NK catalogue that seemed promising, so that almost all of our present effort is confined to dyes that are specially synthesized. These are analogues of the dyes shown in Table 2 of Ross et al (1977), with relatives of the merocyanines illustrated in Figure 1 predominating.

We have made many decisions about which direction to follow in searching for better dyes, without having good information for making the decision. For example, we decided to try extrinsic signals even though only a few of the possible intrinsic signals had been investigated. Then we decided to look at photosensitizing dyes after trying relatively few dyes of other classes, etc. This process of making decisions is analogous to a blind insect climbing a tree and making a random choice at each branch point. Eventually the insect reaches a fine twig and can go no further, but there is no way of knowing whether that twig is at the top or near the bottom of the tree.

Waggoner & Grinvald (1977) have attempted to calculate the maximum possible signal size and have predicted that the maximum fractional change in absorption is only three times larger than those presently available and that the maximum fluorescence signal-to-noise ratio is about 10 times larger than the present absorption signal-to-noise ratio. However it remains to be seen whether they have made correct assumptions in their calculations. The relative advantages of fluorescence and absorption measurements are discussed by Waggoner & Grinvald (1977).

MONITORING ACTIVITY IN INDIVIDUAL NEURONS OF AN INVERTEBRATE CNS

The results illustrated in Figures 2 and 3 show that for 1 msec steps an optical measurement is equivalent to an electrode measurement, and thus it should be possible to use an optical method to monitor activity in a neuron in a central nervous system. Two questions to be answered are:

1. Is the signal big enough to measure in single trials in a 50 μm neuron or will signal averaging be required?
2. Does an optical method have the spatial resolution required to distinguish signals from different neurons?

Since a 50 μm diameter cell body will have a membrane area of only 0.0075 mm^2 (assuming an equivalent sphere), it is expected that the signal-to-noise ratio will be about thirty times smaller than that found in experiments on giant axons. It was indeed difficult to measure an optical signal in single trials in neurons of the segmental ganglion of the leech, *Hirudo medicinalis,* using the dye Merocyanine 540 (Salzberg, Davila & Cohen 1973). When we used the dyes illustrated in Figure 1 it was less difficult, but still not easy (Salzberg et al 1977, Grinvald, Salzberg & Cohen 1977). Figure 4 illustrates optical and electrical signals during just subthreshold and just suprathreshold depolarizations in the supraesophageal ganglion of the barnacle, *Balanus nubilus.* The signal-to-noise ratio for the 60 mV action potential is about 10:1; signal averaging was not used.

With the signal-to-noise ratio of Figure 4, large synaptic potentials should also be monitored by the optical system. Figure 5 illustrates three pairs of simultaneous optical and electrode measurements of synaptic potentials. In the top pair, the relatively large inhibitory synaptic potentials are recorded by both measurements. In the middle pair the first and third excitatory synaptic potentials are recorded satisfactorily, but the second excitatory event is obscured by noise in the optical trace. In the bottom pair, both 4 mV inhibitory potentials are recorded optically, but the signals resulting from synaptic activity are not much larger than the noise in the measurement. Thus, under optimal conditions synaptic potentials of 4 mV could be monitored in single trials. The signal-to-noise ratios obtained during synaptic activity were those expected for an optical signal that depends only on potential.

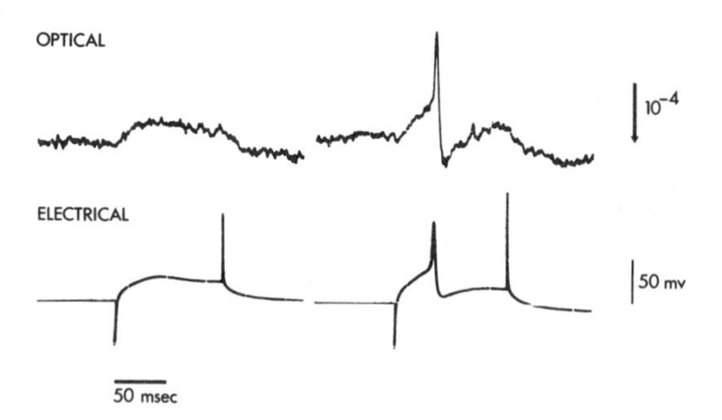

Figure 4 Optical recording of potential changes in a stained barnacle neuron (top traces) during subthreshold and suprathreshold depolarizations. Simultaneous microelectrode recordings are shown in the bottom traces. For both subthreshold and suprathreshold depolarizations the absorption changes have the same shape as the changes in membrane potential. In this and subsequent figures only a single trial was recorded; signal averaging was not used. The response time constant of the light-measuring system was about 600 μsec. (From Salzberg et al 1977. The apparatus used in the barnacle experiments and several of the steps we have taken to eliminate extraneous noise are described in this paper.)

If a presynaptic neuron were stimulated electrically, then signal averaging could be used and averaging n trials would reduce the smallest detectable synaptic potential by a factor of $n^{1/2}$.

A number of different methods might be used to measure the light from individual neurons. One simple way would be to position a light-guide right over the neuron itself. If the light-guide had a diameter that was identical to that of the neuron, then most of the light reaching the light-guide would have passed through (or been emitted from) the neuron. This method has the disadvantage that many small objects would have to be precisely positioned. A second method that would measure light from individual neurons would be to illuminate the preparation with a spot of light that was small compared to the neuron diameter and then rapidly, repeatedly scan the spot over the preparation. In such an apparatus, the time in the scan would specify position. In our experiments, we have used a third method. A microscope objective formed an enlarged image of the ganglion; light guides with diameters similar to the diameters of the neuron images were then used to carry the light to individual photodetectors (Salzberg et al 1977).

In order to show that the optical signal was restricted to the light at the image of an active cell, we stimulated one cell, measured the light from the image of that cell and simultaneously measured the light from the images of four immediately surrounding cells. The signal appeared only in the light from the stimulated cell and not in the light from its neighbors. Thus, in the plane of focus, the optical system had good resolution, and it was easy to determine which cell was active. The localization of the signal to the image of the active neuron implies that an individual

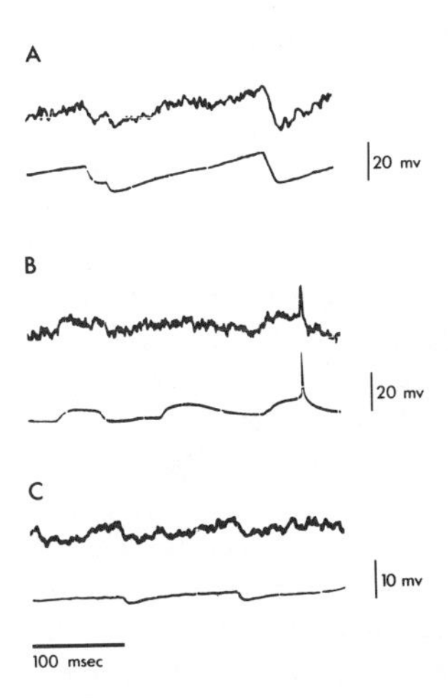

Figure 5 Attempts to use the absorption signal to monitor synaptic potentials in barnacle neurons. The relatively large synaptic potentials in *A* are recorded reliably, but one of the excitatory potentials in *B* was obscured by noise, and in trace *C* the signals in response to the 4 mV inhibitory synaptic potentials are not much larger than the noise in the trace. The response time constant of the light-measuring system was about 1.8 msec. (From Salzberg et al 1977).

light guide that happened to be positioned over parts of the images of two neurons would measure signals from the two different cells, or that two adjacent light guides that were positioned over the image of one neuron would both measure signals from that neuron. Results consistent with this explanation were obtained (Figure 4, Salzberg et al 1977). If a single photodiode detects activity in two neurons simultaneously, then the recording situation is similar to a multi-unit extracellular recording with the difference that the optical record gives information about the time courses of membrane potential changes rather than the time courses of current flow through the extracellular space.

While the resolution of the optical measurement in the focal plane was good, the resolution was poor in the direction perpendicular to that plane (z direction). The focus had to be raised or lowered by 500 μm to reduce the signal size to 10% (Salzberg et al 1977). While the resolution in the z direction can probably be improved by a factor of 5 with improvements in the optical system, it seems likely that in multilayered preparations, the optical recording will be a multi-unit recording. By recording at more than one depth of focus, it should be possible to assign activity to individual neurons.

Pharmacological Effects, Photodynamic Damage, and Dye Bleaching

We were concerned that the binding of dyes to neuronal membranes would have pharmacological effects, and we estimate, in fact, that about a third of the dyes tested did reduce the sodium or potassium currents and/or increase the leakage current in squid axons. However, neither of the dyes illustrated in Figure 1 exhibited these effects at the concentrations used, nor did they have marked effects on barnacle ganglia (Salzberg et al 1977, Grinvald & Cohen 1977). Similarly, no pharmacological effects have been noticed in skeletal muscle (S. M. Baylor & W. K. Chandler, personal communication), cardiac muscle (G. Salama & M. Morad, personal communication), or in tissue-cultured cells from the rat superior cervical ganglion (W. N. Ross & L. Reichardt, personal communication). However, in experiments carried out with D. Livingood there were pharmacological effects when the merocyanine-oxazolone (dye B, Figure 1) was added to lobster cardiac ganglia. In situations where pharmacological effects cause difficulty, they might be obviated by reducing the dye concentration. However, different dyes may have to be used, or an optical method involving extrinsic dyes may be inapplicable.

Although photodynamic damage (damage caused by intense illumination of stained preparations in the presence of oxygen) is found with most dyes, dyes differ by a factor of 100 in the amount of that damage. The molecules illustrated in Figure 1 are among the best in this regard. In experiments on stained squid axons it took about 1000 sec of illumination to cause a reduction of inward current by 50% (Ross et al 1977). In barnacle ganglia we were unable to detect changes in action potential height or width after 300 sec of illumination. Thus, even though earlier experiments using Merocyanine 540 were severely limited by photodynamic damage (Salzberg, Davila & Cohen 1973), this effect has not been a problem with the dyes shown in Figure 1.

With the light intensity used in the experiment shown in Figures 4–6, the signal size was reduced by 50% after three minutes of illumination (Salzberg et al 1977). We presume that this results from light-induced bleaching of the dye. However,

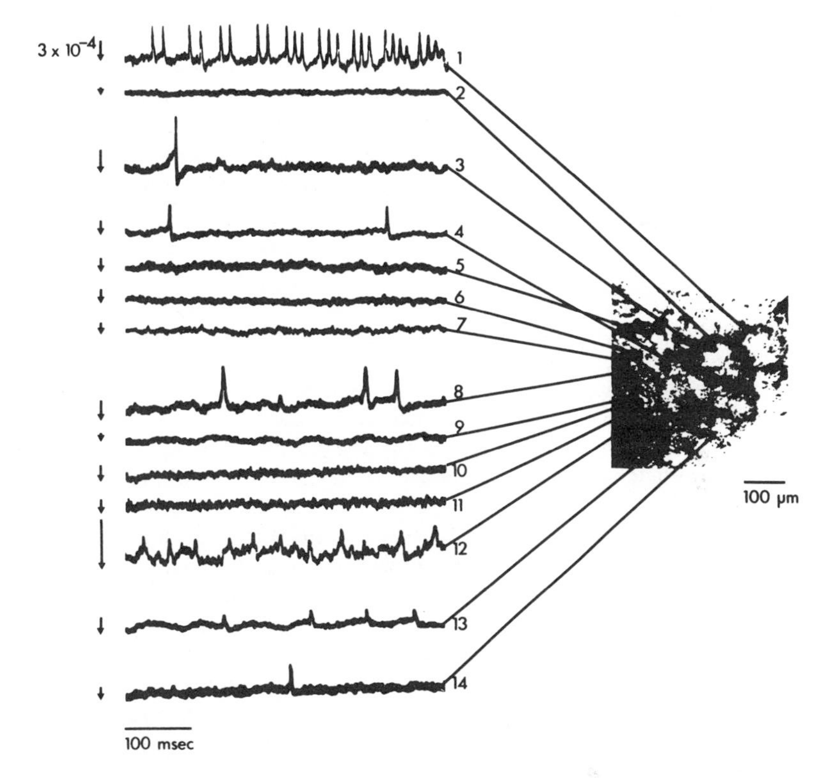

Figure 6 Composite record from a fourteen cell experiment using a barnacle supraesophageal ganglion in which the amount of spontaneous activity was relatively large. Not all neurons were spontaneously active at the same time so four trials measuring the output of fourteen detectors were recorded, and for each detector the trial with the most activity was selected. All four trials were carried out in a barnacle saline with one tenth the normal concentration of calcium and magnesium. The response time constants of the light-measuring systems were about 600 μsec. (From Salzberg et al 1977.)

after bleaching, the ganglion can be restained, and the signal returns to its original size.

SIMULTANEOUS MEASUREMENT OF ACTIVITY IN SEVERAL NEURONS

The apparatus used in the experiments illustrated in Figures 4 and 5 was designed to accomodate several light guide-photodetector combinations so that activity in several neurons could be monitored simultaneously. Because of limitations in the numbers of available amplifiers and oscilloscope traces, the number of cells that could be monitored simultaneously was only 14. It was quickly found that most of the neurons in the isolated supraesophageal ganglion were not spontaneously active, although most will generate action potentials (30–90 mV) when depolarized by passing current. It was also found that most of the neurons can be driven antidromically by stimulating a root (but not the ipsilateral connective). In 10 experiments, between 0 and 4 out of 14 neurons were spontaneously active at some time during four 500 msec trials. However, when the divalent cation concentration was reduced by 90%, the number of spontaneously active neurons increased. For the result shown in Figure 6, we recorded four 500 msec trials and, for each neuron, chose the trial with the most activity. Seven of the fourteen neurons were spontaneously active, some at relatively high rates.

PROGNOSIS

In addition to an optical method for monitoring activity in many cells, attempts have been made to use intracellular and extracellular electrode recordings to achieve the same result. For example, D. Spray, M. Spira, and M. V. L. Bennett (personal communication) have recorded simultaneously from eight neurons in the buccal ganglion of *Navanax* by using intracellular microelectrodes. However, an additional order of magnitude increase in the number of monitored cells will be difficult to attain with individual microelectrodes. Arrays of extracellular electrodes have been fabricated (Thomas et al 1972, Pickard & Welberry 1976) and recordings made from excitable tissue. At present it is not certain whether such a technique can be developed so that individual neurons can be identified and large numbers monitored, but the approach seems promising.

The simple apparatus used for the experiment illustrated in Figure 6 could probably be expanded to monitor 20–30 neurons, although positioning the light guides might become difficult. A larger increase in the number of monitored neurons will probably require some sort of imaging device or fixed detector array at the image plane. Among the many options are videcons (or other imaging tubes), image intensified videcons, charge-coupled devices, or a closely packed bundle of light guides that fans out to individual photodiodes. (If an order-of-magnitude increase in signal size could be attained, then decisions about imaging devices would be less critical; a further order-of-magnitude increase would allow the use of a movie camera to monitor activity). With the signal presently available we expect that it should be possible to construct an optical apparatus that monitors 50–100 neurons. If 100 neurons can be monitored, this would be a large fraction of the total number of neurons in some preparations (e.g. barnacle supraesophageal ganglion or the leech segmental ganglion) and some experiments that are presently too tedious to carry out with microelectrodes might then be feasible. It should be relatively easy to locate all the neurons involved in a particular behavior. Or, simply impaling and stimulating the neurons one at a time and recording from all the other neurons would allow the determination of a relatively complete wiring diagram for the ganglion. Knowledge of the wiring diagram should make it easier to interpret experiments carried out on semi-intact preparations where all the neurons are monitored while the ganglion is controlling behavior. It might then be possible to study changes in the nervous system that occur when behavior is modified.

Although it is easy to suggest possible experiments, it is not clear which preparation and which kind of behavior will provide meaningful results. The ability to record from a large number of neurons could lead to data that are so complicated that they will be impossible to interpret. However, one might be able to find a preparation where most neurons are normally silent and where the simplest behaviors utilize a small number of cells. In such a situation, analysis should be feasible.

Four additional applications of optical methods for monitoring neuron activity are possible. Since the technique might allow a relatively complete determination of the synaptic connections of an identified neuron, it may be possible to study the dynamics and specificity of regeneration in considerable detail. One of us (AG) is planning experiments in this direction.

If a 60 mV action potential in a 50 μm neuron can be monitored with a signal-to-noise ratio of about 10 : 1, then, on the assumption of identical membrane properties, and even without averaging, a signal-to-noise ratio of 1 : 1 would be obtained in a 5 μm neuron and, with averaging, even smaller cells might be monitored. However, this extrapolation to smaller neurons has not yet been checked.

If a sensitive dye were microinjected inside a neuron, and if the dye spread throughout the cell, the time course of the synaptic potential could be monitored at the site of the synapse, along the dendritic tree, and in the cell body. With the ability to monitor membrane potential in dendritic branches, understanding of the integrative function of individual neurons might be greatly improved. Although some dyes do monitor membrane potential when injected into squid axons, we do not know of any efforts to find dyes that would work especially well in this application. One of us (BMS) is planning experiments in this direction.

There is only a preliminary report (Chance, Mayevsky & Smith 1976) of attempts to use optical methods to monitor activity in vertebrate central nervous systems, and no effort was made to record from individual neurons or from a small number of neurons. Although we have no experience in this area, it seems possible that an optical method for monitoring activity might also be useful in vertebrate central nervous systems.

Acknowledgments

We thank Sol Erulkar, Charles Michael, and Dick Orkand for helpful discussion. Our research has been generously supported by grants from the Public Health Service (NS 08437, NS 12253, and MH 27853), by an NIH postdoctoral traineeship, and by a fellowship from EMBO.

NOTE ADDED IN PROOF More stringent tests for pharmacological effects of the dyes in Figure 1 have been carried out on the barnacle supraesophageal ganglion. Using a light stimulus to the median ocellus and recording from high order neurons with suction electrodes on the connective and antennular nerves, it was found that a 20 min incubation at 1 mg ml^{-1} frequently led to an irreversible elimination of the off response. Although lower concentrations or a reduced staining period also blocked the off response, this effect was reversed when the dye solution was replaced with saline (A. Grinvald, S. Lesher, and L. B. Cohen, unpublished results). While it is not yet known whether adequate optical signals will be obtained from preparations that are less stained, we are optimistic that, in the long run, suitable dyes can be found by screening existing and new compounds for both signal size and pharmacologic effect.

Literature Cited

Chance, B., Baltscheffsky, M. 1975. Carotenoid and merocyanine probes in chromatophore membranes. *Biomembranes* 7:33–60

Chance, B., Mayevsky, A., Smith, J. C. 1976. Localized and delocalized potentials in the rat brain cortex. *Neurosci. Abstr.* 2:133 (Abstr.)

Cohen, L. B. 1973. Changes in neuron structure during action potential propagation and synaptic transmission. *Physiol. Rev.* 53:373–418

Cohen, L. B., Keynes, R. D., Hille, B. 1968. Light scattering and birefringence changes during nerve activity. *Nature* 218:438–41

Cohen, L. B., Landowne, D., Shrivastav, B. B., Ritchie, J. M. 1970. Changes in fluorescence of squid axons during activity. *Biol. Bull. Woods Hole* 139: 418–19 (Abstr.)

Cohen, L. B., Salzberg, B. M. 1978. Optical measurement of membrane potential. *Rev. Phys. B.* In press

Cohen, L. B., Salzberg, B. M., Davila, H. V., Ross, W. N., Landowne, D., Waggoner, A. S., Wang, C. H. 1974. Changes in axon fluorescence during activity: molecular probes of membrane potential. *J. Membr. Biol.* 19:1–19

Conti, F. 1975. Fluorescent probes in nerve membranes. *Ann. Rev. Biophys. Bioeng.* 4:287–310

Conti, F., Tasaki, I. 1970. Changes in extrinsic fluorescence in squid axons during voltage clamp. *Science* 169:1322–24

Conti, F., Tasaki, I., Wanke, E. 1971. Fluorescence signals in ANS-stained squid giant axons during voltage clamp. *Biophysik* 8:58–70

Davila, H. V., Cohen, L. B., Salzberg, B. M., Shrivastav, B. B. 1974. Changes in ANS and TNS fluorescence in giant axons from *Loligo. J. Membr. Biol.* 15:29–46

Davila, H. V., Cohen, L. B., Waggoner, A. S. 1972. Changes in axon fluorescence during activity. *Biophys. J.* 12:124a (Abstr.)

Davila, H. V., Salzberg, B. M., Cohen, L. B., Waggoner, A. S. 1973. A large change in axon fluorescence that provides a promising method for measuring membrane potential. *Nature New Biol.* 241: 159–60

Grinvald, A., Cohen, L. B., 1977. Optical monitoring of activity in barnacle neurons in response to light stimulation of the median photoreceptors. *Neurosci. Abstr.* 3:178 (Abstr.)

Grinvald, A., Salzberg, B. M., Cohen, L. B. 1977. Simultaneous recording from several neurons in an invertebrate central nervous system. *Nature* 268:140–42

Patrick, J., Valeur, B., Monnerie, L., Changeux, J.-P. 1971. Changes in extrinsic fluorescence intensity of the electroplax membrane during electrical excitation. *J. Membr. Biol.* 5:102–20

Pickard, R. S., Welberry, T. R. 1976. Printed circuit microelectrodes and their application to honeybee brain. *J. Exp. Biol.* 64:39–44

Ross, W. N., Salzberg, B. M., Cohen, L. B., Grinvald, A., Davila, H. V., Waggoner, A. S., Wang, C. H. 1977. Changes in absorption, fluorescence, dichroism and birefringence in stained giant axons: optical measurement of membrane potential. *J. Membr. Biol.* 33:141–83

Salzberg, B. M., Davila, H. V., Cohen, L. B. 1973. Optical recording of impulses in individual neurons of an invertebrate central nervous system. *Nature* 246: 508–9

Salzberg, B. M., Grinvald, A., Cohen, L. B., Davila, H. V., Ross, W. N. 1977. Optical monitoring of neuronal activity in an invertebrate central nervous system: simultaneous monitoring of several neurons. *J. Neurophysio.* 40:1281–92

Sims, P. J., Waggoner, A. S., Wang, C. H., Hoffman, J. F. 1974. Studies on the mechanism by which cyanine dyes measure membrane potential in red blood cells and phosphatidylcholine vesicles. *Biochemistry* 13:3315–30

Tasaki, I., Carnay, L., Watanabe, A. 1969. Transient changes in extrinsic fluorescence of nerve produced by electric stimulation. *Proc. Natl. Acad. Sci. USA* 64:1362–68

Tasaki, I., Watanabe, A., Hallett, M. 1972. Fluorescence of squid axon membrane labelled with hydrophobic probes. *J. Membr. Biol.* 8:109–32

Tasaki, I., Watanabe, A., Sandlin, R., Carnay, L. 1968. Changes in fluorescence, turbidity, and birefringence associated with nerve excitation. *Proc. Natl. Acad. Sci. USA* 61:883–88

Thomas, C. A. Jr., Springer, P. A., Loeb, G. E., Berwald-Netter, Y., Okun, L. M. 1972. A miniature microelectrode array to monitor the bioelectric activity of cultured cells. *Exp. Cell Res.* 74:61–66

Waggoner, A. S. 1976. Optical probes of membrane potential. *J. Membr. Biol.* 27:317–34

Waggoner, A. S., Grinvald, A. 1977. Mechanisms of rapid optical changes of potential sensitive dyes. *Ann. NY Acad. Sci.* In press

Ann. Rev. Neurosci. 1978. 1:183–214

REGULATION OF AUTONOMIC DEVELOPMENT

◆11508

Ira B. Black

Department of Neurology, Cornell University Medical College, 1300 York Avenue, New York, NY 10021

INTRODUCTION AND OVERVIEW

The autonomic nervous system develops at the interface between the external environment and the basic drives of the individual such as feeding, aggressivity, and sexuality. Autonomic maturation results in the translation of emotional states into appropriate physiologic responses, leading to the integration of internal homeostasis with emotionality and environmental demands. Survival of the species, as well as the individual organism, depends upon the autonomic system, which subserves these critical life functions. It is useful to be aware of these general considerations when one discusses molecular mechanisms underlying development: processes governing autonomic maturation in the periphery undoubtedly occur in the far more complex central nervous system (CNS), and derangement of these processes results in human disease.

The devastating effects of abnormal autonomic development are well documented in man. For example, the genetic disorder, familial dysautonomia is characterized by vasomotor instability, feeding difficulties, relative indifference to pain, cyclic vomiting, cyclic fevers, ataxia, and recurrent pulmonary infections; it may cause death in infancy or childhood (Brunt & McKusick 1970). Autonomic degeneration in adulthood may be associated with postural hypotension resulting in fainting, urinary and fecal incontinence, sexual impotence, loss of sweating, cranial nerve palsies, and a disorder of movement resembling Parkinson's disease (Bradbury & Eggleston 1925, Shy & Drager 1960). The biochemical mechanisms underlying these syndromes (Black & Petito 1976, Siggers et al 1976) may be elucidated by studying autonomic ontogeny, as described below.

The autonomic system in the periphery (Giacobini 1970, Giacobini et al 1970) provides relatively simple models of neuronal development compared to the CNS. Even the simplest brain nuclei contain heterogeneous groupings of cells that differ

0147-006X/78/0325-0183$01.00

morphologically, biochemically, and probably functionally. Although development in the CNS has been documented according to anatomical (Peters & Flexner 1950, Eayrs & Goodhead 1959), ultrastructural (Bunge & Bunge 1965, Aghajanian & Bloom 1967), electrophysiological (Deza & Eidelberg 1967), and biochemical (Hebb 1956, Lognado & Hardy 1967) approaches, studies have been largely descriptive due to the obvious complexity of the system. In contrast, sympathetic ganglia in mouse, rat, and cat contain primarily two neural elements that are in synaptic contact: presynaptic cholinergic nerve terminals and postsynaptic adrenergic neurons (Giacobini 1970, Giacobini et al 1970). More specifically, the well-characterized, relatively simple superior cervical ganglion (SCG) is ideal for the study of neuronal growth and development because the SCG is composed of biochemically distinct, well-defined neural elements consisting primarily of the cholinergic-adrenergic neural unit defined above. In addition, recent studies have indicated the presence of low numbers of small neurons (Williams & Palay 1969), adrenergic fibers (Hamberger et al 1963), and scattered cholinergic cells (Sjoqvist 1962) in sympathetic ganglia. The SCG is anatomically discrete and easily accessible; and its bilaterally symmetric nature allows rigorously controlled experiments within a single animal (Black et al 1971b). Ontogenetically and anatomically there is no fundamental difference between the autonomic and somatic systems; neurons of the latter arise without interruption from the neural crest (Monnier 1968). Hence, while the autonomic system is less complex than central models, data derived from its study may define mechanisms governing growth and development throughout the nervous system.

Specific biochemical markers may be employed to study sympathetic ontogeny. Choline acetyltransferase (ChAc), the enzyme that catalyzes the synthesis of acetylcholine (Fonnum 1970), is highly localized to presynaptic sympathetic terminals in ganglia (Hebb & Waites 1956) and may be used to monitor maturation of these elements (Black & Geen 1973). On the other hand, tyrosine hydroxylase (T–OH), the rate-limiting enzyme in catecholamine biosynthesis (Levitt et al 1965), is localized to postsynaptic adrenergic neurons (Black et al 1971a) and may be employed as an index of development of these cells (Black & Geen 1973). T–OH catalyzes the conversion of tyrosine to L-dopa, the first step in catecholamine synthesis (Levitt et al 1965). Dopa decarboxylase (DDC), which converts L-dopa to dopamine, and dopamine-β-hydroxylase (DBH), which converts dopamine to norepinephrine (NE), are also highly restricted to postsynaptic neurons and may serve as indices of adrenergic ontogeny (Black et al 1971a, Molinoff & Axelrod 1971). Precise biochemical-morphological correlation is made possible by the availability of histofluorescent techniques for the visualization of catecholamines in situ (Falck et al 1962). To a more limited degree, parasympathetic development may be analyzed by analogous methods (Giacobini 1970, Pilar et al 1973, Landmesser & Pilar 1974, Landmesser & Morris 1975).

A formidable array of practical questions and theoretical issues remains to be elucidated by the foregoing approaches. It may be useful to enunciate a number of these questions before proceeding to detailed discussion in the following pages. What are the mechanisms by which cells of the neural plate become progenitors of the neural crest, primordium of the autonomic nervous system? What factor(s) govern neural crest migration; what is the relationship between migration and terminal mitosis; and how does the somitic mesenchyme influence the migrating cells? What intrinsic or extrinsic information guides neuroblasts to their definitive locations, and how does stable association of heterogeneous cellular groups occur at these distant sites? At the final loci, what are the influences of support cells and humoral agents; what factors regulate afferent and efferent synaptogenesis; and how does the neuron acquire information defining the field of innervation? What molecular mechanisms underlie the anterograde and retrograde transsynaptic regulation of development, which occurs at a number of levels of the autonomic neuraxis? What is the role of cell death in normal ontogeny? What regulatory mechanisms, at what critical stage(s) of development, determine the neurotransmitter fate(s) of autonomic neurons?

Partial answers exist to a number of these questions. It is known that (*a*) the neural crest arises from neural plate and neural tube, (*b*) crest cells migrate laterally undergoing critical interactions with somitic mesenchyme, which may influence transmitter fate, and (*c*) neuroblasts arrive at their final sites partially differentiated. It has been demonstrated that autonomic neurons require macromolecules such as nerve growth factor (NGF) for survival, that hormones such as thyroxine are critical for normal differentiation, and that bidirectional regulation at the synapse controls normal differentiation. Recent work has indicated that descending central pathways influence autonomic ontogeny. Cellular interactions between support cells and neurons may also contribute to survival and maturation. In summary, growth and development involves a series of complex interactions between the genetic information encoded within the primitive neuron and the environment.

The present review cannot address all of the foregoing questions. Rather, the sympathetic nervous system is employed as the primary model to examine a number of these issues. We concentrate on mechanisms governing ontogeny in vivo, since other chapters in this volume discuss in vitro systems. However, reference is made to tissue culture work to elucidate in vivo regulation. It is probable that many of the mechanisms defined will be applicable to the nervous system in general. Related insights derived from study of other portions of the nervous system are described. This review approaches autonomic development chronologically, as it occurs in the maturing individual, to define sequential regulatory mechanisms that govern normal ontogeny. After describing the derivation and migration of neural crest cells, regulation within autonomic ganglia is considered, including target organ and endocrine influences, and the effects of drugs. Lastly, autonomic function during adulthood is examined as an extension of the processes governing ontogeny. In this context, transsynaptic regulation, the function and role(s) of NGF, and the phenomenon of regeneration after injury are considered. Several relevant reviews (Guth 1968, Cowan 1970, Molinoff & Axelrod 1971, Zaimis & Knight 1972, Gaze & Keating 1974, Harris 1974, Burnstock & Costa 1975, Purves 1976a, Smith & Kreutzberg 1976, Hendry 1976, Diamond 1976) and excellent texts (Gaze 1970, Jacobson 1970) are recommended.

PRENATAL DEVELOPMENT

The Neural Crest

In pioneering work, His originally described the neural crest in 1868 as a band of material lying between the presumptive epidermis and the neural tube (His 1868). Evidence acquired over the ensuing century has led to general agreement that the neural crest constitutes the progenitor of the peripheral autonomic nervous system. His (1868) and Fusari (1893) initially suggested that the neural crest gives rise to sympathetic neurons, cells of the adrenal medulla, and sensory ganglia. The work of Kuntz and Tello in avian systems supported this contention (Kuntz 1910, Tello 1925), as did the experiments of Müller & Ingvar (1923), Willier (1930), and Brauer (1932). Twenty-five years later, employing *amblystoma* embryos, Detwiler was able to document the development of the neural crest from the neural plate, neural folds, and neural tube (Detwiler 1937). Using a vital stain for crest cells, he definitively demonstrated that the sympathetic nervous system arises from the neural crest (Detwiler 1937). More recently, a series of studies by Hammond and Yntema have supported these contentions. Extirpation of the thoracolumbar neural crest in chick embryos results in absence of peripheral adrenergic neurons in this area and prevents the normal development of chromaffin cells of the adrenal medulla and preaortic areas (Hammond & Yntema 1947). These authors also noted the development of scoliosis, abnormal curvature of the vertebral column, a sign associated with familial dysautonomia in humans (Brunt & McKusick 1970). In other work, Yntema and Hammond demonstrated that removal of cervico-occipital crest prevents the appearance of sympathetic neurons in the cervical area, including the superior cervical ganglion, as well as the esophageal and gastric enteric plexuses (Yntema & Hammond 1945). In virtually simultaneous studies van Campenhout documented the formation of parasympathetic ganglia from neural crest (van Campenhout 1946).

The neural crest itself is first discernible at the neurula stage of development in the edges of the neural folds, where it occurs as a peripheral band on either side of the neural plate (Saxén & Toivonen 1962). As the neural plate closes to form the neural tube, the crest cells assume a dorsal midline position. The mechanisms underlying the differentiation of neural crest cells from neural tube are unknown, although it has been suggested that lateral sectors of the archenteron roof induce crest formation (Raven & Kloos 1945). Raven suggested that all of the ectoderm of the neural plate–neural tube region is presumptive neural crest (Raven 1931), but definitive evidence is lacking.

From the dorsal midline position within the neural tube, the crest cells soon migrate ventrolaterally to form columns on either side of the neural tube (Detwiler 1934). Different longitudinal regions of the crest give rise to different neural and nonneural structures (Hörstadius 1950).

Utilizing an autoradiographic approach, Weston and his colleagues were able to amplify these observations by describing the process of migration of neural crest cells (Weston 1963, Weston & Butler 1966, Weston 1970). Cells labeled with ^{3}H-thymidine migrate in two "sheets," dorsolaterally within the ectoderm and ventrolaterally within the mesoderm (Weston 1963). Grafting of newly condensed crest into progressively older hosts results in progressive attenuation of migration, which suggests that the embryonic microenvironment may be critical in regulating migration patterns (Weston & Butler 1966).

These cellular interactions appear to play a major role in differentiation of the neural crest. It may be recalled that crest gives rise not only to neurons, but also to chromatophores, nonneuronal cells of the peripheral nervous system, chromaffin cells, calcitonin-producing cells, and mesenchymal derivatives of the cephalic region (Coulombre et al 1974). Although the mechanisms governing these various differentiative processes are not fully understood, recent work has begun to provide important clues. The major pathways of *neuronal* differentiation lead to development of sensory neurons (Tennyson 1965) and autonomic neurons (Pick 1963). In turn, autonomic neurons may become cholinergic and/or adrenergic. What factors regulate this critical choice of neurotransmitters? In the chick embryo, appearance of catecholamine histofluorescence in presumptive sympathoblasts is due to interactions with somitic mesenchyme and therefore requires ventral crest migration (Cohen 1972, 1973). Moreover, catecholamine histofluorescence may appear while sympathoblasts are still undergoing mitosis, as judged by ^{3}H-thymidine incorporation. In addition, ablation of ventral neural tube reduces the quantity of nervous tissue formed, which suggests that neural tube, somitic mesenchyme, and sympathoblasts undergo determinative interactions (Cohen 1972, 1973). This contention is supported by in vitro studies of chick neural crest that demonstrate that for catecholamine differentiation crest cells must be contiguous with the somite (Norr 1973). Ventral neural tube may act across a millipore filter to induce appropriate changes in the somite (Norr 1973). In addition, the continued presence of ventral neural tube is required for survival of the differentiating sympathetic neurons. Ventral neural tube may be replaced by NGF in assuring *survival* of the differentiating sympathetic neurons (Norr 1973). These observations suggest that ventral neural tube induces changes in somitic mesenchyme, which then promotes sympathoblast differentiation.

Le Douarin and colleagues have analyzed migration and differentiation by using chimeras in which quail crest cells have been transplanted to chick embryos (Le Douarin & Teillet 1974, Le Douarin et al 1975, Le Douarin & Rival 1975). The quail cells may be conveniently identified by their unique nuclear morphology. Normally, in neural crest the trunk region gives rise to adrenergic sympathetic ganglion and adrenal medullary cells, whereas the "vagal" region (somites 1–7) gives rise to cholinergic enteric ganglion cells. However, if trunk (sympathoadrenal) crest of quail is grafted to the "vagal" region of chick, the crest cells colonize the gut to form cholinergic enteric ganglion cells. If, on the other hand, cephalic crest is transplanted to the "adrenomedullary level," it becomes adrenergic adrenomedullary chromaffin tissue. In contrast, some transplanted cephalic neural crest still differentiates into mesenchymal derivatives even when transplanted to the trunk area (Le Douarin & Teillet 1974). It may be concluded that preferential migratory pathways are located at precise levels of the crest in the embryo and lead cells to their definitive sites. Moreover, the expression of a given phenotype may be regulated by the environment of the definitive site and by cells encountered *en route.* However, the fates of certain derivatives are determined extremely early during development, may be relatively immutable, and may define the pattern of migration.

The problem of neurotransmitter determination may be more complicated than indicated above. In this regard, the work of Ignarro and Shideman with chick embryos is extremely provocative (Ignarro & Shideman 1968). These investigators studied the temporal appearance of the catecholamine-synthesizing enzymes in whole embryo, using embryonic synthesis of transmitters from ^{3}H-tyrosine as an index. The chick embryo converted ^{3}H-tyrosine to ^{3}H-L-dopa on the first day of incubation. This initial, rate-limiting step in catecholamine biosynthesis is catalyzed by T–OH. Thus, T–OH activity is phenotypically expressed from day one of incubation, far before crest cells have reached their definitive sites, and even before migration may have begun. The presence of endogenous L-dopa on the first day of incubation further supports the contention that T–OH is present and functional. These investigators also noted the sequential appearance of T–OH, DDC, DBH, and phenylethanolamine-N-methyltransferase activities on the first, second, fourth, and sixth days of incubation. These enzymes, in sequence, catalyze catecholamine biosynthesis. It would be of extreme interest to define the localization of these enzymes, since recent work suggests that the notochord may synthesize and store catecholamines in chick embryo (Kirby & Gilmore 1972, Lawrence & Burden 1973). Regardless of the cellular localization of the enzyme activities, this work demonstrates that catecholamine enzymes appear sequentially, not in an all-or-none fashion. Moreover, since mature neurotransmitter function is dependent on biosynthetic enzymes, storage mechanisms, physiologic transmitter release, high affinity reuptake processes, and degradative enzymes, it is unlikely that differentiation is a simple all-or-none process. Rather, the work of Ignarro and Shideman suggests that expression of transmitter fate is a progressive process with the appearance of different enzyme gene products at different stages of development. Viewed in this light, it may be extremely difficult to isolate *the* time-frame in which a neuron "becomes adrenergic," or loses "pluripotentiality." It would be extremely interesting to determine when other transmitter enzymes first appear in the chick embryo, and whether enzymes for a number of transmitters normally coexist in the cell at a primitive stage of ontogeny. Such information may help determine whether differentiation simply involves the appearance of critical enzymes and/or the disappearance of others.

Autonomic Ganglia

A large segment of the ventrally migrating crest cells coalesce dorsolateral to the aorta, forming the primary sympathetic chains from which the sympathetic ganglia and adrenal medulla are derived (Yntema & Hammond 1945). Sympathetic ganglia may be visualized in the chick at 3.5 days of incubation, when catecholamine histofluorescence is present (Enemar et al 1965). In the 13- to 15-day rat embryo small fluorescent sympathoblasts with short processes are segmentally arranged dorsally (de Champlain et al 1970, Owman et al 1971), while specific histofluorescence is present in rabbit sympathoblasts at 14 days (Papka 1972). Sympathetic and sensory ganglia are not simply derived from a single progenitor crest cell, since ganglia have recently been shown to consist of two cellular genotypes in allophenic mice, produced from artifical composites of genetically different blastomeres (Dewey et al 1976). Moreover, multiple interactions *within* ganglia are possible: Cell bodies and processes of sympathetic neurons establish extensive contact with each other through soma-somatic, dendro-somatic, and dendro-dendritic contacts (Grillo 1966). In addition, as discussed below, support cell-neuron interactions may be critical for normal development.

The morphologic and biochemical maturation of the embryonic mouse SCG has recently been characterized in vivo and in tissue culture (Coughlin et al 1977). From 13 days of gestation, when the SCG is first visible, to birth at 19 days, T–OH activity increases 30-fold in vivo. Explants of ganglia from 14-day embryos exhibit abundant neurite outgrowth in basal medium without added NGF, and increases in T–OH activity parallel that observed in vivo. Ganglia from 14-day embryos elaborate neurites and exhibit marked increases in enzyme activity over 72 hr in vitro even in the presence of antiserum to NGF (anti-NGF) or NGF + anti-NGF. In direct contrast, ganglia from 18-day fetuses fail to grow without added NGF or in medium containing anti-NGF or NGF + anti-NGF: Virtually no axon outgrowth occurs, and T–OH activity decreases by half. These observations suggest that developmental regulatory mechanisms change radically during embryologic and fetal life of the mammalian SCG (Coughlin et al 1977). NGF may not be an absolute requirement for differentiation of the 14-day ganglion. However, other explanations may be considered. It is possible that in 14-day, but not 18-day, ganglia, support cells produce NGF or an NGF-like substance. This appears unlikely, since sympathetic neurons develop normally even in the presence of anti-NGF, and since previous work has demonstrated that anti-NGF attenuates the influence of support cells (Varon et al 1974). Nevertheless, it is conceivable that support cells directly transfer NGF to neurons in an antibody-resistant form in 14-day, but not 18-day, ganglion cultures. Alternatively, different subpopulations of neurons may develop at different times in ganglia. Thus the 14-day ganglion may contain a subpopulation that differentiates independently of NGF, as well as one that requires NGF. By 18 days of gestation, the entire ganglion population may require NGF. Such a postulate would imply, however, either disappearance of the NGF-independent population by 18 days of gestation or the development of an NGF requirement in this population. It may be more likely that initial biochemical and morphologic development in the ganglion does not require NGF. Regardless of the mechanisms that ultimately prove to be involved, these observations suggest that regulatory influences differ markedly in embryologic and late-fetal ganglia.

The role of support cells in neuronal differentiation has received increasing attention. The effects of NGF on sensory neuron survival and fiber outgrowth in vitro can be reproduced by increasing nonneuronal cell numbers (Burnham et al 1972). In addition, nonneuronal cell numbers may determine whether, and to what extent, NGF can support neurite elaboration (Varon et al 1974), which raises the possibility that support cells also exert a non-NGF function. It is now well established that NGF may be produced by nonneuronal cells, including glia (Monard et al 1973) and fibroblasts (Oger et al 1974, Young et al 1975). Furthermore, the percentage of chick sensory neurons differentiating in culture may be increased by the presence of a layer of glia or heart fibroblasts (Ludueña 1973).

Preganglionic Neurons

The preganglionic cholinergic fibers innervating autonomic ganglia arise from perikarya lying in the intermediolateral column (ILC) and intermediomedial column of the spinal cord (Chung et al 1975). These preganglionic neurons undergo early development in the absence of the neural crest and its embryonic derivatives (Hammond & Yntema 1947): These cells originate from neural tube, not neural crest (Terni 1923). In the chick embryo of up to 4.5 days of incubation, a uniform ventrolateral motor column extends throughout the length of the spinal cord, consisting of all preganglionic visceral and somatic motor neuroblasts (Levi-Montalcini 1950). The thoracolumbar sympathetic outflow segregates by cell migration in a medial and dorsal direction and reaches its definitive location at eight days to form the nucleus of Terni (Levi-Montalcini 1950). The preganglionic cell column and its fibers undergo initial differentiation even in the absence of neural crest derivatives. However, evidence presented more than 25 years ago indicates that fiber elongation is dependent on the presence of peripheral sympathetic ganglia (Yntema & Hammond 1945).

As in many other systems, massive cell death is a normal feature of ILC development (Levi-Montalcini 1950). However, Shieh (1951) demonstrated that transplantation of cervical spinal cord to thoracic levels in embryos leads to increased survival of ILC cells. Many of the neurons sent fibers to peripheral ganglia in that region, which suggests that the field of innervation may regulate survival. Moreover, selective postnatal destruction of postsynaptic adrenergic neurons with 6-hydroxydopamine (6-OHDA) or anti-NGF prevents the normal development of presynaptic ChAc activity (Black et al 1972a). Inhibition of ChAc development after adrenergic destruction is, at least in part, due to decreased survival of ILC neurons (Caserta et al 1976).

Shieh's work implies that the specificity of ganglion innervation by ILC fibers is limited, and that ganglia may be innervated by fibers of different sources. It is clear that some degree of specificity does exist, since spinal cord will innervate rat SCG in tissue culture, whereas cerebral cortex will not (Olson & Bunge 1973). However, since the original studies of vagal anastomoses by Langley (1898), it has been apparent that sympathetic ganglia may be *reinnervated* by preganglionic fibers of various origins. Thus, vagal anastomosis to the denervated SCG results in electrically functional, anatomically demonstrable formation of synapses (Ceccarelli et al 1971) that are as stable as native synapses (Purves 1976b) but do not subserve normal transsynaptic enzyme induction (Östberg et al 1976). In addition to the vagus, hypoglossal nerves (Östberg et al 1976) and somatic nerves (McLachlan 1974) also reinnervate the SCG. Foreign axons form as many synapses per axon as do normal preganglionic axons (Östberg et al 1976).

In summary, ILC neurons probably undergo initial development, and perhaps extension of fibers, even in the absence of postsynaptic neuronal targets. However, fundamentally peripheral event(s), probably dictated by target neurons, determine

ILC cell survival and/or differentiation. Additionally, as will be discussed below, descending central influences regulate ILC neuron differentiation at a later stage of maturation.

POSTNATAL DEVELOPMENT

Anterograde Transsynaptic Regulation in Ganglia

Transsynaptic degeneration is a well-documented phenomenon in the central and peripheral nervous systems (for review see Cowan 1970). Transsynaptic processes are critical for normal function during adulthood as well as normal development during the perinatal period. One approach to the study of developmental transsynaptic mechanisms has involved use of the SCG in the neonatal mouse and rat. ChAc, the enzyme catalyzing the formation of acetylcholine, is highly localized to presynaptic terminals in the ganglion (Hebb & Waites 1956) and may be used to monitor maturation of these elements (Black et al 1971b). T–OH, the rate-limiting enzyme in catecholamine biosynthesis, is localized to postsynaptic adrenergic neurons in the SCG (Black et al 1971b). Studies have indicated that there is a marked postnatal increase in ganglion synapses that is paralleled by a 40- to 50-fold developmental rise of ChAc activity (Black et al 1971b). This suggests that the development of ChAc activity in the SCG reflects synapse formation, which is in agreement with analogous studies in the spinal cord (Burt 1968) and ciliary ganglion (Sorimachi & Kataoka 1974, Burt & Narayanan 1976).

Transection of the preganglionic cholinergic trunk in neonatal mice or rats prevents the normal postnatal six- to ten-fold increase in postsynaptic T–OH activity (Black et al 1971b, 1972b). The effects of ganglion decentralization are reproduced by ganglionic blocking agents, which compete with acetylcholine for postsynaptic nicotinic receptor sites (Black 1973, Hendry 1973, Black & Geen 1973, 1974). However, atropine, a muscarinic antagonist, has no effect on postsynaptic maturation (Black & Geen 1974). These observations suggest that presynaptic terminals, through the mediation of the nicotinic properties of transsynaptic acetylcholine, regulate postsynaptic ontogeny. Although acetylcholine may be necessary for transsynaptic regulation, it may not be sufficient, since acetylcholine or its agonists cannot replace presynaptic terminals in this regulation (Black et al 1972b). Consequently, other transsynaptic factors may also be involved, as has been suggested for the neurotrophic regulation of striated muscle (Smith & Kreutzberg 1976).

The developmental increase of T–OH activity could be due to the activation of preexistent enzyme molecules or the synthesis of new enzyme protein. Immunotitration with a specific antiserum to T–OH was used, to distinguish between these alternatives (Black et al 1974). The ontogenetic increase in T–OH activity in mouse and rat SCG is entirely attributable to accumulation of increased numbers of molecules per adrenergic neuron (Black et al 1974). The kinetics of the developmental rise in T–OH suggest that increased synthesis, and not decreased degradation, is responsible for neonatal development of the enzyme (Black et al 1974). These immunochemical studies, and other physicochemical experiments (Black et al 1974), suggest that there is no alteration in the species of T–OH synthesized during development. Rather, maturation appears to consist of accumulation of the same species of enzyme present in the neonate, in contrast to a number of nonneuronal systems in which cellular ontogeny progresses through sequential appearance and disappearance of different species of enzyme molecules (Paigen 1971).

Ganglion decentralization or administration of ganglionic blocking agents prevents the normal accumulation of postsynaptic T–OH molecules (Black et al 1974). It may be concluded that presynaptic terminals, through the mediation of acetylcholine, regulate the developmental accumulation of T–OH molecules in adrenergic neurons of the SCG. It would be of interest to determine whether the development of DDC and DBH, the other noradrenergic enzymes, is similarly regulated, and whether other postsynaptic cells in the autonomic system are under transsynaptic control. Incomplete evidence suggests that transsynaptic regulation of development is a generalized phenomenon. In the postnatal rat, denervation of the adrenal gland prevents the normal development of postsynaptic T–OH and DBH activities, and the normal increase in norepinephrine content (Patrick & Kirshner 1972). In the avian parasympathetic ciliary ganglion, denervation reduces postsynaptic acetylcholine synthesis (Pilar et al 1973). Deafferentation in the young embryo has more profound effects and may result in almost complete disappearance of cells in the ciliary ganglion (Levi-Montalcini 1947) and in central cochlear nuclei (Levi-Montalcini 1949). Recent work has suggested that development of other central neuron systems is also dependent on anterograde transsynaptic influences. For example, deafferentiation prevents the normal maturation of layer VI pyramid-like neurons of the cerebral cortex (Maeda et al 1974), neurons of the mammalian lateral geniculate body (Weisel & Hubel 1963), and neurons of the superior colliculus (Hess 1958).

Although the definitive mechanisms underlying anterograde transsynaptic regulation in various regions of the nervous system have yet to be defined, several statements are warranted. In the case of the SCG, NGF does not play a role in anterograde transsynaptic regulation since (*a*) NGF cannot prevent or reverse the sequelae of decentralization or ganglionic blockade (Black et al 1972b, Hendry 1973), (*b*) NGF exerts its effects on decentralized ganglia (Black et al 1972b), and (*c*) anti-NGF action does not require intact preganglionic innervation (Hendry 1973). Recent experiments on cultured chick embryo trigeminal ganglia raise interesting questions of general relevance. Exposure to minute direct electrical current (0.00115–11.5 nA mm^{-3}) results in increased neuronal survival and increased fiber outgrowth towards the cathode (Sisken & Smith 1975), which suggests that neuronal depolarization, or an analogue or consequence thereof, enhances development. Perhaps normal activity of a neuron is necessary for normal maturation, and those pathways that fail to perform normally fail to survive.

In a subsequent section the transsynaptic regulation of the development of target innervation by the SCG is discussed in detail.

Central Modulation

The role of central synapses in the development of cholinergic and adrenergic neurons of the rat sympathetic nervous system has recently been characterized (Black et al 1976, Hamill et al 1977). The presynaptic cholinergic ILC neurons of the spinal cord receive afferents descending from suprasegmental levels (Pick 1970), and recent work has documented the existence of direct hypothalamic-ILC projections (Hancock 1976, Saper et al 1976). The transsynaptic regulation of peripheral sympathetic development may be dependent on connections with these descending central fibers and/or may simply depend upon segmental spinal mechanisms similar to those mediating muscle stretch reflexes. One approach to this problem has involved interruption of the descending central pathways that innervate the intermediolateral column cells. However, this cannot be accomplished in the case of the SCG, since its innervation derives mainly from the first three thoracic spinal segments, and lesions rostral to this level result in respiratory paralysis and death. Consequently, the more caudal sixth lumbar (L-6) ganglion was employed. This ganglion is innervated by cells from the eleventh thoracic to the second lumbar spinal segments (Pick 1970).

Transection of the spinal cord at the fifth thoracic segment prevents the normal development of presynaptic terminals and postsynaptic neurons in the L-6 ganglion (Black et al 1976, Hamill et al 1977). Spinal transection blocks the ontogenetic increase of postsynaptic T–OH activity in the L-6 ganglion (Black et al 1976, Hamill et al 1977). However, transection does not alter the ontogeny of T–OH activity in the SCG (Black et al 1976), which derives its innervation from spinal segments rostral to the surgical lesion. Thus, spinal transection interferes with the maturation of sympathetic neurons distal to, but not proximal to, the lesion. Postsynaptic DDC activity, which is regulated differently from T–OH in the adult (Black et al 1971a), also fails to develop normally in L-6 ganglia after transection (Hamill et al 1977).

Since presynaptic ILC neurons regulate the ontogeny of postsynaptic neurons, it is natural to ask whether transection also prevents normal ILC neuron ontogeny. In fact, transection of the spinal cord prevents the normal development of presynaptic ChAc activity in the L-6 ganglion. This effect is not secondary to direct surgical injury to ILC neurons, since histologic examination of the spinal cord does not reveal necrosis, reactive gliosis, or inflammatory infiltrate, and since ILC neurons appear to be normal (Hamill et al 1977). In addition, it is unlikely that these results reflect enzyme inhibition, since naturally occurring inhibitors of ChAc have not been described (Fonnum 1970). Moreover, it is highly improbable that these manipulations alter the proximodistal transport of ChAc, since axonal transport is relatively immutable and is unaffected by nerve interruption (Lubinska 1975), protein synthetic rate (McEwen & Grafstein 1968), or impulse activity (Lubinska 1975, Dahlström 1971). Both of the remaining alternatives, alteration of ChAc turnover in ILC perikarya and abnormal presynaptic terminal maturation in ganglia, are consistent with the observations. The sympathetic nervous system provides a precedent for both possibilities. Transsynaptic factors govern the ontogenetic accumulation of transmitter enzyme molecules in ganglion adrenergic neurons as described above, and also regulate the development of target innervation by adrenergic terminals (see below). It is conceivable, therefore, that spinal transection alters the normal synthesis and/or degradation of ChAc molecules in ILC perikarya *and* the normal development of ganglion innervation by cholinergic terminals.

These observations indicate that descending pathways of suprasegmental origin regulate ILC neuron differentiation. Other work has demonstrated that presynaptic ganglion terminals fail to mature normally after selective postsynaptic destruction

(Black et al 1972a; see below). Consequently, normal development of presynaptic neurons requires intact afferent as well as efferent connections. Thus, regulation by both orthograde and retrograde mechanisms, which has been demonstrated for adrenergic sympathetic neurons (Black et al 1971b, Black & Geen 1973, Dibner & Black 1976a, Dibner et al 1977; see below), also governs presynaptic cholinergic maturation.

In a more general sense, these observations suggest that central pathways regulate maturation of neurons with perikarya in the central nervous system as well as neurons in the periphery. Consequently, central tracts may influence ontogeny of neurons with which efferent contact is made and, also, second order neurons at remote sites. It may be inferred that central lesions can be expressed biochemically at distant loci and result in faulty development of an entire neuronal system. Such a contention may apply to other neuronal pathways as well, since these observations in a visceral motor system parallel mechanisms of development in such sensory systems as acousticovestibular nuclei (Levi-Montalcini 1949), lateral geniculate neurons (Wiesel & Hubel 1963), and striate cortex (Gyllensten et al 1965).

These observations and contentions raise logical questions of the "ultimate" origin of anterograde transsynaptic regulation, and the "ultimate" terminus of transsynaptic regulatory consequences. That is, can a single, critically placed lesion unravel all of neurologic development? Perhaps the answer to this absurd question is that the origin may be the neuronal grouping to which *multiple* afferent inputs converge. The anterograde consequences of a lesion may become progressively more attenuated with each synaptic step, due to interposition of multiple new afferents. A corollary would suggest that the magnitude of transsynaptic effects exerted by a subsystem on a target varies inversely as the number of other subsystem inputs to that target.

Retrograde Transsynaptic Regulation in Ganglia

The work summarized above indicates that presynaptic cholinergic nerve terminals regulate the development of postsynaptic neurons in the SCG. A series of studies has indicated that, conversely, postsynaptic neurons regulate presynaptic maturation through a retrograde process (Black et al 1972a). Selective destruction of adrenergic neurons chemically, with 6-OHDA, or immunologically, with anti-NGF, prevents the normal development of presynaptic ChAc activity in neonatal mouse SCG. After virtually complete destruction of adrenergic neurons in the ganglion, ChAc activity remains at levels observed in the one to three-day-old mouse. This effect is due to the absence of adrenergic neurons and not the presence of degenerating neurons. Since these studies were performed well before the SCG has developed functional innervation of target organs (Iversen et al 1967, de Champlain et al 1970, Black & Mytilineou 1976a), it is unlikely that a complex reflex process involving target organs, their afferent innervation, and the CNS is responsible. Rather, direct retrograde regulation of cholinergic development by adrenergic neurons appears to be critical.

These observations have been confirmed and extended by studies demonstrating that destruction of adrenergic neurons by target extirpation (Dibner et al 1977) or axotomy (Hendry 1975a) also prevents the normal development of presynaptic ChAc activity. Consequently, regardless of the method employed, adrenergic destruction prevents cholinergic maturation. Only severe postsynaptic lesions interfere with presynaptic ontogeny, since ganglionic blockade, which prevents development of adrenergic transmitter enzymes (see above), does not alter ChAc development (Black & Geen 1973, 1974). On the other hand, promotion of abnormally increased adrenergic growth by administration of NGF (Thoenen et al 1972) or increased target organ size (Dibner & Black 1978) results in elevation of ChAc activity above normal.

Perhaps the number of postsynaptic sites in the SCG actually regulates the survival of presynaptic neurons in a manner analogous to that in the somatic motor system (Hamburger 1958, 1975, Prestige 1967, 1970, 1974, Prestige & Wilson 1972, Hollyday & Hamburger 1976). In fact, recent work has demonstrated that destruction of adrenergic neurons with guanethidine in neonatal rats reduces survival of ILC neurons (Caserta et al 1976). Consequently, the failure of ChAc development after adrenergic death may reflect failure of cholinergic survival. Conversely, increased ChAc activity after increased adrenergic growth may actually reflect enhanced survival of ILC neurons.

The mechanisms governing anterograde and retrograde transsynaptic regulation in the SCG are clearly different. However, it is now well-established that reciprocal regulatory relationships between these cholinergic and adrenergic neurons at the synapse are critical for normal development. Such a bidirectional flow of regulatory information may also occur in the CNS and may constitute a major mechanism governing neuronal maturation. Gudden (1870) was perhaps the first to demonstrate retrograde transsynaptic degeneration in the brain, whereas the more recent work of Hamburger (Hamburger 1958, 1975, Hollyday & Hamburger 1976), Cowan & Wenger (1968), and Prestige (Prestige 1967, 1970, 1974, Prestige & Wilson 1972) have stressed the developmental implications of retrograde regulation. Cowan (1970) has observed that the severity of retrograde degeneration in the CNS is a function of the age of the experimental animal. Viewed in conjunction with the work described in previous sections, it is apparent that anterograde and retrograde transsynaptic regulation may be critical in central as well as peripheral development.

Development of Target Innervation

Discussion of retrograde regulation of cholinergic nerves by adrenergic neurons leads naturally to consideration of target organ regulation of adrenergic development. Before addressing this subject it may be helpful to describe the normal process of target innervation by adrenergic neurons. In the rat, adrenergic fibers are visible in various organs at 19–21 days of gestation, although adult levels of endogenous norepinephrine are not approximated until the third to fifth postnatal weeks (de Champlain et al 1970, Owman et al 1971). During the immediate postnatal period, the ganglion sympathoblast contains a well-developed Golgi apparatus, endoplasmic reticulum, mitochondria, and a few large granular vesicles (Eränkö 1972). Within days, cell diameter increases, numerous small granular vesicles (30–60 nm) appear, and catecholamine fluorescence is present in the cytoplasm and the granules

(Eränkö 1972). The uptake of ^{3}H-norepinephrine by targets, an index of innervation by functional axon terminal membrane, increases during the first few postnatal weeks (Iversen et al 1967, Black & Mytilineou 1976a). At birth a faint plexus of fluorescent fibers is present in the sphincter region of the iris, but the adult pattern of innervation does not occur until after the third postnatal week (de Champlain et al 1970). T–OH activity, which is restricted to adrenergic neurons in target organs, increases 50-fold in the iris and 34-fold in pineal nerve terminals of the rat, and adult levels are reached by 1–2 months of age (Black & Mytilineou 1976a). In the pineal gland, adrenergic nerves are present only within the pial covering at birth but invade the parenchyma by 5–6 days and develop the adult appearance at three weeks (Hakanson et al 1967). The circadian rhythm of pineal serotonin N-acetyltransferase, which is regulated by the sympathetic innervation, begins at four days of age (Ellison et al 1972). In the rabbit, maturing autonomic fibers progressively acquire the ability to transmit impulses of high frequency during the first 4–6 postnatal weeks (Schweiler et al 1970).

In a series of critical studies, Olson & Malmfors (1970) demonstrated that the characteristic pattern of innervation of a target is determined by the target itself and not by the innervating neurons. In the iris, for example, the adrenergic innervation consists of a dense plexus in which thick preterminal axons undergo extensive branching to form a dense network of varicose terminals. This pattern differs markedly from that observed in other peripheral organs such as salivary gland or heart, and from central terminal areas such as cerebral and cerebellar cortices. Organs transplanted to the anterior chamber of the eye in rats become normally innervated by adrenergic neurons of the SCG that innervate the iris. Additionally, Purkinje cells in cerebellar grafts also become normally innervated by inhibitory adrenergic fibers of the SCG (Hoffer et al 1976). Conversely, virtually all dopaminergic, noradrenergic, and serotoninergic neuroblasts from central and peripheral tissues are capable of forming a typical varicose ground plexus in the host iris (Olson & Seiger 1972, 1973). Growth of fibers into the target occurs independently of transmitter synthesis, storage, release, reuptake, or receptor blockade (Olson & Malmfors 1970). It is clear that organotypic specificity of innervation resides in the effector organ, while compatible nerves exhibit little specificity. Comparable mechanisms, in which target structures dictate specificity, govern the afferent innervation of adrenergic neurons as discussed above, may facilitate innervation of striated muscle by a variety of cholinergic neurons (Guth 1968), and may be operative in the visual system (Gaze 1970). It is not unreasonable to conclude that this is a generalized phenomenon governing neurologic ontogeny.

A number of studies performed in tissue culture support these contentions. Synaptogenesis and maturation of organotypic synaptic networks can develop in vitro even in the absence of bioelectric activity (Crain & Peterson 1974). Furthermore, sympathetic fiber elongation, catecholamine fluorescence, and perikaryon and nuclear size increase in the presence of targets such as vas deferens (Chamley et al 1973). In the presence of the intact vas deferens, sympathetic neurites appear to be attracted towards the target from a distance of <2 mm (Chamley et al 1973). On the other hand, depletion of NGF in culture retards growth of sympathetic fibers into iris from neurons grown in contact with the iris (Johnson et al 1972). It is not clear whether NGF acts specifically in this context, or whether it plays a permissive role, possibly allowing neuronal survival. In the parasympathetic system, the target salivary gland orients fiber outgrowth from the innervating submaxillary ganglion and exerts this effect across a millipore filter (Coughlin 1975a, 1975b). In addition, formalin-killed salivary gland elicits similar ganglion growth (M. D. Coughlin, M. P. Rathbone, unpublished). Letourneau (1975) has demonstrated that substratum adhesiveness critically affects the migration of growth cones. Lastly, the work of Helfand et al (1976) indicates that heart-conditioned medium stimulates survival and neurite extension from ciliary ganglion through a mechanism independent of NGF.

In addition to determining specificity of the morphologic pattern of innervation, targets regulate *survival* of afferent neurons. Reduction of neuron numbers is a normal development event in sympathetic ganglia (Aguayo et al 1973, Hendry & Campbell 1975), presynaptic ILC neurons (Levi-Montalcini 1950, Shieh 1951), parasympathetic ganglia (Landmesser & Pilar 1974), spinal ventral horn (Hamburger 1958, 1975, Prestige 1967, 1970, 1974, Prestige & Wilson 1972, Hollyday & Hamburger 1976), and the isthmo-optic nucleus (Clarke & Cowan 1975), to cite only some examples. Target extirpation profoundly increases developmental cell death in neurons of the ciliary ganglion (Cowan & Wenger 1968, Landmesser & Pilar 1974), SCG (Gyllensten et al 1965, Hendry & Iversen 1973, Dibner et al 1977), ILC presynaptic neurons (Caserta et al 1976), and spinal ventral horn (Prestige 1967, 1970, 1974, Prestige & Wilson 1972). Conversely, target enlargement increases survival of ILC neurons (Shieh 1951) and of adrenergic neurons (Dibner & Black 1978).

Target influences on cell death probably occur only during a circumscribed period of development. Target extirpation does not appear to affect mitoses or differentiation of neurons in the lateral motor system of the chick (Hamburger 1958, 1975, 1977, Hollyday & Hamburger 1976) or in avian ciliary ganglion (Cowan & Wenger 1968, Landmesser & Pilar 1974). Even those neurons destined to die send axons to appropriate targets from the ciliary ganglion (Cowan & Wenger 1968) and spinal ventral horn (Landmesser & Morris 1975, Hamburger 1977), although cell death is synchronous with formation of peripheral connections (Landmesser & Pilar 1974). Several conclusions may be warranted. First, retrograde control of cell death is in some manner different from regulation of axon outgrowth towards the target. Second, arrival of an axon at the end organ does not, in and of itself, assure neuronal survival. Finally, and more tentatively, the *time* of death may be an expression of information intrinsic to the neuron. It may be helpful to consider these conclusions as we discuss the role(s) of NGF, a factor elaborated by targets, which profoundly alters sympathetic growth and development.

Targets may induce more subtle biochemical changes in afferent neurons, in addition to influencing cell survival. For example, development of ChAc activity in postsynaptic neurons of the ciliary ganglion correlates with formation of synaptic contacts with the target in chicks (Chiappinelli et al 1976). The formation of neuromuscular junctions in culture of spinal cord and muscle is associated with a marked elevation of ChAc activity (Giller et al 1973). In the adult guinea pig, hormonal alteration of the vas deferens and seminal vesicle size influences the content and uptake of norepinephrine by the innervating neurons (Wakade & Kirpekar 1973), which suggests that retrograde mechanisms persist throughout life. These examples are consistent with the aforementioned observation that adrenergic neurons regulate presynaptic cholinergic biochemistry (Black et al 1972a).

In addition to regulation by retrograde factors, anterograde transsynaptic influences in sympathetic ganglia modulate development of target innervation (Black & Mytilineou 1976a). Unilateral deafferentiation of the SCG in 2- to 3-day-old rats prevents the normal development of end organ innervation: T–OH activity, ^{3}H-norepinephrine uptake, innervation density, ground plexus ramification, and fluorescence intensity fail to develop normally in irides innervated by decentralized ganglia (Black & Mytilineou 1976a). This anterograde transsynaptic effect is apparently independent of NGF, since administration of the protein does not prevent or reverse the effects of decentralization (Black & Mytilineou 1976b).

Target Organs and NGF

A considerable body of evidence now indicates that at least some of the influences of target organs are exerted through the mediation of NGF. The great quantity of data amassed over the past 30 years cannot be exhaustively detailed in the present review. Rather, a brief summary is presented so that regulatory mechanisms and outstanding questions may be discussed. The reader is referred to several excellent reviews (Levi-Montalcini & Angeletti 1961, 1968, Levi-Montalcini 1964a, b, 1966, Angeletti 1972, Zaimis & Knight 1972, Hendry 1976).

Thirty years ago, Bueker transplanted mouse sarcoma 180 into chick embryos; he noted that sensory fibers branched into the tumor and that ipsilateral sensory ganglia were enlarged (Bueker 1948). Levi-Montalcini & Hamburger (1951, 1953) extended these observations, noting stimulation of sympathetic and sensory growth by a diffusable agent from specific sarcomas. These insights ultimately led to the isolation of a protein fraction from mesenchymal tumors, snake venom, and male mouse salivary glands that caused profound sympathetic and sensory overgrowth when injected into neonates (see Cohen et al 1954, Levi-Montalcini & Booker 1960a, b, Levi-Montalcini & Angeletti 1961, 1968, Levi-Montalcini 1964a, b, 1966, Olson 1967, Angeletti 1972). In culture, sympathetic and sensory neurons require NGF for survival (Levi-Montalcini & Angeletti 1968). Conversely, treatment of neonates with antiserum to NGF is associated with destruction of adrenergic neurons throughout the animal, resulting in (*a*) decreased cell numbers, (*b*) reduced content of norepinephrine and depressed transmitter enzyme activities in sympathetic ganglia, and (*c*) decreased catecholamine fluorescence, ^{3}H–NE uptake, and NE content of target organs (Cohen 1960, Levi-Montalcini & Booker 1960b, Klingman & Klingman 1965, 1967, Klingman 1965, Hamburger et al 1965, Iversen et al 1966, Hendry & Iversen 1971, Schucker 1972). Treatment of pregnant female mice with anti-NGF results in partial sympathetic destruction in the offspring (Klingman 1966, Klingman & Klingman 1967). Although these observations indicate that NGF plays a critical role during development, the mechanisms involved are not specified.

NGF is secreted by a number of mesenchymal tissues (Oger et al 1974, Young et al 1975), is detectable in a variety of organs (Bueker et al 1960, Waddell et al 1972, Hendry 1972) and is present in highest concentrations in tubules of the submaxillary glands (Bueker et al 1960, Levi-Montalcini & Angeletti 1961, Caramia et al 1962, Schwab et al 1976), which may regulate circulating levels of the protein (Hendry & Iversen 1973). The growth-promoting factor has been purified from snake venom (Cohen & Levi-Montalcini 1956, Cohen 1959) and from mouse sarcoma 180 (Cohen & Levi-Montalcini 1957). Salivary NGF activity has been purified as a 7S protein with mol. wt. of 140,000 (Perez-Polo et al 1972), which may be dissociated to form α, β, and γ subunits. The biological activity is associated with the β subunit (Varon et al 1968, Smith et al 1969), whereas the γ subunit possesses esterase activity that may release the β subunit from the parent complex (Greene et al 1969). Utilizing more vigorous analytic and preparative methods, Bradshaw and colleagues (Angeletti & Bradshaw 1971) have isolated an active 2.5S protein from submaxillary glands that is composed of two identical subunits, each consisting of 118 amino acids with a mol. wt. of 26,518. The 2.5S NGF and the β subunit are most probably identical. Exogenous NGF accumulates in sympathetic ganglia (Angeletti, Angeletti & Levi-Montalcini 1972) and specific binding of NGF to cell surface receptors has been demonstrated for sympathetic ganglia (Herrup & Shooter 1973, Banerjee et al 1973, Frazier, Boyd & Bradshaw 1974, Banerjee et al 1975), dorsal root ganglia (Frazier, Boyd & Bradshaw 1974), and brain (Frazier et al 1974).

NGF exerts a variety of effects on receptive neurons, including activation of the hexose monophosphate shunt (Angeletti et al 1964), increased size of neurons (Banks et al 1975), enhanced incorporation of uridine into RNA (Larrabee 1970), elevation of sympathetic T–OH, DDC, DBH, and MAO activities (Angeletti et al 1972, Phillipson & Sandler 1975), but no alteration in cAMP or adenyl cyclase (Hier et al 1973). However, it is unclear whether any of these effects are fundamental to the biologic action of NGF. Rather, these manifestations may simply be secondary reflections of the critical action(s) of NGF. A number of recent studies, however, have begun to indicate the mode, if not the mechanism, of NGF action.

During postnatal ontogeny there is a normal decrease in the number of adrenergic neurons in the SCG (Hendry & Campbell 1975). The magnitude of this decrease may be markedly enhanced by removal of the salivary glands and iris, targets of some SCG neurons (Hendry & Iversen 1973, Dibner & Black 1976a, Dibner et al 1977). However, target organs can be replaced by a cellulose pellet containing covalently bound NGF (Hendry & Iversen 1973). Additionally, testosterone-induced *hypertrophy* of salivary tubules in neonates, which increases tubular NGF content (Ishii & Shooter 1975), results in increased numbers of adrenergic neurons (Dibner & Black 1978) and elevated transmitter enzyme activities (Dibner & Black 1976b, 1978). Pharmacologically induced hypertrophy of other, non-NGF-containing portions of the salivery glands does not alter SCG neurons (Dibner & Black 1978). It may be inferred that targets regulate sympathetic survival through the mediation of NGF.

NGF may regulate the direction of neurite growth, as well as govern neural survival. In tissue culture, chick embryo sensory ganglia elaborate fibers towards a capillary tube source of NGF (Charlwood et al 1972). This effect is exerted over a distance of up to 2 mm. While these observations are extremely provocative, it may be useful to determine whether the capillary source of NGF favors survival of proximate neurons in a given ganglion, which then develop normal, radially oriented neurites. Regardless of the mechanisms, these results raise the important possibility of chemotactic neuronal responses to NGF.

In addition to reaching responsive neurons through the circulation, NGF appears to undergo retrograde axonal transport from terminals in the target organ to neuronal perikarya (Hendry et al 1974). Injection of NGF into the ocular anterior chamber of mice and rats results in appearance of the protein in ipsilateral SCG perikarya in a biologically active, immunoreactive form (Hendry et al 1974, Hendry, Stach & Herrup 1974, Stöckel et al 1974, Parvicini et al 1975). Moreover, the chromatolysis and neuronal death that may follow axotomy in the neonate is prevented by the administration of NGF (Hendry 1975b, c). The primary source of tissue NGF is not well-defined, although recent evidence suggests that the submaxillary glands may play a pivotal regulatory role (Hendry & Iversen 1973).

In summary, NGF is present in target organs and the circulation, is subject to retrograde axonal transport, interacts with neuron membrane receptors, appears to be critical for neuronal survival in the neonate, may influence the direction of neurite outgrowth, and may cause a multiplicity of metabolic responses in receptive neurons. NGF may cause these effects by (*a*) exerting a *single* metabolic action, enabling responsive developing neurons to survive and express many intrinsic traits; and/or (*b*) directly regulating a variety of neuronal events through separate mechanisms. In either instance it is unlikely that NGF alone accounts for target regulation of neurons in vivo. Although NGF molecules from different targets are identical, neurons reproducibly innervate their respective targets in an *organotypic* pattern. Furthermore, the same neurons innervate different NGF-containing tissues differently (Olson & Malmfors 1970). It may be inferred that complex interactions between target and neuron involve NGF *and* other unidentified factors.

Endocrine Regulation

Considerable evidence indicates that hormonal mechanisms are critical for growth and development of the nervous system (for review see Balazs 1976). Recent work suggests that thyroxine, in particular, is necessary for normal sympathetic maturation. Ablation of the thyroid in neonatal rats with methimazole prevents the normal development of T–OH and ChAc activities in the SCG, and this effect is reversed by thyroxine treatment (Black 1974). Thyroid destruction with ^{131}I also prevents normal sympathetic development (Black 1974). These observations were recently extended by experiments demonstrating that SCG histofluorescence is reduced in hypothyroid neonates and that submaxillary terminals are thinner (Gresik 1976). The mechanisms underlying these effects are unclear, although it is recognized that drugs such as thiourea, which destroy the thyroid, also cause atrophy of the submaxillary gland tubules (Arvy et al 1950). Whether thyroid regulation is exerted via an effect on NGF is unknown.

Glucocorticoids may also play a role in sympathetic ontogeny. Treatment of neonatal rats with dexamethosone results in the appearance of small, intensely fluorescent (SIF) cells in abdominal chromaffin tissue and in sympathetic paravertebral ganglia (Ciaranello et al 1973). In parallel, the activity of phenylethanolamine-N-methyl-transferase, which catalyzes the formation of epinephrine, appears. Thus glucocorticoids may influence differentiation of specific cellular populations in sympathetic ganglia. Although hormone-autonomic relations during maturation have only recently been explored, it appears that these interactions will be increasingly appreciated.

EFFECTS OF DRUGS

Many of the effects that drugs exert on autonomic maturation result from interaction with the normal regulatory mechanisms described above. In the broadest terms, pharmacologic agents may alter anterograde transsynaptic regulation or possibly retrograde transsynaptic regulation by NGF. Undoubtedly, additional drug effects will be defined in the near future and may elucidate other mechanisms governing ontogeny.

Any agents that interfere with orthograde ganglionic transmission may potentially prevent normal development of adrenergic neurons. As discussed in previous sections, administration of the ganglionic blocking agents chlorisondamine or pempidine during the first month of life prevents the normal development of T–OH, DDC, and DBH activities, and total ganglion protein in rats (Hendry 1973, Black & Geen 1973, 1974). Inhibition of development follows nicotinic receptor blockade but not muscarinic blockade with atropine (Black & Geen 1974). The effects of ganglionic blockade are not reversed by NGF administration (Hendry 1973).

Conversely, stimulation of ganglionic transmission results in long-lasting elevation of ganglionic T–OH and DBH activities in neonatal rats (Bartolomé et al 1976). The elevation of enzyme activities persists for at least 45 days in the neonates. Apparently, enhancement of ganglionic transmission, as well as blockade, may cause long-lasting alteration of sympathetic development.

The importance of these observations for treatment of pregnant women and of infants need not be stressed. It may be recalled, for example, that ganglionic blocking agents have been employed to treat hypertension, a hallmark of toxemia (eclampsia) during pregnancy.

Several drugs are capable of destroying adrenergic neurons in neonates, and this effect may result, in part, from interference with NGF function. It is now well recognized that administration of 6-OHDA to neonates results in widespread destruction of adrenergic neurons in the periphery (Angeletti & Levi-Montalcini 1970a, Jaim-Etcheverry & Zieher 1971, Eränkö & Eränkö 1972, Finch et al 1973, Singh & de Champlain 1974, Jonsson et al 1975), and alters development in the CNS (Sachs et al 1974, Kostrzewa & Harper 1974, Jonsson et al 1974; see Jonsson et al 1975 for general review). A number of observations suggest that 6-OHDA antagonizes actions of NGF. First, the developing neuron is sensitive to 6-OHDA and anti-NGF during the same period of development (Angeletti & Levi-Montalcini 1970a, b). Second, the adrenal medulla is resistant to both 6-OHDA and anti-NGF, and the pattern of destruction of sympathetic ganglia is similar to the two agents (de Champlain & Nadeau 1971, Thoenen 1971, Clark et al 1972, Eränkö & Eränkö

1972). Third, NGF administration prevents the effects of 6-OHDA on adrenergic perikarya but not on nerve terminals (Levi-Montalcini et al 1975). These observations are consistent with the contention that chemical axotomy with 6-OHDA prevents retrograde axonal transport of NGF during a critical phase of maturation, resulting in neuronal death. The enhanced increase of T–OH in adrenergic perikarya of rats treated with 6-OHDA and NGF, compared to treatment with NGF alone (Levi-Montalcini et al 1975) may be related to the well-characterized induction of T–OH in adult adrenal medulla after 6-OHDA sympathectomy (Mueller et al 1969). On the other hand, the differences in the cytotoxic pathology elicited by 6-OHDA and anti-NGF, may result from the fact that anti-NGF not only deprives the neuron of NGF but may also form cytotoxic complexes with NGF and complement. It is clear that additional work is required to definitively explain the mechanism of action of 6-OHDA on immature neurons.

Treatment of adult rats with guanethidine also results in apparent postganglionic axotomy (Jensen-Holm & Juul 1970), whereas treatment of neonates results in both loss of target terminals and destruction of ganglion perikarya (Eränkö & Eränkö 1971, 1973). Paradoxically, SIF cell numbers increase in neonates (Eränkö & Eränkö 1971, 1973). Adrenergic cell bodies of neonatal mice and rats treated with guanethidine exhibit mitochondrial swelling, fragmented cristae, and dilated endoplasmic reticulum (Angeletti, Levi-Montalcini & Caramia 1972). However, axon retraction appears to precede perikaryon degeneration (Heath et al 1973), which suggests that interference with NGF transport may, at least in part, mediate the cytotoxic effects. Indeed, in tissue culture, high concentrations of NGF prevent destruction of embryonic chick or rat sympathetic ganglia by guanethidine (Johnson & Aloe 1974). Another drug, bretylium tosylate, has long been known to possess a number of pharmacologic properties similar to guanethidine and also causes mitochondrial lesions in developing sympathetic neurons (Caramia et al 1972). Although it is tempting to conclude that these agents simply interfere with NGF function, it may be recalled that they exert a variety of effects on adrenergic neurons (for summary see Goodman & Gilman 1975). It is not yet clear which, if any, of these actions are specifically related to NGF.

PERSISTENCE OF DEVELOPMENTAL MECHANISMS INTO ADULTHOOD

Many of the mechanisms that govern ontogeny appear to persist, in somewhat altered form, into adulthood. Two examples, anterograde transsynaptic regulation and regulation by NGF, illustrate that adult regulatory processes may have their origins in developmental mechanisms. Transsynaptic factors in sympathetic ganglia and adrenal govern the level of T–OH and DBH in *mature* rats, as well as the development of enzymes in the neonate. Experimental manipulations in adult rats, such as cold exposure, immobilization stress, or reserpine treatment, which reflexly increase nerve impulse frequency, increase T–OH and DBH activities (Mueller et al 1969, Thoenen et al 1969a, b, 1970, Molinoff et al 1970, Molinoff et al 1972, Kvetňanský 1973). As in neonates, the increase in T–OH activity is due to accumulation of increased numbers of enzyme molecules and thus represents biochemical induction (Joh et al 1973). The effects are blocked by deafferentation (Thoenen et al 1969b, Kvetňanský 1973), which indicates that these are transsynaptic processes analogous to that in the newborn. Moreover, pharmacologic ganglionic blockade of nicotinic, but not muscarinic, receptors prevents the increase in enzyme activities (Mueller et al 1970), as in neonates. Lastly, central pathways mediate transsynaptic enzyme induction in adults (Bloom et al 1976), reproducing central function during development. While the increase in enzyme activity governed by transsynaptic mechanisms is permanent in neonates, that in adults is temporary, and activity returns to basal levels when the increased transsynaptic stimulus is removed (Thoenen et al 1969a, b, 1970, Molinoff et al 1970, Molinoff et al 1972, Kvetňanský 1973). Apparently, transsynaptic mechanisms, which are critical for normal development, continue to regulate postsynaptic enzymes throughout life.

In a similar fashion, NGF and anti-NGF affect adrenergic neurons in adults as well as neonates. In adult mice, NGF treatment increases the adrenergic nerve terminal network and NE content of target organs (Bjerre et al 1975). Sialectomy in adults reduces T–OH, DBH, and DDC activities in the SCG, and treatment with NGF reverses this effect (Hendry & Thoenen 1974). Conversely, treatment of adult mice with anti-NGF reduces sympathetic ganglion volume and catecholamine content of targets (Angeletti et al 1971, Bjerre, Wiklund & Edwards 1975), although sympathetic neurons do not appear to undergo necrosis. It is apparent that NGF and anti-NGF exert influences in neonates and adults, although effects in the latter tend to be less profound.

Recent work has suggested that NGF may play a role in regeneration of noradrenergic neurons in adults. Since the early work of Cajal (1928), it has been recognized that sympathetic neurons undergo vigorous regenerative sprouting after injury. Central catecholaminergic neurons also produce sprouts after damage (Katzman et al 1971) and grow into iris and mitral valve transplanted to the rat mesencephalon (Björklund & Stenevi 1971). Intraventricular injection of NGF appears to increase formation and growth of nerve sprouts from transected, ascending noradrenergic axons into irides transplanted to the diencephalon (Bjöklund & Stenevi 1972). Transected peripheral adrenergic neurons also respond to NGF treatment with an apparent increase in regenerative sprouts and acceleration of growth (Björklund et al 1974). Administration of anti-NGF apparently retards regenerative growth of central aminergic neurons into transplanted iris (Bjerre et al 1974).

These examples may indicate that many mechanisms that are critical for normal development continue to regulate autonomic function throughout life. Although the repertoire of neuronal responses may evolve as the organism matures, underlying regulatory stimuli may persist.

CONCLUDING QUESTIONS

The present review has described a rather large number of mechanisms regulating autonomic ontogeny. In the broadest sense, it is apparent that anterograde and retrograde transsynaptic mechanisms, at multiple levels of the neuraxis, regulate autonomic development. A number of humoral controls have also been described. Undoubtedly, additional regulatory processes will be defined in the future. Several theoretical questions arise from these considerations. Why are such a multitude of controls necessary? Does each regulatory agent govern a separate developmental character, or is there a hierarchy of signals to which neurons must respond to "pass the tests" of development? Do multiple regulatory mechanisms tend to preserve only the fittest neurons, liberating the organism from defective nerves and enhancing the probability of survival? What is the molecular basis for the persistence of some developmental mechanisms throughout life? What is the precise relationship of normal development to regeneration after injury? To what extent, and in what manner, is autonomic aging a normal part of ontogeny?

ACKNOWLEDGMENTS

I thank Dr. M. D. Coughlin, Dr. M. D. Dibner, Dr. R. W. Hamill, and Ms. Janet Lindquist Black for helpful review of the manuscript, and Ms. Elise Grossman for excellent technical assistance.

Literature Cited

Aghajanian, G., Bloom, F. E. 1967. The formation of synaptic junctions in developing rat brain: a quantitative electron microscopic study. *Brain Res.* 6:716–27

Aguayo, A. J., Terry, L. C., Bray, G. M. 1973. Spontaneous loss of axons in sympathetic unmyelinated nerve fibers of the rat during development. *Brain Res.* 54:360–64

Angeletti, P. U. 1972. Biological properties of the Nerve Growth Factor. *Adv. Exp. Med. Biol.* 32:83–90

Angeletti, P. U., Levi-Montalcini, R. 1970a. Sympathetic nerve cell destruction in newborn mammals by 6-hydroxydopamine. *Proc. Natl. Acad. Sci. USA* 65:114–21

Angeletti, P. U., Levi-Montalcini, R. 1970b. Specific cytotoxic effects of 6-hydroxydopamine on sympathetic neurons. *Arch. Ital. Biol.* 108:213–21

Angeletti, P. U., Levi-Montalcini, R., Caramia, F. 1971. Analysis of the effects of the antiserum to the Nerve Growth Factor in adult mice. *Brain Res.* 27:343–55

Angeletti, P. U., Levi-Montalcini, R., Caramia, F. 1972. Structural and ultrastructural changes in developing sympathetic ganglia induced by guanethidine. *Brain Res.* 43:515–25

Angeletti, P. U., Levi-Montalcini, R., Kettler, R., Thoenen, H. 1972. Comparative studies on the effect of the Nerve Growth Factor on sympathetic ganglia and adrenal medulla in newborn rats. *Brain Res.* 44:197–206

Angeletti, P. U., Liuzzi, A., Levi-Montalcini, R., Gandini-Attardi, D. 1964. Effect of a Nerve Growth Factor on glucose metabolism by sympathetic and sensory nerve cells. *Biochem. Biophys. Acta* 90:445–50

Angeletti, R. H., Angeletti, P. U., Levi-Montalcini, R. 1972. Selective accumulation of (^{125}I) labelled Nerve Growth Factor in sympathetic ganglia. *Brain Res.* 46:421–25

Angeletti, R. H., Bradshaw, R. A. 1971. Nerve Growth Factor from mouse submaxillary gland: amino acid sequence. *Proc. Natl. Acad. Sci. USA* 68:2417–20

Arvy, L., Debray, P., Gabe, M. 1950. Action de la thiourée sur la glande sous-maxillaire du rat albinos. *C. R. Seances Soc. Biol. Paris* 144:111–13

Balazs, R. 1976. Hormones and brain development. In *Perspectives in Brain Research,* ed. M. A. Corner, D. F. Swaab, 45:139–59. Amsterdam: Elsevier. 489 pp.

Banerjee, S. P., Cuatrecasas, P., Snyder, S. H. 1975. Nerve Growth Factor receptor binding—influence of enzymes, ions and protein reagents. *J. Biol. Chem.* 250:1427–33

Banerjee, S. P., Snyder, S. H., Cuatrecasas, P., Greene, L. A. 1973. Binding of Nerve Growth Factor receptor in sympathetic ganglia. *Proc. Natl. Acad. Sci. USA* 70:2519–23

Banks, B. E. C., Charlwood, K. A., Edwards, D. C., Vernon, C. A., Walter, S. J. 1975. Effects of Nerve Growth Factor from salivary glands and snake venom on the sympathetic ganglia of neonatal and developing mice. *J. Physiol. London* 247:289–98

Bartolomé, J., Seidler, F. J., Anderson, T. R., Slotkin, T. A. 1976. Effects of prenatal reserpine administration on development of the rat adrenal medulla and central nervous system. *J. Pharmacol. Exp. Ther.* 197:293–302

Bjerre, B., Björklund, A., Mobley, W., Rosengren, E. 1975. Short- and long-term effects of Nerve Growth Factor on the sympathetic nervous system in the adult mouse. *Brain Res.* 94:263–77

Bjerre, B., Björklund, A., Stenevi, U. 1974. Inhibition of the regenerative growth of central noradrenergic neurons by cerebrally administered Anti-NGF serum. *Brain Res.* 74:1–18

Bjerre, B., Wiklund, L., Edwards, D. C. 1975. A study of the de- and regenerative changes in the sympathetic nervous system of the adult mouse after treatment with the antiserum to Nerve Growth Factor. *Brain Res.* 92:257–78

Björklund, A., Bjerre, B., Stenevi, U. 1974. Has Nerve Growth Factor a role in the regeneration of central and peripheral catecholamine neurons? In *Dynamics of Degeneration and Growth in Neurons*, ed. K. Fuxe, L. Olson, Y. Zotterman, pp. 389–409. New York: Pergamon. 608 pp.

Björklund, A., Stenevi, U. 1971. Growth of central catecholamine neurones into smooth muscle grafts in the rat mesencephalon. *Brain Res.* 31:1–20

Björklund, A., Stenevi, U. 1972. Nerve Growth Factor: stimulation of regenerative growth of central noradrenergic neurons. *Science* 175:1251–53

Black, I. B. 1973. Development of adrenergic neurons in vivo: inhibition by ganglionic blockade. *J. Neurochem.* 20: 1265–67

Black, I. B. 1974. The role of the thyroid in the growth and development of adrenergic neurons in vivo. *Neurology* 24:377 (Abstr.)

Black, I. B., Bloom, E. M., Hamill, R. W. 1976. Central regulation of sympathetic neuron development. *Proc. Natl. Acad. Sci. USA* 73:3575–78

Black, I. B., Geen, S. C. 1973. Trans-synaptic regulation of adrenergic neuron development: inhibition by ganglionic blockade. *Brain Res.* 63:291–302

Black, I. B., Geen, S. C. 1974. Inhibition of the biochemical and morphological maturation of adrenergic neurons by nicotinic receptor blockade. *J. Neurochem.* 22:301–6

Black, I. B., Hendry, I. A., Iversen, L. L. 1971a. Differences in the regulation of tyrosine hydroxylase and DOPA-decarboxylase in sympathetic ganglia and adrenal. *Nature* 231:27–29

Black, I. B., Hendry, I. A., Iversen, L. L. 1971b. Trans-synaptic regulation of growth and development of adrenergic neurons in a mouse sympathetic ganglion. *Brain Res.* 34:229–40

Black, I. B., Hendry, I. A., Iversen, L. L. 1972a. The role of post-synaptic neurons in the biochemical maturation of presynaptic cholinergic nerve terminals in a mouse sympathetic ganglion. *J. Physiol. London* 221:149–59

Black, I. B., Hendry, I. A., Iversen, L. L. 1972b. Effects of surgical decentralization and Nerve Growth Factor on the maturation of adrenergic neurons in a mouse sympathetic ganglion. *J. Neurochem.* 19:1367–77

Black, I. B., Joh, T. H., Reis, D. J. 1974. Accumulation of tyrosine hydroxylase molecules during growth and development of the superior cervical ganglion. *Brain Res.* 75:133–44

Black, I. B., Mytilineou, C. 1976a. Trans-synaptic regulation of the development of end organ innervation by sympathetic neurons. *Brain Res.* 101:503–21

Black, I. B., Mytilineou, C. 1976b. The interaction of Nerve Growth Factor and trans-synaptic regulation in the development of target organ innervation by sympathetic neurons. *Brain Res.* 108: 199–204

Black, I. B., Petito, C. K. 1976. Catecholamine enzymes in the degenerative neurological disease Idiopathic Orthostatic Hypotension. *Science* 192:910–12

Bloom, E. M., Hamill, R. W., Black, I. B. 1976. Elevation of tyrosine hydroxylase activity in sympathetic neurons after reserpine: the role of the central nervous system. *Brain Res.* 115:525–28

Bradbury, S., Eggleston, M. D. 1925. Postural Hypotension. *Am. Heart J.* 1:73–86

Brauer, A. 1932. A topographical and cytological study of the sympathetic nervous components of the suprarenal of the chick embryo. *J. Morphol.* 53:277–325

Brunt, P. W., McKusick, V. A. 1970. Familial Dysautonomia. *Medicine Baltimore* 49:343–74
Bueker, E. D. 1948. Implantation of tumours in the hind limb field of the embryonic chick and the developmental response of the lumbosacral nervous system. *Anat. Rec.* 102:369–90
Bueker, E. D., Scheinkein, I., Barre, J. L. 1960. Distribution of Nerve Growth Factor specific for spinal and sympathetic ganglia. *Cancer Res.* 20:1220–27
Bueker, E. D., Weis, P., Scheinkein, I. 1965. Sexual dimorphism of mouse submaxillary glands and its relationship to nerve growth stimulating protein. *Proc. Soc. Exp. Biol. Med.* 118:204–7
Bunge, R. P., Bunge, M. B. 1965. Ultrastructural characteristics of synapses forming in cultured spinal cord. *Anat. Rec.* 151:329
Burnham, P. A., Raiborn, C., Varon, S. 1972. Replacement of Nerve Growth Factor by ganglionic non-neuronal cells for the survival in vitro of dissociated ganglionic neurons. *Proc. Natl. Acad. Sci. USA* 69:3556–60
Burnstock, G., Costa, M. 1975. *Adrenergic Neurons.* London: Chapman & Hall. 225 pp.
Burt, A. M. 1968. Acetylcholinesterase and choline acetyltransferase activity in the developing chick spinal cord. *J. Exp. Zool.* 169:107–12
Burt, A. M., Narayanan, C. H. 1976. Choline acetyltransferase, choline kinase and acetylcholinesterase during development of the chick ciliary ganglion. *Exp. Neurol.* 53:703–13
Cajal, S. R. Y. 1928. *Degeneration and Regeneration of the Nervous System,* Vols. 1, 2. London: Oxford Univ. Press. 373 pp., 769 pp.
Caramia, F., Angeletti, P. U., Levi-Montalcini, R. 1962. Experimental analysis of the mouse submaxillary salivary gland in relationship to its Nerve Growth Factor content. *Endocrinology* 70: 915–22
Caramia, F., Angeletti, P. U., Levi-Montalcini, R., Carratelli, L. 1972. Mitochondrial lesions of developing sympathetic neurons induced by bretylium tosylate. *Brain Res.* 40:237–46
Caserta, M. T., Johnson, E. M., Ross, L. L. 1976. Effect of destruction of the postganglionic sympathetic neurons in neonatal rats on development of choline acetyltransferase and survival of preganglionic cholinergic neurons. *Neurosci. Abstr.* II (Pt. I):295 (Abstr.)
Ceccarelli, B., Clementi, F., Mantegazza, P. 1971. Synaptic transmission in the superior cervical ganglion of the cat after reinnervation by the vagus. *J. Physiol. London* 216:87–98
Chamley, J. H., Campbell, G. R., Burnstock, G. 1973. An analysis of the interactions between sympathetic nerve fibers and smooth muscle cells in tissue culture. *Dev. Biol.* 33:344–61
Charlwood, K. A., Lamont, D. M., Banks, B. E. C. 1972. Apparent orientating effects produced by Nerve Growth Factor. See Zaimis & Knight 1972, pp. 102–7
Chiappinelli, V., Giacobini, E., Pilar, G., Uchimura, H. 1976. Induction of cholinergic enzymes in chick ciliary ganglion and iris muscle cells during synapse formation. *J. Physiol. London* 257:749–66
Chung, J. M., Chung, K., Wurster, R. D. 1975. Sympathetic preganglionic neurons of the cat spinal cord: horseradish peroxidase study. *Brain Res.* 91:126–31
Ciaranello, R. D., Jacobowitz, D., Axelrod, J. 1973. Effect of dexamethasone on phenylethanolamine N-methyl-transferase in chromaffin tissue of the neonatal rat. *J. Neurochem.* 20:799–805
Clark, D. W., Laverty, R., Phelan, E. L. 1972. Long-lasting peripheral and central effects of 6-hydroxydopamine in rat. *Br. J. Pharmacol.* 44:233–43
Clarke, P. G. H., Cowan, W. M. 1975. Ectopic and aberrant connections during neural development. *Proc. Natl. Acad. Sci. USA* 72:4455–58
Cohen, A. L. 1972. Expression of sympathetic traits in cells of neural crest origin. *J. Exp. Zool.* 179:167–92
Cohen, A. L. 1973. DNA synthesis and cell division in differentiating avian adrenergic neuroblasts. See Björklund, Bjerre & Stenevi 1974, pp. 359–70
Cohen, S. 1959. Purification and metabolic effects of a nerve growth-promoting protein from snake venom. *J. Biol. Chem.* 234:1129–37
Cohen, S. 1960. Purification of a nerve growth-promoting protein from mouse salivary gland and its neurocytotoxic antiserum. *Proc. Natl. Acad. Sci. USA* 46:302–11
Cohen, S., Levi-Montalcini, R. 1956. Nerve Growth Factor from snake venom. *Proc. Natl. Acad. Sci. USA* 42:571–74
Cohen, S., Levi-Montalcini, R. 1957. Purification and properties of a nerve growth-promoting factor from mouse sarcoma 180. *Cancer Res.* 17:15–20

Cohen, S., Levi-Montalcini, R., Hamburger, V. 1954. A nerve growth-stimulating factor isolated from sarcomas 37 and 180. *Proc. Natl. Acad. Sci. USA* 40:1014–18
Coughlin, M. D. 1975a. Early development of parasympathetic nerves in the mouse submandibular gland. *Dev. Biol.* 43:123–39
Coughlin, M. D. 1975b. Target organ stimulation of parasympathetic nerve growth in the developing mouse submandibular gland. *Dev. Biol.* 43:140–58
Coughlin, M. D., Boyer, D. B., Black, I. B. 1977. Embryologic development of a mouse sympathetic ganglion in vivo and in vitro. *Proc. Natl. Acad. Sci. USA* 74:3438–42
Coulombre, A. J., Johnston, M. C., Weston, J. A. 1974. Conference on neural crest in normal and abnormal embryogenesis. *Dev. Biol.* 36:f1–5
Cowan, W. M. 1970. Anterograde and retrograde transneuronal degeneration in the central and peripheral nervous system. In *Contemporary Research Methods in Neuroanatomy,* ed. W. J. H. Nanta, S. O. E. Ebbeson, pp. 217–51. New York: Springer. 386 pp.
Cowan, W. M., Wenger, E. 1968. Degeneration of the cells of origin of the ciliary ganglion after early removal of the optic vesicle. *J. Exp. Zool.* 168:105–24
Crain, S. M., Peterson, E. R. 1974. Development of neural connections in culture. *Ann. NY Acad. Sci.* 228:6–34
Dahlström, A. 1971. Axoplasmic transport (with particular respect to adrenergic neurons). *Philos. Trans. R. Soc. London Ser. B* 261:325–58
de Champlain, J., Malmfors, T., Olson, L., Sachs, C. 1970. Ontogenesis of peripheral adrenergic neurons in the rat: pre- and post-natal observations. *Acta Physiol. Scand.* 80:276–88
de Champlain, J., Nadeau, R. 1971. 6-hydroxydopamine, 6-hydroxydopa and degeneration of adrenergic nerves. *Fed. Proc.* 30:877–85
Detwiler, S. R. 1934. An experimental study of spinal nerve segmentation in Amblystoma with reference to the plurisegmental contribution to the brachial plexus. *J. Exp. Zool.* 67:395–441
Detwiler, S. R. 1937. Observations upon the migration of neural crest cells and upon the development of the spinal ganglia and vertebral arches in Amblystoma. *Am. J. Anat.* 61:63–94
Dewey, M. J., Yervais, A. G., Mintz, B. 1976. Brain and ganglion development from two genotypic classes of cells in allophenic mice. *Dev. Biol.* 50:68–81
Deza, L., Eidelberg, E. 1967. Development of cortical electrical activity in the rat. *Exp. Neurol.* 17:425–38
Diamond, J. 1976. Trophic regulation of nerve sprouting. *Science* 193:371–77
Dibner, M. D., Black, I. B. 1976a. The effect of target organ removal on the development of sympathetic neurons. *Brain Res.* 103:93–102
Dibner, M. D., Black, I. B. 1976b. Elevation of sympathetic ganglion tyrosine hydroxylase activity in neonatal and adult rats by testosterone treatment. *J. Neurochem.* 27:323–24
Dibner, M. D., Mytilineou, C., Black, I. B. 1977. Target organ regulation of sympathetic neuron development. *Brain Res.* 123:301–10
Dibner, M. D., Black, I. B. 1978. Biochemical and morphological effects of testosterone treatment on developing sympathetic neurons. *J. Neurochem.* In press
Eayrs, J. T., Goodhead, B. 1959. Postnatal development of the cerebral cortex in the rat. *J. Anat.* 93:385–402
Ellison, N., Weller, J. L., Klein, D. C. 1972. Development of a circadian rhythm in the activity of pineal serotonin-N-acetyltransferase. *J. Neurochem.* 19: 1335–41
Enemar, A., Falck, B., Hakanson, R. 1965. Observations on the appearance of norepinephrine in the sympathetic nervous system of the chick embryo. *Dev. Biol.* 11:268–83
Eränkö, L. 1972. Ultrastructure of the developing sympathetic nerve cell and the storage of catecholamines. *Brain Res.* 46:159–75
Eränkö, L., Eränkö, O. 1972. Effect of 6-hydroxydopamine on the ganglion cells and the small intensely fluorescent cells in the superior cervical ganglion of the rat. *Acta Physiol. Scand.* 84:115–24
Eränkö, O., Eränkö, L. 1971. Histochemical evidence of chemical sympathectomy by guanethidine in newborn rats. *Histochem. J.* 3:451–56
Eränkö, O., Eränkö, L. 1973. Effects of drugs on sympathetic nerve cells and small intensely fluorescent (SIF) cells. See Cohen 1973, pp. 181–90
Falck, B., Hillarp, N. A., Thieme, G., Torp, A. 1962. Fluorescence of catecholamines and related compounds condensed with formaldehyde. *J. Histochem. Cytochem.* 10:348–54
Finch, L., Haeusler, G., Thoenen, H. 1973. A comparison of the effects of chemical

sympathectomy by 6-hydroxydopamine in newborn and adult rats. *Br. J. Pharmacol.* 47:249–60

Fonnum, F. 1970. Studies of choline acetyltransferase with particular reference to its subcellular localization. *Norw. Def. Res. Establ. Rep. No. 58.*

Frazier, W. A., Boyd, L. F., Bradshaw, R. A. 1974. Properties of specific binding of ^{125}I-Nerve Growth Factor to responsive peripheral neurons. *J. Biol. Chem.* 249:5513–19

Frazier, W. A., Boyd, L. F., Szutowicz, A., Pulliam, M. W., Bradshaw, R. A. 1974. Specific binding-sites for ^{125}I–Nerve Growth Factor in peripheral tissues and brain. *Biochem. Biophys. Res. Commun.* 57:1096–1103

Fusari, R. 1893. Contribution à l'étude du dévelopment des capsules surrénales et du sympathique chez le poulet et chez les mammifères. *Arch. Ital. Biol.* 18:161–82

Gaze, R. M. 1970. *The Formation of Nerve Connections.* London: Academic. 288 pp.

Gaze, R. M., Keating, M. J., eds. 1974. *Development and Regeneration in the Nervous System—Br. Med. Bull.* Vol. 30. 194 pp.

Giacobini, E. 1970. Biochemistry of synaptic plasticity studies in single neurons. In *Biochemistry of Simple Neuronal Models, Advances in Biochemical Psychopharmacology,* ed. E. Costa, E. Giacobini, 2:9–64. New York: Raven Press. 382 pp.

Giacobini, G., Marchisio, P. C., Giacobini, E., Koslow, S. H. 1970. Developmental changes of cholinesterases and monoamine oxidase in chick embryo spinal and sympathetic ganglia. *J. Neurochem.* 17:1177

Giller, E. L., Schrier, B. K., Shainberg, A., Fisk, H. R., Nelson, P. G. 1973. Choline acetyltransferase activity is increased in combined cultures of spinal cord and muscle cells from mice. *Science* 182:588–89

Goodman, L. S., Gilman, A., eds. 1975. *The Pharmacological Basis of Therapeutics,* p. 434. New York: McMillan. 1704 pp. 5th ed.

Greene, L. A., Shooter, E. M., Varon, S. 1969. Subunit interaction and enzymatic activity of mouse 7s Nerve Growth Factor. *Biochemistry* 8: 3735–41

Gresik, E. W. 1976. Preliminary observations on the effects of chronic hypothyroidism on the development of the superior cervical ganglion of the rat. *Brain Res.* 110:619–22

Grillo, M. A. 1966. Electron microscopy of sympathetic tissues. *Pharmacol. Rev.* 18:387–99

Gudden, B. 1870. Experimental Untersuchungen über das peripherische und centrale Nervensystem. *Arch. Psychiatr. Nervenkr.* 2:693–723

Guth, L. 1968. "Trophic" influences of nerve on muscle. *Physiol. Rev.* 48:645–47

Gyllensten, L., Malmfors, T., Norrlin, M. L. 1965. Effect of visual deprivation on the optic centers of growing and adult mice. *J. Comp. Neurol.* 124:149–60

Hakanson, R., Lombard des Gouttes, M.-N., Owman, C. 1967. Activities of tryptophan hydroxylase, DOPA decarboxylase, and monoamine oxidase as correlated with the appearance of monoamines in developing rat pineal gland. *Life Sci.* 6:2577–85

Hamberger, B., Levi-Montalcini, R., Norberg, K.-A., Sjoqvist, F. 1965. Monoamines in immunosympathectomized rats. *Int. J. Neuropharmacol.* 4:91–95

Hamberger, B., Norberg, K.-A., Sjoqvist, F. 1963. Evidence for adrenergic nerve terminals and synapses in sympathetic ganglia. *Int. J. Neuropharmacol.* 2: 279–82

Hamburger, V. 1958. Regression versus peripheral control of differentiation in motor hypoplasia. *Am. J. Anat.* 102:365–410

Hamburger, V. 1975. Cell death in the development of the lateral motor column of the chick embryo. *J. Comp. Neurol.* 160:535–46

Hamburger, V. 1978. Regulation of cell numbers in neuronal units. In *Neurosci. Res. Program Bull.* In press

Hamill, R. W., Bloom, E. M., Black, I. B. 1977. The effect of spinal cord transection on the development of cholinergic and adrenergic sympathetic neurons. *Brain Res.* 134:269–78

Hammond, W. S., Yntema, C. L. 1947. Depletion in the thoracolumbar sympathetic system following removal of neural crest in the chick. *J. Comp. Neurol.* 85:237–65

Hancock, M. B. 1976. Cells of origin of hypothalamo-spinal projections in the rat. *Neurosci. Lett.* 3:179–84

Harris, A. J. 1974. Inductive functions of the nervous system. *Ann. Rev. Physiol.* 36:251–305

Heath, J. W., Evans, B. K., Burnstock, G. 1973. Axon retraction following guanethidine treatment. *Z. Zellforsch. Mikrosk. Anat.* 146:439–51

Hebb, C. O. 1956. Choline acetylase in the developing nervous system of the rabbit and guinea pig. *J. Physiol. London* 133:566–70

Hebb, C. O., Waites, G. M. H. 1956. Choline acetylase in antero- and retrograde degeneration of a cholinergic nerve. *J. Physiol. London* 132:667–71

Helfand, S. L., Smith, G. A., Wessells, N. K. 1976. Survival and development in culture of dissociated parasympathetic neurons from ciliary ganglia. *Dev. Biol.* 50:541–47

Hendry, I. A. 1972. Developmental changes in tissue and plasma concentrations of the biologically active species of Nerve Growth Factor in the mouse, by using a two site radioimmunoassay. *Biochem. J.* 128:1265–72

Hendry, I. A. 1973. Trans-synaptic regulation of tyrosine hydroxylase activity in a developing mouse sympathetic ganglion: effects of Nerve Growth Factor (NGF), NGF-antiserum and pempidine. *Brain Res.* 56:313–20

Hendry, I. A. 1975a. The retrograde trans-synaptic control of the development of cholinergic terminals in sympathetic ganglia. *Brain Res.* 86:483–87

Hendry, I. A. 1975b. The response of adrenergic neurons to axotomy and Nerve Growth Factor. *Brain Res.* 94:87–97

Hendry, I. A. 1975c. The effects of axotomy on the development of the rat superior cervical ganglion. *Brain Res.* 90:235–44

Hendry, I. A. 1976. Control in the development of the vertebrate sympathetic nervous system. In *Reviews of Neuroscience,* ed. S. Ehrenpreis, I. J. Kopin, 2:149–94. New York: Raven Press. 270 pp.

Hendry, I. A., Campbell, J. 1975. Morphometric analysis of rat superior cervical ganglion after axotomy and Nerve Growth Factor treatment. *J. Neurocytol.* 5:351–60

Hendry, I. A., Iversen, L. L. 1971. Effect of Nerve Growth Factor and its antiserum on tyrosine hydroxylase activity in the mouse superior cervical ganglion. *Brain Res.* 29:159–62

Hendry, I. A., Iversen, L. L. 1973. Changes in tissue and plasma concentrations of Nerve Growth Factor following removal of the submaxillary glands in adult mice and their effects on the sympathetic nervous system. *Nature* 243: 500–4

Hendry, I. A., Stach, R., Herrup, K. 1974. Characteristics of the retrograde axonal transport system for Nerve Growth Factor in the sympathetic nervous system. *Brain Res.* 82:117–28

Hendry, I. A., Stöckel, K., Thoenen, H., Iversen, L. L. 1974. The retrograde axonal transport of Nerve Growth Factor. *Brain Res.* 68:103–21

Hendry, I. A., Thoenen, H. 1974. Changes of enzyme pattern in the sympathetic nervous system of adult mice after submaxillary gland removal: response to exogenous Nerve Growth Factor. *J. Neurochem.* 22:999–1004

Herrup, K., Shooter, E. M. 1973. Properties of the Nerve Growth Factor receptor of avian dorsal root ganglion. *Proc. Natl. Acad. Sci. USA* 70:3884–88

Hess, A. 1958. Optic centers and pathways after eye removal in fetal guinea pigs. *J. Comp. Neurol.* 109:91–115

Hier, D. B., Arnason, B. G. W., Young, M. 1973. Nerve Growth Factor: relationship to the cyclic AMP system of sensory ganglia. *Science* 182:79–81

His, W. 1868. Untersuchungen über die erste Anlage des Wirbeltierleibes. In *Die erste Entwicklung des Hühnchens im Ei.* Leipzig: F. C. W. Vogel. 237 pp.

Hoffer, B. Olson, L., Seiger, A., Bloom, F. 1976. Formation of a functional adrenergic input to intraocular cerebellar grafts: ingrowth of inhibitory sympathetic fibers. *J. Neurobiol.* 6:565–85

Hollyday, M., Hamburger, V. 1975. Reduction of the naturally occurring motor neuron loss by enlargement of the periphery. *J. Comp. Neurol.* 170:311–20

Hörstadius, S. 1950. *The Neural Crest.* London: Oxford Univ. Press. 111 pp.

Ignarro, L. J., Shideman, F. E. 1968. The appearance and concentrations of catecholamines and their biosynthetic enzymes in embryonic and developing chick. *J. Pharmacol. Exp. Ther.* 159: 38–48

Ishii, D. N., Shooter, E. M. 1975. Regulation of Nerve Growth Factor synthesis in mouse submaxillary glands by testosterone. *J. Neurochem.* 25:843–51

Iversen, L. L., de Champlain, J., Glowinski, J., Axelrod, J. 1967. Uptake, storage and metabolism of norepinephrine in tissues of the developing rat. *J. Pharmacol. Exp. Ther.* 157:509–16

Iversen, L. L., Glowinski, J., Axelrod, J. 1966. The physiologic disposition and metabolism of norepinephrine in immunosympathectomized animals. *J. Pharmacol. Exp. Ther.* 151:273–84

Jacobson, M. 1970. *Developmental Neurobiology*. New York: Holt, Rinehart & Winston. 465 pp.

Jacoby, F., Leeson, C. R. 1959. The postnatal development of the rat submaxillary gland. *J. Anat.* 93:201–16

Jaim-Etcheverry, G., Zieher, L. M. 1971. Permanent depletion of peripheral norepinephrine in rats treated at birth with 6-hydroxydopamine. *Eur. J. Pharmacol.* 13:272–76

Jensen-Holm, J., Juul, P. 1970. The effects of guanethidine, pre- and post-ganglionic nerve division of the rat superior cervical ganglion: cholinesterases and catecholamines (histochemistry), and histology. *Acta Pharmacol. Toxicol.* 28: 283–98

Joh, T. H., Geghman, C., Reis, D. J. 1973. Immunochemical demonstration of increased accumulation of tyrosine hydroxylase protein in sympathetic ganglia and adrenal medulla elicited by reserpine. *Proc. Natl. Acad. Sci. USA* 70:2767–71

Johnson, D. G., Silberstein, S. D., Hanbauer, I., Kopin, I. J. 1972. The role of Nerve Growth Factor in the ramification of sympathetic nerve fibres into the rat iris in organ culture. *J. Neurochem.* 19: 2025–29

Johnson, E. M., Aloe, L. 1974. Suppression of the in vitro and in vivo cytotoxic effects of guanethidine in sympathetic neurons by Nerve Growth Factor. *Brain Res.* 81:519–32

Jonsson, G., Malmfors, T., Sachs, C., eds. G., Malmfors, *Chemical Tools in Catecholamine Research I: 6-Hydroxydopamine as a Denervation Tool in Catecholamine Research.* Amsterdam: North Holland. 372 pp.

Jonsson, G., Pycock, C., Fuxe, K., Sachs, C. 1974. Changes in the development of central noradrenaline neurones following neonatal administration of 6-hydroxydopamine. *J. Neurochem.* 22: 419–26

Katzman, R., Björklund, A., Owman, C., Stenevi, U., West, K. A. 1971. Evidence for regenerative axon sprouting of central catecholamine neurons in the rat mesencephalon following electrolytic lesions. *Brain Res.* 25:579–96

Kirby, M. L., Gilmore, S. A. 1972. A fluorescence study on the ability of the notochord to synthesize and store catecholamines in early chick embryos. *Anat. Rec.* 173:469–77

Klingman, G. I. 1965. Catecholamine levels and DOPA-decarboxylase activity in peripheral organs and adrenergic tissue in the rat after immunosympathectomy. *J. Pharmacol. Exp. Ther.* 148:14–21

Klingman, G. I. 1966. In utero immunosympathectomy of mice. *Int. J. Neuropharmacol.* 5:163–70

Klingman, G. I., Klingman, J. D. 1965. Effects of immunosympathectomy on the superior cervical ganglion and other adrenergic tissues of the rat. *Life Sci.* 4:2171–79

Klingman, G. I., Klingman, J. D. 1967. Prenatal and postnatal treatment of mice with antiserum to Nerve Growth Factor. *Int. J. Neuropharmacol.* 6:501–8

Kostrzewa, R. M., Harper, J. W. 1974. Effect of 6-hydroxydopa on catecholamine-containing neurons in brains of newborn rats. *Brain Res.* 69:174–81

Kuntz, A. 1910. Sympathetic system in birds. *J. Comp. Neurol.* 20:283–308

Kvetňansky, R. 1973. Trans-synaptic and humoral regulation of adrenal catecholamine synthesis in stress. In *Frontiers in Catecholamine Research,* ed. E. Usdin, S. Snyder, 223–29. New York: Pergamon Press. 1219 pp.

Landmesser, L., Morris, D. G. 1975. The development of functional innervation in the hind limb of the chick embryo. *J. Physiol. London* 249:301–25

Landmesser, L., Pilar, G. 1974. Synapse formation during embryogenesis on ganglion cells lacking a periphery. *J. Physiol. London* 241:715–36

Langley, J. N. 1898. On the union of cranial autonomic (visceral) fibres with the nerve cells of the superior cervical ganglion. *J. Physiol. London* 23:240–70

Larrabee, M. G. 1970. Metabolism of adult and embryonic sympathetic ganglia. *Fed. Proc.* 29:1919–28

Lawrence, I. E., Burden, H. W. 1973. Catecholamines and morphogenesis of the chick neural tube and notochord. *Am. J. Anat.* 137:199–207

Le Douarin, N. M., Renaud, D., Teillet, M. A., Le Douarin, G. H. 1975. Cholinergic differentiation of presumptive adrenergic neuroblasts in interspecific chimeras after heterotopic transplantations. *Proc. Natl. Acad. Sci. USA* 72:728–32

Le Douarin, N. M., Rival, J. M. 1975. A biological nuclear marker in cell culture: recognition of nuclei in single cells and in heterokaryons. *Dev. Biol.* 47:215–21

Le Douarin, N. M., Teillet, M. A. M. 1974. Experimental analysis of the migration and differentiation of neuroblasts of the autonomic nervous system and of neurectodermal mesenchymal derivatives using a biological cell marking technique. *Dev. Biol.* 41:162–84

Letourneau, P. C. 1975. Cell-to-substratum adhesion and guidance of axonal elongation. *Dev. Biol.* 44:92–101

Levi-Montalcini, R. 1947. Regressione secondaria del ganglio ciliare dopo asportazione della vesciola mesencefalica in embrione di pollo. *R. Acad. Naz. Lincei* 3:144–46

Levi-Montalcini, R. 1949. The development of the acoustico-vestibular centers in the chick embryo in the absence of the afferent root fibers and of descending fiber tracts. *J. Comp. Neurol.* 91:209–41

Levi-Montalcini, R. 1950. The origin and development of the visceral system in the spinal cord of the chick embryo. *J. Morphol.* 86:253–84

Levi-Montalcini, R., 1964a. Growth control of nerve cells. *Science* 143:105–10

Levi-Montalcini, R. 1964b. The Nerve Growth Factor. *Ann. N.Y. Acad. Sci.* 118:149–70

Levi-Montalcini, R. 1966. The Nerve Growth Factor: its mode of action on sensory and sympathetic nerve cells. *Harvey Lect.* 60:217–59

Levi-Montalcini, R., Aloe, L., Mugnaini, E., Oesch, F., Thoenen, H. 1975. Nerve Growth Factor induces volume increase and enhances tyrosine hydroxylase synthesis in chemically axotomized sympathetic ganglia of newborn rats. *Proc. Natl. Acad. Sci. USA* 72:595–99

Levi-Montalcini, R., Angeletti, P. U. 1961. Growth control of the sympathetic system by a specific protein factor. *Q. Rev. Biol.* 36:99–108

Levi-Montalcini, R., Angeletti, P. U. 1968. Nerve Growth Factor. *Physiol. Rev.* 48:534–69

Levi-Montalcini, R., Booker, B. 1960a. Excessive growth of the sympathetic ganglia evoked by a protein isolated from mouse salivary glands. *Proc. Natl. Acad. Sci. USA* 46:373–83

Levi-Montalcini, R., Booker, B. 1960b. Destruction of the sympathetic ganglia in mammals by antiserum to a Nerve Growth Factor. *Proc. Natl. Acad. Sci. USA* 46:384–91

Levi-Montalcini, R., Hamburger, V., 1951. Selective growth stimulating effects of mouse sarcoma on the sensory and sympathetic nervous system of the chick embryo. *J. Exp. Zool.* 116:321–62

Levi-Montalcini, R., Hamburger, V. 1953. A diffusable agent of mouse sarcoma producing hyperplasia of sympathetic ganglia and hyperneurotization of viscera in the chick embryo. *J. Exp. Zool.* 123:233–78

Levitt, M., Spector, S., Sjoerdsman, A., Udenfriend, S. 1965. Elucidation of the rate-limiting step in norepinephrine biosynthesis in the perfused guinea pig heart. *J. Pharmacol. Exp. Ther.* 148: 1–8

Lognado, J. R., Hardy, M. 1967. Brain esterases during development. *Nature* 214: 1207–10

Lubinska, L. 1975. On axoplasmic flow. In *International Review of Neurobiology,* ed. C. C. Pfeiffer, J. R. Smythies, 17:241–96. New York: Academic 358 pp.

Ludueña, M. A. 1973. Nerve cell differentiation in vitro. *Dev. Biol.* 33:268–84

Maeda, T., Tohyama, M., Shimizu, N. 1974. Modification of postnatal development of neocortex in rat brain with experimental deprivation of locus coeruleus. *Brain Res.* 70:515–20

McEwen, B. S., Grafstein, B. 1968. Fast and slow components in axonal transport of protein. *J. Cell Biol.* 38:494–508

McLachlan, E. M. 1974. The formation of synapses in mammalian sympathetic ganglia reinnervated with preganglionic or somatic nerves. *J. Physiol.* 237: 217–42

Molinoff, P. B., Axelrod, J. 1971. Biochemistry of catecholamines. *Ann. Rev. Biochem.* 40:465–500

Molinoff, P. B., Brimijoin, S., Axelrod, J. 1972. Induction of dopamine-β-hydroxylase and tyrosine hydroxylase in rat hearts and sympathetic ganglia. *J. Pharmacol. Exp. Ther.* 182:116–29

Molinoff, P. B., Brimijoin, S., Weinshilboum, R., Axelrod, J. 1970. Neurally mediated increase in dopamine-β-hydroxylase activity. *Proc. Natl. Acad. Sci. USA* 66:453–58

Monard, D., Solomon, F., Rentsch, M., Gysin, R. 1973. Glia-induced morphological differentiation in neuroblastoma cells. *Proc. Natl. Acad. Sci. USA* 70:1894–97

Monnier, M. 1968. Functions of the nervous system. In *General Physiology, Autonomic Functions,* 1:91–129. Amsterdam: Elsevier

Mueller, R. A., Thoenen, H., Axelrod, J. 1969. Compensatory increase in adrenal tyrosine hydroxylase activity after chemical sympathectomy. *Science* 163: 468–69

Mueller, R. A., Thoenen, H., Axelrod, J. 1970. Inhibition of neuronally induced tyrosine hydroxylase by nicotinic receptor blockade. *Eur. J. Pharmacol.* 10:51–56

Müller, E., Ingvar, S. 1923. Über den Ursprung des Sympathicus beim Hühnchen. *Arch. Mikrosk. Anat. Entwicklungsmech.* 99:650–71

Norr, S. C. 1973. In vitro analysis of sympathetic neuron differentiation with chick neural crest cells. *Dev. Biol.* 34:16–38

Oger, J., Arnason, B. G. W., Pantazis, N., Lehrich, J., Young, M. 1974. Synthesis of Nerve Growth Factor by L and 3T3 cells in culture. *Proc. Natl. Acad. Sci. USA* 71:1554–58

Olson, L. 1967. Outgrowth of sympathetic adrenergic neurons in mice treated with Nerve Growth Factor. *Z. Zellforsch. Mikrosk. Anat.* 81:155–73

Olson, L., Malmfors, T. 1970. Growth characteristics of adrenergic nerves in the adult rat. *Acta Physiol. Scand. Suppl.* 348:1–111

Olson, L., Seiger, A. 1972. Brain tissue transplanted to the anterior chamber of the eye: 1, fluorescence histochemistry of immature catecholamine and 5-hydroxytryptamine neurons reinnervating the rat iris. *Z. Zellforsch. Mikrosk. Anat.* 135:175–94

Olson, L., Seiger, A. 1973. See Cohen 1973, pp. 499–508

Olson, M. I., Bunge, R. P. 1973. Anatomical observations on the specificity of synapse formation in tissue culture. *Brain Res.* 59:19–33

Östberg, A.-J. C., Raisman, G., Field, P. M., Iversen, L. L., Zigmond, R. E. 1976. A quantitative comparison of the formation of synapses in the rat superior cervical sympathetic ganglion by its own and by foreign nerve fibers. *Brain Res.* 107:445–70

Owman, C., Sjoberg, N.-O., Swedin, G. 1971. Histochemical and chemical studies on pre- and postnatal development of the different systems of "short" and "long" adrenergic neurones in peripheral organs of the rat. *Z. Zellforsch. Mikrosk. Anat.* 116:319–41

Paigen, K. 1971. The genetics of enzyme realization. In *Enzyme Synthesis and Degradation in Mammalian Systems,* ed. M. Rechcigl Jr., 1–46. Baltimore: Univ. Park Press. 477 pp.

Papka, R. E. 1972. Ultrastructural and fluorescence histochemical studies of developing sympathetic ganglia in the rabbit. *Am. J. Anat.* 134:337–64

Paravicini, U., Stöckel, K., Thoenen, H. 1975. Biological importance of retrograde axonal transport of Nerve Growth Factor in adrenergic neurons. *Brain Res.* 84:279–91

Patrick, R. L., Kirschner, N. 1972. Developmental changes in rat adrenal tyrosine hydroxylase, dopamine-β-hydroxylase and catecholamine levels: effect of denervation. *Dev. Biol.* 29:204–13

Perez-Polo, J. R., De Jong, W. W., Straus, D., Shooter, E. M. 1972. The physical and biological properties of 7s and beta-Nerve Growth Factor from the submaxillary gland. *Adv. Exp. Med. Biol.* 32:91–97

Peters, V. B., Flexner, L. B. 1950. Biochemical and physiological differentiation during morphogenesis: VIII. Quantitative morphologic studies on the developing cerebral cortex of the fetal guinea pig. *Am. J. Anat.* 86:133–57

Phillipson, O. T., Sandler, M. 1975. The influence of Nerve Growth Factor, potassium depolarization and dibutyryl (cyclic) adenosine 3',5'-monophosphate on explant cultures of chick embryo sympathetic ganglia. *Brain Res.* 90:273–81

Pick, J. 1963. The submicroscopic organization of the sympathetic ganglion in the frog (*Rana pipiens*). *J. Comp. Neurol.* 120:409–62

Pick, J. 1970. *The Autonomic Nervous System—Morphological, Comparative, Clinical and Surgical Aspects.* Philadelphia: Lippincott. 483 pp.

Pilar, G., Jenden, D. J., Campbell, B. 1973. Distribution of acetylcholine in the normal and denervated pigeon ciliary ganglion. *Brain Res.* 49:245–56

Prestige, M. C. 1967. The control of cell numbers in the lumbar ventral horns during the development of *Xenopus laevis* tadpoles. *J. Embryol. Exp. Morphol.* 18:359–87

Prestige, M. C. 1970. Differentiation, degeneration and the role of the periphery: quantitative considerations. In *The Neurosciences: Second Study Program,* ed. F. O. Schmitt, 73–82. New York: Rockefeller Univ. Press. 1068 pp.

Prestige, M. C. 1974. Axon and cell numbers in the developing nervous system. *Br. Med. Bull.* 30:107–11

Prestige, M. C., Wilson, M. A. 1972. Loss of axons from ventral roots during development. *Brain Res.* 41:467–70

Purves, D. 1976a. Long-term regulation in the vertebrate peripheral nervous system. *Int. Rev. Physiol. Neurophysiol.* 10:125–77

Purves, D. 1976b. Reinnervation of mammalian sympathetic neurons. *J. Physiol. London* 261:453–75

Raven, C. P. 1931. Zur Entwicklung der Ganglienleiste. I. Die Kinematik der Ganglienleistenentwicklung bei den Urodelen. *Roux Arch. Entw. Mech.* 125:210–92

Raven, C. P., Kloos, J. 1945. Induction by medial and lateral pieces after archenteron roof, with special reference to the determination of the neural crest. *Acta Neerl. Morphol. Norm. Pathol.* 5: 348–62

Sachs, C., Pycock, C., Jonsson, G. 1974. Altered development of central noradrenaline neurons during ontogeny by 6-hydroxydopamine. *Med. Biol.* 52: 55–65

Saper, C. B., Loewy, A. D., Swanson, L. W., Cowan, W. M. 1976. Direct hypothalamo-autonomic connection. *Brain Res.* 117:305–12

Saxén, L., Toivonen, S. 1962. *Primary Embryonic Induction.* Englewood Cliffs, NJ: Prentice-Hall. 271 pp.

Schucker, F. 1972. Effects of Nerve Growth Factor antiserum in sympathetic neurons during early postnatal development. *Exp. Neurol.* 36:59–78

Schwab, M. E., Stöckel, K., Thoenen, H. 1976. Immunocytochemical localization of Nerve Growth Factor in the submandibular gland of adult mice by light and electron microscopy. *Cell. Tissue Res.* 169:289–99

Schweiler, G. H., Douglas, J. S., Bouhys, A. 1970. Postnatal development of the autonomic efferent innervation in the rabbit. *Am. J. Physiol.* 219:391–97

Shieh, P. 1951. The neoformation of cells of preganglionic type in the cervical spinal cord of the chick embryo following its transplantation to the thoracic level. *J. Exp. Zool.* 117:354–95

Shy, G. M., Drager, G. A. 1960. A neurological syndrome associated with Orthostatic Hypotension. *AMA Arch. Neurol.* 2:511–27

Siggers, D. C., Rogers, J. G., Boyer, S. H., Margolet, L., Dorkin, H., Banerjee, S. P., Shooter, E. M. 1976. Increased Nerve Growth Factor β-chain cross-reacting material in Familial Dysautonomia. *N. Engl. J. Med.* 295:629–34

Singh, B., de Champlain, J. 1974. Ontogenesis of sympathetic fibers after neonatal 6-hydroxydopamine treatment in the rat. *Can. J. Physiol. Pharmacol.* 52: 304–18

Sisken, B. F., Smith, S. D. 1975. The effects of minute direct electrical currents on cultured chick embryo trigeminal ganglia. *J. Embryol. Exp. Morphol.* 33:29–41

Sjoqvist, F. 1962. *Cholinergic Sympathetic Ganglion Cells,* Stockholm: P. A. Norstedt & Söner. 41 pp.

Smith, A. P., Greene, L. A., Fisk, H. R., Varon, S., Shooter, E. M. 1969. Subunit equilibria of the 7s Nerve Growth Factor. *Biochemistry* 8:4918–26

Smith, B. H., Kreutzberg, G. W., eds. 1976. *Neurosciences Research Program Bulletin: Neuron–Target Cell Interactions,* Vol. 14. 453 pp.

Sorimachi, M., Kataoka, K. 1974. Developmental change of choline acetyltransferase and acetylcholinesterase in the ciliary and the superior cervical ganglion of the chick. *Brain Res.* 70:123–30

Stöckel, K., Paravicini, U., Thoenen, H. 1974. Specificity of the retrograde axonal transport of Nerve Growth Factor. *Brain Res.* 76:413–22

Tello, J. F. 1925. Sur la formation des chaînes primaire et secondaire du grand sympathique dans l'embryon de poulet. *Trav. Lab. Invest. Biol. Univ. Madrid* 23:1–28

Tennyson, V. 1965. Electron microscopic study of the developing neuroblast of the dorsal root ganglion of the rabbit embryo. *J. Comp. Neurol.* 124:267–317

Terni, T. 1923. Ricerche anatomique sul sistema nervoso autonomie degli uccelli. 1. Il sistema pregangliara spinale. *Arch. Ital. Anat. Embriol.* 20:433–510

Thoenen, H. 1971. Biochemical alterations induced by 6-hydroxydopamine in peripheral adrenergic neurons. In *6-hydroxydopamine and Catecholamine Neurones,* ed. T. Malmfors, H. Thoenen, pp. 75–85. New York: Elsevier. 368 pp.

Thoenen, H., Mueller, R. A., Axelrod, J. 1969a. Increased tyrosine hydroxylase activity after drug-induced alteration of sympathetic transmission. *Nature* 221: 1264

Thoenen, H., Muller, R. A., Axelrod, J. 1969b. Trans-synaptic induction of adrenal tyrosine hydroxylase. *J. Pharmacol. Exp. Ther.* 169:249–54

Thoenen, H., Mueller, R. A., Axelrod, J. 1970. Phase difference in the induction of tyrosine hydroxylase in cell body and nerve terminals of sympathetic neurones. *Proc. Natl. Acad. Sci. USA* 65:58–62

Thoenen, H., Saner, A., Angeletti, P. U., Levi-Montalcini, R. 1972. Increased activity of choline acetyltransferase in sympathetic ganglia after prolonged administration of Nerve Growth Factor. *Nature New Biol.* 236:26–28

van Campenhout, E. 46. The epithelioneural bodies. *Q. Rev. Biol.* 21:327–47

Varon, S., Nomura, J., Shooter, E. M. 1968. Reversible dissociation of mouse Nerve Growth Factor protein into different subunits. *Biochemistry* 7:1296–1303

Varon, S., Raiborn, C., Burnham, P. A. 1974. Implication of a Nerve Growth Factor–like antigen in the support derived by ganglionic neurons from their homologous glia in dissociated cultures. *Neurobiology* 4:317–27

Waddell, W. R., Goldstein, M. N., Bradshaw, R. A., Kirsch, W. M. 1972. Nerve Growth Factor in a patient with liposarcoma. *Lancet* 24:1365–67

Wakade, A. R., Kirpekar, S. M. 1973. "Trophic" influence on the sympathetic nerves of the vas deferens and seminal vesicle of the guinea pig. *J. Pharmacol. Exp. Ther.* 186:528–36

Weston, J. A. 1963. A radioautographic analysis of the migration and localization of trunk neural crest cells in the chick. *Dev. Biol.* 6:279–310

Weston, J. A. 1970. Interaction in neural crest development. *Adv. Morphog.* 8: 41–114

Weston, J. A., Butler, S. L. 1966. Temporary factors affecting localization of neural crest cells in the chick embryo. *Dev. Biol.* 14:246–66

Wiesel, T. N., Hubel, D. H. 1963. The effects of visual deprivation on morphology and physiology of cells in the cat's lateral geniculate body. *J. Neurophysiol.* 26:978–93

Williams, T. H., Palay, S. L. 1969. Ultrastructure of the small neurons in the superior cervical ganglion. *Brain Res.* 15:17–34

Willier, B. H. 1930. A study of the origin and differentiation of the suprarenal gland in the chick embryo by chorio-allantoic grafting. *Physiol. Zool.* 3:201–25

Yntema, C. L., Hammond, W. S. 1945. Depletions and abnormalities in the cervical sympathetic system of the chick following extirpation of the neural crest. *J. Exp. Zool.* 100:237–63

Young, M., Oger, J., Blanchard, M. H., Asdourian, H., Amos, H., Arnason, B. G. W. 1975. Secretion of Nerve Growth Factor by primary chick fibroblast cultures. *Science* 187:361–62

Zaimis, E., Knight, J., eds. 1972. *Nerve Growth Factor and its Antiserum.* London: Athlone Press. 273 pp.

Ann. Rev. Neurosci. 1978. 1:215–96

RECENT ADVANCES IN NEUROANATOMICAL METHODOLOGY

❖11509

E. G. Jones
Department of Anatomy and Neurobiology, Washington University School of Medicine, St. Louis, Missouri 63110

B. K. Hartman
Department of Psychiatry, Washington University School of Medicine, St. Louis, Missouri 63110

INTRODUCTION

The last five years have witnessed a considerable resurgence of interest in fundamental neuroanatomy. This renewed interest has been associated, to a large extent, with the development of new techniques for the analysis of fiber pathways in the central nervous system (CNS), but there have been concomitant advances in techniques for functional and metabolic mapping of neurons and pathways, in methods for intracellular labeling of single neurons, and in certain methods of quantitative analysis. In this article we deal in a rather selective way with certain aspects of these new developments. Because our review is not intended to be comprehensive, reference to certain aspects will inevitably be omitted or mentioned only in passing, but we hope that we have focused upon those that appear to be the most widely applicable.

CONNECTION TRACING METHODS

The rejuvenescence of conventional, fiber-tracing neuroanatomy stems primarily from the introduction of two new methods that are physiologically based, and therefore, obviate some of the problems inherent in the long-established axonal degeneration methods. The well-known capriciousness of the axonal degeneration methods, particularly their relative lack of success in impregnating fine-fibered systems, coupled with the almost inevitable degeneration of fiber systems or cells additional to those under investigation, has considerably reduced their use. Many of the newer methods, particularly those involving autoradiographic and immuno-

0147-006X/78/0325-0215$01.00

logical techniques, probably require no less attention to detail in achieving satisfactory results than do the degeneration methods. (This may account for the much greater proliferation of papers based upon the relatively simpler method of horseradish peroxidase histochemistry.) However, these newer methods have generally come to be regarded as inherently more reliable. Among their most satisfying attributes, the new methods permit a better delineation of the cell types that give rise to a pathway and greater precision in delineating the levels of axonal termination. Identification of cellular origins had previously only been available in the vertebrate central nervous system from painstaking electrophysiological work involving antidromic activation of long tract cells, coupled with intracellular injection of dyes or other markers (e.g. Bryan et al 1972, Trevino et al 1973, see also Snow et al 1976, Jankowska et al 1976). Similarly, although methods for estimating levels of termination were available, their use was restricted by the difficulty in many systems of distinguishing terminal from fiber degeneration, or by the small size of the samples examined electron microscopically. As well as obviating these problems, the newer methods also permit a much more satisfactory correlation of the experimental data with cytoarchitecture, though many workers have paid scant attention to this important detail.

The Autoradiographic Method

The popularity of the autoradiographic method can be traced to the study of Cowan et al (1972), though the groundwork was laid by Lasek et al (1968), Hendrickson (1969), and Schonbach et al (1971), and the understanding of the fundamental process upon which it depends dates to the earlier work of Weiss & Hiscoe (1948), Droz & Leblond (1963), Taylor & Weiss (1965), Grafstein (1967), and Lasek (1968). The method depends upon the new well-documented observation that radioactively labeled amino acids, sugars, or certain other compounds injected in the vicinity of neuronal somata will be taken up by the cells, incorporated into macromolecules by their synthetic machinery and transported at varying velocities down the axons to accumulate at the axon terminals. At any stage after incorporation into larger molecules, the material may be fixed and visualized by conventional light or electron microscopic autoradiographic methods (Rogers 1973, Hendrickson 1975), or its presence detected by liquid scintillation counting. In autonomic ganglion cells, the material reaches the axon from the endoplasmic reticulum and Golgi complex within about 30–90 min of injection (Droz 1965, Droz & Koenig 1970), and the time course is probably similar in other neurons.

BACKGROUND The phenomenon of axoplasmic transport has been reviewed at length by numerous authors (Lasek 1970b, Cowan et al 1972, Lubinska 1975, Kerkut 1975, Jeffrey & Austin 1973, Grafstein 1969, 1975a,b, Droz 1975, Ochs 1974, 1975), and we do not propose to recapitulate the large volume of background data here. Among the more significant contributions of the last two or three years are those that indicate the components transported in the various phases of axoplasmic flow (see review by Grafstein 1975b). The fast phase, moving as a wave at rates of from 100 to 400 mm per day (Lasek 1970, Ochs 1972, Jeffrey & Austin 1973) in most nerves, consists almost entirely of particulate matter: large numbers of different proteins, glycoproteins, glycolipids, and phospholipids are incorporated into membranous structures, particularly the smooth endoplasmic reticulum, certain synaptic vesicles, the axolemma, and those mitochondria that reach the terminals (Schonbach et al 1971, Forman & Ledeen 1972, Grafstein et al 1975, Forman et al 1971, Elam & Agranoff 1971, Droz et al 1973a,b,c, Grafstein 1975a,b). The rapidly transported material seems to be necessary for the maintenance of the axon's terminals since these show early degenerative changes following blockade of fast transport by colchicine, and synaptic transmission is concomitantly depressed (Cuénod et al 1972, see also Cuénod et al 1973). By careful selection of the radiolabeled precursor (see Grafstein 1975b), many of the macromolecules transported in the fast phase can be more or less selectively labeled. Important recent contributions by Droz and his colleagues (Droz et al 1975, Droz 1975), using high-voltage electron microscopic autoradiography on sections that are 0.5–2.0 μm thick, have indicated that the smooth endoplasmic reticulum, which they think extends continuously through the axon from soma to terminals, may play a major role in the transport of rapidly moving materials, including the components of synaptic vesicles. They postulate that separate channels in the reticulum may provide selective routes for different materials. The elegance of this work makes the hypothesis particularly attractive, and the latter part is of some interest in relation to the work of Anderson & McClure (1973), which suggests that different proteins may be transported away from the soma in the central and peripheral branches of dorsal root ganglion cells in cats (see also Ochs & Erdman 1974). At present the relationship between the hypothesis and the well-established observation that microtubule-disrupting agents such as colchicine (Samson 1971) in small doses effectively block fast axoplasmic transport is not clear (Karlsson et al 1971, James et al 1971, McGregor et al 1973, Paulson & McClure 1975; but see Byers 1974).

The slow phase of axoplasmic transport, moving as a front at a rate of ~1–5 mm per day (Taylor & Weiss 1965, Droz & Leblond 1963, Grafstein et al 1970, Karlsson & Sjöstrand 1971a,b, Schonbach et al 1971), consists largely of soluble protein (McEwen & Grafstein 1968), most of which may be associated with structural elements of the axon. Although four or more times as much material is transported in the slow than in the rapid phase of axoplasmic transport, only a small proportion actually reaches the terminals (Hendrickson 1972, Droz et al 1973a,b,c). The number of polypeptides traveling in this phase appears to be limited, five polypeptides accounting for 76% of the transported radioactivity in studies by Hoffman & Lasek (1975). Two of the five have been known for some time to be subunits of tubulin (Grafstein et al 1970; James & Austin 1970, Karlsson & Sjöstrand 1971b, McEwen et al 1971, Hoffman & Lasek 1975). The three others seem to be associated with microfilaments. One of these co-migrates on SDS gels with chick muscle heavy chain myosin. Hoffman and Lasek, on the basis of their results, follow Ochs (1971) in postulating a polarized myosin-actin type interaction between the axolemma and the neurofilamentous-microtubular structural core of the axon, which leads to forces promoting the proximo-distal flow of the slow component. Unlike the fast component, the slower phase of axoplasmic flow ceases in the distal segment of an axon immediately after the axon is cut or ligated or after the application of colchicine to the axon or soma (Frizell et al 1975). Since colchicine does not cause degeneration of the axons studied, the effect cannot simply be due to degeneration and Frizell et al hypothesize that the slow flow may depend upon the existence of a continuously produced, rapidly transported factor.

In addition to materials transported in the fast and slow phases, many others move at intermediate rates in the axon, forming at least two labeled peaks in experiments on the rabbit optic nerve (Karlsson & Sjöstrand 1971, Schonbach & Cuénod 1971, Willard et al 1974). The relevance of this observation to pathway tracing techniques is discussed below.

Certain factors other than microtubule-disrupting drugs (McClure 1972) may affect the rapid, but not the slow, phase of axoplasmic transport. These include temperature (Ochs & Smith 1971, Grafstein et al 1972, Gross 1973, Edström & Hanson 1973b, Brimijoin & Helland 1976) and the ability of the parent cell to continue oxidative metabolism (Ochs 1971, Ochs & Smith 1971, Banks et al 1973). Although to date these factors do not seem to have unduly influenced the results of connection tracing studies, it is possible that under certain circumstances, and in certain species (e.g. cold-blooded animals), they could exert an effect. Factors upon which the fast transport does not seem to depend include the length of the axon (Murray 1974) or the passage of action potentials along the axon (Grafstein et al 1972). The rate seems to be less in young animals than in old (Hendrickson & Cowan 1971, Marchisio & Sjöstrand 1971, Gremo 1974). In immature animals (Hendrickson & Cowan 1971) the rate of slow transport is greater than in the adult, and the rate is said to increase during neuron regeneration, particularly at about the time new connections are being established (Murray & Grafstein 1969, Grafstein 1971), though this has been contested (Frizell & Sjöstrand 1974).

So far as is known, the same fundamental mechanisms discussed above operate equally in myelinated and unmyelinated axons (Dahlström 1971, Edström & Hanson 1973, Gross & Beidler 1973, Ochs 1974, Cancalon & Beidler 1977), and in sensory, motor, and autonomic nerves of all vertebrates (for review see Lasek 1968). Although dendritic transport has been less extensively studied, it appears that similar principles may operate here also (Globus et al 1968, Lux et al 1970, Schubert et al 1972).

METHODS OF APPLICATION A detailed protocol for the preparation of isotope, mode of administration, and conduct of autoradiographic tracing experiments has been published by Cowan et al (1972). This seems to have served as the standard for the majority of ensuing studies. In studies not involving intracellular injections, three methods of application of the labeled amino acid solution seem to be in current usage (some are described by Schubert & Holländer 1975). Hydraulic injection through a small-volume syringe of the Hamilton type is favored by many. Others prefer injection through micropipettes, by iontophoresis or by air pressure or some other compression system. The latter methods have the advantage that they permit recording of evoked or spontaneous electrical activity prior to injection. A new method of application recently described by Davis & Agranoff (1977) involves the implantation of the precursor contained within beads of an ion-exchange resin. This may prove to have the advantage of releasing the precursor over a longer period of time than that over which material introduced in a single injection is available (probably less than one hour, Barondes 1968).

Most workers seem to try to avoid unnecessary damage at the injection site by very slow injection, but there has been no serious attempt to ascertain whether a fast injection necessarily causes more damage than a slow one. Damage is obviously a potential source of spurious results by destroying projection neurons and in certain cases may predispose them to incorporate and transport precursors that they normally do not (Berkley et al 1977).

CHOICE OF PRECURSOR For most straightforward connectional studies, the choice of precursor is probably immaterial, provided that it is incorporated and transported in sufficiently large amounts. Though glycoproteins are transported preferentially in the fast phase of axoplasmic transport, glycoprotein precursors seem to offer no particular advantages over amino acids and in general seem to lead to a good deal of systemic labeling (Droz et al 1973a,b,c, Bennett et al 1973). [^{3}H] leucine has perhaps been the most widely used precursor, and, as a constituent of most proteins, it is an obvious choice (Lasek 1968, Cowan et al 1972). However, there is reason to believe that certain populations of small cells may incorporate and transport relatively small amounts of leucine (Künzle & Cuénod 1973, Berkley et al 1977), and there is one report of failure to label a pathway with [^{3}H] leucine (Sousa-Pinto & Reis 1975).

The [^{3}H]-labeled essential amino acid, proline, has also been used fairly widely (e.g. Hickey & Guillery 1974, Updyke 1975, Jones & Burton 1974, Crossland et al 1974a,b), though less so than leucine. Its introduction in tracing studies seems to have been conditioned by the observation (Elam & Agranoff 1971) that after injection into the eyes of fish, it led to far less, nonspecific (presumably blood-borne) labeling of unconnected parts of the brain than did leucine. Whether [^{3}H] proline leads, in larger animals, in which the ratio of transported material to brain volume is small, to less background labeling than [^{3}H] leucine is perhaps debatable. But [^{3}H] proline seems to have a major disadvantage in that it may not be incorporated in sufficiently large amounts for transport to be visualized in certain systems. Künzle & Cuénod (1973) were unable to label the projection of the lateral reticular nucleus of the medulla oblongata to the cerebellum by using [^{3}H] proline, whereas [^{3}H] leucine labeled the projection quite effectively. In comparative studies on the differential incorporation of [^{3}H] proline and [^{3}H] leucine in the dorsal column nuclei of cats (Berkley 1974, Berkley et al 1977), it was found that the larger neurons of the nuclei incorporated and transported [^{3}H] proline in much smaller amounts than they did [^{3}H] leucine. After injections of [^{3}H] proline and [^{3}H] leucine that contained equal amounts of activity, on opposite sides of the brain, the terminations of medial lemniscal fibers in the thalamic ventrobasal complex were always much more effectively labeled on the side contralateral to the leucine injection. The proline effect seemed to be dose dependent, and it was suggested that the large cells may compete unsuccessfully for the pool of injected proline, over the short time that it is available, with elements in the neuropil that possess a high-affinity uptake system for this amino acid. Since proline may serve as a transmitter agent at certain synapses (Felix & Künzle 1974, Zarzecki et al 1975), a population of synaptic terminals of this type in the dorsal column nuclei would presumably have such a high-affinity uptake system.

The uncertainty surrounding the effectiveness of [^{3}H] leucine or proline in labeling the projections of some neuronal systems has led many workers to use a combination of the two precursors (e.g. Krettek & Price 1977, Edwards & deOlmos 1976, Burton & Jones 1976). Edwards & deOlmos (1976) consider that [^{3}H] proline, though as effective as [^{3}H] leucine for labeling axon terminations, may be less effective at labeling the parent axons. Some workers have included a third labeled amino acid, lysine, in the injected mixture (e.g. Nauta 1974, Graybiel 1975, Saper et al 1976a,b). The popular belief that this precursor labels axonal proteins and, therefore, the trajectory of a fiber pathway more heavily than the other two has not yet been the subject of a controlled investigation (but see Hoffman & Lasek 1975).

Many other labeled amino acid precursors are readily available commercially and have been used from time to time in tracing studies. The most extensive comparative studies in single species are probably those of Elam & Agranoff (1971), of Karlsson & Sjöstrand (1972), and of Hunt & Künzle (1976a). The latter workers found that thirteen different precursors were all transported anterogradely and apparently labeled axon terminal fields with equal efficacy. SOme, however, notably glycine, α-alanine, and serine, presented the potentially confusing effect of also moving retrogradely (from terminals to soma) in certain short axon systems. Although there are certain other indications that labeled amino acids injected into muscles may accumulate by retrograde flow in motoneurons (Kerkut et al 1967, Watson 1968), the synthetic capacities of axon terminals in most sites have generally not been considered sufficient to cause serious contamination of autoradiographic tracing studies by retrograde accumulation in cell somata that project to an injection site (Lasek 1970a,b, Cowan et al 1972, Droz et al 1973a,b,c, Lasek et al 1974).

Precursors of specific macromolecules transported anterogradely in the axoplasmic flow include: [^{3}H] glucosamine for glycolipids and glycoproteins (Forman et al 1971, [^{3}H] glycerol for phospholipids (Grafstein et al 1975), [^{35}S] sodium sulfate for sulfated mucopolysaccharides (Elam et al 1970), [^{3}H] fucose for glycoproteins, (Karlsson & Sjöstrand 1971, Forman et al 1972), and [^{3}H] nucleosides presumably for RNA (Rahmann & Wolburg 1971, Bondy 1971). With the exception of fucose and the nucleosides, which have rather special applications discussed below, at present there seems to be little benefit to be derived from using any of the others in conventional anatomical studies. However, where there are strong grounds for believing that a particular transmitter substance is secreted by a set of axon terminals, advantage has been taken of the fact that these terminals may be selectively demonstrated by injecting the labeled transmitter itself or one of its precursors. This method, has its main application in electron microscope studies, as discussed below.

CHOICE OF ISOTOPE Tritium appears to remain the label of choice in most experimental autoradiographic procedures aimed at neuroanatomical analysis, since its low maximum energy assures the highest resolution in autoradiography. The higher energy β emitters, [^{14}C] and [^{35}S], and the γ emitter [^{125}I] have been used for certain special purposes, some of which are discussed in the course of this review. For a detailed consideration of the principles and fundamental techniques of autoradiography, the work of Rogers (1973) is recommended.

The low-energy, charged β particles have a limited penetrance, and it is probable that in most light microscopic autoradiographs involving tritium only the radioactive label contained in the part of the section lying within 2–3 μm of the emulsion is visualized (Caviness & Barkley 1971). Since counterstained sections of this thickness do not permit a ready delineation of the boundaries of nuclear masses in the central nervous system, a compromise must be sought, and in current publications most workers seem to have chosen sections of 15–25 μm thickness.

INTERPRETATION There are two major areas in which interpretative problems seem to arise regarding the use of the autoradiographic method. One of these relates to the identification of the "effective" injection site, the other to the interpretation of the pattern of transported label at both the light and electron microscopic levels. Each of these is bound up with the question of changes in the appearance of the autoradiograph with time.

The injection site An injection of isotopically labeled precursors made directly into the brain or spinal cord usually consists of a central zone of damage of variable size that seems to depend on both the size of the injection needle or pipette and the rapidity with which the isotope solution is expelled; one μl of solution displaces 1 mm^3 of tissue, so sudden expulsion of a bolus of isotope of this size will predispose to necrosis. Surrounding this at short and medium survival periods (Cowan et al 1972) are two other zones; the extents of these depend primarily on the amount of activity injected, the exposure time, and the development of the autoradiographs. Adjacent to the needle or pipette track is a zone in which, in bright field microscopy, most cells and the intervening neuropil appear homogeneously blackened by overlying silver grains. Where it has been studied, recognizable heavy labeling of the somata occurs within less than 20 min of the injection (e.g. Berkley et al 1977). Like the extent of this central zone, however, the degree of blackening depends on the amount of activity injected and the exposure time. The smallest injections, at exposure times of even 2 or 3 weeks may not be visible at low magnification under bright field conditions.

The zone of intensely blackened cells often ends quite sharply and is replaced by a surrounding zone in which cell labeling rapidly declines and most of the labeling lies over the neuropil. In most cases at relatively short survival times this labeling probably represents direct diffusion from the center of the injection site, with uptake and incorporation by glial cells and neurons, together with transport of macromolecules in the dendrites and short axons of neurons situated in the central zone. It is still uncertain to what extent the spread of this central zone is conditioned by the speed or compressive force of the injection, and no attempt has been made to document the differences between the degree of spread of an electrophoretically applied injection and that of a hydraulic one. Often, however, the outer zone is quite asymmetrical. In these cases, the shape of the injection is commonly determined by the greater resistance to diffusion offered by an adjacent tract of white matter, in comparison with that presented by the gray matter into which the injection is made.

The apparent size of an injection of isotope as seen under constant conditions of exposure and development seems to decline with time as the label is both transported away from the soma and diluted by the continuing metabolic activities of the cells at the injection site. During this time, the rate of turnover of individual proteins probably varies considerably (Lajtha & Marks 1971). Although little attempt has been made to study this in detail, personal experience suggests that significant shrinkage occurs with survival times beyond about 10–14 days. Of course, the injected pool of labeled precursor remains available for incorporation for a much shorter time than this—probably rather less than one hour (Barondes 1968)—but it continues to be transported from the cells for as long as six days after synthesis in dorsal root ganglia (Lasek 1968).

It has sometimes been hinted (e.g. Elam & Agranoff 1971) that new labeled precursor may become available by breakdown of macromolecules labeled by the originally injected material. The degree to which this occurs at the injection site is not known, though it would seem to be a feature of the use of proline (Grafstein et al 1972, Heacock & Agranoff (1977).

Under constant conditions of exposure and development, the peripheral zone of an injection appears to be ineffective in providing detectable amounts of transported material in the terminal fields of axons emanating from cells in the peripheral zone. No systematic study has been done to document this, but in studies of the cortical projections of various nuclei of the thalamus (Burton & Jones 1976) and amygdala (Krettek & Price 1977), spread of the peripheral zone from one nucleus into another was not associated with transported label in the cortical area known from other studies to be associated with the second nucleus. Naturally, the extent of the central zone of the injection will increase with increased exposure times and under different conditions of development, but the same principle seems to hold, provided that autoradiographs of the injection site and terminal regions are subjected to the same treatment. The practice of some workers of exposing the injection site for a short time and the terminal region for a longer time is probably not advisable where questions of topographic organization are important.

Terminal and axonal labeling The initial observations on axoplasmic flow (Weiss & Hiscoe 1948, Droz & Leblond 1963) concerned the transport of materials in the slow phase, which moves in bulk at 1–5 mm per day, progressively filling the whole axon. With the recognition of the fast phase, which travels as a wave at 100–400 mm per day and accumulates within the terminals (Grafstein 1967, Hendrickson 1969, Elam & Agranoff 1971, Karlsson & Sjöstrand 1971, Schonbach et al 1971), it appeared likely that a differential labeling of terminals and of axon trajectories could be obtained in an experimental animal simply by manipulating the survival time (Hendrickson 1972, Cowan et al 1972). In principle, a survival time of 1–2 days should in any long-fibered system label only axon terminals, whereas one of a week or more should label the course of the axons as well. In practice, the difference between terminal and axonal labeling has proved to be far less clear-cut. In most cases, as judged from personal experience and from the evidence of published photomicrographs, a certain proportion of the labeled material transported in the fast phase accumulates along the length of the axon, probably by being incorporated into the axolemma (Droz et al 1975). Moreover, the presence of material transported at two or more intermediate rates (Willard et al 1974), serves to enhance the degree of axonal labeling at short survival times. At virtually all survival times, therefore, the whole course of a pathway is usually laid out before the investigator.

Despite the somewhat negative conclusion of the above paragraph, it is probably fair to say that the shortest survival periods compatible with permitting material to reach the terminals still give the most accurate delineation of a terminal region. This may be of some importance in laminar structures, such as the cerebral cortex, where afferent fibers enter from below and may traverse one or more cellular layers before terminating in another. In many cases, however, where exact levels of termination are not considered important, it may be desirable to allow a relatively long survival time in order to enhance the distinction between labeling of a terminal region and that of the background. In this way more labeled material transported in the fast and intermediate phases will have had time to accumulate in the terminal region. In a selection of recently published connectional studies, survival periods ranged from 6 hr to 30 days (Hickey & Guillery 1974, Rosenquist, Edwards & Palmer 1974, Barber & Raisman 1974, Edwards 1975, Graybiel 1975b, Goldman & Nauta 1977, Saper et al 1976a, LeVay & Gilbert 1976, Conrad & Pfaff 1976a,b, Krettek & Price 1977, Updyke 1977). Even in regions such as the cerebral cortex, where a knowledge of the exact levels of termination is desired, the difference between labeling of axons and their terminal ramifications at relatively long survival periods is often sufficiently distinctive to permit their separate identification (see Jones & Burton 1976).

The relationship between the pattern of autoradiographic labeling and the pattern of distribution of the underlying axonal ramifications is variable. A number of recent studies indicate that the two patterns are related to a certain degree, but that a variety of additional factors can influence grain density and distribution. Since axoplasmically transported labeled material lies in preterminal axons as well as in axon terminals (Schonbach et al 1971, Hendrickson 1969, 1972), and because the number of grains can be affected by the specific activity of the injected material and by development and exposure times, it is obvious that the number of grains cannot indicate, in light microscopic preparations, the absolute number of terminals. There is reason to believe, however, that the differences in the number and density of grains seen in two systems represent differences in the extent and density of the terminal ramifications. In parts of the cerebral cortex (Jones & Burton 1976) known from axonal degeneration and electron microscopic studies to receive many thick thalamocortical fibers that give rise to a dense pattern of terminal ramifications, the density of silver grains in comparable studies conducted with autoradiography is particularly high. In the same section and with approximately equal involvement by the injection of the relevant thalamic nuclei, areas of the cortex known to receive relatively few thin thalamocortical fibers with a sparse pattern of termination, have a far lower density of silver grains. Similar differences are seen in the intense labeling of the terminal ramifications in the striatum of fibers emanating from the thalamic intralaminar nuclei, in comparison with the sparse labeling in the cerebral cortex of axonal ramifications arising from the same nuclei (Jones 1975a).

At the electron microscopic level, the number of silver grains observed over the terminals of large-fibered systems is usually 6–7 times greater than that observed over the terminals of fine-fibered systems (Schonbach et al 1971, Hendrickson 1969, 1972, 1975, Dekker & Kuypers 1976). This probably reflects differences in the amount of material incorporated and transported by large and small parent cells.

The number of silver grains and, to a certain extent, the number of labeled terminals seen in electron microscopic autoradiographs also varies with certain other conditions. Hendrickson (1975) finds that the density of grains observed electron microscopically in the lateral geniculate nucleus, following injections into the eyes of rabbits, is two to three times greater if [^{3}H] lysine is used as the tracer instead of [^{3}H] leucine and/or proline. In this study the percentage of retinal terminals labeled remained constant (65%) after injections of each precursor. In the study of Dekker & Kuypers (1976), however, the number of labeled corticothalamic and mamillothalamic terminals observed in the anterior thalamic nuclei of rats increased markedly with exposure time. These observations indicate that even at the electron microscopic level, the number of grains and the number of labeled terminals are not reliable guides to the proportionate number of terminals contributed to the neuropil of a nucleus by a particular afferent system.

In view of the above, counts of grain density in connection tracing studies probably have little value in absolute terms. Grain counts, however, have a number of useful applications. They can indicate in graphic form where peaks of labeling coincide with particular architectonic subdivisions such as the laminae of the lateral geniculate nucleus (Hickey & Guillery 1974) or of the cerebral cortex (Jacobson & Trojanowski 1975, Jones 1975b, Jones & Burton 1976, Wise & Jones 1976a). They can indicate the position of the leading edge of a wave of labeled growing nerve fibers in relation to cerebral landmarks (Crossland et al 1974b, Wise & Jones 1976a, 1977a, Wise et al 1977, Fricke & Cowan 1977). Finally, they can provide a useful comparative guide to changes in innervation density caused by experimental manipulation (Stanfield & Cowan 1976).

ADVANTAGES OF THE AUTORADIOGRAPHIC TECHNIQUE Probably the major advantage of the autoradiographic technique is its predictability. Provided an appropriate precursor is used, the experimenter can usually be confident that the axonal ramifications of a particular set of cells will be labeled. Moreover, although reports are beginning to appear indicating the possibility of a certain amount of retrograde axoplasmic transport of labeled amino acids (Hunt & Künzle 1976a), in most cases the labeling is undoubtedly the result of anterograde transport. Coupled with this, the functional basis of the technique is clearly understood. One of the obvious advantages of this is the knowledge that fibers of passage traversing the vicinity of an injection cannot be labeled directly. Though they may take up labeled amino acids, axons lack significant protein synthetic capacity and therefore cannot incorporate substantial amounts of precursor into macromolecules (Lasek et al 1976, Droz et al 1973a,b) that would be fixed and, therefore, contaminate the result. This is a clear advantage over both the older axonal degeneration techniques and certain of the newer retrograde labeling methods. Several long-standing uncertainties regarding patterns of connectivity have been resolved because of this particular advantage. The cortical relationships of the intralaminar nuclei of the thalamus have become clearer (Jones 1975a), the existence of descending fibers from the hypothalamus to autonomic nuclei in the spinal cord has been confirmed (Saper et al 1976b), retino-hypothalamic fibers have been positively identified (Moore & Lenn 1972, Hendrickson et al 1972), the connectional relationships of the reticular formation are becoming clearer (Edwards 1975, Edwards & deOlmos 1976). In addition, certain hitherto unsuspected pathways have been identified, such as the substantial one from the cerebellar cortex to the inferior olivary nucleus (Graybiel et al 1973). Others include the projection from certain amygdaloid nuclei to the frontal cortex (Krettek & Price 1974) and that from the subicular region (rather than the hippocampus proper) via the fornix to the hypothalamus (Swanson & Cowan 1975).

The technique seems applicable to all fiber systems. Examples are available of studies on motor (e.g. Droz & Leblond 1963), sensory (e.g. Lasek 1968, Kelts et al 1975, Smith & Mills 1976, Fidone et al 1977), and autonomic (e.g. Droz et al 1973a,b,c, Kelts et al 1975) nerves; unmyelinated fiber systems (e.g. Gross & Beidler 1973, Ochs 1972, Bisby 1976) seem to be just as amenable to study as myelinated. Successful results have been obtained in examples of most vertebrate classes, including many lower forms (e.g. Cook & Whitlock 1975, Barker et al 1975, Guillery & Updyke 1976, Caldwell & Berman 1977), and also in certain invertebrates (Schafer 1973, Sandemann & Denburg 1976). In addition, successful investigations have been carried out in the central nervous system on fiber systems that because of their fine caliber, could not be satisfactorily demonstrated with the older degeneration methods. These include the nigro-striatal system (Carpenter et al 1976), the central noradrenergic system (Pickel et al 1974, McBride & Sutin 1976), the serotoninergic system (Conrad et al 1974, Bobillier et al 1975, Halaris et al 1976), and certain pathways related to the hypothalamus (Saper et al 1976a, Jones et al 1976, Conrad & Pfaff 1976a,b, Swanson 1976). In some fine-fiber systems, such as the corticostriatal, the extent of the terminations emanating from a particular area has been shown to be far greater than previously believed (Künzle 1975, Goldman & Nauta 1977, Jones et al 1977).

The autoradiographic technique also provides a far clearer picture than earlier fiber-tracing methods of the distribution of terminal ramifications in relation to cytoarchitecture. The accuracy of localization is, therefore, greatly enhanced. Light microscopic autoradiographs can be readily counterstained, and the degree of resolution achieved with β particle emission is particularly good, even though one is viewing a marker that lies *over* rather than *in* the section. It is unfortunate that many workers choose to ignore this and, as shown repeatedly in the literature, continue to rely upon unstained or poorly counterstained material. The ability to recognize terminal regions in relation to cytoarchitecture (Figure 1) has led to a better understanding of the differential levels of termination of various sets of afferent fibers in laminated structures such as the lateral geniculate nucleus (Hickey & Guillery 1974, Shatz 1977), cerebral cortex (Rosenquist, Edwards & Palmer 1974, Jones 1975b, Jones et al 1975, Künzle 1976, Jones & Burton 1976, Ribak & Peters 1975, Robson & Hall 1975, LeVay & Gilbert 1976, Wise & Jones 1976a), superior colliculus (Graybiel 1975a, Hubel 1975, Hubel et al 1975, Dräger 1974, Harting & Guillery 1976), and spinal cord (Coulter & Jones 1977). In all of these cases, the enhanced labeling of the terminal regions in comparison with the arriving fibers has also enabled workers to recognize many previously unknown finer organizational aspects of the projections, such as columnar and other grouping of terminations.

Because of the relatively short survival times necessary for obtaining adequate labeling of terminal ramifications and because of the limited amount of trauma caused by an injection of isotope, the autoradiographic method offers considerable advantages for the study of developing connections in immature animals. To date, this advantage has been successfully exploited in studying the developmental specificity of the retino-tectal projection in the chick embryo (Crossland et al 1974a) and

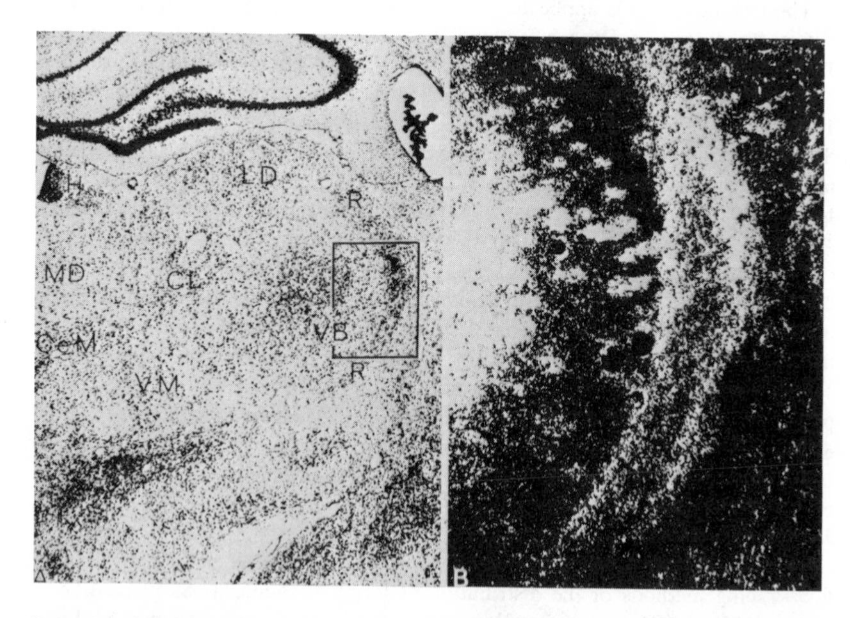

Figure 1 *A.* Counterstained autoradiograph of rat thalamus following an injection of [^{3}H] amino acids in ipsilateral somatic sensory cortex. Area indicated by box is enlarged and shown under darkfield illumination in *B.* There is heavy labeling of terminal ramifications of cortico-thalamic fibers in a sector of the reticular nucleus (*R*) and in a part of the arcuate nucleus of the ventrobasal complex (*VB*). Intervening neuropil and internal capsule show labeling of fiber bundles. *A* X 12; *B* X 60 (Jones 1975d).

in determining the developmental sequence of the ingrowth of various fiber systems into the cerebral cortex of newborn rats (Gottlieb & Cowan 1972, Fricke & Cowan 1977, Wise & Jones 1976a, 1977a). The ability to use short survival times also facilitates the study of developing connections in fetal mammals (Rakic 1976, Wise et al 1977). The latter studies showed that growing thalamocortical axons accumulate in the presumptive white matter beneath the maturing cerebral cortex long before cortical lamination develops and even before the cells of layer IV, upon which the fibers will mainly terminate, have reached the cortex from the ventricular zone. In all of the studies quoted and in several others concerned with the phenomenon of axoplasmic flow (Bondy & Madsen 1971, Marchisio & Sjöstrand 1971, Marchisio et al 1973, 1975, Crossland et al 1974a), labeled material accumulates at what are presumed to be the growing axon tips, though no electron microscopic verification has yet been made.

Another advantage of the autoradiographic technique is that it permits, more readily than the axonal degeneration methods, the examination of the same tissue by both light and electron microscopy. Kopriwa (1973) and Fischer (1975) have outlined the method in general, and Hendrickson (1972, 1975) has published some particularly useful guidelines to the application of electron microscopic autoradiography to the nervous system. In general, the use of the conventional electron microscopic fixative, glutaraldehyde, is to be avoided since it has the capacity to fix free labeled amino acids (Peters & Ashley 1967), i.e. amino acids not incorporated into proteins and axoplasmically transported. As in conventional light microscopic autoradiography, the use of formaldehyde serves to obviate any problems that this may cause.

One of the major drawbacks to the use of electron microscopic autoradiography has been the extended period of time (often several months) required for the exposure of the emulsion. The use of [^{35}S] methionine, which has a very short half-life, can reduce the exposure time considerably, but the resolution suffers. A recent innovation in which part of the radiation is converted to photons promises to be particularly useful in shortening exposure times for both electron and light microscopic autoradiography. Initially, scintillation fluid was simply added to the embedding agent (Fischer et al 1971), but more recently Durie & Salmon (1975) have described a method whereby the scintillation fluid is essentially incorporated into the emulsion layer. The protracted preparation time (which often may be followed by technical failure) for electron microscopic autoradiography will probably limit its widespread usage. It is likely that the labeling of axon terminals for electron microscopy, by causing them to degenerate (Guillery 1970, Heimer 1970), will remain a popular alternative for some years to come.

The autoradiographic method may also be used in combination with other pathway-tracing techniques. Several workers have used combined injections of [^{3}H] amino acids and a retrogradely transported marker (Figure 2), horseradish peroxidase, to label in the same brain the terminal distribution and cells of origin of a particular pathway (Jacobson & Trojanowski 1974, Trojanowski & Jacobson 1975, Wise & Jones 1976a, Steindler & Colwell 1976). By staining alternating autoradiographic sections with an axonal degeneration method, is is also possible to demon-

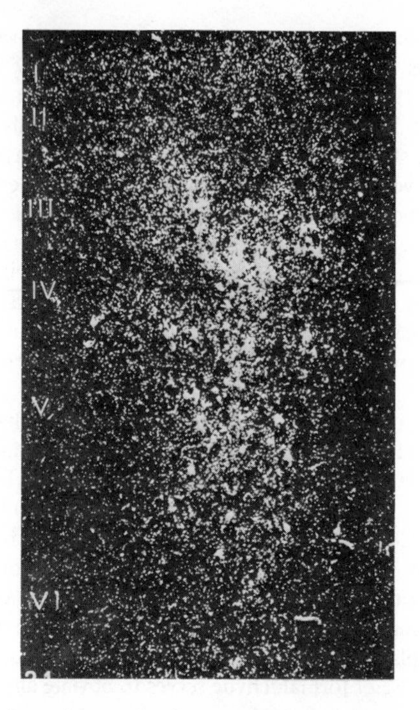

Figure 2 Darkfield photomicrograph of an autoradiograph from a vertical section of rat cerebral cortex following a combined injection of [^{3}H] amino acids and horseradish peroxidase in contralateral cortex. Column of commissural fiber ramifications labeled autoradiographically by anterogradely transported material is superimposed on column of commissurally projecting pyramidal cells retrogradely labeled with horseradish peroxidase. X 65 (Wise & Jones 1976a).

strate a second pathway degenerating as the result of a lesion made in the same brain (Jones & Burton 1974). By making the lesion some days before the injection and killing the animal less than three days after the injection, any incidental degeneration caused by the injection will usually not be stained.

DISADVANTAGES Apart from the somewhat protracted preparation time and the necessity for strict attention to technical details, both of which some workers may find unacceptable, the autoradiographic technique has relatively few disadvantages. Probably the one area where it has limited usefulness is in the analysis of those pathways, such as the corpus callosum and spinothalamic tract, that arise from very widely distributed cells. In these cases it is virtually impossible to label equally every cell contributing to the tract. Therefore the total distribution of such a tract is probably still best studied by means of axonal degeneration techniques (e.g. Jones & Burton 1974, Wise & Jones 1976a). Here, the degeneration methods will probably only be superseded by the introduction of a specific immunofluorescent label that will serve to outline the systems in their entirety.

A potential source of error in autoradiographic studies arises in cases where the labeling of terminal ramifications is particularly intense. In such cases, labeled material apparently leaking out of the axon terminals, either as macromolecules (Grafstein 1971) or as reutilizable precursor (Heacock & Agranoff 1977), can diffuse beyond the confines of the terminal zone into, and even across, the subarachnoid space into overlying parts of the brain (Specht & Grafstein 1973; see also figures of Hubel 1975, Dräger 1974). In this way, the possibility of false positive results may arise.

TRANSNEURONAL LABELING One of the most dramatic uses of the autoradiographic method has been the demonstration of ocular dominance columns of the monkey and cat visual cortex by means of material transported across the retinal synapses in the lateral geniculate nucleus following injection of precursor into an eye (Wiesel, Hubel & Lam 1974, Hubel et al 1975, 1976, Shatz et al 1975). The recognition that [^{3}H] fucose injected into the eyes of mice could lead to the transsynaptic transfer of small amounts of labeled material to thalamic cells and thence to the terminations of thalamocortical fibers in the visual cortex was first made by Grafstein (1971), Grafstein & Laureno (1973), and Specht & Grafstein (1973), though Korr et al (1967) had earlier indicated that labeled material traveling in the axons of motoneurons could enter muscle fibers. The success of the technique as a marker for connectivity depends on the injection of enormous quantities of precursor (up to 3 mCi) commonly repeated several times over several days. Usually, protracted survival times (up to one month) and long exposure of the autoradiographs have also been employed, though in some more recent studies these have tended to be shorter than in the initial studies (Casagrande & Harting 1975, Dräger 1974, Hubel 1975, Kaas et al 1976). The method has also been successfully applied in studying development of the visual pathways in fetal monkeys (Rakic 1976).

There have been no reported studies of the effective use of transneuronal labeling of this type in the central nervous system except after intraocular injections. The eye has the singular advantage that it is external to the remainder of the central nervous system, has a large volume and is surrounded by a dense, fibrous, connective tissue coat. These factors permit the injection of large volumes and concentrations of labeled precursor and restrict diffusion into other parts of the central nervous system. In the brain or spinal cord, it is particularly difficult to inject sufficient label to ensure transneuronal transport, yet at the same time restrict it to the confines of a particular nucleus. In attempts to study transneuronal labeling following injections of the medulla, spinal cord, or deep cerebellar nuclei (S. H. C. Hendry, S. P. Wise, E. G. Jones, unpublished), these difficulties have been further compounded by diffusion of large amounts of material into the subarachnoid space or ventricular system. This is dispersed in the cerebrospinal fluid and leads to large amounts of cellular and neuropil labeling at considerable distances from the injection site.

In the visual system, where transneuronal labeling has been successfully exploited, the most effective precursor has been [^{3}H] fucose, which suggests that glycoproteins or their derivatives are the main molecules being transferred across the retinogeniculate synapse. Most workers studying the phenomenon have also included [^{3}H] proline in their injections, but its effectiveness as a transneuronal label has not been particularly well documented. The nature of the material being transported from the terminals of the retinal ganglion cell axons to the thalamocortical relay cells of the lateral geniculate nucleus is still uncertain. Droz et al (1973a,b,c), studying the turnover of proteins in presynaptic terminals of the chicken ciliary ganglion, considered that most material transferred from cell to cell in this way should be of low molecular weight. In their study, no label could be detected in the postsynaptic cells following local application of an inhibitor of protein synthesis. Specht & Grafstein (1973), on the other hand, favor a transfer of macromolecules, which suggests that these may play a role in trophic interactions between cells. The most recent evidence is that of Heacock & Agranoff (1977), who present evidence that in the goldfish tectum, [^{3}H] proline released by proteolysis of transported protein following an injection in the eye, can be reutilized by neurons of the tectum and lead to significant transneuronal labeling in the contralateral tectum.

The movement of material from retinal axon to lateral geniculate cell does not seem to occur specifically at synaptic sites. As mentioned above, Specht & Grafstein (1973) were able to show that material leaking out of optic nerve terminals in the superior colliculus may lead to labeling of the overlying cerebral cortex. The phenomenon is, therefore, probably best referred to as transneuronal rather than transsynaptic. On the other hand, the restriction of labeling in the rhesus monkey lateral geniculate nucleus essentially to the layers receiving input from the injected eye and the clear segregation of the transneuronally labeled ocular dominance columns in the visual cortex (Wiesel et al 1974, Hubel et al 1975) argues for some degree of specificity. This may, however, be conditioned by no more than the density of the interlaminar plexuses that effectively prevent diffusion to adjacent laminae in the rhesus monkey. In New World monkeys in which these plexuses are largely absent, the restriction is not so obvious, and, as a consequence, ocular dominance columns related to the injected eye are not labeled in isolation (Kaas et al 1976).

Other labeled precursors that, under certain circumstances, appear to move transneuronally, include the nucleosides adenosine and uridine. Uridine, guanosine, and orotic acid were initially used to study what was thought to be the axoplasmic transport of RNA in the visual system of the chick and goldfish (Bray & Austin 1968, Bondy 1972, Ingoglia et al 1973). It has now been shown that although small amounts of RNA may be transported in these experiments, it is probably the free precursor or some derivative (such as ATP) that is mainly transported, and that the labeled RNA seen along the course of the optic nerve after injection of the eye is primarily the result of leakage of the precursor from the axons with subsequent incorporation into glial cells (Autilio-Gambetti et al 1973; but see Bonnet & Bondy 1976). Schubert & Kreutzberg (1974, 1975a,b), first suggested that [^{3}H] adenosine or [^{3}H] uridine might serve as useful transneuronal markers when they showed that, following injection of the visual cortex of rabbits, the axoplasmically transported label accumulated in neuronal somata of the lateral geniculate nucleus. Transneuronal transport of nucleosides has subsequently been shown in a number of sites (Hunt & Künzle 1976b,c, Wise & Jones 1976b, Wise et al 1977, Kruger & Saporta 1977, Schubert et al 1976, Rose & Schubert 1977). The release at the synaptic sites appears to be enhanced by stimulation (Schubert et al 1976), and Rose & Schubert

(1977) propose that this suggests that purinergic mechanisms accompany particularly cholinergic synaptic activity.

In a series of studies, however (Hunt & Künzle 1976b,c, Wise & Jones 1976a,c, Wise et al 1977), it has also been shown that the axoplasmic movement of the labeled nucleoside or its derivatives is bidirectional. The cells of the isthmo-optic and oculomotor nuclei in chickens can be retrogradely labeled, by injections of nucleosides in the eye and extraocular musculature, respectively. The cells of origin of corticothalamic and of reticulo-cerebellar fibers in rats can be respectively retrogradely labeled, by injections in the thalamus and cerebellar cortex.

For reasons that are as yet poorly understood, adenosine seems to provide far more effective retrograde labeling and uridine far more effective anterograde transneuronal labeling (Wise et al 1977).

The transneuronal labeling occurring after [^{3}H] nucleoside injections appears to be nonspecific. Since glial cells are labeled with neurons in regions of termination of a pathway as well as along the length of the pathway, leakage out of the axon and its terminals and subsequent uptake and incorporation by any cells in the vicinity probably accounts for the effect (Wise et al 1977).

There is also some evidence for retrograde transneuronal transport in cells whose axons synapse upon the retrogradely labeled cells and also for secondary anterograde transport in cells labeled by anterograde transneuronal transport (Hunt & Künzle 1976b).

Since the method is to a certain extent unpredictable, it is unlikely that it will achieve widespread usage as either a transneuronal or retrograde marker. It may, however, have some usefulness when used in combination with other markers (see below).

Retrograde Tracing Methods

The phenomenon of retrograde axonal transport had been observed in the 1920s by studies of the movement of stained and unstained particles in nerves growing in vitro (e.g. Matsumoto 1920, Burdwood 1965) and later gained credence from the observation that certain materials such as acetyl cholinesterase and noradrenaline accumulate in azons below a constriction (e.g. Lubinska 1964, Dahlström 1965, 1967, Mayor & Kapeller 1967, Lasek 1967, Geffen & Livett 1971). Studies intimating that certain neurotoxins might also accumulate in neurons by retrograde axoplasmic transport have been reviewed by Kristensson (1975).

The first suggestions that this phenomenon might be used as a retrograde marker for neuronal connectivity were made by Kristensson & Olsson (1971) and Kristensson et al (1971), who used fluorescent labeled albumin and horseradish peroxidase as markers and by LaVail & LaVail (1972), who used horseradish peroxidase alone. Though these initial studies seemed to imply that the technique might only work effectively in immature animals, the retrograde method, particularly that employing horseradish peroxidase, has become one of the most widely applied techniques in Neurobiology. The work of Kristensson & Olsson (1971, 1973a,b, 1974) dealt with retrograde transport in mouse motoneurons and that of LaVail and co-workers (LaVail & LaVail 1972, 1974, LaVail et al 1973) primarily with the chick visual system. Since then, however, the method has been successfully exploited in numerous parts of the central nervous system (see LaVail 1975 for a recent review). It has also been applied in the hypothalamo-hypophysial system (Sherlock et al 1975), in sensory (Furstman et al 1975, Ellison & Clark 1975, Arvidsson 1975) as well as autonomic (Hansson 1973, Schramm et al 1975, Ellison & Clark 1975, DeVito et al 1974, Brownson et al 1977) and motor nerves (Mizuno et al 1975, Gacek 1974), in the developing nervous system (Oppenheim & Heaton 1975, Clarke & Cowan 1976, Clarke et al 1976, Wise & Jones 1976a), in submammalian as well as mammalian forms (Luiten 1975, Parent 1976, Halpern et al 1976, Benowitz & Karten 1976), and in vitro as well as in vivo (Litchy 1973, see also Bunge 1973). Like the autoradiographic method, the retrograde labeling method employing horseradish peroxidase as the marker can be used at both the light and electron microscopic level.

As is reviewed below, the retrograde method, based upon the use of horseradish peroxidase, is not without interpretative difficulties, but its specificity and the facility with which it may be used have made it particularly attractive. As a consequence, the past three years have seen a spate of papers reporting studies conducted with horseradish peroxidase. To be sure, many of these have contributed little that is new, but a considerable number of new observations of fundamental significance has also been made. The laminar origin of most of the efferent pathways of the cerebral cortex has been clearly established for the first time (e.g. Holländer 1974, Jacobson & Trojanowski 1974, Lund et al 1975, Gilbert & Kelly 1975, Wise 1975, Coulter et al 1976, Ebner et al 1976, Jones et al 1977, Wise & Jones 1977a, Berrevoets & Kuypers 1975, Jones & Wise 1977); the cells of origin of the spinothalamic tract have been defined in their entirety for the first time (Trevino & Carstens 1975); separate populations of thalamic cells that correlate closely with functional types defined electrophysiologically have been shown to project differentially upon the visual cortex (LeVay & Ferster 1977); certain hitherto unsuspected pathways, such as that from the basal nucleus of Meynert to the cerebral cortex (Kievit & Kuypers 1975, Divac 1975, Jones et al 1977), have been discovered; the presence of others long suspected, such as the descending pathways from the hypothalamus (Saper et al 1976b) and medullary aminergic nuclei (Kuypers & Maisky 1975, Hancock & Fougerousse 1976, Martin 1977) to the spinal cord, have been confirmed; certain controversies have been resolved, such as that over the cortical relationships of the thalamic intralaminar nuclei (Jones & Leavitt 1974, Strick 1975); with the introduction of other retrograde labels, at least one of which (nerve growth factor; Hendry et al 1974) is specific for certain classes of neuron, the prospects for further advances seem considerable.

BACKGROUND The phenomenon of retrograde transport in axons has been reviewed extensively by LaVail (1975) and by Kristensson (1975). In what follows we shall concentrate on some particular areas only. The greater part of the survey will be concerned with the use of horseradish peroxidase as a marker. Other, alternative labels are considered below.

Since the introduction of the horseradish peroxidase method, three major studies have been devoted to the fine structural correlates of the uptake of horseradish peroxidase (Turner & Harris 1974, LaVail & LaVail 1974, Nauta et al 1975). The latter two studies and one other (Hansson 1973) have also examined the fine structural localization of the transported material. There seems to be general agreement that at the injection site uptake into cell somata, dendrites, presynaptic axons, and axon terminals occurs by a typical endocytotic phenomenon, involving coated vesicles. This occurs within $\sim$ 15 min of the injection, and shortly thereafter the enzyme begins to appear in synaptic and other larger vesicles and in smooth-walled sacs and tubules; it appears in multivesicular bodies within 30 min and, finally, after about 1.5 hr, it appears in dense bodies (i.e. after uptake into the somata; Turner & Harris 1974). The enzyme disappears from organelles in the same order, synaptic vesicles being free by 24 hr, sacs and tubules by 48 hr, and multivesicular bodies by 72–96 hr. In the chick retina (LaVail & LaVail 1974) and the mammalian cerebral cortex (Turner & Harris 1974), all cell somata labeled by direct uptake are free of label 3–4 days after the injection.

The retrograde movement of the enzyme seems to occur primarily in vesicles of varying size that are often intimately associated in the axon with microtubules (LaVail & LaVail 1974). This seems particularly significant in view of the fact that small doses of microtubule-disrupting drugs such as colchicine (Kristensson & Sjöstrand 1972, Kristensson & Olsson 1973, LaVail & LaVail 1974) and vinblastine (Bunt & Lund 1974), serve to block the retrograde transport of the enzyme. Colchicine may also interfere with the uptake process (LaVail & LaVail 1974). In all of the studies quoted above, multivesicular bodies loaded with horseradish peroxidase were also seen to be transported retrogradely. These are presumably assembled in the axon terminals from the endocytotic vesicles. Most reaction product seen in cell somata labeled by retrograde transport is in the form of multivesicular and dense bodies. In most cells the retrogradely transported material disappears within about 4 days (LaVail 1975).

The role of the synaptic vesicle recycling phenomenon postulated by Heuser & Reese (1973) in the packaging of horseradish peroxidase for retrograde transport is uncertain. Heuser & Reese (1973) demonstrated at the frog neuromuscular junction that endocytotic vesicles labeled with horseradish peroxidase became incorporated into the smooth endoplasmic reticulum and were subsequently budded off as new, enzyme-loaded, synaptic vesicles. In the study of Turner & Harris (1974), loaded synaptic vesicles were not often observed budding off from the smooth endoplasmic reticulum. Moreover, in many terminals, most of the synaptic vesicles were loaded, but the smooth endoplasmic reticulum was not. They suggest, therefore, that the material contained in the synaptic vesicles may be retrogradely transported by virtue of the vesicles later coalescing with the smooth endoplasmic reticulum. LaVail & LaVail (1974), in noting similar observations, suggest that only the larger labeled vesicles that they saw, might enter the smooth endoplasmic reticulum and multivesicular bodies to be transported; they suggest that the labeling of synaptic vesicles could remain part of the local recycling effect.

The rate of retrograde transport of horseradish peroxidase is approximately half that of the rapid phase of anterograde axoplasmic transport. It has been estimated to be 80 mm per day in the chick visual system (LaVail & LaVail 1974) and 120 mm per day in the rat visual system (Hansson 1973). In other systems and species the rate seems to be comparable, though it may be expected to increase in regenerating neurons (Kristensson & Sjöstrand 1972, Sjöstrand & Frizell 1975). The amount that is transported seems to be considerable, and, in most cases, quite intense retrograde labeling of cell somata can be obtained with 24 hr of injection. According to Vanegas, Holländer & Distel (1977), most of the transport of the effective label in cat geniculo-cortical neurons occurs within the first eight hours after the injection.

METHODS OF APPLICATION Most of the same methods available for making isotope injections are applicable to the delivery of horseradish peroxidase. In view of the high concentration of enzyme (20–50%) used by most workers for extracellular injections, if a pipette is used, it usually needs to have an opening of at least 20 μm. Electrophoresis, which is favored by some (Graybiel & Devor 1974, Aghajanian & Wang 1977), does not appear to have any particular advantage for larger extracellular injections over hydraulic injections. In view of the differences in the net charges on each of the several components of the commercially available enzyme mixture (Bunt et al 1976), it may be impossible to predict whether an effective injection is being driven electrophoretically.

Because of the regenerative nature of the enzyme-substrate interaction during incubation of the sections, even a small amount of leakage of horseradish peroxidase along an injection track may lead to detectable retrograde labeling in groups of cells not immediately related to the primary region injected. To avoid this, some workers have devised methods whereby the injection needle is introduced through a surrounding sleeve (Graybiel & Devor 1974) and have sought to prevent spread of the injection mass up the needle track by filling the track with oil (Beitz & King 1976). Unfortunately these cause a considerable amount of damage to the tissue, which may be disadvantageous since uptake and retrograde transport clearly seem to occur in damaged axons (see below).

NATURE OF THE MATERIAL TRANSPORTED Commercially available horseradish peroxidase contains three isoenzymes (A, B, and C) in varying proportions depending on the degree of refinement (Shannon et al 1966, Shih et al 1971). Bunt et al (1976) have made a comprehensive study of the relative efficacy of the three isoenzymes in the retrograde labeling of cells in the rat visual system. According to these authors, isoenzymes B and C, which are basic glycoproteins carrying a net positive charge, are readily transported. Isoenzyme A, which differs substantially from the other two in its carbohydrate composition, carries a net negative charge and is not transported. In view of the difference in carbohydrate composition and the fact that experimental oxidative-reductive interference with the carbohydrate moiety of isoenzyme C prior to its injection prevents retrograde labeling, it was suggested that the uptake and the specificity of transport of the three isoenzymes may depend on the carbohydrate component.

CHOICE OF SUBSTRATE The primary substrate for horseradish peroxidase is, of course, hydrogen peroxide, but the oxidative polymerization of chromogens is the effective substrate in anatomical studies of the type discussed here. The readily available chromogens are benzidine and several of its derivatives. In view of impending governmental restrictions on the use of benzidine and related compounds, it is uncertain for how much longer these useful, but potentially hazardous, compounds will remain readily available to workers. A synthetic analogue involving a copolymer of paraphenylene-diamine and catechol recently introduced by Hanker & Yates and their associates (1977 and personal communication, see also Hanker et al 1972) may prove to be a satisfactory alternative. Though the benzidine-blue reaction was first used to demonstrate horseradish peroxidase histochemically (Straus 1964), the chromogen that has been used most widely in neuroanatomical studies is diaminobenzidine, usually as the tetrahydrochloride (Graham & Karnovsky 1966). This and the Hanker-Yates mixture are the only currently available chromogens that are osmiophilic and electron dense, and therefore applicable to electron microscopic studies. For light microscopic studies some workers prefer O-dianisidine (Graham & Karnovsky 1966). This gives a green and much denser reaction product than diaminobenzidine (Adams & Warr 1976, Colman et al 1976), though in comparison with the use of diaminobenzidine, the method is a little more tedious. A useful advantage of another, black, reaction product produced by the interaction of horseradish peroxidase with 4-chloro-1-naphthol is that it may be used in combination with acetylcholinesterase staining (Warr 1975, Adams & Warr 1976). The brown cholinesterase reaction product would be difficult to distinguish from that of diaminobenzidine. An alternative method to differentiate the two reaction products is to convert that due to cholinesterase to a black product by the addition of sodium sulfite (Hardy et al 1976).

Mesulam (1976a) and Mesulam & Van Hoesen (1976) have advocated a return to the use of benzidine as the HRP secondary substrate, claiming that it enables one to detect far lower levels of peroxidase labeling than does diaminobenzidine. The blue reaction product again permits differentiation of the retrogradely transported peroxidase from the cholinesterase reaction product when the two are used in combination (Mesulam & Van Hoesen 1976). Under certain conditions the addition of small amounts of dimethyl sulfoxide may enhance cellular labeling demonstrated with diaminobenzidine (Keefer et al 1976), and prior soaking of sections in cobaltous chloride might also be expected to give some intensification (see below).

EFFECT OF FIXATION A wide variety of fixative mixtures have been used in studies employing horseradish peroxidase. Most, but not all, involve paraformaldehyde and glutaraldehyde in different proportions. Personal experience (Jones & Leavitt 1974) has been that relatively high concentrations of paraformaldehyde tend to destroy a good deal of the enzymatic activity. Others have related similar experiences to us, and Kim & Strick (1976) have recently documented this in a more controlled way. Nevertheless, some workers have still obtained successful results using high concentrations of formaldehyde (Ralston & Sharp 1973, Wong-Riley 1974) and even formaldehyde alone (Bunt et al 1975, Lund et al 1975).

Adequate demonstration of horseradish peroxidase can be obtained with relatively low concentrations of aldehydes, for example 0.5% paraformaldehyde and 1.25% glutaraldehyde (LaVail & LaVail 1974, Halperin & LaVail 1975). But many workers report regularly using higher concentrations, such as 1% paraformaldehyde and 2.5% glutaraldehyde (e.g. Kievit & Kuypers 1975). Sections from material fixed in this way are easier to handle but less satisfactory to counterstain. There is no evidence that moderate lipid extraction of sections in xylene or chloroform, prior to counterstaining, destroys the reaction product, and there is no reason to believe that adequate counterstaining obscures the reaction product (see e.g. Jones et al 1976).

Up to this time the application of horseradish peroxidase has necessitated the use of frozen sections, which may make the processing a little tedious. Recently, however, a modification of the immunoperoxidase method (Sternberger 1973, Vacca et al 1975) and the introduction of [^{3}H]-labeled horseradish peroxidase (Geisert 1976) offer the prospect of applying the technique to paraffin sections, thus speeding up the method considerably and facilitating the handling of serial sections.

THE EFFECTIVE INJECTION SITE A typical injection of horseradish peroxidase consists of a central zone of damage and a surrounding zone that is usually so densely stained as to obscure cellular detail. The size of this zone depends on the amount of enzyme injected. Outside this dense zone is a region of homogenous reaction product of reduced density, in which a Golgi-like impregnation of isolated neurons and their processes often occurs. The size of this zone seems to vary in a rather unpredictable way, ranging from virtual absence to being spread throughout a large proportion of the brain (cf Figure 8 of Jones et al 1977 with Figure 1 of Nauta et al 1974). It is conceivable that this phenomenon may occur due to inadequate fixation of the injected enzyme mass in situ. Dunker et al (1976) have shown that the degree of penetration into the cerebral cortex of horseradish peroxidase injected into the subarachnoid space is increased if the concentration of glutaraldehyde is reduced in the fixative.

Vanegas et al (1977) consider that the peripheral, less dense zone of the injection represents a secondary diffusion that occurs 4–8 hr after injection, after effective uptake at the terminals of the projection neurons, and is associated with the onset of the local inflammatory reaction. This zone diminishes in size again after about 17 hr, but there seems to be no appreciable reduction in size of the dense core of the injection until about the third or fourth day (Meibach & Siegel 1975, Walberg et al 1976, Vanegas et al 1977). Most of the injection in the above studies had disappeared by the 5–7th day.

There are good grounds for believing that the peripheral zone of an injection of horseradish peroxidase does not lead to detectable retrograde labeling of neurons whose axons terminate in it. First, in experiments involving multiple injections of the enzyme in the cerebral cortex of monkeys (Jones & Leavitt 1974), though the peripheral, less dense zones coalesced, neurons in the relevant thalamic relay nucleus were still labeled in columns, each related topographically to an injection site and separated by gaps. Second, expansion of the cortical injection site in the experiments of Vanegas et al (1977) was not associated with an extension of the zone of labeled cells in the thalamus.

The extent of a retrogradely labeled column or cluster of cells in a strongly topographically organized system such as the thalamocortical (Jones & Leavitt 1974, Gilbert & Kelly 1975) or the retino-geniculate (Kelly & Gilbert 1975) is usually directly related to the size of the dense zone of the injection. This seems to indicate that the dense zone constitutes the "effective" injection site. Retrograde transport also seems to occur in the axons whose terminal ramifications are injured in the damaged central zone of an injection. At the center of a column of thalamic cells retrogradely labeled from an injection site in the cortex, there are usually a number of cells that, in addition to containing the usual granules of reaction product, have a diffuse labeling of their perikaryal cytoplasm and processes, which are also somewhat shrunken. It was postulated that these are cells (Jones 1975a) whose axons were damaged by the injection needle or pipette. This effect has recently been investigated in the chick visual system by Halperin & LaVail (1975). These workers found that neurons of the isthmo-optic nucleus, whose terminal ramifications were deliberately damaged at the time of injection into the eye, begin to accumulate retrogradely transported horseradish peroxidase 1–2 hr later than uninjured cells. Thereafter they transiently show a greater accumulation than the normal cells, followed by a more rapid disappearance of the enzyme from their somata. These results point to an increased accumulation of the enzyme by cells whose axons have reconstituted the membranes of their proximal stumps after axotomy. This observation may be compared with the view of Sherlock & Raisman (1975) that retrograde transport of horseradish peroxidase is best demonstrated in those neurons whose somata show the most vigorous reaction to section of their axons. The electron microscopic studies of Kristensson & Olsson (1976) in motor nerves subjected to crush injuries suggest that retrograde transport in the damaged axons occurs in the same manner as in undamaged. In connection with the more rapid clearing of horseradish peroxidase from damaged cells, Kristensson & Olsson (1974) have shown that the rate of retrograde transport in the hypoglossal nerve is sufficiently rapid for material to reach the soma before the onset of chromatolysis due to cutting the axon. Therefore, the more rapid clearing of enzyme from the damaged cells observed by Halperin & LaVail may commence with the onset of the degenerative response in these cells.

UPTAKE BY FIBERS OF PASSAGE? Although Krishnan & Singer (1973) had noted that horseradish peroxidase applied to the surface of nerves could penetrate into the axons, LaVail & LaVail (1974) and Nauta et al (1975) noted in electron microscopic studies that axons traversing zones of diffused horseradish peroxidase did not show evidence of increased endocytotic activity. This suggested that most uptake occurred in terminal and preterminal regions and not in fibers of passage.

There have, however, been numerous reports indicating that crushed or sectioned nerves placed in solutions of the enzyme can result in rapid and effective retrograde labeling of the cells of origin of the damaged axons (Kristensson & Olsson 1974, 1976, DeVito et al 1974, Furstman et al 1975), though the time over which damaged axons can take up the material may be restricted to little more than 10 min after the injury (Kristensson & Olsson 1976). In the spinal cord the application of horseradish peroxidase to the proximal end of the divided cord leads to retrograde labeling in cells at a distance (Hicks & D'Amato 1976), as also do injections associated with damage to central fiber tracts such as the corpus callosum (Lund et al 1975), dorsal acoustic stria (Adams & Warr 1976), and stria medullaris thalami (Herkenham & Nauta 1977). These and many other examples argue strongly for uptake and retrograde transport of horseradish peroxidase by fibers of passage provided that they are damaged by an injection needle or pipette. This, together with the difficulty of assessing the site of the effective injection, constitute the biggest drawbacks to the use of this method. Undoubtedly, many studies claiming to demonstrate projections from the cerebral cortex to various parts of the brainstem will prove to have resulted from interference, by the injection, with fibers of the pyramidal tract.

THE DEGREE OF RETROGRADE LABELING Along with the intense retrograde labeling of damaged cells described above, certain other differences in the degree of retrograde labeling in different cell populations seem to be determined by the nature of the terminal axonal plexus engendered by each cell type. In studies on the intralaminar nuclei of the thalamus (Jones & Leavitt 1974, Jones 1975a), it was found that the cells could only be very lightly labeled by injections in the cerebral cortex but could be very heavily labeled by injections in the striatum. This difference seems to correlate quite well with the observation in concurrent autoradiographic studies that the terminal ramifications of intralaminar cells are sparse and diffusely distributed in the cortex, whereas their terminal ramifications in the striatum are dense and compact. These results suggest that the amount of horseradish peroxidase accumulated by a cell depends upon the amount of terminal membrane available for uptake at the injection site and, probably, on the diameter of the parent axon (Nauta et al 1975, Jones & Leavitt 1974, Jones et al 1977). A converse distinction, however, has recently been made in the spinal cord where horseradish peroxidase injected into muscles leads to dense labeling of γ motoneurons and light labeling of α motoneurons (Strick et al 1976). It is difficult to explain this finding, though it is possible that the enzyme that accumulates within the capsules of muscle spindles remains available to the fusimotor endings for longer than it does to extrafusal motor endings outside the capsule. In principle, these variations in labeling properties might suggest that some cell systems would prove intractable to retrograde labeling. Initially, this seemed to be so in the case of the corticostriatal system of cells (Kuypers et al 1974, Nauta et al 1975, Jones & Leavitt 1974), but these cells have now been identified quite readily (Kitai et al 1976, Jones et al 1977), and there seems to be no known system in which retrograde labeling cannot be demonstrated.

EFFECTS OF NEURAL ACTIVITY ON UPTAKE OF HORSERADISH PEROXIDASE In their study of the synaptic vesicle recycling phenomenon at the frog neuromuscular junction, Heuser & Reese (1975) demonstrated that the uptake and turnover of horseradish peroxidase in synaptic vesicles was greatly enhanced during repetitive stimulation of the muscle nerve. These observations, which were also shown in vitro by Litchy (1973) and in the isolated retina by Schacher et al (1976), have raised the possibility that the amount of horseradish peroxidase taken up and subsequently retrogradely transported, could be enhanced by increased synaptic activity during and after an injection. To date, there have been no reports of success in this direction, but three studies suggest that further investigation may be worthwhile. First, Theodosis et al (1976) have shown that intravenous injections of horseradish peroxidase in rodents that are under conditions that lead to increased posterior pituitary activity result in considerable uptake of the enzyme by pinocytosis in the terminals of neurohypophysial axons. Second, Turner (1977) has found that as the dosage of pentobarbital is increased in anesthetized rabbits, the amount of uptake of horseradish peroxidase into synaptic terminals of the cerebral cortex from the subarachnoid space is markedly reduced. Finally, Singer et al (1977) have shown that the retrograde labeling of thalamocortical relay cells in the cat's lateral geniculate nucleus is markedly diminished if the cells' spontaneous activity is first reduced by removal of an eye. All of these observations suggest that uptake, at least, of horseradish peroxidase for subsequent retrograde transport may be in some way influenced by synaptic activity.

So far as is known, no successful attempt has been made to enhance terminal uptake of horseradish peroxidase by application of substances that increase pinocytotic activity.

UPTAKE OF ADVENTITIOUS HORSERADISH PEROXIDASE Horseradish peroxidase injected into the bloodstream can definitely be taken up by cerebral neurons, especially those such as the area postrema having an intimate relationship to the ventricular system (Broadwell & Brightman 1976). Retrograde cellular labeling can also be seen in these experiments. This is attributable to uptake in neurons that have axons terminating in the circumventricular organs or to uptake from the periphery in sensory, motor, and autonomic nerves.

In each of these situations, relatively large amounts of the enzyme are injected. In the usual experimental situation much smaller amounts are injected into the brain. Some leakage into the vasculature seems to occur, as evidenced by the large amount of uptake in vascular pericytes in the vicinity of an injection. However, it is generally conceded that this is insufficient to result in extraneous retrograde labeling (LaVail & LaVail 1974).

There is no evidence at this time for transneuronal transport of horseradish peroxidase.

ENDOGENOUS PEROXIDATIC ACTIVITY Several workers have reported the presence of endogenous peroxidatic activity that may be demonstrated with diaminobenzidine in several parts of the brainstem and diencephalon. Not all of these reports have been confirmed, and there is reason to believe that some may have been conditioned by a diffuse brown staining that commonly occurs in poorly fixed tissues subjected to diaminobenzidine. This may represent staining of lysed red blood cells. In many sites in which endogenous peroxidatic activity has been reported in neurons or in the neuropil, the reaction is inhibited by inhibitors of heme- and other metalloenzymes (Wong-Riley 1976). In others, however, such as the substantia nigra, the endogenous peroxidatic activity is clearly due to the presence of neuromelanin (Mensah & Finger 1975, Wong-Riley 1976) in the neurons, or as yet uncharacterized dense bodies in astrocytes (Sherlock & Raisman 1975, see also Keefer & Christ 1976). The peroxidatic activity of neuromelanin may be overcome by bleaching the tissue for a prolonged period in 50% hydrogen peroxide (Wong-Riley 1976). The neuromelanin artifact is also said to be reduced if benzidine is used as a marker instead of diaminobenzidine (Mesulam 1976a).

On the whole, the presence of endogenous peroxidase activity in some systems has not presented a problem of interpretation. When due to neuromelanin, as in the substantia nigra and locus coeruleus, it can usually be recognized by the greater homogeneity of the dense granules in comparison with those containing retrogradely transported horseradish peroxidase. The labeling of astrocytes in the hypothalamus is usually clearly distinguishable from neuronal labeling. Only the effect observed in the brainstem by Wong-Riley (1976) remains a source of potential confusion since the stained neurons are said to resemble exactly those labeled by retrograde transport.

THE SEARCH FOR A SECOND LABEL In their initial studies on the phenomenon of retrograde transport, Kristensson & Olsson (1971) noted that bovine serum albumin labeled with the fluorescent dye, Evans blue, was retrogradely transported as successfully as horseradish peroxidase. This suggested at once that the two markers could be used in combination to effect a double retrograde labeling of neurons whose axons branch and project to more than one site. The fact that a successful experiment using the two labels has not yet been reported is indicative of severe practical limitations rather than of the fact that no one has tried the experiment. The major advantage of horseradish peroxidase as a marker is in the high yield of reaction product by small quantities of transported enzyme. The numerous (unreported) failures to effect satisfactory retrograde labeling with fluorescent labeled albumin in the central nervous system probably stem mainly from the fact that, in general, too little fluorophore is transported to be satisfactorily visualized. An interesting use of horseradish peroxidase, itself labeled with two differently colored fluorophores as potential retrograde markers in double labeling studies, has been reported by Hanker et al (1976). To date however, it, too, appears to have met with limited success in the nervous system. The recent successful retrograde labeling of afferent cells to the rat hippocampal formation (Steward & Scoville 1976) with [^{3}H] bovine serum albumin, suggests that there are still prospects for using this in combination with horseradish peroxidase as a double label.

Other potential labels that unfortunately do not seem to be transported retrogradely include alkaline phosphatase, cytochrome C, hemacyanin, and lactoperoxidase (Stoeckel et al 1974, 1975a, Bunt et al 1976). Hansson et al (1975) and Bunt et al (1976), unlike Kristensson & Olsson (1971) and Steward & Scoville (1976), consider that bovine serum albumin is also not transported retrogradely. Other potential second labels that may prove to be applicable in selected systems are fluorescent antibodies bound to enzymes involved in the synthesis of the neurotransmitter characteristic of those systems (Livett et al 1969). For example, an antibody to dopamine β-hydroxylase has recently been shown to be retrogradely transported in sympathetic nerves (Fillenz et al 1976, Nagatsu et al 1976, Ziegler et al 1976). In those systems that are sensitive to nerve growth factor either throughout life (sympathetic ganglion cells) or at some time during their developmental history (dorsal root ganglion cells), [^{125}I] nerve growth factor has been successfully used as a retrogradely transported marker (Hendry et al 1974a,b, Stoeckel et al 1974, Iversen et al 1975, Stoeckel & Thoenen 1975). There seems to be little reason why [^{125}I] nerve growth factor could not be used in combination with horseradish peroxidase, though breakdown of the material into labeled amino acid residues that can be subsequently incorporated into anterogradely transported macromolecules is a potential complication.

Successful double labeling of a population of neurons with branching axons was obtained in the ventrobasal complex of the cat's thalamus by using a combination of retrograde degeneration and retrograde transport (Jones 1975a). In these experiments, the first somatosensory area of the cerebral cortex was ablated, then, six months later and 48 hr before killing the animal, horseradish peroxidase was injected into the second somatic sensory area. The effect was to cause retrograde labeling of cells that had undergone retrograde shrinkage presumably as the result of axon branches to the first somatic sensory area being severed. Unfortunately, this method has limited applications since not all populations of cells undergo such profound, or indeed any retrograde, atrophy after destruction of their axons. In the monkey thalamocortical system we have also been successful in a few cases in double labeling cells in the ventrobasal complex by injecting horseradish peroxidase into the first somatosensory cortex and [^{3}H] adenosine into the second (S. P. Wise & E. G. Jones, unpublished). This method, too, has limited potential, since in any experiment it will always remain uncertain whether the cellular labeling by [^{3}H] nucleosides is by retrograde or by anterograde transneuronal transport (Wise et al 1977).

At the moment, the two most promising candidates for second retrograde labels are [^{125}I] tetanus toxin and [^{3}H] horseradish peroxidase. For obvious reasons, the latter may come to be the more generally acceptable. Unlike nerve growth factor, [^{125}I] tetanus toxin has now been shown to be transported retrogradely in all neuronal systems, central and peripheral, in which it has been applied (Stoeckel et al 1975, Erdmann et al 1975, Price et al 1975, 1976, Schwab et al 1977, Schwab & Thoenen 1976, 1977). The fine structural correlates of the transport process seem to be identical to those already known for horseradish peroxidase (Schwab & Thoenen 1976), and the rate of retrograde transport in central and peripheral nerves appears to be comparable. It has the added advantage of being effective in far lower concentrations than horseradish peroxidase and, as a larger molecule, it does not diffuse as widely (Schwab et al 1977). In addition, the retrogradely transported material can cross from the retrogradely labeled cells into the axon terminals that end on the surface of that cell (Schwab & Thoenen 1976). Moreover, there is some evidence that this may occur specifically at cholinergic synapses (Schwab & Thoenen 1977). There are obvious advantages in this for studies of fine structural localization, though the fact that the tracer is highly toxic and needs to be purified and freshly labeled for each experiment will tend to restrict its widespread use. In principle, however, it could be used in combination with horseradish peroxidase.

Tritiated horseradish peroxidase (Geisert 1976) seems to offer the greatest prospects both as a second label and as a means of avoiding the potential health hazards of benzidine and related compounds. Since the label is demonstrated by autoradiography, [^{3}H] horseradish peroxidase may be used as a straightforward retrograde marker in paraffin sections. According to Geisert, the sensitivity of the tritium-labeled compound is somewhat greater than pure horseradish peroxidase, but it loses much of its enzymatic activity during the tritiating procedure. It may, thus, be used successfully in combination with horseradish peroxidase, because concentrations of the retrogradely transported tritiated compound sufficient to give detectable label in an autoradiograph are apparently too small to produce a recognizable reaction product with diaminobenzidine. Therefore, by injecting a high concentration of unlabeled horseradish peroxidase into area 17 of the cat's visual cortex and a small concentration of [^{3}H] horseradish peroxidase into area 18, Geisert (1976) was able to double label cells of the lateral geniculate nucleus that project by means of axon branches to both areas.

One as yet unresolved problem with the use of [^{3}H] horseradish peroxidase is that the labeled material or labeled breakdown products appear to be transported anterogradely as well, thus leading to relatively heavy labeling of the neuropil in nuclei that receive fibers from neurons involved in an injection site. According to Geisert (1976) no transneuronal labeling of cell somata occurs in these regions, so the phenomenon may not be a potential source of confusion.

COMBINATIONS OF RETROGRADE AND OTHER MARKERS Combinations of horseradish-peroxidase retrograde tracing with an anterograde tracing technique such as autoradiography, or in conjunction with a cytochemical stain for acetylcholinesterase, have already been mentioned. Other techniques with which retrograde horseradish peroxidase labeling has been combined include formaldehyde-induced fluorescence (Pickel et al 1975), axonal degeneration (Blomqvist & Westman 1975), immunohistochemistry (Ljungdahl et al 1975), and [^{3}H] thymidine autoradiography (Nowakowski et al 1975). In the latter case, the method has been successfully used to determine the connectivity of cells generated at a known time in development. Since the horseradish peroxidase method with an appropriate substrate lends itself very successfully to electron microscopic studies (Kristensson & Olsson 1973b, Ralston & Sharp 1973, LaVail & LaVail 1974, Sotelo & Riche 1974, Nauta et al 1975, Winfield et al 1975), many of the combinations mentioned should be as applicable at the electron microscopic as at the light microscopic level.

ANTEROGRADE TRANSPORT OF HORSERADISH PEROXIDASE Several investigators have reported that extracellularly injected horseradish peroxidase may be taken up by cells at the injection site and transported in the anterograde direction. In this way, some particularly striking demonstrations of the terminal ramifications of certain fiber systems have been achieved (e.g. Lynch et al 1974, Repérant 1975, Scalia & Colman 1974, Colman et al 1976, Adams & Warr 1976, Mesulam 1976a). In these cases, the fibers and their terminal ramifications are homogenously filled with dense reaction product. Diffuse filling of axons leaving an injection site has often been reported, and in the initial studies of LaVail & LaVail (1974) it was implied that any anterograde labeling seen was more in the nature of a diffusion process, since it did not extend more than a few mm beyond the injection site and seemed to be primarily conditioned by damage at the injection site. A similar phenomenon in which neurons close to an injection site are diffusely labeled to give a Golgi-like image has also been attributed to damage to their membranes (Jones & Leavitt 1974, Turner & Harris 1974), though this interpretation has been questioned by Vanegas et al (1977).

A number of observations argue against the anterograde transport, in which terminals are labeled, as being necessarily a mechanical process, though it can undoubtedly occur as the result of direct injection into a severed fiber tract (Adams & Warr 1976):

1. Although the failure rate is high, the transported material accumulates in axon terminals over a space of some hours (Scalia & Colman 1974, Colman et al 1976).
2. In these cases, the nature of the reaction product is granular, not diffuse (Repérant 1975, Scalia & Colman 1974).
3. The material appears to be transported in the smooth endoplasmic reticulum and in related tubular organelles, and occasionally as multivesicular or dense bodies (Hansson 1973, Sotelo & Riche 1974, Nauta et al 1975, Winfield et al 1975, Colman et al 1976, Walberg et al 1976).
4. The rate of anterograde transport appears to be comparable to that of retrograde transport—in both mammals and amphibia it is of the order of 80–200 mm per day (Hansson 1973, Colman et al 1976). This rate contrasts markedly with the rate (1–2 mm per day) of anterograde movement of the diffuse labeling described by LaVail & LaVail (1974).

It is possible, therefore, that two phenomena are involved: one a direct diffusion of material into the distal ends of cut axons, not necessarily reaching the terminals; the second a true anterograde transport phenomenon that is not necessarily the result of—though clearly facilitated by—damage, and in which the transported material reaches the axon terminals.

Labeling of Fiber Pathways with Cobalt

Sulfide precipitation of cobalt ions introduced into a cell by direct injection (Pitman et al 1972) or by axonal iontophoresis has proved to be a successful method for delineating the somata and the ramifications of single cells in invertebrates (Mason 1973, Mittenthal & Wine 1973, Winlow & Kandel 1976). The dense, cobalt sulfide precipitate is visible at both the light and electron microscopic level.

Until recently, the technique has had rather limited success in vertebrates (Llinas 1973, Prior & Fuller 1973). However, more promising results have now been reported in the cranial and spinal nerves of several species (Fuller & Prior 1975, Mason 1975, Iles 1976, Martin & Mason 1977). Anterograde filling of axons and retrograde filling of axons and somata have both been observed, and Cunningham (1976) and Cunningham & Freeman (1977) have successfully exploited the method to analyze the branching patterns and developmental plasticity of optic tract axons in rats. It now appears (Mason 1975) that passage of an electrophoretic current during uptake of the cobaltous chloride precursor may no longer be necessary to cause effective labeling, particularly if the Timm's intensification method is applied to the sections (Tyrer & Bell 1974).

The major restrictions on the method seem to be that the cobalt diffuses in the axons only up to a distance of 5–6 mm, so the method is unlikely to prove very useful in the analysis of long pathways. The fact that the application of cobalt must occur over a rather protracted period of time before fixation suggests that the tissue preservation may be poor. However, in Mason's study (1975), fine structural characteristics were fairly well preserved.

LABELING OF INDIVIDUAL CLASSES OF CELLS AND THEIR TERMINALS

Uptake of Specific Materials

It is now possible to identify specific classes of nerve terminal by their uptake of radiolabeled known or putative neurotransmitters (Fuxe et al 1968, Aghajanian et al 1966, Schon & Iversen 1974). In this way, central nervous synapses have been labeled with [^{3}H] serotonin (Descarries et al 1975, Chan-Palay 1976), [^{3}H] noradrenaline (Sotelo 1971, Bloom 1973, Descarries & Lapierre 1973), [^{3}H] dopamine (Fibiger & McGeer 1974), [^{3}H] gamma aminobutyric acid (GABA) (Hökfelt & Ljungdahl 1970, 1972a, Sotelo et al 1972, Hattori et al 1973, Iversen & Bloom 1972, Schon & Iversen 1972, Iversen & Kelly 1975), [^{3}H] glycine (Matus & Dennison 1971, 1972, Iversen & Bloom 1972, Ljungdahl & Hökfelt 1973, Price et al 1976), or [^{3}H] glutamic acid (Storm-Mathisen 1977). In certain instances the labeled transmitter has been associated with a particular class of synaptic vesicle (e.g. Matus & Dennison 1971, Ljungdahl & Hökfelt 1973a,b).

The labeling in this type of study has usually depended upon the existence of a high-affinity uptake system in the terminal for its particular transmitter agent (see Curtis & Johnston 1974, Iversen 1974, Hökfelt & Ljungdahl 1975, and Iversen et al 1975 for recent reviews). But it is also sometimes possible to label axon terminals by axoplasmic transport after injecting the labeled transmitter, or a precursor, into or around the parent cells (Livett et al 1968, Fibiger et al 1972, 1973, McGeer et al 1974, Schwartz et al 1976, Halaris et al 1976, Hunt & Künzle 1976b, Levin & Stolk 1977). These materials may be transported in an unincorporated form, since Hunt and Künzle (1976b) find in the pigeon tectum that [^{3}H] GABA is transported much more rapidly than materials labeled with [^{3}H] proline or leucine. The uptake and transport of transmitters is not always predictable, however, because in some cases, such as the cerebellar Purkinje cells, for undetermined reasons the soma, unlike the terminals, may not accumulate large amounts of the injected material (Hökfelt & Ljungdahl 1975, Iversen & Kelly 1975, Kelly & Dick 1976).

There are two potential sources of confusion in experiments involving uptake of specific transmitters. In the first, and this is especially applicable at the electron microscopic level, care must be taken to avoid injecting concentrations that exceed the K_m value for the specific uptake system, since nonspecific labeling of other terminals will then occur (Bloom 1973). The second, applicable mainly at the light microscope level, lies in the fact that glial cells and certain related cells such as pituicytes and pinealocytes appear also to have sodium dependent, high-affinity uptake systems for GABA (Bowery & Brown 1972, Young et al 1973, Schon & Kelly 1974, Schon et al 1975, Mata et al 1976). In one instance (the crayfish stretch receptor, Orkand & Kravitz 1971), labeled GABA is not taken up in large amounts by the synaptic terminal at which it is known to function as a transmitter agent, but primarily by adjacent glial cells. Iversen, Kelly, and their associates have recently shown that [^{3}H] β-alanine and 2–4, diaminobutyric acid (DABA), which substitute for GABA in the sodium-dependent uptake system for this substance, may be used selectively to label glial cells (in the case of β-alanine) and GABA-containing interneurons (in the case of DABA), in the cerebellum (Iversen et al 1975, Schon & Kelly 1975, Dick & Kelly 1975, Kelly & Dick 1976).

A variant of the labeling of axon terminals by uptake of specific transmitters is that in which axon terminals may be marked by analogues of the transmitter, such as 5,6- or 5,7-dihydroxytryptamine (for serotonin; Baumgarten et al 1971, Baumgarten & Lachenmeyer 1972, Costa et al 1972, Daly et al 1973, 1974), or 6-hydroxydopamine (for noradrenergic or dopamine neurons; Thoenen & Tranzer 1968, 1973). In appropriate doses these cause the relevant cells and their terminals to degenerate, and they thus become visible electron microscopically (e.g. Bloom 1973, Daley et al 1973, Björklund et al 1974, Gerson & Baldessarini 1975, Javoy et al 1976). In the substantia nigra this marking device has been successfully combined with autoradiographic labeling of afferent terminals on the degenerating cells by Hattori et al (1975). In many studies, injections of monoamine oxidase inhibitors alone or prior to the administration of *l*-noradrenaline, serotonin, or analogues such as α-methylnoradrenaline or 5-hydroxydopamine (Tranzer & Thoenen 1967, Richards & Tranzer 1970), have been used successfully to mark the dense core vesicles of monoamine-containing terminals (e.g. Hökfelt 1968, Tranzer et al 1969, Richards & Tranzer 1970, Koda & Bloom 1977, Chan-Palay 1976) (Figure 3). In a comparable manner, peroxidase-antiperoxidase binding to antibodies to substance P has been used to localize this peptide to large, dense core vesicles (Pickel et al 1977). Experimental approaches for cytochemical localization of biogenic amines and for selective destruction of their terminals and cell bodies have been reviewed extensively by Bloom (1973), Iversen & Schon (1973), and Hökfelt & Ljungdahl (1975).

When coupled with electron microscopy, which provides identification of the terminals, and in association with fluorescence or immunohistochemical methods that allow the whole course of a specific transmitter-containing pathway to be outlined, the specific uptake and direct labeling methods now offer the possibility of a biochemical mapping of central nervous pathways.

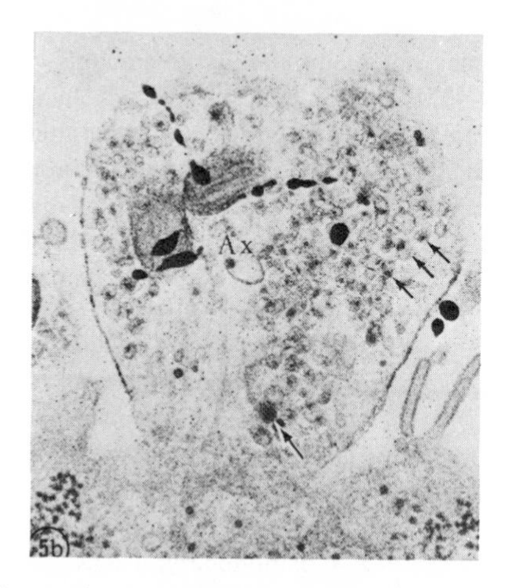

Figure 3 Electron microscopic autoradiograph showing axon terminal (Ax) in ventricular system of monkey, labeled by specific uptake of [^{3}H] serotonin. Arrows indicate large and small dense core vesicles. Prior treatment with a monoamine oxidase inhibitor. X 5500 (Chan-Palay 1976). By courtesy of Dr. V. Chan-Palay, Harvard University.

Work currently being conducted on the binding of putative transmitter substances to their presumed postsynaptic receptor sites also offers the promise of mapping central nervous pathways in terms of function (Snyder 1975). In the peripheral nervous system, labeled α-bungarotoxin (Lee et al 1967) has been used successfully as a marker for cholinergic receptor sites (Frank et al 1976), and there have been some regional studies of the binding of this labeled toxin in the central nervous system (Polz-Tejera et al 1975) and retina (Vogel & Nirenberg 1976). However, in the central nervous system, it has not yet been possible to demonstrate morphologically the binding of a recognizable marker such as this to a particular part of a nerve cell. Regional studies have now been conducted in vitro on the binding of glycine (Young & Snyder 1973), acetylcholine (Yamamura et al 1974), GABA (Zukin et al 1974), and serotonin (Bennett & Aghajanian 1974). Of particular interest are those studies on the binding of opium agonists and antagonists (Terenius 1973, Simon et al 1973, Pert & Snyder 1973, Pert et al 1974a,b), which seems to correlate with the presence of endogenous morphinelike transmitter agents, the enkephalins (Hughes et al 1975, Hughes 1975, Pasternak et al 1975, Terenius & Wahlström 1974, 1975). Up to this time most of these studies have been concerned primarily with the regional distribution of the binding sites for the various substances mentioned, particularly with that of the opiate receptors (Hiller et al 1973, Kuhar et al 1973, Pert et al 1974a,b, LaMotte et al 1976, Atweh & Kuhar 1977). There is good evidence that the binding of opiates is at the cellular (Pert et al 1975) and synaptic level (Pert et al 1974a,b), and antibodies to the naturally occurring enkephalins, when administered to rats, can be localized in axonal plexuses (Elde et al 1976). However, there is still a clear need to localize these putative transmitters and their receptors in terms of specific neuronal pathways and in relation to particular types of synapse.

Intracellular Injection of Single Neurons

The technique of intracellular injection of single neurons with fluorescent dyes has been available for some years (see Kater & Nicholson 1973 for a review). Perhaps the most widely used dye, Procion yellow, has been particularly widely applied in the invertebrate nervous system. The use of Procion brown (Christensen 1973) and of sulfide-precipitated cobalt salts (Pitman et al 1972), both of which are electron dense, permitted application of the technique at the fine structural level. The quality of preservation is rarely optimal with these methods, though a modification of Gillette & Pomeranz (1973), whereby the cobalt is reacted with diaminobenzidine to produce an osmium-binding polymer, has effected a considerable improvement. Procion yellow and Procion brown have been used with limited success to identify the position and, to some extent, the shapes and synaptic organization of neurons identified electrophysiologically in the mammalian central and peripheral nervous system (e.g. Kaneko 1970, 1971, Jankowska & Lindström 1972, Bryan et al 1972, Kelly & Van Essen 1974, Christensen & Ebner 1975, Barker et al 1972, 1976). This is particularly tedious, and the yield of labeled cells is usually very low. More recently, both autoradiography and horseradish peroxidase histochemistry have been adapted to intracellular labeling with striking success.

Intracellular injection of [^{3}H] amino acids, sugars, and nucleosides followed by autoradiography has been used particularly by Schubert, Kreutzberg, and their collaborators to mark individual spinal motoneurons and to study the phenomenon of dendritic transport (Lux et al 1970a,b Globus et al 1968, Schubert et al 1971, 1972, Kreutzberg & Schubert 1975). The method is extremely effective, though it suffers from the necessity of using thin sections to ensure identification of the radioactive tracer. Therefore, serial sections have usually been necessary in order to study the total extent of a labeled cell. There has been too little work done with this technique in mammals to determine whether the finest processes and elements, such as dendritic spines, will be outlined by transported material; in all published studies the axon of a filled cell has not been clearly identified. In the invertebrate, *Aplysia,* however, Schwartz et al (1976) have had considerable success in delineating fine processes and a good deal of the axon of certain serotonin-producing cells by the intracellular injection of [^{3}H] serotonin. In the same animal, Thompson et al (1976) have injected [^{3}H] fucose with similar success. Where cells are coupled electrotonically, the labeled precursor, injected into one cell, passes very rapidly into the other (Globus et al 1973, Rieske et al 1975, Schwartz et al 1976).

Perhaps the most striking success in intracellular labeling has come in the last year with the use of horseradish peroxidase (Figure 4). In several studies the enzyme has been injected into electrophysiologically identified cells, by electrophoresis or air pressure, through micropipettes containing solutions with enzyme concentrations ranging from 4 to 15%. In this way workers have been able to identify the cells of origin and axon collaterals of the cat spino-cervical tract (Jankowska et al 1976, Snow et al 1976) and the collaterals of cat spinal interneurons excited by Group I muscle afferents (Czarkowska et al 1976). The method may be more sensitive than the Procion method since there is usually filling of the finest dendritic processes, spines, and recurrent axon collaterals of cerebellar Purkinje cells (McCrea et al

1976) and of caudate nucleus projection neurons (Kitai et al 1976, McCrea et al 1976). At the electron microscopic level it has been used successfully to identify the terminals of recurrent collaterals of motoneuron axons in the cat (Cullheim & Kellerth 1976) and of sensory and motor neurons in the leech (Muller & McMahan 1976). The material seems to be free in the axoplasm, rather than membrane-bound, so the labeling probably occurs by diffusion.

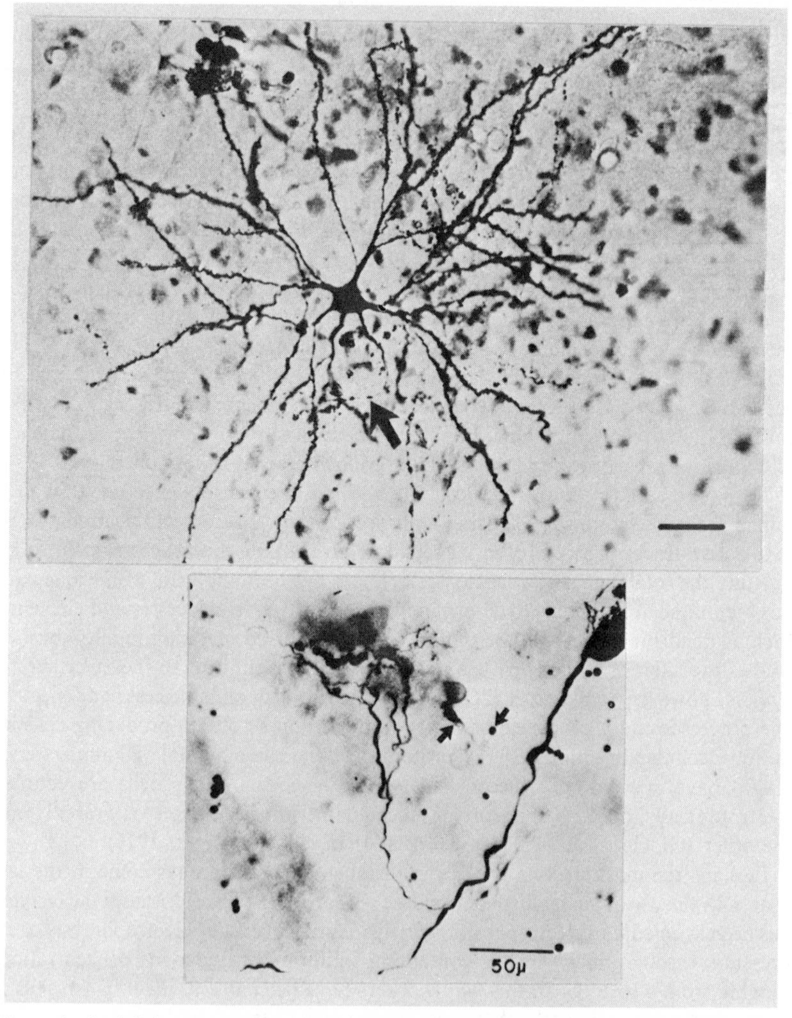

Figure 4 (*top*) Spiny neuron filled with horseradish peroxidase by injection through a recording pipette. Caudate nucleus of cat, arrow indicates axon. X 300 (Kitai et al 1976). (*bottom*) Recurrent collaterals of a Purkinje cell axon labeled by intracellular injection of horseradish peroxidase. Cerebellar cortex of cat. X 600 (McCrea et al 1976). By courtesy of Dr. S. T. Kitai, Wayne State University.

Immunohistochemical Methods

INTRODUCTION The first detailed information concerning the anatomy of a central system containing a particular neurotransmitter agent was brought about by the application of the Falck and Hillarp technique for the histofluorescent localization of catecholamines (Falck et al 1962). The success of this technique kindled an interest on the part of many investigators in obtaining similar anatomical information about other biochemically definable systems. The discovery of a histochemical reaction as specific and sensitive as the formaldehyde-induced fluorescence of catecholamines is rare. In contrast, immunohistochemical techniques theoretically require only that the substance to be localized be antigenic; this technique is, therefore, potentially applicable to the localization of virtually any complex molecule. Although the basic technique has been available for more than 30 years (Coons et al 1942), immunohistochemistry has only recently become an important tool in neuroanatomy. For this reason, methodological innovations and applications to new systems are appearing all the time, and the following discussion is an overview of a rapidly changing field.

ANTIGEN PURIFICATION The preparation of a specific antibody necessitates the use of a pure and homogeneous antigen. In the case of protein antigens, such as enzymes, this may well be the most difficult task in the entire immunohistochemical procedure, for these proteins may represent only a minor fraction of the many, sometimes very similar, proteins in a tissue. At present, the most universally applied criterion for protein purity is the formation of a single protein band on polyacrylamide gel electrophoresis when carried out under several conditions of pH and gel concentration. It is now a relatively standard procedure to utilize acrylamide gel electrophoresis as the final step in purification.

Once the structure of smaller and less complex antigens has been determined (e.g. peptide hormones), the task of obtaining pure antigen may consist of nothing more than buying the pure crystalline compound commercially. In the case of these compounds, however, problems may arise in obtaining a specific antibody.

ANTIBODY PREPARATION The pure antigen is usually homogenized in Freund's adjuvant (in "complete," which contains heat-killed microbacteria, or "incomplete," without bacteria) prior to inoculation. Adjuvants have an extraordinary potentiating effect in the induction of antibodies. A frequently used technique for protein antigens is to excise the protein band from polyacrylamide gels (also an effective adjuvant) after electrophoresis. The gel segment is then pulverized and injected with Freund's adjuvant (Hartman & Udenfriend 1969). Small quantities of protein antigens (~50–100 μg) are now commonly used for inoculation. Although often prompted by the scarcity of the antigen, the use of small amounts may be advantageous. Using a dose greater than that essential to induce a good immunological response does not produce proportionally higher titers of antibodies. Furthermore, if large doses are used, highly antigenic contaminants may be given in sufficient quantities to exceed the threshold necessary for antibody production. Thus, an important principle is that a relatively small inoculating dose of a very pure antigen will generally yield better results than a larger dose of a less pure preparation.

In the case of small molecular weight antigens (e.g. peptides) the preparation of antibody is somewhat different. First, the molecule must be conjugated to a macromolecule (usually a protein such as bovine serum albumin or α-globulin) before it will induce antibody formation [see Parker (1971) for methods of conjugation]. Since the antigen is usually not in short supply, many animals are usually inoculated, perhaps with different doses and schedules in the hope that at least one will produce specific antibody to the hapten antigen.

CRITERIA OF ANTIBODY PURITY AND SPECIFICITY The titer of precipitating antibodies directed against the desired antigen can be determined by precipitation titration (e.g. Kabat & Mayers 1967). It should be possible to quantitatively remove the antigen from the supernatant at equivalence. If this test is not performed, there is substantial risk in assuming that a single band seen on double diffusion in agar is the result of antibody interaction with the specific antigen (i.e. the major immunological response may be to a contaminant). Furthermore, the intensity of the reaction on double diffusion should be commensurate with the precipitin titer determined by titration. After it has been demonstrated that the major immunological response was to the specific antigen, more critical tests for purity and sensitivity should be carried out. The need for careful immunochemical evaluation of antisera prior to the application of immunohistochemistry cannot be overemphasized. None of the various blocking controls done in immunohistochemical localization are adequate to determine purity and specificity.

In precipitating systems one of the most stringent criteria for purity (i.e. freedom from contaminating antibody systems) is to show that *no* precipitin lines are visible when the antiserum is tested in double diffusion experiments against protein fractions eluted from chromatograms during purification either immediately before or after the antigen. These fractions will contain the proteins most likely to have contaminated the inoculating dose.

The most appropriate test for specificity (i.e. lack of antibodies cross-reacting with other similar tissue antigens) is to demonstrate a single immunoprecipitin line on immunoelectrophoresis when the antiserum is run against a *crude* tissue extract. In a crude tissue extract, the concentration of other tissue antigens relative to the specific antigen is substantially retained. Thus, the test system for specificity is made comparable to that encountered during a practical application of immunohistochemistry in a tissue section. The demonstration of a single immunoprecipitin line against the purified antigen is not a satisfactory criterion for antibody specificity.

Another potential problem that arises in the evaluation of antisera to proteins is due to the fact that it is frequently necessary to attempt localization in a heterologous species. Beef tissue (available in large quantities) is frequently used as a source of enzyme antigens, but localization studies are more commonly carried out on smaller laboratory animals. If the heterologous antigen binds quantitatively to antibody-sepharose (Cuatrecasas & Anfinsen 1971), the cross-reactivity between species will probably be sufficient for practical immunohistochemical localization, even though no immunoprecipitin band may form on double diffusion. For example, anti-bovine dopamine β-hydroxylase (DBH) adsorbs greater than 95% of rat DBH activity in a test system (Hartman & Udenfriend 1972) and is sufficiently sensitive to localize even the finest (< 0.5 μm) DBH-containing axons (Hartman 1973a). Although a lower dilution of antiserum will usually be required when applying immunohistochemistry to a heterologous species, immunological specificity will be maintained (i.e. a pure and specific antiserum will be no more likely to have a strong affinity for a random protein in a heterologous species than in the homologous species).

When nonprecipitating antibodies are present (e.g. with hapten antigens), they can be detected using co-precipitating systems (such as adding anti-IgG in a second step) or by equilibrium dialysis against labeled hapten. The principles for testing purity and specificity of small molecular weight antigens are also somewhat different. The same difficulties with antibody purity are not encountered if the inoculating antigen is chemically pure. In immunohistochemical applications, however, the goal is frequently to differentiate the distribution of two chemically related compounds, for example, oxytocin and vasopressin (Vandesande et al 1975c). In this case, the two antigens are so similar that, despite the inoculation of pure vasopressin, antibodies with an affinity for both hormones were induced. Specificity was obtained by adsorbing the antiserum with the undesired antigen, which had been bound to an insoluble matrix (Sepharose 4B) (Vandesande et al 1975c).

The application of immunohistochemical techniques can be divided into two steps. *The primary immunological reaction* involves producing the conditions necessary to permit the reaction of the specific antibody with the tissue antigen in situ. *Localization of the primary antibody* involves those methods that permit visualization of the bound antibody and thereby localize the antigen within the tissue.

The Primary Immunological Reaction

Incubation of a tissue section in an antibody-containing aqueous solution promotes the formation of an immunochemical bond between the specific antibody and the antigen in its original in vivo position. The formation of this bond is influenced by several factors, of which the most important are fixation, retention of the antigenicity, and accessibility of the antigen to antibody. Unfortunately, the optimal conditions relating to one factor frequently exclude the others, and in most cases it is necessary to empirically balance the several factors to obtain optimal overall conditions. Even under the most optimal conditions, however, one never knows when "complete" localization has been accomplished, and overinterpretation of negative results must be avoided.

FIXATION Adequate fixation is important not only to preserve morphology but also to prevent movement of the antigen being localized. It has the negative effect of decreasing the antigenicity of many substances, especially large protein antigens, though antigens vary widely in their ability to withstand fixation. Cross-linking fixatives also tend to make the tissue less easily penetrated by the antibody.

Organic solvents such as ethyl and methyl alcohol, acetone, or chloroform-methanol provide one of the mildest means of fixation that also satisfactorily preserves tissue morphology, at least for light microscopic examination. Because they do not form chemical bonds, they are much more satisfactory in cases where the antigen is bound to other tissue components (e.g. membrane-bound antigens). If an antigen is only partially bound, one runs the risk of localizing only the bound portion. The lipid solvent chloroform-methanol (2 : 1) has been useful for localization of the vesicular enzyme dopamine-β-hydroxylase in the brain (Hartman et al 1972) and has the advantage of increasing penetrability by extracting lipids.

Cross-linking fixatives have been most frequently used for neuroanatomical immunohistochemistry. Formaldehyde solutions are the most common and have been used in a wide variety of concentration, times, pH, and combinations (Parsons et al 1976, Silverman 1975a, Vandesande et al 1975b). Hökfelt et al (1973a) have developed a procedure that has been used for a very large number of different antigens including enzymes and polypeptides. It consists of transcardiac perfusion with cold 4% buffered formaldehyde for 20 min, followed by 90 min to 4 hr of immersion in the same fixative and washing overnight in 5% sucrose before cryostat sectioning. This method seems to offer a good balance between preservation of morphology and antigenicity for light microscopic localization. Glutaraldehyde is usually added for electron microscopic applications (Pickel et al 1976a), although it has a particularly bad reputation for inactivating antigens as well as for interfering with penetration of antibodies. The latter problem may be partially solved by use of other techniques (see below).

There remains a substantial need for improvement in fixation techniques used in immunohistochemistry. Aldehyde fixatives cross-link by forming bridges between side chain amino groups on proteins. Unfortunately, the antigenic determinants frequently include these groups. A step in the right direction is the innovative carbohydrate cross-linking fixative developed by McLean & Nakane (1974). This fixative contains periodate, lysine and paraformaldehyde. The periodate oxidizes the carbohydrate moieties of proteins (which do not contribute to antigenicity) to aldehydes that form bridges with the bifunctional amino acid, lysine. It is then possible to use very low concentrations of formaldehyde and achieve good ultrastructural preservation. These investigators also reported a marked preservation of the antigenic activity of basement membrane antigen over fixation with glutaraldehyde or 2% paraformaldehyde, although antigenicity was not equal to that of the native unfixed antigen.

TISSUE PREPARATION For light microscopy, cryostat sections are commonly used on both fresh (Hartman 1973a) and fixed tissue (Hökfelt et al 1973a), but if the antigen is heat stable, conventional paraffin sections give excellent results (Vandesande et al 1975a–c, Silverman 1975a, Agrawal et al 1977). For electron microscopy, frozen (Tabuchi et al 1976) or Vibratome (Hökfelt et al 1974a, Hökfelt & Ljungdahl 1972b) sections can be treated prior to embedding and then sectioning (Pickel et al 1976a), or tissue may be reacted after thin sectioning (Sternberger

1974). In this case, the embedding medium is etched away with hydrogen peroxide, and penetration distances are reduced, but the antigen must withstand the embedding process, which usually involves heating.

TISSUE PENETRABILITY For antigens present in solution in the cytoplasm of cell bodies, penetration has not been a serious problem (in this case, fixation is more important). On the other hand, for antigens located within small structures, such as fine nerve fibers or vesicles, lack of penetration may completely block the reaction. It should be emphasized that the same antigen may be located in compartments of different accessibility, creating the problem of knowing when localization is "complete."

Several methods are available to increase penetrability of tissue by antibodies. The addition of the detergent Triton X-100 to the antibody solution (Hartman et al 1972, Hartman 1973a) greatly increases penetrability and in well-fixed tissue does not disrupt ultrastructure (Pickel et al 1976a). Rapid freeze-thaw (Nakane 1975), organic solvents (see above), mild trypsin digestion [for paraffin sections (Huang et al 1976)], and the use of the antibody fragment of the immunoglobulin molecule rather than the whole antibody (Nakane 1975) have also been utilized.

PROBLEMS IN SPECIFICITY Appropriate fixation and enhanced tissue penetrability reduce the risk of false negative results, but a number of other difficulties can conspire to create false positives.

Nonspecific binding (i.e. nonimmunochemical binding of the antibody to other elements in the tissue) can be effectively reduced by dilution of the antiserum because these bonds generally have much lower association constants than those of antibodies for their specific antigens. Other means are (*a*) adsorption of the antisera with acetone powders of tissue not containing the antigen; (*b*) preincubation of tissue sections and addition to the reagents of immunoglobulins from another species; (*c*) adjustment of the pH, ionic strength, etc. (e.g. Goldman 1968, Nairn 1969). The best control for nonspecific binding is the substitution of the same antiserum as that used in the reaction, but one which has had the specific antibody removed by immunoadsorption with the antigen. In practice, "normal" serum (i.e. serum from an unimmunized animal) is frequently used for this purpose. Different sera, fixations, tissues, etc, result in a great variety of staining patterns due to nonspecific binding. It is hazardous to interpret a localization as specific, based mainly on a difference in intensity of the staining reaction, when a similar qualitative pattern has been observed with control sera.

Impure antisera result in false positives due to simultaneous localization of the contaminant. Unfortunately, contaminating antibodies are not always detectable by immunochemical tests (either because they are nonprecipitating or below the sensitivity of the tests). Recognition of these false positives is difficult. The frequent suggestion that the lack of reaction with "blocked" antisera is an adequate control for this artifact is generally incorrect because impure antisera may indicate that even the purest available antigen is contaminated. Because of this adsorption may remove both antibodies. Probably the best control is the observation of the effects of serial dilutions of antisera. If a particular type of localization disappears prior to the predominant localization, it may indicate that more than one antigen is being localized.

Cross-reacting antibodies result in false-positive localization of antigenically related compounds. For example, antibodies directed against cyclic-AMP may also bind to ATP. Since these small molecules do not form precipitating antibodies, the presence of this type of problem cannot be detected by immunoelectrophoresis against crude tissue extracts. The presence of cross-reacting antibodies can only be detected by testing for affinity with all the known related compounds and then attempting to remove the cross-reacting antibodies by adsorption. Antisera adsorbed with the specific antigen will not control for this artifact, because the cross-reacting antibodies will also be removed. As in the case of impure antisera, probably the best control is the observation of the effects of serial dilutions. The association constants for related antigens should be sufficiently low that they will not be demonstrated at dilutions that continue to localize the primary antigen. For a general discussion of these problems and their control, see Steiner et al (1976).

Localization of the Primary Antibody

CHOICE OF A VISIBLE MARKER Though a relatively large number of markers are available for visualizing the immunologically bound antigen and antibody, only two have generally been used in neuroanatomical applications.

Fluorescein isothiocyanate (FITC) is the original marker utilized by Coons et al (1942) and is still widely used. It emits an exceptionally high yield of green (517 nm) fluorescence when excited with blue light (peak excitation 490–500 nm). This green fluorescence is in the region of maximum sensitivity for the human retina, and recent advances in the design of fluorescence microscopes (see Ploem 1975) have permitted even greater efficiency in its visualization. Currently it is possible, for example, to view the extremely fine fibers (~0.5 μm) of the adrenergic systems (Hartman 1973a) at the relatively low magnification of 250 times (Figure 5).

Fluorescein is easily conjugated covalently to antibodies by use of the commercially available derivative FITC. Optimal conjugation requires that maximum fluorescein be incorporated per molecule without (*a*) decreasing antibody activity, or (*b*) increasing the tendency for nonspecific binding of the conjugate. Traditionally, the optimal molar ratio is thought to be approximately 2 moles fluorescein per mole antibody (Wood et al 1965). We have found, however, that 6 to 8 mole $mole^{-1}$ can be utilized with proportionally greater sensitivity, without increasing nonspecific fluorescence, if the antibody is purified by affinity chromatography prior to conjugation (Hartman 1973a). A useful nomogram for determining the molar ratio after conjugation from spectrophotometric adsorption measurements at 280 and 495 nm is presented by Wells et al (1966). FITC conjugates are available commercially, but these are often overconjugated and must be tested for nonspecific staining properties prior to use.

Fluorescein (as well as other fluorescent markers) suffers from two serious disadvantages as a marker for immunohistochemistry. The first is lack of permanence.

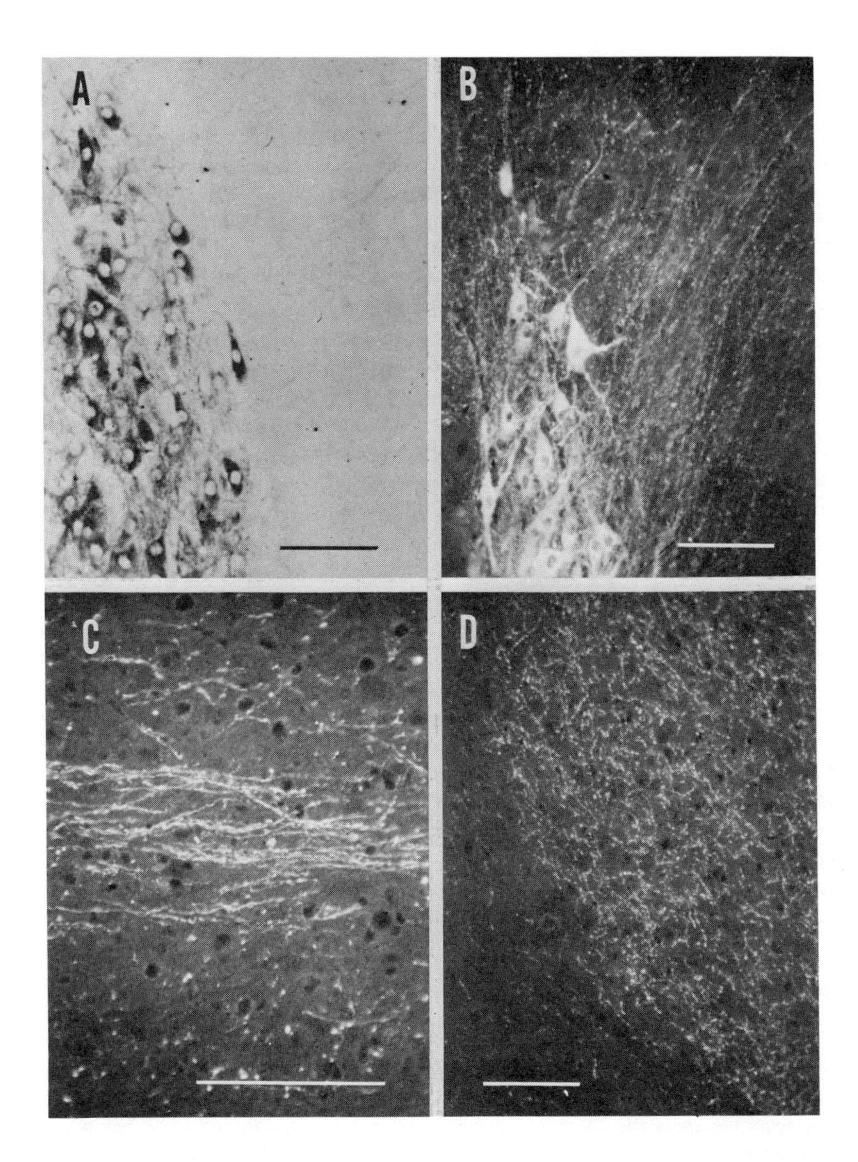

Figure 5 Comparison of peroxidase-antiperoxidase and immunofluorescence localization of DBH. (*A*) PAP localization of DBH in locus coeruleus of rat, horizontal section. Reaction product is restricted to DBH-containing cell bodies and a few proximal processes. In contrast, the indirect immunofluorescence technique demonstrates the presence of DBH in (*B*) locus coeruleus cell bodies and numerous proximal processes, (*C*) nonterminal axon fiber pathways, and (*D*) terminal arborizations, in this case in the anterior ventral n. of the thalamus. The difference is probably due to lack of penetration of the large PAP complex into the vesicles where DBH is localized. Fresh frozen sections fixed in chloroform-methanol (scale bars = 50 μm).

Optimal fluorescence is achieved in basic buffered aqueous mounting media (usually buffered glycerol). In this medium, sections deteriorate usually in 6–8 months. Although synthetic, xylene-soluble mounting media are available for fluorescence work, their use results in some quenching of fluorescence. Furthermore, fluorescent compounds undergo photodecomposition, and the fluorescence fades with prolonged observation. The second major disadvantage of fluorescent markers is that they are not applicable to electron microscopy.

The second marker used in neuroanatomical applications of immunohistochemistry is *horseradish peroxidase* (HRP). The major advantages of immunoperoxidase techniques over immunofluorescence are the relative permanence of the stained sections, especially after postfixation with osmium, and their applicability to electron microscopy. Peroxidase conjugated antibodies were first applied to immunohistochemistry by Nakane & Pierce (1966). Peroxidase activity is revealed in the conventional manner reviewed earlier in this chapter (see also Sternberger 1974, Weir et al 1974).

HRP has been conjugated directly to antibodies by the use of a number of bifunctional cross-linking reagents such as p,p^{1}-difluoro-m,m-dinitro-diphenyl sulfane (Nakane & Pierce 1966) or glutaraldehyde (Avrameas 1968). Two persistent problems have limited the effectiveness of the conjugates for immunohistochemical applications. First, conjugation results in partial inactivation of peroxidase activity. Second, only low yields of usable conjugates result because of the formation of cross-linked polymers of antibody, peroxidase, and antibody-peroxidase conjugates. The new method of conjugation developed by Nakane & Kawaoi (1974) appears to solve these problems by cross-linking between the carbohydrate moiety of peroxidase (preserving its activity) and the amino groups of the antibody. At present, however, the most frequently used procedure avoids chemical conjugation altogether and relies upon the immunochemical binding by an antibody to peroxidase. Peroxidase bound in this way retains most of its enzymatic activity (this procedure is discussed in detail below).

Methods for Localization of the Primary Antibody

THE DIRECT METHOD In the direct method the chosen visible marker is simply conjugated to the primary antibody. Though simple to use, separate conjugates must be prepared for each primary antiserum, and the method is generally significantly less sensitive than the more frequently used indirect techniques (see below). A number of investigators, however, continue to obtain excellent results at both light and EM levels using the direct method, especially in conjunction with Fab-peroxidase conjugates (Nakane 1975, Tabuchi et al 1976). In cases where penetration is the limiting factor this method may, in fact, be the most sensitive.

THE INDIRECT METHOD The indirect method makes use of a second antibody that recognizes and binds to the primary antibody, thereby localizing it. In this case the second antibody is labeled with the visible marker, either FITC or HRP. Immunoglobulin (IgG) molecules themselves are good antigens when injected into another species. For example, goat-antirabbit antibodies have the property of specifically binding to any rabbit antibody (all rabbit antibodies have antigenic sites in common) and labeled goat-antirabbit IgG can therefore be used to localize the primary antibody, regardless of what it is directed against, provided it was prepared in rabbits. Each primary antibody will accommodate multiple secondary antibodies, and each of these, in turn, will have several markers, thus increasing the number of marker molecules per molecule of primary antibody. This amplification by the application of a second antibody layer is one of the major advantages of the indirect technique.

High titer anti-IgG directed against the immunoglobulins of many species as well as FITC-conjugated anti-IgG, are available commercially. The conjugates should be tested before application: no detectable nonspecific staining should result when tissue sections are incubated with labeled anti-IgG alone in the same dilution to be used in localization. It should be emphasized that the anti-IgG has no special specificity for antibodies bound immunochemically and, therefore, detects non-specifically bound primary antibody as well as that bound specifically to antigen.

PEROXIDASE–ANTIPEROXIDASE (PAP) METHOD The PAP variant of the indirect method involves the addition of a third layer of immunochemical reagents (Mason et al 1969, Sternberger & Cuculis 1969). Antibodies are prepared to HRP (antiperoxidase) in the same species as the primary antibody. The function of the antiperoxidase is to bind peroxidase without inactivating it. The antiperoxidase is then bound to the primary antibody by way of a single anti-IgG molecule, which forms a bridge or link between the two. The method makes use of the fact that anti-IgG molecules have two identical binding sites on adjacent arms of the Y-shaped molecule. In the conventional indirect technique one binding site is attached to the primary antibody, and the other is unoccupied. In the PAP technique the second anti-IgG binding site is used to bind the antiperoxidase molecule (itself an IgG) that holds the active HRP. Mason et al (1969) carried out this procedure in four incubation steps: (*a*) primary antibody, (*b*) anti-IgG, (*c*) antiperoxidase, and (*d*) peroxidase. Sternberger & Cuculis (1969) discovered that peroxidase and antiperoxidase could be made to form a relatively stable, soluble complex, "PAP" (Sternberger et al 1970). This complex could then be added in a single incubation and finally reacted with hydrogen peroxide and an appropriate chromogen (normally diaminobenzidine). The details for the preparation of the PAP complex as well as a discussion of its application can be found in Sternberger (1973, 1974).

The impact of this method on immunohistochemistry has been great, especially in the area of electron microscopy. As a detector of a single primary antibody molecule it is extremely sensitive (Moriarty et al 1973) and permits the use of high dilutions of primary antiserum even below antigen saturation (i.e. the point where the maximum number of antibodies are bound to antigen), thus reducing nonspecific binding. The main weakness of the technique lies in the relatively large size of the PAP complex. It has a molecular weight of approximately 420,000 daltons and a pentagonal structure with sides estimated to be 120 Å (Sternberger 1973). This size may produce a penetration problem for some antigens. Furthermore, some of its sensitivity may be sacrificed by a limitation in the number of complex molecules that can be bound to a single antigen simply because of steric considerations. It must be kept in mind that the final sensitivity in detecting an antigen depends not only upon the sensitivity of the marker itself but on the ability of the marker to react with the primary antigen. The localization of fine terminal ramifications of noradrenergic fibers is a case in point. These fibers contain both dopamine-β-hydroxylase and tyrosine hydroxylase as demonstrated by fluorescence immunohistochemistry (Hartman 1973a, Hökfelt 1973a, 1976a). However, it has not been possible to visualize these fibers at either the light or EM level by using PAP-immunoperoxidase (Pickel et al 1976a). This difference is presumably related to the inability of the PAP molecules to penetrate into these particular fibers, since immunofluorescence demonstrates that the primary antibody reaction has taken place, and the PAP technique is able to visualize enzymes both in noradrenergic cell bodies and in the proximal portions of their axons. Figure 5 compares the results of fluorescence and PAP localization of DBH. Almost certainly this particular problem will be solved, but it serves as an example of the potential hazard of assuming that the PAP technique will always be the most sensitive.

APPLICATIONS

Neurotransmitter related enzymes The immunohistochemical localization of enzymes responsible for the synthesis of the various monoaminergic neurotransmitters has served as a prototype for the use of this tool in neuroanatomy. The monoamine systems had the advantage that a great deal was already known about the details of their anatomy from work done using the Falck-Hillarp technique. The feasibility of the immunohistochemical approach was first demonstrated for dopamine-β-hydroxylase (DBH) (Geffen et al 1969, Hartman & Udenfriend 1970, Fuxe et al 1970). Since these initial studies, work in the field has progressed rapidly so that at present all the biosynthetic enzymes of the monoamine systems have now been localized immunohistochemically. One important observation in common to the central monoaminergic terminals localized by electron microscopy to date is that the terminal varicosities do not always have synaptic membrane specializations (Pickel et al 1976a). In this way, they are similar to the varicosities of the peripheral monoaminergic nervous system.

Tyrosine hydroxylase (TH) has been demonstrated immunohistochemically at both the light microscopic [see Pickel et al (1975b, 1976a) Hökfelt et al (1975b, 1976a) for complete references] and electron microscopic (Pickel et al 1975a, 1976a) levels. In terminal areas where tyrosine hydroxylase can be localized electron microscopically (Pickel et al 1976a), it is associated with granular vesicles ranging from 40 to 80 nm in diameter. This appears to resolve the controversy concerning the subcellular localization of this enzyme that has existed since its identification. In axons, the enzyme appears to be associated with microtubular elements, which may indicate the involvement of these structures in axonal transport of the enzyme (Pickel et al 1976a). With immunoperoxidase techniques it has only been possible to localize the enzyme to noradrenergic neurons in the locus coeruleus, but the immunofluorescence methods employed by Hökfelt's group have shown that even the faintly fluorescent cells of the dorsal periventricular system contain it (Hökfelt 1976a).

Dopa decarboxylase has only been localized at the light microscopic level (Hökfelt et al 1973b, 1975a) but was demonstrable in all catecholaminergic neurons, with highest concentration in dopaminergic neurons. Positive immunofluorescence was also observed in serotoninergic (5-HT) neurons of the raphe system and in certain other neuronal groups with unknown neurotransmitters. The localization in 5-HT neurons is consistent with the earlier observation of immunologic similarity or identity between the decarboxylating enzyme for dopa and 5-hydroxytryptophan (Christenson et al 1972).

Dopamine-β-hydroxylase (DBH) is demonstrably absent from dopaminergic neurons and their terminals (Hartman et al 1972, Hökfelt et al 1973a). A complete map of the DBH-containing system in rat brain has recently been published (Swanson & Hartman 1975) DBH immunofluorescence has also demonstrated a close association of centrally originating noradrenergic nerve fibers with the microvasculature of the brain. Based on this evidence, a role for the central adrenergic system in the regulation of cerebral blood flow has been postulated (Hartman & Udenfriend 1972, Hartman et al 1972, Hartman 1973b). Recent physiological evidence (Raichle et al 1975) and electron microscopy (Swanson et al 1977) tend to support the concept of a regulatory effect both on blood flow and capillary permeability. Electron microscopic localization of DBH in the CNS has only been successful in the cell bodies and proximal axons of neurons in the locus coeruleus, where it was associated with a network of interconnecting membranes resembling smooth endoplasmic reticulum (Pickel et al 1976a). It has also been localized in the adrenal gland (Van Orden et all 1977). An interesting observation of Jacobowitz et al (1975) shows that anti-DBH injected into animals appears to be taken up by adrenergic neurons and transported retrogradely, in a manner similar to that described above for peroxidase except that anti-DBH is specific for the peripheral adrenergic system.

The demonstration of phenylethanolamine N-methyl transferase (PNMT), the final enzyme in the catecholamine pathway, in neurons of the CNS(Hökfelt et al 1973c, 1974a) represented the discovery of a new and rather unexpected neuron system that sends fibers to innervate the locus coeruleus and certain hypothalamic nuclei. The PNMT-containing neurons appear to be resistant to the effects of the neurotoxin 6-hydroxydopamine (Jonsson et al 1976). PNMT has only been localized at the EM level in the adrenal medulla (Van Orden et al 1977).

Tryptophan hydroxylase (TrH), the first enzyme in the biosynthetic pathway of the indolamine neurotransmitter serotonin (5-HT), has been localized at the light and electron microscopic level (Joh et al 1975, Pickel et al 1976a), and the neurons containing it have also been shown to send fibers to the locus coeruleus. Electron microscopically, the enzyme appears to be associated with microtubules in a manner comparable to tyrosine hydroxylase.

An additional transmitter-related enzyme that has been localized immunohistochemically is glutamate decarboxylase, which has been demonstrated at both the light and electron microscopic level in Purkinje cell axon terminals (Saito et al 1974, McLaughlin et al 1974). As with the uptake of GABA (see above), the parent cell bodies have proved difficult to visualize, but Ribak & Vaughn (1976) have recently had some success using animals pretreated with colchicine.

Two groups have reported the immunohistochemical localization of choline acetyltransferase (Eng & Uyeda 1974, McGeer et al 1974, Hattori et al 1976), but the purity of the choline acetyltransferase and the specificity of the antibodies prepared by these groups has been disputed by Rossier (1975).

Other brain proteins A number of proteins have been isolated and purified from the CNS on the basis of a unique physical property and subsequently been shown to be specific to the nervous system (i.e. not found in other organs). Although no specific function is presently known for these proteins, their localization has been of interest because of the possibility of using them as markers for particular types or aggregations of cells. S-100 protein is an acidic, small molecular weight, protein that has been extensively localized at the light and electron microscopic level. There is general agreement, both on the basis of biochemical data and immunohistochemical localization, that S-100 is predominantly a glial protein. Controversy exists, however, as to whether the protein is also normally present in small amounts in some neurons. Several laboratories have reported small but significant amounts of S-100 localized to neurons (Tabuchi et al 1976, Hanson et al 1976, Haglid et al 1974), whereas other groups (Ludwin et al 1976, Matus & Mughal 1975) see no reaction in neuronal elements. This controversy is a good example of the kind of problem that may be difficult to solve by immunohistochemical methods alone. Both groups who did not observe S-100 in neurons used tissue perfused with cross-linking fixatives. Those identifying S-100 in neurons used less stringent fixation conditions. S-100 is located in high concentration in glial elements and is small and highly charged. Therefore, the possibility exists that in inadequately fixed tissue it could be displaced from glial cells and appear in neurons. Conversely, excessive fixation may inactivate a small quantity of antigen in neurons so that only the large quantity in glial elements is localized.

Glial fibrillary acidic protein is another brain-specific protein localized to the glial elements [see Ludwin et al (1976) for references]. This protein has been used as a marker for human glial cells grown in culture (Gilden et al 1976) and to study the development of the Bergmann glia in mice with cerebellar malformations (Bignami & Dahl 1974).

Proteolipid protein isolated from myelin has recently been shown immunohistochemically to be a specific protein marker for the myelin sheath in the CNS. This protein was localized to oligodendrocytes during active myelination (Agrawal et al 1977).

Other nervous system–specific proteins that have been localized immunohistochemically are 14-3-2 (Grasso et al 1977), NSP–R (Pickel et al 1976b), olfactory marker protein (Hartman & Margolis 1975), and neurofilament protein (Jorgensen et al 1976).

NEUROENDOCRINOLOGY Immunohistochemistry has been used extensively in attempting to understand the organization of the central systems involved in the regulation of the various pituitary hormones.

Neurophysins, oxytocin and vasopressin The neurophysins (types I and II) are thought to be carrier proteins for the two neurohypophysial hormones oxytocin and vasopressin, respectively. The recent application of immunohistochemical methods has proved much more sensitive than the traditional Gomori stain for localizing the relevant neuronal systems. The localization of neurophysin II was the first application of immunohistochemistry to this system (Livett et al 1971). Since that time both neurophysins and their associated hormones have been extensively studied at the light (Silverman 1975a, Vandesande et al 1975c) and electron microscopic level (see Silverman 1976a). A great many more cell bodies than were recognized with the Gomori stains have been shown to contain these neurosecretory products. In addition to the paraventricular and supraoptic nuclei, many positive magnocellular neurons have been demonstrated between and rostral to these two nuclei (Silverman 1975a, Zimmerman & Antunes 1976). In addition, immunoreactive material has been localized to parvicellular perikarya of the suprachiasmatic nucleus in the rat (Vandesande et al 1975a, Zimmerman et al 1975) and the arcuate nucleus in cattle (De Mey et al 1975) but not of the guinea pig (Silverman 1975a). At least in the major nuclei, cells appear to contain either neurophysin I-oxytocin or neurophysin II-vasopressin, with mixtures of the two types of cell in each nucleus (Vandesande et al 1975c). Although the major projection of these cells is to the neurohypophysis by way of the zona interna of the median eminence, neurophysin-vasopressin positive fibers have also been demonstrated immunohistochemically in the zona externa in close contact with the primary portal plexus (Watkins et al 1974, Vandesande et al 1974, Zimmerman & Antunes 1976, Silverman 1975a, b). The origins of these latter fibers remain uncertain. Recent immunohistochemical studies have shown that the fiber systems derived from the oxytocin- and vasopressin-containing neurons have more widespread distribution than previously supposed. Zimmerman & Antunes (1976) have described positive fibers in the organum vasculosum of the lamina terminalis; a fiber system has been described extending dorsal to the choroid plexus of the ventricular system (Brownfield & Kozlowski 1977); a long, descending neurophysin-containing fiber pathway descends from the hypothalamus and terminates over a wide area in the medulla and spinal cord (Swanson 1977).

Luteinizing hormone-releasing hormone (LHRH), thyrotropin-releasing hormone (TRH), and somatostatin Several of the hypothalamic peptides involved in the regulation of the release of the various anterior pituitary hormones have been isolated, their structures determined, and antibodies prepared to them. Immunohistochemical localization of these peptides has resulted in an explosion of information concerning the organization of the relevant systems. Unfortunately, the results obtained by different laboratories have not been altogether consistent, but as more immunohistochemical data are published, previous controversies are being resolved by the demonstration of actual anatomical differences between the species examined, the stage of the endocrine cycle examined, and certain other factors, rather than by nonspecificity of the immunological reagents used.

The decapeptide LHRH was the first hormone of this type localized by immunohistochemistry (Barry et al 1973). More information is available on its distribution than on that of the other releasing factors, and it is clear that the LHRH system is more widely distributed than previously realized. The greatest concentration of LHRH is in the median eminence, where in most species it is contained in neuronal varicosities (Goldsmith & Ganong 1975, Silverman & Desnoyers 1976) associated with the primary portal plexus (Leonardelli et al 1973; see Pelletier et al 1976). An interesting exception to this generalization appears to be the mouse, where LHRH in the median eminence is said to be in tanycytes rather than neuronal processes (Zimmerman et al 1974, Silverman 1976b). One of the major controversies in the field at present is the location of the perikarya that give rise to these fibers. Their identification has proved difficult because under normal conditions the cell bodies do not contain sufficient hormone to permit adequate immunohistochemical localization. In the study of Barry et al (1973), who used castrated male guinea pigs treated with intraventricular colchicine, immunoactive cell bodies were found in an extensive area of the rostral hypothalamus extending back to, but not including, the arcuate nucleus. Barry and co-workers postulated the existence of a preoptico-infundibular pathway which might be the main route to the median eminence mainly on the basis of the orientation of labeled fiber segments and lack of immunoreactive cell bodies in the arcuate nucleus (see Barry & Dubois 1976a,b). Silverman (1976b), however, has presented strong evidence that the main source is, in fact, the medial basal hypothalamus (which contains the arcuate nucleus) by isolating this region and median eminence from the remainder of the brain with knife cuts and noting no significant decrease of LHRH in the median eminence fibers. Silverman was able to confirm the localization of LHRH in cell bodies of the rostral hypothalamus but also observed reactive perikarya in the arcuate nucleus. Other regions that appear to receive LHRH-containing fibers include the organum vasculosum of the lamina terminalis (Zimmerman et al 1974, Pelletier et al 1976, King & Gerall 1976), the suprachiasmatic, arcuate, and interpenducular nuclei. Silverman (1976b) and Pelletier et al (1976) also reported LHRH-reactive material in a nonneuronal distribution in other periventricular organs.

The peptide TRH has also been demonstrated by immunofluorescence (Hökfelt et al 1975c, d). As in the case of LHRH the major concentration of fibers appeared to be associated with the primary portal plexus of the median eminence. Fibers have, however, been found to be distributed quite widely through many diverse parts of the CNS. Localization of cell bodies has again proved difficult, but a few faintly fluorescent perikarya have been demonstrated in the dorsomedial hypothalamic nucleus.

The tetradecapeptide somatostatin has also been localized immunohistochemically and has been found to be widely distributed in many parts of the brain and spinal cord as well as in several peripheral organs (pancreas, thyroid, gut) [see Parsons et al (1976) for complete references]. Somatostatin is localized predominantly in small subependymal neurons located within the periventricular nucleus and, more rostrally, in the preoptic portion of the suprachiasmatic n. (Hökfelt et al 1975a, Elde & Parsons 1975, Alpert et al 1976). The major terminal areas appear to be the median eminence, the pituitary stalk (Hökfelt et al 1975a) and the organum vasculosum of the lamina terminalis (King et al 1975, Parsons et al 1976), although many other nuclei appear to receive fibers.

Other peptides and cyclic nucleotides Other peptides that have been localized immunohistochemically include certain substances that are thought to be involved either directly or indirectly in synaptic transmission. These include substance P (Hökfelt et al 1975e, f, 1976b), which has also been demonstrated at the electron microscopic level (Pickel et al 1977), and leucine-enkephalin (Elde et al 1976). Each has a fairly widespread distribution though leucine-enkephalin has not yet been demonstrated immunohistochemically in cell somata. The localization of cyclic nucleotides has not yet been studied in great detail (Bloom et al 1972, Siggins et al 1973, Steiner et al 1976), but future application of the immunohistochemical method to these compounds is feasible and should yield information of fundamental significance.

Although the application of immunohistochemistry to neuroendocrine problems is only a few years old, the power of the technique is apparent, for the studies quoted above have permitted a far better localization than that obtained by regional assay. The anatomical findings imply that these compounds may not only regulate pituitary function but also influence many other diverse parts of the CNS. Incomplete localization, especially of cell bodies, appears to be a greater difficulty than false positives due to antibody nonspecificity, although the latter can never be completely ruled out. Methods are rapidly being developed that take advantage of different physiological or pharmacological states to enhance the quantity of antigen within the cell bodies. These procedures have the potential danger of highlighting only one part of the system and could lead to erroneous conclusions. Additional effort in the direction of promoting a more complete primary immunological reaction might be rewarding. Almost all investigators have used some variant of formalin or picric acid formalin in paraffin-embedded sections with little concern about the possible effects of this treatment on the antigen and its accessibility.

"Functional" Mapping Studies

GENERAL There is no easily applied metabolic technique in the CNS comparable to the glycogen depletion method (Edström & Kugelberg 1968) that has been used so successfully to map out the distribution of particular kinds of α-efferent and fusimotor fibers in the peripheral nervous system (Brown & Butler 1973, 1975, Burke et al 1974, Barker et al 1976). It has been possible for some time to study biochemically, in dissected pieces of brain, the regional distribution of certain transmitter-related enzymes such as glutamate decarboxylase and tyrosine hydroxylase (e.g. Storm-Mathisen 1975, Ben-Ari et al 1975, 1977, Brownstein et al 1976, Fahn 1976, Tappaz et al 1976, Massari et al 1976), and histochemical methods for monoamine oxidase, acetylcholinesterase, succinate dehydrogenase, and many other enzymes are well established. The differential distribution of these enzymes offers the possibility of determining a "histochemical profile" for many parts of the brain. In several studies, changes in the distribution of some of these enzymes or of transmitters themselves following lesions have been used successfully to map certain specific pathways (e.g. Kataoka et al 1974, Fonnum et al 1974, Storm-Mathisen 1974, 1976, McGeer & McGeer 1975, Kizer et al 1976). By studying the time course of appearance of succinate dehydrogenase activity in neonatal rats, Killackey & Belford

(1976) have been successful in mapping the maturation of the trigeminal pathways from periphery to cerebral cortex. Similarly, the study of the distribution of heavy metals by the Timm's silver-sulfide technique has provided new insights into the connectivity of the hippocampal formation (Haug 1973).

Another form of functional mapping that has recently been explored by Pfaff and his collaborators involves the identification of cell groups in the hypothalamus and the amygdala on the basis of their ability to accumulate [^{3}H] estradiol (Stumpf 1968, Pfaff 1968, Pfaff & Kener 1974, Attramadal 1970, Morrell et al 1975, Pfaff et al 1976).

One interesting marking technique used by LeVay (1977) in the visual cortex seems to depend upon the apparent reduction in protein synthesis that occurs in neurons subjected to functional deafferentation. In LeVay's study, neurons of layer IV in the visual cortex of growing kittens showed an absence of all polyribosomal aggregations following visual deprivation. The dispersed ribosomes of these cells could be caused to reaggregate by prolonged barbiturate anesthesia, which suggests that the effect is functional rather than degenerative. Whether it may be applicable to other sensory systems remains to be determined.

2-DEOXYGLUCOSE Probably the most promising technique for the mapping of pathways or brain regions that are active under certain behavioral conditions is that which makes use of the regional changes in glucose consumption that occur during central nervous activity. [^{14}C]-2-deoxy-D-glucose is a labeled structural analogue of glucose that was first introduced by Sokoloff and his associates (Sokoloff et al 1974, 1977) in their studies on regional changes in cerebral blood flow. It quickly became apparent, however, that this marker for increased glucose utilization could be used very successfully for neuroanatomical mapping (Figure 6). Up to the present time, the method has been used to indicate regions of the brain and spinal cord that are active under particular stimulus conditions (Kennedy et al 1975, Sharp et al 1975, Sharp 1976a–c, Des Rosiers et al 1974, Sharp 1974, 1976a,b,c, Schwartz et al 1976, Collins et al 1976, Hubel et al 1978, Durham & Woolsey 1977). In these studies the whole active pathway is labeled within a few minutes after administration of an intravenous pulse of the tracer. The labeled material visualized in subsequent autoradiographs is [^{14}C]-2-deoxy-D-glucose-6-phosphate. One of the most dramatic examples of the use of this technique comes from the visual cortex, where it has been possible to dissociate active from inactive ocular dominance columns by injecting [^{14}C] 2-deoxyglucose in monkeys in which one eye had been removed or covered one or more days earlier (Kennedy et al 1975, 1976; L. Sokoloff, personal communication). In this example the label serves to outline not simply the regions of thalamic fiber termination in layer IV but also functionally related vertical zones of cortex lying superficial and deep to layer IV. This confirms that the initial flow of activity in the sensory cortex is in the vertical dimension. Similarly, in the vestibulo-cerebellar cortex alternating active and inactive strips have been demonstrated in rats following rotation of the animal for 45 min. In order to demonstrate some more subtle organizational details, such as the orientation columns of the visual cortex, it may be necessary to suppress other activity by rather deep anesthesia (Hubel et al 1978).

The major disadvantage of the deoxyglucose technique as currently applied is that it is regional in its application, and its resolution is limited. These defects stem to a large extent from the highly diffusible nature of the deoxyglucose. It has not yet been possible to satisfactorily label individual classes of cells or to distinguish separately the labeling of cell somata or specific processes, such as axons or synaptic terminals in the neuropil. Some improvement in resolution has been effected by the use of [^{3}H]-2 deoxyglucose, but identification of active synapses may ultimately depend upon the introduction of a glucose analogue that is converted into a larger molecule that may then be held in the tissue by aldehyde or other fixatives. To date no such compound is in sight. Though [^{14}C] glucose is obviously converted into labeled glycoproteins and other macromolecules that could be fixed by aldehydes, in functional mapping studies of the type discussed here, the increased glucose utilization demonstrated by labeled deoxyglucose appears to be associated with a decreased incorporation of glucose into macromolecules (Sharp 1976b).

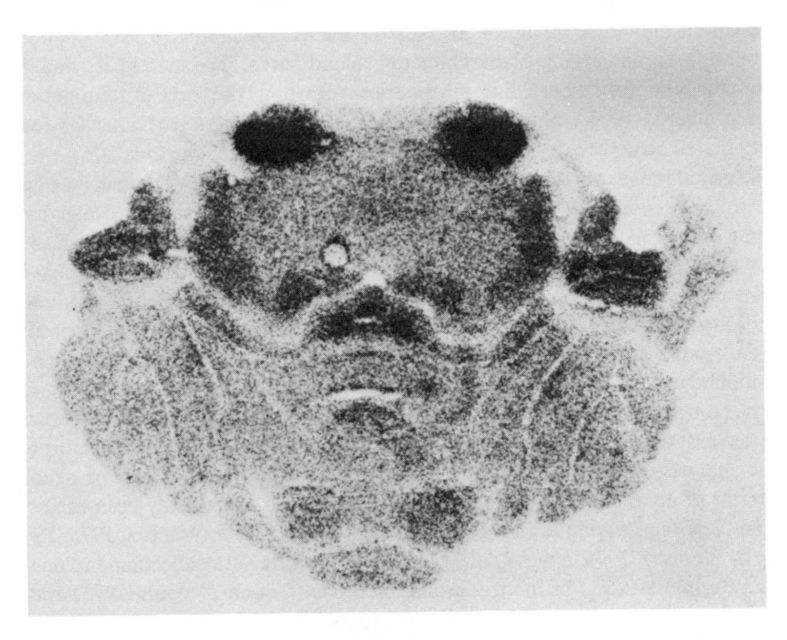

Figure 6 Autoradiograph showing a cross-section of the brainstem and cerebellum from a restrained rat injected with 2-deoxyglucose a short time before sacrifice. Dense zones indicate areas of increased glucose utilization and outline nuclei of auditory, vestibular, and trigeminal pathways as well as vestibulo-cerebellum. X 50. By courtesy of Dr. W. J. Schwartz, National Institutes of Health.

SOME QUANTITATIVE METHODS IN NEUROANATOMY

We do not propose in this brief review to deal with the more traditional methods for data collection and analysis in neuroanatomical studies. Among the more recent developments in a discipline that has traditionally lent itself primarily to qualitative investigations, are those that permit a more quantitative approach and that introduce a certain amount of automation. In the general area of autoradiography, several systems have been developed for determining relative or absolute grain densities in autoradiographs. The earlier systems relied upon photometric methods that analyzed grain density on the basis of the degree of reflection of incident light by the grains under darkfield illumination (Goldstein & Williams 1971, Rogers 1973). The reflectance is proportional to the number of grains. Image analysis by television systems such as the "Quantimat" that scan the microscope field in a linear fashion, recording changes in the intensity of transmitted light, have also been applied to autoradiographs (e.g. Prensky 1971). As pointed out by Price & Wann (1975), the reliability of these systems tends to vary since they are influenced by the size of the silver grains as much as by the actual number of grains. Recently Wann et al (1974) have developed a computer-assisted television scanning system that not only counts individual grains but also scans the autoradiographs in a systematic way in relation to underlying cytoarchitectonic landmarks. The system is, therefore, particularly useful in providing information regarding the level of termination of afferent fiber systems within particular nuclei or in relation to particular laminae of a laminated structure (e.g. Krettek & Price 1977, Jones & Burton 1976).

A second innovation that permits the collection of quantitative data in an efficient and much less tedious way than the traditional one, and less expensively than in some commercially available systems, is the computer-assisted pantagraph of Cowan & Wann (1973). This simple device consists of a commercially available data tablet upon which photo- or electron micrographs or camera lucida outlines are positioned and traced by a pen that emits repetitive electrical sparks. The positions of the sparks are "heard" by a pair of directionally selective strip microphones positioned at right angles to one another. The completion of each outline is signalled to the computer by the reentry of the pen into an acoustical "window" set up when the pen first touches each new profile. The computer program permits the morphometric analysis of cross-sectional profiles of cell somata, nuclei, axons, or dendrites in terms of areas and diameters. This program, which has already been used quite extensively (e.g. Alving & Cowan 1971, Jones 1975b, Lee & Woolsey 1975, Pasternak & Woolsey, 1975, Jones et al 1976, 1977), permits data collection and analysis at a rate far in excess of the traditional methods (Cowan & Wann 1973) and has much to commend it.

The principle of computer reconstruction from serial drawings, or photo- or electron micrographs has been applied by a number of workers. In the system of Levinthal and his colleagues (Levinthal & Ware 1972, Levinthal et al 1973), a 35 mm movie film is first made from the sequence of sections. The X and Y coordinates of a profile are fed into a digitizer, and the Z coordinates are recorded from the frame number of the movie. This system permits the reconstructed item to be rotated in the *X, Y,* or *Z* axes, thus providing a view of the structure from any angle. Quantification consists of the measurement of dendritic or axonal branches and the volume of the dendritic tree. Levinthal and his associates (Levinthal & Ware 1972, Macagno et al 1973, LoPresti et al 1973, Levinthal et al 1976) have used this system to dramatic effect in studying the symmetry in the branching patterns of particular cells in isogenic organisms and the changes in shape that some of these undergo during development, especially in relation to the formation of synaptic connections. A comparable system developed by Rakic et al (1974) uses tracings of electron microscopic profiles drawn onto transparent acetate sheets. The contours of these are then digitized, and, in the subsequent reconstruction, the intensity of each contour is weighted according to its distance from the observer.

Possibly the simplest method for providing three-dimensional visualization of nerve cells is the ingenious technique of McKenzie & Vogt (1976), whereby a modified *X-Y* plotter connected to the microscope stage controls permits drawing the cell as though seen in planes at right angles to the surface of the slide.

Over the years, many authors have sought to quantify the linear dimensions and branching patterns of nerve cells as seen in Golgi-impregnated material (e.g. Bok 1959, Sholl 1953, Ruiz-Marcos & Valverde 1970, Colon & Smit 1970). In the last two years, Berry and his associates, by using network analysis, have provided impressive data regarding the variations that occur in the dendritic fields of normal neurons in the cerebellum and elsewhere and the changes that occur in these during normal and disordered development (Hollingworth & Berry 1975, Berry et al 1975, Berry & Bradley 1976a,b, Bradley & Berry 1976a,b). It is possible that this quantitative approach may find wider application in developmental studies and in cases in which neuronal geometry may be altered in some specific manner by disease (e.g. Purpura & Suzuki 1976).

An early attempt to apply computer technology to the analysis of dendritic lengths and branching patterns was that of Glaser & Van der Loos (1965). These workers used transducers connected to the stage controls of a microscope and a focusing potentiometer to transfer information to an *X-Y* plotter, interfaced to an analogue computer to calculate dendritic lengths. Since then, several systems have been developed for the analysis of Golgi-impregnated or dye-injected nerve cells, making use of digital computer technology (Garvey et al 1973, Wann et al 1973, 1974, Lindsay & Scheibel 1974, Marin-Padilla & Stibitz 1974, Cowan et al 1975, Llinas & Hillman 1975, Valverde 1976). The available systems vary somewhat in their degree of automation. In that of Valverde, for example, the operator must make a drawing, record data points in two steps, and type the data into the computer. In the systems of Llinas & Hillman and of Wann et al, data, though still selected by the operator, are fed directly from the microscope into the computer, the computer to a large extent controlling the focus and the movement of the stage. All of these methods permit sophisticated analyses of nerve cells in terms of dendritic (and axonal) lengths, orientations, and branching patterns. However, up to the present time their main applications have been in the classification of nerve cells in terms of their three-dimensional organization (Figure 7) (Jones 1975c, Woolsey et al 1975, Valverde 1976). For the future, they offer the

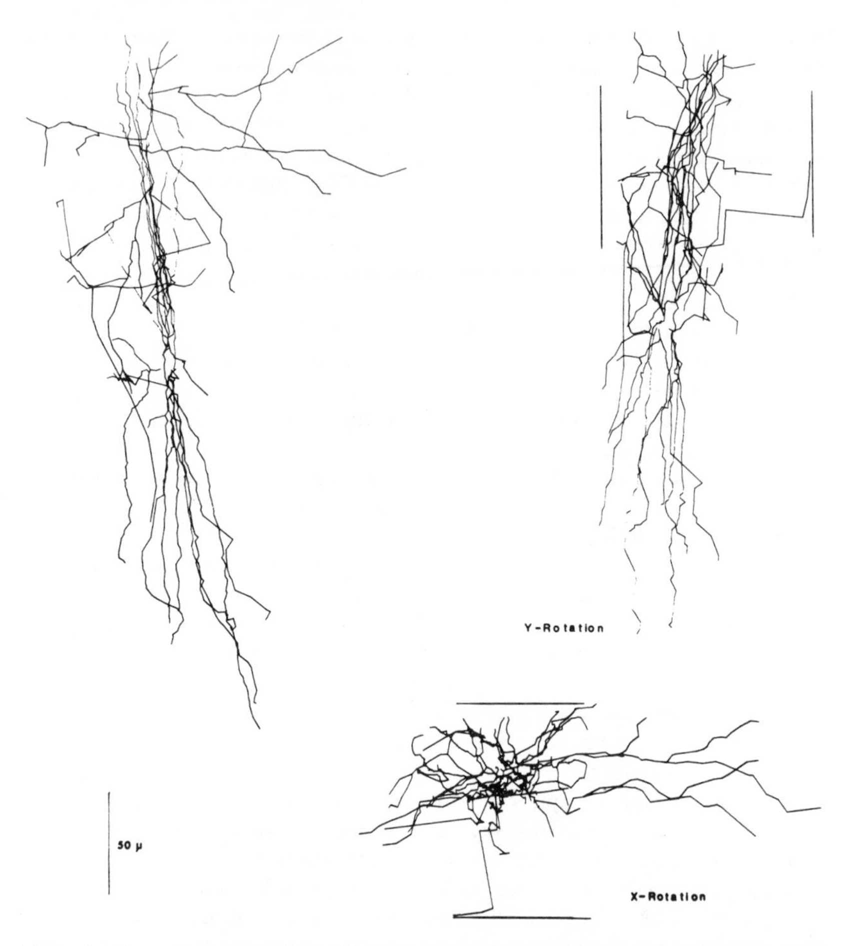

Figure 7 Computer tracing of a Golgi-impregnated, large multipolar cell from the cerebral cortex of a monkey. Cell viewed en face on left, rotated as though seen from side on right, and as though seen from above at bottom. Lines indicate surfaces of section (Jones 1975c).

prospect of providing quantitative parameters for the classification of cells and for assessing the changes that may occur during development and after sensory deprivation or deafferentation.

CONCLUSION

The preceding review has by no means been exhaustive. Several areas have been treated rather cursorily and others not at all. Among the latter, mention should be made of the exciting new observations that have been made with freeze-fracture–etch techniques on the functional morphology of the synapse (e.g. Pfenninger 1973,

Landis & Reese 1974, Heuser et al 1974). The use of tissue culture particularly for the study of synaptic interactions between cells (Bunge 1975) has, similarly, not been mentioned. Important as they are, the observations in these two areas rest upon technical foundations that are already well established, and the methodologies have already been well reviewed in the papers referred to.

There is no way to predict what technical innovations will be forthcoming in the next few years. Since about 1972, after a period of some twenty years during which the majority of anatomical studies were carried out with well-established axonal degeneration techniques, numerous new methods have become available. Therefore, though allowing for the fact that some further fundamental breakthroughs may occur, it is likely that the next few years will be a period of consolidation. That is, a period of exploitation of the new methods for further advancing our understanding of neural organization. It would seem that pathway tracing will continue to occupy a major part of neuroanatomy. Given the interpretative problems associated with the use of horseradish peroxidase, it may be that the popularity of this retrograde tracer will decline in favor of the more predictable autoradiographic method. The autoradiographic method and, particularly, the immunohistochemical methods seem to hold the promise of analyzing neural pathways not simply in terms of their topographical and morphological organization but in terms of the transmitter substances, transmitter-related enzymes, and the types of synaptic receptor that they contain. In this way it may be possible in the not too distant future to provide what has been called a "biochemical anatomy" of the brain and spinal cord at both the light and electron microscopic levels.

Since the majority of the new mapping techniques lend themselves to the study of fetal and immature as well as mature animals, it is likely that there will be an upswing in the number of studies being devoted to the development of neural connectivity. In association with this we would expect to see an increased use of quantitative techniques for assessing the maturation of cell form in relation to the establishment of connections. In addition, there seems little doubt that studies of synaptic induction and of the nature of those substances on the surfaces of growing cells that permit them to recognize one another will continue at the anatomical as well as at the physiological and biochemical levels.

It is nearly a century since the first significant experimental neuroanatomical studies of Gudden, Nissl, and Monakow. In the intervening years, the questions being asked of neuroanatomy and the experimental approach have essentially remained unchanged. The newer methodologies, by bringing together the conventional morphological and the biochemical approaches, seem to have brought us closer than ever before towards correlating neural structure with function.

Acknowledgments

We are indebted to Drs. W. M. Cowan, D. I. Gottlieb, and S. P. Wise for commenting upon the manuscript. Personal work quoted in the text was supported by grants NS10526, EY00092, NS12311, and MH70451 from the National Institutes of Health and the National Institute of Mental Health, United States Public Health Service.

Literature Cited

Adams, J. C., Warr, W. B. 1976. Origins of axons in the cat's acoustic striae determined by injection of horseradish peroxidase into severed tracts. *J. Comp. Neurol.* 170:107–21

Aghajanian, G. K., Bloom, F. E., Lovell, R., Sheand, M., Freedman, D. X. 1966. The uptake of 5-hydroxytryptamine-^{3}H from the cerebral ventricles: autoradiographic localization. *Biochem. Pharmacol.* 15:1401–3

Aghajanian, G. K., Wang, R. Y. 1977. Habenular and other midbrain raphe afferents demonstrated by a modified retrograde tracing technique. *Brain Res.* 122:229–42

Agrawal, H. C., Hartman, B. K., Shearer, W. T., Kalmbach, S., Margolis, F. L. 1977. Purification and immunohistochemical localization of rat brain myelin proteolipid protein. *J. Neurochem.* 28:495–508

Alpert, L. C., Brawer, J. R., Patel, Y. C., Reichlin, S. 1976. Somatostatinergic neurons in anterior hypothalamus: immunohistochemical localization. *Endocrinology* 98:255–61

Alving, B. M., Cowan, W. M. 1971. Some quantitative observations on the cochlear division of the eighth nerve in the squirrel monkey (*Saimiri sciureus*). *Brain Res.* 25:229–39

Anderson, L. E., McClure, W. O. 1973. Differential transport of protein in axons: comparison between the sciatic nerve and dorsal columns of cats. *Proc. Natl. Acad. Sci. USA* 70:1521–25

Arvidsson, J. 1975. Location of cat trigeminal ganglion cells innervating dental pulp of upper and lower canines studied by retrograde transport of horseradish peroxidase. *Brain Res.* 99:135–39

Attramadal, A. 1970. Cellular localization of ^{3}H-oestradiol in the hypothalamus. *Z. Zellforsch.* 104:572–81

Atweh, S. F., Kuhar, M. J. 1977. Autoradiographic localization of opiate receptors in rat brain. I. Spinal cord and lower medulla. *Brain Res.* 124:53–67

Autilio-Gambetti, L., Gambetti, P., Shafer, B. 1973. RNA and axonal flow. Biochemical and autoradiographic study in the rabbit optic system. *Brain Res.* 53:387–98

Avrameas, S. 1968. Détection d'anticorps et d'antigènes à l'aide d'enzymes. *Bull. Soc. Chim. Biol.* 50:1169–78

Banks, P., Mayor, D., Mraz, P. 1973. Metabolic aspects of the synthesis and intra-axonal transport of noradrenaline storage vesicles. *J. Physiol.* 229:383–94

Barber, P. C., Raisman, G. 1974. An autoradiographic investigation of the projection of the vomeronasal organ to the accessory olfactory bulb in the mouse. *Brain Res.* 81:21–30

Barker, D., Bessou, P.,Jankowska, E., Pagès, B., Stacey, M. J. 1972. Distribution des axones fusimoteurs statiques et dynamiques aux fibres musculaires intrafusales chez le chat. *C. R. Acad. Sci., Ser. D.* 275:2527–30

Barker, D., Bessou, P., Jankowska, E., Pagès, B., Stacey, M. J. 1975a. Distribution of static and dynamic γ axons to cat intrafusal muscle fibres. *J. Anat.* 119:199

Barker, D., Emonet-Denand, F., Harker, D. W., Jami, L., Laporte, Y. 1976. Distribution of fusimotor axons to intrafusal muscle fibres in cat tenuissimus spindles as determined by the glycogen-depletion method. *J. Physiol. London* 261:49–69

Barker, J. L., Hoffman, P. N., Gainer, H., Lasek, R. J. 1975b. Rapid transport of proteins in the sonic motor system of the toadfish. *Brain Res.* 97:291–302

Barondes, S. H. 1968. Further studies on the transport of protein to nerve endings. *J. Neurochem.* 15:343–50

Barrett, J., Cook, M., Ledbury, P., Zucker, R., Whitlock, D. 1973. Axoplasmic transport from rat dorsal root ganglia. *Anat. Rec.* 175:269

Barry, J., Dubois, M. P. 1976a. Immunofluorescence study of LRF-producing neurons in the cat and the dog. *Neuroendocrinology* 18:290–98

Barry, J., Dubois, M. P. 1976b. Immunoreactive LRF neurosecretory pathways in mammals. *Acta Anat.* 94:497–503

Barry, J., Dubois, M. P., Poulain, P. 1973. LRF-producing cells of the mammalian hypothalamus. A fluorescent antibody study. *Z. Zellforsch.* 146:351–66

Baumgarten, H. G., Björklund, A., Lachenmeyer, L., Nobin, A., Stenevi, U. 1971. Long-lasting selective depletion of brain serotonin by 5,6,-dihydroxytryptamine. *Acta Physiol. Scand. Suppl.* 373:1–16

Baumgarten, H. G., Lachenmeyer, L. 1972. 5,7-Dihydroxytryptamine: improvement in chemical lesioning of indoleamine neurons in mammalian brain. *Z. Zellforsch.* 135:399–414

Beitz, A. J., King, G. W. 1976. An improved technique for the microinjection of

horseradish peroxidase. *Brain Res.* 108:175–79
Ben-Ari, Y., Zigmond, R. E., Moore, K. E. 1975. Regional distribution of tyrosine hydroxylase, norepinephrine and dopamine within the amygdaloid complex of the rat. *Brain Res.* 87:96–101
Ben-Ari, Y., Zigmond, R. E., Shute, C. C. D., Lewis, P. R. 1977. Regional distribution of choline acetyltransferase and acetylcholinesterase within the amygdaloid complex and stria terminalis system. *Brain Res.* 120:435–45
Bennett, G., DiGiamberardino, L., Koenig, H. L., Droz, B. 1973. Axonal migration of protein and glycoprotein to nerve endings. II. Radioautographic analysis of the renewal of glycoproteins in nerve endings of chicken ciliary ganglion after intracerebral injection of [^{3}H] fucose and [^{3}H] glucosamine. *Brain Res.* 60: 129–46
Bennett, J. L., Aghajanian, G. K. 1974. D-LSD-binding to brain homogenates: possible relationship to serotonin receptors. *Life Sci.* 15:1935–44
Benowitz, L. I., Karten, H. J. 1976. Organization of the tectofugal visual pathway in the pigeon: a retrograde transport study. *J. Comp. Neurol.* 167: 503–20
Berkley, K. 1974. Differential labeling of neural pathways converging on the ventrobasal complex of cat thalamus. *Brain Res.* 66:342–48
Berkley, K. J., Graham, J., Jones, E. G. 1977. Differential incorporation of tritiated proline and leucine by neurons of the dorsal column nuclei in the cat. *Brain Res.* 132:485–505
Berrevoets, C. E., Kuypers, H. G. J. M. 1975. Pericruciate cortical neurons projecting to brain stem reticular formation, dorsal column nuclei and spinal cord in the cat. *Neurosci. Lett.* 1:257–62
Berry, M., Bradley, P. M. 1976a. The application of network analysis to the study of branching patterns of large dendritic fields. *Brain Res.* 109:111–32
Berry, M., Bradley, P. 1976b. The growth of the dendritic trees of Purkinje cells in irradiated agranular cerebellar cortex. *Brain Res.* 116:361–87
Berry, M., Hollingworth, T., Anderson, E. M., Flinn, R. M. 1975. Application of network analysis to the study of the branching patterns of dendritic fields. In *Physiology and Pathology of Dendrites, Advances in Neurology,* ed. G. S. Kreutzberg, vol. 12, pp. 217–45. N.Y: Raven Press
Bignami, A., Dahl, D. 1974. The development of Bergmann glia in mutant mice with cerebellar malformations: reeler, staggerer and weaver. Immunofluorescence study with antibodies to the glial fibrillary acidic protein. *J. Comp. Neurol.* 155:219–29
Bisby, M. A. 1976. Orthograde and retrograde axonal transport of labeled protein in motoneurons *Exp. Neurol.* 50:628–40
Björklund, A., Baumgarten, H. G., Nobin, A. 1974. Chemical lesioning of central monoamine axons by means of 5,6-dihydroxytryptamine and 5,7-dihydroxytryptamine. *Adv. Biochem. Psychopharmacol.* 10:13–33
Bloom, F. E. 1973. Ultrastructural identification of catecholamine-containing central synaptic terminals. *J. Histochem. Cytochem.* 21:333–48
Blomqvist, A., Westman, J. 1975. Combined HRP and Fink-Heimer staining applied on the gracile nucleus in the cat. *Brain Res.* 99:339–42
Bloom, F. E., Hoffer, B. J., Battenberg, E. R., Siggins, G. R., Steiner, A. L., Parker, C. W., Wedner, H. J. 1972. Adenosine 3',5'-monophosphate is localized in cerebellar neurons: immunofluorescence evidence. *Science* 177:436–38
Bobillier, P., Pettijean, F., Salvent, D., Ligier, M., Sequin, S. 1975. Differential projections of the nucleus raphe dorsalis and nucleus centralis as revealed by autoradiography. *Brain Res.* 85:205–10
Bok, S. T. 1959. *Histonomy of the Cerebral Cortex.* Amsterdam: Elsevier
Bondy, S. C. 1971. Axonal transport of macromolecules. II. Nucleic acid migration in the central nervous system. *Exp. Brain Res.* 13:135–39
Bondy, S. C. 1972. Axonal migration of various ribonucleic acid species along the optic tract of the chick. *J. Neurochem.* 19:1769–76.
Bondy, S. C., Madsen, C. J. 1971. Development of rapid axonal flow in the chick embryo. *J. Neurobiol.* 2:279–86
Bonnet, K. A., Bondy, S. C. 1976. Transport of RNA along the optic pathway of the chick: an autoradiographic study. *Exp. Brain Res.* 26:185–91
Bowery, N. G., Brown, D. A. 1972. γ-aminobutyric acid uptake by sympathetic ganglia. *Nature New Biol.* 238: 89–91
Bradley, P., Berry, M. 1976a. The effects of reduced climbing and parallel fibre input on Purkinje cell dendritic growth. *Brain Res.* 190:133–51

Bradley, P., Berry, M. 1976b. Quantitative effects of climbing fibre deafferentation on the adult Purkinje cell dendritic tree. *Brain Res.* 112:133–40

Bray, J. J., Austin, L. 1968. Flow of protein and ribonucleic acid in peripheral nerve. *J. Neurochem.* 15:731–40

Brimijoin, S., Helland, L. 1976. Rapid retrograde transport of dopamine-β-hydroxylase as examined by the stop-flow technique. *Brain Res.* 102:217–28

Broadwell, R. D., Brightman, M. W. 1976. Entry of peroxidase into neurons of the central and peripheral nervous systems from extracerebral and cerebral blood. *J. Comp. Neurol.* 166:257–83

Brown, M. C., Butler, R. G. 1973. Studies on the site of termination of static and dynamic fusimotor fibres within spindles of the tenuissimus muscle of the cat. *J. Physiol. London* 233:000–573

Brown, M. C., Butler, R. G. 1975. An investigation into the site of termination of static gamma fibres within muscle spindles of the cat peroneus longus muscle. *J. Physiol. London* 247:131–43

Brownfield, M. S., Kozlowski, G. P. 1977. The hypothalamic-choroidal tract. I. Immunohistochemical evidence demonstrating neurophysin pathways to telencephalin choroid plexuses and cerebral spinal fluid. *Cell Tissue Res.* 178:111–27

Brownstein, M. J., Mroz, E. A., Kizer, J. S., Palkovits, M., Leeman, S. E. 1976. Regional distribution of substance P in the brain of the rat. *Brain Res.* 116:299–305

Brownson, R. H., Uusitalo, R., Palkama, A. 1977. Intraaxonal transport of horseradish peroxidase in the sympathetic nervous system. *Brain Res.* 120:407–22

Bryan, R. N., Trevino, D. L., Willis, W. D. 1972. Evidence for a common location of alpha and gamma motoneurons. *Brain Res.* 38:193–96

Bunge, M. B. 1973. Uptake of peroxidase by growth cones of cultured neurons. *Anat. Rec.* 175:280

Bunge, R. P. 1975. Changing uses of nerve tissue culture 1950–1975. In *The Nervous System. Vol. I: The Basic Neurosciences,* ed. D. B. Tower, pp. 31–42. New York: Raven Press

Bunt, A. H., Haschke, R. H., Lund, R. D., Calkins, D. F. 1976. Factors affecting retrograde axonal transport of horseradish peroxidase in the visual system. *Brain Res.* 102:152–55

Bunt, A. H., Hendrickson, A. E., Lund, J. S., Lund, R. D., Fuchs, A. F. 1975. Monkey retinal ganglion cells: morphometric analysis and tracing of axonal projections, with a consideration of the peroxidase technique. *J. Comp. Neurol.* 164:265–86

Bunt, A. H., Lund, R. D. 1974. Blockage by vinblastine of ortho- and retrograde axonal transport in the retinal ganglion cells. *Anat. Rec.* 178:507–8

Burdwood, W. O. 1965. Rapid bidirectional particle movement in neurons. *J. Cell Biol.* 27:115A

Burke, R. E., Levine, D. N., Salcman, M., Tsairis, P. 1974. Motor units in cat soleus muscle: physiological, histochemical and morphological characteristics. *J. Physiol. London* 238:503–14

Burton, H., Jones, E. G. 1976. The posterior thalamic region and its cortical projection in New World and Old World monkeys. *J. Comp. Neurol.* 168:249–302

Byers, M. R. 1974. Structural correlates of rapid axonal transport: evidence that microtubules may not be directly involved. *Brain Res.* 75:97–113

Caldwell, J. H., Berman, N. 1977. The central projection in the retina in *Necturus maculosus. J. Comp. Neurol.* 171:455–63

Cancalon, P., Beidler, L. M. 1977. Differences in the composition of the polypeptides deposited in the axon and the nerve terminals by fast axonal transport in the garfish olfactory nerve. *Brain Res.* 121:215–28

Carpenter, M. B., Nakano, K., Kim, R. 1976. Nigrothalamic projections in the monkey demonstrated by autoradiographic techniques. *J. Comp. Neurol.* 165:401–15

Casagrande, V. A., Harting, J. K. 1975. Transneuronal transport of tritiated fucose and proline in the visual pathways of tree shrew *Tupaia glis. Brain Res.* 96:367–72

Caviness, V. S. Jr., Barkley, D. S. 1971. Section thickness and grain count variation in tritium autoradiography. *Stain Technol.* 46:131–35

Chan-Palay, V. 1976. Serotonin axons in the supra- and subependymal plexuses and in the leptomeninges; their roles in local alterations of cerebrospinal fluid and vasomotor activity. *Brain Res.* 102:103–30

Christensen, B. N. 1973. Procion brown: An intracellular dye for light and electron microscopy. *Science* 182:1255

Christensen, B. N., Ebner, F. F. 1975. The ultrastructure of physiologically studied cortical cells. *Neurosci. Abstr.* 1:127

Christenson, J., Dairman, W., Udenfriend, S. 1972. On the identity of dopa decarboxylase and 5-hydroxytryptophan decarboxylase. *Proc. Natl. Acad. Sci. USA* 69:343–47

Clarke, P. G. H., Cowan, W. M. 1976. The development of the isthmo-optic tract in the chick, with special reference to the occurrence and correction of developmental errors in the location and connections of isthmo-optic neurons. *J. Comp. Neurol.* 167:143–63

Clarke, P. G. H., Rogers, L. A., Cowan, W. M. 1976. The time of origin and the pattern of survival of neurons in the isthmo-optic nucleus of the chick. *J. Comp. Neurol.* 167:125–41

Coleman, P. D., West, M. J., Wyss, U. R. 1973. Computer-aided quantitative neuroanatomy. In *Digital Computers in the Behavioral Laboratory,* ed. B. Weiss, pp. 370–426. New York: Appleton-Century-Crofts

Collins, R. C., Kennedy, C., Sokoloff, L., Plum, F. 1976. Metabolic anatomy of focal motor seizures. *Arch. Neurol.* 33:536–42

Colman, D. R., Scalia, F., Cabrales, E. 1976. Light and electron microscopic observations on the anterograde transport of horseradish peroxidase in the optic pathway in the mouse and rat. *Brain Res.* 102:156–63

Colon, E. J., Smit, G. J. 1970. Quantitative analysis of the cerebral cortex. II. A method for analyzing basal dendritic plexuses. *Brain Res.* 22:363–80

Conrad, L. C., Leonard, C. M., Pfaff, D. W. 1974. Connections of the median and dorsal raphe nuclei in the rat. An autoradiographic and degeneration study. *J. Comp. Neurol.* 156:179–206

Conrad, L. C. A., Pfaff, D. W. 1976a. Efferents from medial basal forebrain and hypothalamus in the rat. I. An autoradiographic study of the medial preoptic area. *J. Comp. Neurol.* 169: 185–219

Conrad, L. C. A., Pfaff, D. W. 1976b. Efferents from medial basal forebrain and hypothalamus in the rat. II. An autoradiographic study of the anterior hypothalamus. *J. Comp. Neurol.* 169: 221–61

Cook, M. L., Whitlock, D. G. 1975. Axoplasmic transport in the toad *Bufo marinus. Brain Res.* 96:247–66

Coons, A. H., Creech, H. J., Jones, R. N., Berliner, E. 1942. The demonstration of pneumococcal antigen in tissues by the use of fluorescent antibody. *J. Immunol.* 45:159–70

Costa, E., Daly, J., Lefevre, H., Meek, J., Revuelta, A., Sparro, F., Strada, S. 1972. Serotonin and catecholamine concentrations in brain of rats injected intracerebrally with 5,6-dihydroxytryptamine. *Brain Res.* 44:304–8

Coulter, J. D., Ewing, L., Carter, C. 1976. Origin of primary sensorimotor cortical projections to lumbar spinal cord of cat and monkey. *Brain Res.* 103:366–72

Coulter, J. D., Jones, E. G. 1977. Differential laminar distribution of corticospinal fibers arising in cytoarchitectonic fields of the sensory-motor cortex in monkeys. *Brain Res.* 129:335–40

Cowan, W. M., Gottlieb, D. I., Hendrickson, A. E., Price, J. L., Woolsey, T. A. 1972. The autoradiographic demonstration of axonal connections in the central nervous system. *Brain Res.* 37:21–51

Cowan, W. M., Wann, D. F. 1973. A computer system for the measurement of cell and nuclear sizes. *J. Microscopy* 99:331–48

Cowan, W. M., Woolsey, T. A., Wann, D. F., Dierker, M. L. 1975. The computer analysis of Golgi-impregnated neurons. In *Golgi Centennial Symposium. Proceedings,* ed. M. Santini, pp. 81–85. New York: Raven Press

Crossland, W. J., Cowan, W. M., Rogers, L. A., Kelly, J. P. 1974a. The specification of the retino-tectal projection in the chick. *J. Comp. Neurol.* 155:127–64

Crossland, W. J., Currie, J. R., Rogers, L. A., Cowan, W. M. 1974b. Evidence for a rapid phase of axoplasmic transport at early stages in the development of the visual system of the chick and frog. *Brain Res.* 78:483–89

Cuatrecasas, P., Anfinsen, C. B. 1971. Affinity chromatography. In *Methods in Enzymology,* ed. W. B. Jakoby, vol. 22, p. 345. New York: Academic

Cuénod, M., Boesch, J., Marko, P., Perisic, M., Sandri, C., Schonbach, J. 1972. Contributions of axoplasmic transport to synaptic structures and functions. *Int. J. Neurosci.* 4:77–87

Cuénod, M., Marko, P., Niederer, E. 1973. Disappearance of particulate tectal protein during optic nerve degeneration in the pigeon. *Brain Res.* 49:422–26

Cullheim, S., Kellerth, J. -O. 1976. Combined light and electron microscopic tracing of neurones, including axons and synaptic terminals after intracellular injection of horseradish peroxidase. *Neurosci. Lett.* 2:307–13

Cunningham, T. J. 1976. Early eye removal produces excessive bilateral branching in the rat: application of the cobalt filling method. *Science* 194:857–58

Cunningham, T. J., Freeman, J. A. 1977. Bilateral ganglion cell branches in the normal rat: a demonstration with electro-physiological collision and cobalt tracing methods. *J. Comp. Neurol.* 172:165–75

Curtis, D. R., Johnston, G. A. R. 1974. Amino acid transmitters in the mammalian central nervous system. *Rev. Physiol.* 69:97–188

Czarkowska, J., Jankowska, E., Sybirska, E. 1976. Axonal projections of spinal interneurones excited by group I afferents in the cat, revealed by intracellular staining with horseradish peroxidase. *Brain Res.* 118:115–18

Dahlström, A. 1965. Observations on the accumulation of noradrenaline in the proximal and distal parts of peripheral adrenergic nerves after compression. *J. Anat.* 99:677–89

Dahlström, A. 1967. The transport of noradrenaline between two simultaneously performed ligations of the sciatic nerves of a rat and cat. *Acta Physiol. Scand.* 69:158–66

Dahlström, A. 1971. Axoplasmic transport (with particular respect to adrenergic neurons). *Phil. Trans. Soc. London Ser. B.* 261:325–58

Daly, J., Fuxe, K., Jonsson, G. 1973. Effects of intracerebral injections of 5,6-dihydroxytryptamine on central monoamine neurons: evidence for selective degeneration of central 5-hydroxytryptamine neurons. *Brain Res.* 49:476–82

Daly, J., Fuxe, K., Jonsson, G. 1974. 5,7-Dihydroxytryptamine as a tool for the morphological and functional analysis of central 5-hydroxytryptamine neurons. *Res. Commun. Chem. Pathol. Pharmacol.* 7:175–87

Davis, R. E., Agranoff, B. W. 1977. Microimplantation of [^{3}H] proline on a single bead of ion-exchange resin. *Brain Res.* 124:341–46

Dekker, J. J., Kuypers, H. G. J. M. 1976. Quantitative EM study of projection terminals in the rat's AV thalamic nucleus. Autoradiographic and degeneration techniques compared. *Brain Res.* 117:399–422

De Mey, J., Dierickx, K., Vandesande, F. 1975. Identification of neurophysin producing cells. III. Immunohistochemical demonstration of neurophysin I–producing neurons in the bovine infundibular nucleus. *Cell Tissue Res.* 161:219–24

Descarries, L., Lapierre, Y. 1973. Noradrenergic axon terminals in the cerebral cortex of rat. I. Radioautographic visualization after topical application of DL-[^{3}H] norepinephrine. *Brain Res.* 51:141–60

Descarries, L., Beaudet, A., Watkins, K. C. 1975. Serotonin nerve terminals in adult rat neocortex. *Brain Res.* 100:563–88

Des Rosiers, M. H., Kennedy, C., Patlak, C. S., Pettigrew, K. D., Sokoloff, L., Reivich, M. 1974. Relationship between local cerebral blood flow and glucose utilization in the rat. *Neurology* 24:389

DeVito, J. L., Clausing, K. W., Smith, O. A. 1974. Uptake and transport of horseradish peroxidase by cut end of the vagus nerve. *Brain Res.* 82:269–71

Dick, F., Kelly, J. S. 1975. L-2,4-diaminobutyric acid (L-DABA) as a selective marker for inhibitory nerve terminals in rat brain. *Br. J. Pharmacol.* 53:439–40

Divac, I. 1975. Magnocellular nuclei of the basal forebrain project to neocortex, brain stem, and olfactory bulb. Review of some functional correlates. *Brain Res.* 93:385–98

Dräger, U. C. 1974. Autoradiography of tritiated proline and fucose transported transneuronally from the eye to the visual cortex in pigmented and albino mice. *Brain Res.* 82:284–93

Droz, B. 1965. Synthèse et transfert des protéines cellulaires dans les neurones ganglionnaires: Étude radioautographique quantitative en microscopie électronique. *J. Microsc. Paris* 6:201–28

Droz, B. 1969. Protein metabolism in nerve cells. *Int. Rev. Cytol.* 25:363–90

Droz, B., Koenig, H. L. 1970. Localization of protein metabolism in neurons. In *Protein Metabolism of the Nervous System,* ed. A. Lajtha, pp. 93–108. New York: Plenum

Droz, B., Koenig, H. L., DiGiamberadino, L. 1973a. Axonal migration of protein and glycoprotein to nerve endings. I. Radioautographic analysis of the renewal of protein in nerve endings of chicken ciliary ganglion after intracerebral injection of [^{3}H]lysine. *Brain Res.* 60:93–127

Droz, B., Koenig, H. L., DiGiamberadino, L. 1973b. Axonal migration of protein and glycoprotein to nerve endings. II. Radioautographic analysis of the renewal of glycoproteins in nerve endings of chicken ciliary ganglion after intracerebral injection of [^{3}H]fucose and [^{3}H]-glucosamine. *Brain Res.* 60:129–46

Droz, B., Koenig, H. L., DiGiamberadino, L. 1973c. Axonal migration of protein and glycoprotein to nerve endings. III. Cell fraction analysis of chicken ciliary ganglion after intracerebral injection of labeled precursors of proteins and glycoproteins. *Brain Res.* 60:147–59.

Droz, B. 1975. Synthetic machinery and axoplasmic transport: maintenance of neuronal connectivity. In *The Nervous System. Vol. I. The Basic Neurosciences,* ed. D. B. Tower, pp. 111–27. New York: Raven Press

Droz, B., Leblond, C. P. 1963. Axonal migration of proteins in the central nervous system and peripheral nerves as shown by radioautography. *J. Comp. Neurol.* 121:325–46

Droz, B., Rambourg, A., Koenig, H. L. 1975. The smooth endoplasmic reticulum: structure and role in the renewal of axonal membrane and synaptic vesicles by fast axonal transport. *Brain Res.* 93: 1–13

Dunker, R. O., Harris, A. B., Jenkins, D. P. 1976. Kinetics of horseradish peroxidase migration through cerebral cortex. *Brain Res.* 118:199–217

Durham, D., Woolsey, T. A. 1977. Barrels and columnar cortical organization: evidence from 2-deoxy-glucose experiments. *Anat. Rec.* 187:570

Durie, B. G. M., Salmon, S. E. 1975. High speed scintillation autoradiography. *Science* 190:1093–95

Ebner, F. F., Donaghue, J. P., Foster, R., Christensen, B. 1976. The organization of opossum somatic sensory-motor cortex. *Neurosci. Abstr.* 2:135

Edström, A., Hanson, M. 1973a. Retrograde axonal transport of proteins *in vitro* in frog sciatic nerves. *Brain Res.* 61: 311–20

Edström, A., Hanson, M. 1973b. Temperature effects on fast axonal transport of proteins *in vitro* in frog sciatic nerves. *Brain Res.* 58:345–54

Edström, L., Kugelberg, E. 1968. Histochemical composition, distribution of fibres and fatiguability of single motor units. *J. Neurol. Neurosurg. Psychiat.* 31:424–33

Edwards, S. B. 1975. Autoradiographic studies of the projections of the midbrain reticular formation: descending projections of nucleus cuneiformis. *J. Comp. Neurol.* 161:341–58

Edwards, S. B., deOlmos, J. S. 1976. Autoradiographic studies of the projections of the midbrain reticular formation: ascending projections of nucleus cuneiformis. *J. Comp. Neurol.* 165: 417–31

Elam, J. S., Agranoff, B. W. 1971. Rapid transport of protein in the optic system of the goldfish. *J. Neurochem.* 18: 375–87

Elam, J. S., Goldberg, M., Radin, N. S., Agranoff, B. W. 1970. Rapid axonal transport of sulfated mucopolysaccharide proteins. *Science* 170:458–59

Elde, R., Hökfelt, T., Johansson, O., Terenius, L. 1976. Immunohistochemical studies using antibodies to leucine-enkephalin: initial observations on the nervous system of the rat. *Neuroscience* 1:349–51

Elde, R. P., Parsons, J. A. 1975. Immunocytochemical localization of somatostatin in cell bodies of the rat hypothalamus. *Am. J. Anat.* 144:541–48

Ellison, J. P., Clark, G. M. 1975. Retrograde axonal transport of horseradish peroxidase in peripheral autonomic nerves. *J. Comp. Neurol.* 161:103–13

Eng, L. F., Uyeda, C. T. 1974. Antibody to bovine choline acetyltransferase and immunofluorescent localization of the enzyme in neurones. *Nature* 250: 243–45

Erdmann, G., Wiegand, H., Wellhoner, H. H. 1975. Intraaxonal and extraaxonal transport of ^{125}I-tetanus toxin in early local tetanus. *Naunyn-Schmiedeberg's Arch. Exp. Path. Pharmakol.* 290: 357–73

Fahn, S. 1976. Regional distribution studies of GABA and other putative neurotransmitters and their enzymes. In *GABA in Nervous System Function,* ed. E. Roberts, T. N. Chase, D. B. Tower, pp. 169–86. New York: Raven Press

Falck, B., Hillarp, N. A., Thieme, G., Torp, A. 1962. Fluorescence of catecholamines and related compounds condensed with formaldehyde. *J. Histochem. Cytochem.* 10:348–54

Felix, D., Künzle, H. 1974. Iontophoretic and autoradiographic studies on the role of proline in nervous transmission. *Pfluegers Arch.* 350:135–44

Fibiger, H. C., McGeer, E. G. 1974. Accumulation and axoplasmic transport of dopamine but not of amino acids by axons of the nigro-neostriatal projection. *Brain Res.* 72:366–69

Fibiger, H. C., McGeer, E. G., Atmadja, S. 1973. Axoplasmic transport of dopamine in nigro-striatal neurons. *J. Neurochem.* 21:373–85

Fibiger, H. C., Pudritz, R. E., McGeer, P. L., McGeer, E. G. 1972. Axonal transport in nigro-striatal and nigro-thalamic neurons: effects of medial forebrain bundle lesions and 6-hydroxydopamine. *J. Neurochem.* 19:1697–1708

Fidone, S. J., Zapata, P., Stensaas, L. J. 1977. Axonal transport of labeled material into sensory nerve endings of cat carotid body. *Brain Res.* 124:9–28

Fillenz, M., Gagnon, C., Stoeckel, K., Thoenen, H. 1976. Selective uptake and retrograde axonal transport of dopamine-β-hydroxylase antibodies in peripheral adrenergic neurons. *Brain Res.* 114: 293–304

Fischer, H. A. 1975. Autoradiography of specimens for electron microscopy—presentation of a method with special remarks on the use of samples from the nervous system. *Brain Res.* 85:237–40

Fischer, H. A., Korr, H., Thiele, H., Werner, G. 1971. Kürzere autoradiographische Exposition elektronenmikroskopischer Präparate durch Szintillatoren. *Naturwissenschaften* 58:101–2

Fonnum, F., Grofová, I., Rinvik, E., Storm-Mathisen, J. 1974. Origin and distribution of glutamate decarboxylase in substantia nigra of the cat. *Brain Res.* 71:77–92

Forman, D. S., Grafstein, B., McEwen, B. S. 1972. Rapid axonal transport of [^{3}H]fucosyl glycoproteins in the goldfish optic system. *Brain Res.* 48: 327–42

Forman, D. S., Ledeen, R. W. 1972. Axonal transport of gangliosides in the goldfish optic nerve. *Science* 177:630–33

Forman, D. S., McEwen, B. S., Grafstein, B. 1971. Rapid transport of radioactivity in goldfish optic nerve following injections of labeled glucosamine. *Brain Res.* 28:119–30

Frank, E., Gautvik, K., Sommerschild, H. 1976. Persistence of junctional acetylcholine receptors following denervation. *Cold Spring Harbor Symp. Quant. Biol.* 40:275–81

Fricke, R., Cowan, W. M. 1977. An autoradiographic study of the development of the entorhinal commissural afferents to the dentate gyrus of the rat. *J. Comp. Neurol.* 173:231–50

Frizell, M., Sjöstrand, J. 1974. The axonal transport of slowly migrating ^{3}H leucine labeled proteins and the regeneration rate in regenerating hypoglossal and vagus nerves of the rabbit. *Brain Res.* 81:267–83

Frizell, M., McLean, W. G., Sjöstrand, J. 1975. Slow axonal transport of proteins; blockade by interruption of contact between cell body and axon. *Brain Res.* 86:67–73

Fuller, P. M., Prior, D. J. 1975. Cobalt iontophoresis techniques for tracing afferent and efferent connections in the vertebrate CNS. *Brain Res.* 88:211–20

Furstman, L., Saporta, S., Kruger, L. 1975. Retrograde axonal transport of horseradish peroxidase in sensory nerves and ganglion cells of the rat. *Brain Res.* 84:320–24

Fuxe, K., Goldstein, M., Hökfelt, T., Joh, T. H. 1970. Immunohistochemical localization of dopamine-β-hydroxylase in the peripheral and central nervous system. *Res. Commun. Chem. Pathol. Pharmacol.* 1:627–36

Fuxe, K., Hökfelt, T., Ritzen, M., Ungerstedt, U. 1968. Studies on uptake of intraventricularly administered tritiated noradrenaline and 5-hydroxytryptamine with combined fluorescence and autoradiographic techniques. *Histochemie* 16:186–94

Gacek, R. R. 1974. Localization of neurons supplying the extraocular muscles in the kitten using horseradish peroxidase. *Exp. Neurol.* 44:381–403

Garvey, C. F., Young, J. H. Jr., Coleman, P. D., Simon, W. 1973. Automated three-dimensional dendrite tracking system. *Electroencephalogr. Clin. Neurophysiol.* 35:199–204

Garvey, C., Young, J., Simon, W. 1972. Semi-automatic dendritic tracking and focusing by computer. *Anat. Rec.* 172:314

Geffen, L. B., Livett, B. G. 1971. Synaptic vesicles in sympathetic neurons. *Physiol. Rev.* 51:98–157

Geffen, L. B., Livett, B. G., Rush, R. A. 1969. Immunohistochemical localization of protein components of catecholamine storage vesicles. *J. Physiol. London* 204:593–605

Geisert, E. E. Jr. 1976. The use of tritiated horseradish peroxidase for defining neuronal pathways: a new application. *Brain Res.* 117:130–35

Gerson, S., Baldessarini, R. J. 1975. Selective destruction of serotonin terminals in rat forebrain by high doses of 5,7-dihydroxytryptamine. *Brain Res.* 85:140–45

Gilbert, C. D., Kelly, J. P. 1975. The projections of cells in different layers of the cat's visual cortex. *J. Comp. Neurol.* 163:81–105

Gilden, D. H., Wroblewska, Z., Eng, L. F., Rorke, L. B. 1976. Human brain in tissue culture. V. Identification of glial cells by immunofluorescence. *J. Neurol. Sci.* 29:177–84

Gillette, R., Pomeranz, B. 1973. Neuron geometry and circuitry in the electron microscope: intracellular staining with osmiophilic polymer. *Science* 182: 1256–58

Glaser, E. M., Van der Loos, H. 1965. A semi-automatic computer-microscope for the analysis of neuronal morphology. *IEEE Trans. Biomed. Eng.* 12:22–31

Globus, A., Lux, H. D., Schubert, P. 1968. Somadendritic spread of intracellularly injected tritiated glycine in cat spinal motoneurons. *Brain Res.* 11:440–45

Globus, A., Lux, H. D., Schubert, P. 1973. Transfer of amino acids between neuroglial cells and neurons in the leech ganglion. *Exp. Neurol.* 40:104–13

Goldman, M. 1968. *Fluorescent Antibody Methods.* New York: Academic. 303 pp.

Goldman, P. S., Nauta, W. J. H. 1977. An intricately patterned prefrontocaudate projection in the rhesus monkey. *J. Comp. Neurol.* 171:369–86

Goldsmith, P. C., Ganong, W. F. 1975. Ultrastructural localization of luteinizing hormone releasing hormone in the median eminence of the rat. *Brain Res.* 97:181–93

Goldstein, D. J., Williams, M. A. 1971. Quantitative autoradiography: an evaluation of visual grain counting, reflectance microscopy, growth adsorbance measurements and flying-spot microdensitometry. *J. Microscopy Oxford* 94:215–240

Gottlieb, D. I., Cowan, W. M. 1972. Evidence for a temporal factor in the occupation of available synaptic sites during the development of the dentate gyrus. *Brain Res.* 41:452–56

Grafstein, B. 1967. Transport of protein by goldfish optic nerve fibers. *Science* 157:196–98

Grafstein, B. 1969. Axonal transport: communication between soma and synapse. In *Advances in Biochemical Psychopharmacology,* ed. E. Costa, R. Greengard, pp. 11–25. New York: Raven

Grafstein, B. 1971. Transneuronal transfer of radioactivity in the central nervous system. *Science* 172:177–79

Grafstein, B. 1975a. The eyes have it: axonal transport and regeneration in the optic nerve. In *The Nervous System. Vol. 1: The Basic Neurosciences,* ed. D. B. Tower, pp. 147–51. New York: Raven

Grafstein, B. 1975b. Principles of anterograde axonal transport in relation to studies of neuronal connectivity. In *The Use of Axonal Transport for Studies of Neuronal Connectivity,* ed. W. M. Cowan, M. Cuénod, pp. 47–68. Amsterdam: Elsevier

Grafstein, B., Laureno, R. 1973. Transport of radioactivity from eye to visual cortex in the mouse. *Exp. Neurol.* 39:44–57

Grafstein, B., McEwen, B. S., Shelanski, M. 1970. Axonal transport of neurotubule protein. *Nature* 227:289–90

Grafstein, B., Miller, J. A., Ledeen, R. W., Haley, J. H., Specht, S. C. 1975. Axonal transport of phospholipid in goldfish optic system. *Exp. Neurol.* 46:261–81

Grafstein, B., Murray, M. 1969. Transport of protein in goldfish optic nerve during regeneration. *Exp. Neurol.* 25:494–508

Grafstein, B., Murray, M., Ingoglia, N. A. 1972. Protein synthesis and axonal transport in retinal ganglion cells of mice lacking visual receptors. *Brain Res.* 44:37–48

Graham, R. C., Karnovsky, M. J. 1966. The early stages of absorption of injected horseradish peroxidase in the proximal tubules of mouse kidney in ultrastructural cytochemistry by a new technique. *J. Histochem. Cytochem.* 14:291–302

Grasso, A., Haglid, K. G., Hansson, H. A., Persson, L., Rönnbäck, L. 1977. Localization of 14-3-2 protein in the rat brain by immunoelectron-microscopy. *Brain Res.* 122:582–85

Graybiel, A. M. 1975a. Anatomical organization of retinotectal afferents in the cat: an autoradiographic study. *Brain Res.* 96:1–24

Graybiel, A. M. 1975b. Wallerian degeneration and anterograde tracer methods. In *The Use of Axonal Transport for Studies of Neuronal Connectivity,* ed. W. M. Cowan, M. Cuénod, pp. 173–247. Amsterdam: Elsevier

Graybiel, A. M., Devor, M. 1974. A microelectrophoretic delivery technique for use with horseradish peroxidase. *Brain Res.* 68:167–73

Graybiel, A. M., Nauta, H. J. W., Lasek, R. J., Nauta, W. J. H. 1973. A cerebello-olivary pathway in the cat: an experimental study using autoradiographic tracing techniques. *Brain Res.* 58: 205–11

Gremo, F. 1974. Radioautographic analysis of ^{3}H-fucose labeled glycoproteins transported along the optic pathway of chick embryos. *Cell Tissue Res.* 153: 465–76

Gross, G. W. 1973. The effect of temperature on the rapid axoplasmic transport in C-fibers. *Brain Res.* 56:359–63

Gross, G. W., Beidler, L. M. 1973. Fast axonal transport in the C-fibers of the garfish olfactory nerve. *J. Neurobiol.* 4:413–28

Guillery, R. W. 1970. Light and electron microscopical studies of normal and degenerating axons. In *Contemporary Research Methods in Neuroanatomy,* ed. W. J. H. Nauta, S. O. E. Ebbesson, pp. 77–105. Berlin: Springer

Guillery, R. W., Updyke, B. V. 1976. Retinofugal pathways in normal and albino axolotes. *Brain Res.* 109:235–44

Haglid, K. G., Hamberger, A., Hansson, H. A., Hydén, H., Persson, L., Rönnbäck, L. 1974. S-100 protein in synapses of the central nervous system. *Nature* 251–34

Halaris, A. E., Jones, B. E., Moore, R. Y. 1976. Axonal transport in serotonin neurons of the midbrain raphe. *Brain Res.* 107:555–74

Halperin, J. J., LaVail, J. H. 1975. A study of the dynamics of retrograde transport and accumulation of horseradish peroxidase in injured neurons. *Brain Res.* 100:253–70

Halpern, M., Want, R. T., Colman, D. R. 1976. Centrifugal fibers to the eye in a nonavian vertebrate: source revealed by horseradish peroxidase studies. *Science* 194:1185–87

Hancock, M. B., Fougerousse, C. L. 1976. Spinal projections from the nucleus locus coeruleus and nucleus subcoeruleus in the cat and monkey as demonstrated by the retrograde transport of horseradish peroxidase. *Brain Res. Bull.* 1: 229–34

Hanker, J. S., Anderson, W. A., Bloom, F. E. 1972. Osmiophilic polymer generation: catalysis by transition metal compounds in ultrastructural cytochemistry. *Science* 175:991–93

Hanker, J. S., Norden, J. J., Oppenheim, R. W., Diamond, I. T. 1976. Design and characterization of fluorochrome-conjugated horseradish peroxidases for neuronal tracing. *Neurosci. Abstr.* 2:37

Hanker, J. S., Yates, P. E., Metz, C. B., Carson, K. A., Light, A., Rustioni, A. 1977. A new specific, sensitive and noncarcinogenic reagent for the demonstration of horseradish peroxidase (HRP). *Neurosci. Abstr.* 3:30

Hanson, M., Tonge, D., Edström, A. 1976. Retrograde axonal transport of exogenous protein in frog nerves. *Brain Res.* 100:458–61

Hansson, H.-A. 1973. Uptake and intracellular bidirectional transport of horseradish peroxidase in retinal ganglion cells. *Exp. Eye Res.* 16:377–88

Hansson, H.-A., Hydén, H., Rönnbäck, L. 1975. Localization of S-100 protein in isolated nerve cells by immunoelectron microscopy. *Brain Res.* 93:349–52

Hardy, H., Heimer, L., Switzer, R., Watkins, D. 1976. Simultaneous demonstration of horseradish peroxidase and acetylcholinesterase. *Neurosci. Lett.* 3:1–5

Harting, J. K., Guillery, R. W. 1976. Organization of retinocollicular pathways in the cat. *J. Comp. Neurol.* 166:133–44

Hartman, B. K. 1973a. Immunofluorescence of dopamine-β-hydroxylase. Application of improved methodology to the localization of the peripheral and central noradrenergic nervous system. *J. Histochem. Cytochem.* 21:312–32

Hartman, B. K. 1973b. The innervation of cerebral blood vessels by central noradrenergic neurons. In *Frontiers in Catecholamine Research,* ed. E. Usden, S. H. Snyder, pp. 91–96. New York: Pergamon

Hartman, B. K., Margolis, F. L. 1975. Immunofluorescence localization of the olfactory marker protein. *Brain Res.* 96:176–80

Hartman, B. K., Udenfriend, S. 1969. A method for immediate visualization of proteins in acrylamide gels and its use for preparation of antibodies to enzymes. *Annal. Biochem.* 30:391–94

Hartman, B. K., Udenfriend, S. 1970. Immunofluorescent localization of dopamine-β-hydroxylase in tissues. *Mol. Pharmacol.* 6:85–94

Hartman, B. K., Udenfriend, S. 1972. The application of immunological techniques to the study of enzymes regulating catecholamine synthesis and degradation. *Pharmacol. Rev.* 24:311–30

Hartman, B. K., Zide, D., Udenfriend, S. 1972. The use of dopamine-β-hydroxylase as a marker for the noradrenergic pathway of the central nervous system in rat. *Proc. Natl. Acad. Sci. USA* 69:2722–26

Hattori, T., Fibiger, H. C., McGeer, P. L. 1975. Demonstration of a pallido-nigral projection innervating dopaminergic neurons. *J. Comp. Neurol.* 162:487–504

Hattori, T., McGeer, P. L., Fibiger, H. C., McGeer, E. G. 1973. On the source of GABA-containing terminals in the substantia nigra. Electron microscopic autoradiographic and biochemical studies. *Brain Res.* 54:103–14

Hattori, T., Singh, V. K., McGeer, E. G., McGeer, P. L. 1976. Immunohistochemical localization of choline acetyltransferase containing neostriatal neurons and their relationship with dopaminergic synapses. *Brain Res.* 102: 164–73

Haug, F.-M. S. 1973. Heavy metals in the brain. A light microscope study of the rat with Timms sulphide silver method. Methodological considerations and cytological and regional staining patterns. *Ergeb. Anat. Entwicklungsgesch.* 47:1–69

Heacock, A. M., Agranoff, B. W. 1977. Reutilization of precursor following axonal transport of [^{3}H]proline-labeled protein. *Brain Res.* 122:243–54

Heimer, L. 1970. Selective silver-impregnation of degenerating axoplasm. In *Contemporary Research Methods in Neuroanatomy*, ed. W. J. H. Nauta, S. O. E. Ebbesson, pp. 106–31. Berlin: Springer

Hendrickson, A. E. 1969. Electron microscopic radioautography: identification of origin of synaptic terminals in normal nervous tissue. *Science* 165:194–96

Hendrickson, A. E. 1972. Electron microscopic distribution of axoplasmic transport. *J. Comp. Neurol.* 144:381–98

Hendrickson, A. 1975. Technical modifications to facilitate tracing synapses by electron microscopic autoradiography. *Brain Res.* 85:241–48

Hendrickson, A. E., Cowan, W. M. 1971. Changes in the rate of axoplasmic transport during postnatal development of the rabbit's optic nerve and tract. *Exp. Neurol.* 30:403–22

Hendrickson, A. E., Wagoner, N., Cowan, W. M. 1972. An autoradiographic and electron microscopic study of retinohypothalamic connections. *Z. Zellforsch.* 135:1–26

Hendry, I. A., Stach, R., Herrup, K. 1974. Characteristics of the retrograde axonal transport system for nerve growth factor in the sympathetic nervous system. *Brain Res.* 82:117–28

Hendry, I. A., Stöckel, K., Thoenen, H., Iversen, L. L. 1974. Retrograde axonal transport of nerve growth factor. *Brain Res.* 68:103–21

Herkenham, M., Nauta, W. J. H. 1977. Afferent connections of the habenular nuclei in the rat. A horseradish peroxidase study, with a note on the fiber-of-passage problem. *J. Comp. Neurol.* 173: 123–45

Heuser, J. E., Reese, T. S. 1973. Evidence for recycling of synaptic vesicle membrane during transmitter release at the frog neuromuscular junction. *J. Cell. Biol.* 57:315–44

Heuser, J. E., Reese, T. S., Landis, D. M. D. 1974. Functional changes in frog neuromuscular junctions studied with freeze-fracture. *J. Neurocytol.* 3:109–31

Hickey, T. L., Guillery, R. W. 1974. An autoradiographic study of retinogeniculate pathways in the cat and the fox. *J. Comp. Neurol.* 156:239–54

Hicks, S. P., D'Amato, C. J. 1976. Locating corticospinal neurons in mature, infant and malformed (radiation) rats with retrograde axonal transport of horseradish peroxidase. *Neurosci. Abstr.* 2:123

Hiller, J. M., Pearson, J., Simon, E. J. 1973. Distribution of stereospecific binding of the potent narcotic analgesic etorphine in the human brain: predominance in the limbic system. *Res. Commun. Chem. Pathol. Pharmacol.* 6:1052–61

Hoffman, P. N., Lasek, R. J. 1975. The slow component of axonal transport. Identification of major structural polypeptides of the axon and their generality among mammalian neurons. *J. Cell. Biol.* 66:351–66

Hökfelt, T. 1968. *In vitro* studies on central and peripheral monoamine neurons at the ultrastructural level. *Z. Zellforsch.* 91:1–74

Hökfelt, T., Efendic, S., Hellerstrom, C., Johansson, O., Luft, R., Arimura, A. 1975a. Cellular localization of somatostatin in endocrine-like cells and neurons of the rat with special references to the A_1 cells of the pancreatic islets and to the hypothalamus. *Acta Endocrinol. Suppl.* 200:5–41

Hökfelt, T., Fuxe, K., Goldstein, M. 1973a. Immunohistochemical localization of aromatic L-amino acid decarboxylase (DOPA decarboxylase) in central dopamine and 5-hydroxytryptamine nerve cell bodies of the rat. *Brain Res.* 53:175–80

Hökfelt, T., Fuxe, K., Goldstein, M. 1975b. Applications of immunochemistry to studies on monoamine cell systems with special reference to nervous tissues. *Ann. NY Acad. Sci.* 254:407–32

Hökfelt, T., Fuxe, K., Goldstein, M., Joh, T. H. 1973b. Immunohistochemical studies of three catecholamine synthesizing enzymes: aspects on methodology. *Histochemie* 33:231–54

Hökfelt, T., Fuxe, K., Goldstein, M., Johansson, O. 1973c. Evidence for adrenaline neurons in the rat brain. *Acta Physiol. Scand.* 89:286–88

Hökfelt, T., Fuxe, K., Goldstein, M., Johansson, O. 1974a. Immunohistochemical evidence for the existence of adrenaline in the rat brain. *Brain Res.* 66:235–51

Hökfelt, T., Fuxe, K., Johansson, O., Jeffcoate, S., White, N. 1975c. Distribution of thyrotropin-releasing hormone (TRH) in the central nervous system as revealed with immunohistochemistry. *Eur. J. Pharmacol.* 34:389–92

Hökfelt, T., Fuxe, K., Johansson, O., Jeffcoate, S., White, N. 1975d. Thyrotropin-releasing hormone (TRH) containing nerve terminals in certain brain stem nuclei and in the spinal cord. *Neurosci. Lett.* 1:133–39

Hökfelt, T., Johansson, O., Fuxe, K., Goldstein, M., Park, D. 1976a. Immunohistochemical studies on the localization and distribution of monoamine neuron systems in the rat brain. I. Tyrosine hydroxylase in the mes- and diencephalon. *Med. Biol.* 54:427–53

Hökfelt, T., Kellerth, J. O., Nilsson, G., Pernow, B. 1975e. Substance P: localization in the central nervous system and in some primary sensory neurons. *Science* 190:899–90

Hökfelt, T., Kellerth, J. -O., Nilsson, G., Pernow, B. 1975f. Experimental immunohistochemical studies on the localization and distribution of substance P in cat primary sensory neurons. *Brain Res.* 100:235–52

Hökfelt, T., Ljungdhal, Å. 1970. Cellular localization of gamma-aminobutyric acid (^{3}H-GABA) in rat cerebellar cortex: an autoradiographic study. *Brain Res.* 22:391

Hökfelt, T., Ljungdhal, Å. 1972a. Autoradiographic identification of cerebral and cerebellar cortical neurones accumulating labeled gamma-aminobutyric acid (^{3}H-GABA). *Exp. Brain Res.* 14:354

Hökfelt, T., Ljungdahl, Å. 1972b. Modification of the Falck-Hillarp formaldehyde fluorescence method using the Vibratome: simple, rapid and sensitive localization of catecholamine in sections of unfixed or fixed brain tissue. *Histochemie* 29:324–39

Hökfelt, T., Ljungdahl, Å. 1975. Uptake mechanisms as a basis for the histochemical identification and tracing of transmitter-specific neuron populations. In *The Use of Axonal Transport for Studies of Neuronal Connectivity,* ed. W. M. Cowan, M. Cuénod, pp. 249–86. Amsterdam: Elsevier

Hökfelt, T., Ljungdahl, Å., Johansson, O., Lindblom, D. 1974b. The Vibratome: a useful tool in transmitter histochemistry. In *Amine Fluorescence Histochemistry,* ed. M. Fujiwara, C. Tanaka, pp. 1–12. Tokyo: Igaku Shoin

Hökfelt, T., Meyerson, B., Nilsson, G., Pernow, B., Sachs, C. 1976b. Immunohistochemical evidence for substance P–containing nerve endings in the human cortex. *Brain Res.* 104:181–86

Holländer, H. 1974. On the origin of the corticotectal projections in the cat. *Exp. Brain Res.* 21:433–39

Hollingworth, T., Berry, M. 1975. Network analysis of dendritic fields of pyramidal cells in the neocortex and Purkinje cells in the cerebellum of the rat. *Phil. Trans. R. Soc. London Ser. B.* 270:227–62

Huang, S.-N., Minassian, H., More, J. D. 1976. Application of immunofluorescent staining on paraffin sections improved by trypsin digestion. *Lab. Invest.* 35:383–90

Hubel, D. H. 1975. An autoradiographic study of the retino-cortical projections in the tree shrew (*Tupaia glis*). *Brain Res.* 96:41–50

Hubel, D. H., LeVay, S., Wiesel, T. N. 1975. Mode of termination of retinotectal fibers in macaque monkey: an autoradiographic study. *Brain Res.* 96:25–40

Hubel, D. H., Wiesel, T. N., LeVay, S. 1976. Functional architecture of area 17 in normal and monocularly deprived macaque monkeys. *Cold Springs Harbor Symp. Quant. Biol.* 40:581–89

Hubel, D. H., Wiesel, T. N., Stryker, M. 1978. Anatomical demonstration of orientation columns in macaque monkey. *J. Comp. Neurol.* 177:361–79

Hughes, J. 1975. Isolation of an endogenous compound from the brain with pharmacological properties similar to morphine. *Brain Res.* 88:295–308

Hughes, J., Smith, T., Kosterlitz, H. W., Fothergill, L. A., Morgan, B. A., Morris, H. R. 1975. Identification of two related pentapeptides from the brain with potent opiate agonist activity. *Nature* 258:577–79

Hunt, S. P., Künzle, H. 1976a. Bidirectional movement of label and transneuronal transport phenomena after injection of [^{3}H]-adenosine into the central nervous system. *Brain Res.* 112:127–32

Hunt, S. P., Künzle, H. 1976b. Observations on the projections and intrinsic organization of the pigeon optic tectum: an autoradiographic study based on anterograde and retrograde, axonal and dendritic flow. *J. Comp. Neurol.* 170: 153–72

Hunt, S. P., Künzle, H. 1976c. Selective uptake and transport of label within three identified neuronal systems after injection of ^{3}H-GABA into the pigeon optic tectum: an autoradiographic Golgi study. *J. Comp. Neurol.* 170:173–90

Iles, J. F. 1976. Central terminations of muscle afferents on motoneurones in the cat spinal cord. *J. Physiol. London* 262:91–117

Ingoglia, N. A., Grafstein, B., McEwen, B. S., McQuarrie, I. G. 1973. Axonal transport of radioactivity in the goldfish optic systems following intraocular injection of labeled RNA precursors. *J. Neurochem.* 20:1605–15

Iversen, L. L. 1974. Biochemical aspects of synaptic modulation. *The Neurosciences, Third Study Program,* ed. F. O. Schmitt, F. G. Worden, pp. 905–15. Cambridge, Mass.: MIT Press

Iversen, L. L., Bloom, F. E. 1972. Studies of the uptake of ^{3}H-GABA and ^{3}H-glycine in slices and homogenates of rat brain and spinal cord by electron microscopic autoradiography. *Brain Res.* 41:131–43

Iversen, L. L., Dick, F., Kelly, J. S., Schon, F. 1975a. Uptake and localization of transmitter amino acids in the nervous system. *Metabolic Compartmentation in CNS,* ed. S. Berl, D. Schneider, p. 65. New York: Plenum Publishing

Iversen, L. L., Kelly, J. S. 1975. Uptake and metabolism of γ-aminobutyric acid by neurones and glial cells. *Biochem. Pharmacol.* 24:933–38

Iversen, L. L., Schon, F. E. 1973. The use of autoradiographic techniques for the identification and mapping of transmitter-specific neurones in CNS. In *New Concepts in Neurotransmitter Regulation,* ed. A. J. Mandell, pp. 153–93. New York: Plenum

Iversen, L. L., Stöckel, K., Thoenen, H. 1975b. Autoradiographic studies of the retrograde axonal transport of nerve growth factor in mouse sympathetic neurones. *Brain Res.* 88:37–43

Jacobowitz, D. M., Ziegler, M. G., Thomas, J. A. 1975. *In vivo* uptake of antibody to dopamine-β-hydroxylase into sympathetic elements. *Brain Res.* 91:165–70

Jacobson, S., Trojanowski, J. Q. 1974. The cells of origin of the corpus callosum in rat, cat and rhesus monkey. *Brain Res.* 74:149–55

Jacobson, S., Trojanowski, J. Q. 1975. Corticothalamic neurons and thalamocortical terminal fields: an investigation in rat using horseradish peroxidase and autoradiography. *Brain Res.* 85:385–401

James, K. A. C., Austin, L. 1970. The binding *in vitro* of colchicine to axoplasmic protein from chicken sciatic nerve. *Biochem. J.* 117:773–77

James, K. A. C., Bray, J. J., Morgan, I. G., Austin, L. 1971. The effect of colchicine on the transport of axonal protein in the chicken. *Biochem. J.* 117:767–71

Jankowska, E., Lindström, S. 1972. Morphology of interneurones mediating Ia reciprocal inhibition of motoneurones in the spinal cord of the cat. *J. Physiol. London* 226:805–23

Jankowska, E., Rastad, J., Westman, J. 1976. Intracellular application of horseradish peroxidase and its light and electron microscopical appearance in spinocervical tract cells. *Brain Res.* 105:557–62

Javoy, F., Sotelo, C., Herbet, A., Agid, Y. 1976. Specificity of dopaminergic neuronal degeneration induced by intracerebral injections of 6-hydroxydopamine in the nigrostriatal dopamine system. *Brain Res.* 102:201–16

Jeffrey, P. L., Austin, L. 1973. Axoplasmic transport. *Prog. Neurobiol.* 2:207–55

Joh, T. H., Shikimi, T., Pickel, V. M., Reis, D. J. 1975. Brain tryptophan hydroxylase: purification of production of antibodies to and cellular and ultrastructural localization in serotonergic neurons of rat midbrain. *Proc. Natl. Acad. Sci. USA* 72:3575–79

Jones, E. G. 1975a. Possible determinants of the degree of retrograde neuronal labeling with horseradish peroxidase. *Brain Res.* 85:249–53

Jones, E. G. 1975b. Lamination and differential distribution of thalamic afferents in the sensory-motor cortex of the squirrel monkey. *J. Comp. Neurol.* 160:167–204

Jones, E. G. 1975c. Varieties and distribution of non-pyramidal cells in the somatic sensory cortex of the squirrel monkey. *J. Comp. Neurol.* 160:205–68

Jones, E. G. 1975d. Some aspects of the organization of the thalamic reticular complex. *J. Comp. Neurol.* 162:285–308

Jones, E. G., Burton, H. 1974. Cytoarchitecture and somatic sensory connectivity of thalamic nuclei other than the ventrobasal complex in the cat. *J. Comp. Neurol.* 154:395–432

Jones, E. G., Burton, H. 1976. Areal differences in the laminar distribution of thalamic afferents in cortical fields of the insular, parietal and temporal regions of primates. *J. Comp. Neurol.* 168:197–247

Jones, E. G., Burton, H., Porter, R. 1975. Commissural and cortico-cortical "columns" in the somatic sensory cortex of primates. *Science* 190:572–74

Jones, E. G., Burton, H., Saper, C. B., Swanson, L. W. 1976. Midbrain, diencephalic and cortical relationships of the basal nucleus of Meynert and related structures in primates. *J. Comp. Neurol.* 167:385–420

Jones, E. G., Coulter, J. D., Burton, H., Porter, R. 1977. Cells of origin and terminal distribution of corticostriatal fibers arising in the sensory-motor cortex of monkeys. *J. Comp. Neurol.* 173:53–80

Jones, E. G., Leavitt, R. Y. 1974. Retrograde axonal transport and the demonstration of non-specific projections to the cerebral cortex and striatum from thalamic intralaminar nuclei in the rat, cat and monkey. *J. Comp. Neurol.* 154:349–78

Jones, E. G., Wise, S. P. 1977. Size, laminar and columnar distribution of efferent cells in the sensory-motor cortex of monkeys. *J. Comp. Neurol.* 175:391–438

Jonsson, G., Fuxe, K., Hökfelt, T., Goldstein, M. 1976. Resistance of central phenylethanolamine-N-methyl transferase containing neurons to 6-hydroxydopamine. *Med. Biol.* 54:421–26

Jorgensen, A. O., Subrahmanyan, L., Turnbull, C., Kalnins, V. I. 1976. Localization of the neurofilament protein in neuroblastoma cells by immunofluorescent staining. *Proc. Natl. Acad. Sci. USA* 73:3192–96

Kaas, J. H., Lin, C.-S., Casagrande, V. A. 1976. The relay of ipsilateral and contralateral retinal input from the lateral geniculate nucleus to striate cortex in the owl monkey: a transneuronal transport study. *Brain Res.* 106:371–78

Kabat, E. A., Mayers, M. M. 1967. *Experimental Immunochemistry.* Springfield, Ill: Thomas

Kaneko, A. 1970. A physiological and morphological identification of horizontal, amacrine and bipolar cells in goldfish retina. *J. Physiol. London* 207:623–33

Kaneko, A. 1971. Electrical connexions between horizontal cells in the dogfish retina. *J. Physiol. London* 213:95–105

Kapeller, K., Mayor, D. 1967. The accumulation of noradrenaline in constricted sympathetic nerves as studied by fluorescence and electron microscopy. *Proc. R. Soc. London Ser. B.* 167:282–92

Karlsson, J.-O., Hansson, H.-A., Sjöstrand, J. 1971. Effect of colchicine on axonal transport and morphology of retinal ganglion cells. *Z. Zellforsch.* 115:265–83

Karlsson, J.-O., Sjöstrand, J. 1971a. Rapid intracellular transport of fucose-containing glycoproteins in retinal ganglion cells. *J. Neurochem.* 18:2209–16

Karlsson, J.-O., Sjöstrand, J. 1971b. Synthesis, migration and turnover of protein in retinal ganglion cells. *J. Neurochem.* 18:749–67

Karlsson, J.-O., Sjöstrand, J. 1971c. Transport of microtubular protein in axons of retinal ganglion cells. *J. Neurochem.* 18:975–82

Karlsson, J.-O., Sjöstrand, J. 1972. Axonal transport of proteins in retinal ganglion cells. Amino acid incorporation into rapidly transported proteins and distribution of radioactivity to the lateral geniculate body and the superior colliculus. *Brain Res.* 37:279–85

Kataoka, K., Bak, I. J., Hassler, R., Kim, J. S., Wagner, A. 1974. L-Glutamate decarboxylase and choline acetyltransferase activity in the substantia nigra and the striatum after surgical interruption of the strio-nigral fibres of the baboon. *Exp. Brain Res.* 19:217–27

Kater, S. B., Nicholson, C. 1973. *Intracellular Staining in Neurobiology.* New York: Springer

Keefer, D. A., Christ, J. F. 1976. Distribution of endogenous diaminobenzidine-staining cells in the normal rat brain. *Brain Res.* 116:312–16

Keefer, D. A., Spatz, W. B., Misgeld, U. 1976. Golgi-like staining of neocortical neurons using retrogradely transported horseradish peroxidase. *Neurosci. Lett.* 3:233–37

Kelly, J. P., Van Essen, D. C. 1974. Cell structure and function in the visual cortex of the cat. *J. Physiol. London* 238:515–47

Kelly, J. P., Gilbert, C. D. 1975. The projections of different morphological types of ganglion cells in the cat retina. *J. Comp. Neurol.* 163:65–80

Kelly, J. S., Dick, F. 1976. Differential labelling of glial cells and GABA-inhibitory interneurons and nerve terminals following microinjection of [B–^{3}H]alanine, [^{3}H]DABA, and [^{3}H]GABA into single folia of the cerebellum. *Cold Spring Harbor Symp., Quant. Biol.* 40:93–106

Kelts, K. A., Shepperdson, F. T., Land, L. J., Whitlock, D. G. 1975. Postganglionic connections between autonomic ganglia demonstrated autoradiographically. *Neurosci. Abstr.* 1:321

Kennedy, C., Des Rosiers, M. H., Jehle, J. W., Reivich, M., Sharpe, F., Sokoloff, L. 1975. Mapping of functional neural pathways by autoradiographic survey of local metabolic rate with [^{14}C]deoxyglucose. *Science* 187:850–53

Kennedy, C., Des Rosiers, M. H., Sakurada, O., Shinohara, M., Reivich, M., Jehle, J. W., Sokoloff, L. 1976. Metabolic mapping of the primary visual system of the monkey by means of the autoradiographic [^{14}C]deoxyglucose technique. *Proc. Natl. Acad. Sci. USA* 73:4230–34

Kerkut, G. A. 1975. Axoplasmic transport. *Comp. Biochem. Physiol.* 51A:701–4

Kerkut, G. A., Shapira, A., Walker, R. J. 1967. The transport of ^{14}C-labelled material from CNS $\rightleftharpoons$ muscle along a nerve trunk. *Comp. Biochem. Physiol.* 23: 729–48

Kievit, J., Kuypers, H. G. J. M. 1975. Basal forebrain and hypothalamic connections to frontal and parietal cortex in the rhesus monkey. *Science* 187:660–62.

Killackey, H. P., Belford, G. 1976. Discrete afferent terminations in the trigeminal pathway of the neonatal rat. *Anat. Rec.* 184:446

Kim, C. C., Strick, P. L. 1976. Critical factors involved in the demonstration of horseradish peroxidase retrograde transport. *Brain Res.* 103:356–61

King, J. C., Arimura, A., Gerall, A. A., Fishback, J. B., El Kind, K. E. 1975. Growth hormone-release inhibiting hormone (GH-RIH) pathway of the rat hypothalamus revealed by the unlabeled antibody peroxidase-anti-peroxidase method. *Cell Tissue Res.* 160: 423–30.

King, J. C., Gerall, A. A. 1976. Localization of luteinizing hormone-releasing hormone. *J. Histochem. Cytochem.* 24: 829–45

Kitai, S. T., Kocsis, J. D., Preston, R. J., Sugimori, M. 1976. Monosynaptic inputs to caudate neurons identified by intracellular injection of horseradish peroxidase. *Brain Res.* 109:601–6

Kizer, J. S., Palkovits, M., Brownstein, M. J. 1976. The projections of the A8, A9 and A10 dopaminergic cell bodies: evidence for a nigral-hypothalamic–median eminence dopaminergic pathway. *Brain Res.* 108:363–70

Koda, L. Y., Bloom, F. E. 1977. A light and electron microscopic study of noradrenergic terminals in the rat dentate gyrus. *Brain Res.* 120:327–35

Kopriwa, B. M. 1973. A reliable, standardized method for ultrastructural electron microscopic radioautography. *Histochemie* 37:1–17

Korr, I. M., Wilkinson, P. N., Chornock, F. W. 1967. Axonal delivery of neuroplasmic components to muscle cells. *Science* 155:342–45

Kreutzberg, G. W., Schubert, P. 1975. The cellular dynamics of intraneuronal transport. In *The Use of Axonal Transport for Studies of Neuronal Connectivity,* ed. W. M. Cowan, M. Cuénod, pp. 83–111. Amsterdam: Elsevier

Krettek, J. E., Price, J. L. 1974. A direct input from the amygdala to the thalamus and the cerebral cortex. *Brain Res.* 67:169–74

Krettek, J. E., Price, J. L. 1977. The cortical projections of the mediodorsal nucleus and adjacent thalamic nuclei in the rat. *J. Comp. Neurol.* 171:157–91

Krishnan, N., Singer, M. 1973. Penetration of peroxidase into peripheral nerve fibers. *Am. J. Anat.* 136:1–14

Kristensson, K. 1975. Retrograde axonal transport of protein tracers. In *The Use of Axonal Transport for Studies of Neuronal Connectivity,* ed. W. M. Cowan, M. Cuénod, pp. 69–81. Amsterdam: Elsevier

Kristensson, K., Olsson, Y. 1971. Retrograde axonal transport of protein. *Brain Res.* 29:363–65

Kristensson, K., Olsson, Y. 1973a. Diffusion pathways and retrograde axonal transport of protein tracers in peripheral nerves. *Prog. Neurobiol.* 1:85–109

Kristensson, K., Olsson, Y. 1973b. Uptake and retrograde transport of protein tracers in hypoglossal neurons. Fate of a tracer and reaction of the nerve cell bodies. *Acta Neuropathol.* 23:43–47

Kristensson, K., Olsson, Y. 1974. Retrograde axonal transport of horseradish peroxidase in transected axons. I. Time relationships between transport and induction of chromatolysis. *Brain Res.* 79: 101–9

Kristensson, K., Olsson, Y. 1976. Retrograde transport of horseradish peroxidase in transected axons. 3. Entry into injured axons and subsequent localization in perikaryon. *Brain Res.* 115:201–13

Kristensson, K., Olsson, Y., Sjöstrand, J. 1971. Axonal uptake and retrograde transport of exogenous proteins in the hypoglossal nerve. *Brain Res.* 32:399–406

Kristensson, K., Sjöstrand, J. 1972. Retrograde transport of protein tracer in the rabbit hypoglossal nerve during regeneration. *Brain Res.* 45:175–81

Kruger, L., Saporta, S. 1977. Axonal transport of [^{3}H] adenosine in visual and somatosensory pathways. *Brain Res.* 122:132–36

Kuhar, M. J., Pert, C. B., Snyder, S. H. 1973. Regional distribution of opiate receptor binding in monkey and human brain. *Nature* 245:447–50

Künzle, H. 1975. Bilateral projections from precentral motor cortex to the putamen and other parts of the basal ganglia. An autoradiographic study in *Macaca fascicularis. Brain Res.* 88:195–210

Künzle, H. 1976. Alternating afferent zones of high and low axon terminal density within the macaque motor cortex. *Brain Res.* 106:365–70

Künzle, H., Cuénod, M. 1973. Differential uptake of [^{3}H]proline and [^{3}H]leucine by neurons: its importance for the autoradiographic tracing of pathways. *Brain Res.* 62:213–17

Kuypers, H. G. J. M., Kievit, J., Groen-Klevant, A. C. 1974. Retrograde axonal transport of horseradish peroxidase in rat's forebrain. *Brain Res.* 67:211–18

Kuypers, H. G. J. M., Maisky, V. A. 1975. Retrograde axonal transport of horseradish peroxidase from spinal cord to brain stem cell groups in the cat. *Neurosci. Lett.* 1:9–14

Labedsky, L., Lierse, W. 1968. Die Entwicklung der Succinodehydrogenaseactivität im Gehirn der Maus während der Postnatalzeit. *Histochemie* 12:130–51

Lajtha, A., Marks, N. 1971. Protein turnover. In *Handbook of Neurochemistry, Vol. 5B, Metabolic Turnover in the Nervous System,* ed. A. Lajtha, pp. 551–629. New York: Plenum

LaMotte, C., Pert, C. B., Snyder, S. H. 1976. Opiate receptor binding in primate spinal cord: distribution and changes after dorsal root section. *Brain Res.* 112: 407–12

Landis, D. M. D., Reese, T. S. 1974. Differences in membrane structure between excitatory and inhibitory synapses in the cerebellar cortex. *J. Comp. Neurol.* 155:93–126

Lasek, R. J. 1967. Bidirectional transport of radioactively labelled axoplasmic components. *Nature* 216:1212–14

Lasek, R. J. 1968. Axoplasmic transport in cat dorsal root ganglion cells: as studied with L-[^{3}H]-leucine. *Brain Res.* 7: 360–77

Lasek, R. J. 1970a. Axonal transport of proteins in dorsal root ganglion cells of the growing cat: a comparison of growing and mature neurons. *Brain Res.* 26: 121–26

Lasek, R. J. 1970b. Protein transport in neurons. *Int. Rev. Neurobiol.* 13:289–321

Lasek, R. J., Gainer, H., Przybylski, R. J. 1974. Transfer of newly synthesized proteins from Schwann cells to the squid giant axon. *Proc. Natl. Acad. Sci. USA* 71:1188–92

Lasek, R. J., Joseph, B. S., Whitlock, D. G. 1968. Evaluation of a radioautographic neuronanatomical tracing method. *Brain Res.* 8:319–36

LaVail, J. H. 1975. Retrograde cell degeneration and retrograde transport techniques. In *The Use of Axonal Transport for Studies of Neuronal Connectivity,* ed. W. M. Cowan, M. Cuénod, pp. 217–23. Amsterdam: Elsevier

LaVail, J. H., LaVail, M. M. 1972. Retrograde axonal transport in the central nervous system. *Science* 176:1416–17

LaVail, J. H., LaVail, M. M. 1974. The retrograde intraaxonal transport of horseradish peroxidase in the chick visual system: a light and electron microscopic study. *J. Comp. Neurol.* 157:303–57

LaVail, J. H., Winston, K. R., Tish, A. 1973. A method based on retrograde intraaxonal transport of protein for identification of cell bodies of origin of axons terminating within the CNS. *Brain Res.* 58:470–77

Lee, C. Y., Tseng, L. F., Chiu, T. H. 1967. Influence of denervation on localization of neurotoxins from elapid venoms in rat diaphragm. *Nature* 215:1177–79

Lee, K. J., Woolsey, T. A. 1975. A proportional relationship between innervation density and cortical neuron number in the somatosensory system of the mouse. *Brain Res.* 99:349–53

Leonardelli, J., Barry, J., Dubois, M. P. 1973. Mis en évidence par immunofluoresence d'un constituant immunologiquement apparenté au LHRF dans l'hypothalamus et l'eminence médiane chez les mammifères. *C. R. Acad. Sci.* 276: 2043–46

Leontovich, T. A. 1975. Quantitative analysis and classification of subcortical forebrain neurons. In *Golgi Centennial Symposium: Perspectives in Neurobiology,* ed. M. Santini, pp. 101–22. New York: Raven Press

LeVay, S. 1977. Effects of visual deprivation on polyribosome aggregation in visual cortex of the cat. *Brain Res.* 119:73–86

LeVay, S., Ferster, D. 1977. Relay cell classes in the lateral geniculate nucleus of the cat and the effects of visual deprivation. *J. Comp. Neurol.* 172:563–84

LeVay, S., Gilbert, C. D. 1976. Laminar patterns of geniculocortical projection in the cat. *Brain Res.* 113:1–20

Levin, B. E., Stolk, J. M. 1977. Axoplasmic transport of norepinephrine in the locus coeruleus-hypothalamic system in the rat. *Brain Res.* 120:303–15

Levinthal, F., Macagno, E., Levinthal, C. 1976. Anatomy and development of identified cells in isogenic organisms. *Cold Spring Harbor Symp. Quant. Biol.* 40:321–31

Levinthal, C., Macagno, E., Tountas, C. 1973. Computer-aided reconstruction from serial sections. *Fed. Proc.* 33: 2336–40

Levinthal, C., Ware, R. 1972. Three-dimensional reconstruction from serial sections. *Nature* 236:207–10

Lindsay, R. D., Scheibel, A. B. 1974. Quantitative analysis of the dendritic branching pattern of small pyramidal cells from adult rat somesthetic and visual cortex. *Exp. Neurol.* 45:424–34

Litchy, W. J. 1973. Uptake and retrograde transport of horseradish peroxidase in frog sartorius nerve *in vitro*. *Brain Res.* 56:377–81

Livett, B. G., Geffen, L. B., Austin, L. 1968. Proximo-distal transport of [^{14}C] noradrenaline and protein in sympathetic nerves. *J. Neurochem.* 15:931–39

Livett, B. G., Geffen, L. B., Rush, R. A. 1969. Immunohistochemical evidence for the transport of dopamine-β-hydroxylase and a catecholamine binding protein in sympathetic nerves. *Biochem. Pharmacol.* 18:923–24

Livett, B. G., Uttenthal, L. O., Hipe, D. B. 1971. Localization of neurophysin II in the hypothalamo-neurohypophysial system of the pig by immunofluorescence histology. *Phil. Trans. R. Soc. London B. Ser.* 261–371

Ljungdahl, A., Hökfelt, T. 1973a. Accumulation of ^{3}H-glycine in interneurons of the cat spinal cord. *Histochemie* 33:277–80

Ljungdahl, A., Hökfelt, T. 1973b. Autoradiographic uptake patterns of [^{3}H] GABA and [^{3}H] glycine in central nervous tissues with special reference to the cat spinal cord. *Brain Res.* 62:587–95

Ljungdahl, A., Hökfelt, T., Goldstein, M., Park, D. 1975. Retrograde peroxidase tracing of neurons combined with transmitter histochemistry. *Brain Res.* 84: 313–19

Llinás, R. 1973. Procion yellow and cobalt as tools for the study of structure-function relationships in vertebrate central nervous systems. In *Intracellular Staining in Neurobiology*, ed. S. B. Kater, C. Nicholson, pp. 211–26. New York: Springer

Llinás, R., Hillman, D. E. 1975. A multipurpose tridimensional reconstruction computer system for neuroanatomy. In *Golgi Centennial Symposium Proceedings*, ed. M. Santini, pp. 71–79. New York: Raven Press

LoPresti, V., Macagno, E. R., Levinthal, C. 1973. Structure and development of neuronal connections in isogenic organisms: cellular interactions in the development of the optic lamina in *Daphnia*. *Proc. Natl. Acad. Sci. USA* 70: 433–37

Lubinska, L. 1964. Axoplasmic streaming in regenerating and in normal nerve fibers. *Prog. Brain Res.* 13:1–71

Lubinska, L. 1975. On axoplasmic flow. *Int. Rev. Neurobiol.* 17:241–95

Ludwin, S. K., Kosek, J. C., Eng, L. F. 1976. The topographical distribution of S-100 and GFA proteins in the adult rat brain: an immunohistochemical study using horseradish peroxidase–labelled antibodies. *J. Comp. Neurol.* 165:197–208

Luiten, P. G. M. 1975. The horseradish peroxidase technique applied to the teleostean nervous system. *Brain Res.* 89: 181–86

Lund, J. S., Lund, R. D., Hendrickson, A. E., Bunt, A. H., Fuchs, A. F. 1975. The origin of efferent pathways from the primary visual cortex, Area 17, of the macaque monkey as shown by retrograde trasnport of horseradish peroxidase. *J. Comp. Neurol.* 164:287–304

Lux, H. D., Schubert, P., Kreutzberg, G. W. 1970a. Direct matching of morphological and electrophysiological data in cat spinal motoneurons. In *Excitatory Synaptic Mechanisms*, ed. P. Anderson, J. K. S. Jansen, pp. 189–98. Oslo: Universitetsforlaget

Lux, H. D., Schubert, P., Kreutzberg, G. W., Globus, A. 1970b. Excitation and axonal flow: autoradiographic study on motoneurons intracellularly injected with a ^{3}H-amino acid. *Exp. Brain Res.* 10:197–204

Lynch, G., Gall, C., Mensah, P., Cotman, C. W. 1974. Horseradish peroxidase histochemistry: a new method for tracing efferent projections in the central nervous system. *Brain Res.* 65:373–80

Macagno, E. R., LoPresti, V., Levinthal, C. 1973. Structure and development of neuronal connections in isogenic organisms: variations and similarities in the optic system of *Daphnia magna*. *Proc. Natl. Acad. Sci. USA* 70:57–61

Marchisio, P. C., Gremo, F., Sjöstrand, J. 1975. Axonal transport in embryonic neurons. The possibility of a proximodistal axolemmal transfer of glycoproteins. *Brain Res.* 85:281–86

Marchisio, P. C., Sjöstrand, J. 1971. Axonal transport in avian optic pathway during development. *Brain Res.* 26:204–11

Marchisio, P. C., Sjöstrand, J., Aglietta, M., Karlsson, J. O. 1973. The development of axonal transport of proteins and glycoproteins in the optic pathway of chick embryos. *Brain Res.* 63:273–84

Marín-Padilla, M., Stibitz, G. R. 1974. Three-dimensional reconstruction of the basket cell of the human motor cortex. *Brain Res.* 70:511–14

Martin, M. R., Mason, C. A. 1977. The seventh cranial nerve of the rat. Visualization of efferent and afferent pathways by cobalt precipitation. *Brain Res.* 121: 21–41

Martin, R. 1977. Retrograde labeling of medullary raphe nuclei neurons in the monkey following spinal cord injections of horseradish peroxidase. *Anat. Rec.* 187:645–46

Mason, C. A. 1973. New features of the brain-retrocerebral neuroendocrine complex of the locust, *Schistocera vaga* (Scudder). *Z. Zellforsch.* 141:19–32

Mason, C. A. 1975. Delineation of the rat visual system by the axonal iontophoresis-cobalt sulfide precipitation technique. *Brain Res.* 85:287–94

Mason, T. E., Phifer, R. F., Spicer, S. S., Swallow, R. A., Dreskin, R. B. 1969. An immunoglobulin-enzyme bridge method for localizing tissue antigens. *J. Histochem. Cytochem.* 17:563–69

Massari, V. J., Gottesfeld, Z., Jacobowitz, D. M. 1976. Distribution of glutamic acid decarboxylase in certain rhombencephalic and thalamic nuclei of the rat. *Brain Res.* 118:147–51

Mata, M. M., Schrier, B. K. Klein, D. C., Weller, J. L., Chiou, C. Y. 1976. On GABA function and physiology in the pineal gland. *Brain Res.* 118:383–94

Matsumoto, T. 1920. The granules, vacuoles and mitochondria in the sympathetic nerve fibres cultivated *in vitro*. *Bull. Johns Hopkins Hosp.* 31:91–93

Matus, A. I., Dennison, M. E. 1971. Autoradiographic localization of tritiated glycine at "flat-vesicle" synapses in spinal cord. *Brain Res.* 32:195–97

Matus, A. I., Dennison, M. E. 1972. Autoradiographic study of exogenous glycine by vertebrate spinal cord slices *in vitro*. *J. Neurocytol.* 1:27–44

Matus, A., Mughal, S. 1975. Immunohistochemical localization of S–100 protein in brain. *Nature* 258:746–48

Mayor, D., Kapeller, K. 1967. Fluorescence microscopy and electron microscopy of adrenergic nerves after constriction at two points. *J. R. Microsc. Soc.* 87: 277–94

McBride, R. L., Sutin, J. 1976. Projections of the locus coeruleus and adjacent pontine tegmentum in the cat. *J. Comp. Neurol.* 165:265–84

McClure, W. O. 1972. Effect of drugs upon axoplasmic transport. *Adv. Pharmacol. Chemother.* 10:185–220

McCrea, R. A., Bishop, G. A., Kitai, S. T. 1976. Intracellular staining of Purkinje cells and their axons with horseradish peroxidase. *Brain Res.* 118:132–36

McEwen, B. S., Forman, D. S., Grafstein, B. 1971. Components of fast and slow axonal transport in the goldfish optic nerve. *J. Neurobiol.* 2:361–77

McEwen, B. S., Grafstein, B. 1968. Fast and slow components in axonal transport of protein. *J. Cell Biol.* 38:494–507

McGeer, P., McGeer, E. 1975. Neurotransmitter synthetic enzymes. *Prog. Neurobiol.* 1:71–118

McGeer, P. L., McGeer, E. G., Singh, V. K., Chase, W. H. 1974. Choline acetyltransferase localization in the central nervous system by immunohistochemistry. *Brain Res.* 81:373–79

McGregor, A. M., Komiya, Y., Kidman, A. D., Austin, L. 1973. The blockage of axoplasmic flow of proteins by colchicine and cytochalasins A and B. *J. Neurochem.* 21:1059–66

McKenzie, J. D. Jr., Vogt, B. A. 1976. An instrument for light microscopic analysis of three-dimensional neuronal morphology. *Brain Res.* 111:411–15

McLaughlin, B. J., Wood, J. G., Saito, K., Barker, R., Vaughn, J. E., Roberts, E., Wu, J.-Y. 1974. The fine structural localization of glutamate decarboxylase in synaptic terminals of rodent cerebellum. *Brain Res.* 76:377–91

McLean, I. W., Nakane, P. K. 1974. Periodate-lysine-paraformaldehyde fixative. A new fixative for immuno-electron microscopy. *J. Histochem. Cytochem.* 22: 1077–83

Meibach, R. C., Siegel, A. 1975. The origin of fornix fibers which project to the mammillary bodies in the rat: a horseradish peroxidase study. *Brain Res.* 88:508–12

Mensah, P., Finger, T. 1975. Neuromelanin: a source of possible error in HRP material. *Brain Res.* 98:183–88

Mesulam, M.-M. 1976a. The blue reaction product in horseradish peroxidase neurohistochemistry: incubation parameters and visibility. *J. Histochem. Cytochem.* 24:1273–80

Mesulam, M.-M. 1976b. A horseradish peroxidase method for the identification of the efferents of acetyl cholinesterase-containing neurons. *J. Histochem. Cytochem.* 24:1281–86

Mesulam, M.-M., Van Hoesen, G. W. 1976. Acetylcholinesterase-rich projections from the basal forebrain of the rhesus monkey to neocortex. *Brain Res.* 109: 152–57

Mittenthal, J. E., Wine, J. J. 1973. Connectivity patterns of crayfish giant interneurons: visualization of synaptic regions with cobalt dye. *Science* 179: 182–84

Mizuno, N., Konishi, A., Sato, M. 1975. Localization of masticatory motoneurons in the cat and rat by means of retrograde axonal transport of horseradish peroxidase. *J. Comp. Neurol.* 164: 105–15

Moore, R. Y., Lenn, N. J. 1972. A retinohypothalamic projection in the rat. *J. Comp. Neurol.* 146:1–14

Moriarty, G. C., Moriarty, C. M., Sternberger, L. A. 1973. Ultrastructural immunocytochemistry with unlabeled antibodies and the peroxidase-antiperoxidase complex. A technique more sensitive than radioimmunoassay. *J. Histochem. Cytochem.* 21:825–33

Morrell, J. I., Kelly, D. B., Pfaff, D. W. 1975. Sex steroid binding in the brains of vertebrates. Studies with light-microscopic autoradiography. In *Brain-Endocrine Interaction. II. The Ventricular System,* ed. K. M. Knigge, D. E. Scott, pp. 230–56. Basel: Karger

Muller, K. J., McMahan, U. J. 1976. The shapes of sensory and motor neurones and the distribution of their synapses in ganglia of the leech: a study using intracellular injection of horseradish peroxidase. *Proc. R. Soc. London Ser. B.* 194:481–99

Murray, M. 1974. Axonal transport in the asymmetric optic axons of flatfish. *Exp. Neurol.* 42:636–46

Murray, M., Grafstein, B. 1969. Changes in the morphology and amino acid incorporation of regenerating goldfish optic neurones. *Exp. Neurol.* 23:544–60

Nagatsu, I., Kondo, Y., Kato, T., Nagatsu, T. 1976. Retrograde axoplasmic transport of inactive dopamine-β-hydroxylase in sciatic nerves. *Brain Res.* 116: 277–85

Nairn, R. C. 1969. *Fluorescent Protein Tracing.* Edinburgh: Livingston. 3rd ed.

Nakane, P. K. 1975. Recent progress in the peroxidase-labeled antibody method. *Ann. NY Acad. Sci.* 254:203–11

Nakane, P. K., Kawaoi, A. 1974. Peroxidase-labeled antibody a new method of conjugation. *J. Histochem. Cytochem.* 22: 1084–91

Nakane, P. K., Pierce, G. S. 1966. Enzyme-labeled antibodies: a new preparation and application for localization of antigens. *J. Histochem. Cytochem.* 14: 929–31

Nauta, H. J. W. 1974. Evidence of a pallidohabenular pathway in the cat. *J. Comp. Neurol.* 156:19–27

Nauta, H. J. W., Kaiserman-Abramof, I. R., Lasek, R. J. 1975. Electron microscopic observations of horseradish peroxidase transported from the caudoputamen to the substantia nigra in the rat: possible involvement of the agranular reticulum. *Brain Res.* 85:373–84

Nauta, H. J. W., Pritz, M. B., Lasek, R. J. 1974. Afferents to the rat caudatoputamen studied with horseradish peroxidase. An evaluation of a retrograde neuroanatomical research method. *Brain Res.* 67:219–38

Nowakowski, R. S., LaVail, J. H., Rakic, P. 1975. The correlation of the time of origin of neurons with axonal projection: the combined use of [^{3}H]-thymidine autoradiography and horseradish peroxidase histochemistry. *Brain Res.* 99:343–48

Ochs, S. 1971. Local supply of energy to the fast axoplasmic transport mechanism. *Proc. Natl. Acad. Sci. USA* 68:1279–82

Ochs, S. 1972. Fast transport of material in mammalian nerve fibers. *Science* 176: 252–60

Ochs, S. 1974. Systems of material transport in nerve fibers (axoplasmic transport) related to nerve function and trophic control. *Ann. NY Acad. Sci.* 228: 202–23

Ochs, S. 1975. Axoplasmic transport. In *The Nervous System, Vol. 1: The Basic Neurosciences,* ed. D. B. Tower, pp. 137–46. New York: Raven Press

Ochs, S., Erdman, J. 1974. "Routing" of fast transported materials in nerve fibers. *Abstr. Soc. Neurosci.* 4:359

Ochs, S., Smith, C. B. 1971. Effect of temperature and rate of stimulation on fast axoplasmic transport in mammalian nerve fibers. *Fed. Proc.* 30:665

Oppenheim, R. W., Heaton, M. B. 1975. The retrograde transport of horseradish peroxidase from the developing limb of the chick embryo. *Brain Res.* 98:291–302

Orkand, P. M., Kravitz, E. A. 1971. Localization of the sites of γ-aminobutyric acid (GABA) uptake in lobster nerve-muscle preparations. *J. Cell. Biol.* 49: 75–89

Parent, A. 1976. Striatal afferent connections in the turtle (*Chrysemys picta*) as revealed by retrograde axonal transport of horseradish peroxidase. *Brain Res.* 108: 25–36

Parker, C. W. 1971. Nature of immunological responses and antigen-antibody interaction. In *Principles of Competitive Protein-Binding Assays,* ed. W. D. Odell, W. M. Daughaday. London: Lippincott

Parsons, J. A., Erlandsen, S. L., Hegre, O. D., McEvoy, R. C., Elde, R. P. 1976. Central and peripheral localization of somatostatin. Immunoenzyme immunocytochemical studies. *J. Histochem. Cytochem.* 24:872–82

Pasternak, G. W., Goodman, R., Snyder, S. H. 1975. An endogenous morphine-like factor in mammalian brain. *Life Sci.* 16: 1765–69

Pasternak, J. F., Woolsey, T. A. 1975. The number, size and spatial distribution of neurons in lamina IV of mouse SmI neocortex. *J. Comp. Neurol.* 160:291–306

Paulson, J. C., McClure, W. O. 1975. Microtubules and axoplasmic transport. Inhibition of transport by podophyllotoxin: an interaction with microtubule protein. *J. Cell. Biol.* 67:461–67

Pelletier, G., Leclerc, R., Duke, D. 1976. Immunohistochemical localization of hypothalamic hormones. *J. Histochem. Cytochem.* 24:864–71

Pert, C. B., Aposhian, D., Snyder, S. H. 1974a. Phylogenetic distribution of opiate receptor binding. *Brain Res.* 75:356–61

Pert, C. B., Kuhar, M. J., Snyder, S. H. 1975. Autoradiographic localization of the opiate receptor in rat brain. *Life Sci.* 16: 1849–54

Pert, C. B., Snowman, A. M., Snyder, S. H. 1974b. Localization of opiate receptor binding in synaptic membranes of rat brain. *Brain Res.* 70:184–88

Pert, C. B., Snyder, S. H. 1973. Opiate receptor: demonstration in nervous tissue. *Science* 174:1011–14

Peters, T. Jr., Ashley, C. A. 1967. An artefact in radioautography due to binding of free amino acids to tissue by fixatives. *J. Cell. Biol.* 33:53–60

Pfaff, D. W., Gerlach, J. L., McEwen, B. S., Ferin, M., Carmel, P., Zimmerman, E. A. 1976. Autoradiographic localization of hormone-concentrating cells in the brain of the female rhesus monkey. *J. Comp. Neurol.* 170:279–93

Pfaff, D., Keiner, M. 1974. Atlas of estradiol-concentrating cells in the central nervous system of the female rat. *J. Comp. Neurol.* 151:121–58

Pfaff, D. W. 1968. Uptake of estradiol-17β-H^3 in the female rat brain. An autoradiographic study. *Endocrinology* 82:1149–55

Pfenninger, K. H. 1973. Synaptic morphology and cytochemistry. *Prog. Histochem. Cytochem.* 5:11–81

Pickel, V. M., Joh, T. H., Reis, D. J. 1975a. Ultrastructural localization of tyrosine hydroxylase in noradrenergic neurons of brain. *Proc. Natl. Acad. Sci. USA* 72: 659–63

Pickel, V. M., Joh, T. H., Field, P. M., Becker, C. G., Reis, D. J. 1975b. Cellular localization of tyrosine hydroxylase by immunocytochemistry. *J. Histochem. Cytochem.* 23:1–12

Pickel, V. M., Joh, T. H., Reis, D. J. 1976a. Monamine-synthesizing enzymes in central dopaminergic noradrenergic and serotonergic neurons. Immunocytochemical localization by light and electron microscopy. *J. Histochem. Cytochem.* 24:792–806

Pickel, V. M., Reis, D. J., Leeman, S. E. 1977. Ultrastructural localization of substance P in neurons of rat spinal cord. *Brain Res.* 122:534–40

Pickel, V. M., Reis, D. J., Masangos, P. J., Zomzely-Neurath, C. 1976b. Immunocytochemical localization of nervous system specific protein (NSP-R) in rat brain. *Brain Res.* 105:184–87

Pickel, V. M., Segal, M., Bloom, F. E. 1974. A radioautographic study of the efferent pathways of the nucleus locus coeruleus. *J. Comp. Neurol.* 55:15–42

Pickel, V. M., Joh, T. H., Reis, D. J. 1975c. Ultrastructural localization of tyrosine hydroxylase in noradrenergic neurons of brain. *Proc. Natl. Acad. Sci. USA* 72: 659–63

Pitman, R. M., Tweedle, C. D., Cohen, M. J. 1972. Branching of central neurons: intracellular cobalt injection for light and electron microscopy. *Science* 176: 412–14

Ploem, J. S. 1975. Immunofluorescence and related staining techniques. General Introduction. *Ann. NY Acad. Sci.* 254: 4–20

Polz-Tejera, G., Schmidt, J., Karten, H. J. 1975. Autoradiographic localization of α-bungarotoxin-binding sites in the central nervous system. *Nature* 258: 349–51

Prensky, W. 1971. Automated image analysis in autoradiography. *Exp. Cell Res.* 68: 388–94

Price, D. L., Griffin, J., Young, A., Peck, K., Stocks, A. 1975. Tetanus toxin: direct evidence for retrograde intra-axonal transport. *Science* 188:945–47

Price, D. L., Stocks, A., Griffin, J., Young, A., Peck, K. 1976. Glycine-specific synapses in rat spinal cord: identification by electron microscopic autoradiography. *J. Cell. Biol.* 68:389–95

Price, J. L., Wann, D. F. 1975. The use of quantitative autoradiography for axonal tracing experiments and an automated system for grain counting. In *The Use of Axonal Transport for Studies of Neuronal Connectivity*, ed. W. M. Cowan, M. Cuénod, pp. 155–72. Amsterdam: Elsevier

Prior, D. J., Fuller, P. M. 1973. The use of a cobalt iontophoresis technique for identification of the mesencephalic trigeminal nucleus. *Brain Res.* 64:472–75

Pritz, M. B., Northcutt, R. G. 1977. Succinate dehydrogenase activity in the telencephalon of crocodiles correlates with the projection areas of sensory thalamic nuclei. *Brain Res.* 124:357–60

Purpura, D. P., Suzuki, K. 1976. Distortion of neuronal geometry and formation of aberrant synapses in neuronal storage disease. *Brain Res.* 116:1–22

Rahmann, H., Wolburg, H. 1971. Intraaxonaler Transport von ^{3}H-Uridin-Verbindungen im Tractus opticus von Teleosteern. *Experientia* 27:903–4

Raichle, M. E., Hartman, B. K., Eichling, J. O., Sharpe, L. G. 1975. Central noradrenergic regulation of cerebral blood flow and vascular permeability. *Proc. Natl. Acad. Sci. USA* 72:3726–30

Rakic, P. 1976. Prenatal genesis of connections subserving ocular dominance in the rhesus monkey. *Nature* 261:467–71

Rakic, P., Stensaas, L. J., Sayre, E. P., Sidman, R. L. 1974. Computer-aided three-dimensional reconstruction and quantitative analysis of cells from serial electron microscopic montages of foetal monkey brain. *Nature* 250:31–34

Ralston, H. J., Sharp, R. V. 1973. The identification of thalamocortical relay cells in the adult cat by means of retrograde axonal transport of horseradish peroxidase. *Brain Res.* 62:273–78

Repérant, J. 1975. The orthograde transport of horseradish peroxidase in the visual system. *Brain Res.* 85:307–12

Ribak, C. E., Peters, A. 1975. An autoradiographic study of the projections from the lateral geniculate body of the rat. *Brain Res.* 92:341–68

Ribak, C. E., Vaughn, J. E. 1976. Immunocytochemical localization of GAD in somata and dendrites of GABA-ergic neurons following colchicine treatment. *Neurosci. Abstr.* 2:796

Richards, J. G., Tranzer, J. P. 1970. The ultrastructural localization of amine storage sites in the central nervous system with the aid of a specific marker, 5-hydroxydopamine. *Brain Res.* 17: 463–69

Rieske, E., Schubert, P., Kreutzberg, G. W. 1975. Transfer of radioactive material between electrotonically coupled neurons of the leech central nervous system. *Brain Res.* 84:365–82

Robson, J. A., Hall, W. C. 1975. Connections of layer VI in striate cortex of the grey squirrel (*Sciureus carolinensis*). *Brain Res.* 93:133–39

Rogers, A. W. 1973. *Techniques of Autoradiography.* Amsterdam: Elsevier. 372 pp. 2nd ed.

Rose, G., Schubert, P. 1977. Release and transfer of [^{3}H] adenosine derivatives in the cholinergic septal system. *Brain Res.* 121:353–57

Rosenquist, A. C., Edwards, S. B., Palmer, L. A. 1974. An autoradiographic study of the dorsal lateral geniculate nucleus and the posterior nucleus in the cat. *Brain Res.* 80:71–93

Rossier, J. 1975. Immunohistochemical localization of choline acetyltransferase: real or artefact? *Brain Res.* 98:619–22

Ruiz-Marcos, A., Valverde, F. 1970. Dynamic architecture of the visual cortex. *Brain Res.* 19:25–39

Saito, K., Barber, R., Wu, J.-Y., Matsuda, T., Roberts, E., Vaughn, J. E. 1974. Immunohistochemical localization of glutamate decarboxylase in rat cerebellum. *Proc. Natl. Acad. Sci. USA* 71: 269–73

Samson, F. E. 1971. Mechanism of axoplasmic transport. *J. Neurobiol.* 2:347–60

Sandeman, D. C., Denburg, J. L. 1976. The central projections of chemoreceptor axons in the crayfish revealed by axoplasmic transport. *Brain Res.* 115: 492–96

Saper, C. B., Loewy, A. D., Swanson, L. W., Cowan, W. M. 1976a. Direct hypothalamo-autonomic connections. *Brain Res.* 117:305–12

Saper, C. B., Swanson, L. W., Cowan, W. M. 1976b. The efferent connections of the ventromedial nucleus of the hypothalamus of the rat. *J. Comp. Neurol.* 169:409–42

Scalia, F., Colman, D. R. 1974. Aspects of the central projection of the optic nerve in the frog as revealed by anterograde migration of horseradish peroxidase. *Brain Res.* 79:496–504

Schacher, S., Holtzman, E., Hood, D. C. 1976. Synaptic activity of frog retinal photoreceptors. A peroxidase uptake study. *J. Cell. Biol.* 70:178–92

Schafer, R. 1973. Acetylcholine: fast axoplasmic transport in insect chemoreceptor fibers. *Science* 180:315–16

Schon, F., Beart, P. M., Chapman, D., Kelly, J. S. 1975. On GABA metabolism in the gliocyte cells of the rat pineal gland. *Brain Res.* 85:479–90

Schon, F., Iversen, L. L. 1972. Selective accumulation of [^{3}H]-GABA by stellate cells in rat cerebellar cortex *in vivo. Brain Res.* 42:503–7

Schon, F., Iversen, L. L. 1974. The use of autoradiographic techniques for the identification and mapping of transmitter-specific neurones in the brain. *Life Sci.* 15:157–75

Schon, F., Kelly, J. S. 1974. Autoradiographic localization of [^{3}H]-GABA and [^{3}H]-glutamate over satellite glial cells. *Brain Res.* 66:275–88

Schon, F., Kelly, J. S. 1975. Selective uptake of ^{3}H-β-alanine by glia: association with glial uptake system for GABA. *Brain Res.* 86:243–57

Schonbach, J., Cuénod, M. 1971. Axoplasmic migration of protein. A light microscopic autoradiographic study in the avian retino-tectal pathway. *Exp. Brain Res.* 12:275–82

Schonbach, J., Schonbach, C., Cuénod, M. 1971. Rapid phase of axoplasmic flow and synaptic proteins: an electron microscopical study. *J. Comp. Neurol.* 141:485–98

Schramm, L. P., Adair, J. R., Stribling, J. M., Gray, L. P. 1975. Preganglionic innervation of the adrenal gland of the rat: a study using horseradish peroxidase. *Exp. Neurol.* 49:540–53

Schubert, P., Holländer, H. 1975. Methods for the delivery of tracers to the central nervous system. In *The Use of Axonal Transport for Studies of Neuronal Connectivity,* ed. W. M. Cowan, M. Cuénod, pp. 113–25. Amsterdam: Elsevier

Schubert, P., Kreutzberg, G. W. 1974. Axonal transport of adenosine and uridine derivatives and transfer to postsynaptic neurons. *Brain Res.* 76:526–30

Schubert, P., Kreutzberg, G. W. 1975a. Dendritic and axonal transport of nucleoside derivatives in single motoneurons and release from dendrites. *Brain Res.* 90:319–23

Schubert, P., Kreutzberg, G. W. 1975b. [^{3}H] adenosine, a tracer for neuronal connectivity. *Brain Res.* 85:317–20

Schubert, P., Kreutzberg, G. W., Lux, H. D. 1972. Neuroplasmic transport in dendrites: effect of colchicine on morphology and physiology of motoneurones in the cat. *Brain Res.* 47:331–43

Schubert, P., Lee, K., West, M., Deadwyler, S., Lynch, G. 1976. Stimulation-dependent release of ^{3}H-adenosine derivatives from central axon terminals to target neurons. *Nature* 260:541–42

Schubert, P., Lux, H. D., Kreutzberg, G. W. 1971. Single cell isotope injection technique, a tool for studying axonal and dendritic transport. *Acta Neuropathol.* 5:179–86

Schwab, M., Agid, Y., Glowinski, J., Thoenen, H. 1977. Retrograde axonal transport of ^{125}I-tetanus toxin as a tool for tracing fiber connections in the central nervous system: connections of the rostral part of the rat neostriatum. *Brain Res.* 126:211–24

Schwab, M. E., Thoenen, H. 1976. Electron microscopic evidence for a trans-synaptic migration of tetanus toxin in spinal cord motoneurons: an autoradiographic study. *Brain Res.* 105:213–27

Schwab, M., Thoenen, H. 1977. Selective trans-synaptic migration of tetanus toxin after retrograde axonal transport in peripheral sympathetic nerves: a comparison with nerve growth factor. *Brain Res.* 122:459–74

Schwartz, J. H., Goldman, J. E., Ambron, R. T., Goldberg, D. J. 1976. Axonal transport of vesicles carrying [^{3}H] serotonin in the metacerebral neuron of *Aplysia californica. Cold Spring Harbor Symp. Quant. Biol.* 40:83–92

Shannon, L. M., Kay, E., Lew, J. Y. 1966. Peroxidase isozymes from horseradish roots. I. Isolation and physical properties. *J. Biol. Chem.* 241:2166–72

Sharp, F. R. 1974. Activity related 2-deoxy-D-glucose uptake in the central nervous system of the rat. In *Program and Abstracts, Society for Neuroscience, Fourth Annual Meeting, St. Louis, Missouri,* p. 422

Sharp, F. R. 1976a. Activity-related increases of glucose utilization associated with reduced incorporation of glucose into its derivatives. *Brain Res.* 107:663–66

Sharp, F. R. 1976b. Relative cerebral glucose consumption of neuronal perikarya and neuropil determined with 2-deoxyglucose in resting and swimming rat. *Brain Res.* 110:127–39

Sharp, F. R. 1976c. Rotation induced increases of glucose metabolism in rat vestibular nuclei and vestibulocerebellum. *Brain Res.* 110:141–51

Sharp, F. R., Kauer, J. S., Shepherd, G. M. 1975. Local sites of activity-related glucose metabolism in rat olfactory bulb during olfactory stimulation. *Brain Res.* 98:596–600

Shatz, C. 1977. A comparison of visual pathways in Boston and midwestern Siamese cats. *J. Comp. Neurol.* 171:205–28

Shatz, C., Lindstrom, S., Wiesel, T. 1975. Ocular dominance columns in the cat's visual cortex. *Neurosci. Abstr.* 1:56

Sherlock, D. A., Field, P. M., Raisman, G. 1975. Retrograde transport of horseradish peroxidase in the magnocellular neurosecretory system of the rat. *Brain Res.* 88:403–14

Sherlock, D. A., Raisman, G. 1975. A comparison of anterograde and retrograde axonal transport of horseradish peroxidase in the connections of the mammillary nuclei in the rat. *Brain Res.* 85:321–24

Shih, J. H. C., Shannon, L. M., Kay, E., Lew, J. Y. 1971. Peroxidase isoenzymes from horseradish roots. IV. Structural relationships. *J. Biol. Chem* 246:4546–51

Sholl, D. A. 1953. Dendritic organization in the neurons of the visual and motor cortices of the cat. *J. Anat.* 87:387–406

Siggins, G. R., Battenberg, E. R., Hoffer, B. J., Bloom, F. E., Steiner, A. L. 1973. Noradrenergic stimulation of cyclic adenosine monophosphate in rat Purkinje neurons: an immunocytochemical study. *Science* 179:585–88

Silverman, A. J. 1975a. The hypothalamic magnocellular neurosecretory system of the guinea pig. I. Immunohistochemical localization of neurophysin in the adult. *Am. J. Anat.* 144:433–44

Silverman, A. J. 1975b. The hypothalamic magnocellular neurosecretory system of the guinea pig. II. Immunohistochemical localization of neurophysin and vasopressin in the fetus. *Am. J. Anat.* 144:445–60

Silverman, A. J. 1976a. Ultrastructural studies on the localization of neurophypophysial hormones and their Carrier proteins. *J. Histochem. Cytochem.* 24:816–27

Silverman, A. J. 1976b. Distribution of luteinizing hormone–releasing hormone (LHRH) in the guinea pig brain. *Endocrinology* 99:30–41

Silverman, A. J., Desnoyers, P. 1976. Ultrastructural localization of luteinizing hormone–releasing hormone (LHRF) in median eminence of guinea pig. *Cell Tissue Res.* 169:157–66

Simon, E. J., Hiller, J. M., Edelman, I. 1973. Stereospecific binding of the potent narcotic analgesic [^{3}H] etorphine to rat brain homogenate. *Proc. Natl. Acad. Sci. USA* 70:1947–49

Singer, W., Holländer, H., Vanegas, H. 1977. Decreased peroxidase labeling of lateral geniculate neurons following deafferentation. *Brain Res.* 120:133–37

Sjöstrand, J., Frizell, M. 1975. Retrograde axonal transport of rapidly migrating proteins in peripheral nerves. *Brain Res.* 85:325–30

Smith, P. G., Mills, E. 1976. Autoradiographic identification of the terminations of petrosal ganglion neurons in the cat carotid body. *Brain Res.* 113: 174–78

Snow, P. J., Rose, P. K., Brown, A. G. 1976. Tracing axons and axon collaterals of spinal neurons using intracellular injection of horseradish peroxidase. *Science* 191:312–13

Snyder, S. H. 1975. The opiate receptor in normal and drug altered brain function. *Nature* 257:185–89

Sokoloff, L., Reivich, M., Kennedy, C., Des Rosiers, M. H., Patlak, C. S., Pettigrew, K. D., Sakurada, O., Shinohara, M. 1977. The [^{14}C] deoxyglucose method for the measurement of local cerebral glucose utilization: theory, procedure, and normal values in the conscious and anesthetized albino rat. *J. Neurochem.* 28:897–916

Sokoloff, L., Reivich, M., Patlak, C. S., Pettigrew, K. D., Des Rosiers, M., Kennedy, C. 1974. The [^{14}C] deoxyglucose method for the quantitative determination of local cerebral glucose consumption. *Proc. 5th Meet. Am. Soc. Neurochem.* p. 86.

Sotelo, C. 1971. The fine structural localization of [^{3}H] norepinephrine in the substantia nigra and area postrema of the rat. An autoradiographic study. *J. Ultrastruct. Res.* 36:824–41

Sotelo, C., Privat, A., Drian, M. J. 1972. Localization of ^{3}H GABA in tissue culture of rat cerebellum using electron microscopy radioautography. *Brain Res.* 45: 302–8

Sotelo, C., Riche, D. 1974. The smooth endoplasmic reticulum and the retrograde and fast orthograde transport of horseradish peroxidase in the nigrostriato-nigral loop. *Anat. Embryol.* 146:209–18

Sousa-Pinto, A., Reis, F. F. 1975. Selective uptake of [^{3}H] leucine by projection neurons of the cat auditory cortex. *Brain Res.* 85:331–36

Specht, S. C., Grafstein, B. 1973. Accumulation of radioactive protein in mouse cerebral cortex after injection of ^{3}H-fucose into the eye. *Exp. Neurol.* 41:705–22

Specht, S. C., Grafstein, B, 1977. Axonal transport and transneuronal transfer in mouse visual system following injection of [^{3}H] fucose into the eye. *Exp. Neurol.* 54:352–68

Stanfield, B., Cowan, W. M. 1976. Evidence for a change in the retinohypothalamic projection in the rat following early removal of one eye. *Brain Res.* 104: 129–36

Steinberger, L. A. 1973. Enzyme immunocytochemistry. In *Electron Microscopy of Enzymes: Principles and Methods,* Vol. 1, ed. M. A. Hyatt, pp. 150–91. New York: Van Nostrand Reinhold

Steindler, D. A., Colwell, S. A. 1976. Reeler mutant mouse: maintenance of appropriate and reciprocal connections in the cerebral cortex and thalamus. *Brain Res.* 113:386–93

Steiner, A. L., Ong, S.-H., Wedner, H. J. 1976. Cyclic nucleotide immunocytochemistry. In *Advances in Cyclic Nucleotide Research,* Vol. 7, ed. P. Greengard, G. A. Robison. New York: Raven Press

Sternberger, L. A. 1974. *Immunocytochemistry.* Englewood Cliffs, N. J.: Prentice-Hall

Sternberger, L. A., Cuculis, J. T. 1969. Methods for enzymatic intensification of the immunocytochemical reaction without use of labeled antibodies. *J. Histochem. Cytochem* 17:190

Sternberger, L. A. 1973. Enzyme immunocytochemistry. In *Electron Microscopy of Enzymes Principles and Methods,* Vol. 1, ed. M. A. Hayatt, pp. 150–91. New York: Van Nostrand Reinhold

Sternberger, L. A., Hardy, P. H. Jr., Cuculis, J. J., Meyer, H. G. 1970. The unlabeled antibody enzyme method of immunohistochemistry: preparation and properties of soluble antigen-antibody complex (horseradish peroxidase–antihorseradish peroxidase) and its use in identification of spirochetes. *J. Histochem. Cytochem* 18:315–33

Steward, O., Scoville, S. A. 1976. Retrograde labeling of central nervous pathways with tritiated or Evans blue–labeled bovine serum albumin. *Neurosci. Lett.* 3:191–96

Stoeckel, K., Paravicini, U., Thoenen, H. 1974. Specificity of the retrograde axonal transport of nerve growth factor. *Brain Res.* 76:413–21

Stoeckel, K., Schwab, M., Thoenen, H. 1975a. Comparison between the retrograde axonal transport of nerve growth factor and tetanus toxin in motor, sensory and adrenergic neurons. *Brain Res.* 99:1–16

Stoeckel, K., Schwab, M., Thoenen, H. 1975b. Specificity of retrograde transport of nerve growth factor (NGF) in sensory neurons: a biochemical and morphological study. *Brain Res.* 89: 1–14

Stoeckel, K., Thoenen, H. 1975. Retrograde axonal transport of nerve growth factor: specificity and biological importance. *Brain Res.* 85:337–42

Storm-Mathisen, J. 1974. Choline acetyltransferase and acetylcholinesterase in fascia dentata following lesions of entorhinal afferents. *Brain Res.* 80:181–97

Storm-Mathisen, J. 1975. Accumulation of glutamic acid decarboxylase in the proximal parts of presumed GABA-ergic neurones after axotomy. *Brain Res.* 87:107–9

Storm-Mathisen, J. 1976. Distribution of the components of the GABA system in neuronal tissue: cerebellum and hippocampus—effects of axotomy. In *GABA in Nervous System Function,* ed.

E. Roberts, T. N. Chase, D. B. Tower, pp. 149–68. New York: Raven Press
Storm-Mathisen, J. 1977. Glutamic acid and excitatory nerve endings: reduction of glutamic acid uptake after axotomy. *Brain Res.* 120:379–86
Straus, W. 1964. Factors affecting the cytochemical reaction of peroxidase with benzidine and the stability of the blue reaction produce. *J. Histochem. Cytochem.* 12:462–69
Strick, P. L. 1975. Multiple sources of thalamic input to the primate motor cortex. *Brain Res.* 88:372–77
Strick, P. L., Burke, R. E., Kanda, K., Kim, C. C., Walmsley, B. 1976. Differences between alpha and gamma motoneurons labeled with horseradish peroxidase by retrograde transport. *Brain Res.* 113:582–88
Stumpf, W. E. 1968. Estradiol-concentrating neurons: topography in the hypothalamus by dry-mount autoradiography. *Science* 162:1001–3
Swanson, L. W. 1976. An autoradiographic study of the efferent connections of the preoptic region in the rat. *J. Comp. Neurol.* 167:227–56
Swanson, L. W. 1977. Immunohistochemical evidence for a neurophysin-containing autonomic pathway arising in the paraventricular nucleus of the hypothalamus. *Brain Res.* 128:346–53
Swanson, L. W., Connelly, M. A., Hartman, B. K. 1977. Ultrastructural evidence for central monoaminergic innervation of blood vessels in the paraventricular nucleus of the hypothalamus. *Brain Research.* 136:166–73
Swanson, L. W., Cowan, W. M. 1975. Hippocampo-hypothalamic connections: origin in subicular cortex, not Ammon's horn. *Science* 189:303–4
Swanson, L. W., Hartman, B. K. 1975. The central adrenergic system: an immunofluorescence study of the localization of cell bodies and their efferent connections in the rat, utilizing dopamine-β-hydroxylase as a marker. *J. Comp. Neurol.* 163:467
Tabuchi, K., Kirsch, W. M., Nakane, P. K. 1976. The fine structural localization of S-100 protein in rodent cerebellum. *J. Neurol. Sci.* 28:65–76
Tappaz, M. L., Brownstein, M. J., Palkovits, M. 1976. Distribution of glutamate decarboxylase in discrete brain nuclei. *Brain Res.* 108:371–80
Taylor, A. C., Weiss, P. 1965. Demonstration of axonal flow by the movement of tritium-labeled protein in mature optic nerve fibres. *Proc. Natl. Acad. Sci. USA* 54:1521–27
Terenius, L. 1973. Characterization of receptor for narcotic analgesics in synaptic membrane fractions from rat brain. *Acta Pharmacol (Kbh.)* 33:377–84
Terenius, L., Wahlström, A. 1974. Inhibitor(s) of narcotic receptor binding in brain extracts and cerebrospinal fluid. *Acta Pharmacol. (Kbh.)* 35:55
Terenius, L., Wahlström, A. 1975. Morphine-like ligand for opiate receptors in human CSF. *Life Sci.* 16:1759–64
Theodosis, D. T., Dreifuss, J. J., Harris, M. C., Orci, L. 1976. Secretion-related uptake of horseradish peroxidase in neurophypophysial axons. *J. Cell Biol.* 70:294–303
Thoenen, H., Tranzer, J. P. 1968. Chemical sympathectomy by selective destruction of adrenergic nerve endings with 6-hydroxydopamine. *Naunyn-Schmiedeberg's Arch. Exp. Pathol. Pharmakol.* 261:271–88
Thoenen, H., Tranzer, J. P. 1973. The pharmacology of 6-hydroxydopamine. *Ann. Rev. Pharmacol.* 13:169–80
Thompson, E. B., Schwartz, J. H., Kandel, E. R. 1976. A Radioautographic analysis in the light and electron microscope of identified *Aplysia* neurons and their processes after intra-somatic injection of L-^{3}H fucose. *Brain Res.* 112:251–81
Tranzer, J. P., Thoenen, H. 1967. Electromicroscopic localization of 5-hydroxydopamine (3,4,5-trihydroxy-phenylethylamine), a new "false" sympathetic transmitter. *Experientia* 23:743–45
Tranzer, J. P., Thoenen, H. 1968. An electron microscopic study of selective acute degeneration of sympathetic nerve terminals after administration of 6-hydroxydopamine. *Experientia* 24: 155–56
Tranzer, J. P., Thoenen, H., Snipes, R. L., Richards, J. G. 1969. Recent developments on the ultrastructural aspect of adrenergic nerve endings in various experimental conditions. *Prog. Brain Res.* 31:33–46
Trevino, D. L., Carstens, E. 1975. Confirmation of the location of spinothalamic neurons in the cat and monkey by the retrograde transport of horseradish peroxidase. *Brain Res.* 98:177–82
Trevino, D. L., Coulter, J. D., Willis, W. D. 1973. Location of cells of origin of spinothalamic tract in lumbar enlargement of the monkey. *J. Neurophysiol.* 36:750–61

Trojanowski, J. Q., Jacobson, S. 1975. A combined horseradish peroxidase–autoradiographic investigation of reciprocal connections between superior temporal gyrus and pulvinar in squirrel monkey. *Brain Res.* 85:347–53

Turner, P. T. 1977. Effect of pentobarbital on uptake of horseradish peroxidase by rabbit cortical synapses. *Exp. Neurol.* 54:24–32

Turner, P. T., Harris, A. B. 1974. Ultrastructure of exogenous peroxidase in cerebral cortex. *Brain Res.* 74:305–26

Tyrer, N. M., Bell, E. M. 1974. The intensification of cobalt-filled neurone profiles using a modification of Timm's sulphide-silver method. *Brain Res.* 73: 151–55

Updyke, B. V. 1975. The patterns of projection of cortical areas 17, 18, and 19 onto the laminae of the dorsal lateral geniculate nucleus in the cat. *J. Comp. Neurol.* 163:377–95

Updyke, B. V. 1977. Topographic organization of the projections from cortical areas 17, 18 and 19 onto the thalamus, pretectum and superior colliculus in the cat. *J. Comp. Neurol.* 173:81–122

Vacca, L. L., Rosario, S. L., Zimmerman, E. A., Tomashefsky, P., Ng, P.-Y., Hsu, K. C. 1975. Application of immunoperoxidase techniques to localize horseradish peroxidase in the central nervous system. *J. Histochem. Cytochem.* 23:208–15

Valverde, F. 1976. Aspects of cortical organization related to the geometry of neurons with intra-cortical axons. *J. Neurocytol.* 5:509–29

Vandesande, F., DeMay, J., Dierickx, K. 1974. Identification of neurophysin producing cells. I. The origin of the neurophysin-like substance–containing nerve fibers of the external region of the median eminence of the rat. *Cell Tissue Res.* 151:187–200

Vandesande, F., Dierickx, K., DeMay, J. 1975a. Identification of the vasopressin-neurophysin producing neurons of the rat suprachiasmatic nuclei. *Cell Tissue Res.* 156:377–80

Vandesande, F., Dierickx, K., DeMay, J. 1975b. Identification of separate vasopressin-neurophysin II and oxytocin-neurophysin I containing nerve fibers in the external region of the bovine median eminence. *Cell Tissue Res.* 158:509–16

Vandesande, F., Dierickx, K., DeMey, J. 1975c. Identification of the vasopressin-neurophysin II and the oxytocin-neurophysin I producing neurons in the bovine hypothalamus. *Cell Tissue Res.* 156:189–200

Vanegas, H., Holländer, H., Distel, H. J. 1977. Early stages of uptake and transport of horseradish peroxidase by cortical structures, and its use for the study of local neurons and their processes. *J. Comp. Neurol.* 177:193–211

Van Orden, L. S., Burke, J. P., Redick, J. A., Rybarczyk, K. E., Van Orden, D. E., Baker, H. A., Hartman, B. K. 1977. Immunocytochemical evidence for particulate localization of phenylethanolanine-N-methyltransferase in adrenal medulla. *Neuropharmacol.* 16:129–33

Vogel, Z., Nirenberg, M. 1976. Localization of acetylcholine receptors during synaptogenesis in retina. *Proc. Natl. Acad. Sci. USA* 73:1806–10

Walberg, F., Brodal, A., Hoddevik, G. H. 1976. A note on the method of retrograde transport of horseradish peroxidase as a tool in studies of afferent cerebellar connections, particularly those from the inferior olive; with comments on the orthograde transport in Purkinje cell axons. *Exp. Brain Res.* 24:383–401

Wann, D. F., Price, J. L., Cowan, W. M., Agulnek, M. A. 1974. An automated system for counting silver grains in autoradiographs. *Brain Res.* 81:31–58

Wann, D. F., Woolsey, T. A., Dierker, M. L., Cowan, W. M. 1973. An on-line digital-computer system for the semiautomatic analysis of Golgi-impregnated neurons. *IEEE Trans. Biomed. Eng.* 20:233–47

Warr, W. B. 1975. Olivocochlear and vestibular efferent neurons of the feline brain stem: their location, morphology and number determined by retrograde axonal transport and acetylcholinesterase histochemistry. *J. Comp. Neurol.* 161:159–82

Watkins, W. B., Schwabedal, P., Bock, R. 1974. Immunohistochemical demonstration of a CRF-associated neurophysin in the external zone of the rat median eminence. *Cell Tissue Res.* 152: 411–21

Watson, W. E. 1968. Centripetal passage of labeled molecules along mammalian motor axons. *J. Physiol. London* 196: 122–23

Weir, E. E., Pretlow, T. G., Pitts, A., Williams, E. E. 1974. A more sensitive and specific histochemical peroxidase stain for the localization of cellular antigen by the enzyme-antibody conjugate

method. *J. Histochem. Cytochem.* 22: 1135–40
Weiss, P., Hiscoe, H. B. 1948. Experiments on the mechanism of nerve growth. *J. Exp. Zool.* 107:315–95
Wells, A. F., Miller, C. E., Nadel, M. K. 1966. Rapid fluorescein and protein assay method for fluorescent-antibody conjugates. *Appl. Microbiol.* 14:271–75
Wiesel, T. N., Hubel, D. H., Lam, D. 1974. Autoradiographic demonstration of ocular dominance columns in the monkey by means of transneuronal transport. *Brain Res.* 79:273–79
Willard, M., Cowan, W. M., Vagelos, P. R. 1974. The polypeptide composition of intra-axonally transported proteins: evidence for four transport velocities. *Proc. Natl. Acad. Sci. USA* 71:2183–87
Winfield, D. A., Gatter, K. C., Powell, T. P. S. 1975. An electron microscopic study of retrograde and orthograde transport of horseradish peroxidase to the lateral geniculate nucleus of the monkey. *Brain Res.* 92:462–67
Winlow, W., Kandel, E. R. 1976. The morphology of identified neurons in the abdominal ganglion of *Aplysia californica*. *Brain Res.* 112:221–50
Wise, S. P. 1975. The laminar organization of certain afferent and efferent fiber systems in the rat somatosensory cortex. *Brain Res.* 90:139–42
Wise, S. P., Hendry, S. H. C., Jones, E. G. 1977. Prenatal development of sensory-motor cortical projections in cats. *Brain Res.* In press
Wise, S. P., Jones, E. G. 1976a. The organization and postnatal development of the commissural system of the somatic sensory cortex in the rat. *J. Comp. Neurol.* 168:313–43
Wise, S. P., Jones, E. G. 1976b. Transneuronal or retrograde transport of [^{3}H] adenosine in the rat somatic sensory system. *Brain Res.* 107:127–31
Wise, S. P., Jones, E. G. 1977a. Developmental specificity in the thalamocortical system of the rat. *J. Comp. Neurol.* In press
Wise, S. P., Jones, E. G. 1977b. Cells of origin and terminal distribution of corticofugal pathways from the rat somatic sensory cortex. *J. Comp. Neurol.* 175: 129–58
Wise, S. P., Jones, E. G., Berman, N. 1977. Direction and specificity of the axonal and transcellular transport of nucleosides. *Brain Res.* In press
Wong-Riley, M. T. T. 1974. Demonstration of geniculocortical and callosal projection neurons in the squirrel monkey by means of retrograde axonal transport of horseradish peroxidase. *Brain Res.* 79:267–72
Wong-Riley, M. T. T. 1976. Endogenous peroxidatic activity in brain stem neurons as demonstrated by their staining with diaminobenzidine in normal squirrel monkeys. *Brain Res.* 108:257–78
Wood, B. T., Thompson, S. H., Goldstein, G. 1965. Fluorescent antibody staining. III. Preparation of fluorescein-isothiocyanate-labeled antibodies. *J. Immunol.* 95:225–29
Woolsey, T. A., Dierker, M. L., Wann, D. F. 1975. Mouse SmI cortex: qualitative and quantitative classification of Golgi-impregnated barrel neurons. *Proc. Natl. Acad. Sci. USA* 72:2165–69
Yamamura, H. I., Kuhar, M. J., Greenberg, D., Snyder, S. H. 1974. Muscarinic cholinergic receptor binding: regional distribution in monkey brain. *Brain Res.* 76:541–46
Young, A. B., Snyder, S. H. 1973. Strychnine binding associated with glycine receptors of the central nervous system. *Proc. Natl. Acad. Sci. USA* 70:2832–36
Young, J. A. C., Brown, D. A., Kelly, J. S., Schon, F. 1973. Autoradiographic localization of sites of [^{3}H] γ-aminobutyric acid accumulation in peripheral autonomic ganglia. *Brain Res.* 63: 479–86
Zarzecki, P., Blum, P. S., Cordingley, G. E., Somjen, G. G. 1975. Microiontophoretic studies of the effects of L-proline on neurons in the mammalian central nervous system. *Brain Res.* 89:187–91
Zeki, S. M., Sandeman, D. R. 1976. Combined anatomical and electrophysiological studies on the boundary between the second and third visual areas of rhesus monkey cortex. *Proc. R. Soc. London Ser. B* 194:555–62
Ziegler, M. G., Thomas, J. A., Jacobowitz, D. M. 1976. Retrograde axonal transport of antibody to dopamine-β-hydroxylase. *Brain Res.* 104:390–95
Zimmerman, E. A., Antunes, J. L. 1976. Organization of the hypothalamic-pituitary system: current concepts from immunohistochemical studies. *J. Histochem. Cytochem.* 24:807–15
Zimmerman, E. A., Defendini, R., Sokol, H. W., Robinson, A. G. 1975. The distribution of neurophysin-secreting pathways in the mammalian brain. Light microscopic studies using the immunoperoxidase technique. *Ann. NY Acad. Sci.* 248:92–111

Zimmerman, E. A., Hsu, K. C., Ferin, M., Kozlowski, G. P. 1974. Localization of gonadotropin-releasing hormone (Gn.Rh) in the hypothalamus of the mouse by immunoperoxidase technique. *Endocrinology* 5:1–8

Zukin, S. R., Young, A. B., Snyder, S. H. 1974. Gamma-aminobutyric acid binding to receptor sites in the rat central nervous system. *Proc. Natl. Acad. Sci. USA* 71:4802–7

Ann. Rev. Neurosci. 1978. 1:297–326

MECHANISMS OF CORTICAL DEVELOPMENT: A VIEW FROM MUTATIONS IN MICE[1]

❖11510

Verne S. Caviness, Jr., and Pasko Rakic

Departments of Neurology and Neuropathology, Massachusetts General Hospital and Harvard Medical School, and Eunice Kennedy Shriver Center for Mental Retardation, Inc., Waverley, Massachusetts 02179, and the Children's Hospital Medical Center, Boston, Massachusetts 02115

INTRODUCTION

Cortical structures in the mammalian central nervous system are remarkable among biological tissues for the variety and elegance of their cellular forms, for the geometric regularity in the spatial arrangements of like and unlike neuronal elements, and for the heterogeneity in the types of interrelationships among cells. Complex modifications in the structure and synaptic connectivity of the cortex may result from mutations at single genetic loci. Structural modifications induced by a mutation may be specific in character and often independent of variations in genetic background (e.g. Caviness, So & Sidman 1972, Rakic & Sidman 1973c, Rakic 1976b, Sotelo 1975a, Mariani et al 1977). Various aspects of the effects of single gene mutations on the structure and function of the nervous system have been recently reviewed (Sidman 1972, 1974, Rakic 1974b, 1976b, Pak & Pinto 1976, Goldman 1976, Caviness 1977, Dräger 1977).

The present review is concerned specifically with cellular modifications induced in cortical structures of the central nervous system by mutations in mice. Only a few single gene mutations that lead to cortical maldevelopment are known in other mammalian species including man (e.g. Volpe & Adams 1972, Guillery, Casagrande & Oberdorfer 1974, Shatz 1977). Even in mice, where more than 140 mutations are known to affect the developing nervous system (Sidman, Green & Appel 1965), only 7 are recognized that selectively affect the development and structure of the cortex (Table 1). All 7 of these "cortical mutations" modify the structure of the cerebellar cortex; only one, the reeler mutation, also affects the structure of the cerebral cortex. Cortical malformation in all of these mutations, with the exception of the cerebral cortical malformation in the reeler, is associated with the death of one or more

[1]This study was supported in part by USPHS Grant 1 R01 NS12005–02 and NS11223, National Institutes of Health, Bethesda, Maryland 20015

0147-006X/78/0325-0297$01.00

Table 1 Cortical mutants in mice

Name and gene symbol	Inheritance and chromosome	Salient behavioral and morphological characteristics; earliest and critical recent references
Leaner (tg^{1a})	autosomal recessive (8)	Ataxia and hypertonia at two weeks of age. Particularly in the anterior and nodular lobes of cerebellum, there is degeneration of granule cells, most before completing migrations, and degeneration of a few Purkinje cells (Meier & MacPike 1971, Sidman 1968, Sidman and Herndon, personal communication)
Lurcher (Lc)	autosomal semi-dominant (6)	Homozygous (Lc/Lc) dies in perinatal period. Heterozygote is ataxic with hesitant, reeling gait and has seizures when startled. Its cerebellum is half normal size. Purkinje cells degenerate; granule cells are reduced in numbers (Philips 1960, Caddy & Briscoe 1975, Swisher & Wilson 1977)
Nervous (nr)	autosomal recessive (8)	Hyperactivity and ataxia develop by 3½ weeks. Mitochondria of all Purkinje cells become spherical and enlarged. Ninety percent of Purkinje cells die between three and six weeks of age. Surviving Purkinje cells recover normal cytology. Most photoreceptors also degenerate by the 7th week (Sidman & Green 1970, Landis 1973, Mullen & LaVail 1975)
Purkinje cell degeneration (pcd)	autosomal recessive (13)	Moderate ataxia of gait. Degeneration of all Purkinje cells between the fifteenth day and third month of age. Photoreceptors in the retina degenerate at a slower rate (Mullen et al 1976, Landis & Mullen 1977)
Reeler (rl)	autosomal recessive (5)	Reeling ataxia of gait, dystonic postures, and tremors. Systematic malposition of neuron classes in forebrain and cerebellum. Small cerebellum with reduced complement of granule cells (Falconer 1951, Rakic 1976b, Mariani et al 1977; for cerebral cortex, Caviness 1977)
Staggerer (sg)	autosomal recessive (9)	Ataxia with tremors. Dendritic arbor of Purkinje cell is dysplastic with few spines. No Purkinje cell–parallel fiber synapses. Eventual degeneration of granule cells (Sidman et al 1962, Sotelo 1975a, Landis & Reese 1977)
Weaver (wv)	autosomal recessive (?)	Homozygote: ataxia, hypotonia, and tremor. Cerebellar cortex reduced in volume. Most cells of external granular layer degenerate prior to migration. Bergmann-glial fibers hypoplastic. Heterozygote: no behavioral abnormality. Cerebellum slightly reduced in size. Incomplete migration of some granule cells (Lane 1964, Rezai & Yoon 1972, Rakic & Sidman 1973b,c, Sotelo 1975b)

classes of neurons. In the reeler, neurons of all classes survive even though they may be in abnormal positions and some neuron classes may be diminished in number.

The primary objective of this review is to survey variations in cell position, in cell form, and in cell-cell interconnections in three of the cortical mutants: reeler, weaver, and staggerer (Figure 1). Certain structural variations observed in these three mutants are systematically related and hence of considerable interest since they point to specific cellular interactions in development that may be critical determinants of form and synaptic organization in normal cortical structures.

The cortical malformations of these three mutants have been analyzed by a number of current anatomical methods, and the observations that can be drawn from them are complementary. The usefulness of the other mutants listed in Table 1 for the study of cortical development is not apparent at the present time and must await more detailed anatomical analysis. An abbreviated description of the behavioral and morphological characteristics of each mutant is provided with an introductory bibliography in Table 1 together with some notes on the pathogenesis of the cortical malformation.

ATTRIBUTES OF CELL CLASS AS THEMES FOR ANALYSIS

Cortical structures in the mammalian central nervous system are formed of literally billions of neurons and glial elements, divisible into multiple cell classes (Ramón y Cajal 1911, Peters, Palay & Webster 1976). As reviewed elsewhere (Sidman & Rakic 1973, Rakic 1975a), during development neurons of a given class achieve their characteristic positions, both through migration and through rearrangements in cell position subsequent to migration. Once the cells achieve their definitive position and begin to grow, the cytologic features characteristic of each cell class emerge (e.g. Ramón y Cajal 1911, Morest 1969a,b, Hinds 1972a,b). The developing young neuron deploys its dendritic and axonal processes so as to achieve class-characteristic patterns of afferent and efferent connections with neurons of the same and other classes. Thus, these three attributes of cell class: position, form, and patterns of connectivity may be taken as the principal morphological manifestations of a neuron's developmental history (Caviness 1977). In the present review, these three attributes of cell class will form the basis of a comparative description of normal neural structure and certain mutationally induced abnormalities.

DEVELOPMENTAL EVENTS LEADING TO NEURONAL POSITION IN THE CEREBRAL CORTEX

The neurons of most cortical structures in the mammalian brain are formed in generative zones that are remote from the definitive locations that the cells occupy in the adult animal (Sidman & Rakic 1973, Rakic 1975a). For example, neurons destined for the neocortex, for olfactory cortical structures, and for Ammon's horn are generated in the ventricular and subventricular zones of the forebrain, while Purkinje cells in the cerebellar cortex are formed in the ventricular zone along the dorsal margin of the fourth ventricle. On the other hand, some of the granule cells of the dentate fascia of the hippocampal formation, are generated in situ (Schlessinger, Cowan & Gottlieb 1975, Nowakowski & Rakic 1978), and the different classes

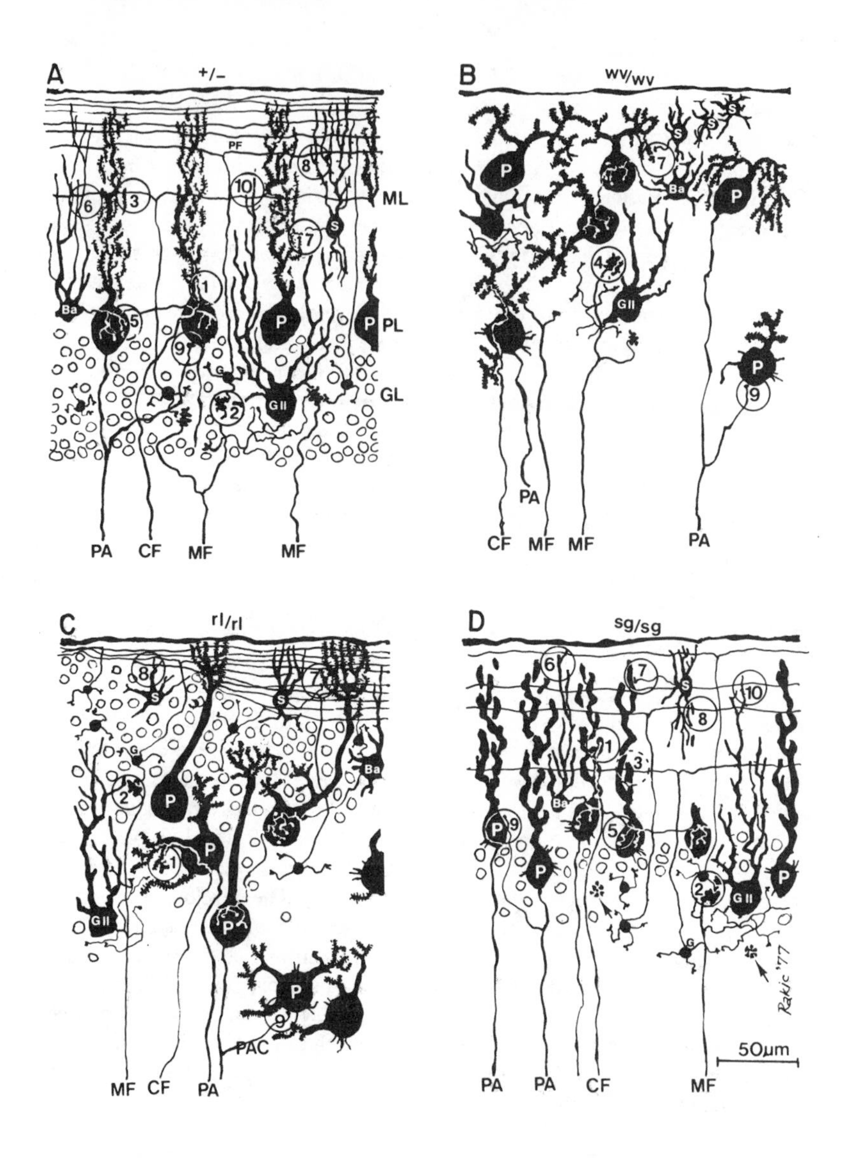

Figure 1 Composite semi-schematic drawing of the neuronal arrangement and synaptic circuitry of the normal (*A*), homozygous weaver (*B*), reeler (*C*), and staggerer (*D*) cerebellum (based on Rakic 1976b, and D. Landis, personal communication). The shapes of neuronal silhouettes are drawn from Golgi preparations, and the positions of unimpregnated granule cells are outlined. All sections are oriented longitudinal to the folium and drawn at approximately the same magnification. *A.* Normal cerebellum of a 3-week-old mouse (C57BL/6J +/+). *B.* Cerebellar cortex of a 3-week-old homozygous weaver mouse (C57BL/6J wv/wv) in a parasagittal plane where granule cells are absent. *C.* Midsagittal outline of the 3-week-old reeler mouse (C57BL/6J rl/rl). The area represented in the drawing is in an area of transition between the relatively well-organized cortex where the molecular layer contains properly oriented parallel fibers (right side) and an abnormal segment of cortex with numerous granule cells situated close to the pia above the Purkinje cells (left side). *D.* Cerebellar cortex of

of stellate cell in the cerebellar cortex are generated at the interface between the external granular layer and the developing molecular layer (Rakic 1972b, 1973).

On completing their final mitoses, most cortical neurons migrate substantial distances across complex terrains in order to reach their destinations within the cortex (Rakic 1971a,b, 1972a). Here the neurons become segregated by class into laminae at successively more superficial levels (Caviness 1976a,b). Analysis of two mutants, the reeler and the weaver, has led to the identification of at least two independent mechanisms that govern the final positions of cortical neurons: (*a*) cell migration that delivers neurons to the cortex and (*b*) the segregation of neurons into class-specific laminae (Caviness 1977).

Cell Migration

Neurons in several forebrain cortical structures and the granule cells of the cerebellar cortex have been observed, during their phase of migration, to remain tightly apposed to the radially disposed processes of glial cells (Rakic 1971a,b, 1972a). In the cerebellar cortex of reeler, these glial fibers are often deployed at anomalous inclinations that may be quite oblique (see Figure 6*B* in Rakic 1976b). Nevertheless, electron micrographs indicate that in reelers, as in their normal littermates, the migrating granule cells are always apposed to glial fibers (Figure 2). This observation strongly implies that this type of apposition is not an incidental association and that in normal animals it is not merely a coincidence of common radial alignment. Rather, it seems to be due to a unique cellular interrelationship that is essential for neuronal guidance across an extended and complex terrain through which the cell might otherwise not be able to navigate (Rakic 1971a, 1972a, 1975a).

The cellular mechanisms by which apposition of migrating neuron and glial fiber is achieved and by which the young neuron is impelled to "scale" the glial fiber are unknown. Whatever their nature, one, or both, of these mechanisms is disrupted in weaver mice. In this mutation the migration of a single neuronal population, the cerebellar granule cells, is affected. In the heterozygous condition of this mutation, granule cell to glial fiber apposition is achieved, and the migration of granule cells from the external granular layer toward the internal granular layer proceeds, though at a retarded rate (Rezai & Yoon 1972, Rakic & Sidman 1973a,b). However, many granule cells terminate in heterotopic positions in the molecular layer and never reach the internal granular layer. Despite their heterotopic positions, the granule

2½-week-old staggerer mouse (C57BL). Many granule cells are still present at this age, although many are in the process of degeneration (arrows). Note the presence of all classes of synapses except the class between parallel fiber and Purkinje cell dendritic spines (broken circle marked by number 3). Abbreviations: Ba, basket cell; CF, climbing fiber; G, granule cell; GII, Golgi type II cell; MF, mossy fiber; P, Purkinje cell; PA, Purkinje cell axon; PF, parallel fiber; S, stellate cell. The major classes of synapses, all identified ultrastructurally, are encircled and numbered: *1,* climbing fiber to Purkinje cell dendrite; *2,* mossy fiber to granule cell dendrite; *3,* granule cell axon (parallel fiber) to Purkinje cell dendrite; *4,* mossy fiber to Golgi type II cell dendrite; *5,* basket cell axon to Purkinje cell soma; *6,* parallel fiber to basket cell dendrite; *7,* stellate cell axon to Purkinje cell dendrite; *8,* parallel fiber to stellate cell dendrite; *9,* Purkinje cell axon collateral to Purkinje cell soma; *10,* parallel fiber to Golgi type II cell dendrite.

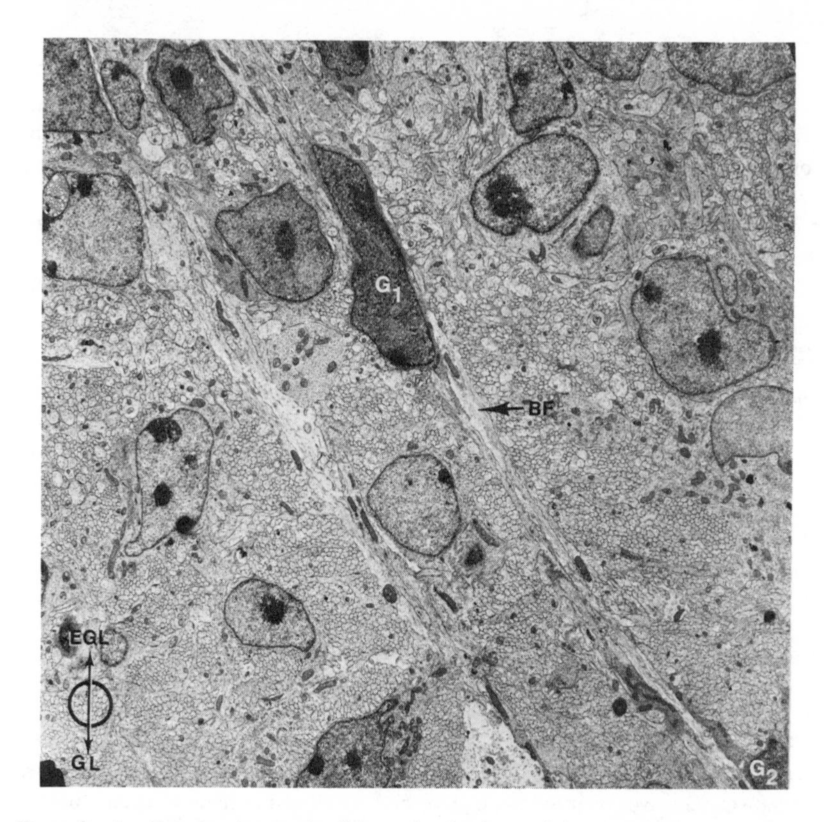

Figure 2 An electron micrograph of the molecular layer of the cerebellum in a 9-day-old reeler (C57BL/6J rl/rl). Two postmitotic granule cells (G_1 and G_2) migrate from the external granular layer situated above to the granular layer situated below. The arrows in the left lower corner indicate the radial axis between external granular layer (EGL) and granular layer (GL). In their descent, migrating granule cells are closely apposed to an obliquely inclined Bergmann glial fiber. X 4,500.

cells are viable, and they develop their class-specific cytologic features and their characteristic afferent and efferent connections (Rakic & Sidman 1973c). In the homozygous condition only an exceptional granule cell migrates to the internal granular layer (Rakic & Sidman 1973b, Sotelo & Changeux 1974b, Sotelo 1975a); the majority die at the interface of the molecular and external granular layers without emitting an axon (an event which normally precedes migration). The few cells that successfully migrate through the molecular layer into the internal granular layer, usually die (Sotelo & Changeux 1974b). It is highly significant that structural abnormalities are also observed in the Bergmann glial fibers in these animals (Rakic & Sidman 1973b, Sotelo & Changeux 1974b). In both the heterozygous and homozygous condition, the usual lamellate expansions seen on the radial glial fibers appear not to be fully developed, the cytoplasmic matrix and organelles are altered or immature, and there are irregularities in the contour of the radial processes.

These observations have led to two quite different views of the mechanism by which the weaver mutation disrupts the migration of granule cells. The first places the greatest emphasis upon the atypical morphological features of the radial glial fibers (Rakic & Sidman 1973b). It has been suggested that these abnormalities in the glial cells impede or disrupt their interaction with the young neurons and that in consequence cellular migration is impaired or aborted. According to the second view it is the neurons themselves that are directly affected by the mutation (Sotelo & Changeux 1974b). When minimally expressed in the heterozygous state, it is only the migratory capacity of the granule cells that is impaired. In the more severe condition, in the homozygous state, there is a more general disturbance of granule cell development as indicated by the failure of the cells to form axons and to migrate along the glial fibers. Ultimately, the disturbance leads to neuronal death.

It is possible that neither view completely accounts for the disruption of granule cell migration and development in the weaver. For example, when dissociated cells from the external granular layer of homozygous weaver animals are reaggregated in tissue culture, the normal tendency of separate aggregates to contact each other by fasiculated glial processes is not observed (Trenkner, Hatten & Sidman 1976) and neuronal death proceeds at an abnormally rapid rate. The serum of homozygous weaver animals appears to contain elevated concentrations of cholesterol and other lipids (Trenkner, Hatten & Sidman 1976). When the concentrations of these substances are reduced by lipid extraction, the viability of the dissociated granule cells is substantially increased. It thus seems possible that the effect of the weaver mutation upon the granule cells may be indirect, and mediated by toxic effects of circulating metabolites upon the glial fibers and/or granule cells.

Postmigratory Neuronal Distribution

RELATIVE POSITIONS OF NEURONAL CLASSES The cortical malformations in the reeler mouse place in focus the cellular events through which neuronal position is achieved at the termination of migration (Caviness 1977). A majority of cortical neurons in the reeler (in contrast to those in the weaver and other cortical mutants) appear to be healthy and remain viable throughout the life of the organism; however, in all cortical structures, with the exception of the olfactory bulb, there is a systematic malposition of neuronal classes in this mutant.

Within the neocortex, and the olfactory and entorhinal cortices, there is an inversion in the relative positions of the polymorphic and pyramidal cell classes (Caviness & Sidman 1972, 1973a,b, Devor, Caviness & Derer 1975, Caviness 1976a,b). Whereas in normal animals polymorphic cells occupy the deepest stratum, while pyramidal cells are found in most of the more superficial layers, the reverse is true in the reeler (e.g. Figure 4*A,B*). This is particularly noticeable in the neocortex and in the piriform cortex, where normally large pyramidal cells are located deeply, while medium-sized and small pyramids are more superficially located; in the reeler the largest pyramids are concentrated superficially, while medium-sized and small pyramids are concentrated in the depths of the cortex (Figure 4*B*). Interestingly, in the neocortex of the reeler, the small stellate (or granule) cells are intercalated between the large and medium-sized pyramidal cells at the midcortical level, as in normal animals.

In Ammon's horn, which consists predominantly of pyramidal cells, there is in the reeler an abnormal degree of radial dispersion of the cells rather than the compact cellular lamina seen in normal animals. Similarly, in the dentate gyrus, the granule cells are abnormally diffused throughout the hilar region (Figure 4*D*). In the cerebellum of the normal animal, the Purkinje cells are arranged in a monolayer at the interface between the molecular and granule cell layers (Figure 1*A*). In the reeler, the majority of Purkinje cells lie at an intermediate depth between the granule cells and the cells of the deep cerebellar nuclei. (Figures 2*C* and 3) (Rakic & Sidman 1972, Rakic 1976b, Mariani et al 1977, So and Caviness, unpublished observation).

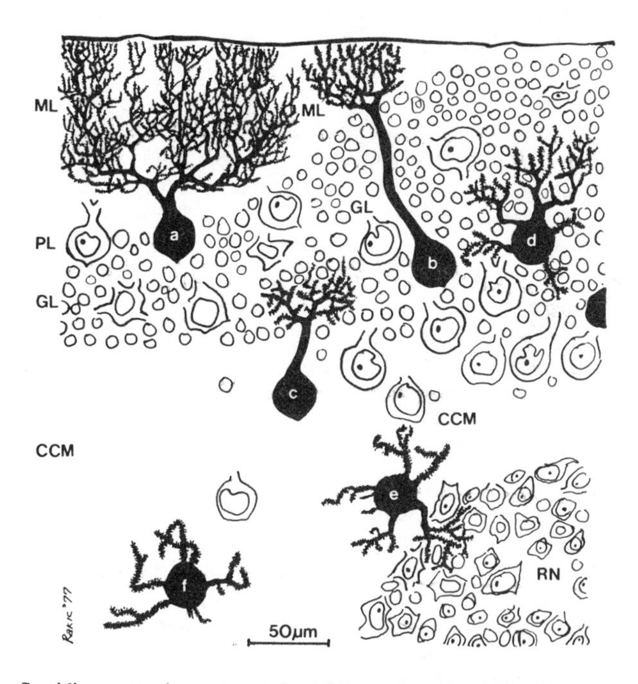

Figure 3 Semidiagrammatic representation of the systematic relationship between the shape and size of the Purkinje cell dendritic arbor and the position of the cell in the reeler cerebellum as visualized in sections cut transversely to the folium. To the left, a few Purkinje cells are in normal positions (PL) with respect to the molecular layer (ML) and granular layer (GL). On the right, most Purkinje cells are situated within or below the granule cells in the central cerebellar mass (CCM). Deep or roof nuclei (RN) lie below the Purkinje cells in the central cerebellar mass. The range of Purkinje cell shapes, based on the extensive study by Mariani et al (1977) is supplemented from our own Golgi impregnations: *a,* Purkinje cell in the normal position; *b,* Purkinje cell with its soma situated in the depth of the granular layer, its single elongated dendritic shaft branches in the molecular layer; *c,* Purkinje cell with soma situated in the central cerebellar mass, its elongated dendritic shaft branches in the granular layer; *d,* Purkinje cell situated among granular cells; *e* and *f,* Purkinje cells with their somata located within the central cerebellar mass. Cells *d, e,* and *f* have multiple primary dendrites. For further explanation, see text.

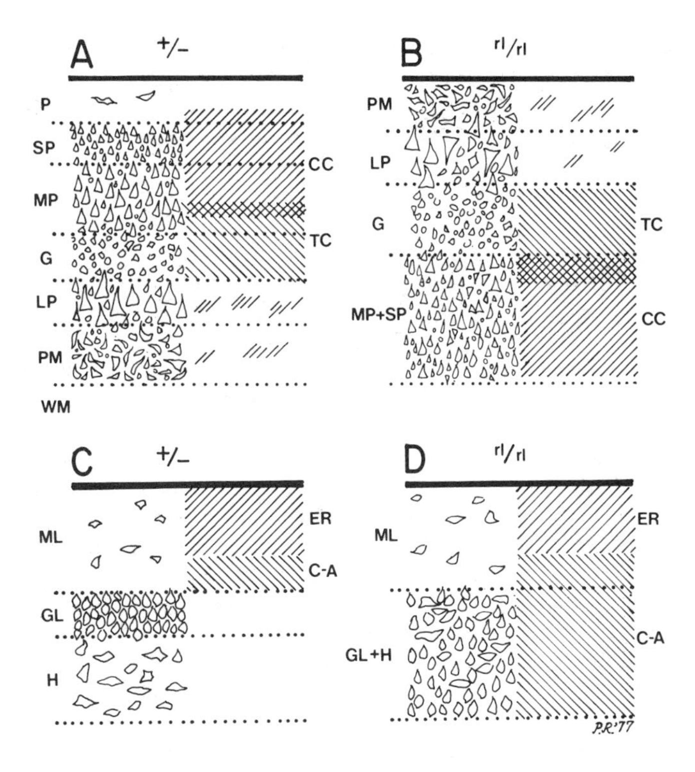

Figure 4 Schematic representation of the distribution of basic neuronal classes and their major afferent systems in normal (+/–) neocortex (*A*) and dentate gyrus (*C*) and in corresponding cortical areas of the reeler (rl/rl) mouse (*B, D*). (Based on studies of Caviness 1976a,b Caviness & Yorke 1976, Stanfield et al 1975, Bliss et al 1976). The left half of column *A* displays the normal positions of plexiform layer (P); layer of small pyramids (SP); layer of medium pyramids (MP); layer of granule (stellate) cells (G); layer of large pyramids (LP); layer of polymorphous cells (PM); and the subcortical white matter (WM). The left half of column *B* shows the relative distributions of the homologous classes of neurons in the corresponding cortical area in the reeler mouse. Note the absence of plexiform layer and misplacement and/or reversal of position of some cell classes. The right half of columns *A* and *B* indicates the territories occupied by callosal (CC) and by thalamo-cortical terminals (TC) in the normal and the reeler mouse, respectively. The left half of column *C* displays the position of the molecular layer (ML), granule cell layer (GL), and hilus (H) of the fascia dentate of the normal mouse. The left half of column *D* shows abnormal mixing of granule cells and neurons of the hilus (GL + H) in the fascia dentata of the reeler mouse. The right half of columns *C* and *D* indicates territories occupied by afferents from the entorhinal area (ER) and of the commissural-association systems systems (C–A) in the normal and the reeler mouse, respectively. The (C–A) in the normal and the reeler mouse, respectively. The distributions of terminals of the TC and CC afferent systems to the neocortex and of the C–A system to the dentate gyrus vary in normal and mutant cortex, depending on the position of their target cells. The position of terminals of the ER afferent system is identical in mutant and normal dentate gyrus and is independent of the position of its target cells.

Although there is a segregation of neurons by class at specific laminar levels within the various cortical structures in the reeler, the compactness, or degree, of segregation of cells in any given class is less than in normal animals. Further, the degree of overlap of adjacent classes is greater; this is particularly conspicuous in the case of the stellate cells of the neocortex, which are abnormally dispersed and overlap the pyramidal cells more widely than normal at midcortical levels (Caviness 1976a,b). Similarly, in the cerebellum a small number of Purkinje cells overlap the granule cell population, and a rare Purkinje cell may even be found in a "normal" position between the granule cell layer and molecular layer (Mariani et al 1977).

SEGREGATION OF NEURONAL CLASSES Neither the events of cell division nor of cell migration appear to contribute to the neuronal malposition found in the reeler. Thus, the cells of each class, in all cortical structures, are generated at the same times in normal and reeler littermates (Caviness & Sidman 1973b, Caviness 1973, Devor, Caviness & Derer 1975, Stanfield 1977, So and Caviness, unpublished observation), and the cells of each class migrate in the sequence in which they are formed and at an essentially normal rate (Sidman 1968). Moreover, the radical glial fibers in the forebrain of the mutant are normally deployed (Caviness 1977, Pinto and Caviness, unpublished observation). In the cerebellar cortex, as mentioned earlier and illustrated in Figure 2, there appears to be no abnormality in the capacity of granule cells to migrate along radial glial fibers despite the obliquity of these fibers (Rakic 1976b).

Anomalies in the patterns of cell segregation first become manifest in the reeler after the neurons have completed their migrations to the presumptive cortical zone. Three different patterns of anomalous cell segregation have been observed: one typical of olfactory and neocortex, one typical of Ammon's horn, and one typical of the cerebellar cortex. In the neocortex and piriform cortex, the pattern of segregation of the earliest formed polymorphic cells is radically different in mutant and normal littermates (Caviness 1977). The first neurons of this class to be generated become post-mitotic on, or before, the eleventh day of gestation (E11). These earliest-formed cells complete their migrations by E13 in both normal and mutant animals, and at this time their post-migratory pattern of arrangement is indistinguishable in the two groups of animals. Over the succeeding 24–48 hr, as later-generated cells migrate into the cortical zone, the so-called cortical plate emerges, between the cell-sparse plexiform zone, above, and the subplate zone, below. Simultaneously, in normal animals, the early-formed polymorphic neurons that have completed their migrations by E13 separate into two subpopulations: a small number of cells retain their superficial position in the plexiform zone, while a larger number come to lie at the level of the "subplate" of Kostovic & Molliver (1974). Thus, the ultimate disposition of this apparently homogeneous cell group forms a two-tiered "scaffold" with tiers of cells at the most superficial and at the deepest levels of the developing cortex. In the reeler mutant, on the other hand, an anomalous plexiform zone emerges at the midcortical level during the same E13–E15 interval. Most of the polymorphic cells that have completed their migrations by E13 remain superficial to this anomalous plexiform zone and because only a rare, early-formed cell comes to lie at the deepest level of the cortex, a two-tiered scaffold does not appear.

A different sequence of events is characteristic of the development of the cytologically simpler Ammon's horn in this mutant (Caviness 1973, Stanfield 1977). In normal animals, the somata of the majority of neurons in this structure lie in a compact single lamina, only a few are scattered in the plexiform zone overlying the principal lamina. In the reeler, the majority of pyramidal cell somata are also positioned in a fairly well-ordered lamina, but the lamina is interrupted and irregular in outline (Caviness & Sidman 1973a). In addition an excessive number of neurons are scattered throughout the molecular layer superficial to the principal cellular lamina. As in normal animals, the superficially scattered neurons in the reeler are the earliest formed, and those in the compact lamina are formed later. However, in the mutant, the period of formation of the cells lying in the molecular layer continues for two days longer than in normal animals (Caviness 1973).

A complex sequence of events leads to the segregation of the Purkinje cells into a monolayer at the interface of the molecular and granule cell layers in the normal cerebellum (Rakic & Sidman 1973b). These events seem to occur in two separate phases. The first phase involves active neuronal migration. The Purkinje cells and neurons destined for the deep cerebellar nuclei undergo their final mitoses during the latter half of the second week of gestation and migrate together into the cerebellar anlage during the third week. Their migrations completed, the Purkinje cells form a diffuse multicellular layer external to the deep nuclei and separated from them by bundles of fibers of unknown origin. The second phase of Purkinje cell segregation occurs later, during the first postnatal week, while granule cell migration is proceeding rapidly. At this time, the Purkinje cells become redistributed into a characteristic monolayer without active cell migration (Rakic & Sidman 1973b).

The primary segregation of Purkinje cells from neurons of the deep nuclei is clearly abnormal in E13-E15 reeler fetuses (So and Caviness, unpublished observations). The Purkinje cells remain closely superimposed upon the deep nuclei as a continuous undulating mass, many cells in depth. A fiber zone appears at a level superficial to the mass of Purkinje cells and not between them and the neurons in the deep nuclei as in normal animals. The majority of the granule cells migrate into positions superficial to this layer of fibers. Therefore, only a relatively small number of Purkinje cells actually come into contact with granule cells and undergo the normal postmigratory redistribution.

THE MECHANISM OF SEGREGATION—AN HYPOTHESIS Our understanding of intracortical neuron segregation in the intact animal has been substantially extended by observations in normal-reeler chimeric preparations (Mullen 1977). In the cerebella of such animals, patches of cortex with the reeler abnormality are distributed as a mosaic in a larger field of cytoarchitectonically normal cortex. The critical observation in these chimeric cerebella is that Purkinje cells of both the normal and the mutant genotype are found in patches of cortex where cell pattern is abnormal. Similarly, Purkinje cells of both genotypes are found in regions of the cortex where cell pattern is normal. These observations imply that "positional instructions," both normal and abnormal, reside outside the Purkinje cells themselves (Mullen 1977); Purkinje cells of both normal and reeler genotypes respond in the same way, whether these instructions are normal or abnormal.

During migration cells move with respect to each other in the radial dimension and the segregation of neurons, by class, occurs along this dimension as the cells reach their definitive positions. The mechanisms that determine the relative position of each class of cell must, therefore, be sequentially distributed along the radial dimension of the developing cortex.

The segregation patterns of different neuronal classes of cortical neurons in normal animals and in reeler mutants might reasonably be viewed as expressions of differential adhesive affinities of the cells. Such a mechanism might be similar in principle to that thought to govern the segregation of heterogeneous cell populations in tissue culture (Sidman 1972, Steinberg & Gepner 1973, Moscona 1974, Caviness 1977). Only the radial glial fibers span the full radial dimension of the cortex during the period when cell position is established. It is thus plausible that differential cellular binding affinities along these fibers determine neuronal class positions. In the hippocampus of the reeler, those neurons with anomalous "positional" behavior are the first to be formed, while those with more "normal" behavior are formed last (Caviness 1973). The anomalous cell positions may, therefore, be due to a delay in the emergence of adhesive mechanisms that designate neuronal class position, rather than absence of such "adhesive clues."

DEVELOPMENT OF NEURONAL MORPHOLOGY

A cell of any given neuronal class in the normal animal shares in common with other cells of the same class certain distinctive morphological features. These include dendritic topology (Hollingworth & Berry 1975, Berry & Bradley 1976), the presence and morphology of dendritic spines, the general shape of the soma, and the general pattern of distribution of the axon. Each neuron of a given class receives the same types of afferent connections (including both inhibitory and excitatory inputs). Each afferent system has a characteristic distribution upon the cell, terminating upon either dendritic spines or smooth portions of the dendrites; on the terminal segments, branch points, or main shafts of the dendrites; upon the cell soma; or even upon initial segments of the axon. Although all neurons within a class share these *qualitative* features, each cell is probably unique with respect to the actual number and lengths of higher order dendritic branches, the number and density of distribution of its dendritic spines, its somatic size, and the exact number and pattern of arrangement of its afferent and efferent synapses. This simple generalization applies not only to the mammalian nervous system, where large numbers of neurons may play equivalent roles in multineuronal assemblies, but also to "relatively simple" invertebrate nervous systems, such as that of isogenic, Daphnia magna, where each neuron has a unique identity (Levinthal, Macagno & Levinthal 1976).

Substantially greater variability is encountered in the morphology of cortical neurons of mutant mice than in normal animals. In general, however, the variations in cell form do not obscure the essential cellular features upon which neuronal taxonomy is based. There may be excessive variation in the *quantitative* aspects of cell form, particularly in the number and length of dendritic branches, and in the density of dendritic spines. In addition, there may be radical alterations in certain aspects of cellular morphology that in normal animals vary relatively little. For example, neurons that normally have only a single primary dendrite may, in the mutants, be multipolar. The disposition of the dendritic arbor, the shape of the dendritic spines, and even the pattern of distribution of "postsynaptic" densities may be grossly abnormal in the cortex of these mutant animals. These abnormal features are illustrated, in different respects, by the morphology of three cell classes in the mutants: the Purkinje cells of the cerebellar cortex, the pyramidal cells of neocortical and allocortical formations, and the granule cells of the dentate gyrus of the hippocampal formation.

The Purkinje Cell of the Cerebellar Cortex

As outlined in an earlier section, and illustrated in Figures 1 and 3, the majority of Purkinje cells in the reeler mouse form a compact mass of cells closely apposed to the deep nuclei. A relatively smaller number are scattered more superficially among the overlying granule cells (So and Caviness, unpublished observations). The variable relationships of the Purkinje cells to the parallel fibers in the molecular layer, or to the granule cell layer itself, appear to be major determinants of Purkinje cell form in the reeler.

PATTERN OF DENDRITIC ARBORIZATION The dendritic arbor of the Purkinje cell, which is strictly planar in normal animals, may in the reeler have several geometric configurations, planar, cylindrical to conical, or spherical to elliptical (Mariani et al 1977). The normal *planar* configuration is seen in the relatively small number of Purkinje cells that are located in the "normal" position at the interface between the granule cell zone and the molecular layer (cell *a* in Figure 3). Second order dendritic branches of a single apical dendrite extend tangentially, in a normal orthogonal relationship to the deepest-lying parallel fibers, and higher order branches ascend through the overlying parallel fibers in a characteristic planar configuration. The *cylindrical* or *conical* (*en parapluie,* Mariani et al 1977) dendritic configuration is characteristic of those Purkinje cells whose dendritic arbors span two compartments. The soma may lie within or below the granule cell layer (cells *b–e* in Figure 3). In either case, a single, apical dendrite generally ascends to the suprajacent compartment where most branching occurs. If the soma lies within the granule cell layer, the dendrite ascends to branch among the parallel fibers (cell *b* in Figure 3); if it is below the granule cell layer, the primary dendrite ascends and branches among the granule cells (cell *c* in Figure 3). A *spherical* or *elliptical* configuration is generated when the Purkinje cell and its dendrites lie wholly within, or completely below, the granule cell layer (cell *d* in Figure 3 and cell *f* in Figure 3). In the latter case, the dendrites may penetrate the deep nuclei (cell *e* in Figure 3). Typically, such radially symmetric cells within the granule cell layer have two or three primary dendrites, while those below may have up to seven (cell *f* in Figure 3). In neither instance is there any indication of preferential growth toward another compartment—neither towards the parallel fibers in the case of the cells within the granule cell layer, nor towards the granule cell layer by those that lie below it.

VOLUME–DENSITY OF DENDRITIC BRANCHING Whether the dendritic configuration is planar or cylindrical, the length of the distal dendritic segments is usually least, and the volume-density (Berry & Bradley 1976) of dendritic branches greatest, when the dendritic arbor lies within the molecular layer. Dendritic segments appear somewhat longer, and the volume-density of branches is somewhat less where the dendritic arbor lies within the granule cell layer. Dendritic segments are very much longer, and the volume-density of dendritic branches is relatively low when the dendrites are subjacent to the granule cell layer.

GRANULE CELLS AS DETERMINANTS OF PURKINJE CELL FORM Those observations, which derive from the study of Mariani et al (1977), identify at least two specific consequences of Purkinje cell–granule cell interactions upon the form of the Purkinje cell dendrites. First, it is evident that the axons of the granule cells provide the optimum stimulus to growth and branching of terminal dendritic segments of the Purkinje cell (see also Berry & Bradley 1976). The stimulus provided by the granule cell layer itself, though important, seems somewhat less potent. Second, the spatial pattern of interaction with the granule cells and their axons appears also to influence the number of primary dendrities on the Purkinje cell. Perhaps because of the differing stimulus potencies that these two elements have upon Purkinje cell dendritic growth, dendrite development is consolidated into a single primary dendrite if the dendritic arbor extends among the parallel fibers from a soma lying within the granule cell layer. This also occurs if the dendrites extend into the granule cell layer from a soma lying below (cells *b* and *c* in Figure 3). Where the cell encounters no such differential in the stimulus to dendritic growth (as when the entire cell lies within, or below, the granule cell layer) multiple primary dendritic processes are formed, and maintained, in the mature cell (cells *d* and *f* in Figure 3).

It is probably significant that when the Purkinje cells lie fully within or fully below the granule cell layer, there is little evidence for a preferred direction of growth towards the parallel fibers or towards the granule cell layer, respectively (compare cells *d* and *f* in Figure 3). Evidently, the cellular interactions that provide the stimulus to growth also require cellular propinquity or actual cell-cell contact. Whatever the nature of the stimulus, it probably does not require synaptic interactions since the formation of synapses generally occurs after the elaboration of the dendritic arbor.

There has, as yet, been no systematic study of the dendritic configurations of other neuronal classes in the cerebella of mutant mice. The overall size of the dendritic arbors of the interneurons of the molecular layer has been observed to be reduced in the homozygous condition of the weaver mutation (Rakic & Sidman 1973c, Sotelo 1975a,b), which suggests that interaction of these neurons with the parallel fibers also serves as a stimulus for the growth of the dendrites of these interneurons (Rakic 1972b).

DENDRITIC SPINES AND THEIR SUBMEMBRANOUS DENSITIES Three different types of spine have been observed on Purkinje cells during the course of their development in the mouse. Late in gestation, the young Purkinje cells bristle with "perisomatic spines." These appendages, that were described originally by Ramón y Cajal (1890), are postsynaptic to the developing climbing fibers. In normal mice these spines are transient and disappear completely during the first two postnatal weeks (Larramendi 1969). As the perisomatic spines disappear two additional types of spine emerge upon the developing dendrites: (*a*) stubby spines that are concentrated upon the primary dendrite and the lower order dendritic segments—these are the definitive sites of termination of climbing fibers in the adult animal; (*b*) small spines that eventually become the more numerous type and are usually distributed on higher order dendritic branches—these spines form asymmetric synaptic contacts with parallel fiber axons. Both of these types of spine persist on the mature cells (Larramendi 1970, Mugnaini 1972, Palay & Chan-Palay 1974).

The Purkinje cells in the weaver and reeler mutants are innervated by climbing fibers. However, all of these neurons in the weaver, and many in the reeler, lack parallel fiber innervation. This results in two synaptic abnormalities. First, a few perisomatic spines persist beyond the first few weeks of postnatal life (Rakic & Sidman 1973c, Sotelo 1975b, Mariani et al 1977); the persistence of these spines cannot be directly attributed to the incomplete parallel fiber innervation, but appears rather to be related to the decreased size of the Purkinje cell somata and their dendritic arbors. Ramón y Cajal (1911) was the first to suggest that the perisomatic spines become incorporated into the somata during the course of normal cell growth. Second, the small spines, replete with normal "postsynaptic densities" (Hanna, Hirano & Papas 1976, Landis & Reese 1977) develop even in the absence of parallel fiber innervation in both the weaver (Hirano & Dembitzer 1973, Rakic & Sidman 1973b, Sotelo 1973, 1975a,b) and reeler mutants (Rakic 1976b, Mariani et al 1977), as they do in a variety of experimental conditions in which the granule cell population is destroyed, or severely depleted (Altman & Anderson 1972, Hirano, Dembitzer & Jones 1972, Herndon, Margolis & Kilham 1971, Llinás, Hillman & Precht 1973). Such "non-innervated spines" and the associated membrane densities are stable and have been observed, enveloped in astroglial processes, in hybrid weaver mice two years of age (Rakic 1976b). The shape of these "naked" spines is extremely variable and often bizarre (Sotelo 1975b, Mariani et al 1977, Landis & Reese 1977). They may be elongated and filamentous, or short and stubby; they may be bifid, or have secondary or even tertiary order bifurcations. Evidently, contact with parallel fibers significantly modulates spine form, even though it is not necessary for spine formation.

SUBMEMBRANOUS DENSITIES WITHOUT SPINES The dendritic spines of Purkinje cells are always associated with submembranous densities not only in normal animals but also in the reeler and weaver mutants. The staggerer and the *pcd* mutants provide evidence that the formation of these submembranous densities is due to a mechanism that is independent of spine formation in the Purkinje cells. In these two mutants there is either an absence or a marked decrease in spine formation. In the staggerer, small spines do not appear until at least the third week of postnatal life and are relatively few in number (Landis & Sidman 1974, Hirano & Dembitzer 1975, Sotelo & Changeux 1974a, Sotelo 1975a, Yoon 1976). In the *pcd* mutant, the development of spines proceeds normally during the first two postnatal weeks (Landis & Mullen 1978); during the third week, as most of the distal dendritic segments are developing, degeneration of the Purkinje cells becomes apparent: spines do not develop on these terminal dendritic segments. In both the staggerer and the *pcd* mutant, the abnormal spine-free dendritic segments are surrounded by parallel fibers. In both mutants, submembranous densities, similar to those normally found on the dendritic spines, appear on portions of the spine-free dendrites. These submembranous densities resemble those found on dendritic shafts of interneurons, that are normally spine-free. However, unlike the densities on interneurons, those on dendritic membranes of the staggerer and the *pcd* mutant are not sites of synaptic engagement with parallel fibers.

Pyramidal Cells in the Cerebral Cortex and Granule Cells in the Hippocampal Formation

CELL ALIGNMENT The full range of pyramidal (and nonpyramidal) cells that is normally found in Golgi preparations of the cerebral cortex is also present in the disordered forebrain cortices of the reeler (Caviness 1976b, 1977). However, there are dramatic anomalies in the alignment, the polarity, and various aspects of dendritic configuration of many of the pyramidal cells, and similar abnormalities are seen in the granule cells of the reeler dentate gyrus.

Many pyramidal neurons in the mutant, particularly those whose apical dendrites span the mid-portion of the cortex, are radially aligned. However, a substantial number (and especially those with somata in the outer half of the neocortex and in Ammon's horn) are inverted (Figure 5*c*). Those with extremely superficial or deep

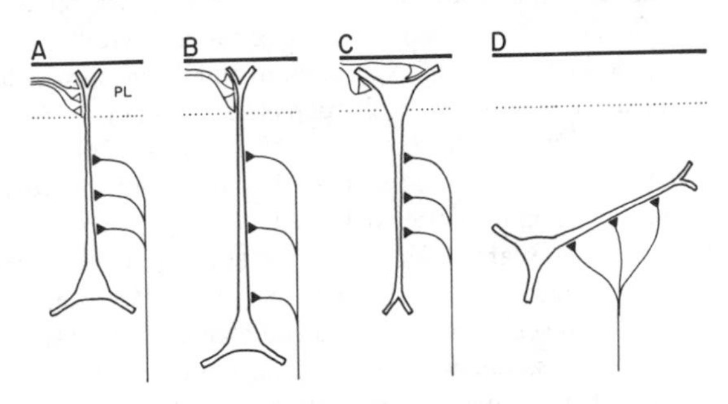

Figure 5 Schematic representation of synaptic distribution of the principal afferent system to the pyramidal cells in the allocortex in the normal and the reeler mouse, based upon observations drawn independently from Golgi impregnations and hodologic studies. Although grossly simplified, the diagram illustrates some of the patterns of afferent mismatching that may occur as a result of cell malposition in the reeler cortex. In each of the four examples, the position of the pia is indicated by a thick horizontal line. The terminal zone of the lateral olfactory tract in the plexiform layer (PL) is delineated by a dashed line below. *A.* Normal relationship in which axons of the lateral olfactory tract innervate the distal segment of the apical dendrite, situated in the plexiform layer (PL), whereas ascending afferents of the ipsilateral association system are distributed uniformly along the proximal segment of the apical shaft. *B.* In the reeler, the cell with its soma lying more deeply than normal in the cortex has an abnormally elongated apical dendritic shaft. Ascending terminals are more widely distributed but at a lower density. In the reeler, a reduced number of axons of the lateral olfactory tract distribute abnormally large boutons to the terminal dendritic branches in the plexiform layer. *C.* An inverted pyramidal cell in the reeler, with the soma situated near the pial surface, receives the same two inputs, but in reverse order: basal dendritic branches synapse with axons from the lateral olfactory tract, whereas apical branches receive input from the ascending association system. *D.* The apical dendrite of a tangentially aligned, deeply situated pyramidal cell is isolated from axons of the lateral olfactory tract. Other class(es) of afferents may be distributed to the terminal segment of its apical dendrite.

somata, on the other hand, frequently deviate to varying degrees from radial alignment; this is particularly true of the more distal segments of the apical processes which usually become increasingly tangential in alignment. Where the pyramidal cells are radially aligned, collateral branches radiate from the apical dendrites, and basal dendrites extend from the cell somata in normal symmetric fashion, and this pattern appears to be independent of the polarity of the cells. At the deepest and at the most superficial levels of the cortex, the smaller dendritic branches of the neuron tend to be deflected asymmetrically in the same direction as the apical process.

From material prepared with general cellular stains, it appears that pyramidal neurons in the reeler, like their homologues in the normal animal, are radially aligned with an ascending polarity as they migrate into the cortex (Pinto and Caviness, unpublished). Statistically, the later the cell is generated in the mutant, the deeper in the cortex will be its final position (Caviness & Sidman 1973b, Devor, Caviness & Derer 1975). This relationship between the time of cell origin and the relative position occupied by the cells in the adult cortex is the reverse of that found in the normal cerebral cortex, where the earliest formed cells assume the deepest positions, while the last-formed cells are the most superficially located (Angevine & Sidman 1961, Sidman & Rakic 1973, Rakic 1974a, 1975b).

Deviations in pyramidal cell alignment and inversions in cell polarity might occur by either one, or possibly a combination of two, different mechanisms. Both would involve interactions between the pyramidal cell and afferent axons that are known to penetrate the cortex at abnormally oblique inclinations, and in some instances even to enter the cortex from the opposite direction to that seen in normal animals (reviewed in a later section). On the one hand, dendritic growth might be aberrantly directed in response to the abnormal direction of the afferent fibers (e.g. see Vaughn, Henrickson & Grieshaber 1974). On the other hand, the entire cell might be pivoted from the normal radial orientation as the cortex expands and the mechanical stresses developing between afferent axons and the dendritic-somatic axis of the cell may bring about a realignment of the cell.

DIRECTED DENDRITIC GROWTH The majority of pyramidal cells in the piriform cortex, and the majority of granule cells in the dentate gyrus are radially aligned and normally polarized in the reeler. However, the length of the apical processes of these cells is very variable and may be much longer than normal (Figure 5*B*). The apical dendrites of most of these cells ascend to the most superficial level of the cortex. The terminal dendritic branches of these apical processes ramify normally in a laminar zone occupied by terminals of axons from the olfactory bulb or from the entorhinal cortex, respectively, even though the distance of the cell somata from this superficial zone of afferent terminals is very much greater than normal for some cells in the mutant (Figure 5). The axons of both afferent systems are already in their normal superficial positions before many of their target cells are generated, or before they achieve their definitive positions (Derer, Caviness & Sidman 1977, Pinto and Caviness, unpublished, Stanfield 1977). The final positions of the cells may be abnormally deep in the mutant and therefore the somata lie at a greater than normal distance from the afferent system. Conjunction of the dendrites and their afferent inputs must come about, then, by this directed growth of the dendrite towards the axons of these superficially located afferent systems, which appear never to penetrate the cortex from their superficial location.

It is possible that these afferent systems are the source of the stimulus to directed dendritic growth. The dendrites of a few granule cells of the dentate gyrus (Stanfield, Caviness & Cowan 1975) and certain pyramidal cells in the piriform cortex (Devor, Caviness & Derer 1975) do not reach their respective superficial afferent system (Figure 5). Such cells are aligned tangentially, deep within the cortex, and there is no evident direction of their dendrites towards the superficially lying afferent system. It is possible that such deeply lying neurons are beyond the effective range of the dendrite directing stimulus from the superficial afferent fiber systems.

FORMATION AND MAINTENANCE OF DENDRITIC SPINES Although the matter has not yet been quantitatively verified, the pattern of distribution of spines upon most cortical pyramids in the cerebral cortex of the mutants appears to be normal. The spines on the granule cells of the dentate fascia, on the other hand, are distributed anomalously along the length of their dendrites and also appear on the cell somata. This is true, even when the somata lie deeply within the hilus of the dentate gyrus (Bliss, Chung & Stirling 1976). Thus, in the reeler spines are not confined to those segments of the dendritic arbor lying within the molecular layer—as is the case in normal animals (Stanfield, Caviness & Cowan 1975, Bliss, Chung & Stirling 1976). Commissural and ipsilateral association afferents to the granule cells invade the hilus in an anomalous fashion in the reeler. Evidently, the spines develop and are maintained upon the intrahilar dendritic segments and somata of these cells in the reeler in response to innervation by commissural and ipsilateral association, inputs, because they disappear if these afferents are surgically interrupted (Bliss, Chung & Stirling 1976).

DEVELOPMENT OF CONNECTIONS

Two obstacles to the development of normal cortical circuitry are posed by these genetic mutations. First, there is the malposition of the principal cortical neuronal classes in both the forebrain and cerebellum in the reeler; this tests the developmental interdependence of normal circuitry and normal spatial arrangement of cellular elements. Second, in the cerebella of the reeler, weaver, and staggerer, an entire afferent system, the parallel fiber system, may fail to form synapses with many Purkinje cells. These cellular abnormalities examine the potential for heterologous synaptic rearrangements among cerebellar cortical neurons.

Topology of Organization of Afferent Systems

Regional variations in cortical width, cell packing density, and the relative prominence of homologous cell and fiber elements are closely parallel in the cortical structures of the forebrain in normal and reeler mice (Caviness & Sidman 1972, 1973a, Caviness 1975, 1976a). That is, by cytoarchitectonic and fiberarchitectonic criteria, normal and mutant cortical structures appear to be homologous and topologically identical. The organization of the major extrinsic cortical connections in the reeler forebrain also appear to be topologically normal. The evidence for this comes from extensive surveys of neocortical (Caviness 1977), hippocampal (Stanfield, Caviness & Cowan 1975, Bliss & Chung, 1974, Bliss, Chung & Stirling 1976), olfactory (Devor, Caviness & Derer 1975), and cerebellar (Steindler 1975, So and Caviness, unpublished) afferent connections in the reeler mutant. The precision of the normal topologic organization of the various cortical afferent systems in the reeler is convincingly illustrated by the patterns of distribution of the callosal and thalamo-cortical systems. The distribution of callosal terminals appears to be identical in the normal and mutant neocortex; connections between the two hemispheres are homotopic and symmetric (Yorke & Caviness 1975, Caviness & Yorke 1976). Their distribution is heavy and homogeneous in the frontal fields, extremely sparse in the central regions of the postero-medial barrel subfield and in area 17. Callosal terminals are densely concentrated at the boundary between areas 17 and 18a at the 3–4 boundary, and at the boundary between field 1 and the medial, smaller limb of field 3.

Similarly, the distribution of thalamic afferents appears to be normal in the reeler (Frost & Caviness 1974, Colwell 1976, Steindler 1975, 1976). For example, the dorsal lateral geniculate nucleus (LGd) projects to area 17 (Frost & Caviness 1974). The rostral part of the LGd projects caudally, whereas the medial aspect projects laterally in field 17. Physiological observations (Dräger 1976, Mangini and Pearlman, personal communication) have established that the visual field is represented normally within area 17 of the mutant. Similarly, the ventrobasal nucleus projects to the barrel field within the somatosensory cortex (Caviness, Frost & Hayes 1976, Welt & Steindler 1977, Steindler 1977). Within the barrel field of both the mutant and the normal mouse, this projection is distributed as a tangential mosaic with clusters of terminals that are coextensive with the cellular barrels (Caviness, Frost & Hayes 1976). It is of note that studies of the parietal (Colwell 1976, Steindler 1975, Steindler 1977) and occipital (Caviness 1977) cortical areas indicate that these regions in the mutant, as in normal animals, are reciprocally connected with their principal thalamic projection nuclei.

Comparative Cellular Organization of Connections in the Reeler and Normal Cortex

The relative position of neurons in the radial dimension of the cortex is found to be disordered in the reeler. Nevertheless, the cells of origin of the various cortical efferent systems show a characteristic laminar arrangement, as do the target cells of the major extrinsic afferent systems.

CELLS OF ORIGIN OF CORTICAL EFFERENT SYSTEMS The laminar segregation of neurons by class is matched by a corresponding laminar segregation of the neurons of origin of the various cortical efferent systems in the reeler (Caviness 1977), just as in the normal animal (Wise 1975). For example, the medium-sized pyramids that are superficially placed in the normal mouse cortex, but in the depths of the cortex in the reeler, are the principal neurons of origin of callosal axons

(Yorke & Caviness 1975, Caviness 1976b). Polymorphic and large pyramidal neurons, which lie in the deepest strata of the normal cortex but are situated superficially in the reeler, provide the major subcortical projections such as those from the visual cortex to the LGd and to the superior colliculus (Caviness 1977).

CELLULAR TARGETS OF CORTICAL AFFERENT SYSTEMS The axon terminals of the extrinsic afferent systems in the reeler cortex, as in the normal animal, are concentrated in laminar fashion at characteristic radial levels. However, there are important similarities and differences in the patterns of distribution of axon terminals in the normal and reeler cortex. These suggest that two quite different mechanisms govern the distribution of these different afferent systems during development (Devor, Caviness & Derer 1975, Caviness & Yorke 1976). One mechanism appears to deliver *axons* to the appropriate radial level of the mutant and normal cortex independently of the positions of the target cells in the two genotypes (ER in Figure 4*C,D*). The other mechanism appears to deliver the *terminals* of the afferent axons to different laminar levels of the mutant and normal cortex. This second mechanism is significantly influenced by the position of the target cells that are different in the two genotypes (Figure 4*A,B*).

AFFERENTS WITH TERMINALS AT IDENTICAL CORTICAL LEVELS Terminals of the following afferent systems are found at identical laminar levels in the reeler and normal cortex, despite dramatic differences in the positions of their target cells in the two genotypes: the olfactory bulb and entorhinal projections in the forebrain and the parallel fibers in the cerebellar cortex. Thus in both reeler and normal animals, afferent fibers from the olfactory bulb are gathered in the lateral olfactory tract and are distributed within the most superficial strata of the various olfactory cortical structures. Short terminal branches of the olfactory bulb axons are rigidly confined to an immediately subjacent laminar zone. Similarly, perforant tract fibers emanating from the entorhinal cortex enter the molecular layer of the dentate gyrus and are distributed in laminar fashion to the most superficial two-thirds of the molecular layer of this structure in normal animals and across its entire thickness in reelers (Figure 4*D*). In the cerebellum the parallel fiber axons are deployed exclusively in the molecular layer. In each of these three afferent systems, the spatial domain of the axon terminals is rigidly fixed with respect to the cortical surface (Figure 1, 4*C,D*), and it has been suggested that contact with investing mesenchymal structures or the glial external limiting membrane determines the plane of growth of the lateral olfactory tract (Devor, Caviness & Derer 1975).

AFFERENTS WITH TERMINALS AT DIFFERENT CORTICAL LEVELS The majority of cortical afferent systems is distributed at variable radial levels in both the mutant and normal cortex (Figure 4*A,B*). Axons of these systems penetrate the relevant cortical structures, and their terminals are distributed in strata whose positions in the cortex correspond to the laminar positions of their target cells. Examples of afferent systems belonging to this group include the thalamo-cortical and callosal afferent systems of the neocortex (Figure 4*A,B*), the commissural and ipsilateral association systems of the piriform cortex and of the dentate gyrus, and the mossy and climbing fiber systems of the cerebellar cortex.

In the neocortex, for example, terminals of the callosal system, though widely distributed, are concentrated most densely among small and medium-sized pyramidal cells in the outer half of the normal cortex but in the depths of the reeler neocortex (Figure 4*A,B*) (Yorke & Caviness 1975, Caviness & Yorke 1976). Terminals of the thalamo-cortical afferent systems are most densely concentrated among stellate cells at the midcortical level in both genotypes. They also extend among the medium-sized pyramids that lie in a supragranular position in the normal animal but in an infragranular position in the mutant (Figures 4*A,B*) (Yorke & Caviness 1975, Caviness & Yorke 1976).

The commissural and ipsilateral association fiber systems of both the piriform cortex (Devor, Caviness & Derer 1975) and hippocampal (Figure 4*C,D*) formation (Stanfield, Caviness & Cowan 1975, Bliss, Chung & Stirling 1976) in the mutant have an expanded laminar distribution through the zone traversed by the ascending dendrites of the respective target cells. Their distribution corresponds to the position of the homologous segments of their target cells; these are substantially more elongated in the mutant than in the normal animal.

Finally, the axons of the different mossy fiber systems to the cerebellar cortex terminate most densely in the laminar zone occupied by granule cells, and in the reeler they ramify among the heterotopic granule cells in the molecular layer (Rakic & Sidman 1973c, Rakic 1976b, Steindler 1975, So and Caviness, unpublished observations). During normal development, the climbing fibers terminate most densely within the zone occupied by Purkinje cells at the interface of the molecular and granule cell layers (Larramendi 1969, Rakic & Sidman 1973b). Subsequently they ascend along the growing Purkinje cell dendrites (Ramón y Cajal 1911, Larramendi 1969). In the reeler mouse the majority of the Purkinje cells lies subjacent to the granule cell layer and some of their dendrites descend away from the pial surface (Figure 1*C*) (Rakic 1976b, Mariani et al 1977). Nevertheless, climbing fibers grow along these dendrites. Such a course followed by a climbing fiber axon to a Purkinje cell dendrite is illustrated in Figure 1*C* (encircled synapse marked by arabic numeral 1; see also Rakic 1976b).

AXON TRAJECTORIES OF CORTICAL AFFERENT SYSTEMS Perhaps even more remarkable than the varied intracortical distributions of the terminals of the extrinsic afferent systems are the variations seen in the trajectories by which axons reach their target fields in the reeler. On the one hand, there are afferent systems, like the callosal system of the neocortex, that follow essentially normal trajectories in the reeler mouse. Axons descend radially from their cells of origin to the subcortical white matter, cross the midline in the corpus callosum and continue towards their respective target cytoarchittectonic fields. There they ascend directly to their target cells in a normal radial fashion. Whatever mechanism provides for the guidance of axons of the callosal system, it evidently acts in the same way in both genotypes.

Further, it appears to be effective in the mutant despite the malpositioning of the neurons from which the axons originate, and of the target cells in the opposite hemisphere (Caviness & Yorke 1976).

The trajectories of thalamo-cortical axons and those of the anterior limb of the anterior commissure, unlike those of the callosal system, are dramatically anomalous in the reeler. In both normal and reeler mice, the thalamo-cortical axons ascend from the thalamus in large fiber bundles. In both genotypes, these fiber bundles cross the central white matter and deliver the thalamo-cortical axons into a fiber stratum that courses tangentially in the zone of polymorphic cells. In normal animals, the zone of polymorphic cells and the tangential fiber stratum lie in the depths of the cortex so that the ascending trajectory of the axons is relatively short. In the mutant, the polymorphic cell zone and the associated tangential fiber stratum lie at the most superficial level of the cortex. The ascending fiber fascicles thus have to traverse the full width of the cortex in order to deliver their axons into the fiber stratum. In both the mutant and normal neocortex, the thalamo-cortical axons course in the tangential fiber stratum until they reach their target cytoarchitectonic field; there they penetrate the cortex to reach their target cells by *ascending* in the normal animal and by *descending* in the reeler (Frost & Caviness 1974, Caviness, Frost & Hayes 1976). Similarly, the anterior limb of the anterior commissure courses through a fiber plane lying in the zone of polymorphic cells. Again, this fiber system lies at the most superficial cortical level in the mutant, whereas in the normal animal it is found at the deepest cortical level. At the base of the olfactory peduncle these fibers, in the reeler, traverse the full width of the cortex to decussate in the anterior commissure. The transcortical trajectories of the thalamo-cortical and anterior commissure axons in the reeler have no homologue in the normal animal. These anomalous trajectories might be established by two quite different developmental mechanisms. One mechanism might compel active growth of axons from subcortical levels to the polymorphic cell zone as the first in a succession of specific intracortical objectives. Given the positional rearrangements of cortical neurons that are evident from the earliest hours of cortical plate formation in the reeler (Caviness 1977, Pinto and Caviness, unpublished observations), axon growth along a gradient of diffusible substances might provide for the anomalous transcortical trajectory.

A trajectory through the polymorphic cell layer [which may be a part of the subplate (Kostovic & Molliver 1974)] in the developing normal cortex might govern the distribution of axons to the appropriate cytoarchitectonic field. Though situated at opposite cortical levels, the neurons and synapses in this zone develop precociously but, interestingly, at the same time in normal and reeler embryos (Pinto and Caviness, unpublished observations). Evidence for this derives from the observation that thalamo-cortical axons extend through the polymorphic cell zone before the stellate cells of layer IV (which are their principal target cells) complete their migrations (Lund 1976, Rakic 1976a, Rakic 1977).

There is a plausible alternative mechanism that might account for the growth of thalamo-cortical axons into, and through, the zone of polymorphic cells. Golgi impregnations of the cortex of E15 mouse embryos indicate that some superficially placed, intrinsic cortical neurons have already directed their axons centrally. The initial course of these axons is through the subplate in the normal animal and through the superficial polymorphic cell zone in the mutant (Pinto and Caviness, unpublished observations). The initial deployment of these axons may have occurred even before there was a separation between the inner and outer levels of the cortex of the mutant. It is thus possible that thalamo-cortical topologic equivalence and the bridging axon trajectories are established by these early corticofugal axons. Bundles of thalamo-cortical axons that arise somewhat later in development might simply fasciculate with these cortical efferents.

Local Circuitry—Intrinsic Connections of Visual Cortex

The organization of intrinsic cortical circuitry is less well-known than the organization of extrinsic afferent systems. However, from Golgi impregnations of the visual cortex it seems that the axons of local circuit neurons (Rakic 1976c) and the collaterals of axons of pyramidal neurons are deployed in patterns that are similar in normal and mutant animals despite cell malposition in the latter (Caviness 1976b). For example, the axons of spiny stellate cells lying at midcortical levels in both the reeler and normal cortex have similar distributions in the two genotypes; they ramify mainly within the region of their dendritic arbors, but some axonal branches may continue for short distances among the adjacent pyramidal cell classes. Similarly, the axons of large pyramidal cells in both reeler and normal animals extend out of the cortex. In normal animals they pass directly into the subcortical white matter, giving off a few collaterals in the region of the cell body. In reeler, these cells are located superficially in the cortex, but their axons descend directly to the subcortical white matter and give off few axon collaterals among other cell classes during their transcortical trajectory. On the other hand, the axons of medium-sized and small pyramidal cells in both normal and mutant animals have rich collateral plexuses at the level of the cell body and an extensive collateral system among the cells at other cortical levels. In normal mice, this intracortical circuitry is achieved as the axon descends through the subjacent cortical layers. Since the same cell types lie in the depths of the cortex in the mutant their axons may take an initially ascending course. Eventually they give off collaterals that ascend further to ramify among the overlying cells. However, it should be emphasized that the axon itself eventually makes a hairpin turn to leave the cortex and descend into the subcortical white matter.

This analysis of Golgi material of normal and reeler visual cortex is consonant with physiological studies that have demonstrated that the receptive field properties of many neurons in the primary visual cortex of the mutant are normal with respect to size, shape, binocularity and orientation (Dräger 1976, Mangini and Pearlman, personal communication). These physiological responses clearly depend upon complex sequences of specific connections among the various cortical neurons.

Local Circuitry—Synaptic Arrangements

The neurological mutants considered in this review all show impaired locomotor behavior: ataxic gait, dystonic postures, and tremors. Each has structural abnormalities of the cerebellum, and the abnormal motor functions are presumably related to the disordered cerebellar circuitry. Only the reeler has recognizable structural abnormalities of the forebrain, and the only published study of the behavioral functions of the cerebral cortex in the reeler (Bliss & Errington 1977) indicates that these animals make random responses on alternate trials in a "T-maze," whereas normal animals choose opposite directions on 75–80% of alternate trials.

Abnormalities in neural circuitry, at the synaptic level are presumably the ultimate basis for disordered function in these mutants. Three general types of synaptic abnormality might be proposed. These are: (*a*) afferent fiber–target cell mismatch; (*b*) absence of a specific class of afferent system–target neuron synaptic interrelationships; and (*c*) heterologous synapses, i.e. synapses between an afferent system and a class of target neurons that do not occur in the normal nervous system.

AFFERENT FIBER–TARGET CELL MISMATCH Such mismatches are illustrated diagrammatically in Figure 5. In the forebrain and in the granule cell layer of the reeler cerebellum, afferent system to target cell connections appear to be both homotopic and topologically normal. However, abnormalities in the position, configuration, orientation, or alignment of the target cells may modify the patterns of distribution, or certain quantitative relationships, of a specific afferent system upon individual neurons. For example, the commissural and ipsilateral association afferent systems in the dentate gyrus and piriform cortex are evidently distributed at a reduced density and to abnormally long segments of the ascending dendrites of the relevant cells (Figure 5*B*) (Devor, Caviness & Derer 1975, Caviness and Korde, unpublished, Stanfield, Caviness & Cowan 1975, Bliss, Chung & Stirling 1976).

Where allocortical or neocortical pyramidal cell alignment deviates from the normal radial arrangement or where the cells are frankly reversed in polarity, the cells may either have a reduced or exaggerated synaptic input from a given afferent system (Figure 5*C*). For example, an axon system that is normally distributed to the apical dendrites of pyramidal cells, when they are normally aligned, might be more densely distributed to the basal dendrites of the same cell if they are inverted (Figure 5*C*).

At a higher level of resolution, there might be several types of synaptic anomalies; the two most obvious are errors in the number of cells contacted by a single axon or the number of synapses formed by a single axon with a given neuron. The first of these anomalies has been documented, physiologically, in the cerebella of weaver (Crepel & Mariani 1975, 1976) and reeler mice (Mariani et al 1977). Multiple climbing fiber innervation of Purkinje cells is a transient arrangement in the normal, developing cerebellum; in the adult, the innervation ratio is reduced to 1:1 (Ramón y Cajal 1911, Changeux & Danchin 1976). Superficially, in the adult reeler cerebellum—within and above the granule cell layer where Purkinje cells may be richly innervated by parallel fibers—the climbing fiber–Purkinje cell ratio may also be 1:1. However, Purkinje cells in the agranular weaver cerebellum, and those lying below the granule cell layer in the reeler, retain the "immature" arrangement of multiple climbing fiber innervation.

The second synaptic abnormality in innervation ratio has been documented by electron microscopy in the piriform cortex of the reeler. In both mutant and normal animals projecting axons from the olfactory bulb terminate exclusively upon dendritic spines or terminal dendritic segments, where they form type 1, asymmetric synapses (Caviness 1977, Caviness and Korde, unpublished observations). In the mutant cortex the corresponding axon terminals may be two to three times larger than normal (Figure 5*B,C*) and may contact as many as three to five dendritic spines rather than one or two as in normal animals.

It is important to note that these anomalous features are not unique to the reeler. They are also found in the reconstituted neuropil of sublamina Ia of the piriform cortex some weeks after subtotal olfactory bulb ablation in normal animals (Caviness, Korde & Williams 1977). This suggests that the aberrant synaptic interrelationships seen in the reeler may be either due to a delay in primary innervation or to less efficient synaptogenesis. It is also possible that there is a dynamic equilibrium between the formation and degeneration of synaptic contacts (see also Watson 1976) which in the reeler may be shifted toward synaptic degeneration in response to imbalanced strenghts of the multiple afferent systems that converge on the target cells.

ABSENCE OF A CLASS OF AXON–CELL SYNAPTIC INTERRELATIONSHIPS The complete absence of an entire class of afferent system–neuron interaction has only been unequivocally established in the cerebella of the three mutants considered in this review (Figure 1). In the homozygous condition of the weaver mutation, where all the granule cells of the vermal and paravermal regions degenerate (Rakic & Sidman 1973c, Sotelo 1973, 1975a), and at the infragranular level of the reeler cerebellar cortex (Rakic 1976b) there are no parallel fiber–Purkinje cell synapses. In the cerebellum of the staggerer mouse, the Purkinje cells do not form spines on their tertiary dendritic branches, and no synapses are formed on the smooth dendrites by parallel fibers despite close dendrite–parallel fiber apposition (Landis & Sidman 1974, Sotelo & Changeux 1974a, Sotelo 1975a).

Evidence for the absence of a specific class of afferent-cell synapses is more circumstantial in the disordered forebrain cortical structures of the reeler. A relatively small number of dentate granule cells, and of pyramidal cells in the piriform cortex, are tangentially aligned in the depths of their respective cortices (Figure 5 *D*) (Stanfield, Caviness & Cowan 1975, Devor, Caviness & Derer 1975). The distal segments of the ascending dendrites of these neurons apparently do not achieve appropriate synaptic contact with the entorhinal or olfactory bulb afferent systems, respectively. Nevertheless, the distal segments of their dendrites are richly invested with spines. If spine development in these cells is contingent upon innervation (Globus & Schiebel 1967, Valverde & Estaban 1968, Ryugo, Ryugo & Killackey 1975), it is possible that the relevant synaptic sites have been assumed by some other, as yet unidentified, afferent systems.

HETEROLOGOUS SYNAPSES As reviewed earlier, the Purkinje cells that develop when the granule cells are reduced in number, as in the cerebella of the reeler and weaver mice, form dendritic spines, and these have typical submembranous "postsynaptic" densities. Despite the fact that other excitatory afferent systems, including the climbing and mossy fiber systems, and the various inhibitory systems, converge upon these cells, these Purkinje cell spines generally do not receive presynaptic inputs (Rakic & Sidman 1973c, Sotelo 1975a,b, Rakic 1976b, Mariani et al 1977. Occasionally, a mossy fiber synapse may be observed on a Purkinje cell (Sotelo 1975a, Mariani et al 1977).

Presumably, all classes of synaptic anomaly, including mismatch, absence, and heterologous junctions, contribute to the behavioral disorders seen in the mutant animals. It is perhaps surprising that the bioelectric properties of "abnormally wired" Purkinje cells in weaver and reeler cerebella are remarkably normal (Crepel & Mariani 1975, 1976, Mariani et al 1977). This seeming paradox may also have important implications for the mechanisms that govern the development of neural circuitry. The mechanisms that guide axons through their trajectory and determine the relative positions of their terminals may provide only a rough sketch of the ultimate wiring diagram of the system. The final spatial arrangement, and quantitative details of excitatory and inhibitory synaptic influences upon individual cells (e.g. Llinás 1975) may reflect a trial-and-error "selection" of afferents by the target neurons. This selection is made from a great number of available choices in the local cellular milieu (e.g. see Changeux & Danchin 1976). The criteria for the final "selection" are unknown, but under conditions of abnormal development the synapses selected may lead to a malfunction of the neural system of which the cell is a part.

Acknowledgments

We wish to thank Ursula Dräger, Douglas Frost, Story Landis, Robert Herndon, Richard Mullen, Cecilia Pinto, Alan Pearlman, Constantino Sotelo, Kwok-fai So, Brent Stanfield, Dennis Steindler, Carol Welt, and their associates for access to their unpublished work. We owe a special debt to Carla Shatz, Roger Williams, Richard Nowakowski, Richard Mullen, Kwok-fai So, and Sonal Jhaveri for their constructive comments during the preparation of the manuscript.

Literature Cited

Altman, J., Anderson, W. J. 1972. Experimental reorganization of the cerebellar cortex. I. Morphological effects of elimination of all microneurons with prolonged x-irradiation started at birth. *J. Comp. Neurol.* 146:355–406

Angevine, J. B. Jr., Sidman, R. L. 1961. Autoradiographic study of cell migration during histogenesis of cerebral cortex in the mouse. *Nature* 192:766–68

Berry, M., Bradley, P. 1976. The growth of the dendritic trees of the Purkinje cells in the cerebellum of the rat. *Brain Res.* 112:1–35

Bliss, T. V. P., Chung, S. H. 1974. An electrophysiological study of the hippocampus of the "reeler" mutant mouse. *Nature* 252:153–55

Bliss, T. V. P., Chung, S. H., Stirling, R. V. 1976. Electrophysiological and anatomical observations on the hippocampus of the reeler mouse. *Exp. Brain Res.* 1:235–36

Bliss, T. V. P., Errington, M. L. 1977. "Reeler" mutant mice fail to show spontaneous alternation. *Brain Res.* 124:168–70

Caddy, K. W. T., Biscoe, T. J. 1975. Preliminary observations on the cerebellum in the mutant mouse Lurcher. *Brain Res.* 91:276–80

Caviness, V. S. Jr. 1973. Time of neuron origin in the hippocampus and dentate gyrus of normal and reeler mutant mice: an autoradiographic analysis. *J. Comp. Neurol.* 151:113–20

Caviness, V. S. Jr. 1975. Structure of the visual cortex of the reeler mutant mouse: a Golgi analysis. *Neurosci. Abstr.* 1:101

Caviness, V. S. Jr. 1976a. Patterns of cell and fiber distribution in the neocortex of the reeler mutant mouse. *J. Comp. Neurol.* 170:435–48

Caviness, V. S. Jr. 1976b. Reeler mutant mice and laminar distribution of afferents in the neocortex. *Exp. Brain Res. Suppl.* 1:267–73

Caviness, V. S. Jr. 1977. The reeler mutant mouse: a genetic experiment in developing mammalian cortex. *Soc. Neurosci. Symp.* 2:27–46

Caviness, V. S. Jr., Frost, D. D., Hayes, N. L. 1976. Barrels in somatosensory cortex of normal and reeler mutant mice. *Neurosci. Lett.* 3:7–14

Caviness, V. S. Jr., Korde, M. G., Williams, R. S. 1977. Cellular events induced by ablation of the olfactory bulb in the molecular layer of the piriform cortex of the mouse. *Brain Res.* 134:13–34

Caviness, V. S. Jr., Sidman, R. L. 1972. Olfactory structures of the forebrain in the reeler mutant mouse. *J. Comp. Neurol.* 145:85–104

Caviness, V. S. Jr., Sidman, R. L. 1973a. Retrohippocampal, hippocampal, and related structures of the forebrain in the reeler mutant mouse. *J. Comp. Neurol.* 147:235–54

Caviness, V. S. Jr., Sidman, R. L. 1973b. Time of origin of corresponding cell classes in the cerebral cortex of normal and reeler mutant mice: An autoradiographic analysis. *J. Comp. Neurol.* 148:141–52

Caviness, V. S. Jr., So, D. K., Sidman, R. L. 1972. The hybrid reeler mouse. *J. Hered.* 63:241–46

Caviness, V. S. Jr., Yorke, C. H. Jr. 1976. Interhemispheric neocortical connections of the corpus callosum in the reeler mutant mouse: A study based on anterograde and retrograde methods. *J. Comp. Neurol.* 170:449–60

Changeux, J.-P., Danchin, A. 1976. Selective stabilization of developing synapses as mechanism for the specification of neuronal networks. *Nature* 264:705–12

Colwell, S. A. 1976. Combined anterograde-retrograde tracing of the connections of reeler mouse cortex. II. Thalamocortical-corticothalamic reciprocity. *Anat. Rec.* 184:380

Crepel, F., Mariani, J. 1975. Anatomical, physiological and biochemical studies of the cerebellum from mutant mice. I. Electrophysiological analysis of cerebellar cortical neurons in the staggerer mouse. *Brain Res.* 98:135–47

Crepel, F., Mariani, J. 1976. Multiple innervation of Purkinje cells by climbing fibers in the cerebellum of the weaver mutant mouse. *J. Neurobiol.* 7:579–82

Derer, P., Caviness, V. S. Jr., Sidman, R. L. 1977. Early cortical histogenesis in the primary olfactory cortex of the mouse. *Brain Res.* 123:27–40

Devor, M., Caviness, V. S. Jr., Derer, P. 1975. A normally laminated afferent projection to an abnormally laminated cortex: some olfactory connections in the reeler mouse. *J. Comp. Neurol.* 164:471–82

Dräger, U. C. 1976. Reeler mutant mice: physiology in primary visual cortex. *Exp. Brain Res. Suppl.* 1:274–76

Dräger, U. C. 1977. In *Function and Formation of Neural Systems,* ed. G. S. Stent, Berlin: Dahlem Konferenzen. pp. 111–38

Falconer, D. S. 1951. Two new mutants "trembler" and "reeler," with neurological actions in the house mouse (*Mus* musculus). *J. Genet.* 50:192–201

Frost, D. O., Caviness, V. S. Jr. 1974. *Progr. Abstr. Fourth Annu. Meet. Soc. Neurosci.* p. 217

Globus, A., Schiebel, A. B. 1967. The effect of visual deprivation on cortical neurons: a Golgi study. *Exp. Neurol.* 19:331–45

Goldman, P. S. 1976. Maturation of mammalian nervous system and the ontogeny of behavior. In *Advances in the Study of Behavior,* ed. J. S. Rosenblatt, R. A. Hinde, E. Shaw, C. Beer, 7:1–90. New York: Academic

Guillery, R. W., Casagrande, V. A., Oberdorfer, M. D. 1974. Congenitally abnormal vision in Siamese cats. *Nature* 252:195–99

Hanna, R. B., Hirano, A., Pappas, G. D. 1976. Membrane specializations of dendritic spines and glia in the weaver mouse cerebellum: a freeze fracture study. *J. Cell Biol.* 68:403–10

Hatten, M. E., Trenkner, E., Sidman, R. L. 1976. Cell migration and cell-cell interactions in primary cultures of embryonic mouse cerebellum. *Neurosci. Abstr.* 2:1023

Herndon, R. M., Margolis, G., Kilham, L. 1971. The synaptic organization of the malformed cerebellum induced by perinatal infection with the feline panleukopenia virus. *J. Neuropathol. Exp. Neurol.* 30:557–70

Hinds, J. W. 1972a. Early neuron differentiation in the mouse olfactory bulb. I. Light microscopy. *J. Comp. Neurol.* 146:233–52

Hinds, J. W. 1972b. Early neuron differentiation in the mouse olfactory bulb. II. Electron microscopy. *J. Comp. Neurol.* 146:253–76

Hirano, A., Dembitzer, H. M. 1973. Cerebellar alteration in the weaver mouse. *J. Cell Biol.* 56:478–86

Hirano, A., Dembitzer, H. M. 1975. Fine structure of staggerer cerebellum. *J. Neuropathol. Exp. Neurol.* 34:1–11

Hirano, A., Dembitzer, H. M., Jones, M. 1972. An electron microscopic study of cycasin-induced cerebellar alteration. *J. Neuropathol. Exp. Neurol.* 31:113–25

Hollingworth, T., Berry, M. 1975. Network analysis of dendritic fields of pyramidal cells in neocortex and Purkinje cells in the cerebellum of the cat. *Philos. Trans. R. Soc. London* 270:227–63

Kostovic, I., Molliver, M. E. 1974. *Anat. Rec.* 178:395 (Abstr.)

Landis, D. M. D., Reese, T. S. 1977. Structure of the Purkinje cell membrane in staggerer and weaver mutant mouse. *J. Comp. Neurol.* 171:247–60

Landis, D. M. D., Sidman, R. L. 1974. Cerebellar cortical development in the staggerer mouse. *J. Neuropathol. Exp. Neurol.* 33:180

Landis, S. 1973. Ultrastructural changes in the mitochondria of cerebellar Purkinje cells of nervous mutant mice. *J. Cell Biol.* 57:782–97

Landis, S. C., Mullen, R. J. 1978. The development and degeneration of Purkinje cells in *pcd* mutant mice. *J. Comp. Neurol.* 177:125–43

Lane, P. 1964. Personal communication in *Mouse News Letter* 30:32

Larramendi, L. M. H. 1969. In *Neurobiology of Cerebellar Evolution and Development,* ed. R. Llinas, pp. 803–43. Chicago: Am. Med. Assoc. Educ. Res. Fdn.

Larramendi, L. M. H. 1970. In *Cerebellum in Health and Disease,* ed. W. S. Fields, W. D. Willis, pp. 63–110. St. Louis: Green

Levinthal, F., Macagno, E., Levinthal, C. 1976. In *The Synapse, Cold Spring Harbor Symp. Quant. Biol.* 40:321–31

Llinás, R. 1975. Electroresponsive properties of dendrites in central neurons. *Adv. Neurol.* 12:1–13

Llinás, R., Hillman, D. E., Precht, W. 1973. Neuronal circuit reorganization in mammalian agranular cerebellar cortex. *J. Neurobiol.* 4:69–94

Lund, R. D. 1976. The development of laminar connections in the mammalian visual cortex. *Exp. Brain Res. Suppl.* 1:255–58

Mariani, J., Crepel, F., Mikoshiba, K., Changeux, J.-P., Sotelo, C. 1977. Anatomical, physiological and biochemical studies of the cerebellum from reeler mutant mouse. *Philos. Trans. R. Soc. London* 281:1–28

Meier, H., MacPike, A. D. 1971. Three syndromes produced by two mutant genes in the mouse. *J. Hered.* 62:297–302

Moscona, A. A. 1974. In *The Cell Surface in Development,* ed. A. A. Moscona, pp. 67–99. New York: Wiley

Mugnaini, E. 1972. In *The Comparative Anatomy and Histology of the Cerebellum. Vol. III. The Human Cerebellum, Cerebellar Connections, and Cerebellar Cortex,* ed. O. Larsell, J. Jansen, pp. 201–64. Minneapolis: Univ. Minnesota Press

Mullen, R. J. 1977. Genetic dissection of the CNS with mutant–normal mouse and rat chimeras. *Soc. Neurosci. Symp.* 2:47–65

Mullen, R. J., Eicher, E. M., Sidman, R. L. 1976. Purkinje cell degeneration, a new neurological mutation in the mouse. *Proc. Natl. Acad. Sci. USA* 73:208–12

Mullen, R. J., LaVail, M. M. 1975. Two new types of retinal degeneration in cerebellar mutant mice. *Nature* 258:528–30

Morest, D. K. 1969a. The differentiation of cerebral dendrites: a study of postmigratory neuroblasts in the medial nucleus of the trapezoid body. *Z. Anat. Entwicklungsgesch.* 128:271–89

Morest, D. K. 1969b. The growth of dendrites in the mammalian brain. *Z. Anat. Entwicklungsgesch.* 128:290–317

Nowakowski, R. S., Rakic, P. 1978. Site of origin and route of migration of neurons in the hippocampal region of the rhesus monkey. Submitted for publication.

Pak, W. L., Pinto, L. H. 1976. Genetic approach to the study of the nervous system. *Ann. Rev. Biophys. Bioeng.* 5:397–448

Palay, S. L., Chan-Palay, V. 1974. *Cerebellar Cortex: Cytology and Organization.* Berlin, Heidelberg, New York: Springer. 348 pp.

Peters, A., Palay, S. L., Webster, S. L. 1976. *The Fine Structure of the Nervous System: The Neurons and Supporting Cells.* Philadelphia: Sanders. 406 pp.

Phillips, R. J. S. 1960. "Lurcher." A New gene in linkage group XI of the house mouse. *J. Genet.* 57:35–42
Rakic, P. 1971a. Neuron-glia relationship during granule cell migration in developing cerebellar cortex. A Golgi and electron microscopic study in Macacus rhesus. *J. Comp. Neurol.* 141:283–312
Rakic, P. 1971b. Guidance of neurons migrating to the fetal monkey neocortex. *Brain Res.* 33:471–76
Rakic, P. 1972a. Mode of cell migration to the superficial layers of fetal monkey neocortex. *J. Comp. Neurol.* 145:61–84
Rakic, P. 1972b. Extrinsic cytological determinants of basket and stellate cell dendritic pattern in the cerebellar molecular layer. *J. Comp. Neurol.* 146:335–54
Rakic, P. 1973. Kinetics of proliferation and latency between final division and onset of differentiation of the cerebellar stellate and basket neurons. *J. Comp. Neurol.* 147:523–46
Rakic, P. 1974a. Neurons in rhesus monkey visual cortex: systematic relation between time of origin and eventual disposition. *Science* 183:425–27
Rakic, P. 1974b. In *Frontiers in Neurology and Neuroscience Research,* ed. P. Seeman, G. M. Brown, pp. 112–32. Toronto: Univ. of Toronto Press
Rakic, P. 1975a. In *Morphogenesis and Malformation of the Face and Brain,* ed. D. Bergsma. 9:95–129. New York: Liss
Rakic, P. 1975b. In *Brain Mechanisms in Mental Retardation,* ed. N. A. Buchwald, M. Brazier, pp. 3–40. New York: Academic
Rakic, P. 1976a. Prenatal genesis of connections subserving ocular dominance in the rhesus monkey. *Nature* 261:467–71
Rakic, P. 1976b. In *The Synapse, Cold Spring Harbor Symp. Quant. Biol.* 40:333–46
Rakic, P. 1976c. *Local Circuit Neurons.* Cambridge: MIT Press. 161 pp.
Rakic, P. 1977. Prenatal development of the visual system in the rhesus monkey. *Philos. Trans. R. Soc. London Ser. B.* 278:245–60
Rakic, P., Sidman, R. L. 1972. Synaptic organization of displaced and disoriented cerebellar cortical neurons in reeler mice. *J. Neuropath. Exp. Neurol.* 31:192 (Abstr.)
Rakic, P., Sidman, R. L. 1973a. Weaver mutant mouse cerebellum: defective neuronal migration secondary to specific abnormality of Bergmann glia. *Proc. Nat. Acad. Sci. (Wash.)* 70:240–44
Rakic, P., Sidman, R. L. 1973b. Sequence of developmental abnormalities leading to granule cell deficit in cerebellar cortex of weaver mutant mice. *J. Comp. Neurol.* 152:103–32
Rakic, P., Sidman, R. L. 1973c. Organization of cerebellar cortex secondary to deficit of granule cells in weaver mutant mice. *J. Comp. Neurol.* 152:133–62
Ramón y Cajal, S. 1890. Sur les fibres nerveuses de la couche granuleuse du cervelet et sur l'évolution des éléments cerebelleux. *Int. Monatsch. Anat. Physiol. (Leipzig)* 7:12–31
Ramón y Cajal, S. 1911. *Histologie du Système Nerveux de l'Homme et des Vertébrés.* Paris: Maloine. Reprinted by Consejo Superior de Investigaciones Cientificas, Madrid, 1955
Rezai, Z., Yoon, C. H. 1972. Abnormal rate of granule cell migration in the cerebellum of "weaver" mutant mice. *Dev. Biol.* 29:17–26
Ryugo, D. K., Ryugo, R., Killackey, H. P. 1975. Changes in pyramidal cell density consequent to vibrissae removal in the newborn rat. *Brain Res.* 96:82–87
Schlessinger, A. R., Cowan, W. M., Gottlieb, D. I. 1975. An autoradiographic study of the time of origin and the pattern of granule cell migration in the dentate gyrus of the rat. *J. Comp. Neurol.* 159:149–76
Schneider, G. E. 1969. Two visual systems. *Science* 163:895–902
Shatz, C. 1977. Abnormal connections in the visual system of Boston Siamese cats: a physiological study. *J. Comp. Neurol.* 171:229–46
Sidman, R. L. 1968. In *Physiological and Biochemical Aspect of Nervous Integration,* ed. F. D. Carlson, pp. 163–93. Englewood Cliffs, N.J.: Prentice-Hall
Sidman, R. L. 1970. In *Cell Interactions, Proc. Third Lepetit Colloq.,* ed. L. G. Silvestri, pp. 1–13. Amsterdam: North-Holland
Sidman, R. L. 1974. In *The Cell Surface in Development,* ed. A. A. Moscona, pp. 221–53. New York: Wiley
Sidman, R. L., Green, M. C. 1970. In *Les Mutants Pathologiques chez l'Animal, Leur Interet pour la Recherche Biomedicale,* ed. M. Sabourdy, pp. 69–79. Paris: Editions du Centre National de La Recherche Scientifique
Sidman, R. L., Green, M. C., Appel, S. H. 1965. *Catalog of the Neurological Mutants of the Mouse.* Cambridge: Harvard Univ. Press 82 pp.
Sidman, R. L., Lane, P. W., Dickie, M. M. 1962. Staggerer, a new mutation in the mouse cerebellum. *Science* 137:610–12

Sidman, R. L., Rakic, P. 1973. Neuronal migration, with special reference to developing human brain: a review. *Brain Res.* 62:1–35

Sotelo, C. 1973. Permanence and fate of paramembranous synaptic specializations in "mutants" and experimental animals. *Brain Res.* 62:345–51

Sotelo, C. 1975a. Dendritic abnormalities of Purkinje cells in cerebellum of neurological mutant mice (weaver and staggerer). *Adv. Neurol.* 12:335–51

Sotelo, C. 1975b. Anatomical, physiological and biochemical studies of cerebellum from mutant mice. II. Morphological study of cerebellar cortical neurons and circuits in the weaver mouse. *Brain Res.* 94:19–44

Sotelo, C., Changeux, J.-P. 1974a. Transsynaptic degeneration "en cascade" in the cerebellar cortex of staggerer mutant mice. *Brain Res.* 67:519–26

Sotelo, C., Changeux, J.-P. 1974b. Bergmann fibers and granular cell migration in the cerebellum of homozygous weaver mutant mouse. *Brain Res.* 77:484–91

Stanfield, B. 1977. The morphogenesis of the hippocampus in the normal and reeler mutant mouse. *Anat. Rec.* 187:721

Stanfield, B. B., Caviness, V. S. Jr., Cowan, W. M. 1975. The organization of the afferents to the dentate gyrus of the hippocampal formation in the reeler mutant mouse. *Neurosci. Abstr.* 1:774

Steinberg, M. S., Gepner, I. A. 1973. Are concanavalin A receptor sites mediators of cell-cell adhesion? *Nature New Biol.* 241:249–51

Steindler, D. A. 1975. Distribution of spinal afferents in the cerebellum of the reeler mutant mouse. *Neurosci. Abstr.* 1:516

Steindler, D. A. 1976. Combined anterograde-retrograde tracing of the connections of reeler mouse cortex (I): on neuronal specificity in the mutant nervous system. *Anat. Rec.* 184:540

Steindler, D. A. 1977. Cortical barrels, thalamic barreloids, and trigeminocerebellar projections in normal and reeler mutant mice. *Anat. Rec.* 187:722–23

Swisher, D. A., Wilson, D. B. 1977. Cerebellar histogenesis in the Lurcher (*La*) mutant mouse. *J. Comp. Neurol.* 173:205–17

Trenkner, E., Hatten, M. E., Sidman, R. L. 1976. Histogenesis of normal and mutant mouse cerebellar cells in a microculture system. *Neurosci. Abstr.* 1:1028

Valverde, F., Estéban, M. E. 1968. Prestriate cortex of mouse: location and the effects of enucleation of the number of dendritic spines. *Brain Res.* 9:145–48

Vaughn, J. E., Henrickson, C. K., Grieshaber, J. A. 1974. A quantitative study of synapses on motor neuron dendritic growth cones in developing mouse spinal cord. *J. Cell Biol.* 60:664–72

Volpe, J. J., Adams, R. D. 1972. Cerebro-hepato-renal syndrome of Zellweger: an inherited disorder of neuronal migration. *Acta Neuropathol.* 20:175–98

Watson, W. E. 1976. *Cell Biology of Brain.* New York: Wiley 527 pp.

Welt, C., Steindler, D. A. 1977. Somatosensory cortical barrels and thalamic barreloids in reeler mutant mice. *J. Neurosci.* 2:755–66

Wise, S. P. 1975. The laminar organization of certain afferent and efferent fiber systems in the rat somatosensory cortex. *Brain Res.* 90:139–42

Yoon, C. H. 1976. Pleiotropic effect of the staggerer gene. *Brain Res.* 109:206–15

Yorke, C. H. Jr., Caviness, V. S. Jr. 1975. Interhemispheric neocortical connections of the corpus callosum in the normal mouse: a study based on anterograde and retrograde methods. *J. Comp. Neurol.* 164:232–45

Ann. Rev. Neurosci. 1978. 1:327–61

TROPHIC MECHANISMS IN THE PERIPHERAL NERVOUS SYSTEM

❖11511

Silvio S. Varon
Department of Biology, School of Medicine, University of California, San Diego, La Jolla, California 92093

Richard P. Bunge
Department of Anatomy, School of Medicine, Washington University, St. Louis, Missouri 63110

INTRODUCTION

The word "trophic" implies nutrition, but as generally used in neurobiology it may include any relatively long-term influence that passes from one cell or tissue to another, either during development or in the mature state. Trophic interactions can involve components of the peripheral nervous system (PNS) in two distinct manners, either as *sources* of trophic influences to non-neural tissues or as *recipients* of trophic influences from other cells. The term "neurotrophic" has been most often used in the first sense, to describe the consequences of neuronal activity on the state of peripheral targets (two examples being skeletal muscle and sensory receptors). This review stresses the converse orientation, namely trophic influences that are exerted by either neural or non-neural tissues, but which are *directed to* neurons and supporting cells.

Figure 1 offers a schematic view of the neurons and related tissues which may be concerned with trophic influences in the PNS. The sites where trophic influences are thought to act are indicated with numbers to facilitate reference. Included are the *sensory neuron* interacting with: (*a*) satellite and Schwann cells, herein referred to as "peripheral glia," at sites 7 and 8, (*b*) peripheral tissue at site 2, and (*c*) central neuronal targets at site 1. Certain aspects of the provision of trophic influences *to* this neuronal type are stressed rather than its own complex influence on sensory receptors in its peripheral field of innervation (e.g. Zalewski & Silvers 1973). Also excluded from detailed consideration are the influences that the periphery may exert during early stages of development of this type of neuron in determining the patterns of its connectivity within the central nervous system (for review see Jacobson 1970).

0147-006X/78/0325-0327$01.00

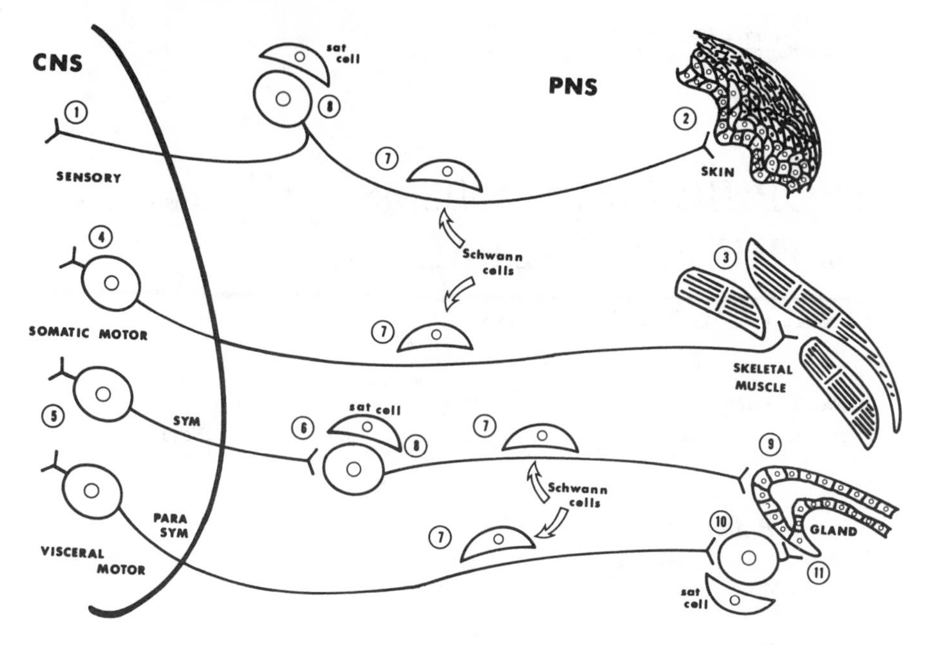

Figure 1 This diagram depicts the many sites (numbers 1–11) at which trophic interactions may occur in the peripheral nervous system. Neurons, their supporting cells (satellite [*sat*] cells and Schwann cells), and targets of the various neuronal types present are shown. Both sympathetic (*sym*) and parasympathetic (*parasym*) portions of the peripheral autonomic nervous system are diagrammed. For details see text.

The *somatic motor neuron* is involved in both anterograde and retrograde trophic interactions (at sites 3, 4, and 7 of Figure 1) that are complex and incompletely understood. Only some of the retrograde relationships are considered here. Motor neuron influences on striated muscle have been extensively reviewed elsewhere (e.g. Harris 1974, Purves 1976a, Fischbach et al 1976, among others). The *visceral motor system,* including both sympathetic and parasympathetic components, is one of our primary points of emphasis. As trophic sources for the ganglionic neurons, one must consider presynaptic neurons as well as peripheral glia and the visceral targets. Because trophic interactions between pre- and postsynaptic neurons in the sympathetic system (site 6) are the subject of a separate review in this volume (Black 1978), our emphasis is on the trophic influences from peripheral tissue and glia (sites 7, 8, 9, and 11). This emphasis derives from the availability of considerable new data and from the lack of any recent systematic review of the interactions at these particular sites.

Biological Situations Involving Trophic Interactions

One of the most dramatic processes in neural morphogenesis is the natural occurrence of neuronal cell death (Cowan 1973). Within a circumscribed period of embryonic life, 50 to 80% of all the postmitotic neurons of a particular region die. While the time at which this occurs varies for different regions and different animal species, its wide occurrence has been verified in amphibian ventral spinal cord (Prestige 1970), chick dorsal root ganglion (DRG) (Hamburger & Levi-Montalcini 1949), cervical visceromotor column (Levi-Montalcini 1950), somatic motor cord (Hamburger 1975), isthmo-optic nucleus (Cowan & Wenger 1968), mesencephalic nucleus of trigeminal nerve (Rogers & Cowan 1973), and ciliary ganglia (Landmesser & Pilar 1974 a,b), among others. Generally, death of the neurons coincides with the time of arrival of their axons in their target territory. In several instances this neuronal death has been demonstrated to be increased by removal of the periphery before it was reached by the nerve fibers and decreased by implantation of tissue, such as supranumerary limb buds, to serve as additional periphery. The lethal effects of peripheral deprivation are less dramatic when the latter is imposed after neural connections are made and are progressively delayed as removal of the periphery is further delayed (Prestige 1970). These repeatedly observed features support the view that death occurs in those neurons whose processes, on reaching their peripheral territory, fail to make permanent connections. This numerical remodeling to fit functional need has been aptly called *neurothanasia* (Hollyday & Hamburger 1976).

By considering these observations on the timing of neuron loss during embryogenesis and extrapolating from the definition of several early stages of neuron development proposed by Prestige (1970), we make reference in this review to four stages of differentiation in neurons comprising the PNS. During *Stage 1* of neuronal development, axonal growth proceeds toward the prescribed periphery, but cell survival is *independent* of the potential availability of target tissue (discussion by Cowan & Wenger 1967). *Stage 2* is characterized by a state of *vulnerability* of the cell, leading to death unless connection is made. *Stage 3* is reached after neuron/target connections have been made; during this period cell *stability* may persist indefinitely. A *Stage 4* of *new vulnerability* will be imposed following disconnection of the neuronal soma from its peripheral target, e.g. by axotomy (Grafstein 1975). At present the most widely supported hypothesis to explain the stability of Stage 3 presumes (*a*) the occurrence of a trophic agent supplied by the periphery, (*b*) the reception of the agent by nerve terminals, (*c*) the transmission of a signal (the trophic agent itself, or a "second message" triggered by it) from the nerve terminal to the neuronal soma, and (*d*) the translation of the signal into the appropriate survival process. It is useful, both in assessing their applicability and in exploring their potential implications, to examine how these concepts would fit several of the biological situations discussed above.

Survival and neurite production by a neuron in Stage 1 may reflect availability of trophic factors *directed to* the neuron, a "neuronotrophic" factor (NTF), at levels adequate for growth but from sources not directly involving the peripheral territory. Transition to the vulnerability of Stage 2 may imply the loss of trophic support to below maintenance level. Effective trophic support requires the concurrence of several components, any one of which could be responsible for the cell vulnerability: (*a*) the earlier source of NTF may have ceased to be available, either through declining production or through reduced accessibility (e.g. following the establishment of a blood-nerve barrier); (*b*) an extrinsic situation has arisen (e.g. arrival of a hormone; see Prestige 1970), which "shifts" the cells to increased consumption activities without compensatory rise of the NTF supply and, as in the previous case, converts the positive production/consumption balance to a negative one; (*c*) a similar increase in the "need" for NTF, even with its supply continuing at the earlier level, may reflect the extension of the neuronal fiber beyond the ability of the soma to sustain it under such a limited trophic drive; (*d*) the increased need for NTF may be imposed on the cell by its encounter with the periphery and, possibly, the very attempt to connect with target cells (Landmesser & Pilar 1976).

Stabilization (Stage 3) requires an interaction between nerve terminal and periphery, which may go beyond mere propinquity. Consolidation of a connection may be required to activate the target cell as an NTF producer or to ensure adequate transfer of NTF to the nerve terminal. A suggested mechanism by which substances may be exchanged between target cell and neuron is provided by the observation (Rees, Bunge & Bunge 1976, among others) that a population of coated vesicles can be observed in certain targets, apparently in transit from Golgi apparatus to the cell surface, specifically to those areas of the plasmalemma where the presynaptic fiber is in contact with the target cell. If these vesicles contained trophic materials their apparent transit route would deliver this material directly to the ingrowing processes of the presynaptic neuron, where it would be available for uptake by endocytosis.

The vulnerability following disconnection from the periphery (Stage 4) may reflect an altered set of influences causing a shift from lower to higher NTF requirements. More simply, it may reveal an altered availability of NTF to the cells (i.e. interruption of its supply route from the periphery, loss of periphery, loss of peripheral glia, or failure of appropriate interactions with these components).

Cell survival, however, has not been the only feature that has been ascribed to the presence of trophic influences. In several cases "trophic" influences have been invoked to explain specific behavioral changes in one cell following its interaction with another. Examples of such situations are: (*a*) peripheral specification or respecification of intracentral connections (reviewed by Jacobson 1970); (*b*) the effect of neuronal contact on the distribution pattern of acetylcholine receptors on muscle cells (Fischbach et al 1976, Purves 1976a); (*c*) the anterograde transsynaptic regulation of transmitter-synthesizing enzymes in sympathetic neurons (Black 1978); (*d*) the altered association of glial cells and presynaptic boutons with the dendrites and somata of axotomized neurons (Grafstein 1975); (*e*) the shift of transmitter synthesis recently observed in cultured autonomic neurons (Patterson 1978). Several of these examples are examined in detail below.

Also relevant to the definition of trophic influences are the "guidance" interactions among neural and nonneural cells. Guidance influences are assumed, throughout and beyond neural development, to underlie cell migration, axonal direction, and selective connectivity. Guidance influences may depend on trophic interactions, for example, if selective pathways are defined by humoral or surface gradients of a trophic agent. Alternatively, guidance influences could themselves constitute a trophic mechanism since changes in intercellular adhesiveness may have serious consequences for the survival and/or selective behavior of a cell (Varon & Tayrien 1977).

In all the situations discussed thus far the object of trophic influences has been the neuron. It has long been appreciated, however, that glial cells also are affected in their maintenance, proliferation, and differentiated behavior by interaction with their neighboring neurons in ways that are equally deserving of the "trophic" appellation. The two best-known situations involving glia-directed signals are the reactive gliosis observed in PNS and CNS injury, and myelinogenesis. We present below a review of the new evidence that a potent mitogenic signal passes from the axon to the proliferating Schwann cell population during PNS development. Because these mechanisms have generally received very little attention until recently, their discussion is taken up in some detail below.

The Concept of Trophic Versus Specifying Influences

"Trophic" influences have thus been invoked in conjunction with cell proliferation, maintenance, death, process elongation, directional guidance, acquisition of functions, and functional activity, whether occurring during development, in the adult state, under pathological conditions, in the course of regeneration, or in vitro. How can one reconcile a single concept, i.e. the trophic influence, with as many such effects as possible, without reducing it to a useless term?

Figure 2 illustrates a schema to define general trophic influences as distinct from influences which specify genetic programs for use within the cell. The balance between anabolic and catabolic activities of a cell (production/consumption ratio) determines whether the cell will be in a deficit, a steady-state, or a growth situation. *Trophic* influences may be defined as those directed to the regulation of this anabo-

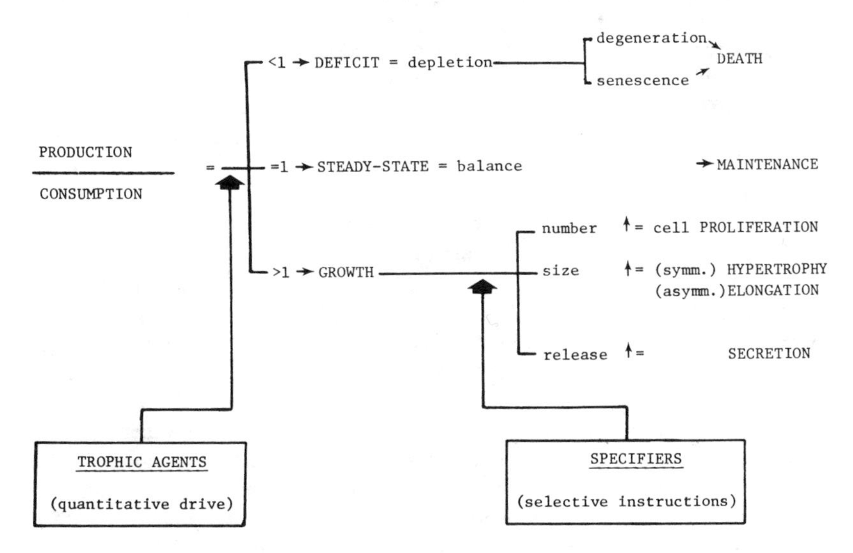

Figure 2 Trophic and specifying influences. A conceptional framework (Varon 1977a).

lic/catabolic balance, mainly through a regulation of production activities. What specific programs are to benefit from a trophic drive is, in turn, determined by *specifying* agents or conditions. According to this scheme, cell death, maintenance, and growth are to be viewed as quantitatively different responses to the same trophic agent when it is available in different amounts. Selection among the various growth modalities (proliferation, elongation, or function-related secretion), acquisition of new functions, or the expression of different behaviors, will not reflect trophic influences except as a source of sufficient drive. Rather, they will result from specifications imposed by stable genetic changes (differentiation) or variable extrinsic influences (modulation). This does not rule out the possibility that a given trophic agent may be instrumental in the expression of selected programs as well. For example, raising anabolic activities above certain "thresholds" could call forth apparently new cell performances. In the course of this review influences which can be defined as either "trophic" or "specifying" are discussed; certain interactions must remain poorly categorized, however, until further experimentation permits their proper classification.

NERVE GROWTH FACTOR (NGF) AS A MODEL

General Features of the NGF Phenomenon

Nerve Growth Factor (NGF) is the term for a particular protein (or closely related group of proteins) discovered by Levi-Montalcini & Hamburger (1951) and found to elicit neurite outgrowth and other responses selectively from neurons of the sympathetic and embryonic dorsal root ganglia. The history of NGF and much of the research on this phenomenon have been periodically reviewed (Levi-Montalcini 1966, Levi-Montalcini & Angeletti 1968, Varon 1975a, Hendry 1976, Bradshaw & Young 1976).

The main source of NGF for experimental work has been the adult male mouse submandibular gland, whose content of other regulatory factors and enzymes is still only partially explored (Levi-Montalcini & Angeletti 1968, Greene, Tomita & Varon 1971). In the gland extract, the NGF agent occurs in a complex, 7S NGF, which contains three different subunits (Varon, Nomura & Shooter 1967, 1968, Smith et al 1969). Only one of these, *beta,* is responsible for the NGF activity, whereas different activities have been described for the other two subunits, *gamma* (Greene, Shooter & Varon 1968, 1969, Greene, Tomita & Varon 1971) and *alpha* (Varon & Raiborn 1972a,b). The individual activity of each subunit is affected, at least quantitatively, by the association with the other two (Varon 1975a). Nevertheless, the functional significance of the 7S complex for the NGF or the other activities remains to be elucidated (Greene, Shooter & Varon 1969, Varon & Raiborn 1972b, Ishii & Shooter 1976). A shorter procedure for the isolation of mouse submandibular NGF (Bocchini & Angeletti 1969) yields a protein (2.5S NGF) almost identical to the beta-NGF (Greene et al 1971, Moore, Mobley & Shooter 1974). Both forms are dimers. The amino acid sequence of the 13,250 dalton monomer chain was found to display numerous analogies with the proinsulin/insulin polypeptides (Frazier,

Angeletti & Bradshaw 1972). The beta-NGF protein has recently been crystallized (Wlodawer, Hogson & Shooter 1975).

The traditionally recognized *target tissues* for NGF are sympathetic ganglia and embryonic dorsal root ganglia (DRG). This restrictive view of NGF targets has been challenged by reports that NGF also elicits responses from cells in the CNS, from catecholamine-containing neurons in vivo (Björklund & Stenevi 1972, Berger, Wise & Stein 1973, Bjerre, Björklund & Stenevi 1974) and in vitro (Bjerre & Björklund 1973), as well as from embryonic optic tectum cells in vitro (Merrell et al 1975). A clonal cell line derived from rat pheochromocytoma has recently been reported to respond to NGF (Greene & Tischler 1976). In addition, DRG neurons have been shown to respond to NGF at postnatal as well as embryonic ages, under appropriate in vitro conditions (Varon, Raiborn & Tyszka 1973, see also below). Finally, NGF also appears to elicit responses in certain nonneural tissues, i.e. embryonic cartilage rudiments (Eisenbarth, Drezner & Lebovitz 1975). It is possible that several of these responsive tissues are not "physiological" targets of NGF but, rather, recognize it as an "analog" of their own trophic agents. On the other hand, a case could be made that NGF is an agent generally directed to cells derived from neural crest (Varon 1968, 1975a)—a category that may encompass most of the responsive cells listed thus far.

The *actual sources* of NGF in vivo and at different stages of development have yet to be firmly identified. There are several potential candidates for such a role (Varon 1975a). Circulating blood and body fluids contain demonstrable traces of NGF (Hendry & Iversen 1973). Peripheral tissues have also been shown to contain minute amounts of NGF by biological or by immunochemical criteria (Angeletti & Vigneti 1971, Johnson, Gorden & Kopin 1971, Hendry & Iversen 1973, Young et al 1976), although it has not always been determined whether they represent production sites or simply storage depots. Implantation into chick embryos of mesenchymal tumors was, in fact, the starting point for the discovery and investigation of NGF (Bueker 1948, Levi-Montalcini & Hamburger 1951), and cultures of normal as well as tumor-forming fibroblasts have more recently been shown to produce and release NGF (Young et al 1976). Glial cells from peripheral ganglia have been proposed as a source of NGF, and NGF production has been reported from glial tumors and cultured glial cell lines, as will be discussed in detail below. Intriguingly, C1300 neuroblastoma cultures also appear to produce and release NGF (Young et al 1976).

Despite these advances in the study of several features of NGF, the mode of action of this factor continues to be poorly understood (Varon 1975a,c). The physical association between NGF and target tissues has come under considerable investigation in the past four years, but it has not been resolved whether NGF acts only on the outside of cells, only inside cells, or in both locations. Information is almost entirely lacking on transduction mechanisms that link the NGF-target cell association with the earliest observable responses of the cells. Finally, target cell responses to the administration of NGF have not yet been properly categorized; to accomplish this it will be necessary to distinguish between behaviors uniquely elicited by the factor and performances intrinsic to the target cell and only quantitatively regulated by NGF.

NGF as a Trophic Agent

One major contribution of the study of neural cells in vitro has been to draw attention to the need for distinguishing between the inherent capabilities of these cells and the extent to which they become expressed in different cellular and humoral environments (for discussion see Varon 1975b, Varon & Saier 1975). Conversely, one must distinguish between extrinsic influences imparting new programs to the target cells (differentiation agents) and those regulating the quantitative expression of existing cell programs (modulation agents).

The key point is that NGF is required for the *survival* of the target neuron. In vitro this has been amply demonstrated both with sympathetic and DRG neurons, and with explant as well as dissociated cell cultures (Levi-Montalcini & Angeletti 1968, 1971, Weis 1970, Burnham, Raiborn & Varon 1972, Varon & Raiborn 1971, 1972c). In vivo a similar demonstration would require all NGF sources to be shut down. Such a situation may occur, at least in some NGF target tissues, on administration of antiserum against NGF, which causes massive destruction of sympathetic neurons—the so-called immunosympathectomy (Steiner & Schonbaum 1972, Zaimis & Knight 1972). In the trophic concept illustrated in Figure 2, death or survival, as well as maintenance and growth, are reduced to a quantitative difference in the availability of trophic factor. In vivo, the same neuronotrophic factor, i.e. NGF, has been reported to promote (*a*) proliferation of neuroblasts if applied at early enough (embryonic) stages; (*b*) hypertrophy and neurite elongation if applied at more advanced developmental stages; and (*c*) hypertrophy if applied to an adult animal (Varon 1975a). The converse situation in the newborn rodent, i.e. a decrease in function-related anabolism, has been illustrated by Hendry (1975), who caused a reduction in tyrosine hydroxylase in the homolateral superior cervical ganglion by unilateral removal of the submandibular gland (an important target tissue for this ganglion) and prevented this reduction by replacing the ablated gland with an artificial NGF depot.

The recognition of the encompassing role of NGF has important conceptual correlates. Experimental investigation of the NGF effects is hampered by the absence of a true control, since the low end of the graded NGF action is death itself. At the upper end of this graded action, and under appropriate permissive circumstances, the target cell will display maximal expression of all its available programs and will therefore be indifferent to the administration of additional (excess) NGF, a situation which may incorrectly lead to the classification of the cell as a nontarget. At intermediate ranges of NGF availability, the target cell will exhibit intermediate, or "immature," behaviors, or even fail to execute programs for which a threshold trophic drive is required. It is, in fact, possible to argue that postmitotic development of certain NGF-dependent neurons might reflect temporal changes in NGF availability rather than a progressive expression of new programs of differentiation.

The Periphery as a Source of NGF

The submandibular gland of adult male mice has the property of storing uniquely large amounts of the factor in its tubular portions. Synthesis of NGF within the gland tissue has been repeatedly reported (Varon 1975a), most recently by Ishii & Shooter (1976). No data are available to support or deny the occurrence of NGF synthesis in the submandibular glands of other species, even though the amount of NGF that they store is very low. In fact, the previously cited study by Hendry (1975) indicates that the rat submandibular gland provides a meaningful supply of NGF to the sympathetic neurons that innervate it. Thus, this gland can be viewed, more generally, as an example of a peripheral source of NGF. The acquisition of NGF from such sources by the innervating neurons presumably involves selective uptake of the protein by the nerve terminals, retrograde axonal transport of NGF or of a "second messenger," and accumulation within the neuronal soma—much as has been postulated above for the protection against naturally occurring neuronal death that is offered by a peripheral territory during development.

A direct demonstration of such a mechanism has been provided for sympathetic neurons upon injection of labeled NGF in the anterior chamber of the eye (e.g. Stoeckel et al 1976), as well as for DRG neurons upon injection of radioactive NGF in the corresponding dermotome (Stoeckel & Thoenen 1975). It should be noted that, in the latter case, the experiments were carried out on adult rats, at an age when DRG are supposedly no longer sensitive to NGF administration. Thus, the ability of NGF target neurons to acquire the factor from the periphery appears to be retained beyond the age at which it may have represented a necessary mechanism. Also noteworthy is the observation by Hendry (1975) that a depot preparation of NGF bound to cellulose was effective in substituting for an ablated submandibular gland as a peripheral source of NGF, which indicates that the retrograde axonal route does not require synaptic connections within the peripheral territory. Finally, treatment with 6-hydroxydopamine or surgical axotomy causes the death of superior cervical ganglionic neurons in postnatal rodents, thus demonstrating their ongoing dependence on peripheral trophic support. Levi-Montalcini (1975) pointed out that, in both situations, NGF administration insures survival of the neurons, elicits massive overgrowth of their neurites, and enhances the syntheses of one of their characteristic enzymes (tyrosine hydroxylase). In addition to providing yet another illustration of the trophic spectrum already discussed, these observations indicate the ability of the sympathetic neurons to receive NGF by means other than uptake at established nerve terminals.

Experiments concerned with sympathetic innervation of the iris offer additional information on the neuronotrophic performance of peripheral tissue. Johnson et al (1972) demonstrated the presence of NGF and described its release by iris in tissue culture. Iris reinnervation by catecholaminergic central axons may occur after iris transplantation into the CNS (Björklund & Stenevi 1972). More recently, Olson et al (1977) and Hoffer et al (1977) have reported that hippocampal tissue, transplanted into the anterior chamber of the eye, will undergo normal morphological and functional development and will become innervated by peripheral sympathetic fibers from the superior cervical ganglion via the host iris. Together, these data offer a reasonable case for (*a*) production of NGF in the iris tissue, (*b*) a tactic role of iris NGF in attracting and distributing sympathetic fibers within the iris territory, and (*c*) some similarity between trophic and recognition mechanisms operating in PNS and those that operate in CNS.

Peripheral Glia as a Source of NGF

DRG neurons have been traditionally regarded as responding to NGF only within restricted prenatal periods. Failure to respond to exogenous NGF need not exclude postnatal DRG neurons from the list of NGF target cells, since the neurons may already be maximally supplied with NGF from other sources. DRG explant cultures also appear not to require exogenous NGF, thus ruling out a dependence on peripheral sources. However, when postnatal mouse DRG were dissociated and seeded in surface cultures, neuronal survival was increased several-fold by the addition of NGF to the medium (Varon, Raiborn & Tyszka 1973). The newly expressed requirement for exogenous NGF was traced to the loss of ganglionic nonneuronal cells incurred during the manipulation. Addition to DRG neurons of increasing numbers of ganglionic nonneuronal cells, even in the absence of exogenous NGF, led to increased neuronal attachment, neurite production, and long-term survival (Burnham, Raiborn & Varon 1972). Quantitatively comparable behavior was elicited by graded concentrations of NGF in the medium, by graded nonneuronal populations on the floor of the culture dish, or by a combination of both (Varon, Raiborn & Burnham 1974a). The NGF-like competence of DRG nonneuronal cells, studied in detail with postnatal mouse cells, was also displayed by DRG cells from fetal mice, chick embryos, and neonatal rats but was not observed with a variety of nonganglionic cell populations (Varon, Raiborn & Burnham 1947b). Purified antibody against beta-NGF blocked the NGF-like activity of ganglionic nonneurons, even when the immune treatment was applied before presentation of the cells to their target neurons (Varon, Raiborn & Burnham 1974c, Varon, Raiborn & Norr 1974). These several observations are best explained in terms of production and delivery by ganglionic nonneuronal cells of a trophic agent that is immunochemically as well as biologically similar to beta-NGF.

Direct demonstrations that glial cells can produce NGF-like proteins have been reported more recently by a number of investigators. Longo & Penhoet (1974) extracted and partially purified protein from a rat glioma grown in vivo, which they described as having biological, electrophoretic, and immunochemical properties characteristic of NGF. Arnason et al (1974) reported a similar factor produced and released by human glioma cells in vitro. Clonal cultures derived from the C6 line of rat astrocytoma cells have been found to release into their medium an agent with NGF-like biological properties (Schwartz, Chuang & Costa 1977, Murphy et al 1977). C6 cells also release a trophic factor that is active on C1300 neuroblastoma cultures but not on normal DRG explants (Monard et al 1975). Finally, Ebendal & Jacobson (1975) have reported that neurite outgrowth from chick spinal and sympathetic ganglia is strongly elicited by monolayers of normal human glia and by the medium conditioned over such cultures. It is interesting to note that, in all these cases, the source of NGF-like agents was glia from central, rather than peripheral, neural tissue, thus encouraging the speculation of a possible role of NGF or NGF-like factors in the CNS as well as in the PNS (Björklund & Stenevi 1972, Schwartz, Chuang & Costa 1977).

Based on the initial study with mouse DRG cell cultures, the hypothesis has been proposed that glia may act as a NGF source in vivo for ganglionic and possibly other neural tissues (Burnham, Raiborn & Varon 1972, Varon 1975a). In more general terms, both peripheral and central glia can be considered as sources of neuronotrophic factors, of which NGF is but an example. In the PNS, connection of neurons with peripheral target tissues takes place before a full complement of glial cells is available to the neurons, and peripheral tissues could be the prevalent source of trophic factors at that time. Later in development, satellite cells become very prominent in both DRG and sympathetic ganglia, and their postulated neuronotrophic effectiveness would increase accordingly. Axotomy, beside interrupting the peripheral connections, is also known to alter the association between glial cells and neuronal somata (Grafstein 1975), and both consequences could concur to deprive the neuron of its needed trophic support (see below). Finally, in the CNS, neurons lack a connection to peripheral tissues, and their only potential source of trophic factors is represented by surrounding glial cells, as well as by other neurons (Varon & Saier 1975).

Delivery Routes and Sites of Action

The occurrence of multiple sources of NGF comes as no surprise if one considers the trophic importance of the factor for its target neurons and the extended span of cell history over which NGF acts. One may tentatively speculate that each source plays a predominant role at a different stage of development—at very early stages before blood-neural barriers develop circulating NGF may serve as a major source; later peripheral targets may become the prime source as connections are made and consolidated; and finally neighboring glia may become more important as neural growth comes to completion and close glial association is established with neuronal somata and neurites. However, the "outdated" sources and supply mechanisms are likely to persist, perhaps even continue to operate at modest levels, and may be ready to resume a predominant role under abnormal demand. This view requires that the site and mode of action of the factor at the target cell level be the same, whatever the sources and delivery routes of NGF.

Figure 3 summarizes the three main putative sources of NGF and their physical relationships to the target neuron. Peripheral NGF, taken up at terminals and transported retrogradely, will reach the neuronal perikaryon, still membrane-bound in endocytotic vesicles (Schwab & Thoenen 1976). In contrast, blood-borne NGF would arrive in the extracellular space in a soluble form, most likely in complex with other circulating proteins (Almon & Varon 1978). Finally, glial NGF will either be released into the extracellular space, like circulating NGF, or will occur immobilized on the glial surface to act by direct contact with the target cell surface. The occurrence of NGF-specific, high affinity binding sites on target tissue membrane has been widely documented (Banerjee et al 1973, Herrup & Shooter 1973, Frazier et al 1974), although several problems with such binding studies have been recognized (e.g. Almon & Varon 1978). Suggestive evidence has been presented that NGF could act after being "immobilized" by covalent coupling to sepharose beads (Frazier, Boyd & Bradshaw 1973), bacteriophage (Oger et al 1974), or red blood cells (Revoltella, Bertolini & Pediconi 1974). DRG cells have been found capable in vitro of both endocytotic uptake and exocytotic release of active NGF, with an involvement of NGF-specific surface binding sites (Burnham & Varon 1973,

Norr & Varon 1975). Attempts to reconcile these several aspects can be made along a number of lines (Varon 1975c), the ultimate choice depending on a better understanding of molecular sites and mechanisms of NGF action than is currently available. The most likely alternatives are:

1. There is only one site of action, and it is intracellular. In this case, NGF-specific binding sites on the cell surface would be merely acting as "selective gates" for the entry of NGF delivered extracellularly. The internalized NGF (by terminal or somal endocytosis) would have to be released from inside the endocytotic vesicle. Intriguing information in this direction has been recently reported (Andres, Jeng & Bradshaw 1977), which suggests that the NGF-containing vesicle might fuse with the nuclear envelope, and either alter properties of the nuclear envelope or release the factor for action on intranuclear molecular targets.
2. The single site of action is extracellular, namely a true NGF "receptor" on the surface of the target cell. In this case, endocytotic uptake of NGF, coupled with its exocytotic release into the perisomal space, would act as an insurance route to guarantee external availability of the factor or as a regulatory mechanism to control its external concentration. The in vitro studies already cited (e.g. Norr & Varon 1975) provide support for the plausibility of such an interpretation.
3. The single site of action is the surface membrane receptor, regardless of its extra- or intracellular location. NGF is likely to be bound to cell surface sites, whether such sites are still facing extracellular fluid at the outer surface of the plasma membrane or at the inner surface of the vesicular membrane. This view, like the previous one, requires that the action of NGF be restricted to the membrane and coupled with transduction mechanisms that will transmit the signal—or its consequences—to the intracellular machinery.

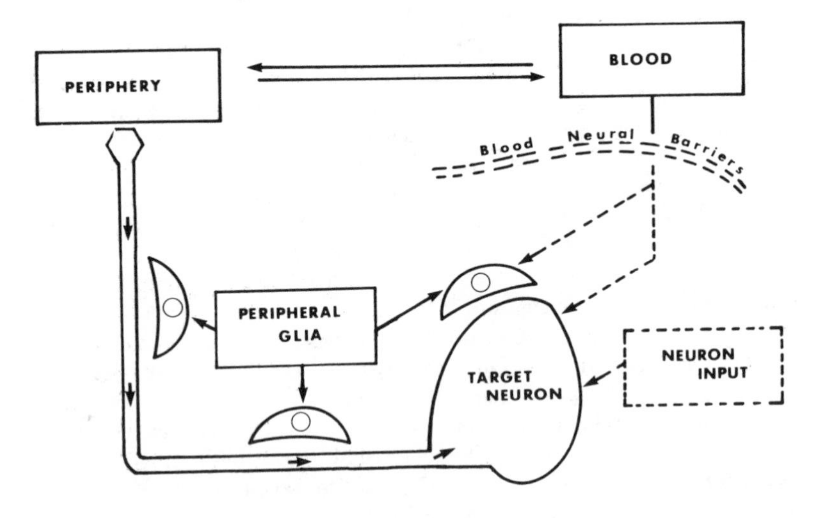

Figure 3 Sources and delivery routes of Nerve Growth Factor.

In all three cases, interaction of NGF with target cell membranes is likely to be the first step of a sequence of events leading to the gross responses currently recognized, i.e. survival and hypertrophy, neurite extension, metabolic changes, production of transmitter enzymes (Levi-Montalcini & Angeletti 1968, Varon 1975a). Given the trophic role of NGF, one must presume that an early consequence of this interaction would be to affect some central function of the cell, which in turn could radiate to and control a variety of cellular activities. Two recent studies (Horii & Varon 1975, 1977) draw attention to rapid changes in membrane permeability elicited by NGF with the shortest latency observed thus far. Other candidates have been suggested for the "central function" that NGF might crucially regulate, although none has yet been adequately substantiated. Microtubule proteins have been proposed to play a role, both as surface receptors for NGF binding and as transducing agents of the NGF-receptor interaction (Levi-Montalcini, Revoltella & Calissano 1974). That cAMP is involved in the action of NGF has been suggested by some investigators (Roisen, Murphy & Braden 1972, Haas et al 1972), refuted by others (Frazier et al 1973, Mizel & Bamburg 1975), and, more recently, reproposed as a very early, transient occurrence (Nikodijevic et al 1975). Cell-cell and cell-substratum adhesion are implicated in the regulation of a variety of cellular activities (e.g. Letourneau 1975), and changes in surface adhesiveness may prove to be an early, primary effect of NGF (Varon, Raiborn & Burnham 1974a,b, Varon & Tayrien 1977). Finally, one should note that permeation properties, cytoskeletal components, cyclic nucleotides, and cell adhesiveness, as well as ionic fluxes and intracellular distribution, are clearly interrelated features; more knowledge of these phenomena will be required before one could identify which of them may be the primary site of NGF action (Varon 1975a).

NGF Function in Neuronal Response to Injury

Recent studies permit a more precise definition of the role of NGF in the steady-state condition of the mature noradrenergic autonomic neuron. Purves (1975, 1976b) has shown that interruption of peripheral signals to a sympathetic neuron, either by axotomy or colchicine application, results in the loss of preganglionic synaptic contacts. These physiological observations extend earlier anatomical observations on synapse loss from axotomized noradrenergic neurons by Matthews & Nelson (1975). Recovery of these contacts takes place if regeneration of the peripheral connection is allowed. Of particular interest to the present discussion is the observation that loss of the preganglionic synaptic contacts is partially prevented by topical administration of NGF to the axotomized ganglion (Purves & Njå 1976), and that systemic administration of antiserum to NGF causes the loss of some ganglionic synapses (Njå & Purves 1977). These findings generally support the view that a trophic influence from the periphery is mediated by an axonally transported supply of an NGF-like substance and further extend the roster of NGF effects to include the regulation of synaptic input.

Chromatolysis (which includes an eccentric positioning of the neuronal nucleus and a redistribution of Nissl material) is another prominent reaction of axotomized noradrenergic neurons, particularly when axon section occurs relatively close to the cell body. West & Bunge (1976) have reported that the chromatolytic reaction in axotomized neurons of the adult rat superior cervical ganglion can largely be prevented by providing substantial amounts of NGF at the site of axonal transection. A single application of 50 μg of 2.5*s* NGF to the region where the severed axons are presumably beginning a process of regrowth prevented the onset of the chromatolytic reaction that normally occurs 3–4 days after axotomy. A biologically inactive analog was ineffective. These authors suggest that isolation of the neuronal soma from its peripheral sources of NGF may be a dominant component in the initiation of the chromatolytic response in adrenergic neurons.

It has generally been observed (Cragg 1970) that section near the peripheral termination of the axon may not result in chromatolysis, and that the severity of the chromatolytic response is proportional to the amount of axon amputated. If the major source of trophic factor were the target itself, the site of axonal section should not be critical. If, on the other hand, the axon receives a trophic supply from sources along its entire length, e.g. the peripheral glial sheath, a direct relationship between the degree of chromatolysis and the amount of axon amputated would be expected. This review has discussed earlier the ability of cultured glia to substitute for NGF in the support of NGF-dependent neurons. It seems reasonable to suggest that the degree of chromatolytic response may reflect the amount of axon amputated because the number of Schwann cells no longer contributing to the NGF supply is changed. Studies of sympathetic neurons allowed to mature in culture with and without accompanying Schwann cells have established the following: (*a*) noradrenergic neurons grown in isolation for 3 weeks in culture will degenerate within 3 days if the NGF component in their medium is withdrawn (Lazarus et al 1976), and (*b*) these neurons, grown in the same culture conditions but provided with Schwann cells, do not die when exogenous NGF is withdrawn; instead, they first undergo a chromatolytic change and then become atrophic (West & Bunge 1977). This type of experiment makes it possible to demonstrate directly that NGF withdrawal may cause a chromatolytic change in the noradrenergic neuron. Observations of this kind are consistent with the already discussed concept of cell death, maintenance, and growth as graded responses to different levels of trophic factor.

NEURONOTROPHIC FACTORS FOR CHOLINERGIC NEURONS

Neuronotrophic Influences Other than NGF

NGF has long been viewed as the one recognized member of a *family* of neuron-directed trophic factors (NTFs), each member of which may be selectively competent to support only certain categories of nerve cells. A successful search for other members of this postulated family of NTFs will depend on the fortunate choice of (*a*) a readily identifiable class of neurons as the target, (*b*) a workable test system, and (*c*) a putative source of the NTF. *Cholinergic neurons* fit the first qualification, in that they are identifiable by acetylcholine-related properties (e.g. choline acetyltransferase, choline uptake, and acetylcholine storage) and can be collected from selected neural regions (e.g. ciliary ganglia, ventral and ventrolateral spinal cord). *In vitro systems* using these neurons could be investigated for response to a putative

NTF in terms both of neuronal survival and/or growth and of neuronal function (transmitter properties). *Peripheral targets* suitable for the chosen category of neurons (e.g. skeletal muscle for ventral cord neurons, striated and smooth muscle for ciliary neurons) represent probable sources of NTF, as do glial cells related to those neurons.

Several recent studies have begun to build a case for the occurrence of cholinergic neuronotrophic factor(s) and/or cholinergic specifying factors and provide approaches to their investigation. The following subsections examine in detail a number of such systems.

The Submandibular Ganglion

One of the most dramatic examples of the direct influence of target tissues on the development of a nerve cell is provided by observations on the early development of the parasympathetic neurons of the mouse submandibular ganglion both in vivo and in vitro. Coughlin (1975a) studied the early axonal outgrowth from the small group of neurons that make up this ganglion in vivo and in vitro, using electron microscopy, vital staining, and histochemical localization of acetylcholinesterase. He demonstrated a direct relationship between the development of an axonal outgrowth from the ganglion and the development of the epithelium of the related submandibular salivary gland. Axonal growth and branching occurs in parallel with the growth and branching of the lobules of the gland. It seems likely that parasympathetic nerve growth in this instance is stimulated and directed by the glandular epithelium. Coughlin (1975b) then extended these observations to an in vitro system that allowed the direct demonstration of the dependence of nerve growth on gland tissue. When cultured alone, the submandibular ganglion showed little axonal extension, but in the presence of the salivary gland epithelium, a dramatic stimulation of axon outgrowth occurred, again patterned according to glandular growth. It was further observed that stimulation of axonal growth could occur if the glandular epithelium and neurons were cultured on opposite sides of a filter, which suggests that direct contact between the epithelium and the nerve was not necessary. The capacity to stimulate axonal outgrowth was to a large extent, but not completely, restricted to epithelium from the salivary gland. Another autonomic ganglion known to contain cholinergic neurons, the parasympathetic ganglion of the pelvic plexus, also showed directed nerve outgrowth in relation to salivary gland epithelium.

These observations indicate that the target tissue for parasympathetic nerve fibers in the body provides some factor that stimulates and directs axonal growth within that target. It is clear from observations in a number of laboratories that this factor is not NGF, for NGF has been shown to have no effect on the growth of parasympathetic nerve fibers (discussed by Coughlin 1975b).

The Ciliary Ganglion

The ciliary ganglia are one of the four pairs of autonomic ganglia topographically associated with the cranial nerves. In the chick, each ciliary ganglion contains some 6000–7000 neurons in two subpopulations: (*a*) larger ciliary neurons innervating striated intrinsic muscles of the eye and smaller choroid neurons innervating smooth muscles in the vascular choroid. Both classes provide cholinergic synapses to their targets. The neurons receive synapses (also cholinergic) from two separate sets of preganglionic fibers that arise from the accessory oculomotor nuclei (Edinger-Westfall) in the midbrain and provide the motor roots of the ganglion. During development preganglionic synapse formation takes place at 5–8 days, before any neuronal death is apparent. Axonal outgrowth from the ganglion occurs along two distinct nerves (ciliary and choroid) and reaches the periphery at about day 8. Between days 8 and 12, as peripheral synapses appear, 50% of the neurons of the ciliary ganglion die. Peripheral deprivation raises the toll to over 90% of the neuronal population but has no effect on the earlier acquisition of preganglionic synapses or on the elongation of axons. Thus, the ciliary ganglion is a classic example of the naturally occurring neuronal death already discussed.

Landmesser & Pilar (1976, Pilar & Landmesser 1976) have provided an excellent analysis of the ultrastructural changes that take place in ciliary ganglion cells during the several phases of their development. At day 8, regardless of the presence or absence of the periphery, the neurons have ovoid somata with eccentric nuclei, a developed Golgi apparatus, sparse smooth endoplasmic reticulum (ER), and scant rough ER. Glial cells are already present. Between 10 and 14 days, all neurons, whether or not fated to die, undergo drastic changes. The nucleus loses its finger-like indentations, free ribosomes aggregate, rough ER increases and organizes into subsurface Nissl-like layers. Peripheral synapses are formed, and production of transmitter enzymes increases. None of these changes take place if the periphery has been removed. Thus, the neuronal switch from an "elongation" to a "secretory" mode depends on contact with the periphery (specifying influence?) and occurs regardless of the fate of neurons and synaptic connections—a possible base for the acquisition of vulnerability that marks Stage 2 as already discussed. The reverse switch, from secretory to elongation modes, is observed if axotomy occurs, and a return to the "secretory" pattern follows successful regeneration of the axotomized axons. In the absence of axotomy, however, the cells naturally fated to die do not undergo the reversion from secretory to elongation patterns. Rather, a gradual dilation of the rough ER (accumulation of secretory products because of nonfunctional synapses?) leads to cytoplasmic disruption, then to nuclear disruption. These data raise the possibility, not previously considered, that programmed cell death may result, not from an insufficient supply of a trophic factor from the periphery, but from the concurrent presence of both trophic drive and secretory-specifying influence, in a situation where the cell is not allowed to use the products of such instructions.

Few studies have yet been carried out with ciliary ganglionic *cultures.* Explant cultures of chick ganglia have been shown to develop a vigorous neuritic outgrowth and to have the ability to form synapses on skeletal muscle cells (Hooisma et al 1975). Dissociated cell cultures, however, have been successfully achieved only recently with chick embryo ciliary ganglia (Helfand, Smith & Wessells 1976). Neuronal survival required both a highly adhesive substratum (polyornithine) and the presence of medium conditioned over heart muscle cell cultures. The heart-conditioned medium supplied a nondialyzable factor, the presence of which was continually required. Even under such conditions, only 50% of the neuronal population of the source ganglia (8 day chick embryo) was retained in culture, and of these more than 90% underwent massive death at about one week. It remains to be ascertained whether the initial loss of neurons and the later death derived from the same or different causes. It is also possible that too little of the factor provided by the heart-conditioned medium was available.

The Spinal Cord

Spinal cord cell cultures from chick or mouse have repeatedly been reported to contain neurons capable of forming synapses on skeletal muscle cells (Shimada, Fischman & Moscona 1969, Fischbach 1972, Peacock, Nelson & Goldstone 1973, among others). Giller et al (1973) reported that co-cultures of fetal mouse spinal cord and skeletal muscle cells exhibited a substantial increase with time in choline acetyltransferase (CAT) activity in addition to an increase in synaptic interactions. This observation is susceptible to several interpretations. Formation of cholinergic synapses could result in increased accumulation of CAT in the presynaptic terminals, with no particular stimulation of CAT synthesis. Alternatively, synapse formation may elicit an increased production of the enzyme by the cholinergic neurons of the culture. Lastly, increased production of CAT may be stimulated by the muscle cells independently of synapse formation. In the light of the concept proposed in Figure 2, a stimulation of CAT production could, in turn, reflect a trophic influence directed to cholinergic neurons or a specifying influence committing cord neurons to a cholinergic output. Additional information provided in the report cited was inadequate to discriminate between these two possibilities, since DNA and protein contents of the cultures, as well as specific activities of the several other enzymes measured, were presumably heavily influenced by the overwhelming presence of noncholinergic elements.

A more detailed study, recently carried out in the same laboratory (Giller et al 1977), adds considerable information on some of these points. CAT stimulation could be obtained in co-cultures treated with α-bungarotoxin, thus ruling out the need for functional neuromuscular synapses. Conditioned medium from skeletal muscle cell cultures was also effective in eliciting a CAT increase from spinal cord cell cultures, which demonstrates that the interaction between muscle and nerve cells was mediated by humoral factors. The active constituents of the conditioned medium were temperature stable, nondialyzable, and of molecular weight(s) greater than 50,000. The increase in CAT activity was linearly related both to the concentration of conditioned medium provided to the cord culture and to the length of the treatment. The response of cord cultures to either conditioned medium or to intact muscle cells could not be saturated with increasing supply of the stimulus. However, the time spent in vitro by the cord culture before treatment was applied was critical since a delay of several days resulted in little, if any, stimulation of CAT activity. This last feature could reflect a progressive decline in the susceptibility of cord neurons to the stimulus or a progressive disappearance from the culture of those neurons that were a target for the stimulus. Future investigations on this question should be directed to culture systems where the majority of the neurons present are potential targets for cholinergic trophic or specifying factors. One such system may be the monolayer culture of ciliary ganglion cells (see preceding subsection). Another such system might be provided by sympathetic ganglionic neurons, as is made clear in the next section of this review.

NEUROTRANSMITTER SPECIFYING INFLUENCES

At some stage in development, a determination must be made regarding the type of neurotransmitter to be synthesized, stored and released by each particular type of neuron. Many questions remain unanswered as to how this determination is made. Among these questions are: How late in its development does the neuron retain the option to choose alternate neurotransmitter types? Is the final decision made at the time of the last mitosis? Is the selection influenced by neuronal interaction with target tissue?

Recent experiments on the development of the autonomic neuron are beginning to provide suggestive answers. These new observations derive from experiments with autonomic neurons in culture, as well as from experimental embryology, and lead to at least one common conclusion: The autonomic neuron may respond to influences from surrounding tissue in making the decision to employ a specific neurotransmitter.

Observations from Experimental Embryology

It now seems generally agreed that the neurons of the peripheral autonomic nervous system, at least in the trunk region, are exclusively of neural crest origin (Weston 1970). Their development is marked by extensive migrations to sites of residence among the peripheral tissues of the body. During this migration the neural crest cells continue to divide and, thus, increase their number. In the early part of their migration these cells pass between the developing neural tube and the adjacent somites. It has been demonstrated that exposure to both neural tube and somite influences the subsequent ability of these cells to express adrenergicity (Cohen 1972, Norr 1973). During migration to the more ventral regions of the embryo, chick neural crest cells must interact with both ventral neural tube and somitic mesenchyme in order to differentiate into neuroblasts exhibiting specific catecholamine fluorescence. These observations demonstrate the important influence of the migratory pathway in directing the expression of neuronal characteristics. In the light of subsequent observations from experimental embryology and tissue culture (discussed below) there remains some question of how firm a commitment to "adrenergicity" the neuron makes on the basis of its experience during migration.

Of direct interest to this discussion are the experiments of Le Douarin & Teillet (1974) that utilize the technique of implanting neural crest from quail into chick embryos. Structural characteristics of the quail interphase nucleus allow recognition of quail cells among chick tissues. When midtrunk neural crest precursor cells normally destined to become the adrenergic cells of the adrenal medulla are transplanted to the "vagal" region, they follow the migratory course of the cells that normally colonize the gut wall, take up the position of cholinergic neurons in the gut wall, and now fail to exhibit their presumptive adrenergic characteristics. Conversely, if precursor cells from the upper cervical region that are normally destined to provide cholinergic parasympathetic neurons for the gut wall are transplanted to the midtrunk region, they will migrate to the region of the adrenal medulla and there exhibit the characteristic fluorescence of adrenergic cells. From these and related experiments these workers conclude that differentiation of autonomic neuroblasts is ultimately controlled by the environment in which these neural crest cells are localized at the end of their migration. Because these neurons are believed to become postmitotic at the time they arrive at their tissue destination, and because the site of this final destination seems to these workers more critical in influencing neuronal differentiation than does the migratory pathway, the postmitotic neuron appears to "decide" which neurotransmitter to employ after interaction with its immediate postmigratory environment. The experimental system provides no evidence regarding the type of local factors that may be involved and whether these factors may be related to the vital maintenance of these neuronal types (trophic function), as well as influencing their differentiation (specifying function). One possible conclusion is that migrating neural crest cells, while specified as autonomic, are not committed to become cholinergic or adrenergic. The latter specification occurs secondarily in response to their final local environment. It is not known whether other neurons have this capability to make a secondary adjustment of their transmitter synthesis in response to local environment. Nevertheless, this phenomenon could be a common one in nervous system development and be undetected, because the nature of the transmitter for most neurons, developing or mature, is not known.

Observations from Experiments in Tissue Culture

Principal neurons of the rat superior cervical ganglion (SCG) can be established as dissociated cells in long-term cultures without the survival of the small, intensely fluorescent (presumably dopaminergic) "interneuron" normally found in the animal (Bray 1970). These isolated sympathetic neurons exhibit expected noradrenergic characteristics such as the synthesis and accumulation of catecholamines (Mains & Patterson 1973). They also specifically take up exogenous noradrenaline and demonstrate CA^{2+}-dependent release of this transmitter during induced depolarization (Burton & Bunge 1975, Patterson et al 1976). When these neurons are co-cultured with thoracic spinal cord explants, they receive a typical cholinergic preganglionic input (Ko, Burton & Bunge 1976). When cultured with the appropriate target of brown fat, they will provide endings on the fat cells that have the cytochemical characteristics of noradrenergic contacts (Ko et al 1976).

The interpretation of these generally expected results is considerably complicated by the following observations:

1. Cultures of dissociated sympathetic neurons form numerous synaptic contacts with each other (Claude 1973, Bunge et al 1974, Rees & Bunge 1974).

2. Initial cytochemical studies of these synapses showed that they contain predominantly dense-cored vesicles and thus appear to be noradrenergic (Claude 1973, Bunge et al 1974, Rees & Bunge 1974, O'Lague et al 1974).
3. Physiological studies clearly indicate that synaptic signals between these cultured neurons are nicotinic-cholinergic (O'Lague et al 1974, Ko et al 1976). In addition, dissociated SCGN will provide cholinergic innervation to striated muscle in culture (Nurse & O'Lague 1975).

These observations raised the question of whether only one type of neuron was present in the cultures or whether cholinergic mechanisms were developed in a subpopulation of undifferentiated cells present in the neonatal rat at the time the culture was prepared. This question has now been addressed by systematic analysis of synaptic vesicle cytochemistry (Johnson et al 1976, Landis 1976) and by direct physiological analysis of single neurons grown on microislands of heart muscle cells (Furshpan et al 1976). The latter observations clearly demonstrate that at least during one stage of their development in culture these neurons are capable of releasing both acetylcholine and norepinephrine. The systematic analysis of synaptic vesicle cytochemistry (Johnson et al 1976) indicated that neurons expressing adrenergic functions at the time they were put into culture gradually "shift" to take on cholinergic characteristics. It has further been shown that these cholinergic characteristics can be induced by the use of media conditioned over heart cells, as well as over certain other types of nonneuronal cells (Patterson & Chun 1974, 1977). Components in chick embryo extract and in human placental serum, when present in culture media in substantial concentrations, may also act as inducing agents (Ross & Bunge 1976, and unpublished). These experiments leave little doubt that at least certain neurons are able to shift production and utilization from one transmitter to another or to acquire the ability to produce and utilize a second transmitter (see also Bird & James 1975).

This concept of "transmitter-shift" capability should be clearly distinguished from the concept championed by Burn & Rand (1965) that acetylcholine might play a routine part in the release of norepinephrine from autonomic neurons. The Burn and Rand hypothesis requires that acetylcholine and norepinephrine production regularly occur in the same neuron. We now know that adrenergic neurons cultured under conditions conducive to choline acetyltransferase (CAT) accrual initially reveal no evidence of cholinergic activity, although later this activity is substantial (Johnson et al 1976); this is opposite to a developmental sequence that would support the Burn-Rand concept (Burn 1971). Although the tissue culture data indicate that both transmitters may be released from the same neuron under certain conditions, the data seem to best support the view that this occurs only during a certain developmental stage; the cultured neuron will eventually make a long-term commitment to adrenergicity or cholinergicity.

Most recently data have been presented that indicate that this capability for shifting transmitters may be expressed in the sympathetic neuron only during a limited period in its development (Ross, Johnson & Bunge 1977). Comparison of the amount of CAT that can be induced in sympathetic neurons put into culture at various ages indicated that the neurons from 2-day-old rat pups were maximally inducible, and that neurons from rats more than three weeks of age are not inducible at all under the same conditions. The definition of a "critical age" for the "transmitter-shift" phenomenon suggests that it may be a reflection of a developmental mechanism normally operative in the animal. The possible significance of this "transmitter selection" plasticity in the normal developmental sequence of the autonomic nervous system has recently been discussed in detail elsewhere (Bunge, Johnson & Ross 1978).

The aspect of this still-developing saga that is of particular interest to the present discussion is the clear demonstration that some factor capable of inducing the cholinergic shift in rat noradrenergic neurons is released into tissue culture medium by several types of nonneuronal cells, including one of the target cells of the autonomic neuron, the heart cell (Patterson & Chun 1977). Other cells capable of conditioning medium with a similar CAT-inducing activity include vascular elements, skeletal muscle, and embryonic fibroblasts. Fibroblasts from species other than rat are much less effective. It is clear from this and previous work that the inductive factor is not NGF (Chun & Patterson 1976). The inducing factor influences differentiation, whereas NGF has its influence on survival.

Among the many questions that remain unanswered regarding this inductive phenomenon are: What is the nature of the factor? How is it taken up by the neuron? At what level (cytoplasmic or nuclear) does it influence neuronal transmitter synthesis? Why is it present in such diverse tissues and fluids? Is the factor active in influencing neuronal differentiation in vivo?

Until these and other questions can be answered, this factor remains partially characterized, not as "trophic" in the sense discussed originally above, but as "specifying." The general developmental implication of these observations is discussed in more detail in Patterson (1978) and Bunge, Johnson & Ross (1978).

SATELLITE AND SCHWANN CELLS AS TARGETS OF TROPHIC INFLUENCES

It has been determined by direct observation through the thin skin of the developing tadpole (reviewed by Speidel 1964) that the leading front of the developing sensory nerves invading the body tissues consists of actively advancing neuritic growth cones. The addition of Schwann cells to the developing nerve occurs proximal to this front; here proliferation gradually provides a full complement of ensheathing cells so that the mature ensheathment of both unmyelinated and myelinated nerve fibers may occur. Recent evidence suggests that the cellular proliferation necessary to accomplish this ensheathment is controlled by a signal passing from neuron to Schwann cell. We also review here new evidence that shows that the signal for myelinogenesis (i.e. the signal that determines which Schwann cells will not merely ensheath axons but will form myelin sheaths) is of axonal origin and not a function of subsets of the Schwann cell population.

The Mitogenic Signal Provided by Axons to Schwann Cells

Tissue culture procedures that allow the separation of neurons of the peripheral nervous system from their supporting cell population (Schwann or connective tissue cells) have provided an opportunity to demonstrate directly the strict control by neurons (and their axonal processes) of Schwann cell proliferation. Before these experiments, the question had not been addressed of how the numbers of supporting cells might be controlled as the peripheral nerves formed and grew, although there was some indication from in vivo observations that Schwann cell numbers adapt to excessive neuronal loss during development by a curtailment of proliferation (Aguayo et al 1976). It should be noted that satellite cell proliferation may be influenced, under certain conditions, by changes in neuronal activity such as increased electrical stimulation (Schwyn 1967), environmental stress (Dropp & Sodetz 1971), or axotomy (Friede & Johnstone 1967, among others). This proliferation is seen in the region of the cell body of the affected neuron; the nature of the signal is not known.

Methods for the culture of normal Schwann cells alone have been reported recently (Wood & Bunge 1975, Wood 1976). Explants of sensory ganglia from fetal rats are allowed to spawn a rich halo of rapidly outgrowing axons; with the judicious use of antimitotic agents it is possible to suppress the fibroblasts but not the Schwann cells that normally accompany this axonal outgrowth. After several weeks there occurs a substantial halo of outgrowth consisting entirely of Schwann cells and neurites; the explant containing the neuronal cell bodies is then excised. As the neurites degenerate over several days a cellular halo remains that consists entirely of Schwann cells.

After several days in culture these Schwann cell "beds" become mitotically quiescent. Fewer than 1% of cells show incorporation of DNA precursors, and the number of cells in the culture dish does not increase (or significantly decrease), even after many weeks in culture. To demonstrate that Schwann cell numbers are under direct control of the axon it is necessary to prepare a sensory ganglion explant that has been exposed to extensive treatment with antimitotic agents so that an outgrowth of axons entirely free of Schwann cell components occurs. When a ganglion prepared in this manner is transferred to the region of a culture dish just beyond the periphery of the Schwann cell "bed" its bare axons grow into only a portion of the Schwann cell population. After allowing the axons and Schwann cells to interact in the presence of H^3 thymidine, autoradiographic analysis demonstrates that (*a*) the labeling index (cells labeled/over total cells) in regions of the Schwann cell population invaded by neurites is generally over 90%, and (*b*) the labeling index in remaining portions of the Schwann cell field is less than 5%. These experiments suggest that a potent mitogenic signal is delivered by direct or close membrane contact from the axon to the Schwann cell. In the tissue culture situation described, Schwann cell proliferation continues until sufficient cells are provided to populate the entire neuritic field of the implanted ganglion. In time, myelin will form in relation to the larger axons of this sensory neurite population. A variety of axonal types (sensory, autonomic, spinal somatic motor, and certain CNS fibers) are now known, by the use of similar techniques, to provide this type of mitogenic signal to Schwann cells (P. Wood, unpublished). The nature of the signal provided is not yet known, but because mitogenic stimulation is not seen in portions of the Schwann cell population not directly in contact with growing neurites (present in the same culture dish) it appears that contact between neurite and Schwann cell is necessary for provision of the signal (Wood & Bunge 1975). Related information from chick sensory ganglia has also been reported recently (Varon 1977b).

Similar results have been reported from a third laboratory (McCarthy & Partlow 1976a,b), which used a very different experimental system (chick embryo tissues rather than perinatal rat; autonomic neurons rather than sensory; and initial rather than delayed cell separation techniques). This method utilizes differences between neuronal and nonneuronal cell adhesiveness to obtain cultures in which 99% of the cells were chick autonomic neurons. The authors believe that the majority of their nonneuronal cells are either satellite or Schwann cells. By using various methods of recombining the neuronal and nonneuronal cell populations, these authors demonstrated that the presence of neurons increased thymidine incorporation into nonneuronal cells by 230–370%. In cases where neurons and nonneuronal cells were not grown in direct physical contact but were allowed to communicate with one another via the culture medium, no stimulation of thymidine incorporation was observed. More recently, this laboratory reported (Hanson & Partlow 1977) that homogenates of ganglionic neurons, as well as other cells, increased thymidine incorporation into ganglionic nonneuronal cells. We have recently obtained evidence that the particulate fraction of homogenates of sensory neurites provides a mitogenic signal when applied to pure Schwann cell populations prepared as described above, and furthermore that this signal is not present in particulate fractions from neurites exposed to trypsin digestion prior to homogenization (Salzer, Glaser & Bunge 1977).

The Axonal Signal for Myelinogenesis

From the time of the experiments of Simpson & Young (1945), it has been generally believed that control of the formation of myelin around certain axons was a property of the axon and not the Schwann cell. These early experiments involved the study by light microscopy of cross-anastomoses created between myelinated and unmyelinated nerves. Recently (Weinberg & Spencer 1975, Aguayo, Charron & Bray 1976), light and electron microscopic studies of cross-anastomosed nerves have confirmed that fibers from a myelinated nerve, forced to grow into the distal stump of an unmyelinated nerve, acquire myelin over a period of several weeks. In addition, such new studies have provided information on both migratory and proliferative activities of Schwann cells.

The cross-anastomosis experiments convincingly demonstrate a respecification of Schwann cells by myelinogenic axons only if one can rule out the alternate possibility that myelinated nerve fibers growing into the unmyelinated nerve stump were not bringing with them, by migration, already specified Schwann cells. Aguayo, Charron & Bray (1976) studied specifically the propensity of Schwann cells from injured nerves to migrate into adjacent regions. They accomplished this by heavily labeling the Schwann cells of a largely unmyelinated nerve and transplanting this labeled segment to a gap in a largely myelinated nerve of an unlabeled host mouse. As the host axons grew into the transplant, they became myelinated only by the Schwann cells originally resident in the transplant and not by cells migrating in from the proximal or distal host stumps. Similarly, there was little migration of labeled Schwann cells out of the transplant in either direction (see also Romine, Aguayo & Bray 1975).

The nature of the signal for myelinogenesis provided by certain axons to the Schwann cells is not known. The signal appears much more complex than the presentation of an adequate axonal diameter, as suggested earlier (for discussion, see Weinberg & Spencer 1975, Aguayo et al 1976), and is more likely to reside in some specific surface characteristic of the "competent" axon. Another question is whether Schwann cells are susceptible at all times to the myelinogenic signal when it is presented to them. Aguayo and co-workers (Aguayo et al 1976, Aguayo, Charron & Bray 1976) have suggested that Schwann cells may regain the required multipotentiality only after undergoing mitosis. They studied Schwann cell proliferation that occurred in the first few days after nerve anastomoses and established that it constituted an early local response to injury that was independent of axonal regrowth into the region. Because of the substantial amount of Schwann cell proliferation that occurs as part of this early local response and as the proximal axons grow into the distal stump, these workers suggest that the multipotentiality of the Schwann cell in relating to either a myelinated or an unmyelinated axon may be made possible by the fact that Schwann cells invariably divide after nerve injury and may thus reacquire the multipotentiality that permits them to respond to specific axonal influences (see also Hall & Gregson 1975).

The Axonal Signal for the Production of Basal Lamina by the Schwann Cell

Use of the special tissue culture preparations described above (Wood 1976) have permitted new observations on the source of the prominent basal lamina seen in relation to Schwann cells throughout the peripheral nervous system. These in vitro experiments utilize initial preparations that contain (*a*) Schwann cells only, (*b*) Schwann cells in physical contact with axons, or (*c*) an outgrowth containing axons without Schwann cells (Williams, Wood & Bunge 1976). The normal functioning of both Schwann cells and neurons in these cultures is indicated by normal ensheathment of smaller axons by Schwann cells and by myelin formation around larger axons when these two elements are cultured together for a period of several weeks. Schwann cells in contact with axons for several days exhibit a basal lamina (Figure 4). After removal of axons from these young cultures, the basal lamina begins to degenerate and is largely absent after one week. Schwann cells cultivated for extensive periods of time in the absence of axons appear markedly atrophic and, unlike the normal cell, contain substantial amounts of ribosome-free endoplasmic reticulum and few ribosomes in rosettes (Figure 5). When the axon–Schwann cell relationship is restored by the addition of a ganglion to engender a new axonal outgrowth, Schwann cell processes ensheath the bare axons, the Schwann cell cytoplasm again regains its normal appearance, and within several days a basal lamina reappears in

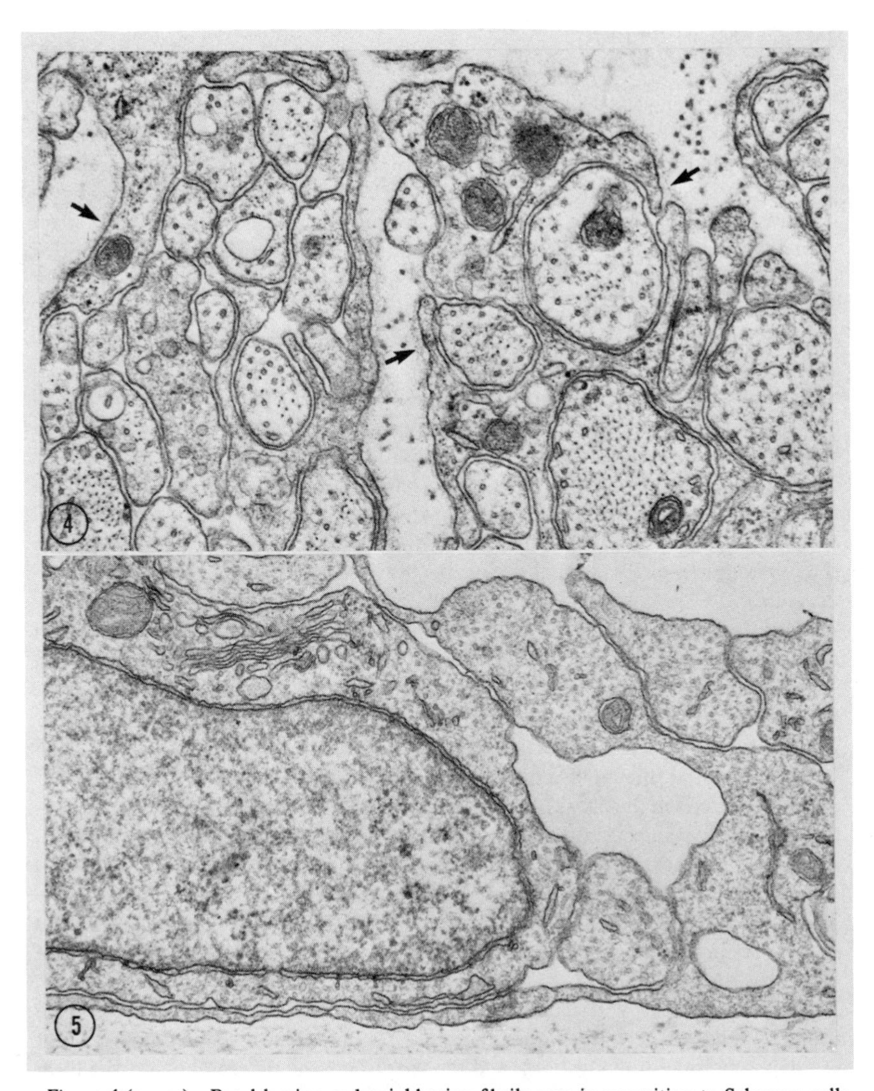

Figure 4 (*upper*) Basal lamina and neighboring fibrils seen in apposition to Schwann cells related to axons. This tissue culture preparation contained only sensory neuronal cell bodies, their axons, and related Schwann cells. Under these conditions Schwann cells produce basal lamina material (arrows) and closely related, small, fibrillar elements lacking the prominent banding of the typical collagen fibril. Electron micrograph courtesy of M. B. Bunge and P. M. Wood. X 47,000.

Figure 5 (*lower*) Lack of basal lamina formation by Schwann cells in the absence of axons. In this tissue culture preparation the neuronal cell bodies (and related axons) have been removed and the remaining Schwann cells subsequently cultured for several weeks. Under these conditions the Schwann cell exhibits no adjacent basal lamina or fibrillar material. The cells appear atrophic, with a decrease in rough endoplasmic reticulum, but their numbers remain stable. The fibrillar material at the base of the picture is the reconstituted collagen layer provided for cell adhesion and growth. Electron micrograph courtesy of M. B. Bunge, A. K. Williams, and P. M. Wood. X 47,000.

relation to the Schwann cells. These observations indicate that normal Schwann cell function, especially the initiation of basal lamina production, is reversibly dependent upon a normal axon–Schwann cell relationship and presumably depends upon a signal from the axon to the Schwann cell. These studies correlate well with in vivo observations indicating that migrating Schwann cells temporarily out of contact with an axonal element are not surrounded by the basal lamina so characteristic of this cell type in the normal mature nerve (Billings-Gagliardi, Webster & O'Connell 1974).

It seems reasonable to consider the possibility that the secretory activity engendered by axonal contact with Schwann cells may be instrumental in turning off the mitogenic signal discussed above. Whereas the initial contact of newly formed axolemma with the Schwann cell may serve as a mitogenic signal, the Schwann cell may in time respond to this contact by the secretion of material that provides (among other functions) a covering for certain axolemmal components, thereby, a masking of the mitogenic signal. In this way both the stimulation of Schwann cell numbers and the eventual stabilization of these numbers may be accomplished in the development of the normal peripheral nerve. An alternate explanation can be offered for the termination of Schwann cell proliferative activity as a nerve matures. As continuing cell division provides a complete mosaic of ensheathment, Schwann cells come into close contact at their lateral borders with their nearest neighbors. They could then be influenced by the "density-dependent regulation of growth" (see discussion by Holley 1974), a well-established mechanism in the control of cellular proliferation observed in normal cells in tissue culture.

The substantial amount of Schwann cell proliferation that is known to occur as the axon degenerates in the distal stump of a severed nerve may result from (*a*) unmasking of the mitogenic signal during the course of axonal breakdown in the injured nerve, or (*b*) loss of the contact inhibition established with maturation of the normal nerve when injury causes a disturbance of cellular architecture in the distal stump. There thus may be a temporary reexpression of the mitogenic signal that engenders a transient phase of Schwann cell proliferation during the degenerative phase of Wallerian degeneration.

The discussion above establishes that trophic and specifying signals pass from axons to Schwann cells for the regulation of proliferation, myelination (a form of cellular expansion or elongation), and for the stimulation of secretion (in the form of basal lamina). These interactions correlate with the scheme shown in Figure 2, where trophic factors are postulated to control cell numbers, cell size, and cellular release mechanisms (secretion), whereas "specifying" influences dictate which particular type of cellular activity will be expressed.

These interactions between nerve and axon appear to be expressed in an abnormal manner in a variety of peripheral nerve diseases. A particularly instructive example is provided in the recent study by Aguayo et al (1977) on the "trembler" mouse. The Schwann cells of this mouse mutant (*a*) are present in abnormally large numbers, (*b*) respond poorly to the myelinogenic signal, even from normal axons, and (*c*) appear to provide inadequate trophic support to their partner axons, as expressed by smaller axonal diameter (A. Aguayo, personal communication). Further studies of such mutants should provide better understanding of the two-way trophic relationships between neuron and Schwann cells and the way in which the expression of different glial functions are controlled.

CONCLUDING REMARKS

This review considers several of the humoral or cellular influences known to act on the neurons and glia of the peripheral nervous system. An effort has been made to distinguish, at both the conceptual and the informational levels, between trophic and specifying influences—that is, between extrinsic modulation of the overall metabolic activity of a cell and extrinsic agencies that somehow influence selection of distinct programs to be expressed from within the cell. These efforts have led to a generalized concept of the possible mechanism and action of trophic influences; in this concept the graded availability of trophic factors, whether to neurons or glia, determines whether survival, maintenance, or growth (in a variety of different modalities) will occur. In the particular case of trophic influences directed to glial cells from their partner neurons, some speculative considerations have been addressed to possible relationships between the different response modalities and, hence, to the specifying influences that may dictate their selection.

The mechanisms by which trophic influences operate on their target cells remain largely a matter of speculation at present. One notable exception has been the discussion on the mode of action of Nerve Growth Factor, currently the only available model for a neuron-directed trophic agent. Transfer of trophic signals from one cell to another cell may occur by a physical passage of humoral substances. The feasibility of the transfer of materials from Schwann cell to axon has now been directly shown by Lasek, Gainer & Przbylski (1974) with the demonstration of transfer of protein produced in the Schwann cell to the interior of the squid axon. Endocytotic uptake of NGF, from peripheral sources in vivo or from surrounding fluid in vitro, has also been documented. Nevertheless, the authors of this review tend to favor the notion that many (if not all) trophic agents, and possibly some specifying agents as well, interact with their target cell at the cell surface and exert their primary action on the cell membrane.

Another important emphasis of this review has been on possible sources of trophic substances. In the past, the sources of trophic substances for peripheral neurons have generally been considered to be the target tissue of the peripheral nerves. In our review we have stressed the equal importance of the supporting tissues of the nerve itself, particularly the satellite cells and the cells of Schwann (i.e. the peripheral glia). This extension in our concept of possible trophic sources may help to explain why peripheral neurons can be maintained in culture without their target tissues, or even without the exogenous supply of their putative trophic products (e.g. NGF for sensory and sympathetic ganglion cells) as long as the culture conditions favor close association (explants) or numerical abundance (proliferating outgrowth regions, monolayer cultures) of supporting cells, both Schwann cells and connective tissue components.

These changes in our concept of possible trophic sources have implications for mechanisms of nerve regeneration that should not be overlooked. Peripheral nerves regenerate most effectively when they are able to enter the distal stumps from which they have been separated by injury. Cajal (1928) and many other workers have stressed the importance of the Schwann cell within these nerve trunks in promoting the prolonged axonal growth necessary for the accomplishment of functional restoration. If trophic substances are available continually from cells within the nerve trunk, these substances would be able to guide and stimulate nerve growth during the prolonged period of regeneration, then no time limit is set during which the nerve must reach its peripheral target to obtain a source of trophic support. The failure of regeneration in CNS tissues of higher vertebrates may derive in large part from lack of a comparable sustaining terrain for axonal regrowth.

ACKNOWLEDGMENTS

We wish to thank Mrs. Gloria Finot for patient and helpful secretarial assistance, and the Editors of the *Annual Review of Neuroscience* for useful suggestions. Work in the authors' laboratories is supported by USPHS-NIH grants NS 07606, NS 09923, NS 11888, and grant RG 1118 from the National Multiple Sclerosis Society.

Literature Cited

Aguayo, A., Attiwell, M., Trecarten, J., Perkins, S., Bray, G. 1977. Abnormal myelination in transplanted trembler mouse Schwann cells. *Nature* 265: 73–75

Aguayo, A., Charron, L., Bray, G. 1976. Potential of Schwann cells from unmyelinated nerves to produce myelin: a quantative ultrastructural and radiographic study. *J. Neurocytol.* 5:565–73

Aguayo, A., Peyronnard, J., Terry, L., Romine, J., Bray, G. 1976. Neonatal neuronal loss in rat superior cervical ganglia: retrograde effects on developing preganglionic axons and Schwann cells. *J. Neurocytol.* 5:137–55

Almon, R. A., Varon, S. 1978. Associations of Beta Nerve Growth Factor with the Alpha and Gamma subunits of the 7S macromolecule. *J. Neurochem.* In press

Andres, R. Y., Jeng, I., Bradshaw, R. A. 1977. Nerve Growth Factor receptors: Identification of distinct classes in plasma membranes and nuclei of embryonic dorsal root neurons. *Proc. Natl. Acad. Sci. USA* 74:2785–89

Angeletti, P. U., Vigneti, E. 1971. Assay of Nerve Growth Factor (NGF) in subcellular fractions of peripheral tissues by microcomplement fixation. *Brain Res.* 33:601–4

Arnason, B. G. W., Oger, J., Pantazis, N. J., Young, M. 1974. Secretion of nerve cell growth factor by cancer cells. *J. Clin. Invest.* 53:2a (Abstr.)

Banerjee, S. P., Snyder, S. H., Cuatrecasas, P., Greene, L. A. 1973. Binding of Nerve Growth Factor receptor in sympathetic ganglia. *Proc. Natl. Acad. Sci. USA* 70:2519–23

Berger, B. D., Wise, C. B., Stein L. 1973. Nerve Growth Factor: enhanced recovery of feeding after hypothalamic damage. *Science* 180:506–8

Billings-Gagliardi, S., Webster, H., O'Connell, M. 1974. In vivo and electron microscopic observations on Schwann cells in developing tadpole nerve fibers. *Am. J. Anat.* 141:375–92

Bird, M., James, D. 1975. The cultures of previously dissociated embryonic chick spinal cord cells in feeder layers of liver and kidney, and the development of paraformaldehyde induced fluorescence on the former. *J. Neurocytol.* 4:633–46

Bjerre, B., Björklund, A. 1973. The production of catecholamine-containing cells in vitro by young chick embryos: effects of "Nerve Growth Factor" (NGF) and its antiserum. *Neurobiology* 3:140–61

Bjerre, B., Björklund, A., Stenevi, U. 1974. Inhibition of the regenerative growth of central noradrenergic neurons by intracerebrally administered anti-NGF serum. *Brain Res.* 74:1–18

Björklund, A., Stenevi, U. 1972. Nerve Growth Factor: stimulation of regenerative growth of central noradrenergic neurons. *Science* 175:1251–53

Black, I. B. 1978. Regulation of autonomic development. *Ann. Rev. Neurosci.* 1: 183–214

Bocchini, V., Angeletti, P. U. 1969. The Nerve Growth Factor: purification as a 30,000 molecular weight protein. *Proc. Natl. Acad. Sci. USA* 64:787–94

Bradshaw, R. A., Young, M. 1976. Nerve Growth Factor—recent developments and perspectives. *Biochem. Pharmacol.* 25:1445–49

Bray, D. 1970. Surface movements during the growth of single explanted neurons. *Proc. Natl. Acad. Sci. USA* 65:905–10

Bueker, E. D. 1948. Implantation of tumors in the hind limb field of the embryonic chick and the developmental response of the lumbo-sacral nervous system. *Anat. Rec.* 102:369–90

Bunge, R., Johnson, M., Ross, D. 1978. Nature and nurture in the development of the autonomic neuron. *Science.* In press

Bunge, R., Rees, R., Wood, P., Burton, H., Ko, C.-P. 1974. Anatomical and physiological observations on synapses formed on isolated autonomic neurons in tissue culture. *Brain Res.* 66:401–12

Burn, J. H. 1971. *The Autonomic Nervous System.* Oxford: Blackwell

Burn, J. H., Rand, M. J. 1965. Acetylcholine in adrenergic transmission. *Ann. Rev. Pharmacol.* 5:163–82

Burnham, P. A., Raiborn, C., Varon, S. 1972. Replacement of Nerve Growth Factor by ganglionic nonneuronal cells for the survival in vitro of dissociated ganglionic neurons. *Proc. Natl. Acad. Sci. USA* 69:3556–60

Burnham, P. A., Varon, S. 1973. In vitro uptake of active Nerve Growth Factor in dorsal root ganglia of embryonic chick. *Neurobiology* 3:232–45

Burton, H., Bunge, R. 1975. A comparison of the uptake and release of H^3 norepinephrine in rat autonomic and sensory ganglia in tissue culture. *Brain Res.* 97:157–62

Cajal, R. 1928. *Degeneration and Regeneration of the Nervous System.* Vol. I. New York: Hafner

Changeux, J.-P., Danchin, A. 1976. Selective stabilisation of developing synapses as a mechanism for the specification of neuronal networks. *Nature* 264:705–12

Chun, L., Patterson, P. 1976. The role of NGF in the development of rat sympathetic neurons in vitro. *6th Annu. Meet. Soc. Neurosci.* 2:191 (Abstr.)

Claude, P. 1973. Electron microscopy of dissociated rat sympathetic neurons in vitro. *J. Cell Biol.* 59:57a

Cohen, A. 1972. Factors directing the expression of sympathetic nerve traits in cells of neural crest origin. *J. Exp. Zool.* 179:167–82

Coughlin, M. 1975a. Early development of parasympathetic nerves in the mouse submandibular gland. *Dev. Biol.* 43: 123–39

Coughlin, M. 1975b. Target organ stimulation of parasympathetic nerve growth in the developing mouse submandibular gland. *Dev. Biol.* 43:140–58

Cowan, W. M. 1973. Neuronal death as a regulative mechanism in the control of cell number in the nervous system. In *Development and Aging in the Nervous System.* New York: Academic. pp. 19–41

Cowan, W. M., Wenger, E. 1967. Cell loss in the trochlear nucleus of the normal development and after radical extirpation of the optic vesicle. *J. Exp. Zool.* 164:267–80

Cowan, M., Wenger, E. 1968. Degeneration in the nucleus of origin of the preganglionic fibers of the chick ciliary ganglion following early removal of the optic vesicle. *J. Exp. Zool.* 168:105–24

Cragg, B. 1970. What is the signal for chromatolysis? *Brain Res.* 23:1–21

Dropp, J., Sodetz, F. 1971. Autoradiographic study of neurons and neuroglia in autonomic ganglia of behaviorally stressed rats. *Brain Res.* 33:419–30

Ebendal, T., Jacobson, C. A. 1975. Human glial cells stimulating outgrowth of axons in cultured chick embryo ganglia. *Zoon* 3:169–72

Eisenbarth, G. S., Drezner, M. K., Lebovitz, H. E. 1975. Inhibition of chondromucoprotein synthesis: an extraneuronal effect of Nerve Growth Factor. *J. Pharmacol. Exp. Ther.* 192:630–34

Fischbach, G. D. 1972. Synapse formation between dissociated nerve and muscle cells in low density cell cultures. *Dev. Biol.* 28:407–29

Fischbach, G., Berg, D., Cohen, S., Frank, E. 1976. Enrichment of nerve-muscle synapses in spinal cord–muscle cultures and identification of relative peaks of ACh sensitivity at sites of transmitter release. In *The Synapse, Cold Spring Harbor Symp. Quant. Biol.* 40:347–57

Frazier, W. A., Angeletti, R. H., Bradshaw, R. A. 1972. Nerve Growth Factor and insulin. *Science* 176:482–88

Frazier, W. A., Boyd, L. F., Bradshaw, R. A. 1973. Interaction of Nerve Growth Factor with surface membranes: biological competence of insolubilized Nerve Growth Factor. *Proc. Natl. Acad. Sci. USA* 70:2931–35

Frazier, W. A., Boyd, L. F., Szutowicz, A., Pulliam, M. W., Bradshaw, R. A. 1974. Specific binding sites for ^{125}I-Nerve Growth Factor in peripheral tissues and brain. *Biochem. Biophys. Res. Commun.* 57:1096–1103

Frazier, W. A., Ohlendorf, C. E., Boyd, L. F., Aloe, L., Johnson, E. M., Ferrendelli, J. A., Bradshaw, R. 1973. Mechanism of action of Nerve Growth Factor and cyclic AMP on neurite outgrowth in embryonic chick sensory ganglia: demonstration of independent pathways of stimulation. *Proc. Natl. Acad. Sci. USA* 70:2448–52

Friede, R., Johnstone, M. 1967. Responses of thymidine labeling of nuclei in gray matter and nerve following sciatic transection. *Acta Neuropathol.* 70:218–31

Furshpan, E., Macleish, P., O'Lague, P., Potter, D. 1976. Chemical transmission between rat sympathetic neurons and cardiac myocytes developing in microcultures: evidence for cholinergic, adrenergic, and dual function neurons. *Proc. Natl. Acad. Sci. USA* 73:4225–29

Giller, E. L., Neale, J. H., Bullock, P. N., Schrier, B. K., Nelson, P. G. 1977. Choline acetyltransferase activity of spinal cord cell cultures increased by co-culture with muscle and by muscle-conditioned medium. *J. Cell Biol.* 74: 16–29

Giller, E. L., Schrier, B. K., Shainberg, A., Fisk, R., Nelson, P. 1973. Choline acetyltransferase activity is increased in combined cultures of spinal cord and muscle cells in mice. *Science* 182:588–89

Grafstein, B. 1975. The nerve cell body response to axotomy. *Exp. Neurol.* 48: 32–51

Green, L. A., Shooter, E. M., Varon, S. 1968. Enzymatic activities of mouse Nerve Growth Factor and its subunits. *Proc. Natl. Acad. Sci. USA* 60:1383–88

Greene, L. A., Shooter, E. M., Varon, S. 1969. Subunit interaction and enzymatic activity of mouse 7S Nerve Growth Factor. *Biochemistry* 8: 3735–41

Greene, L. A., Tischler, A. S. 1976. Establishment of a noradrenergic clonal line of rat adrenal pheochromocytoma cells which respond to nerve growth factor. *Proc. Natl. Acad. Sci. USA.* 73:2424–28

Greene, L. A., Tomita, J. T., Varon, S. 1971. Growth-stimulating activities of mouse submaxillary esteropeptidases on chick embryo fibroblasts in vitro. *Exp. Cell Res.* 64:387–95

Greene, L. A., Varon, S., Piltch, A., Shooter, E. M. 1971. Substructure of the β subunit of mouse 7S Nerve Growth Factor. *Neurobiology* 1:37–48

Haas, D. C., Hier, D. B., Arnason, B. G. W., Young, M. 1972. On a possible relationship of cyclic AMP to the mechanism of action of Nerve Growth Factor. *Proc. Soc. Exp. Biol. Med.* 140:45–47

Hall, S., Gregson, N. 1975. The effects of mitomycin C on the process of remyelination in the mammalian peripheral nervous system. *Neuropathol. Appl. Neurobiol.* 1:149–70

Hamburger, V. 1975. Cell death in the development of the lateral motor column of the chick embryo. *J. Comp. Neurol.* 160:535–46

Hamburger, V., Levi-Montalcini, R. 1949. Proliferation, differentiation and degeneration in the spinal ganglion of the chick embryo under normal and experimental conditions. *J. Exp. Zool.* 111: 457–501

Hanson, G., Partlow, L. 1977. Stimulation of thymidine incorporation in glia by cell sonicates. *Trans. Am. Soc. Neurochem.* 8:142

Harris, A. 1974. Inductive functions of the nervous system. *Ann. Rev. Physiol.* 36: 251–305

Helfand, S. L., Smith, G. A., Wessels, N. K. 1976. Survival and development in culture of dissociated parasympathetic neurons from ciliary ganglia. *Dev. Biol.* 50:541–47

Hendry, I. A. 1975. Response of the adrenergic system to changes in tissue and plasma levels of nerve growth factor (NGF). In *Neurotransmission, Vol. II of Proc. 6th Int. Congr. Pharmacol.*, ed. L. Ahtee, pp. 249–58. Forssa, Finland: Forssen Kirjapamo Oy.

Hendry, I. A. 1976. Control in the development of the vertebrate sympathetic nervous system. *Rev. Neurosci.* 2:149–93

Hendry, I. A., Iversen, L. L. 1973. Changes in tissue and plasma concentrations of Nerve Growth Factor following the removal of the submaxillary glands in adult mice and their effects on the sympathetic nervous system. *Nature* 243: 500–4

Herrup, K., Shooter, E. M. 1973. Properties of the β Nerve Growth Factor receptor of avian dorsal root ganglia. *Proc. Natl. Acad. Sci. USA* 70:3884–88

Hoffer, B., Seiger, A., Freedman, R., Olson, L., Taylor, D. 1977. Electrophysiology and cytology of hippocampal formation transplants in the anterior chamber of the eye. II. Cholinergic mechanisms. *Brain Res.* 119:109–32

Holley, R. 1974. Serum factors and growth control. In *Control of Proliferation in Animal Cells.* ed. B. Clarkson, R. Baserga. Cold Spring Harbor, NY: Cold Spring Harbor Lab.

Hollyday, M., Hamburger, V. 1976. Reduction of the naturally occurring motor neuron loss by enlargement of the periphery. *J. Comp. Neurol.* 170:311–20

Hooisma, J., Slaff, D. W., Meeter, E., Stevens, W. F. 1975. The innervation of chick striated muscle by the chick ciliary ganglion in tissue culture. *Brain Res.* 85:79–85

Horii, Z. I., Varon, S. 1975. Nerve Growth Factor—induced rapid activation of RNA labeling in dorsal root ganglionic dissociates from the chick embryo. *J. Neurosci. Res.* 1:361–75

Horii, Z. I., Varon, S. 1977. Nerve Growth Factor action on membrane permeation to exogenous substrates in dorsal root ganglionic dissociates from the chick embryo. *Brain Res.* 124:121–33

Ishii, D. N., Shooter, E. M. 1976. Regulation of Nerve Growth Factor synthesis in mouse submaxillary glands by testosterone. *J. Neurochem.* 25:843–51

Jacobson, M. 1970. Development, specification and diversification of neuronal connections. In *The Neurosciences: Second Study Program,* ed. F. O. Schmitt, pp. 116–29. New York: Rockefeller Univ. Press

Johnson, D. G., Gorden, P., Kopin, I. J. 1971. A sensitive radioimmuno assay for 7S Nerve Growth Factor antigens in serum and tissue. *J. Neurochem.* 18: 2355–62

Johnson, M., Ross, D., Meyers, M., Rees, R., Bunge, R., Wakshull, E., Burton, H. 1976. Synaptic vesicle cytochemistry changes when cultured sympathetic neurones develop cholinergic interactions. *Nature* 262:308–10

Johnson, D. G., Silberstein, S., Hanbauer, I., Kopin, I. J. 1972. The role of Nerve Growth Factor in the ramification of sympathetic nerve fibers into the rat iris in organ culture. *J. Neurochem.* 19:2025–29

Ko, C.-P., Burton, H., Bunge, R. P. 1976. Synaptic transmission between rat spinal cord explants and dissociated superior cervical ganglion neurons in tissue culture. *Brain Res.* 117:437–60

Ko, C.-P., Burton, H., Johnson, M., Bunge, R. 1976. Synaptic transmission between rat superior cervical ganglion neurons in dissociated cell cultures. *Brain Res.* 117:461–85

Landis, S. 1976. Rat sympathetic neurons and cardiac myocytes developing in microcultures: correlation of the fine structure of endings with neurotransmitter function in single neurons. *Proc. Natl. Acad. Sci. USA* 73:4220–24

Landmesser, L., Pilar, G. 1974a. Synapse formation during embryogenesis on ganglion cells lacking a periphery. *J. Physiol. London* 241:715–36

Landmesser, L., Pilar, G. 1974b, Synaptic transmission and cell death during normal ganglionic development. *J. Physiol. London* 241:737–49

Landmesser, L., Pilar, G. 1976. Fate of ganglionic synapses and ganglion cell axons during normal and induced cell death. *J. Cell. Biol.* 68:357–73

Lasek, R., Gainer, H., Pryzbylski, R. 1974. Transfer of newly synthesized proteins from Schwann cells to the squid giant axon. *Proc. Natl. Acad. Sci. USA* 71: 1188–92

Lasher, R. A., Zagon, I. S. 1972. The effect of potassium on neuronal differentiation in cultures of dissociated newborn rat cerebellum. *Brain Res.* 41:482–88

Lazarus, K., Bradshaw, R., West, N., Bunge, R. 1976. Adaptive survival of rat sympathetic neurons cultured without supporting cells on exogenous nerve growth factor. *Brain Res.* 113:159–64

Le Douarin, N. M., Teillet, M.-A. M. 1974. Experimental analysis of the migration and differentiation of neuroblasts of the autonomic nervous system and of neurectodermal mesenchymal derivatives, using a biological cell marking technique. *Dev. Biol.* 41:162–83

Letourneau, P. C. 1975. Possible roles for cell-to-substratum adhesion in neuronal morphogenesis. *Dev. Biol.* 44:77–91

Levi-Montalcini, R. 1950. The origin and development of the visceral system in the spinal cord of the chick embryo. *J. Morphol.* 86:253–83

Levi-Montalcini, R. 1966. The Nerve Growth Factor: its mode of action on sensory and sympathetic nerve cells. *Harvey Lect.* 60:217–59

Levi-Montalcini, R. 1975. Milestones, unanswered questions and current studies on Nerve Growth Factor. See Hendry 1975, pp. 221–30

Levi-Montalcini, R., Angeletti, P. U. 1968. Nerve Growth Factor. *Physiol. Rev.* 48:534–69

Levi-Montalcini, R., Angeletti, P. U. 1971. Ultrastructure and metabolic studies on sensory and sympathetic nerve cells treated with the Nerve Growth Factor and its antiserum. In *Hormones in Development,* ed. M. Hamburgh, E. J. W. Barrington, pp. 719–30. New York: Appleton

Levi-Montalcini, R., Hamburger, V. 1951. Selective growth stimulating effects of mouse sarcoma on sensory and sympathetic nervous system of the chick embryo. *J. Exp. Zool.* 116:321–62

Levi-Montalcini, R., Revoltella, R., Calissano, P. 1974. Microtubule proteins in the Nerve Growth Factor mediated response. Interaction between the Nerve Growth Factor and its target cells. *Recent Prog. Horm. Res.* 30:635–69

Longo, A. M., Penhoet, E. E. 1974. Nerve Growth Factor in rat glioma cells. *Proc. Natl. Acad. Sci. USA* 71:2347–49

Mains, R. E., Patterson, P. H. 1973. Primary cultures of dissociated sympathetic neurons. *J. Cell Biol.* 59:329–66 (3 parts)

Matthews, M., Nelson, V. 1975. Detachment of structurally intact nerve endings from chromatolytic neurons of rat superior cervical ganglion during the depression of synaptic transmission induced by post-ganglionic axotomy. *J. Physiol. London* 245:91–135

McCarthy, K., Partlow, L. 1976a. Preparation of pure neuronal and non-neuronal cultures from embryonic chick sympathetic ganglia: a new method based on both differential cell adhesiveness and the formation of homotypic neuronal aggregates. *Brain Res.* 114:391–414

McCarthy, K., Partlow, L. 1976b. Neuronal stimulation of ^{3}H thymidine incorporation by primary cultures of highly purified non-neuronal cells. *Brain Res.* 114:415–26

Merrell, R., Pulliam, M. W., Randono, L., Boyd, L. F., Bradshaw, R., Glaser, L. 1975. *Proc. Natl. Acad. Sci. USA* 72:4270–74

Mizel, S. B., Bamburg, J. R. 1975. Studies on the action of Nerve Growth Factor III. Differential requirement of chick embryonic dorsal root ganglia neurons for NGF. *Dev. Biol.* 49:20–28

Monard, D., Stockel, K., Goodman, R., Thoenen, H. 1975. Distinction between nerve growth factor and glial factor. *Nature* 258:444–45

Moore, J. B., Mobley, W. C., Shooter, E. M. 1974. Proteolytic modification of the β Nerve Growth Factor protein. *Biochemistry* 13:833–40

Murphy, R. A., Oger, J., Saide, J. D., Blanchard, M. H., Arnason, B. G. W., Hogan, E., Pantazis, N. J., Young, M. 1977. Secretion of Nerve Growth Factor by central nervous system glioma cells in culture. *J. Cell Biol.* 72:769–73

Nelson, P. G., Peacock, J. H. 1973. Electrical activity in dissociated cell cultures from fetal mouse cerebellum. *Brain Res.* 61:163–74

Nikodijevic, B., Nikodijevic, O., Yu, M-Y. W., Pollard, H., Guroff, G. 1975. The effect of Nerve Growth Factor on cyclic AMP levels in superior cervical ganglia of the rat. *Proc. Natl. Acad. Sci. USA* 72:4769–71

Njå, A., Purves, D. 1977. The effects of NGF and its antiserum on synapses in the superior cervical ganglion of the guinea pig. *J. Physiol.* In press

Norr, S. 1973. *In vitro* analysis of sympathetic neuron differentiation of chick neural crest cells. *Dev. Biol.* 34:16–38

Norr, S., Varon, S. 1975. Dynamic, temperature-sensitive association of ^{125}I-Nerve Growth Factor in vitro with ganglionic and non-ganglionic cells from embryonic chick. *Neurobiology* 5:101–18

Nurse, C. A., O'Lague, P. H. 1975. Formation of cholinergic synapses between dissociated sympathetic neurons and skeletal muscles of the rat in cell culture. *Proc. Natl. Acad. Sci. USA* 72:1[illegible]55–59

Oger, J., Arnason, B. G. W., Pantazis, N., Lehrich, J., Young, M. 1974. Synthesis of Nerve Growth Factor by L and 3T3 cells in culture. *Proc. Natl. Acad. Sci. USA* 71:1554–58

O'Lague, P. H., Obata, K., Claude, P., Furshpan, E. J., Potter, D. D. 1974. Evidence for cholinergic synapses between dissociated rat sympathetic neurons in cell culture. *Proc. Natl. Acad. Sci. USA* 71:3602–6

Olson, L., Freedman, R., Seiger, A., Hoffer, B. 1977. Electrophysiology and cytology of hippocampal formation transplants in the anterior chamber of the eye. I. Intrinsic organization. *Brain Res.* 119:87–106

Patterson, P. H. 1978. Environmental determination of autonomic neurotransmitter functions. *Ann. Rev. Neurosci.* 1: 1–17

Patterson, P., Chun, L. 1974. The influence of non-neuronal cells on catecholamine and acetylcholine synthesis and accumulation in cultures of dissociated sympathetic neurons. *Proc. Natl. Acad. Sci. USA* 71:3607–10

Patterson, P., Chun, L. 1977. The induction of acetylcholine synthesis in primary cultures of dissociated rat sympathetic neurons. *Dev. Biol.* 56:263–80

Patterson, P. H., Reichardt, L. F., Chun, L. L. Y. 1976. Biochemical studies on the development of primary sympathetic neurons in cell culture. *Cold Spring Harbor Symp. Quant. Biol.* 40:389–97

Peacock, J. H., Nelson, P. G., Goldstone, M. W. 1973. Electrophysiological study of cultured neurons dissociated from spinal cords and dorsal root ganglia of fetal mice. *Dev. Biol.* 30:137–52

Pilar, G., Landmesser, L. 1976. Ultrastructural differences during embryonic cell death in normal and peripherally deprived ciliary ganglia. *J. Cell Biol.* 68:339–56

Prestige, M. C. 1970. Differentiation, degeneration, and the role of the periphery: quantitative considerations. See Jacobson 1970, pp. 73–82

Purves, D. 1975. Functional and structural changes in mammalian sympathetic neurones following interruption of the axons. *J. Physiol. London* 252:429–63

Purves, D. 1976a. Long-term regulation in the vertebrate peripheral nervous system. *Int. Rev. Physiol.* 10:125–77

Purves, D. 1976b. Functional and structural changes in mammalian sympathetic neurons following colchicine application to post-ganglionic nerves. *J. Physiol. London* 259:159–75

Purves, D., Njå, A. 1976. Effect of nerve growth factor on synaptic depression following axotomy. *Nature* 260:535–36

Rees, R., Bunge, R. 1974. Morphological and cytochemical studies of synapses formed in culture between isolated rat superior cervical ganglion neurons. *J. Comp. Neurol.* 157:1–12

Rees, R., Bunge, M., Bunge, R. 1976. Morphological changes in the neuritic growth cone and target neuron during synaptic junction development in culture. *J. Cell Biol.* 68:240–63

Revoltella, R., Bertolini, L., Pediconi, M. 1974. Unmasking of Nerve Growth Factor membrane-specific binding sites in synchronized murine C1300 neuroblastoma cells. *Exp. Cell Res.* 85:89–94

Rogers, L. A., Cowan, W. M. 1973. The development of the mesencephalic nucleus of the trigeminal nerve in the chick. *J. Comp. Neurol.* 147:291–320

Roisen, F. J., Murphy, R. A., Branden, W. G. 1972. Neurite development in vitro. I. The effects of adenosine 3'5'-cyclic monophosphate (cyclic AMP). *J. Neurobiol.* 3:347–68

Romine, J., Aguayo, A., Bray, G. 1975. Absence of Schwann cell migration along regenerating unmyelinated nerves. *Brain Res.* 98:601–6

Ross, D., Bunge, R. 1976. Choline acetyltransferase in cultures of rat superior cervical ganglion. *Abstr. 6th Ann. Meet. Soc. Neurosci.* 2:769

Ross, D., Johnson, M., Bunge, R. 1977. Evidence that development of cholinergic characteristics in adrenergic neurons is age dependent. *Nature* 267:536–39

Salzer, J. L., Glaser, L., Bunge, R. P. 1977. Stimulation of Schwann cell proliferation by a neurite membrane fraction. *J. Cell Biol.* 75:118a (Abstr.)

Schwab, M., Thoenen, H. 1976. Electron microscopic autoradiographic and cytochemical localization of retrogradely transported nerve growth factor (NGF) in the rat sympathetic ganglion. *J. Cell Biol.* 70:299a

Schwartz, J. P., Chuang, D.-M., Costa, E. 1977. Regulation by isoproterenol of Nerve Growth Factor levels in C_6 glioma. *Trans Am. Soc. Neurochem.* 8:140

Schwyn, R. 1967. An autoradiographic study of satellite cells in autonomic ganglia. *Am. J. Anat.* 121:727–40

Shimada, Y., Fishman, D. A., Moscona, A. A. 1969. The development of nerve-muscle junctions in monolayer cultures of embryonic spinal cord and skeletal muscle cells. *J. Cell Biol.* 43:382–87

Simpson, S., Young, J. 1945. Regeneration of fibre diameter after cross-union of visceral and somatic nerves. *J. Anat.* 79:48–65

Smith, A. P., Greene, L. A., Risk, H. R., Varon, S., Shooter, E. M. 1969. Subunit equilibria of the 7S Nerve Growth Factor protein. *Biochemistry* 8:4918–26

Speidel, C. 1964. *In vivo* studies of myelinated nerve fibers. *Int. Rev. Cytol.* 16:173–231

Steiner, G., Schonbaum, E., eds. 1972. *Immunosympathectomy.* New York: Elsevier

Stoeckel, K., Guroff, G., Schwab, M., Thoenen, H. 1976. The significance of retrograde axonal transport for the accumulation of systemically administered nerve growth factor (NGF) in the rat superior cervical ganglion. *Brain Res.* 109:271–84

Stoeckel, K., Thoenen, H. 1975. Retrograde axonal transport of nerve growth factor: specificity and biological importance. *Brain Res.* 85:337–42

Varon, S. 1968. Nerve Growth Factors: a selective review. Conference on the Biology of Neuroblastoma (Seattle, 1967). *J. Pediatr. Surg.* 3:120–24

Varon, S. 1975a. Nerve Growth Factor and its mode of action. *Exp. Neurol.* 48:75–92

Varon, S. 1975b. Neurons and glia in neural cultures. *Exp. Neurol.* 48:93–134

Varon, S. 1975c. Glia, Nerve Growth Factor and ganglionic metabolism. See Hendry 1975, pp. 275–84

Varon, S. 1977a. Neural growth and regeneration: a cellular perspective. *Exp. Neurol.* 54:1–6

Varon, S. 1977b. Neural cell isolation and identification. In *Cell, Tissue and Organ Cultures in Neurobiology,* ed. S. Fedoroff and L. Hertz. New York: Academic. In press

Varon, S., Nomura, J., Shooter, E. M. 1967. Subunit structure of a high molecular weight form of the Nerve Growth Factor from mouse submaxillary gland. *Proc. Natl. Acad. Sci. USA* 57:1782–89

Varon, S., Nomura, J., Shooter, E. M. 1968. Reversible dissociation of the mouse Nerve Growth Factor into different subunits. *Biochemistry* 7:1296–1303

Varon, S., Raiborn, C. W. Jr. 1971. Excitability and conduction in neurons of dissociated ganglionic cell cultures. *Brain Res.* 30:83–98

Varon, S., Raiborn, C. W. Jr. 1972a. Dissociation of chick embryo spinal ganglia and effects on cell yields by the mouse 7S Nerve Growth Factor protein. *Neurobiology* 2:196–212

Varon, S., Raiborn, C. W. Jr. 1972b, Protective effect of mouse 7S Nerve Growth Factor protein and its Alpha Subunit on embryonic sensory ganglionic cells during dissociation. *Neurobiology* 2:183–96

Varon, S., Raiborn, C. W. Jr. 1972c. Dissociation, fractionation and culture of chick embryo sympathetic ganglionic cells. *J. Neurocytol.* 1:211–21

Varon, S., Raiborn C. W. Jr. Burnham, P. A. 1974a. Comparative effects of Nerve Growth Factor and ganglionic non-neuronal cells on purified mouse ganglionic neurons in culture. *J. Neurobiol.* 5:355–71

Varon, S., Raiborn, C. W. Jr., Burnham, P. A. 1974b. Selective potency of homologous ganglionic non-neuronal cells for the support of dissociated ganglionic neurons in culture. *Neurobiology* 4:231–52

Varon, S., Raiborn, C. W. Jr., Burnham, P. A. 1974c. Implication of a Nerve Growth Factor-like antigen in the support derived by ganglionic neurons from their homologous glia in dissociated cultures. *Neurobiology* 4:317–27

Varon, S., Raiborn, C. W. Jr., Norr, S. 1974. Association of antibody to Nerve Growth Factor with ganglionic non-neurons (glia) and consequent interference with their neuron-supportive action. *Exp. Cell Res.* 88:247–56

Varon, S., Raiborn, C. W. Jr., Tyszka, E. 1973. In vitro studies of dissociated cells from newborn mouse Dorsal Root Ganglia. *Brain Res.* 54:51–63

Varon, S., Saier, M. 1975. Culture techniques and glial-neuronal interrelationships in vitro. *Exp. Neurol.* 48:135–62

Varon, S., Tayrien, M. 1977. Influence of substratum on ganglionic cells in surface culture. *Trans Am. Soc. Neurochem.* 8:90

Weineberg, H., Spencer, P. 1975. Studies on the control of myelinogenesis. I. Myelination of regenerating axons after entry into a foreign unmyelinated nerve. *J. Neurocytol.* 4:395–418

Weis, P. 1970. The in vitro effect of the Nerve Growth Factor on chick embryo spinal ganglia:—a light-microscopic evaluation. *J. Embryol. Exp. Morphol.* 24:381–92

West, N., Bunge, R. 1976. Prevention of the chromatolysis response in rat superior cervical ganglion neurons by nerve growth factor. *Abstr. 6th Ann. Meet. Soc. Neurosci.* 2:1038

West, N. Bunge, R. 1977. Observations on the role of nerve growth factor (NGF) and Schwann cells in the chromatolytic response of sympathetic neurons. *Anat. Rec.* 187:747

Weston, J. 1970. The migration and differentiation of neural crest cells. *Adv. Morphog.* 8:41–114

Williams, A., Wood, P., Bunge, R. 1976. Evidence that the presence of Schwann cell basal lamina depends upon interaction with neurons. *J. Cell Biol.* 70:138a

Wlodawer, A., Hogson, K. O., Shooter, E. M. 1975. Crystallization of Nerve

Growth Factor from mouse submaxillary glands. *Proc. Natl. Acad. Sci. USA* 72:777–79

Wood, P. M. 1976. Separation of functional Schwann cells and neurons from normal peripheral nerve tissue. *Brain Res.* 115:361–75

Wood, P., Bunge, R. 1975. Evidence that sensory axons are mitogenic for Schwann cells. *Nature* 256:662–64

Young, M., Murphy, R. A., Saide, J. D., Pantazis, N. J., Blanchard, M. H., Arnason, B. G. W. 1976. Studies on the molecular properties of nerve growth factor and its cellular biosynthesis and secretion. In *Surface Membrane Receptors,* ed. R. A. Bradshaw, W. A. Frazier, R. C. Merrell, G. I. Gottlieb, pp. 247–68. New York: Plenum

Zaimis, E., Knight, J., eds. 1972. *Nerve Growth Factor and Its Antiserum.* London: Athlone Press

Zalewski, A. A., Silvers, W. K. 1973. Trophic function of neurons in homografts of ganglia in immunologically tolerant rats. *Exp. Neurol.* 41:777–81

Ann. Rev. Neurosci. 1978. 1:363–94

AUDITORY MECHANISMS OF THE LOWER BRAINSTEM

❖11512

John F. Brugge and C. Daniel Geisler

Department of Neurophysiology, Department of Electrical and Computer Engineering, and Waisman Center on Mental Retardation and Human Development, University of Wisconsin, Madison, Wisconsin 53706

INTRODUCTION

Brainstem centers of the mammalian auditory system have been under intensive investigation over the past few years. Anatomical, electrophysiological, and biochemical techniques have all been brought to bear on the complex mechanisms by which these centers process acoustic information. In this article, we review work on two closely related cell groups in the auditory pathway, the cochlear nuclei and the superior olivary complex of the medulla and pons, which form major links in the synaptic chain between auditory nerve fibers, on the one hand, and higher auditory centers on the other. The cochlear nuclear complex is the first synaptic station in the auditory pathway. It is within this region that all information encoded in primary auditory nerve fibers is received and recoded. Neurons of the superior olivary complex are principal targets for cells located throughout the cochlear nuclei and are among the first cells to receive and process binaural input. Reciprocal connections between these two structures establishes a basis for functional feedback loops, while a pathway from the superior olivary complex to the cochlear hair cells provides a route for possible central control of primary afferent input. Thus, understanding the mechanisms by which information is processed within these brainstem areas is critical to understanding the function of the entire auditory system. We have selected material that we believe represents the current state of understanding in this rapidly moving field of mammalian sensory physiology. There are other reviews of some of this work and related studies (Hawkins 1964, Eldredge & Miller 1971, Erulkar 1972, Møller 1972, Harrison & Howe 1974a,b, Evans 1975, Goldberg 1975, Webster & Aitkin 1975).

AFFERENT INPUT TO THE CENTRAL AUDITORY SYSTEM

The auditory nerve is comprised of an array of fibers, most of which link inner hair cells of the organ of Corti with second order neurons of the cochlear nuclear

0147-006X/78/0325-0363$01.00

complex (CN) of the brainstem. In material stained by Nissl methods the cochlear complex can be divided on cytoarchitectural grounds into three easily recognizable divisions: the anteroventral (AVCN), posteroventral (PVCN) and dorsal cochlear nuclei (DCN). These divisions have been recognized in over 100 mammalian species, although there are great variations among species in the size and cellular structure of the three subdivisions (Merzenich 1970, Merzenich et al 1973). Early anatomists, using the Golgi method, clearly showed that primary fibers bifurcate upon entering the CN; each fiber sends an ascending branch to the AVCN and a descending branch to both the PVCN and DCN. It is now generally agreed that all auditory nerve fibers terminate within the cochlear nuclear complex (Powell & Cowan 1962, Osen 1970).

Extensive electrophysiological studies of single auditory nerve fibers have shown that many characteristics of their discharge patterns are common to all fibers. In response to a sustained tone, a single fiber exhibits a continuous train of action potentials that adapts slowly over time. The average rate of discharge increases monotonically with increasing sound pressure level over an intensity range of about two orders of magnitude. When stimulated by high frequency tones, the temporal ordering of the discharges can be described as nearly random; the interspike interval distribution has a short mode (less than 12 msec) and a long tail that can be fitted rather well by an exponential distribution. In the absence of stimulation, most fibers display a spontaneous discharge activity that has these same general interval characteristics. For signals with frequency content below about 4 kHz, the instantaneous probability of discharge reflects the waveshape of the stimulating waveform that arises on the basilar membrane in the small region innervated by that fiber. Presumably, the same information recorded in a distal segment of a primary afferent nerve is transmitted along each limb of the bifurcating fiber and made available to each of its terminals within the cochlear complex.

Each auditory nerve fiber is responsive to tones within a limited range of frequencies and intensities called the *response area.* For each fiber, the response area is sharply tuned to one particular frequency called the *best* or *characteristic* frequency. These response characteristics presumably reflect the mechanical tuning properties of the region of the basilar membrane innervated by that fiber. Because the tuning properties, and hence the frequency selectivity, of the basilar membrane change systematically as a function of distance along the membrane, a relationship between best frequency and cochlear place is established. This tonotopic organization of the cochlear partition is preserved in the principal synaptic stations of the auditory pathway.

COCHLEAR NUCLEAR COMPLEX

Cellular Architecture of the Cochlear Nuclear Complex

Osen (1969a) used both the Nissl and Glees methods to arrive at a simple scheme of organization that embraced earlier cytoarchitectonic models of the CN based on Golgi (Lorente de Nó 1933) and protargol silver stained (Harrison & Warr 1962, Harrison & Irving 1965, 1966b) material. Nine cell classes can be identified in the cat CN in Nissl material; the entire complex can be divided into a corresponding number of cell areas according to the distribution of cell types. A more detailed cytoarchitectonic map, based on Nissl, reduced silver, and Golgi material has since appeared (Brawer, Morest & Kane 1974) that also draws together many of the cytoarchitectonic parcellations of the cochlear nuclei proposed by earlier workers. By using criteria similar to those described by Osen (1969a), Kiang and his coworkers (1975) have constructed, from Nissl-stained serial sections, a three-dimensional model of the cat CN built up of cubes, each 80 μm on a side. Thus, neurons can be assigned not only to a particular subdivision of the CN but to a particular cube within it. If adjustments can be made to account for the inevitable interanimal variations that occur in the CN, such a model may provide a useful common frame of reference for detailed reporting of anatomical, biochemical, and physiological data from different animals and different laboratories.

Each of the three subdivisions of the cat CN contains a complete and orderly representation of the tonal spectrum (Rose, Galambos & Hughes 1960). Although the CN has not been mapped completely, these data suggest that auditory nerve fibers innervating a restricted region of the basilar membrane branch, as a roughly horizontal layer within each of the three CN subdivisions, to form isofrequency sheets. This suggested layering is in harmony with the known orderly distribution of primary afferent fibers seen in normal Golgi material and in Nauta-stained sections following lesions in restricted regions of the cochlea (Powell & Cowan 1962, Sando 1965, Osen 1970, Pirsig, Noda & Lehmann 1972). It follows from the morphological diversity of cell populations within each subdivision that an isofrequency sheet may comprise several cell classes. Furthermore, cells that share the same best frequency, but are reached by different branches of eighth-nerve fibers, may be contacted by very different terminal structures. This morphological heterogeneity is reflected in the variety of complex response patterns that can be recorded from CN neurons (Kiang et al 1973, Morest et al 1973, Kiang 1975).

ANTEROVENTRAL COCHLEAR NUCLEUS The AVCN is separated from the PVCN and DCN by entering rootlets of the auditory nerve and a narrow band of granule cells. Within the AVCN of the cat, three regions can be distinguished according to the morphology of the primary afferent terminals seen in Golgi material (Brawer, Morest & Kane 1974, Brawer & Morest 1975). The rostral region encompasses Osen's (1969a) large spherical cell area. The caudal region, which has been divided into a ventral and dorsal part, includes Osen's areas of small spherical, globular, and multipolar cells.

There is good circumstantial evidence that the spherical cells seen in Nissl preparations correspond to the spheroid-shaped bushy cells seen in Golgi material. Bushy cells are found throughout the AVCN but are most heavily concentrated in rostral areas and, like the large spherical cells, are innervated by large end bulbs of Held. Bushy cells in the caudal region of the AVCN receive generally smaller end bulbs of Held. Ellipsoid-shaped stellate cells, a second category of large neurons in the CN of the cat, are scattered throughout the AVCN. They do not receive end bulbs but are contacted by auditory nerve terminals or collaterals in perineuronal nests. Some multipolar and globular cells that are distributed throughout the central part of the

AVCN may correspond to these stellate cells. Lorente de Nó (1976), on the basis of his studies of Golgi material, has suggested that the central branches of those spiral ganglion cells which establish contacts with a very few neighboring internal hair cells, innervate the large spherical cells in the oral pole of the AVCN. The main branches of other types of spiral ganglion cells are said to reach only the caudal regions of the AVCN.

The fine structure of the large spherical cells and the afferents that end upon them have been described in several animals including the rat, guinea pig, chinchilla, and cat (Lenn & Reese 1966, Pirsig, Reinecke & Lehmann 1969, McDonald & Rasmussen 1971, Gentschev & Sotelo 1973, Ibata & Pappas 1976). Four types of axosomatic and axodendritic contacts have been identified (Figure 1). The first type is a large elongated structure, which is identical to the calyciform ending seen in Golgi material. Each terminal makes multiple synaptic contacts with the large spherical cell soma and dendrites. Some contacts are made on dendritic spines. The fine structure of these contacts is similar to that seen in most other axon terminals within the central nervous system that presumably release chemical transmitters. A second type of axosomatic ending is a small bouton containing spherical vesicles that are slightly smaller than those found in the calyciform endings. As a rule, they make

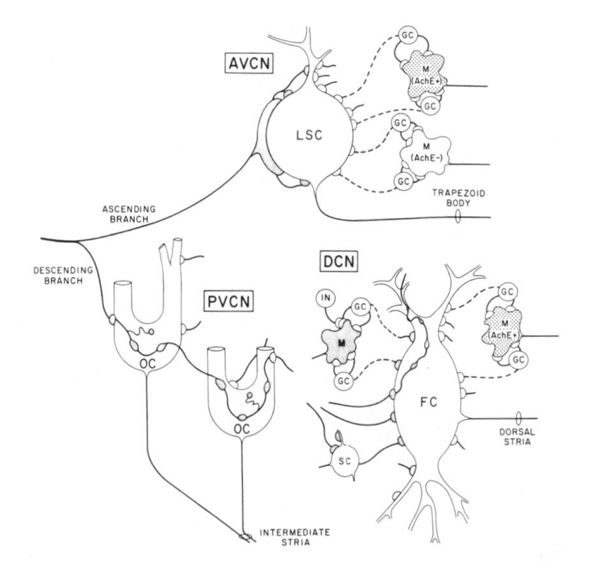

Figure 1 Organization of terminal plexus within regions of large spherical cells (LSC) of AVCN, octopus cells (OC) of PVCN, and fusiform cells (FC) of DCN. Terminals are of cochlear (fine stipples) and central origin. Some mossy-like endings (M) of central origin may contain acetylcholinesterase (coarse stipples). Mossy endings, including those of primary afferents in DCN, contact granule cell (G) dendrites, which in turn may contact principal cells within the respective subdivision. Interneurons (IN) of DCN contact granule cell dendrites. Small cells (SC) of DCN are intermediate links between primary afferents and fusiform cells.

a single contact with the cell soma or dendrite. Some boutons are connected to the calyceal process by a narrow stalk, an observation made earlier in Golgi material (Cajal 1909). The calyceal endings and many of the small boutons with round vesicles disappear after sectioning the eighth nerve. The fact that there are only two or three calyceal endings that contact a single large spherical cell suggests a convergence of primary afferent activity from only a few hair cells in the cochlea. A third type of ending is a bouton containing predominantly flattened vesicles. These endings, which do not degenerate after lesions of the cochlea, are found on both somata and dendrites. They may be terminals of centrifugal axons (Rasmussen 1960), some of which mediate inhibitory activity in the AVCN (Starr & Wernick 1968, Comis 1970). In addition to calyciform processes and boutons containing flattened vesicles, dendrites are contacted by a small number of terminals containing many dense-core vesicles, which suggests that the AVCN is reached by nerve fibers containing catecholamines (Ibata & Pappas 1976). Thus, spherical cells in the AVCN receive, in addition to the end bulbs of Held, large numbers of small synaptic knobs which in part are supplied by collaterals of cochlear fibers and in part by fibers that have a variety of other origins (Lorente de Nó 1976).

The granule cell region of the AVCN contains several different types of synaptic endings. The most common endings contain boutons with round vesicles; flattened vesicles are rarely seen. A small fraction of these endings contains acetylcholinesterase (AChE). A conspicuous component of the superficial layer is the mossy ending that contains round vesicles and makes synaptic contact with several granule cell dendrites in a way reminiscent of cerebellar glomeruli. The sources of AChE-positive mossy-like endings, and perhaps much of the other AChE activity of the molecular and granule cell layers of the cochlear nuclei, appear to be fibers that arise in large cells of the superior olivary complex and reach the CN via the olivocochlear bundle as either collaterals or terminal axons (Osen & Roth 1969). Not all of the mossy endings are AChE-positive. None of them appear to degenerate after surgical section of the eighth nerve. This variety of centrifugal afferent supply suggests that several different populations of SOC neurons contribute to these pathways. The detailed synaptic structure of the other types of cell within the AVCN has not been well studied.

POSTEROVENTRAL COCHLEAR NUCLEUS The descending branch of the auditory nerve projects to the caudal regions of the cochlear complex that includes the PVCN and DCN. It presumably contacts all cell types within the PVCN, although the mode of termination varies greatly from one cell type to the next (Harrison & Irving 1966b, Pirsig 1968, Feldman & Harrison 1969, Osen 1970, Kane 1973, 1974a). The PVCN is composed of a rich variety of cell types recognizable in both Nissl- and Golgi-stained material (Lorente de Nó 1933, Pirsig 1968, Osen 1969a, Brawer, Morest & Kane 1974). Brawer and his associates (1974) have subdivided the PVCN into eight regions by correlating Nissl cytoarchitecture with maps of cell types seen in Golgi preparations.

One of the divisions in the central part of the PVCN is composed almost exclusively of the octopus cells described by Osen (1969a). These cells are the only ones in the PVCN for which the ultrastructure and connections have been well worked out (Kane 1973, 1974a, 1976a,b 1977). The surface of each octopus cell soma is covered with spicules. Usually 4–6 stout dendritic trunks emerge from the perikaryon perpendicular to the caudally coursing auditory nerve fibers. These primary dendrites give off secondary branches, which in turn divide into distal branches. The tertiary dendrites are oriented nearly parallel to the auditory nerve fibers. Massive preterminal degeneration occurs in this region four days after destruction of the cochlea, indicating that it receives an enormous primary afferent input. Of the four cytologically distinct types of terminals that can be identified in electron micrographs of normal and experimental material, two are of cochlear origin. One type, which covers about 50% of the somal surface and 70% of the dendritic surface, appears to be the large boutons of primary afferent fibers identified in Golgi preparations (Figure 1). A second type of bouton ending is seen on somata and dendrites in continuity with small unmyelinated fibers that may be collaterals of primary fibers. Most terminals of these two types degenerate following cochlear ablation, indicating, for the first time in the CN, that two morphologically distinct endings of auditory nerve fibers contact a single morphological cell type. A third type of terminal emerges from unmyelinated fibers to form a long synaptic complex that occupies a large proportion of the postsynaptic membrane. It occurs less frequently than the other two and, in contrast to them, contains flattened vesicles. These endings degenerate after large lesions of the contralateral superior olivary complex (SOC). Small endings, located on distal dendrites, degenerate after ipsilateral SOC lesions and may have their origins in the peri-olivary region.

DORSAL COCHLEAR NUCLEUS The DCN comprises three distinct neuronal layers beneath a layer of ependymal cells. The outer molecular layer contains small cells and granule cells embedded in a fine neuropil. The fusiform-cell layer, which gives the DCN its strikingly laminar appearance, is composed of the somata of fusiform cells and a high concentration of granule cells. The deep layer contains small cells, granule cells, and giant cells. The descending branch of the auditory nerve ramifies widely in the DCN. Although some earlier experimental studies did not include the granule or small cells of the molecular layer, or the apical dendrites of the fusiform cells as targets of primary afferent fibers (Powell & Cowan 1962, Osen 1970), more recent work leaves little doubt that these elements are contacted by first-order fibers (Cohen, Brawer & Morest 1972, Kane 1974a,b).

The ultrastructure of several of these neurons, and the terminals that impinge upon them, have recently been correlated with the morphology of this region as seen in Golgi material (Cohen, Brawer & Morest 1972, Kane 1974a,b). Within the DCN six types of axons can be recognized in rapid Golgi impregnations, two of which are auditory nerve fibers. Primary afferents terminate upon the somata and dendrites of fusiform cells as *boutons en passant* and *boutons termineaux* (Figure 1). They also end upon the somata of small cells; these cells project in turn upon fusiform cells and upon the somata and dendrites of granule cells. Primary afferents also reach the giant cells in the deeper layers. Primary collaterals extend into both the fusiform cell and molecular layers to terminate as irregular-shaped swellings, reminiscent of mossy fiber endings in the cerebellum. Mossy terminals form the core of glomeruli, which also include granule cell dendrites. Axons of granule cells may project upon fusiform cells to provide still another indirect afferent path to this cell type. Finally, eighth-nerve fibers contribute to axonal nests that are found within the fusiform layer. Axonal nests are distinct from glomeruli in that they are not isolated by glial envelopes but consist of tightly packed neuronal profiles including distinct types of dendrites. Some of the mossy terminals in this region are AChE-positive and do not degenerate after cochlear nerve destruction, which suggests that they arise from fibers of the olivocochlear bundle (McDonald & Rasmussen 1971). Thus, it looks as though mossy endings in the DCN may originate from several sources, as they do in the cerebellar cortex. Kane (1974a,b) has described three other types of axon terminals in the DCN. One type contains pleomorphic vesicles and usually occurs upon the dendrites of the outer fusiform cell and molecular layers. In glomeruli, endings of this type contact the outer surface of granule cell dendrites. Another type, which contains flattened vesicles, is never seen in glomeruli but contacts proximal dendrites and the somata of fusiform and small cells. Endings of a third type are seen only in the axonal nests where they make contact with several types of dendrites. The sources of these latter terminals are unknown.

Neuropharmacology of the Cochlear Nuclear Complex

AMINO ACIDS There is considerable evidence that amino acids and small peptides serve as neurotransmitters in the mammalian central nervous system (Iverson et al 1973). Because of the widespread inhibitory activity that exists within the CN, studies have been conducted on two transmitter candidates, gamma-aminobutyric acid (GABA) and glycine, which act as inhibitory neurotransmitters in some other regions of the nervous system. The concentrations of GABA and the activity of the related enzymes, glutamate decarboxylase (GAD) and GABA-transaminase (GABA–T), in the CN as a whole, are less than those in the inferior colliculus and about the same as those in the cerebral cortex (Tachibana & Kuriyama 1974, Fisher & Davies 1976, Fex & Wenthold 1976). However, these compounds are not uniformly distributed throughout the CN. Although it constitutes only about one third of the total weight of the nucleus, the DCN contains nearly 50% of the total GABA, GAD, and GABA–T (Tachibana & Kuriyama 1974, Fisher & Davies 1976). Godfrey & Matschinsky (1976) have developed a quantitative histochemical method for constructing a three-dimensional biochemical map of the cochlear complex and have plotted their results on the CN block model of Kiang et al (1975). They show that the molecular and fusiform layers of the DCN contain the highest levels of GABA in the cochlear nucleus. In addition, GABA is present in the AVCN with the highest levels in the more dorsal and rostral parts. It is also found within the deep layers of the DCN (Lowry et al 1974, Godfrey & Matschinsky 1976, Godfrey et al 1977). In the cat, Davies (1975) has found the highest levels of GABA–T in the fusiform and giant cells of the DCN and in the spherical and multipolar cells of the AVCN. In the guinea pig CN, globular, multipolar, and giant cells have appreciable GABA–T activity (Davies 1973).

The levels of glycine have also been shown to be high in the superficial layers of the DCN, although not as high as the GABA levels. Furthermore, the distribution patterns for glycine and GABA do not coincide completely (Lowry et al 1974,

Godfrey et al 1977). GABA is not likely to be a transmitter for primary afferent fibers, since it is not found in the auditory nerve (Godfrey et al 1977), nor is GAD localized in the eighth nerve terminals (Fex & Wenthold 1976). Moreover, cochlear lesions do not substantially change the level of GAD in the cochlear nucleus (Fisher & Davies 1976). Glycine levels within the auditory nerve are also not impressive (Godfrey et al 1977). Glutamate is rather evenly distributed over the cochlear nucleus, wheras aspartate levels are slightly higher in the ventral cochlear nucleus as compared with the DCN or the granule cell fields (Godfrey et al 1977).

ACETYLCHOLINE The presence of high concentrations of AChE within the CN and some of the centrifugal fibers that reach it, together with the differential effects on CN neurons of locally applied pharmacological agents, provides strong evidence that ACh may be a transmitter substance for two of the afferent systems of central origin that terminate in the CN.

By using histochemical methods in the cat and chinchilla, Rasmussen (1967) traced AChE-positive fibers from the lateral nucleus of the superior olivary complex (LSO) to the AVCN, where they were seen to disperse diffusely among dendrites and somata. In electron micrographs, AChE-positive endings have been localized within the AVCN of both species (McDonald & Rasmussen 1971). The greatest concentration of reactive terminals is in the granular layer, although AChE-positive synapses are seen making contact with large neurons in the AVCN. AChE-positive terminals survive section of the auditory nerve, although not all the endings that remain after cochlear deafferentation are AChE-positive. Further evidence for the cholinergic nature of this pathway comes from electrophysiological experiments. Direct-current stimulation of the medial aspect of the ipsilateral LSO results in increased excitability, and often a reduction in threshold to air-borne sounds, in cells of the AVCN (Comis & Whitfield 1968). Local injection of ACh into the AVCN produces similar effects (Whitfield 1968). Moreover, the effect of current stimulation can be partially blocked by the local application of cholinergic blocking agents. When hemicholinium, a blocker of ACh synthesis, is injected into the brain ventricles, the level of ACh within the cochlear nucleus can be reduced by some 85% following electrical stimulation of this pathway. Finally, when hemicholinium is applied locally to the CN and the superior olive stimulated electrically to deplete presynaptic stores of transmitter, the electrical excitability of neurons in the CN is reduced and the sound threshold increased (Comis & Davies 1969).

Comis (1970) has found that electrical stimulation of neurons within, or dorsal to, the lateral part of the LSO produces inhibition in the majority of neurons sampled in all parts of the cochlear nucleus. Neurons inhibited by this olivary stimulation are also inhibited by local application of noradrenaline but are excited by ACh. Under normal circumstances some of these neurons are likely to be inhibited indirectly via the cochlea, by activation of the olivocochlear bundle, while others are probably affected directly over the pathway from the LSO to the CN. Comis (1970) also made the interesting observation that when excitation of a CN neuron to electrical stimulation of the LSO was abolished by the local application of hemicholinium, subsequent electrical stimulation of the same site in the olivary

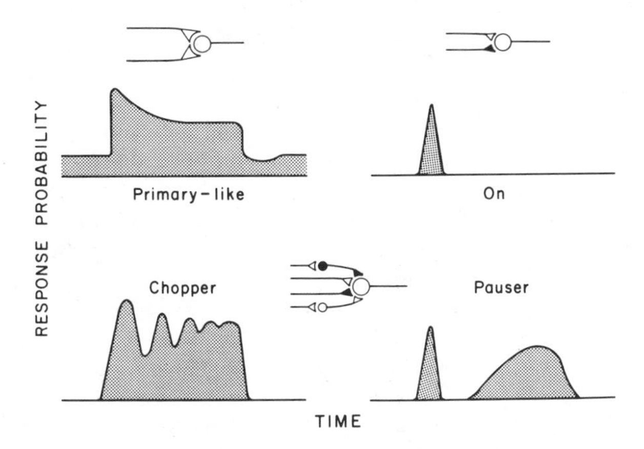

Figure 2 Schematic drawing of poststimulus time histograms showing four categories of temporal discharge patterns that can be recorded from cells in CN (after Pfeiffer 1966b). Each inset indicates a model of spatial arrangements of excitatory synapses (open symbols) and inhibitory synapses (shaded symbols) due to primary afferent inputs that could produce that response.

There is strong, but indirect, evidence that at least some primary-like neurons in the AVCN are spherical cells. Primary-like responses have been obtained from many neurons in the spherical cell area of the AVCN. Most convincing, however, is the fact that the extracellularly recorded action potentials of many AVCN neurons show a positive deflection just prior to the usual diphasic waveshape (Pfeiffer 1966a, Goldberg & Brownell 1973). This initial positive deflection has properties of a presynaptic potential and appears to reflect a large, synchronized synaptic input such as would be expected to occur at a terminal end bulb. The extremely short click latency of these neurons, which is only about 0.5 msec longer than that of auditory nerve fibers (Kiang, 1975), also suggests a very secure synaptic connection. Furthermore, primary-like discharges occur in those fibers of the trapezoid body that originate in the spherical cell region of the AVCN (Brownell 1975). The overall picture that emerges from these studies is that within AVCN there exists a population of cells, probably large spherical cells contacted by end bulbs of Held, that relay with high fidelity information they receive from a very few primary afferent fibers. Thus, the extensive studies of the markedly nonlinear response of primary-like AVCN neurons to multicomponent stimuli (Rose et al 1974, Smoorenburg et al 1976, Greenwood, Merzenich & Roth 1976, Gibson et al 1977) are of considerable importance. Not only do the results describe stimulus-response relationships within a major component of the AVCN, but they also cast a great deal of light on the response characteristics of the less-accessible auditory nerve fibers and, hence, indirectly upon the vibration patterns of the basilar membrane.

Not all cells within the AVCN that might be classified as primary-like show all of the discharge properties of auditory nerve fibers. For example, some of these cells display irregular spontaneous activity that has a somewhat different probability distribution than that seen in the auditory nerve (Kiang et al 1965, Pfeiffer 1966b). An intriguing model, which is compatible with limited convergent input, can mimic this departure from an exponential distribution. The model requires that (*a*) a second-order cell should receive inputs from several spontaneously active primary fibers; (*b*) incoming interspike intervals should occur randomly with an approximately exponential distribution; and (*c*) each incoming pulse should trigger an outgoing one (Molnar & Pfeiffer 1968). Some AVCN cells respond to low-frequency tonal stimuli with a sustained discharge, but, unlike primary-fiber patterns, the discharges tend to be synchronized to the stimulus wave form only at very low frequencies; other cells show no phase synchrony at any audible frequency (Rupert & Moushegian 1970, Rose et al 1974, Britt 1976, Smoorenburg et al 1976). The inability of these cells to relay low-frequency information has not yet been accounted for.

Neurons that display primary-like discharge patterns to best-frequency stimuli have been seen in the PVCN and the DCN as well. Some of those in the PVCN have been tentatively identified as globular cells (Kiang et al 1973). Unlike most primary-like cells in the AVCN, these neurons often have a very brief pause in the firing pattern immediately following the initial discharge at tonal onset, and they usually show inhibitory regions surrounding their excitatory response area (Kiang et al 1975, Brownell 1975).

Responses that occur chiefly at the onset of a best-frequency stimulus were classified by Pfeiffer (1966b) as *on* patterns. In these cells, onset responses are obtained at all frequencies and intensities within the response area (Britt & Starr 1976a). As a rule, these neurons are not active spontaneously. Godfrey, Kiang & Norris (1975a) subdivided this class into two subclasses. They call neurons that respond only at sound onset *on-type I;* those that, in addition to generating onset spikes, display low levels of activity throughout the stimulation period are referred to as *on-type L.* While on-responses are recorded throughout the CN of the cat, they are the only type of response recorded in the octopus cell region of the PVCN (Godfrey, Kiang & Norris 1975a). Fibers in the intermediate stria, which are believed to be largely axons of octopus cells (Warr 1972), also display on-responses (Kiang et al 1973, Adams 1976). An interesting property of cells exhibiting on-responses is their ability to encode the occurrence of rapidly repeating stimuli. Møller (1969a) reports that neurons displaying on-patterns in the rat CN discharge to each repetition of a stimulus, whether it is a click or a brief tone pulse, at repetition rates reaching 700 per sec. The length of the silent period between stimuli, rather than stimulus rate *per se,* seems to be the critical determinant of firing probability. Godfrey, Kiang & Norris (1975a) found that on-type I neurons in the cat CN not only display click-evoked patterns similar to those recorded in the rat, but also encode periodicities of short duration low-frequency tones. As Møller (1969a) observes, these neurons carry considerable information about stimulus periodicity and perhaps play a role in detecting periodicity pitch. Identification of some of these on-neurons with octopus cells (Kiang et al 1973) is strengthened by the observation that on-neurons display the broadest response areas of any cell type in the CN (Møller 1969a, Godfrey, Kiang & Norris 1975a). This broadness would be expected for those cells which have long dendrites oriented so as to intercept primary fibers from widely spaced segments of the basilar membrane. Intracellular recording from these cells gives little clue as to the synaptic mechanisms involved in generating the onset response. Britt & Starr (1976a) showed that membrane potentials recorded from these cells have an initial large amplitude depolarization corresponding to the onset burst of spikes that may be followed by a lower amplitude sustained depolarization. Rarely is hyperpolarization following stimulus onset observed, which suggests that inhibitory synaptic activity is generated at sites distant from the recording electrode.

The remaining two response categories defined by Pfeiffer are much more difficult to identify with any particular cell type since the response patterns are quite variable and can change from one category to another when stimulus conditions are changed (Godfrey, Kiang & Norris 1975b, Britt & Starr 1976a). The *chopper* pattern shows, within the first 25–50 sec after stimulus onset, a response periodicity that is unrelated to stimulus frequency. An interesting model that accounts for these discharge periodicities assumes that the generating cells receive many randomly occurring inputs, each of which causes an EPSP of a magnitude that is small compared to that needed to generate a discharge (Molnar & Pfeiffer 1968). The time required for the summation of the large number of EPSPs necessary to reach threshold determines the periodicity of the output discharges. Van Gisbergen and his co-workers (1975b) found that neurons displaying chopper patterns have long onset latencies, which is consistent with the concept of the gradual buildup of membrane potential. The particular time course associated with the postulated increase in cell potential may well be an individual characteristic of each cell. This possibility is suggested by a neuron model that uses a changing threshold voltage having a specified time constant that is unique to any one cell (Geisler & Goldberg 1966). This model produces interspike interval patterns with statistical properties that are very similar to those of certain CN cells studied by Goldberg & Greenwood (1966). Cells generating the chopper type of pattern are found throughout the CN. They constitute the predominant type of cell in the PVCN, although they can vary considerably in their response characteristics (Godfrey, Kiang & Norris 1975a). Within the DCN, chopper patterns are primarily recorded in the deep layer (Godfrey, Kiang & Norris 1975b).

A fourth class of response pattern shows an initial burst of spikes followed by a pause and then a resumption of activity. Like the chopper pattern, this *pauser* pattern can change with stimulus parameters. Neurons displaying this pattern are encountered primarily in the molecular and fusiform-cell layers of the DCN (Godfrey, Kiang & Norris 1975b), although there is evidence that they occur elsewhere in the CN (Caspary 1972). Membrane events that accompany the pauser pattern are variable and reveal little of the inhibitory synaptic activity that causes this particular type of response activity (Britt & Starr 1976a).

An additional class that is often added to Pfeiffer's scheme includes the *buildup* patterns. These patterns vary a great deal in their time course but have in common a period of time, beginning shortly after stimulus onset, during which there is a slowly increasing discharge rate. This response augmentation period may be preceded by a large onset response, a mild onset response, or no onset response at all (Britt & Starr 1976a). The buildup patterns are not a general characteristic of the cells displaying them, for the pattern changes with stimulus conditions. Like pauser patterns, buildup responses are largely encountered in the molecular and fusiform-cell layers of the DCN. Britt & Starr (1976a) have evidence that in cells displaying this pattern, active inhibition, manifested by membrane hyperpolarization, plays a role both in generating the discharge patterns evoked by best-frequency stimulation and in inhibiting spontaneous activity when certain other tones are sounded.

Evans & Nelson (1973a) recently added another dimension to neuron classification. Their scheme is based on the relative balance between a cell's excitatory and inhibitory inputs. They divide neurons into five classes, depending on the response patterns evoked by tone bursts over a wide range of frequencies and intensities. At one end of the scale are neurons that are only excited by tonal stimuli. These cells, found most in the VCN, probably generate the primary-like discharge patterns. At the other end of the scale are neurons that exhibit only inhibitory responses. The other three classes are intermediate between these extremes. Figure 3 gives a particularly lucid representation of the response patterns that can be obtained from a neuron that is excited or inhibited according to the frequency and intensity of the tone burst. Neurons exhibiting the largest inhibitory influences are found mostly in

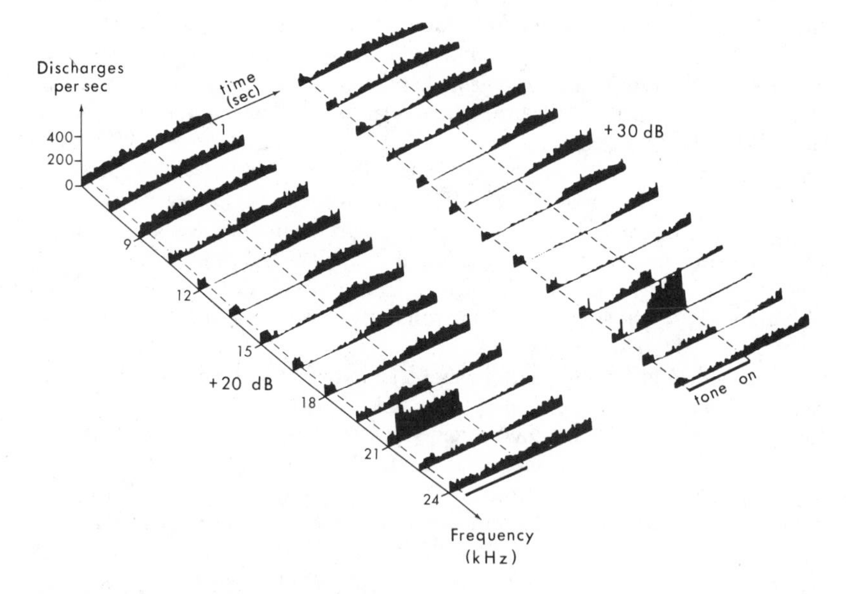

Figure 3 Array of poststimulus time histograms obtained from a single CN neuron by using pure tones of 13 different frequencies and 2 intensities. Notice the differences in the temporal discharge patterns. Intensity level refers to the electrical input to the earphones, relative to threshold intensity for best-frequency tones (From Evans & Nelson 1973a).

the DCN, and, as might be expected for that location, are extremely sensitive to barbiturate anesthesia (Evans & Nelson 1973a, Young & Brownell 1976). At least some of the inhibitory influences exerted in varying degrees on CN cells have been shown to be basically independent of inputs of central origin. In a surgical *tour de force,* Evans & Nelson (1973b) isolated the cochlear nucleus from the brainstem and still observed inhibitory responses in CN neurons. In the same experiment, inhibitory activity was also obtained in the DCN by electrical stimulation of the AVCN.

PROCESSING OF COMPLEX STIMULI The use of sinusoidal stimuli has been essential in delimiting the frequency ranges to which individual CN neurons are sensitive. The success of signal-analysis techniques in other fields has prompted some investigators to use a neuron's tonal response properties to account for its responses to more complex stimuli. For some CN neurons, van Gisbergen et al (1975a) successfully employed linear theory to predict, from responses to tonal stimuli, certain temporal properties of the responses evoked by wide-band noise. In an earlier study, Greenwood & Goldberg (1970) found that the effects of increasing the bandwidth of a noise signal could be correlated with the way in which a neuron responded to increases in tonal intensity. Neurons that responded to intensity increases with monotonically increasing discharge rates were excited by modest increases in the noise bandwidth. By contrast, neurons that showed nonmonotonic rate-intensity curves—an indication of significant inhibitory inputs—almost always responded to increasing noise bandwidth with a reduction in response rate. In another interesting noise study, Møller (1970) showed, by a comparision of threshold intensities, that the effective bandwidth over which these neurons integrated input was very closely related to the bandwidth of their pure-tone response areas. In yet another study, Bilsen et al (1975) were able to show that certain aspects of the responses of some CN neurons to noise with a cosine-modulated spectrum could be predicted from the noise spectrum by using spectrum-weighting functions that closely resembled the frequency sensitivities of the neurons.

Often, however, it is unclear from a response area just how the cell will integrate neural input generated by various frequency components of a complex sound. For example, Møller (1976) described one particular CN neuron that responded in an excitatory manner to band-pass noise presented alone but was inhibited by that same noise sounded with a best-frequency tone. Greenwood & Goldberg (1970) also report complex interactions between tone and noise stimuli. In one case they found that adding noise to a tonal stimulus could either augment or diminish the cell's response rate, depending upon the intensity level of the *tone.* Finally, there are even rare neurons that are not excited by tones but that respond to the very broad frequency spectrum of click stimulation (Mast 1970b, 1973, Britt & Starr 1976a).

Perhaps the chief difficulty in using response-area representations to predict neural responses to complex stimuli is the reliance on the frequency-domain representation of acoustic signals. Nowhere is this better illustrated than with frequency-modulated (FM) signals. The frequency spectrum of an FM signal is, by definition of spectra, not a function of time. Nevertheless, the output of a tuned filter excited by a slowly modulated tone is large only during those time periods in the modulation cycle when the instantaneous frequency is approximately equal to the best frequency of the filter (e.g. see Møller 1969b). Clearly FM stimulation of an array of tuned filters would produce quite complex temporal output patterns. It is not surprising therefore that the degree of success in predicting responses of a CN cell to FM sinusoidal stimuli is related to the complexity of innervation. By using tones with frequency-modulating wave forms that were triangular in shape, Britt & Starr (1976b) obtained a wide variety of response patterns from CN neurons. As expected from their primarily excitatory input, primary-like CN neurons respond well to all FM stimuli. Also as expected, on-neurons, that are able to respond to each of a series of separate stimuli at high repetition rates, are virtually silent in response to FM signals with very low-frequency modulating waveforms (20 sec stimulus period), whereas they respond as vigorously as primary-type neurons to rapidly modulated stimuli (20 msec stimulus period). In sharp contrast, neurons displaying buildup and pause response patterns react to FM stimuli in completely unpredictable ways. Stimulus waveforms generated by use of the fastest and slowest modulation rates are quite effective for these neurons, whereas lower response rates are obtained at intermediate modulation rates. Further complexities of the responses of cat CN neurons to FM stimuli have been reported by Erulkar, Butler, & Gerstein (1968), who separated response patterns into three groups, according to the similarity between the patterns obtained during the periods of rising instantaneous frequency and those recorded during the falling frequency phase. The extreme example of one response type, the asymmetrical, is illustrated by neurons that discharge only during one direction of frequency sweep (Britt & Starr 1976b). Depending on the particular distribution of dendritic inputs that is assumed, any one of these response patterns can be reproduced by current neuron models (Fernald 1971).

Responses to naturally produced vowel sounds have been obtained from CN neurons (Moore & Cashin 1974). For many neurons, the general level of responsiveness to any particular vowel may be inferred from the relative amounts of sound energy that fall into the neuron's excitatory and inhibitory frequency bands. However, curious exceptions to the general rule of balancing excitatory and inhibitory inputs have been observed. In some neurons, the response to a vowel can be augmented by adding a tone that is inhibitory when presented alone. Different types of CN neurons seem to respond to the vowel sounds in characteristic ways. For example, the output of primary-like neurons is similar to that of primary afferents, whereas on-neurons seem to emphasize the temporal peaks observed in the primary responses.

Other types of complex stimuli have been used to activate CN neurons. However, as the information handling aspects of CN neurons are not well understood, it is difficult to relate those results to basic neuronal processes.

SUPERIOR OLIVARY COMPLEX

Cellular Architecture of the Superior Olivary Complex

The superior olivary complex comprises a number of closely grouped nuclei that span the ventral region of the pons. Four of the nuclei are easily recognized in

Nissl-stained material in the cat (Berman 1968). The most conspicuous of the four, in coronal sections, is the S-shaped superior olivary nucleus proper, usually referred to as the lateral superior olivary nucleus (LSO). The accessory, or medial, superior olivary nucleus (MSO) is a curved band of spindle-shaped cells that borders the LSO ventrally, rostrally, and medially. A third cell group, lying within the fibers of the trapezoid body, ventromedial to the MSO, is the nucleus of the trapezoid body (NTB), [also referred to as the medial nucleus of the trapezoid body (Morest 1968, 1973)]. Finally a broad band of cells, collectively referred to as the preolivary nuclei (PON), curves around the MSO and the LSO, and borders these structures ventrally, laterally, and rostrally. Cajal (1909) recognized two nuclei within the PON, and termed them the external and internal preolivary nuclei, respectively. These groups have since been referred to as the lateral (LPO) and medial (MPO) preolivary nuclei (Papez 1930, Stotler 1953) or the lateral and ventral nuclei of the trapezoid body, respectively (Morest 1968, 1973). Goldberg & Brown (1968) located these nuclei in the dog, and Harrison & Warr (1962) have recognized the homologous groups of cells in the rat. Scattered groups of cells form a cap laterally, dorsally, and medially over MSO and LSO; in the cat these cell groups are poorly defined in Nissl material, but have been referred to collectively as the peri-olivary nuclei.

Irving & Harrison (1967) and Harrison & Feldman (1970), in addition to reviewing the comparative anatomical literature on the superior olivary complex, measured the sizes of the LSO, MSO, and NTB in a number of mammalian species. There is a wide variation in size and shape among the homologous nuclear groups, but there is no evidence that the absolute sizes of these nuclei are related to taxonomic order. Of particular interest, however, is the close relationship between eye diameter, size of the abducens nucleus, and size of MSO in all animals studied. Since the MSO may be involved in mechanisms of sound localization, it has been suggested that these mechanisms may include a visual reflex component.

LATERAL SUPERIOR OLIVARY NUCLEUS Scheibel & Scheibel (1974) have analyzed the structure of the cells and the afferent neuropils within both the LSO and MSO in Golgi material of the cat. The LSO is composed of cells, which in cross section appear to have bi-tufted dendrites that extend more-or-less at right angles to the axis of curvature of the nucleus. However, in sagittal section, the dendrites of the LSO neurons are seen to branch extensively, often running for considerable distances in the rostrocaudal direction. These neurons are, in fact, multipolar cells with two-dimensional dendritic fields lying in a rostro-caudal plane. Afferent terminals arising from fibers of the trapezoid body distribute between these flattened and extensively interwoven dendritic trees as elongated sheets of neuropil over many hundred microns, again in a rostrocaudal direction (Figure 4). The afferent terminal plexus contains a variety of presynaptic specializations along its length, and the fibers contact distal and proximal dendrites as well as cell somata.

The LSO receives bilateral innervation. The ipsilateral input comes via the trapezoid body from the rostral spherical cell area of the AVCN (Stotler 1953, Harrison & Warr 1962, Warr 1966, Osen 1969b). Within the LSO, synaptic endings consist of ringlike boutons distributed over the somata and proximal dendrites (Harrison

& Warr 1962). Osen (1969b) suggests that in the cat, small spherical cells are the major contributors to this path, since both the small spherical cell area and the LSO represent the entire acoustic spectrum; large spherical cells, which receive predominantly low-frequency inputs, would then project to the predominantly low-frequency MSO. Harrison & Irving (1966a), on the other hand, have traced fibers to the LSO, in the rat, from an area in the rostral part of the AVCN (region III) which contains large spherical cells.

The ipsilateral input to the cells of the LSO is almost exclusively excitatory and is organized tonotopically (Tsuchitani & Boudreau 1966, Guinan, Norris & Guinan 1972, Tsuchitani 1977). Neurons in the tip of the dorsolateral limb of the nucleus have the lowest best frequencies. Neurons located further along the tilted *S* have successively higher best frequencies, and the highest frequencies are represented in the ventromedial tip. This organization agrees with that of the afferent input from the ipsilateral cochlear complex (Warr 1966, van Noort 1969). An interesting aspect of the tonotopic organization of the LSO is the relatively large volume of tissue devoted to high-frequency representation. The available best-frequency maps indicate that within the range of nearly eight octaves represented within the LSO, well over 50% of the cells in a cross-sectional area have best frequencies restricted to the three highest octaves, between 5 kHz and 44 kHz.

Stimulation of the contralateral ear inhibits the response to ipsilateral ear stimulation in the great majority of neurons in the LSO (Boudreau & Tsuchitani 1968, 1970, Guinan, Norris & Guinan 1972, Tsuchitani 1977). Anatomical and physiological

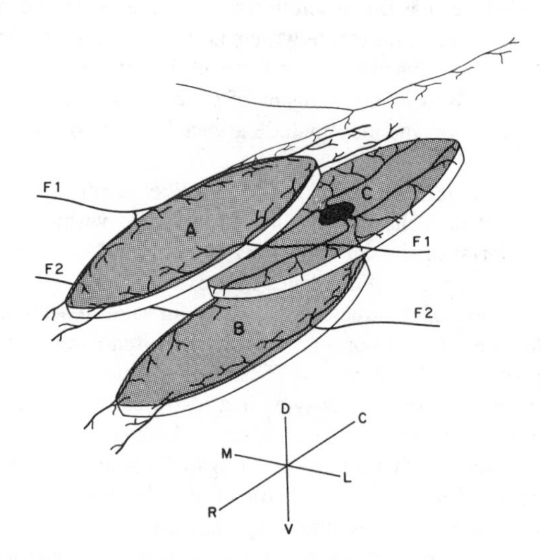

Figure 4 Schematic drawing of three (A–C) interleaving disc-shaped neurons in LSO or MSO. Dendritic organization is shown only in neuron C. The flattened afferent neuropil between dendritic fields in shown as it would appear from two cells of the left and right cochlear nuclei, respectively, which have two different best frequencies (*F1* and *F2*).

evidence suggests that this inhibitory input to the LSO originates from contralateral globular cells in the caudal AVCN and the rostral PVCN, and reaches the LSO indirectly via the principal cells of the NTB. Despite the fact that afferent inputs from the two sides are derived from different cell types, they encode sound stimuli in almost the same way. For any one LSO cell, the best frequency for contralateral (inhibitory) stimulation is almost identical to that for ipsilateral (excitatory) stimulation; the inhibitory response area may be only slightly broader than the excitatory response area (Boudreau & Tsuchitani 1970). Moreover, response thresholds of a neuron at best frequency are similar for the two sides. Total conduction time from the contralateral ear is about equal to that from the ipsilateral ear, because the entire ipsilateral response, including the initial spike, can be inhibited by simultaneous contralateral stimulation. Apparently the high conduction velocity and secure, short-latency synaptic transmission involved in the contralateral path (see below) compensates for its longer path length and the intervening synapse. Thus, to the first approximation, the number of discharges evoked in a neuron in the LSO is due to the difference between the level of excitatory drive from the ipsilateral ear and the level of inhibitory drive from the contralateral ear (Boudreau & Tsuchitani 1970). The few neurons in the LSO that do not receive inhibitory input from the contralateral ear are apparently confined to the dorsolateral and ventromedial tips of the nucleus, where the neurons have best frequencies below 1.0 kHz or above 30 kHz, respectively.

The discharge patterns of LSO neurons in response to tones are very regular (Boudreau & Tsuchitani 1970, Guinan, Norris & Guinan 1972, Tsuchitani 1977) and, using Pfeiffer's scheme, would be classified as chopper patterns. As previously suggested, such patterns may indicate that these neurons receive many different inputs, a conclusion that is compatible with their large dendritic fields. Yet these large numbers of inputs must have very similar properties, for the response areas of LSO neurons are very similar to those of sharply tuned CN neurons.

MEDIAL SUPERIOR OLIVARY NUCLEUS Degeneration studies at the light microscopic level in the cat (Stotler 1953, Warr 1966, van Noort 1969), rat (Harrison & Irving 1966a), dog (Goldberg & Brown 1968), and rhesus monkey (Strominger & Strominger 1971) have shown that axons from the contralateral and ipsilateral AVCN terminate on the medial and lateral dendrites, respectively, of the cells in the MSO. A small afferent input to the medial side of the ipsilateral MSO has been noted in the cat (Warr 1966), dog (Goldberg & Brown 1968) and chinchilla (Perkins 1973). Preterminal axons branch predominantly within horizontal layers between the elongated and flattened dendritic fields of MSO neurons, similar to the arrangement seen in LSO (Morest 1973, Scheibel & Scheibel 1974). Since each sheet of neuropil represents only a small segment of the tonotopically organized afferent input, the dendritic field of each neuron is brought into contact with axon terminals representing, at low stimulus levels, a narrow segment of the frequency spectrum. This narrow frequency range is represented along the entire rostrocaudal extent of the nucleus. Thus, neurons in both the MSO and LSO display sharply tuned threshold curves (Boudreau & Tsuchitani 1970, Guinan, Norris & Guinan 1972).

It is difficult to record from MSO neurons in the cat, since the lateral-to-medial thickness of the nucleus is not great, and large amplitude sound-evoked slow potentials make it hard to discriminate action potentials. Moreover, some investigators of this area have depended upon the slow-potential profile for localizing recording sites rather than on more accurate histological methods. Thus, the number of systematically studied cells that has been positively identified within the cat's MSO is relatively small (Guinan, Norris & Guinan 1972). Only in the dog, with its folded MSO, has a sizeable number of these cells been studied (Goldberg & Brown 1968, 1969). Comparing the data from these two species, there are some generalizations that can be made.

First, the large majority of MSO neurons are responsive to stimulation of either ear. Most of these binaurally responsive neurons are excited by stimulation of both ears, but an appreciable fraction are excited by stimulation of one ear and inhibited by stimulation of the other. The predominantly excitatory synaptic activity within this nucleus is reflected in the behavior of the remarkably large slow potentials that are recorded in response to monaural click stimulation. The amplitude and sign of the evoked wave form are systematically related to the ear stimulated and to the electrode's position in the nucleus. This spatial distribution can be accounted for only if one assumes that the predominant synaptic activity contributed by each ear is depolarizing (Galambos et al 1959, Goldberg & Brown 1968). Second, the MSO is tonotopically organized. Neurons with low best frequencies are located dorsally, and those with high best frequencies are located ventrally. For most of the binaurally responsive neurons, the best frequencies and response areas determined for each ear are similar. Third, MSO cells respond in a tonic manner to long tonal stimuli. However, unlike those of the LSO, the discharge patterns of MSO cells are irregular and resemble in many ways the response patterns of primary-like cells of the AVCN.

Electron microscopic studies in cat (Clark 1969, Schwartz 1972, Lindsey 1975) and chinchilla (Perkins 1973) have revealed some of the complex synaptic organization in the MSO. Two types of endings on the somata and dendrites of MSO neurons have been identified. One type of terminal contains spherical vesicles and makes multiple synaptic contacts on the somata and dendrites of the cells that constitute the main band of the MSO and of the cells that lie at the margins of the nucleus. Some of these endings appear to be terminal boutons, whereas others are nodal synapses making contact *en passage*. Similar profiles have been identified in the MSO of the chinchilla, although here they occur exclusively on proximal dendrites, where they make up about 85% of the synaptic population. These endings degenerate following lesions of the AVCN. A second type of ending seen in the cat, which also contacts the somata and dendrites of all MSO neurons, contains flattened vesicles; it is unaffected by cochlear nucleus deafferentation. Synapses containing flattened vesicles have not been identified in the MSO of the chinchilla. However, in this animal boutons containing small or intermediate-size round vesicles, stemming from unmyelinated axons and covering the cell somata and axon hillock regions, also survive CN lesions. The source of these spared afferents is not known. There is no evidence that Golgi type II cells are present in the MSO. It is also unlikely that these incoming fibers are from regions in the CN other than the

AVCN. It is possible that some of them are terminals of recurrent axons. Probably they represent inputs from sources located within the auditory system, possibly from higher auditory centers as part of the descending auditory pathway.

NUCLEUS OF THE TRAPEZOID BODY The neuronal architecture and synaptic organization of this cell group, which lies just medial to MSO, have been examined in greatest detail in the cat (Lenn & Reese 1966, Morest 1968, 1973, Jean-Baptiste & Morest 1975). Morest (1968) has described three neuronal types within the NTB that he has called *principal cells, stellate cells,* and *elongate cells.* Each is contacted by different axonal endings.

The principal cells, which are the most numerous type, receive the large axosomatic calyces of Held stemming from large fibers within the trapezoid body. The origin of these fibers appears to be the globular cells of the contralateral CN (Harrison & Warr 1962, Harrison & Feldman 1970, Warr 1972). The projection to the NTB is tonotopically organized. The orientation of the isofrequency contours corresponds roughly to the orientation of the laminae formed by the flattened and elongated dendritic fields of the principal cells and the arborizations of their afferent terminals (Guinan, Norris & Guinan 1972). Each principal cell receives one calyx, and each calyx embraces but a single principal cell. From electron micrographs the calyces appear to be chemically transmitting synapses (Lenn & Reese 1966, Morest 1973, Jean-Baptiste & Morest 1975). The interface between the calyceal process and the postsynaptic membrane surface shows multiple synaptic complexes. The electrophysiology of this region indicates a secure synaptic coupling between the calyceal ending and the principal cell. In both the dog (Goldberg & Brown 1968) and cat (Guinan, Guinan & Norris 1972), cells in this region display first-spike latencies in response to acoustic stimulation that are uniformly shorter than the latencies of neurons in other regions of the superior olivary complex. In a systematic study of the response properties of 218 neurons in the NTB of the cat, Guinan, Norris & Guinan (1972) found that about 65 percent of the cells sampled in this region showed complex waveforms, similar to those seen in AVCN cells presumed to receive end bulbs of Held (Pfeiffer 1966a). These neurons (which were identified as principal cells) are excited only by contralateral monaural stimulation and are the only NTB cells exhibiting a tonotopic organization. Consistent with this identification are the observations of Guinan, Norris & Guinan (1972) that cells with complex waveforms have response area properties and discharge patterns that closely resemble those of the large-diameter fibers of the trapezoid body (Brownell 1975) and of neurons in the caudal part of the AVCN and the rostral PVCN (Kiang et al 1973, Godfrey, Kiang & Norris 1975a). The destination of the axons of the principal cells has not been established in Golgi material (Morest 1968). However, there is some experimental anatomical evidence to suggest that they project to the LSO of the same side (Harrison & Warr 1962, Rasmussen 1967). Morest (1968, 1973) has shown that there are other inputs to the principal cells that could account for the transformation of the incoming signals and give rise to the complex discharge patterns that are recorded from some other neurons in the NTB (Guinan, Norris & Guinan 1972). In addition to the calyceal endings, a large number of smaller axosomatic and axodendritic contacts are seen arising from fibers that originate in the contralateral CN and other, as yet unknown, sources (Jean-Baptiste & Morest 1975). The calyx-bearing fibers give off various types of collaterals, one of which forms small boutons on the somata of the nearby principal cells. Another type of collateral also ramifies within the NTB, and may contact the dendrites of principal cells.

In both the cat (Morest 1968) and rat (Harrison & Feldman 1970), collaterals arising from the preterminal regions of the calyx-bearing axons, as well as from axons of principal cells, project to a group of cells adjacent to the NTB. In the cat, one of these cell groups is the dorsomedial peri-olivary nucleus, from which axons in the olivocochlear bundle arise. Neurons giving rise to this recurrent fiber pathway may thus sample both the input and output volleys of NTB principal neurons.

The stellate and elongate cells are consistently associated with pericellular nests and endings other than calyces. They have long dendrites that extend at right angles to the isofrequency planes and, thus, can come into contact with afferent bundles representing a wide range of frequencies. These may be the NTB cells that have broadly tuned response areas (Guinan, Norris & Guinan 1972). The identification of stellate and elongate cells by other physiological criteria has been difficult. Guinan and his colleagues suggested that *off-neurons,* i.e. those that respond briefly to the termination of a stimulus, might be stellate cells, as their response patterns are recorded primarily in more lateral regions of the NTB, where the stellate cells are more heavily concentrated.

Another type of neuron seen in the NTB responds only at the termination of a stimulus but, unlike the brief responses of off-neurons, discharges over a relatively long period of time (Guinan, Norris & Guinan 1972).

PREOLIVARY NUCLEI Afferent fibers to the medial (MPO) and lateral (LPO) preolivary nuclei originate within the ventral cochlear nuclei. They travel within the trapezoid body and end primarily upon cell bodies by means of boutons and modified end bulbs of Held, which resemble those seen in the ventral cochlear nucleus. The exact source of these afferents has not yet been determined (Stotler 1953; Harrison & Feldman 1970). Stotler (1953) originally reported that, in the cat, lesions in the ventral cochlear nuclei resulted in widespread degeneration of terminals within both ipsilateral preolivary nuclei. Later, in a Nauta study in the cat, Warr (1966) found that after lesions in the rostral AVCN, preterminal degeneration appeared in the anterior part of the ipsilateral LPO but little, if any, showed up in the posterior portion of the LPO or in the MPO. It would appear, therefore, that cells in the ventral cochlear nucleus, other than those in the rostral part of the AVCN, are the sources of the ipsilateral input to these nuclei. Goldberg & Brown (1968) reported that in the dog and cat the topographically organized afferent input to the LPO arises almost exclusively within the ipsilateral cochlear complex. These observations are in good agreement with electrophysiological findings in this area (Guinan, Norris & Guinan 1972, Tsuchitani 1977), in that the great majority of LPO cells are activated only by stimulation of the ipsilateral ear. In the cat, fibers from both the contralateral and ipsilateral CN terminate in the MPO, but in the dog only a contralateral projection has been identified with certainty (Goldberg & Brown 1968). In the cat, fibers from the contralateral CN terminate throughout the MPO but are more heavily concentrated in the medial part of the nucleus, whereas the ipsilateral fibers terminate mainly in its lateral part. In both the dog (Goldberg & Brown 1968) and cat (Guinan, Norris & Guinan 1972), many of the cells in the MPO are driven only by stimulation of the contralateral ear, as might be expected from its known connections. A smaller percentage of MPO neurons are excited only by stimulation of the ipsilateral ear or by binaural stimulation. It is of interest that the MPO receives a descending input from the inferior colliculus and perhaps the nuclei of the lateral lemniscus (Rasmussen 1964, Moore & Goldberg 1966). Some neurons in the MPO of the dog and cat are characterized by regular firing at low rates (Goldberg, Adrian & Smith 1964, Goldberg & Brown 1968, Guinan, Norris & Guinan 1972). In this respect, these cells resemble olivocochlear fibers, the uncrossed component of which may originate in this region.

PERI-OLIVARY NUCLEI Certain neurons that are distributed as a continuous arc over the LSO and MSO have been divided into as many as six separate cell groups according to their cellular architecture, afferent input, or position relative to major nuclei of the SOC (Warr 1966, 1969, Morest 1968, 1973, Guinan, Norris & Guinan 1972, Tsuchitani 1977). These groups include the anterolateral (ALPO), posterior (PPO), ventromedial (VMPO), dorsomedial (DMPO), dorsal (DPO), and dorsolateral (DLPO) peri-olivary cell groups. The latter two, the DPO and DLPO, include the retro-olivary mass described by Cajal (1909). Goldberg & Brown (1968) identified a portion of this area in the dog as the retro-olivary region that forms part of the adjacent reticular formation. The peri-olivary nuclei receive some of their afferent input, via the three acoustic striae, from cells in both the DCN and PVCN (Fernandez & Karapas 1967, Warr 1969, 1972). Specific cell types within the PVCN send their axons via either the stria of Held or the trapezoid body to overlapping target fields in the peri-olivary nuclei (Warr 1969, 1972). Guinan, Guinan & Norris (1972) and Tsuchitani (1977) have carried out systematic studies of the response patterns of neurons in some of these cell groups. There is a great variety of response patterns, even within a given area, that is perhaps due to the overlapping afferent input from different cell types within the cochlear complex. Many of the cells exhibit response characteristics similar to those of the olivocochlear bundle; they produce regular discharge patterns, low discharge rates, and long latencies.

The DMPO has been studied in some detail by Morest (1968, 1973). It is continuous ventrally with the nucleus of the trapezoid body from which it also receives a collateral afferent input. Two types of neurons are seen in the DMPO in the cat. The more common type, the *elongate cell,* extends dendrites parallel to the laminae of the afferent axonal plexus arising from the NTB. This pattern of organization suggests that the tonotopic organization in the NTB is preserved in the DMPO, at least among the elongate cells. Cells within the DMPO that have sharply tuned threshold curves may be the elongate neurons (Guinan, Norris & Guinan 1972). *Radiate cells,* on the other hand, have long, radiating dendrites that often extend into the surrounding reticular formation. Cells of this type may include those from which Guinan and his associates obtained broadly tuned responses. Response patterns of DMPO neurons vary greatly, reflecting the varied afferent input they receive not only from the NTB and CN but also from the inferior colliculus (Rasmussen 1960). The DMPO is considered a major source of the crossed olivocochlear bundle (Rasmussen 1967, Warr 1975). Indeed, the regular discharge patterns of some DMPO neurons resemble those obtained from fibers of the olivocochlear bundle. Further evidence on this point was obtained by Osen & Roth (1969), who traced AChE-positive fibers from large cells of the DMPO, as well as the DPO and MPO, through the olivocochlear bundle to the CN. Osen & Roth identified the large cells in the DMPO as the multipolar (radiate) neurons described by Morest (1968). Elongate cells, on the other hand, are not AChE-positive (Osen & Roth 1969, Warr 1975) and are not labeled after the cochlea or the sectioned olivocochlear bundle are injected with HRP (Warr 1975). The discharge properties of cells in lateral peri-olivary groups (DPO and DLPO) have been studied by Guinan, Norris & Guinan (1972) and Tsuchitani (1977). The majority of cells in these areas display chopper patterns and are either excited by ipsilateral monaural or binaural stimulation. Cells with low best frequencies are located laterally in this region and high best-frequency cells are located more medially. Recently, Kane (1976b) has shown that elongated lateral peri-olivary neurons on the ipsilateral side, and multipolar peri-olivary cells on the contralateral side, send their axons to the octopus cells of the cochlear complex.

Information Processing in the Superior Olivary Complex

Despite the fact that the SOC cells are at least one synapse removed from CN neurons, the discharge characteristics of the cells in the two groups are similar in many ways. For example, most of the SOC neurons respond in a sustained manner to tonal stimuli, and their discharge patterns generally look like those recorded in the CN. The interspike interval distributions of spontaneously occurring discharges for SOC neurons can be put into the same few categories as those used for CN neurons (Guinan, Norris & Guinan 1972). Furthermore, the discharge properties of some medially located SOC neurons that have been carefully studied under conditions of sound stimulation as well as silence (Goldberg, Adrian & Smith 1964) very closely resembly those of CN neurons. In fact, one type of model originally developed to account for the response patterns of these SOC neurons (Goldberg, Adrian & Smith 1964, Geisler & Goldberg 1966) can be applied equally well to the interval statistics of CN neurons. Thus, it is not surprising that the broad classes of response patterns to tonal stimuli used to categorize CN neurons have been fruitfully employed for SOC neurons as well (Guinan, Guinan & Norris 1972, Tsuchitani 1977). To encompass all of the discharge types, a few new categories have had to be created (e.g. *after* response), and the chopper category has been subdivided by Tsuchitani (1977) into two subclasses. Nevertheless, it would be difficult to identify the location of most CN or SOC neurons just from their monaural discharge patterns.

The main distinction of many SOC neurons is that they receive short-latency information from both CN complexes. Consequently, they very likely play a crucial role in the integration of binaural stimuli. Unfortunately, the mechanisms involved in binaural integration are largely unknown. In fact, there have been only a few studies in which SOC responses have been obtained for systematically varied binaural stimuli.

The principal study in this area has been done on cells in the medial regions of the dog's SOC by Goldberg & Brown (1969). Using a sample drawn almost equally from the MSO and MPO, they found that a very interesting distinction could be made between different classes of binaurally responsive neurons. Neurons that are excited by stimulation of each ear (excitatory-excitatory neurons) are relatively insensitive to interaural intensity change, as long as the average intensity remains constant. A decrease of responsiveness to the attenuated sound at one ear is partially compensated for by the increase in responsiveness to the augmented sound of the other. Because the discharges of most excitatory-excitatory neurons are dominated by sounds from one side or the other, the compensation is not perfect. Nevertheless the discharge rate of these neurons is primarily determined by *average binaural intensity level.*

Neurons that are excited by sound to one ear but inhibited by stimulation of the other ear present a complementary picture. These so-called excitatory-inhibitory neurons are relatively more sensitive to interaural intensity changes than to average binaural intensity changes. When sound levels are increased binaurally, the increase in excitatory input from one side is partially offset by the increased inhibitory input from the other side (see also Moushegian, Rupert & Whitcomb 1964). In contrast, interaural intensity changes cause large response changes, as each side tends to augment the effects of the other. Thus, the discharge rate of an excitatory-inhibitory neuron is principally determined by the *interaural intensity difference.* This sound cue is known to be important to listeners in localizing the source of sounds containing energy at high frequencies. Since the LSO in the cat is composed almost exclusively of excitatory-inhibitory cells, most of which have fairly high best frequencies, it would appear that, at least in the cat, this nucleus plays a major role in high-frequency sound localization.

If the source of a sound is to one side of the head, the sound reaches the farther ear later than it reaches the closer one. Listeners use this disparity to localize low-frequency sounds. In the dog, Goldberg & Brown (1969) recorded from neurons in the MSO and MPO that encode these small interaural time cues; the cell discharge rate is a function of the difference in time of arrival of the tones at the two ears. Recall that some regions of the SOC receive, from the AVCN of each side, input that typically tends to be synchronized with the waveform of an effective low-frequency tone. It is not surprising, therefore, that some SOC neurons integrate the temporal patterns of discharges coming from each side. The work of Goldberg & Brown (1969) indicates that when interaural time delay is adjusted so that discharges from the two CN converge on an MSO cell at nearly the same instant, the probability of a discharge from that cell is maximal. When the interaural time difference is shifted, as usually happens when the sound source moves, the converging discharges arrive at a different time, and the output of the neuron is consequently reduced. At least for best-frequency stimuli, the most effective interaural delay time is not greatly altered either by interaural or binaural changes in intensity. For some neurons in the SOC this delay time remains fixed for many frequencies within the response area (Moushegian, Stillman & Rupert 1971). Rose and his colleagues (1966) have called this delay the *characteristic delay* of the neuron. Neurons that have a characteristic delay are clearly capable of encoding information about the location of a multicomponent low-frequency signal source. The presence of a population of delay-sensitive neurons implies that a place mechanism is acting to encode the location of a low-frequency sound (Rose et al 1966, Goldberg & Brown 1969). It is a common finding in sensory systems that the topography of the sensory surface is *preserved* in successive populations of neurons along the relevant central pathway. The central representation of low-frequency auditory space, on the other hand, may be *created* by temporal and spatial neural interactions that depend critically upon high-fidelity synaptic transmission of time information from the cochlea to the binaural centers of the brainstem.

In further work on the kangaroo rat, Moushegian and his co-workers (1975) have shown that delay-sensitive MSO cells, including both the binaurally excitable cells and those exhibiting excitatory-inhibitory interactions, can be affected by interaural time delay in several ways. For some cells the response rate, but not the degree of synchronization, is a function of interaural time delay. In others, the two measures co-vary with interaural time delay. On occasion, the intensity level can also have an important effect upon the degree of synchronization to a binaurally presented tone.

Finally, one must be rather careful when characterizing as solely excitatory or inhibitory the effects on an SOC neuron of stimulating a particular ear. Mixed inhibitory and excitatory effects from one side can be clearly demonstrated in some cells (Moushegian, Rupert & Whitcomb 1964, Guinan, Norris & Guinan 1972). However, even neurons that do not overtly display such mixed effects may be subject to them. For example, Goldberg & Brown (1969) recorded from several neurons that showed either excitatory or inhibitory interactions depending upon the particular value of interaural time delay. The discharge rates that occurred at the most favorable time delays were greater than the sum of the monaural response rates and clearly showed mutually excitatory responses. The rates that occurred at the least favorable delays were less than either of the monaural rates and just as clearly showed mutually inhibitory responses. Therefore, it would appear that the input to some SOC neurons from each ear may be a sequence of both inhibitory and excitatory events. The discharge of the SOC cell is, thus, dependent upon the spatial and temporal interactions of these inputs.

SUMMARY AND CONCLUSIONS

Encouraged by new methods and multidisciplinary approaches, the study of lower auditory brainstem centers has, in recent years, advanced rapidly on several fronts, and a great deal of new information about the way these centers process acoustic information has been obtained. Correlated light and electron microscopic work is giving us a structural basis for the various temporal response patterns that can be recorded from single neurons with extracellular microelectrodes. Intracellular recordings, from neurons subsequently identified by dye injection, have begun to reveal synaptic mechanisms that underly these complex excitatory and inhibitory interactions. Some candidate neurotransmitters that may be operating in the CN and SOC have been identified through combined biochemical, electrophysiological, and pharmacological studies. Sensitive histochemical methods coupled with cytoarchitectonics have begun to show the distribution of these substances. Modern anatomical tracer methods, coupled with electrophysiological recording, are opening new doors for working out the complex auditory brainstem connectivity patterns. The latter approaches give promise of being especially rewarding in the SOC, which until recently has been difficult to study with traditional degeneration methods. Sophisticated studies of primary afferent fibers, made possible by advanced computer technology, continue to add new information about the inputs to the central auditory system. With all of this ongoing work, progress should be rapid in our understanding of how the different centers in the mammalian brainstem function during auditory communication.

Literature Cited

Adams, J. C. 1976. Single unit studies on the dorsal and intermediate acoustic striae. *J. Comp. Neurol.* 170:97–106

Berman, A. L. 1968. *The Brain Stem of the Cat. A Cytoarchitectonic Atlas with Stereotaxic Coordinates.* Madison: Univ. Wis. Press. 175 pp.

Bilsen, F. A., ten Kate, J. H., Buunen, T. J. F., Raatgever, J. 1975. Responses of single units in the cochlear nucleus of the cat to cosine noise. *J. Acoust. Soc. Am.* 58:858–66

Boudreau, J. C., Tsuchitani, C. 1968. Binaural interaction in the cat superior olive S segment. *J. Neurophysiol.* 31:442–54

Boudreau, J. C., Tsuchitani, C. 1970. Cat superior olive S-segment cell discharge to tonal stimulation. In *Contributions to Sensory Physiology,* Vol. 4, ed. W. D. Neff, pp. 143–213. New York: Academic

Brawer, J. R., Morest, D. K. 1975. Relations between auditory nerve endings and cell types in the cat's anteroventral cochlear nucleus seen with the Golgi method and Nomarski optics. *J. Comp. Neurol.* 160:491–506

Brawer, J. R., Morest, D. K., Kane, E. C. 1974. The neuronal architecture of the cochlear nucleus of the cat. *J. Comp. Neurol.* 155:251–99

Britt, R. H. 1976. Intracellular study of synaptic events related to phase-locking responses of cat cochlear nucleus cells to low frequency tones. *Brain Res.* 112:313–27

Britt, R., Starr, A. 1976a. Synaptic events and discharge patterns of cochlear nucleus cells. I. Steady-frequency tone bursts. *J. Neurophysiol.* 39:162–78

Britt, R., Starr, A. 1976b. Synaptic events and discharge patterns of cochlear nucleus cells. II. Frequency-modulated tones. *J. Neurophysiol.* 39:179–94

Brownell, W. E. 1975. Organization of the cat trapezoid body and the discharge characteristics of its fibers. *Brain Res.* 94:413–33

Cajal, S. R. 1909. *Histologie du Système Nerveux de L'Homme et des Vertébrés.* Madrid: Instituto Ramón y Cajal. 986 pp.

Caspary, D. 1972. Classification of subpopulations of neurons in the cochlear nuclei of the kangaroo rat. *Exp. Neurol.* 37:131–51

Clark, G. M. 1969. The ultrastructure of nerve endings in the medial superior olive of the cat. *Brain Res.* 14:293–305

Cohen, E. S., Brawer, J. R., Morest, D. K. 1972. Projections of the cochlea to the dorsal cochlear nucleus in the cat. *Exp. Neurol.* 35:470–79

Comis, S. D. 1970. Centrifugal inhibitory processes affecting neurones in the cat cochlear nucleus. *J. Physiol. London* 210:751–60

Comis, S. D., Guth, P. S. 1974. The release of acetylcholine from the cochlear nucleus upon stimulation of the crossed olivo-cochlear bundle. *Neuropharmacology* 13:633–41

Comis, S. D., Davies, W. E. 1969. Acetylcholine as a transmitter in the cat auditory system. *J. Neurochem.* 16:423–29

Comis, S. D., Whitfield, I. C. 1967. Centrifugal excitation and inhibition in the cochlear nucleus. *J. Physiol. London* 188:34P-35P

Comis, S. D., Whitfield, I. C. 1968. Influence of centrifugal pathways on unit activity in the cochlear nucleus. *J. Neurophysiol.* 31:62–68

Davies, W. E. 1973. The role of 4-aminobutyrate in the lower auditory system of the guinea pig. *Biochem. Soc. Trans.* 1:134–36

Davies, W. E. 1975. The distribution of GABA transaminase-containing neurones in the cat cochlear nucleus. *Brain Res.* 83:27–33

Eldredge, D. H., Miller, J. D. 1971. Physiology of hearing. *Ann. Rev. Physiol.* 33:281–310

Erulkar, S. D. 1972. Comparative aspects of spatial localization of sound. *Physiol. Rev.* 52:237–360

Erulkar, S. D., Butler, R. A. Gerstein, G. L. 1968. Excitation and inhibition in cochlear nucleus. II. Frequency-modulated tones. *J. Neurophysiol.* 31: 537–48

Evans, E. F. 1975. Cochlear nerve and cochlear nucleus. In *Handbook of Sensory Physiology,* Vol. V/2, ed. W. D. Keidel, W. D. Neff, pp. 1–108. New York: Academic

Evans, E. F., Nelson, P. G. 1973a. The responses of single neurones in the cochlear nucleus of the cat as a function of their location and the anaesthetic state. *Exp. Brain Res.* 17:402–27

Evans, E. F., Nelson, P. G. 1973b. On the functional relationship between the dorsal and ventral divisions of the cochlear nucleus of the cat. *Exp. Brain Res.* 17:428–42

Falck, B., Owman, C. 1965. A detailed methodological description of the fluorescence method for the cellular demonstration of biogenic monoamines. *Acta Univ. Lund. Sect. 2,* No. 7. pp. 5–23

Feldman, M. L., Harrison, J. M. 1969. The projection of the acoustic nerve to the ventral cochlear nucleus of the rat. A Golgi study. *J. Comp. Neurol.* 137: 267–94

Fernald, R. D. 1971. A neuron model with spatially distributed synaptic input. *Biophys. J.* 11:323–40

Fernandez, C., Karapas, F. 1967. The course and termination of the striae of Monakow and Held in the cat. *J. Comp. Neurol.* 131:371–86

Fex, J., Fuxe, K., Lennerstrand, G. 1965. Absence of monoamines in olivo-cochlear fibres in cat. *Acta Physiol. Scand.* 64:259–62

Fex, J., Wenthold, R. J. 1976. Choline acetyltransferase, glutamate decarboxylase and tyrosine hydroxylase in the cochlea and cochlear nucleus of the guinea pig. *Brain Res.* 109:575–85

Fisher, S. K., Davies, W. E. 1976. GABA and its related enzymes in the lower auditory system of the guinea pig. *J. Neurochem.* 27:1145–55

Fuxe, K. 1965. Evidence for the existence of monoamine neurons in the central nervous system. IV. The distribution of monoamine nerve terminals in the central nervous system. *Acta Physiol. Scand.* 247 (suppl. 64): 37–85

Galambos, R., Schwartzkopff, J., Rupert, A. 1959. Microelectrode study of superior olivary nuclei. *Am. J. Physiol.* 197: 527–36

Geisler, C. D., Goldberg, J. M. 1966. A stochastic model of the repetitive activity of neurons. *Biophys. J.* 6:53–69

Gentschev, T., Sotelo, C. 1973. Degenerative patterns in the ventral cochlear nucleus of the rat after primary deafferentation. An ultrastructural study. *Brain Res.* 62:37–60

Gibson, M. M., Hind, J. E., Kitzes, L. M., Rose, J. E. 1977. Estimation of traveling wave parameters from the response properties of cat AVCN neurons. In *Psychophysics and Physiology of Hearing,* ed. E. F. Evans, J. P. Wilson, pp. 57–88. London: Academic

Godfrey, D. A., Carter, J. A., Berger, S. J., Lowry, O. H., Matschinsky, F. M. 1977. Quantitative histochemical mapping of candidate transmitter amino acids in cat cochlear nucleus. *J. Histochem. Cytochem.* 25:417–31

Godfrey, D. A., Kiang, N. Y. S., Norris, B. E. 1975a. Single unit activity in the posteroventral cochlear nucleus of the cat. *J. Comp. Neurol.* 162: 247–68

Godfrey, D. A., Kiang, N. Y. S., Norris, B. E. 1975b. Single unit activity in the dorsal cochlear nucleus of the cat. *J. Comp. Neurol.* 162:269–84

Godfrey, D. A., Matschinsky, F. M. 1976. Approach to three-dimensional mapping of quantitative histochemical measurements applied to studies of the cochlear nucleus. *J. Histochem. Cytochem.* 24:697–712

Godfrey, D. A., Williams, A. D., Matschinsky, F. M. 1977. Quantitative histochemical mapping of enzymes of the cholinergic system in cat cochlear nucleus. *J. Histochem. Cytochem.* 25:397–416

Goldberg, J. M. 1975. Physiological studies of auditory nuclei of the pons. In *Handbook of Sensory Physiology,* Vol. V/2, ed. W. D. Keidel, W. D. Neff, pp. 109–44. New York: Academic

Goldberg, J. M., Adrian, H. O., Smith, F. D. 1964. Response of neurons of the superior olivary complex of the cat to acoustic stimuli of long duration. *J. Neurophysiol.* 27:706–49

Goldberg, J. M., Brown, P. B. 1968. Functional organization of the dog superior olivary complex: an anatomical and electrophysiological study. *J. Neurophysiol.* 31:639–56

Goldberg, J. M., Brown, P. B. 1969. Response of binaural neurons of dog superior olivary complex to dichotic tonal stimuli: some physiological mechanisms of sound localization. *J. Neurophysiol.* 32:613–36

Goldberg, J. M., Brownell, W. E. 1973. Discharge characteristics of neurons in anteroventral and dorsal cochlear nuclei of cat. *Brain Res.* 64:35–54

Goldberg, J. M., Greenwood, D. D. 1966. Response of neurons of the dorsal and posteroventral cochlear nuclei of the cat to acoustic stimuli of long duration. *J. Neurophysiol.* 29:72–93

Greenwood, D. D., Goldberg, J. M. 1970. Response of neurons in the cochlear nuclei to variations in noise bandwidth and to tone-noise combinations. *J. Acoust. Soc. Am.* 47:1022–40

Greenwood, D. D., Maruyama, N. 1965. Excitatory and inhibitory response areas of auditory neurons in the cochlear nucleus. *J. Neurophysiol.* 28:863–92

Greenwood, D. D., Merzenich, M. M., Roth, G. L. 1976. Some preliminary observations on the interrelations between two-tone suppression and combination-tone driving in the anteroventral cochlear nucleus of the cat. *J. Acoust. Soc. Am.* 59:607–33

Guinan, J. J. Jr., Guinan, S. S., Norris, B. E. 1972. Single auditory units in the superior olivary complex. I: Responses to sounds and classifications based on physiological properties. *Int. J. Neurosci.* 4:101–20

Guinan, J. J. Jr., Norris, B. E., Guinan, S. S. 1972. Single auditory units in the superior olivary complex. II: Locations of unit categories and tonotopic organization. *Int. J. Neurosci.* 4:147–66

Harrison, J. M., Feldman, M. L. 1970. Anatomical aspects of the cochlear nucleus and superior olivary complex. See Boudreau & Tsuchitani 1970, pp. 95–142

Harrison, J. M., Warr, W. B. 1962. A study of the cochlear nuclei and ascending auditory pathways of the medulla. *J. Comp. Neurol.* 119:341–80

Harrison, J. M., Howe, M. E. 1974a. Anatomy of the afferent auditory system of mammals. In *Handbook of Sensory Physiology,* Vol. V/1, ed. W. D. Keidel, W. D. Neff, pp. 284–336. New York: Academic

Harrison, J. M., Howe, M. E. 1974b. Anatomy of the descending auditory system (mammalian). See Harrison & Howe 1974a, pp. 365–88

Harrison, J. M., Irving, R. 1965. The anterior ventral cochlear nucleus. *J. Comp. Neurol.* 124:15–42

Harrison, J. M., Irving, R. 1966a. Ascending connections of the anterior ventral cochlear nucleus in the rat. *J. Comp. Neurol.* 126:51–64

Harrison, J. M., Irving, R. 1966b. The organization of the posterior ventral cochlear nucleus in the rat. *J. Comp. Neurol.* 126:391–402

Hawkins, J. E. Jr. 1964. Hearing. *Ann. Rev. Physiol.* 26:453–80

Ibata, Y., Pappas, G. D. 1976. The fine structure of synapses in relation to the large spherical neurons in the anterior ventral cochlear nucleus of the cat. *J. Neurocytol.* 5:395–406

Irving, R., Harrison, J. M. 1967. The superior olivary complex and audition: a comparative study. *J. Comp. Neurol.* 130:77–86

Iversen, L. L., Kelly, J. S., Minchin, M., Schon, F., Snodgrass, S. R. 1973. Role of amino acids and peptides in synaptic transmission. *Brain Res.* 62:567–76

Jean-Baptiste, M., Morest, D. K. 1975. Transneuronal changes of synaptic endings and nuclear chromatin in the trapezoid body following cochlear ablations in cats. *J. Comp. Neurol.* 162: 111–34

Kane, E. C. 1973. Octopus cells in the cochlear nucleus of the cat: heterotypic synapses upon homeotypic neurons. *Int. J. Neurosci.* 5:251–79

Kane, E. C. 1974a. Patterns of degeneration in the caudal cochlear nucleus of the cat after cochlear ablation. *Anat. Rec.* 179:67–91

Kane, E. C. 1974b. Synaptic organization in the dorsal cochlear nucleus of the cat: a light and electron microscopic study. *J. Comp. Neurol.* 155:301–29

Kane, E. S. C. 1976a. Descending projections to specific regions of cat cochlear nucleus: a light microscopic study. *Exp. Neurol.* 52:372–88

Kane, E. S. 1976b. Descending inputs to caudal cochlear nucleus in cats: a horseradish peroxidase (HRP) study. *Am. J. Anat.* 146:433–41

Kane, E. S. 1977. Descending inputs to the octopus cell area of the cat cochlear nucleus: an electron microscopic study. *J. Comp. Neurol.* 173:337–54

Kiang, N. Y. S. 1975. Stimulus representation in the discharge patterns of auditory neurons. In *The Nervous System, Vol. 3: Human Communication and Its Disorders,* ed. D. B. Tower, pp. 81–96. New York: Raven

Kiang, N. Y. S., Godfrey, D. A., Norris, B. E., Moxon, S. E. 1975. A block model of the cat cochlear nucleus. *J. Comp. Neurol.* 162:221–46

Kiang, N. Y. S., Morest, D. K., Godfrey, D. A., Guinan, J. J. Jr., Kane, E. C. 1973. Stimulus coding at caudal levels of the cat's auditory nervous system: I. Response characteristics of single units. In *Basic Mechanisms In Hearing,* ed. A. R. Møller, pp. 455–78. New York: Academic. 941 pp.

Kiang, N. Y. S., Pfeiffer, R. R., Warr, W. B., Backus, A. S. N. 1965. Stimulus coding in the cochlear nucleus. *Ann. Otol. Rhinol. Laryngol.* 74:463–85

Koerber, K. C., Pfeiffer, R. R., Warr, W. B., Kiang, N. Y. S. 1966. Spontaneous spike discharges from single units in the cochlear nucleus after destruction of the cochlea. *Exp. Neurol.* 16: 119–30

Lavine, R. A. 1971. Phase-locking in response of single neurons in cochlear complex of the cat to low-frequency tonal stimuli. *J. Neurophysiol.* 34: 467–83

Lenn, N. J., Reese, T. S. 1966. The fine structure of nerve endings in the nucleus of the trapezoid body and the ventral cochlear nucleus. *Am. J. Anat.* 118:375–89

Lindsey, B. G. 1975. Fine structure and distribution of axon terminals from the cochlear nucleus on neurons in the medial superior olivary nucleus of the cat. *J. Comp. Neurol.* 160:81–103

Lorente de Nó, R. 1933. Anatomy of the eighth nerve: III. General plan of structure of the primary cochlear nuclei. *Laryngoscope* 43:327–50

Lorente de Nó, R. 1976. Some unresolved problems concerning the cochlear nerve. *Ann. Otol. Rhinol. Laryngol.* Suppl. 34, 85:1–28

Lowry, O. H., Godfrey, D. A., Carter, J. A., Matschinsky, F. M. 1974. Quantitative distribution of alleged transmitter amino acids in the cochlear nucleus. *J. Acoust. Soc. Am.* 56:S19

Mast, T. E. 1970a. Binaural interaction and contralateral inhibition in dorsal cochlear nucleus of the chinchilla. *J. Neurophysiol.* 33:108–15

Mast, T. E. 1970b. Study of single units of the cochlear nucleus of the chinchilla. *J. Acoust. Soc. Am.* 48:505–12

Mast, T. E. 1973. Dorsal cochlear nucleus of the chinchilla: excitation by contralateral sound. *Brain Res.* 62:61–70

McDonald, D. M., Rasmussen, G. L. 1971. Ultrastructural characteristics of synaptic endings in the cochlear nucleus having acetylcholinesterase activity. *Brain Res.* 28:1–18

Merzenich, M. M. 1970. Morphological specialization of the cochlear nuclear complex in certain mammals. *Anat. Rec.* 166:347

Merzenich, M. M., Kitzes, L., Aitkin, L. 1973. Anatomical and physiological evidence for auditory specialization in the mountain beaver (*Aplodontia rufa*). *Brain Res.* 58:331–44

Møller, A. R. 1969a. Unit responses in the rat cochlear nucleus to repetitive, transient sounds. *Acta Physiol. Scand.* 75:542–51

Møller, A. R. 1969b. Unit responses in the cochlear nucleus of the rat to sweep tones. *Acta Physiol. Scand.* 76:503–12

Møller, A. R. 1970. Unit responses in the cochlear nucleus of the rat to noise and tones. *Acta Physiol. Scand.* 78:289–98

Møller, A. R. 1972. Coding of sounds in lower levels of the auditory system. *Q. Rev. Biophys.* 5:59–155

Møller, A. R. 1976. Dynamic properties of excitation and two-tone inhibition in the cochlear nucleus studied using amplitude-modulated tones. *Exp. Brain Res.* 25:307–21

Molnar, C. E., Pfeiffer, R. R. 1968. Interpretation of spontaneous spike discharge patterns of neurons in the cochlear nucleus. *Proc. IEEE* 56:993–1004

Moore, R. Y., Goldberg, J. M. 1966. Projections of the inferior colliculus in the monkey. *Exp. Neurol.* 14:429–38

Moore, T. J., Cashin, J. L. 1974. Response patterns of cochlear nucleus neurons to

excerpts from sustained vowels. *J. Acoust. Soc. Am.* 56:1565–76
Morest, D. K. 1968. The collateral system of the medial nucleus of the trapezoid body of the cat, its neuronal architecture and relation to the olivo-cochlear bundle. *Brain Res.* 9:288–311
Morest, D. K. 1973. Auditory neurons of the brain stem. *Adv. Oto-Rhino-Laryngol.* 20:337–56
Morest, D. K., Kiang, N. Y. S., Kane, E. C., Guinan, J. J. Jr., Godfrey, D. A. 1973. Stimulus coding at caudal levels of the cat's auditory nervous system: II. Patterns of synaptic organization. In *Basic Mechanisms In Hearing,* ed. A. R. Møller, pp. 479–509. New York: Academic. 941 pp.
Moushegian, G., Rupert, A. L., Gidda, J. S. 1975. Functional characteristics of superior olivary neurons to binaural stimuli. *J. Neurophysiol.* 38:1037–48
Moushegian, G., Rupert, A., Whitcomb, M. A. 1964. Brain-stem neuronal response patterns to monaural and binaural tones. *J. Neurophysiol.* 27:1174–91
Moushegian, G., Stillman, R. D., Rupert, A. L., 1971. Characteristic delays in superior olive and inferior colliculus. In *Physiology of the Auditory System,* ed. M. B. Sachs, pp. 245–54. Baltimore: Nat. Educ. Consultants. 393 pp.
Osen, K. K. 1969a. Cytoarchitecture of the cochlear nuclei in the cat. *J. Comp. Neurol.* 136:453–84
Osen, K. K. 1969b. The intrinsic organization of the cochlear nuclei in the cat. *Acta Oto-Laryngol.* 67:352–59
Osen, K. K. 1970. Course and termination of the primary afferents in the cochlear nuclei of the cat. An experimental study. *Arch. Ital. Biol.* 108:21–51
Osen, K. K., Roth, K. 1969. Histochemical localization of cholinesterases in the cochlear nuclei of the cat, with notes on the origin of acetylcholinesterase-positive afferents and the superior olive. *Brain Res.* 16:165–85
Papez, J. W. 1930. Superior olivary nucleus: its fiber connections. *Arch. Neurol. Psychiatry* 24:1–20
Perkins, R. E. 1973. An electron microscopic study of synaptic organization in the medial superior olive of normal and experimental chinchillas. *J. Comp. Neurol.* 148:387–416
Pfeiffer, R. R. 1966a. Anteroventral cochlear nucleus: wave forms of extracellularly recorded spike potentials. *Science* 154:667–68
Pfeiffer, R. R. 1966b. Classification of response patterns of spike discharges for units in the cochlear nucleus: toneburst stimulation. *Exp. Brain Res.* 1: 220–35
Pfeiffer, R. R., Kiang, N. Y. S. 1965. Spike discharge patterns of spontaneous and continuously stimulated activity in the cochlear nucleus of anesthetized cats. *Biophys. J.* 5:301–16
Pickles, J. O. 1976a. Role of centrifugal pathways to cochlear nucleus in determination of critical bandwith. *J. Neurophysiol.* 39:394–400
Pickles, J. O. 1976b. The noradrenaline-containing innervation of the cochlear nucleus and the detection of signals in noise. *Brain Res.* 105:591–96
Pickles, J. O., Comis, S. D. 1973. Role of centrifugal pathways to cochlear nucleus in detection of signals in noise. *J. Neurophysiol.* 36:1131–37
Pirsig, W. 1968. Regionen, Zelltypen und Synapsen im ventralen Nucleus cochlearis des Meerschweinchens, *Arch. Klin. Exp. Ohren Nasen Kehlkopfheilkd.* 192:333–50
Pirsig, W., Noda, Y., Lehmann, I. 1972. Tonotope Abbildung der Cochlea im Nucleus cochlearis ventralis des Meerschweinchens. *Arch. Klin. Exp. Ohren Nasen Kehlkopfheilkd.* 202:494–500
Pirsig, W., Reinecke, M., Lehmann, I. 1969. Die Ultrastruktur der Synapsen im Ventralen Nucleus cochlearis des Meerschweichens. I. Mitteilung. *Arch. Klin. Exp. Ohren Nasen Kehlkopfheilkd.* 195:152–68
Powell, T. P. S., Cowan, W. M. 1962. An experimental study of the projection of the cochlea. *J. Anat.* 96:269–84
Rasmussen, G. L. 1960. Efferent fibers of the cochlear nerve and cochlear nucleus. In *Neural Mechanisms of the Auditory and Vestibular Systems,* ed. G. L. Rasmussen, W. F. Windle, pp. 105–15. Springfield: Thomas. 422 pp.
Rasmussen, G. L. 1964. Anatomic relationships of the ascending and descending auditory systems. In *Neurological Aspects of Auditory and Vestibular Disorders,* ed. W. S. Field, B. R. Alford, pp. 5–19. Springfield: Thomas
Rasmussen, G. L. 1967. Efferent connections of the cochlear nucleus. In *Sensorineural Hearing Processes and Disorders,* ed. A. B. Graham, pp. 61–75. Boston: Little, Brown. 543 pp.
Rose, J. E., Galambos, R., Hughes, J. 1960. Organization of frequency sensitive neurons in the cochlear nuclear com-

plex of the cat. See Rasmussen 1960, pp. 116–36

Rose, J. E., Gross, N. B., Geisler, C. D., Hind, J. E. 1966. Some neural mechanisms in the inferior colliculus of the cat which may be relevant to localization of a sound source. *J. Neurophysiol.* 29:288–314

Rose, J. E., Kitzes, L. M., Gibson, M. M., Hind, J. E. 1974. Observations on phase-sensitive neurons of anteroventral cochlear nucleus of the cat: nonlinearity of cochlear output. *J. Neurophysiol.* 37:218–53

Rupert, A. L., Moushegian, G. 1970. Neuronal responses of kangaroo rat ventral cochlear nucleus to low-frequency tones. *Exp. Neurol.* 26:84–102

Sando, I. 1965. The anatomical interrelationships of the cochlear nerve fibers. *Acta Oto-Laryngol.* 59:417–36

Scheibel, M. E., Scheibel, A. B. 1974. Neuropil organization in the superior olive of the cat. *Exp. Neurol.* 43:339–48

Schuknecht, H. F., Churchill, J. A., Doran, R. 1959. The localization of acetylcholinesterase in the cochlea. *AMA Arch. Otolaryngol.* 69:549–59

Schwartz, I. R. 1972. Axonal endings in the cat medial superior olive: coated vesicles and intercellular substance. *Brain Res.* 46:187–202

Shute, C. C. D., Lewis, P. R. 1965. Cholinesterase-containing pathways of the hindbrain: afferent cerebellar and centrifugal cochlear fibers. *Nature* 205: 242–46

Smoorenburg, G. F., Gibson, M. M., Kitzes, L. M., Rose, J. E., Hind, J. E. 1976. Correlates of combination tones observed in the response of neurons in the anteroventral cochlear nucleus of the cat. *J. Acoust. Soc. Am.* 59:945–62

Starr, A., Wernick, J. S. 1968. Olivocochlear bundle stimulation: effects on spontaneous and tone-evoked activities of single units in cat cochlear nucleus. *J. Neurophysiol.* 31:549–64

Stotler, W. A. 1953. An experimental study of the cells and connections of the superior olivary complex of the cat. *J. Comp. Neurol.* 98:401–31

Strominger, N. L., Strominger, A. I. 1971. Ascending brain stem projections of the anteroventral cochlear nucleus in the rhesus monkey. *J. Comp. Neurol.* 143:217–42

Swanson, L. W., Hartman, B. K. 1975. The central adrenergic system. An immunofluorescence study of the location of cell bodies and their efferent connections in the rat utilizing dopamine-β-hydroxylase as a marker. *J. Comp. Neurol.* 163:467–506

Tachibana, M., Kuriyama, K. 1974. Gamma-aminobutyric acid in the lower auditory pathway of the guinea pig. *Brain Res.* 69:370–74

Tsuchitani, C. 1977. Functional organization of lateral cell groups of cat superior olivary complex. *J. Neurophysiol.* 40:296–318

Tsuchitani, C., Boudreau, J. C. 1966. Single unit analysis of cat superior olive S segment with tonal stimuli. *J. Neurophysiol.* 29:684–97

van Gisbergen, J. A. M., Grashuis, J. L., Johannesma, P. I. M., Vendrik, A. J. H. 1975a. Neurons in the cochlear nucleus investigated with tone and noise stimuli. *Exp. Brain Res.* 23:387–406

van Gisbergen, J. A. M., Grashuis, J. L., Johannesma, P. I. M., Vendrik, A. J. H. 1975b. Statistical analysis and interpretation of the initial response of cochlear nucleus neurons to tone bursts. *Exp. Brain Res.* 23:407–423

van Noort, J. 1969. *The structure and connections of the inferior colliculus. An investigation of the lower auditory system.* Assen: Van Gorcum. 108 pp.

Warr, W. B. 1966. Fiber degeneration following lesions of the anterior ventral cochlear nucleus of the cat. *Exp. Neurol.* 14:453–74

Warr, W. B. 1969. Fiber degeneration following lesions in the posteroventral cochlear nucleus of the cat. *Exp. Neurol.* 23:140–55

Warr, W. B. 1972. Fiber degeneration following lesions in the multipolar and globular cell areas in the ventral cochlear nucleus of the cat. *Brain Res.* 40:247–70

Warr, W. B. 1975. Olivocochlear and vestibular efferent neurons of the feline brain stem: their location, morphology and number determined by retrograde axonal transport and acetylcholinesterase histochemistry. *J. Comp. Neurol.* 161:159–82

Webster, W. R., Aitkin, L. M. 1975. Central auditory processing. In *Handbook of Psychobiology,* ed. M. S. Gazzaniga, C. Blakemore, pp. 325–64. New York: Academic. 639 pp.

Whitfield, I. C. 1968. The pharmacological behaviour of the cochlear nucleus. In *Drugs and Sensory Functions,* ed. A. Herxheimer, pp. 167–74. London: Churchill. 338 pp.

Young, E. D., Brownell, W. E. 1976. Responses to tones and noise of single cells in dorsal cochlear nucleus of unanesthetized cats. *J. Neurophysiol.* 39:282–300

Ann. Rev. Neurosci. 1978. 1:395–415

NEUROPHYSIOLOGY OF EPILEPSY

❖11513

David A. Prince
Department of Neurology, Stanford University School of Medicine,
Stanford, California 94305

For Irving Wagman, my friend and teacher

INTRODUCTION

Epilepsy is a disorder of brain function characterized by paroxysmal stereotyped alterations in behavior associated with synchronous excessive discharge in large aggregates of neurons. It is estimated that 0.5–2% of the world's population is affected by epilepsy—perhaps as many as 80 million people. The pathogenesis of this disorder is not well understood. Because of its high incidence relative to other forms, and the ease with which it can be produced in experimental animals, focal epilepsy which results from abnormal neuronal activities in a localized brain region has been most widely studied. In this review I will examine several current hypotheses regarding the cellular neurophysiological mechanisms of focal epileptogenesis, in light of new data which require that we substantially modify previously held views. For a more extensive treatment of various aspects of this problem, the reader is referred to recent monographs and reviews (Ajmone-Marsan 1969, Jasper 1969, Jasper, Ward & Pope 1969, Purpura 1969a,b, Spencer & Kandel 1969, Ward 1969, Purpura et al 1972, Ayala et al 1973, Prince 1974, 1976).

The rather narrow scope of this review inevitably does injustice to the challenge and complexity of epilepsy research, and to the significant insights into basic aspects of brain function that it has provided. As Ward and colleagues have pointed out, "An adequate model cannot be provided by examining the phenomenon of epilepsy from the parochial viewpoint of a single discipline in the neurosciences. . . . Our goal is to view . . . epilepsy from the broad base of neurobiology" (Ward, Jasper & Pope 1969).

0147-006X/78/0325-0395$01.00

NEURONAL BEHAVIOR IN PENICILLIN FOCI

The most extensive intracellular studies of experimental epilepsy have been done in foci produced by application of penicillin to cat neocortex (Matsumoto 1964, Matsumoto & Ajmone-Marsan 1964a,b, Prince 1966, Prince & Wilder 1967, Prince 1968a,b, Matsumoto, Ayala & Gumnit 1969, Prince 1969a, Ayala, Matsumoto & Gumnit 1970, Prince 1971, Prince & Gutnick 1972) and hippocampus (Dichter, Herman & Selzer 1972, Dichter & Spencer 1969a,b, for reviews see Ajmone-Marsan 1969, Prince 1969b, Prince 1972, Ayala et al 1973, Prince 1974, 1976). Data obtained from these experiments from the bulk of our knowledge of cellular events in focal epileptogenesis.

Matsumoto & Ajmone-Marsan (1964a,b) and Matsumoto (1964) first described the large amplitude (ca. 20–30 mV) prolonged (ca. 50–100 msec) membrane depolarizations and associated high frequency bursts of spikes that occur in neocortical neurons of penicillin epileptiform foci, coincident with interictal paroxysmal discharges in the electroencephalogram (EEG). These depolarization shifts have been described in neurons of several other acute epileptiform foci (Goldensohn & Purpura 1963, Li 1959, Hardy 1970, Schwartzkroin, Bromley & Shimada 1977, and others).

There has been considerable discussion concerning the mechanisms of depolarization shift generation. The most widely held hypothesis is that the depolarization shift recorded in the neuronal cell body is actually a sum of synchronous excitatory postsynaptic potentials generated by complex interactions in large groups of neurons in which augmentation of excitatory and/or depression of inhibitory synaptic activity has taken place (see Ayala et al 1973, Prince 1974, 1976, for reviews). As an alternative to this "giant synaptic potential" hypothesis, it has been proposed that depolarization shift generation is an intrinsic property of a given cell due to active membrane processes, and that synaptic events serve as a mechanism for synchronizing populations of neurons, rather than as *generators* for depolarization shifts (Schwartzkroin & Prince 1977a,b; see also Prince 1968a). It is obvious that the paroxysmal discharges seen in the EEG and the behavioral abnormalities in the animal must be produced by activities in large groups of neurons. It is equally obvious that there must be abnormalities in elements of a group if the group functions in an abnormal way. Recent data, reviewed below, suggest that both changes in synaptic input and intrinsic active spike generating responses contribute to depolarization shift generation, with the former serving as a trigger for the latter.

The Depolarization Shift as a Giant Synaptic Potential

The observation that depolarizing shifts may be readily evoked in neocortex or hippocampus by orthodromic but not intracellular stimuli (Dichter & Spencer 1969a, Matsumoto, Ayala & Gumnit 1969, Prince 1968a) has been used to support the hypothesis that depolarization shifts are giant synaptic potentials. Other evidence favoring a synaptic origin for depolarization shifts includes the finding that they may be graded in amplitude when evoked at short intervals (Matsumoto 1964,

Prince 1966) or during development or waning of an epileptiform focus (Matsumoto & Ajmone-Marsan 1964a, Prince 1968a). In recordings from immature cortex where synaptic connectivity is not well developed, depolarization shifts resemble augmented EPSPs (Prince & Gutnick 1972). Also in support of the "giant synaptic potential" hypothesis is the finding that, in some cells, depolarization shifts tend to increase and decrease in amplitude when triggered during hyperpolarizing and depolarizing intracellular current pulses, respectively, as would EPSPs associated with increased conductance (Prince 1968a, Dichter & Spencer 1969b, Matsumoto, Ayala & Gumnit 1969). This indicates that at least part of the process responsible for depolarization shift generation has an equilibrium potential positive to the resting membrane potential, but does not rule out a contribution by nonsynaptic (active) membrane events. One example of such a nonsynaptic process is the voltage-dependent depolarization produced by inward calcium currents in molluscan neurons (Eckert & Lux 1976).

Attempts to demonstrate a reversal potential for the depolarization shift which would support the "giant synaptic potential" hypothesis have been generally unsuccessful. Depolarization shift inversion with current has been reported only in badly depolarized neurons, and the published figures (e.g. Figure 6 of Prince 1968a or Figure 7 of Matsumoto, Ayala & Gumnit 1969) are difficult to interpret. In these examples, other effects—such as contamination with IPSPs of increased amplitude; significant decreases in membrane resistance associated with passage of intense depolarizing currents in already damaged neurons; and increased field potential contributions to the intracellular signals secondary to low membrane resistance—may have led to erroneous conclusions (Prince 1968a).

Further support for the "giant synaptic potential" hypothesis of depolarization shift generation has come from studies of effects of penicillin on EPSPs and IPSPs in a number of preparations. Although data on effects of penicillin on EPSPs in cortex are scanty and hard to interpret (e.g. Walsh 1971), there is evidence for facilitation of EPSPs in invertebrate systems (Futamachi & Prince 1975) and in the hippocampal slice (Schwartzkroin & Prince 1977b). One hypothesis proposes that specific increases in recurrent excitation are responsible for depolarization shift generation (Ayala et al 1973); however, some of its authors have also recognized that the long multisynaptic pathways that might be involved in depolarization shift generation in neocortex are "certainly different from a simple recurrent excitatory system" (Matsumoto, Ayala & Gumnit 1969). The experiments which form the basis for the "recurrent excitation" hypothesis relied upon stimulation of the deafferented fornix (Dichter & Spencer 1969b) and deafferented corpus callosum (Ayala & Vasconetto 1972) to activate recurrent excitatory circuits. Although recurrent excitatory inputs would be expected to be paucisynaptic, long latencies were required to trigger depolarization shifts. In neither instance were antidromically activated neurons that generated recurrent EPSPs illustrated. In the case of callosal stimulation (Ayala & Vasconetto 1972), spread of stimulus current to orthodromic pathways was not ruled out. To date it has not been possible to evoke depolarization shifts in cells of cat penicillin foci using pyramidal tract stimulation, even though recurrent excitatory synaptic circuits are present (Takahashi, Kubota & Uno 1967) and active, judging by responses containing EPSPs and multiple spikes (Matsumoto, Ayala & Gumnit 1969).

Earlier conclusions regarding the lack of penicillin effects on IPSPs (Matsumoto & Ajmone-Marsan 1964a, Prince 1968b, Matsumoto, Ayala & Gumnit 1969, Dichter & Spencer 1969b) are probably incorrect in light of more recent reports that penicillin suppresses GABA-mediated inhibition (Curtis et al 1972, Clarke & Hill 1972, Van Duijn, Schwartzkroin & Prince 1973, Meyer & Prince 1973, Hochner, Spira & Werman 1976). The large hyperpolarizations that follow depolarization shifts in penicillin foci may be inverted with intracellular chloride iontophoresis (Prince 1968b), which suggests that IPSP components are present; however, other mechanisms most probably contribute, since depolarization shifts evoked in hippocampal slice pyramidal neurons treated with low chloride medium are also followed by hyperpolarizations even when IPSPs are blocked (Yamamoto 1972). A calcium-activated potassium conductance evoked by depolarization, similar to that present in molluscan (Meech 1972) and other neurons, has recently been found in mammalian hippocampal CA1 pyramidal cells (Hotson, Schwartzkroin & Prince 1977) and is a likely contributor to the postdepolarization shift hyperpolarization.

These data on the synaptic effects of penicillin suggest that the drug could produce a net increase in excitation within cortical neuronal aggregates by depressing IPSPs and increasing EPSPs. In addition to its effects on PSPs, the drug might also produce repetitive firing in cortical presynaptic terminals (Noebels & Prince 1977a) which could add to its epileptogenic action. It has also been proposed that penicillin increases excitability in part by significantly decreasing resting chloride conductance (Futamachi & Prince 1975, Hochner, Spira & Werman 1976). This conclusion was derived from experiments in crustacean muscle where chloride probably contributes a large portion of the normal membrane conductance. No such action can be detected in recordings from cell bodies of hippocampal CA1 neurons at penicillin concentrations sufficient to generate epileptiform discharges (Schwartzkroin & Prince 1976, 1977b), making this an unlikely possibility unless such conductance changes occur on distant processes where they are undetectable (Noebels & Prince 1977a).

The Depolarization Shift as an Active Neuronal Response

The possibility that active spike generating responses contribute to depolarization shift generation was raised in early studies of intracellular activities in penicillin foci, where it was noted that depolarization shift behavior was rather atypical for that of a synaptically generated potential (Matsumoto & Ajmone-Marsan 1964a, Prince 1968a). For example, as the intensity of an orthodromic input to the focus is gradually increased, an interictal EEG discharge and an associated full-amplitude depolarization shift are abruptly triggered at threshold. During long trains or pairs of stimuli, it can be shown that depolarization shifts are followed by periods of refractoriness for depolarization shift generation, whereas synaptic potentials are not (Matsumoto 1964, Prince 1966, 1968b). In some experiments, variations in latency for depolarization shift triggering of over 100 msec may occur (e.g. see

Figure 13 of Prince 1966); however, the depolarization shift ultimately evoked has a very stereotyped appearance from stimulus to stimulus. It has been suggested that these long delays are related to involvement of extensive interneuronal pathways in cortex, which ultimately provide an abrupt synchronous input to the depolarization shift-generating neurons (Prince 1965, 1966, 1968a, Dichter & Spencer 1969b); however, the likelihood that depolarization shifts of identical amplitude and duration would be produced with widely fluctuating latencies in the same neuron by such circuitry seems small. On the other hand, stereotyped responses might be expected if depolarization shift generation resulted from intrinsic neuronal activities evoked by a synaptic input.

Support for the concept that depolarization shift generation represents an intrinsic cell response has been provided by the finding that depolarization shift-like potentials may be triggered by intracellular current pulses in motoneurons of cat spinal cord exposed to penicillin (Kao & Crill 1972); however, this observation has not been confirmed by others (Lothman & Somjen 1976b; H. D. Lux, personal communication). Depolarization shift-like potentials have also been evoked by intracellular stimulation in hippocampal cells of unspecified type in tissue culture (Zipser, Crain & Bornstein 1973); hippocampal pyramidal neurons are known to generate such bursts under normal (nonepileptogenic) conditions (Kandel & Spencer 1961; Wong & Prince 1977). Depolarization shift-like potentials also occur in convulsant-treated molluscan neurons isolated from their synaptic inputs (Speckmann & Caspers 1973; Williamson & Crill 1976). Other evidence supporting the view that the depolarization shift is an active neuronal response to depolarization is discussed below.

On the basis of the relatively small effects of injected current on depolarization shift versus EPSP amplitude in some cells, and the observation that the first portion of the depolarization shift might consist of spikes arising abruptly without obvious prepotentials, this reviewer concluded that depolarization shifts might originate from distant dendritic membrane in some cells (Prince 1968a, 1971). A similar conclusion was reached by Matsumoto, Ayala & Gumnit (1969), who also noted that the effects of current on the depolarization shift were small. Data from laminar analysis of interictal discharges are compatible with a large contribution to depolarization shift generation from dendritic electrogenesis (Gumnit & Takahashi 1965, Gleason 1971). Other explanations for the insensitivity of depolarization shifts to current are possible, however (e.g. Klee 1975).

NONSYNAPTIC FACTORS IN EPILEPTOGENESIS

Over the past few years applications of new techniques have significantly added to our knowledge about the cellular mechanisms in focal epilepsy. This new information emphasizes important nonsynaptic mechanisms in epileptogenesis and requires us to modify some of the hypotheses and conclusions drawn from earlier data. The remainder of this review deals with three such areas: (*a*) generation of repetitive discharges in axonal terminals; (*b*) changes in ionic microenvironment during epileptogenesis, which may have influences on cellular activities; and (*c*) nonsynaptic regenerative cellular activities which may contribute to generation of depolarization shifts.

Antidromic Bursts in Cortical Axons During Epileptogenesis

Although it has been suggested that nonsynaptic mechanisms, such as electrical interactions or effects of chemical substances and ions released by neuronal activities, might have a role in epileptogenesis, proof that such processes are important in cortical foci has been difficult to obtain (see Jasper 1969, Purpura 1969b for reviews). Recently, evidence for nonsynaptic interactions has been provided by studies of thalamocortical relay (TCR) cells whose axons project to cortical penicillin foci (Gutnick & Prince 1972, 1974). The hypothesis that abnormalities in activities of presynaptic terminals contribute to penicillin epileptogenesis (Prince 1969b) was tested by making recordings of spontaneous activities of thalamocortical relay cells during cortical penicillin discharges. It was found that, coincident with interictal EEG discharges, spike bursts might arise from below the usual firing level (Figure 1*A1*). The cortical origin of these bursts was proven by demonstrating that known antidromic spikes did not collide with spikes of the burst (Figure 1*A2*). Similar antidromic spike bursts may be detected during cortical penicillin interictal discharges in thalamocortical relay cells of lateral geniculate (Rosen, Vastola & Hildebrand 1973, Scobey & Gabor 1975), and during both penicillin and strychnine discharges in n. ventralis lateralis (Schwartzkroin, Mutani & Prince 1975). It is not known whether antidromic bursts occur in chronic foci. The phenomenon has been reported in callosal axons (Schwartzkroin et al 1975) and dorsal root fibers of cat spinal cord treated with penicillin (Lothman & Somjen 1976a). Axo-axonal synapses on thalamocortical relay cell axons have not been described in the neocortex, so that the bursts are most likely generated by nonsynaptic means.

The potential importance of the antidromic burst phenomenon during the period of transition from interictal to ictal discharge has been emphasized (Gutnick & Prince 1975). In recent experiments we have examined the transition period in more detail in neocortical penicillin foci of cats not exposed to barbiturates (Noebels & Prince 1977b). Cortical stimuli (1 sec^{-1}) initially trigger brief EEG afterdischarges during which neurons generate orthodromic bursts of high frequency decrementing spikes with irregular intervals (Figure 1*B1*). During electrographic ictal episodes, rhythmically recurring orthodromic bursts are gradually replaced by long trains of regular interval antidromic spikes (1*B2*). If axonal bursts propagate orthodromically into even a portion of the extensive axonal arborization of TCR cells, they would represent a potent stimulus for excessive transmitter release and a powerful synchronizing mechanism in cortical neuronal populations.

Since penicillin produces antidromic bursts originating in axon terminals of a mammalian nerve-muscle preparation (Noebels & Prince 1977a), it is important to ask whether the antidromic burst phenomenon in cat cortex is related to the convulsant drug, or to epileptogenesis per se. Gabor & Scobey (1975) suggested that non–penicillin-treated axons of geniculo-cortical relay cells participated in antidromic burst generation; however, penicillin spread through the cortex was not entirely ruled out in their experiments. We have obtained data relevant to this issue

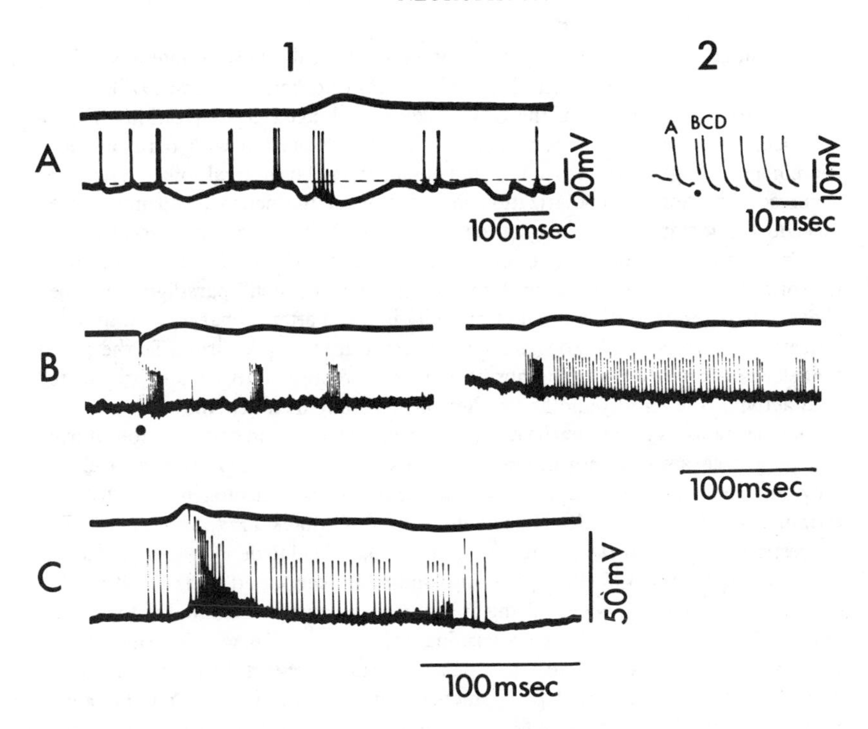

Figure 1 Antidromic bursting originating in cortical axons. *A1*: Intracellular recording showing spike burst arising below firing level (dotted line) during IPSP and cortical interictal EEG discharge in relay cell of n. VPL whose axon projects to penicillin focus in cortex. *A2*: Interposition of cortical stimulus during spontaneous burst similar to that in *A1* evokes an antidromic spike (C) in another relay cell. Interval between spikes B and C is less than 2 times the latency for invasion of antidromic spikes triggered in the cortex, proving that bursts spikes originate in cortex.

B: Antidromic bursting in neuron during transition from interictal to ictal discharge. *B1*: During brief rhythmic EEG afterdischarge evoked by single cortical shock in penicillin focus, VPL relay neuron generates bursts of orthodromic spikes. *B2*: Shortly after onset of ictal episode provoked by several cortical stimuli, cell generates single orthodromic burst followed by train of regular interval spikes shown to be antidromic by collision technique.

C: Spontaneous paroxysmal EEG and intracellular recording of VPL relay cell discharge during tonic phase of electrically-evoked cortical seizure. Gaps in the antidromic spike train (midportion of segment) are mulitples of the regulator interspike interval in the antidromic burst. See text for further description.

Dots: cortical stimuli. Upper traces in *A1, B, C*: EEG from cortical surface. Time calibration in *B2* for *B1* and *B2*. [*A* from (Gutnick & Prince 1972); *B, C* from J. L. Noebels & D. A. Prince, unpublished).]

by examining activities in relay neurons of n. ventralis postero-lateralis (VPL) during electrically evoked cortical afterdischarges (Noebels & Prince 1977b). During the early (tonic) phase of the EEG seizure, prolonged depolarizations (Figure 1*C*) occurred in a high proportion of TCR cells, associated with both bursts of decrementing, high-frequency spikes and bursts of regular interval spikes. The latter occurred even during depolarization inactivation of the neuron, and might even precede the membrane depolarization (Figure 1*C*). The regular interval spikes, which were the predominant event during portions of the afterdischarge, were shown to be antidromic in origin by using the "noncollision" paradigm described above. These results indicate that electrically-induced afterdischarges lead to depolarization of intracortical axons and repetitive antidromic spike firing. To the extent that electrical afterdischarges in normal cortex are representative of any propagated ictal activity, the antidromic firing phenomenon may be a general one.

The mechanisms that underlie repetitive spike generation in cortical axons during epileptogenesis are not known. One possibility is that changes in the extracellular ionic environment, which are known to occur during epileptogenesis, affect the terminals (Pedley et al 1976, Heinemann, Lux & Gutnick 1977, and others). The suggestion that intracortical increases in $[K^+]_o$ might lead to depolarization of axons and bursting (Gutnick & Prince 1972) is indirectly supported by the finding that large increases in $[K^+]_o$ occur during epileptogenesis and are maximal in the deeper cortical lamina where TCR axons terminate (Moody, Futamachi & Prince 1974). Also, in the spinal cord exposed to penicillin there appears to be a coincidence between the site of maximal $[K^+]_o$ increase in the ventral horn and the site of termination of the muscular afferents that are primarily involved in antidromic bursting during spinal cord seizures (Lothman & Somjen 1976a). The decreases in $[Ca^{++}]_o$ that occur in cortex during seizures (Heinemann, Lux & Gutnick 1977) may also increase axonal excitability (Frankenhaeuser & Hodgkin 1957) and act cooperatively with the changes in $[K^+]_o$. Depolarizations produced by large extracellular field potentials seem less likely to be the cause of antidromic bursting since antidromic bursts follow the onset of interictal EEG discharges by up to 50 msec, at a time considerably later than the peak inward transmembrane currents generated in deeper cortical layers (Gleason 1971). Also, bursts may occur only at certain phases of the ictal episode and may even precede the cortical paroxysm (e.g. Figure 1*C*). However an interaction between extracellular currents and ionically depolarized terminals could give rise to axonal bursting. Alternatively the terminals might be depolarized by other substances released by neuronal activity.

Epileptogenesis and the Ionic Microenvironment

A contribution of alterations in extracellular ionic environment to epileptogenesis was suggested by Green (1964) to explain the tendency for generation of epileptiform activities in the hippocampus. The hypothesis that K^+ released by neurons might reach sufficient concentrations in the restricted extracellular space to influence neuronal activities and ultimately lead to development of seizures received support from findings that K^+ was released during epileptogenesis (Fertziger & Ranck 1970) and that seizures might be provoked in hippocampus by superfusion with solutions containing increased $[K^+]$ (Zuckermann & Glaser 1968). Recordings from glial cells, known to behave as potassium electrodes in lower animals (Kuffler & Nicholls 1966), also lead to predictions of substantial increases in $[K^+]_o$ during focal epileptogenesis (Prince 1971, Sypert & Ward 1971, Dichter, Herman & Selzer 1972, Ransom 1974). The application of K^+ ion-sensitive microelectrodes to measurements of cortical $[K^+]_o$ (Vyskočil, Kříž & Bureš 1972, Prince, Lux & Neher 1973) provided the first direct evidence that significant alterations in ionic microenvironment occur during cellular activities and therefore might have a role in the nonsynaptic modulation of neuronal excitability. Increases in $[K^+]_o$ lasting seconds occur during interictal spikes in neocortex (Prince, Lux & Neher 1973) and hippocampus (Figure 2*A;* Fisher et al 1976) and sustained elevations of up to 10–12 mM occur during ictal episodes (Moody, Futamachi & Prince 1974, Futamachi, Mutani & Prince 1974, Sypert & Ward 1974). Important features of these changes in $[K^+]_o$ include their laminar distribution through the cortex with distinct maxima at particular depths, their long duration, and their tendency to peak at about 10–12 mM (Moody et al 1974, Futamachi et al 1974, Futamachi & Pedley 1976, Fisher et al 1976).

Despite the well-documented increase in $[K]_o$ with activity the question remains as to whether these changes in $[K^+]_o$ affect the activity of cortical neurons. Superfusion of solutions with high $[K^+]_o$ can produce epileptiform discharges in hippocampus (Zuckermann & Glaser 1968); however, the resulting concentration and distribution of K^+ in brain are different from those reached by endogenously released K^+ during epileptiform activity, so that the results of such experiments are hard to interpret (see Pedley et al 1976 and Fisher et al 1976 for discussion). One report (Sypert & Ward 1974) suggests that there is a "critical" level of $[K^+]_o$ associated with the onset of propagated seizures; however, in these experiments $[K^+]_o$ was not measured at the site of onset of ictal activity, nor was it possible to analyze the effects of raised $[K^+]_o$ independent of variations in the stimulus parameters that elicited the ictal discharge. In other experiments no correlation has been found between baseline $[K^+]_o$, or changes in $[K^+]_o$, and the transition between interictal and ictal activity (Futamachi, Mutani & Prince 1974, Moody, Futamachi & Prince 1974, Mutani, Futamachi & Prince 1974). The proposal that the low seizure threshold in hippocampus versus neocortex is related to higher buildups of $[K^+]_o$ in the former has not been substantiated with direct K^+ measurements (Fisher et al 1976) although it has been shown that increases in $[K^+]_o$ can increase excitability in hippocampal slices studied in vitro (Ogata 1975, Ogata et al 1976; see below). Studies in injured cortex where gliosis is a prominent feature have not shown significant alterations in resting $[K^+]_o$ or in the ease with which iatrogenic increases in $[K^+]_o$ are handled (Pedley, Fisher & Prince 1976) and thus do not support the hypothesis of Pollen & Trachtenberg (1970) that K^+ accumulation occurs in such areas because of decreased glial K^+ buffering capacity. Intracellular recordings from glia in such cortex have shown a decrease in input resistance (Glötzner 1973) that would tend to make glial cells more, rather than less effective as buffers for increased $[K^+]_o$. Simultaneous recordings of $[K^+]_o$ and glial membrane potentials in acute epileptiform foci (Futamachi & Pedley 1976) have provided strong evidence for a functional glial syncytium in neocortex that might be effective in regulating focal increases in $[K^+]_o$.

One of the difficulties in examining effects of endogenously generated changes in extracellular ionic concentrations on neuronal excitability has been that the measured alterations are generated by activities of the very neuronal aggregates being examined. We have attempted to resolve this problem in part by exposing neurons

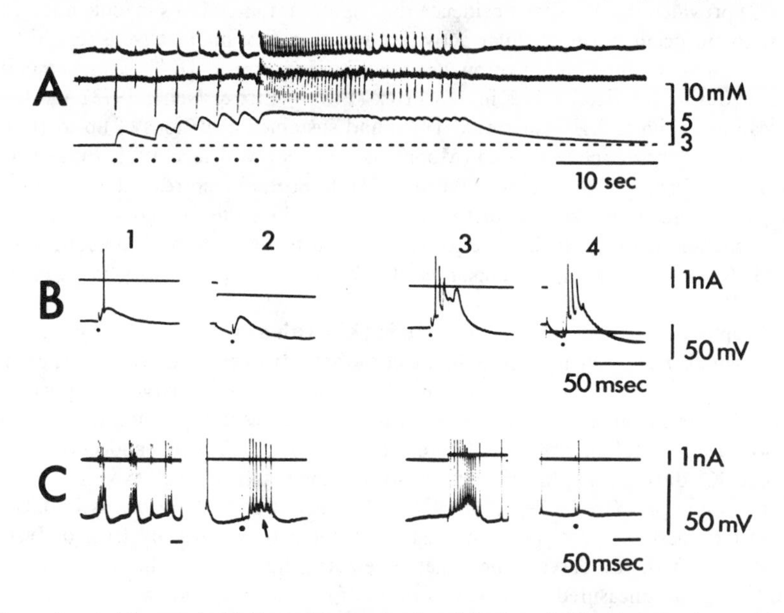

Figure 2 *A*: Changes in $[K^+]_o$ during interictal and ictal penicillin discharge in cat hippocampus. Upper trace: EEG from hippocampal surface; middle trace: field potential from reference micropipette located 50 μ from K^+-sensitive electrode; bottom trace: K^+ activity signal derived from differential recording between reference electrode and ion-sensitive electrode. Dots: stimuli to fornix. Dashed line in bottom trace: 3mM baseline level. Polarity: negativity up in surface EEG and down in reference trace.

B: Recordings from hippocampal CA1 neurons in vitro. *B1* and *2*: Typical responses to Schaffer collateral stimulation (dots) in medium containing 3 mM $[K^+]$ and 2 mM penicillin. *B3* and *4*: Responses of another cell in same slice after perfusing with medium containing 5 mM $[K^+]$ and 2 mM penicillin. The same stimulus as in *B1–2* evokes typical depolarization shifts whose generation is dependent on the increased $[K^+]_o$. Upper trace: current monitor.

C: Recordings from CA1 neuron in solution containing 0.75 mM Ca^{++} (*C1–2*) and shortly after onset of exchange with solution containing 7.5 mM Ca^{++} (*C3–4*). Bursting occurs spontaneously (*C1*) and is evoked by both orthodromic (*C2*) and intracellular (*C3*) stimuli. Generation of spikes during hyperpolarizations in *C1* and *C3* and small spikes of *C2* (arrow) suggest the presence of more than one spike initiating zone. After 5 min of exchange with medium containing 7.5 mM Ca^{++}, orthodromic stimulus in the Schaffer collaterals (dot) evokes normal response—a single spike followed by a small depolarizing afterpotential. Spikes in *B–C* retouched for clarity. Time mark in *C4* for *C2–4.* 50 msec time mark under *C1.* [*A* from (Fisher et al 1976); *B* from P. A. Schwartzkroin & D. A. Prince, unpublished; *C* from J. R. Hotson & D. A. Prince, unpublished).]

of the in vitro brain slice preparation to concentrations of extracellular ions in the range known to occur during epileptogenesis in cortex in vivo, as measured with ion-sensitive electrodes. Although this approach does not allow us to mimic the intracortical ionic distributions, or the activity state of neurons in vivo, it does provide direct evidence that ionic changes of the magnitude measured in neocortex or hippocampus can produce significant net increases in excitability of a neuronal aggregate. In the hippocampal slice preparation, even modest changes in $[K^+]$ of the bathing medium, well within the range recorded in vivo, have dramatic effects on penicillin-induced epileptiform activities (Figure 2*B*). Larger increments to levels of 10–12 mM may lead to generation of epileptiform bursts in such slices (Ogata 1975, Ogata, Hori & Katsuda 1976, Prince & Schwartzkroin, unpublished) without exposure to a convulsant drug. These findings appear to be at odds with earlier conclusions based on in vivo experiments (Fisher et al 1976) discussed above. It is possible that the complexity of events present during epileptogenesis in vivo obscured these effects of $[K^+]_o$ increases. For example, the particular intracortical distribution of increased $[K^+]_o$ and the other associated ionic shifts that are generated in the epileptiform focus, may offset the expected excitability increases which would result from uniform exposure of a cellular population to increased $[K^+]_o$, alone.

Although attention regarding effects of altered ionic microenvironment in epileptogenesis has been focused largely on potassium, other ions should not be neglected. For example, calcium is involved in transmitter release (Katz & Miledi 1969), in the generation of dendritic spikes (Llinas & Hess 1976), and the regulation of spike generation and membrane conductance following depolarization (Meech 1972, Barrett & Barrett 1976), as well as in the modulation of membrane excitability (Frankenhaeuser & Hodgkin 1957). Although a decrease in $[Ca^{++}]_o$ might be expected during intense neuronal activity, it is difficult to predict the net effect of such a change on cortical excitability. The development of a Ca^{++}-sensitive microelectrode (Oehme, Kessler & Simon 1976) has allowed measurements of Ca^{++} activity during evoked cortical potentials and epileptogenesis (Heinemann, Lux & Gutnick 1977). Substantial and long-lasting decreases in $[Ca^{++}]_o$ to levels as low as 0.9–0.5 mM l^{-1} do occur during seizures, presumably secondary to Ca^{++} entry into presynaptic terminals, dendrites, and depolarized neurons. When solutions with $[Ca^{++}]$ in this range are applied to the hippocampal slice, spontaneous and evoked bursts of action potentials occur in CA_3 and CA_1 pyramidal neurons as well as repetitive small spikes resembling the fast prepotentials of Spencer & Kandel (1961) (Figure 2*C*). The implication is that changes in $[Ca^{++}]_o$ of the magnitude demonstrated in cortex during seizures may have profound effects on nerve cell excitability.

The results obtained with ion-sensitive microelectrodes have forced revisions in the concept of close homeostatic control of the extracellular ionic environment in the mammalian brain. Potent effects may arise from ionic shifts occurring during intense neuronal activities, and these may in turn contribute to epileptogenesis. It is possible that increases in $[K^+]_o$ together with decreases in $[Ca^{++}]_o$ and $[Cl^-]_o$ account for the phenomenon of antidromic bursting in cortical axons, as well as for certain long-duration changes in excitability known to occur during interictal epileptogenesis (Prince 1971, Heinemann, Lux & Gutnick 1977) and in the transition from interictal to ictal activities.

Epileptogenesis In Vitro

The finding that epileptiform discharges can be produced in hippocampal slices maintained in vitro, where cellular activities may be examined in detail while the preparation is manipulated in ways hitherto not possible in mammalian CNS (Yamamoto & Kawai 1967, Yamamoto 1972) has lead to increased use of in vitro brain slice preparations in studies of epileptogenesis (Ter Keurs, Voskuyl & Meinardi 1973, Voskuyl, Ter Keurs & Meinardi 1975, Ogata 1975, Ogata, Hori & Katsuda 1976, Schwartzkroin & Prince 1976, 1977, Courtney & Prince 1977, Hotson, Schwartzkroin & Prince 1977a,b Wong & Prince 1977). Yamamoto (1972) demonstrated that depolarization shift-like potentials can be induced in hippocampal CA_3 pyramidal neurons treated with strychnine, with solutions containing low chloride concentration, or with other agents. The large majority of CA_3 pyramidal neurons generate depolarization shift-like bursts without exposure to an epileptogenic medium (Wong & Prince 1977; see Figure 5 below), so that an additional criterion is required to determine whether epileptogenesis is induced by a particular maneuver. The characteristics of extracellular field potentials are useful in this respect. We have recently studied the effects of penicillin on neuronal activities in the hippocampal slice preparation (Schwartzkroin & Prince 1976, 1977a, b). Several findings from these experiments provide important new information regarding mechanisms of depolarization shift generation.

Perfusion of the hippocampal slice with medium containing as little as 1.7 mM penicillin (1,000 U cc^{-1}) changes the potential evoked orthodromically in the CA_1 pyramidal region to a multipeaked field potential that has a rhythmic character (cf Figure 3*A1* and 3*A2*). Similar potentials occur spontaneously in CA_1 and CA_3 regions, but not in the dentate granule cell area. These abnormal field potentials are virtually identical to those recorded in the hippocampus in vivo after penicillin application (Dichter & Spencer 1969a, Dichter, Herman & Selzer 1972). They are not seen normally in extracellular recordings from the CA_3 region, even though spontaneously bursting neurons are present. This suggests that generation of field potential bursts requires synchronization of neuronal activities. Of considerable interest relative to intracellular data described below is the finding that the CA_3 region serves as a pacemaker in this preparation for the generation of epileptiform activity. Field potential bursts in CA_3 always lead those in CA_1 (Figure 3*A3*). Cutting the Schaffer collateral connections between CA_1 and CA_3 eliminates burst activity in CA_1; however, it persists in CA_3 (Figure 3*A4*). In fact, CA_3 can generate rhythmic bursting even when it is completely isolated from both dentate granule cell input and CA_1. Thus, the three major neuronal populations in the in vitro hippocampus have different susceptibilities to the development of epileptiform activity following penicillin exposure.

CA_1 neurons bathed in penicillin-containing medium generate potentials similar to the depolarization shifts recorded in vivo during spontaneous or orthodromically evoked epileptiform field potentials. Bursts of spikes of varying amplitudes riding on underlying slow depolarizations occur in some cells, usually followed by hyperpolarizations (e.g. Figure 3*B1–2*). In other neurons, bursts are generated in which spikes appear to arise abruptly from the baseline without significant underlying

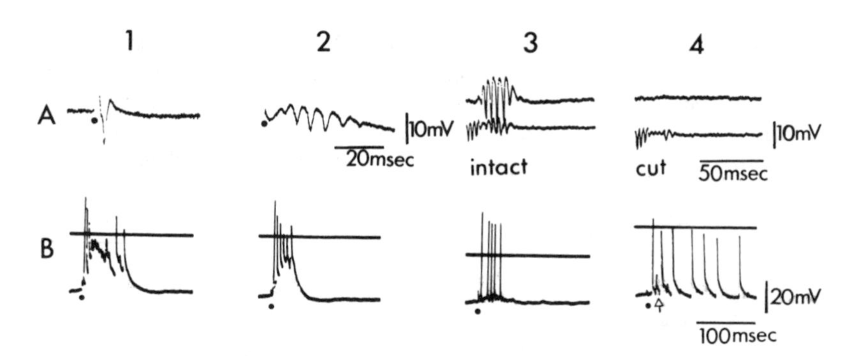

Figure 3 Penicillin epileptogenesis in the hippocampal slice preparation. *A1*: Normal evoked field potential in CA_1 region after stimulation of Schaffer collaterals (dot). *A2*: Same preparation exposed to medium containing 2 mM penicillin; less intense stimulation evokes a field potential with multiple peaks. *A3*: Spontaneous field potential bursts in CA_1 (upper trace) and CA_3 (lower) in preparation exposed to penicillin. *A4*: Persistence of spontaneous burst in CA_3 but disappearance in CA_1 after cutting connections between these areas.

B: Evoked intracellular activities in CA_1 neurons after exposure to penicillin. Dots: Schaffer collateral stimulation. Cells of C1 and C2 show DS-like potentials. Cells of C3 and C4 show spike bursts with minimal underlying depolarization and small partial spikes (arrow in *B4*). [*A3–4* from Schwartzkroin & Prince, 1977b; *A1–2* and *B* from (Schwartzkroin & Prince 1977a).]

depolarization (Figure 3*B3–4*). Small partial spikes are frequently seen within these bursts (arrow in Figure 3*B4*). In contrast to the expected behavior of hypothesized "giant EPSPs," depolarization shifts in some hippocampal CA_1 neurons show delayed onset or even complete blockade when they are evoked during hyperpolarizing current pulses (Figure 4*A*). In other cells, with increasing levels of DC intracellular hyperpolarization, a progressive blockade of burst spiking may occur, associated with attenuation of the underlying depolarization shift (Figure 4*B*). At the same time partial spikes resembling the "fast prepotentials" of Spencer & Kandel (1961) are uncovered (arrows in Figure 4*B2–4*). These arise from below the normal spike threshold and appear to underlie the generation of some burst spikes.

These findings are inconsistent with "giant synaptic potential" hypothesis of depolarization shift generation. They indicate that under some circumstances depolarization shift-like potentials recorded in the soma during epileptogenesis may be the result of summations of all-or-none events (e.g. spike afterpotentials in the neuron of Figure 4*A*). In cases where hyperpolarizing current blocks spike generation without uncovering an EPSP, the generator potential for spike bursts must be at a distant site, perhaps on distal dendrites. To date we have been unable to uncover a "giant EPSP" in any depolarization shift-generating hippocampal CA_1 neuron, although orthodromic EPSPs are certainly present in some. This is not to suggest that EPSPs are unimportant for depolarization shift generation, but rather that they serve as triggering potentials as opposed to generators for the depolarization shift(s) in these cells. In fact, penicillin epileptogenesis is blocked in the hippocampal slice during perfusion of solutions containing low [Ca^{++}] and high [Mg^{++}] that also block

block synaptic activity (Schwartzkroin & Prince 1976, 1977b). The role of hypothesized penicillin-augmented EPSPs (Ayala et al 1973, Futamachi & Prince 1975) may be to provide an orthodromic synchronous depolarization to a population of neurons which in turn generate "their own" depolarization shift(s) and the resulting characteristic epileptiform field potentials. Depolarizing current pulses can evoke

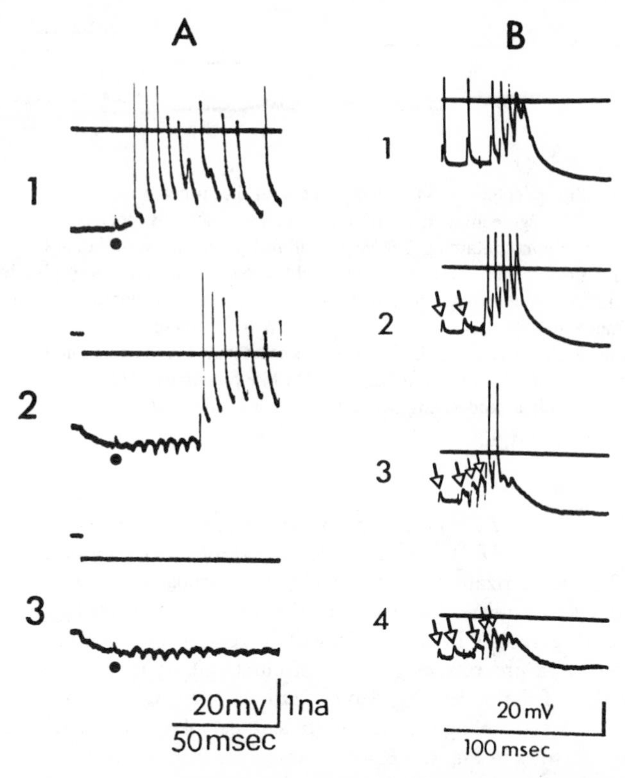

Figure 4 Effects of hyperpolarizing current pulses on penicillin-induced DS generation in CA1 cells. *A1*: DS with multiple sized burst spikes evoked by Schaffer collateral stimulation (dot). Hyperpolarizing current pulses of increasing intensity in *A2* and *A3* delay the onset (*A2*) and completely block (*A3*) the DS, although epileptiform field potential seen intracellularly is present.

B: Spontaneous DS generation at resting membrane potential (*B1*) and at increasing levels of DC intracellular hyperpolarization (*B2–4*). As hyperpolarization increases, spontaneous spikes are blocked revealing underlying fast prepotentials (arrows in *B2*). Blockade of burst spikes (*B3–B4*) causes attenuation of DS slow envelope and reveals underlying rhythmic fast prepotentials (3rd and 4th arrows of *B3*; 3rd, 4th and 5th arrows of *B4*). Bursts of *B3–B4* contain some reflection of extracellular field potentials. Spikes in *B1* cut off at peaks. Upper traces: current monitor with zero level in *A1, B1*. Calibration under *B4*: 20 mV and 1 nA for sweeps of *B*. [*A* from (Schwartzkroin & Prince 1977a); *B* from P. A. Schwartzkroin & D. Prince, 1977b.]

zation shifts in a small number of uninjured CA_1 neurons of the penicillin-treated slice, whereas orthodromic stimuli are invariably effective in triggering these potentials. This difference may be related to the distribution of the orthodromic input onto dendritic membrane.

The propensity for CA_3 versus CA_1 pyramidal cells to act as pacemakers for penicillin epileptiform discharges may be related to fundamental differences in properties of these neurons in the "normal" state. In comparison to CA_1 cells, a high proportion of CA_3 cells normally generate spontaneous bursts that may be evoked and reset using brief intracellular current pulses [proving that the bursts are intrinsic to the cell under study and independent of synaptic input (Wong & Prince 1977; see also Kandel & Spencer 1961)]. Similar bursts are evoked by orthodromic stimuli. Examples of spontaneous and intracellularly evoked bursts in CA_3 neurons are shown in Figure 5*A* and 5*B1–2*. These burst discharges are similar in appearance to those generated by orthodromic stimulation in CA_1 cells of the penicillin-treated preparation (cf. depolarization shifts in Figures 2*B*, 3*B*, and 4 with bursts in Figure 5*A*) although they are not associated with similar field potentials. Bursts evoked by intracellular stimuli share other properties with depolarization shifts including (*a*) occurrence at full amplitude when the stimulus reaches threshold; (*b*) post-burst hyperpolarizations with increased conductance; (*c*) post-burst excitability cycles of 1–2 sec during which a second burst cannot be triggered with the same stimulus; (*d*) a tendency to show spike inactivation and small broad spikes at their peaks; and (*e*) a latency that decreases as stimulus intensity increases. CA_3 pyramidal cell bursts are divisible into an early phase consisting of 3 or 4 large-amplitude, fast spikes and a later phase of several slow spikes with multiple peaks and humps generated at a higher threshold of depolarization. The late phase tends to be triggered in an all-or-none fashion during spontaneous or evoked bursts (Figure 5*A2*; cf. Figure 5*A3* and 5*A4*). Analysis of these bursts suggests that the underlying depolarizing envelope is generated by a summation of depolarizing spike afterpotentials. The late-phase spikes and probably the depolarizing afterpotentials are blocked by bath perfusion with 1 mM Mn^{++}, which suggests that these potentials are mediated by Ca^{++} entry. Support for this conclusion comes from the finding that CA_3 neurons can generate repetitive spikes when depolarized in the presence of tetrodotoxin (TTX) (Figure 5*B3–4*; 5*C*). TTX-resistant spikes resemble the small broad spikes of the late phase of spontaneous or evoked bursts, occur at a high threshold (cf. spikes of Figure 5*B1* and B*3*) and are also reversibly blocked by manganese application (cf. Figure 5*C3* and 4). In recordings from dendrites of identified CA_3 pyramidal neurons, TTX-resistant spikes have amplitudes as large as 60 mV and faster rise times than those recorded in the cell body, suggesting that they are generated on dendrites (Wong, Prince & Basbaum, unpublished). Taken together, this evidence suggests that CA_3 pyramidal cells possess the capacity for repetitive dendritic calcium spike electrogenesis, which normally plays a role in burst generation. By analogy with data from avian Purkinje cells (Llinas & Hess 1976), the calcium spikes may be arising at multiple dendritic sites. This would account for their tendency to assume a variety of shapes (Figure 5*B4*), their occasional multi-peaked appearance (Figure 5*C3*), and their high threshold relative to normal (sodium) spikes. Calcium spike generation also cccurs in hippocampal CA_1

pyramidal cells (Schwartzkroin & Slawsky 1977), but not during normal spontaneous activity. We have thus far been unable to elicit TTX-resistant spikes in granule cells.

The relationship between the mechanisms of spontaneous bursting in CA_3 neurons and depolarization shift generation in hippocampal (or other) neurons is unclear. There does appear to be a parallelism among the three major cell groups in hippocampus between the capacity to generate bursts under normal conditions and the degree of involvement in penicillin epileptogenesis. Penicillin may have an effect on the intrinsic neuronal events leading to burst generation; or it may only serve

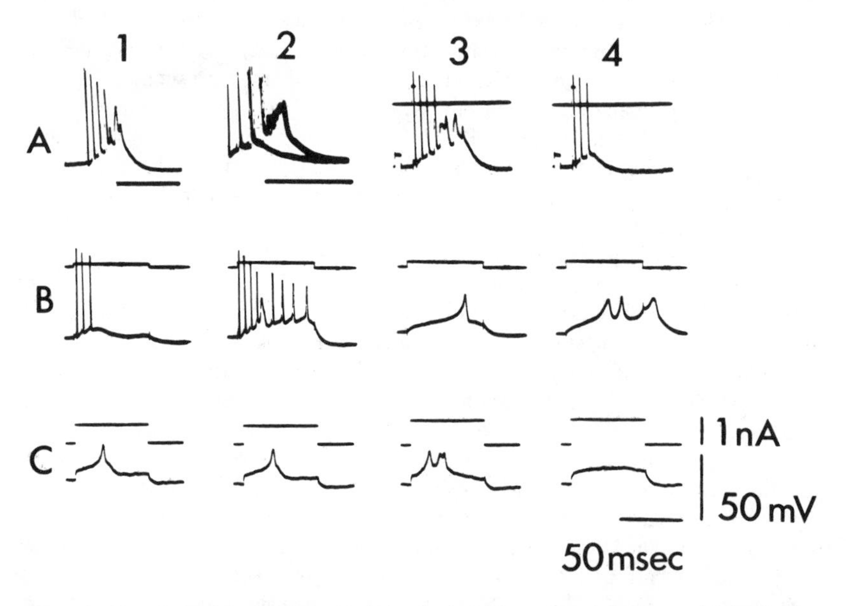

Figure 5 Depolarizations and bursting behavior in 3 hippocampal CA_3 pyramidal cells from slice in normal medium. In cell of *A*, spontaneous bursts (*A1–2*) and some triggered by 2 msec intracellular current pulses (*A3*) show initial phase with high amplitude, regular interval rapid spiking, followed by second phase with small amplitude slower spikes occurring at a higher threshold. *A2*: Superimposed sweeps of spontaneous burst activity triggered from first spike of the burst show that late phase is generated in all-or-none fashion. Similar behavior occurs in bursts triggered by intracellular depolarizing current pulses (cf *A3* and *A4*).

B: Another neuron showing brief burst during weak intracellular depolarizing current pulse (*B1*) and prolonged decrementing burst with an intermixed slow spike during stronger depolarization (*B2*). *B3* and *B4*: Single and repetitive high threshold, slow spikes evoked by intracellular depolarizations after bath perfusion with TTX (5×10^{-6} g ml^{-1}). Current pulses in *B2* and *B3* are identical; current increased in *B4*.

C: Single (*C1–C3*) and repetitive (*C3*) TTX-resistant spikes in another neuron. Behavior prior to TTX was similar to that seen in *A1* and *A3*. *C4*: TTX-resistant spikes blocked after local application of 20 mM Mg^{++} to CA_3 region. Calibrations in *C4* for all sweeps except for *A1–A2* where 50 msec time marks shown. Some spikes retouched for clarity. 1st and 2nd spikes of bursts in *A2* cut off by scope face. (From R. K. S. Wong & D. A. Prince, unpublished.)

to increase the synchrony and intensity of excitation through its known effects on EPSPs and IPSPs, bringing populations of cells to threshold to generate intrinsic depolarization shifts. The functional role of calcium spikes in terms of the output of hippocampal pyramidal neurons under normal or epileptogenic conditions requires further definition. They might serve to amplify distant dendritic depolarizations and result in increased spike output from the neuron, or, through activation of a delayed potassium conductance and associated hyperpolarization, act to limit bursts of sodium spikes and decrease responsiveness to excitatory inputs.

CONCLUSIONS

Data from a variety of experimental preparations emphasize the importance of depolarization shift generation as a characteristic event in neurons of acute epileptiform foci. The hypothesis that these depolarization shifts represent "giant" excitatory postsynaptic potentials requires substantial revision in light of more recent data. Evidence from the hippocampal in vitro slice preparation suggests that different populations of neurons have different susceptibilities for generation of epileptogenic discharges and that intrinsic regenerative neuronal events play an important role in depolarization shift generation. Similarities are noted between the characteristics of bursts normally generated in CA_3 neurons, which are in part dependent on dendritic calcium spike generation, and those of penicillin-evoked depolarization shifts. These and other data suggest that penicillin may either affect the intrinsic neuronal events leading to burst generation or (more likely) serve to increase the intensity of excitation through effects on EPSPs and IPSPs, thus bringing populations of neurons to threshold to generate their own (intrinsic) depolarization shifts. Thus, at least in hippocampal neurons, depolarization shifts are not "giant" synaptic potentials. Other significant nonsynaptic factors have been found that probably affect the excitability of neurons in epileptogenic foci. Alterations in the ionic microenvironment that occur in vivo, such as increased $[K^+]_o$ and decreased $[Ca^{++}]_o$, are of sufficient magnitude to induce depolarization shifts and epileptiform field potentials in vitro. Nonsynaptic generation of bursts in intracortical axons occurs during epileptogenesis, perhaps as a consequence of these ionic alterations, and may be an important factor that amplifies excitation and leads to transitions from interictal to ictal discharges.

ACKNOWLEDGMENTS

I am grateful to Ms. Geraldine Chase for secretarial assistance and to Erica Prince who helped edit the manuscript. Drs. John Hotson, Jeffrey Noebels, Philip Schwartzkroin, Robert Wong, Robert Fisher, Michael Gutnick, Kin Futamachi, and Timothy Pedley made invaluable contributions to much of the work described here. I thank Reva Prince for her encouragement and helpful suggestions.

This work was supported by USPHS Grants NS06477 and NS12151 from the NINCDS, by a grant from the California Community Foundation, and by the Morris Research Fund.

Literature Cited

Ajmone-Marsan, C. 1969. Acute effects of topical epileptogenic agents. In *Basic Mechanisms of the Epilepsies,* ed. H. H. Jasper, A. A. Ward, Jr., A. Pope, pp. 299–328. Boston: Little, Brown. 835 pp.

Ayala, G. F., Dichter, M., Gumnit, R. J., Matsumoto, H., Spencer, W. A. 1973. Genesis of epileptic interictal spikes: New knowledge of cortical feedback systems suggests a neurophysiological explanation of brief paroxysms. *Brain Res.* 52:1–17

Ayala, G. F., Matsumoto, H., Gumnit, R. J. 1970. Excitability changes and inhibitory mechanisms in neocortical neurons during seizures. *J. Neurophysiol.* 33: 73–85

Ayala, G. F., Vasconetto, C. 1972. Role of recurrent excitatory pathways in epileptogenesis. *Electroencephalogr. Clin. Neurophysiol.* 33:96–98

Barrett, E. F., Barrett, J. N. 1976. Separation of two voltage-sensitive potassium currents, and demonstration of a tetrodotoxin-resistant calcium current in frog motoneurones. *J. Physiol. London* 255:737–74

Clarke, G., Hill, R. G. 1972. Effects of a focal penicillin lesion on responses of rabbit cortical neurones to putative neurotransmitters. *Br. J. Pharmacol.* 44: 435–41

Courtney, K. R., Prince, D. A. 1977. Epileptogenesis in neocortical slices. *Brain Res.* 127:191–96

Curtis, D. R., Game, C. J. A., Johnston, G. A. R., McCulloch, R. M., Maclachlan, R. M. 1972. Convulsive action of penicillin. *Brain Res.* 43:242–45

Dichter, M. A., Herman, C. J., Selzer, M. 1972. Silent cells during interictal discharges and seizures in hippocampal penicillin foci. Evidence for the role of extracellular K^+ in the transition from the interictal state to seizures. *Brain Res.* 48:173–83

Dichter, M., Spencer, W. A. 1969a. Penicillin-induced interictal discharges from the cat hippocampus. I. Characteristics and topographical features. *J. Neurophysiol.* 32:649–62

Dichter, M., Spencer, W. A. 1969b. Penicillin-induced interictal discharges from the cat hippocampus. II. Mechanisms underlying origin and restriction. *J. Neurophysiol.* 32:663–87

Eckert, R., Lux, H. D. 1976. A voltage-sensitive persistent calcium conductance in neuronal somata of *Helix. J. Physiol.* 254:129–51

Fertziger, A. P., Ranck, J. B. 1970. Potassium accumulation in interstitial space during epileptiform seizures. *Exp. Neurol.* 26:571–85

Fisher, R. S., Pedley, T. A., Moody, W. J. Jr., Prince, D. A. 1976. The role of extracellular potassium in hippocampal epilepsy. *Arch. Neurol. Chicago* 33:76–83

Frankenhaeuser, B., Hodgkin, A. L. 1957. The action of calcium on the electrical properties of squid axons. *J. Physiol. London* 137:218–44

Futamachi, K. J., Mutani, R., Prince, D. A. 1974. Potassium activity in rabbit cortex. *Brain Res.* 75:5–25

Futamachi, K. J., Pedley, T. A. 1976. Glial cells and extracellular potassium: their relationship in mammalian cortex. *Brain Res.* 109:311–22

Futamachi, K. J., Prince, D. A. 1975. Effect of penicillin on an excitatory synapse. *Brain Res.* 100:589–97

Gabor, A. J., Scobey, R. P. 1975. Spatial limits of epileptogenic cortex: its relationship to ectopic spike generation. *J. Neurophysiol.* 38:395–404

Gleason, C. A. 1971. *The effects of applied direct current fields on electrical activity in cat cortex.* PhD thesis. Stanford Univ. 135 pp.

Glötzner, F. L. 1973. Membrane properties of neuroglia in epileptogenic gliosis. *Brain Res.* 55:159–71

Goldensohn, E. S., Purpura, D. P. 1963. Intracellular potentials of cortical neurons during focal epileptogenic discharges. *Science* 139:840–42

Green, J. D. 1964. The hippocampus. *Physiol. Rev.* 44:561–608

Gumnit, R. J., Takahashi, T. 1965. Changes in direct current activity during experimental focal seizures. *Electroencephalogr. Clin. Neurophysiol.* 19:63–74

Gutnick, M. J., Prince, D. A. 1972. Thalamocortical relay neurons: antidromic invasion of spikes from a cortical epileptogenic focus. *Science* 176:424–26

Gutnick, M. J., Prince, D. A. 1974. Effects of projected cortical epileptiform discharges on neuronal activities in cat VPL. I. Interictal discharges. *J. Neurophysiol.* 37:1310–27

Gutnick, M. J., Prince, D. A. 1975. Effects of projected cortical epileptiform discharges on neuronal activities in ventrobasal thalamus of the cat: Ictal discharge. *Exp. Neurol.* 46:418–31

Hardy, R. W. 1970. Unit activity in Premarin®-induced cortical epileptiform foci. *Epilepsia* 11:179–86

Heinemann, U., Lux, H. D., Gutnick, M. J. 1977. Extracellular free calcium and potassium during paroxysmal activity in the cerebral cortex of the cat. *Exp. Brain Res.* 27:237–43

Hochner, B., Spira, M. E., Werman, R. 1976. Penicillin decreases chloride conductance in crustacean muscle: a model for the epileptic neuron. *Brain Res.* 107: 85–103

Hotson, J. R., Schwartzkroin, P. A., Prince, D. A. 1977. Calcium activated after hyperpolarization in hippocampal slices maintained in vitro. *Neurosci. Abstr.* 3:218 (Abstr. 680)

Jasper, H. H. 1969. Mechanisms of propagation: extracellular studies. See Ajmone-Marsan 1969, pp. 421–38

Jasper, H. H., Ward, A. A. Jr., Pope, A., eds. 1969. *Basic Mechanisms of the Epilepsies.* Boston: Little, Brown. 835 pp.

Kandel, E. R., Spencer, W. A. 1961. Electrophysiology of hippocampal neurons. II. Afterpotentials and repetitive firing. *J. Neurophysiol.* 24:243–59

Kao, L. I., Crill, W. E. 1972. Penicillin-induced segmental myoclonus. I. Motor responses and intracellular recording from motoneurons. *Arch. Neurol. Chicago* 26:156–61

Katz, B., Miledi, R. 1969. Tetrodotoxin-resistant electric activity in presynaptic terminals. *J. Physiol. London* 203: 459–87

Klee, M. R. 1975. Differences between monosynaptic and polysynaptic excitatory postsynaptic potentials in cat motoneurons. In *Golgi Centennial Symposium: Perspectives in Neurobiology,* ed. M. Santini, pp. 261–71. New York: Raven Press

Kuffler, S. W., Nicholls, J. G. 1966. Physiology of neuroglial cells. *Ergeb. Physiol. Biol. Chem. Exp. Pharmakol.* 57:1–90

Li, C.-L. 1959. Cortical intracellular potentials and their responses to strychnine. *J. Neurophysiol.* 22:436–50

Llinas, R., Hess, R. 1976. Tetrodotoxin-resistant dendritic spikes in avian Purkinje cells. *Proc. Natl. Acad. Sci. USA* 73:2520–23

Lothman, E. W., Somjen, G. G. 1976a. Functions of primary afferents and responses of extracellular K^+ during spinal epileptiform seizures. *Electroencephalogr. Clin. Neurophysiol.* 41:253–67

Lothman, E. W., Somjen, G. G. 1976b. Reflex effects and postsynaptic membrane potential changes during epileptiform activity induced by penicillin in decapitate spinal cords. *Electroenceph-alogr. Clin. Neurophysiol.* 41:337–47

Matsumoto, H. 1964. Intracellular events during the activation of cortical epileptiform discharges. *Electroencephalogr. Clin. Neurophysiol.* 17:294–307

Matsumoto, H., Ajmone-Marsan, C. 1964a. Cortical cellular phenomena in experimental epilepsy: Interictal manifestations. *Exp. Neurol.* 9:286–304

Matsumoto, H., Ajmone-Marsan, C. 1964b. Cortical cellular phenomena in experimental epilepsy: Ictal manifestations. *Exp. Neurol.* 9:305–26

Matsumoto, H., Ayala, G. F., Gumnit, R. J. 1969. Neuronal behavior and triggering mechanisms in cortical epileptic focus. *J. Neurophysiol.* 32:688–703

Meech, R. W. 1972. Intracellular calcium injection causes increased potassium conductance in *Aplysia* nerve cells. *Comp. Biochem. Physiol. A* 42:493–99

Meyer, H., Prince, D. A. 1973. Convulsant actions of penicillin: effects on inhibitory mechanisms. *Brain Res.* 53:477–82

Moody, W. J. Jr., Futamachi, K. J., Prince, D. A. 1974. Extracellular potassium activity during epileptogenesis. *Exp. Neurol.* 42:248–63

Mutani, R., Futamachi, K. J., Prince, D. A. 1974. Potassium activity in immature cortex. *Brain Res.* 75:27–39

Noebels, J. L., Prince, D. A. 1977a. Presynaptic origin of penicillin afterdischarges at mammalian nerve terminals. *Brain Res.* 138:59–74

Noebels, J. L., Prince, D. A. 1977b. Synchronous discharges in neocortex alter the excitability of intracortical axon terminals. *Neurosci. Abstr.* 3:143 (Abstr. 443)

Oehme, M., Kessler, M., Simon, W. 1976. Neutral carrier Ca^{2+}-microelectrode. *Chimia* 30:204–6

Ogata, N. 1975. Ionic mechanisms of the depolarization shift in thin hippocampal slices. *Exp. Neurol.* 46:147–55

Ogata, N., Hori, N., Katsuda, N. 1976. The correlation between extracellular potassium concentration and hippocampal epileptic activity in vitro. *Brain Res.* 110:371–75

Pedley, T. A., Fisher, R. S., Futamachi, K. J., Prince, D. A. 1976. Regulation of extracellular potassium concentration in epileptogenesis. *Fed. Proc.* 35:1254–59

Pedley, T. A., Fisher, R. S., Prince, D. A. 1976. Focal gliosis and potassium movement in mammalian cortex. *Exp. Neurol.* 50:346–61

Pollen, D. C., Trachtenberg, M. C. 1970. Neuroglia: gliosis and focal epilepsy. *Science* 167:1253–58

Prince, D. A. 1965. Cyclical spike driving in chronically isolated cortex. *Epilepsia* 6:226–42

Prince, D. A. 1966. Modification of focal cortical epileptogenic discharge by afferent influences. *Epilepsia* 7:181–201

Prince, D. A. 1968a. The depolarization shift in "epileptic" neurons. *Exp. Neurol.* 21:467–85

Prince, D. A. 1968b. Inhibition in "epileptic" neurons. *Exp. Neurol.* 21:307–21

Prince, D. A. 1969a. Electrophysiology of "epileptic" neurons: spike generation. *Electroencephalogr. Clin. Neurophysiol.* 26:476–87

Prince, D. A. 1969b. Discussion: Microelectrode studies of penicillin foci. See Ajmone-Marsan 1969, pp. 320–28

Prince, D. A. 1971. Cortical cellular activities during cyclically occurring inter-ictal epileptiform discharges. *Electroencephalogr. Clin. Neurophysiol.* 31:469–84

Prince, D. A. 1972. Topical convulsant drugs and metabolic antagonists. In *Experimental Models of Epilepsy,* ed. D. P. Purpura, J. K. Penry, D. Tower, D. M. Woodbury, R. Walter, pp. 51–83. New York: Raven Press. 615 pp.

Prince, D. A. 1974. Neuronal correlates of epileptiform discharges and cortical DC potentials. In *Handbook of Electroencephalography and Clinical Neurophysiology,* ed. O. Creutzfeldt, 2C:56–70. Amsterdam: Elsevier. 157 pp.

Prince, D. A. 1976. Cellular activities in focal epilepsy. In *Brain Dysfunction in Infantile Febrile Convulsions,* ed. M. A. B. Brazier, F. Coceani, pp. 187–212. New York: Raven Press. 370 pp.

Prince, D. A., Gutnick, M. J. 1972. Neuronal activities in epileptogenic foci of immature cortex. *Brain Res.* 45:455–68

Prince, D. A., Lux, H. D., Neher, E. 1973. Measurement of extracellular potassium activity in cat cortex. *Brain Res.* 50:489–95

Prince, D. A., Wilder, B. J. 1967. Control mechanisms in cortical epileptogenic foci: "Surround" inhibition. *Arch. Neurol. Chicago* 16:194–202

Purpura, D. P. 1969a. Stability and seizure susceptibility of immature brain. See Ajmone-Marsan 1969, pp. 481–505

Purpura, D. P. 1969b. Discussion: Mechanisms of propagation: Intracellular studies. See Ajmone-Marsan 1969, pp. 441–51

Purpura, D. P., Penry, J. K., Tower, D. B., Woodbury, D. M., Walter, R. D., eds. 1972. *Experimental Models of Epilepsy.* New York: Raven Press. 615 pp.

Ransom, B. R. 1974. The behavior of presumed glial cells during seizure discharge in cat. *Brain Res.* 69:83–99

Rosen, A. D., Vastola, E. F., Hildebrand, Z. J. M. 1973. Visual radiation activity during a cortical penicillin discharge. *Exp. Neurol.* 40:1–12

Schwartzkroin, P. A., Bromley, B., Shimada, Y. 1977. Recordings from cortical epileptogenic foci induced by cobalt iontophoresis. *Exp. Neurol.* 55:353–67

Schwartzkroin, P. A., Futamachi, K. J., Noebels, J. L., Prince, D. A. 1975. Transcallosal effects of a cortical epileptiform focus. *Brain Res.* 99:59–68

Schwartzkroin, P. A., Mutani, R., Prince, D. A. 1975. Orthodromic and antidromic effects of a cortical epileptiform focus on ventrolateral nucleus of the cat. *J. Neurophysiol.* 38:795–811

Schwartzkroin, P. A., Prince, D. A. 1976. Penicillin-induced activity in hippocampal slices maintained in vitro. *Neurosci. Abstr.* 2(Abstr. 380):266

Schwartzkroin, P. A., Prince, D. A. 1977a. Penicillin-induced epileptiform activity in the hippocampal in vitro preparation. *Ann. Neurol.* 1:463–69

Schwartzkroin, P. A., Prince, D. A. 1977b. Cellular and field potential properties of epileptogenic hippocampal slices. *Brain Res.* In press

Schwartzkroin, P. A., Slawsky, M. 1977. Probable calcium spikes in hippocampal neurons. *Brain Res.* 135:157–61

Scobey, R. P., Gabor, A. J. 1975. Ectopic action potential generation in epileptogenic cortex. *J. Neurophysiol.* 38:383–94

Speckmann, E. J., Caspers, H. 1973. Paroxysmal depolarization and changes in action potentials induced by pentylenetetrazol in isolated neurons of *Helix pomatia. Epilepsia* 14:397–408

Spencer, W. A., Kandel, E. R. 1961. Electrophysiology of hippocampal neurons. IV. Fast prepotentials. *J. Neurophysiol.* 24:272–85

Spencer, W. A., Kandel, E. R. 1969. See Ajmone-Marsan 1969, pp. 575–604

Sypert, G. W., Ward, A. A. Jr. 1971. Unidentified neuroglia potentials during propagated seizures in neocortex. *Exp. Neurol.* 28:308–25

Sypert, G. W., Ward, A. A. Jr. 1974. Changes in extracellular potassium activity dur-

ing neocortical propagated seizures. *Exp. Neurol.* 45:19–41
Takahashi, K., Kubota, K., Uno, M. 1967. Recurrent facilitation in cat pyramidal tract cells. *J. Neurophysiol.* 30:22–34
Ter Keurs, W. J., Voskuyl, R. A., Meinardi, H. 1973. Effects of penicillin on evoked potentials of excised prepiriform cortex of guinea pig. *Epilepsia* 14:261–71
Van Duijn, H., Schwartzkroin, P. A., Prince, D. A. 1973. Action of penicillin on inhibitory processes in the cat's cortex. *Brain Res.* 53:470–76
Voskuyl, R. A., Ter Keurs, H. E. D. J., Meinardi, H. 1975. Actions and interactions of dipropylacetate and penicillin on evoked potentials of excised prepiriform cortex of guinea pig. *Epilepsia* 16:583–92
Vyskočil, F., Kříž, N., Bureš, J. 1972. Potassium-selective microelectrodes used for measuring the extracellular brain potassium during spreading depression and anoxic depolarization in rats. *Brain Res.* 39:255–59
Walsh, G. O. 1971. Penicillin iontophoresis in neocortex of cat: effects on the spontaneous and induced activity of single neurons. *Epilepsia* 12:1–11
Ward, A. A. Jr. 1969. The epileptic neuron: chronic foci in animals and man. See Ajmone-Marsan 1969, pp. 263–88
Ward, A. A. Jr., Jasper, H. H., Pope, A. 1969. Clinical and experimental challenges of the epilepsies. See Ajmone-Marsan 1969, pp. 1–12
Williamson, T. L., Crill, W. E. 1976. The effects of pentylenetetrazol on molluscan neurons. I. Intracellular recordings and stimulation. *Brain Res.* 116:217–24
Wong, R. K. S., Prince, D. A. 1977. Burst generation and calcium mediated spikes in hippocampal CA3 neurons. *Neurosci. Abstr.* 3:148 (Abstr. 465)
Yamamoto, C. 1972. Intracellular study of seizure-like afterdischarges elicited in thin hippocampal sections in vitro. *Exp. Neurol.* 35:154–64
Yamamoto, C., Kawai, N. 1967. Seizure discharges evoked in vitro in thin section from guinea pig hippocampus. *Science* 155:341–42
Yamamoto, C., Kawai, N. 1968. Generation of the seizure discharge in thin sections from the guinea pig brain in chloride-free medium in vitro. Jpn. J. Physiol. 18:620–31
Zipser, B., Crain, S. M., Bornstein, M. B. 1973. Directly evoked "paroxysmal" depolarizations of mouse hippocampal neurons in synaptically organized explants in long-term culture. *Brain Res.* 60:489–95
Zuckermann, E. C., Glaser, G. H. 1968. Hippocampal epileptic activity induced by localized ventricular perfusion with high-potassium cerebrospinal fluid. *Exp. Neurol.* 20:87–110.

Ann. Rev. Neurosci. 1978. 1:417–43

BIOLOGY OF CULTURED NERVE AND MUSCLE

♦11514

J. Patrick, S. Heinemann, and D. Schubert
The Neurobiology Department, The Salk Institute, San Diego, California 92112

In 1907 Ross Harrison used explant cultures of amphibian embryos to distinguish between two alternatives for the formation of nerve fibers (Harrison 1907). By culturing isolated pieces of neural epithelium in a controlled environment, he successfully ruled out the idea that the nerve fiber resulted from anastomosis between a number of cells; he also provided convincing support for the Cajal hypothesis of neurite extension. These experiments not only provide the first use of tissue culture for the study of the nervous system, but also exemplify the potential inherent in cell culture. The study of nervous system development and function must at some point focus on the development and function of the individual cells. Although this is difficult in vivo, tissue culture permits the study of individual cell types in an environment that minimizes the multiple interactive events characteristic in the animal. Purified populations of nerve and muscle, or homogeneous populations derived from clonal cell lines, have provided preparations suitable for biochemical and electrophysiological analysis. In addition, there is now evidence that these preparations are not dead-end simplifications but ones in which cellular interactions similar to those occurring in vivo can be studied.

We review some aspects of the development and function of the nervous system in which cell culture has provided a unique access to the problem. Since we cannot be comprehensive we have chosen to deal either with areas in which a substantial body of information has accrued, or areas in which experiments in tissue culture have successfully distinguished between alternative explanations for observations made in vivo. Since primary cultures are dealt with elsewhere (Patterson 1978, Varon & Bunge), this review emphasizes clonal cell lines.

THE SPECTRUM OF CLONAL CELL LINES

To exploit clonal cell lines for the study of neurobiological problems, a collection of cells is required that is representative of the in vivo cell types. Although this task is far from complete, a number of lines have been described. The majority of these were derived from spontaneous or chemically induced neoplasms; some were ob-

0147-006X/78/0325-0417$01.00

tained by selectively cultivating embryonic tissue. The initial cell lines were of glial origin, followed by the popular C1300 neuroblastoma clones. More recently, a large number of nerve and glial lines were isolated from rat brain, and a clone of sympathetic ganglion-like cells was isolated from a rat pheochromocytoma. A brief description of these nerve and glial cell lines will be presented, followed by some characteristics of a few cell lines of muscle origin.

Cell Lines from the Nervous System

The C1300 mouse neuroblastoma probably originates from sympathetic ganglion cells; the properties of the numerous clones derived from the original tumor have been amply reviewed (Breakefield 1976, Haffke & Seeds 1975, Schubert et al 1973). Clones with widely different characteristics have been isolated. This variation, which is also reflected in chromosome number, is probably a consequence of unselected mutations during the large number of in vivo tumor generations. Clones have been isolated that synthesize different neurotransmitters (Amano et al 1972, Breakefield 1976), have different pharmacological properties (Gilman & Nirenberg 1971), and express different morphologies. In addition, selection techniques have been applied to obtain clones with desired characteristics (Breakefield & Nirenberg 1974). Finally, to generate a larger spectrum of cell lines with neuronal properties, C1300 neuroblastoma cells were fused with other cell lines and hybrid cells were selected (Minna et al 1972). These hybrid lines have been successfully used to study the opiate receptor (Klee & Nirenberg 1974) and cell-cell interactions (Nelson et al 1976).

At the time the first cell lines derived from the C1300 neuroblastoma were described, it was generally assumed that it was difficult, if not impossible, to chemically induce brain tumors of neuronal origin. However, the procedure described by Druckrey (Druckrey et al 1967) for transplacentally inducing tumors with nitrosoethylurea was successfully used to obtain tumors with neuronal properties (Schubert et al 1974a). Of over 100 clonal cell lines derived from independent tumors, approximately 5% had the characteristics expected of nerve cells; the majority were probably of glial origin since they contained one or more nervous system-specific proteins but were not electrically excitable (Kidokoro et al 1975, West et al 1977).

Another approach to obtaining nerve cell lines that divide permanently, particularly from selected tissues, is to transform embryonic cells that have been placed in primary culture. Using a viral transforming agent, a nerve cell line that synthesizes neurophysin and vasopressin was selected from the hypothalamus (de Vitry et al 1974). Another hypothalamic cell line, derived from a chemically induced tumor, synthesizes thyrotropin releasing factor (Grimm-Jorgensen et al 1976).

Greene & Tischler adapted cells from a rat pheochromocytoma to culture and isolated a clone designated PC12 (Greene & Tischler 1976). These cells synthesize catecholamines, store them, and respond to nerve growth factor (NGF) with the extension of neurites and the cessation of growth. In addition, PC12 synthesizes acetylcholine and synapses with skeletal muscle (Schubert et al 1977). Although the cells were derived from a tumor of the adrenal medulla, the resultant clone is phenotypically similar to primary cultures of sympathetic ganglion cells.

Myoblast Cultures

In addition to nerve and glial lines, an extensive effort has been made to obtain clonal cultures of muscle that are representative of the many cell types in vivo. The first myoblast line was isolated from rat skeletal muscle by Yaffe (Yaffe 1968). Subsequently a cell line that shares many properties with smooth muscle was derived from a brain neoplasm (Schubert et al 1974b), and cell lines have been obtained from embryonic rat aorta (Kimes & Brandt 1976a) and heart (Kimes & Brandt 1976b). All of these lines are maintained as dividing mononucleated myoblasts that are relatively undifferentiated when compared with stationary cells. In most respects the latter are similar to their in vivo counterparts. Electrophysiological, pharmacological, and biochemical changes are temporally associated with myoblast differentiation in stationary phase cultures (Kidokoro et al 1975). Thus a defined set of clonal target tissues is available for in vitro innervation studies.

BIOCHEMISTRY AND ELECTROPHYSIOLOGY OF CELL LINES

One of the major contributions of clonal cell lines is to provide a basis for the biochemical description of individual cell types. Although primary cultures are useful for electrophysiological studies, their heterogeneity limits the information that can be obtained about the chemical characteristics of individual cells.

Nervous System-Specific Proteins

It has been argued that S100 and 14-3-2 proteins are specifically localized in glia and nerve, respectively (Cicero et al 1970, Perez et al 1970). An examination of over 50 clonal cell lines of nerve and glia has shown that S100 and 14-3-2 proteins are found in both cell types, although all nerve cells contain 14-3-2. Not all glia synthesize detectable amounts of S100. Other biochemical markers are required to sort out these cell types.

Extracellular Macromolecules

If most cell types are labeled for several hours with a radioactive amino acid, approximately 95% of the total protein synthesized is associated with the cell pellet following centrifugation. About 4% of the protein synthesized is released into the medium as soluble macromolecules. Some are identified extracellular proteins such as collagen precursors, but the majority are glycoproteins derived from the turnover of cell surface components (Doljanski & Kapeller 1976, Schubert 1976). The remaining 1% of the extracellular protein is tightly associated with the culture dish in which the cells were grown, and can be quantitatively removed by ionic detergent. In fibroblastlike cells this material, designated substrate attached material (SAM), consists of a variety of proteins, including glycosoaminoglycans, the LETS glycoprotein, and actin (Culp 1976). In contrast, the SAM derived from nerve, glia, and muscle, but not other cell types, consists primarily of an apparently homogeneous protein of approximately 55,000 mol wt. This protein, designated SAM B, is chemically distinct from actin, tubulin, and a collagen fragment (Schubert 1977). Although SAM has been implicated in the adhesion of fibroblastlike cells to the growth surface (Culp & Buniel 1976), its function in the nervous system remains to be defined. It is likely, however, that both secreted and substrate attached extracellular molecules are of importance for cell-cell recognition (McClay & Moscona 1974) and such esoteric processes as the establishment of gradients during development. A detailed description of their cell associations and chemistry is most accessible through the use of clonal cell lines.

Neurotransmitter Metabolism

Of the five clonal CNS nerve cell lines examined, at least three are capable of synthesizing two or more neurotransmitters (Schubert et al 1974a). For example, B65 makes ACh, GABA, and dopamine, whereas B103 synthesizes both ACh and GABA. Clonal analysis has verified that a single cell can make multiple transmitters (Kimes et al 1974). In both cases the concentration of GABA is the highest of the transmitters present, probably reflecting the fact that many CNS cells secrete GABA as a transmitter.

In addition to the transmitter synthesis outlined above, all nerve, glia, and other cell types derived from neuroectoderm (e.g. melanocytes) are able to synthesize and store GABA and β-alanine at levels representing between 0.1% and 2% of their free amino acid pools (Schubert et al 1975). This low level of GABA and β-alanine synthesis may reflect the common embryological origin of these cells and, since it is not found in cells from other primordia, may serve as a useful marker for this group.

Electrophysiology: Nerve or Glia?

Nerve and muscle can be distinguished from most other cell types on the basis of their electrical excitability, detected either electrophysiologically or by veratridine-stimulated sodium flux (Catterall 1975). The initial classification of clonal cell lines derived from brain was made on the basis of the electrophysiological method; the electrically excitable cells containing either S100 or 14-3-2 proteins were classed as nerve and the remainder as glia. However, when the CNS cell lines were examined in detail by both methods, a number of ambiguities arose. First, B65 was excitable by microelectrode techniques but negative in the sodium-flux assay. The possibility that the cell line has a calcium action potential has not been tested. Second, a number of cells that were electrophysiologically negative, such as B82 and B108, gave reproducible responses to veratridine (Stallcup & Cohn 1976). These lines also express surface antigens characteristic of both nerve and glial cells (W. Stallcup, personal communication). It thus appears that among the collection of CNS cell lines there are cells whose characteristics represent a spectrum of properties between the "classical" nerve cell and glial cell. This is also true of neurotransmitters, for some glial cell lines synthesize ACh. In fact, Schwann cells can release ACh in a quantal fashion following denervation (Bevan et al 1973).

A number of alternatives could account for these intermediate cell types; none has been ruled out.

1. They represent a primordial cell that can give rise to either nerve or glia by, for example, amplification or suppression of the action potential mechanism.
2. They are partially differentiated neurons that have been transformed and developmentally frozen at an early (or incomplete) stage of development.
3. They represent a class of mature cells that have not yet been described in vivo.
4. They are not representative of a normal embryonic nerve cell population but are an abnormal phenotype resulting from transformation.

DIFFERENTIATION: MORPHOLOGICAL ASPECTS

The morphological differentiation of nerve and glia in culture has received much attention since the initial in vitro observations of Harrison. Although the tissue culture paradigm offers a geometrically simplified version of in vivo differentiation, perhaps some requirements and constraints can be described for such processes as neurite outgrowth and morphological differentiation.

Mode of Differentiation

It has long been recognized that, in vitro, nerve cells acquire neurites by extension of the limiting membrane; the tip of the neurite is invariably associated with an active growth cone. Recently, another mode of process formation has been described in clonal cells that involves the retraction of parts of an extended cytoplasm; this process has been termed shrinkage (Steinbach & Schubert 1975). Neurite formation via both extension and shrinkage can be induced by dibutyryladenosine 3'5'-monophosphate, DBcAMP, a reagent known to cause morphological changes in many cell types. All nerve cell lines examined formed neurites via elongation, whereas cell lines of glial origin formed processes by shrinkage, showing that process formation can be achieved by different mechanisms in the two cell types.

Although it is generally assumed that the microtubule organelle is required for neurite extension, semi-serial sections show that microtubules are not present in newly formed neurites up to 100 microns in length in C1300 (Schubert et al 1971) and CNS (Klier *et al* 1975) cells. Since the vinca alkaloids vinblastine and colchicine inhibit neurite formation, it is possible that the primary site of action is at the plasma membrane. Colchicine-binding activity has been demonstrated in purified membranes (Stadler & Frank 1974, Altstiel & Landsberger 1977). Finally, DBcAMP, which induces neurite formation, and colchicine, which inhibits it, may ultimately affect the same structure, for cAMP protects cells from colchicine-induced neurite retraction (Schubert 1974) and DBcAMP reverses the effect of colcemid in dorsal root ganglion cell cultures (Roisen & Murphy 1973).

A Possible Mechanism

If the morphological differentiation of nerve cells is defined in terms of increased morphological and ultrastructural complexity, a wide variety of compounds and growth conditions are able to differentiate this group of cells. Serum deprivation (Seeds et al 1970, Schubert et al 1971, Luduena 1973), BUdR (Schubert & Jacob 1970), growth-conditioned medium (Schubert et al 1971, Monard et al 1973),

DBcAMP (Furmanski et al 1971), organic acids (Schubert et al 1971), and dimethylsulfoxide (DMSO) (Kimhi et al 1976) all induce morphological and ultrastructural changes in some clones of the C1300 neuroblastoma, some of the CNS nerve cell lines, and primary cultures of nervous tissue. Finally, both NGF and DBcAMP induce neurite formation in a sympathetic nerve cell line (Schubert & Whitlock 1977.)

Given the fact that the above conditions are able to generate a more complex ultrastructure in cultured nerve cells, is there a common mechanism? There are two broad classes of alternatives, selective and instructive. The first requires that the inducers alter a preexisting structure, possibly by shifting a two-state process into the previously less preferred state. For example, nerve cells in culture are constantly extending and retracting processes. Reagents that stabilize the elongated state would induce neurite formation. The instructive models require a reprogramming of cellular function. For example, NGF could induce the synthesis of a new class of proteins that direct neurite formation. The available data favor a model involving selective neurite stabilization by increased cell-substratum adhesion.

There are a number of arguments against instructive mechanisms. First, it is hard to envision how such diverse reagents as BUdR and DMSO can alter gene expression *directly*. Second, de novo protein synthesis is not required for at least the first 5 hr of neurite outgrowth (Seeds et al 1970, Schubert 1974). Finally, there is no qualitative change in protein synthesis during NGF induced neurite formation (Garrels & Schubert 1978).

A recent example of a selective developmental mechanism has come from studies with clonal Friend cell lines, where DMSO and short chain fatty acids induce erythropoiesis; they also induce neurite extension in the C1300 neuroblastoma. In Friend cells, these compounds increase the phase transition temperature of the membrane, indicating a decreased fluidity and thus an increased stability (Lyman et al 1976). It was suggested that induction of differentiation may be a direct result of the interaction of these compounds with cell membranes. It has similarly been argued that membrane-mediated events are responsible for triggering nerve cell differentiation (Schubert et al 1973). Thus the initial event in nerve differentiation may be an alteration of the plasma membrane, which could, in turn, trigger further steps in development.

Recent studies with the PC12 sympathetic cell line have shown that NGF increases the rate of cell-substratum adhesion and suggested that this enhanced adhesion is ultimately responsible for neurite outgrowth (Schubert & Whitlock 1977, Schubert et al 1978).

The following evidence suggests that cAMP is a mediator of the NGF induced responses of PC12 cells (Schubert & Whitlock 1977, Schubert et al 1977, Schubert et al 1978). 1. NGF elevates the level of endogenous cAMP. 2. Both NGF and cAMP increase cell-cell and cell-substratum adhesiveness. 3. They also increase the specific activity of choline acetyltransferase and intracellular acetylcholine accumulation. 4. Theophylline, a phosphodiesterase inhibitor, potentiates the effect of cAMP and NGF. 5. Cholera toxin, which increases endogenous cAMP accumulation, also promotes neurite outgrowth and adhesiveness. 6. NGF and cAMP lead to similar changes in total protein synthesis (Garrels & Schubert 1978). These biochemical data, plus the morphological and ultrastructural similarities of neurite outgrowth caused by DBcAMP and NGF, strongly favor the role of cAMP as a "second message" in the NGF responsive cell. Additional experiments have indicated that increased mobilization of calcium ions may be responsible for the restructuring of the cell surface, defined in terms of increased lectin agglutinability, which is correlated with enhanced cell-substratum adhesion and neurite extension (Schubert et al 1978).

DIFFERENTIATION: BIOCHEMICAL AND ELECTROPHYSIOLOGICAL ASPECTS

The macromolecular changes in protein synthesis associated with the differentiation of cells can be grouped in two categories. The first is that group of proteins that are undetectable in one state and present in another. These apparently qualitative changes in macromolecular synthesis are exemplified by changes in the type of actin synthesized during myogenesis (Gruenstein & Rich 1975, Garrels & Gibson 1976). Another group consists of proteins that are present in both states but in quantitatively different amounts. These quantitative changes have been termed modulation (Grobstein 1959) and are usually associated with reversible cellular events. To date all of the changes in macromolecular synthesis associated with nerve differentiation in culture are of the latter type; no qualitative changes have been reproducibly detected, and the morphological and biochemical differentiation of all clonal nerve cell systems is reversible.

Perhaps the clonal nerve cell line studied in most detail besides the C1300 clones is the sympathetic clone PC12 (Greene & Tischler 1976). This cell line responds to NGF by the extension of neurites in a manner analogous to the response of C1300 to serum deprivation or DBcAMP. When the pattern of total protein synthesis was studied using a two-dimensional gel system (Garrels & Gibson 1976), no qualitative differences in protein synthesis were detected between control cultures and cells exposed to NGF (Garrels & Schubert 1978). This two-dimensional gel system has detected many qualitative changes during the myogenesis of a skeletal muscle cell line (Garrels & Gibson 1976). It thus appears that at least in PC12, and probably the rest of the clonal nerve lines, the dividing precursor cells and the cells with extensive neurites are not qualitatively distinct in terms of their macromolecular synthesis.

Data concerning the differentiation of the electrical properties of nerve cell membranes are analogous to those of their overall biochemical properties. A qualitative change in membrane properties from the exponentially dividing cell to the morphologically differentiated cell has not been rigorously established. In the case of the C1300, exponentially dividing cells grown in suspension culture are electrically excitable (see Schubert et al 1973 for discussion). A recently published exception to this conclusion is the claim that the PC12 clone of sympathetic ganglion-like cells is not electrically excitable in the absence of NGF, but acquires excitability after exposure to NGF (Dichter, Tischler & Greene 1977a). This conclusion is probably not valid, for other laboratories have found that the same clone of PC12 is excitable in the absence of NGF by electrophysiological methods (A. Ritchie, personal communication) and by the sodium flux assay (Stallcup 1978). The reason for the inability of Dichter, Tischler and Greene to demonstrate excitability in the non-NGF treated cells is probably due to technical difficulties associated with penetrating the small non-NGF exposed cells relative to the very large cells found after exposure to NGF. Analogous false negative electrophysiological results have been obtained with the C1300 neuroblastoma (see Schubert et al 1973 for discussion).

Finally, alterations in enzyme metabolism during morphological differentiation of the CNS nerve lines and PC12 will be described; similar changes associated with the C1300 clones have been reported (Breakefield 1976). Two clones of CNS nerve have been examined in detail (Kimes et al 1974); B65 synthesizes ACh, dopamine, and GABA, while B103 makes both ACh and GABA. An analysis of subclones of the lines shows that they all make two or more transmitters. In the case of catecholamine and GABA synthesis, transmitter synthesis is highest and maximal tyrosine hydroxylase and glutamic acid decarboxylase activities occur in stationary phase cells. However, in B65 and its subclones, ACh is synthesized only in exponentially dividing cells. These experiments established multiple neurotransmitter synthesis by a single nerve clone and demonstrated that this synthesis was related to the state of growth.

A similar study of the PC12 clone of sympathetic ganglionlike cells has shown that the specific activity of choline acetyltransferase (CAT) and ACh synthesis is altered by a number of factors, whereas the amount of catecholamines synthesized per unit of cellular protein remains relatively constant (Schubert et al 1977). The specific activity of CAT and the accumulation of ACh increases as a function of cell density. If NGF is added to the cultures, there is a precocious increase in ACh synthesis. cAMP can also induce an increase in the specific activity of CAT. The stimulation achieved by cAMP and NGF is nonadditive, which suggests that the cyclic nucleotide may mediate the NGF effect. Finally, conditioned medium from essentially all cell types can induce an increase in CAT activity. This conditioned medium effect is like that observed in primary cultures of sympathetic cells (Patterson 1978).

CHOLINERGIC NEUROTRANSMITTER RECEPTORS

The availability of elapid neurotoxins and various neurotransmitter analogues that bind to and inhibit the function of neurotransmitter receptors has greatly expedited their biochemical analysis. This section is concerned with the new insights into receptor structure, function, and regulation that have derived from studies that used these probes in conjunction with excitable cells in tissue culture.

Development of Muscle Nicotinic Acetylcholine Receptors

Myotubes formed from primary cultures of rat (Hartzell & Fambrough 1973) or chick (Vogel et al 1972) myoblasts bind elapid neurotoxins, as do myotubes derived from the fusion of the clonal muscle cell line L6 (Patrick et al 1972) and differentiated cells of the nonfusing muscle cell line BC_3H-1 (Patrick et al 1977). The conclusion that elapid neurotoxins bind to acetylcholine receptors on the surface of these cells in drawn from the following observations:

1. Binding of toxin is inhibited by cholinergic ligands, and the ability of these ligands to inhibit binding parallels their ability to affect receptor function.
2. Binding of toxin blocks acetylcholine-induced receptor activation, and the rate of inhibition parallels the rate of binding.
3. The membrane component that binds toxin sediments in sucrose gradients with a sedimentation coefficient that is the same as that determined for acetylcholine receptor purified from the eel *Electrophorus electricus.*
4. The membrane component that binds toxin carries antigenic determinants that are recognized by antibodies raised against acetylcholine receptor purified from the eel *E. electricus.*

These observations provide convincing evidence that elapid neurotoxins bind to and inhibit function of acetylcholine receptor on muscle cells in culture. The portion of bound toxin that is actually bound to acetylcholine receptor is a function of the concentration of toxin used in the binding reaction and of the length of incubation. For this reason the use of neurotoxins to quantitate acetylcholine receptors usually takes into account only that binding that is inhibited by cholinergic ligands. In fact, a component with an S value about half that of the receptor has been found to bind neurotoxin even in the presence of d-tubocurarine (Devreotes & Fambrough 1975).

The appearance of acetylcholine receptors in cultures of skeletal muscle cells normally parallels the formation of myotubes through fusion of myoblasts (Patrick et al 1972, Vogel et al 1972), although the fusion process is not required. Calcium deprivation inhibits fusion but affects neither the time course of appearance nor the final level of receptors in the culture (Patterson & Prives 1973). Primary cultures that are undergoing fusion have elongated, spindle-shaped, mononucleate cells that are sensitive to ionophoretically applied acetylcholine (Fambrough & Rash 1971) and in some instances have been shown to bind neurotoxin (Sytkowsky et al 1973). Receptor appearance is thus not causally related to cell fusion and may even normally precede it. That acetylcholine receptor synthesis occurs during the time at which receptors appear on the cell surface was demonstrated by incorporation into receptor of radiolabeled methionine that was present during, but not before, fusion (Merlie et al 1976). It is not yet clear, however, whether (*de novo*) synthesis accounts quantitatively for the receptors that appear on the surface during this period.

Distribution of Nicotinic Acetylcholine Receptors on the Muscle Cell Surface

The distribution of acetylcholine receptors on the surface of myotubes can be determined either by assay of receptor function or by localization of neurotoxin bound to receptor. Myotubes formed from chick (Fischbach & Cohen 1973) and rat (Hartzell & Fambrough 1973) have regions that are more sensitive to ionophoretically applied acetylcholine than the rest of the cell surface. Autoradiography of chick (Sytkowsky et al 1973) and rat (Hartzell & Fambrough 1973) primary cultures labeled with iodinated neurotoxin reveals clusters of acetylcholine receptors that in at least one case (Hartzell & Fambrough 1973) were found to correspond to regions with elevated sensitivity to acetylcholine. In one report these receptor clusters were said to correspond to clusters of nuclei (Fischbach & Cohen 1973), a second noted that they were most often found at the cell periphery (Sytkowsky et al 1973), while a third could find no distinctive cytological correlation (Hartzell & Fambrough 1973). Similar experiments using neurotoxin labeled with a fluorescent dye have revealed clusters of acetylcholine receptors on myotubes in primary cultures of frog muscle (Anderson et al 1976) and rat muscle (Axelrod et al 1976).

The structural basis and biological significance of these receptor clusters is unknown. They probably represent areas of high local concentration of receptor per unit membrane, rather than areas of extensive membrane folding (Vogel & Daniels 1976). The clusters are found on both the upper and lower surfaces of the cell (Anderson et al 1976, Axelrod et al 1976), do not occur as a consequence of binding neurotoxin (Sytkowsky et al 1973), and are mobile (Axelrod et al 1976). The distribution of receptors within the cluster is not uniform and the lifetime of a cluster is longer than the lifetime of the receptors it contains (Axelrod et al 1976). Acetylcholine receptors that occur in clusters differ in their mobility from those that are dispersed over the cell surface. By use of a fluorescent photobleaching recovery technique, Axelrod et al (1976) have shown that most disperse receptors are mobile with diffusion constants of 160×10^{-12} cm^2 sec^{-1} at 35°, whereas receptors in clusters are essentially immobile with diffusion coefficients less than 10^{-12} cm^2 sec^{-1} at 35°. All these results suggest that receptors in clusters are anchored in some way, but attempts to convert immobile receptors to mobile receptors by means of a variety of reagents, including those known to disrupt cytoskeletal structures, were unsuccessful (Axelrod, personal communication). The intriguing possibility remains that the mechanisms responsible for cluster formation are related to those that localize receptors at the neuromuscular junction.

The proportion of acetylcholine receptors present in clusters changes with the age of the culture. In one report receptor clusters on myotubes formed from embryonic chick leg muscle increase in frequency as the culture ages up to eleven days post-fusion (Sytkowsky et al 1973). In another instance, both the number of clusters and the total receptor per fiber in chick myotubes were found to decline in older cultures (Prives et al 1976). Total receptor decreased by 50% within eight days of fusion, while clusters effectively disappeared. Since myotubes formed in culture become spontaneously contractile and often detach from the dish it is difficult to separate developmental events from pleiotropic consequences of maturation.

Metabolism of Muscle Nicotinic Acetylcholine Receptor

The number of acetylcholine receptors on the surface of a muscle fiber represents a steady state determined by the rates of synthesis and degradation (Devreotes & Fambrough 1975, Patrick et al 1977). When cultures of muscle are labeled to saturation with iodinated neurotoxin and returned to 37°C the amount of toxin remaining bound to cells decreases exponentially, and the radioactive iodine appears in the culture medium as a low-molecular-weight metabolite with properties of iodinated tyrosine. The rate of loss of bound iodinated toxin is very close to the rate of loss of receptors from the cell surface when protein synthesis is inhibited. The fact that toxin bound to receptor is degraded, and the fact that receptors disappear at the same rate suggest that receptors are also degraded. Consonant with this idea are the observations that both processes are energy and temperature dependent.

Receptor synthesis can be assessed by measurement of the rate of appearance of unoccupied toxin-binding sites after saturating the surface receptors with unlabeled toxin. In differentiated cultures the appearance of unlabeled receptors occurs at a rate equal to the rate of degradation. For a period of about two hours this insertion of new receptors is insensitive to inhibitors of protein synthesis, during which time there appears on the surface a number of receptors equal to 10–15% of the total surface receptors. The appearance of unoccupied toxin-binding sites in the absence of protein synthesis is probably not due to surface receptors that had not bound unlabeled toxin, since overnight incubation in unlabeled toxin fails to eliminate this component. Thus there exists within the cell a class of receptors that are inaccessible to neurotoxin applied to intact cells. These hidden receptors are composed of two populations, one that serves as a precursor to receptors on the cell surface, and one that does not (Devreotes & Fambrough 1975, Patrick et al 1977). Under conditions in which protein synthesis has been inhibited, one of the hidden receptor populations decreases at a rate and to an extent predicted from the appearance of new receptors on the cell surface. The significance of a second nonprecursor population of hidden receptors is not clear. In chick primary cultures this population can be depleted by growing the cells overnight in unlabeled toxin, presumably through chasing labeled surface receptors into this hidden population (Devreotes & Fambrough 1975). In the clonal muscle cell lines that have been tested, the nonprecursor population does not decrease in size after growth in the presence of neurotoxins (Patrick et al 1977).

The evidence for degradation of acetylcholine receptors was initially based on the observation that neurotoxin bound to receptors appears in the medium as a product of degradation (Berg & Hall, 1974), while the evidence for synthesis was based on inhibitor and kinetic experiments. These conclusions have received support from more direct experiments. In one case the specific radioactivity of acetylcholine receptor, synthesized in the presence of radiolabeled methionine, was found to decrease exponentially at a rate very close to that found for degradation of receptor by indirect methods (Merlie et al 1976). Acetylcholine receptors synthesized by chick primary cultures maintained in the presence of amino acids composed of the heavy isotopes of hydrogen, carbon, and nitrogen have a different buoyant density than those synthesized in the presence of naturally occurring amino acids and can be resolved by equilibrium sedimentation in metrizamide–deuterium oxide gradients (Devreotes & Fambrough 1976). They can also be resolved by velocity sedimentation on sucrose–deuterium oxide gradients. By use of these techniques it was found that hidden receptors assume a higher density more rapidly than surface receptors and that there is a lag of about 3 hr before receptors containing the heavy isotopes appear on the surface (Devreotes et al 1977). Quantitatively, these results supported the conclusion that a portion of the hidden receptors are precursors to the surface receptor population and that they undergo an intracellular transit time of about 2 hr before their appearance in the limiting cell membrane. Thus the steady state level of receptors on the cell membrane is attained by a balance between the rates of synthesis and degradation.

Most experiments designed to distinguish biochemically between hidden and surface receptors have been unsuccessful. Both contain carbohydrate and are indistinguishable when antisera directed against eel acetylcholine receptor are used. However, receptors labeled on the cell surface and extracted with detergent have a higher sedimentation coefficient than do hidden receptors extracted with detergent and then labeled with neurotoxin (Patrick et al 1977).

Regulation of Nicotinic Acetylcholine Receptor Metabolism

Receptors localized at the neuromuscular junction are degraded more slowly than receptors that appear in extrajunctional membrane following denervation (Berg & Hall 1974, 1975, Chang & Huang 1975). This observation, coupled with the evidence that the level of extrajunctional receptors appears to be a function of muscle activity (Lomo & Rosenthal 1972), suggests that the electrical activity of cultured muscle cells may regulate the density of extrajunctional receptors on the cell surface. Primary cultures of chick muscle that have been stimulated electrically over almost three days have lower sensitivity to acetylcholine than do cultures that were maintained in either tetrodotoxin or lidocaine (Cohen & Fischbach 1973). These results might have arisen due to differences in input impedance of the stimulated fibers rather than as a consequence of differences in receptor density, although an effort was made to select fibers of comparable size. However, similar experiments that used iodinated neurotoxin to quantitate receptors also revealed higher levels in tetrodotoxin-treated cultures and lower levels in stimulated cultures, compared to controls (Shainberg & Burstein 1976). These differences might arise from alterations in either the rates of synthesis or degradation of receptors. In fact, the rate of synthesis is greater in tetrodotoxin-treated cultures and lower in stimulated cultures, whereas the rate of degradation of receptor in tetrodotoxin-treated cultures is unchanged (Shainberg et al 1976). These results are consistent with the idea that the appearance of extrajunctional receptors following denervation and their disappearance after reinnervation is a consequence of effects of muscle activity on receptor synthesis.

The rate of degradation of acetylcholine receptors is increased following reaction of antiacetylcholine receptor antibody with muscle cells in culture (Heinemann et al 1977). It is known that immunization with purified eel acetylcholine receptor induces a paralysis with the characteristics of muscular blockade (Patrick & Lindstrom 1972) and that antibodies from immunized animals block agonist-induced receptor activation in eel electroplax and, in some cases, on mammalian muscle (Patrick et al 1973). Thus the blockade of neuromuscular transmission in the immunized animal might result from antibody-induced inactivation or from a decrease in the total number of receptors. In fact, application of antireceptor antisera to either primary cultures of rat muscle or to cultures of BC_3H-1, a nonfusing muscle cell line, result in increased rates of degradation of toxin-receptor complexes with only neglibigle decreases in either single channel conductance or mean channel open time. Antibody-induced degradation, or modulation, is temperature and energy dependent and results in a decreased density of receptors on the muscle surface (Heinemann et al 1977).

Non-Muscle Nicotinic Acetylcholine Receptors

Although attention has largely been focused upon the muscle nicotinic receptors, both ganglionic nicotinic acetylcholine receptors and muscarinic acetylcholine receptors are now accessible in tissue culture. Primary cultures of sympathetic ganglion cells have been shown to bind neurotoxin (Greene et al 1973) at a site that can be protected with a variety of cholinergic ligands. The ability of these ligands to inhibit binding of neurotoxin to sympathetic neurons in culture closely parallels their ability to inhibit binding of neurotoxin to membrane fragments prepared from sympathetic ganglia (Greene 1976) or brain (Eterovic & Bennett 1974). Since the ability of these ligands to affect the function of ganglionic or CNS nicotinic receptors is not yet quantitated, it is not possible to compare their efficiencies in these two assay systems. It is clear from a number of studies in vivo, however, that neurotoxins have no effect on agonist-induced activation of this class of nicotinic acetylcholine receptors (Chou & Lee 1969, Nurse & O'Lague 1975, Duggan et al 1976). Thus there remain the possibilities that neurotoxin binds to the ganglionic nicotinic receptor without affecting function or that it binds to a site on a different membrane component. The rat pheochromocytoma adapted to culture by Greene & Tischler (1976) binds neurotoxin and has a functional acetylcholine receptor that is insensitive to neurotoxin. The membrane component in these cells that binds neurotoxin is immunologically distinct from the functional acetylcholine receptor (Patrick & Stallcup 1977a,b).

Muscarinic Acetylcholine Receptors

A wide variety of cells respond to cholinergic stimulation through activation of a muscarinic acetylcholine receptor that differs from either muscle or nerve nicotinic acetylcholine receptor. Little information concerning this receptor has come from studies in tissue culture, largely due to difficulties in growing cells that show muscarinic receptor activity. A prime candidate for such a cell should, in principle, be smooth muscle, but available clonal cell lines whose origin is probably smooth muscle express nicotinic rather than muscarinic receptor activity (Schubert et al 1973, Kimes & Brandt 1976a), as does a cell line isolated from embryonic rat heart (Kimes & Brandt 1976b). Purves (1974) used primary cultures of newborn guinea-pig taenia coli to demonstrate muscarinic responses to ionophoretically applied acetylcholine. The resultant depolarization exhibited a long latency that was not attributable to diffusion. The sensitivity to acetylcholine appears dispersed with no indication of the receptor clusters found in cultures of skeletal muscle. A hyperpolarizing response to iontophoretically applied acetylcholine was found on a rat endothelial cell line (Venter et al 1976). This response was slow (seconds) and was blocked by atropine.

There is now considerable evidence that some muscarinic responses are mediated by cyclic nucleotides (Greengard 1976). Prasad et al (1974) reported that application of acetylcholine to homogenates of C1300 neuroblastoma increased adenylate cyclase activity. Noradrenaline and dopamine also increase the activity, and the effect of each of these three compounds was inhibited by atropine, which was an unexpected finding. Application of acetylcholine to intact neuroblastoma clone N15 resulted in elevated levels of cyclic GMP (Richelson 1977). This effect of carbamylcholine on cyclic GMP levels was inhibited by a variety of antipsychotic drugs, thus making available a homogeneous biological preparation suitable for the study of various antipsychotic drugs. Analogous results were found in neuroblastoma clone N1E-115. (Matsuzawa & Nirenberg, 1975). Carbamylcholine resulted in elevated levels of cyclic GMP and decreased levels of cyclic AMP. In addition, prostaglandin E_1 induced increases in the levels of both cyclic AMP and cyclic GMP, and carbamylcholine inhibited the prostaglandin induced increase in cyclic AMP. A similar but more pronounced effect of acetylcholine on prostaglandin E_1 induced cyclic AMP accumulation was found on a neuroblastoma X glioma hybrid cell line (Traber et al 1975). This inhibitory action of acetylcholine was absent in the presence of atropine and present even when the elevated cyclic AMP levels were obtained by inhibition of phosphodiesterase activity, which suggests that the effect was mediated through a muscarinic receptor and affected cyclase rather than phosphodiesterase activity. These results are consistent with the idea that acetylcholine-induced activation of muscarinic receptors results in elevation of cyclic GMP and inhibits adenylcyclase through some mechanism other than elevated cyclic GMP.

SYNAPSE FORMATION

Systems Studied

There are many reports of chemical transmission between nerve and nerve and between nerve and muscle in tissue culture. Synapses form in explant cultures and in disaggregated primary cultures (see recent reviews, Shimada & Fischman 1973, Fischbach et al 1974, Steinbach 1974a, Nelson 1975). Recently, chemical transmission has been demonstrated by use of clonal nerve cell lines, which have the advantage that they are homogeneous cell populations (Kidokoro & Heinemann 1974, Kidokoro et al 1975, Christian et al 1976, Nelson, Christian & Nirenberg 1976, Schubert, Heinemann & Kidokoro 1977). Since this literature has been reviewed, we do not attempt to list all the systems that have been developed. Instead, experiments will be examined that were designed to ask specific questions about development and function of the neuromuscular junction.

Minimal Requirements for Chemical Transmission

The adult neuromuscular junction has been studied in great detail in terms of its biophysical, biochemical, and morphological properties. In addition it is known that innervation has profound long-term effects upon both the muscle (for reviews see Guth 1968, Harris 1974) and the motoneuron (Kuno 1976). These trophic effects are clearly important for normal muscle activity.

There are three minimal requirements for chemical transmission:

1. A cell must release a neurotransmitter.
2. The released neurotransmitter must match the muscle receptor.
3. The site of transmitter release must be near enough to the muscle to establish a neurotransmitter concentration sufficiently high to activate a significant portion of the receptors.

In principle there need be no other interaction between nerve and muscle at a functional synapse. The chemical transmission from nerve to muscle documented in tissue culture may not reflect any other interaction. In fact Obata (1976) has shown that chemical transmission is established in less than 22 min after the nerve contacts the muscle in culture, which leaves open the possibility that synapses can form without complicated changes in cellular chemistry.

The fact that a nerve-muscle contact meets the minimal requirements for chemical transmission does not necessarily imply that synaptogenesis has occurred. For instance chemical transmission could arise from the random juxtaposition of nerve and muscle. On the other hand such a contact could be a developmental precursor to a synapse and in some instances possess properties of mature synapses. Many of the studies described below are attempts to demonstrate changes in muscle membrane properties as a consequence of innervation in culture.

Distribution of Muscle Acetylcholine Receptor

The acetylcholine receptor on innervated skeletal muscle is confined to the tips of the subsynaptic folds (Fertuck & Salpeter 1976). When the nerve is cut, ACh receptor appears on the extrajunctional membrane (Axelsson & Thesleff 1959) and the extrajunctional membrane becomes susceptible to innervation by implanted nerves. Direct electrical stimulation of denervated muscle decreases the density of extrajunctional ACh receptors but does not affect junctional ACh receptors. The original ACh receptor distribution is restored, and the muscle is largely refractory to innervation at extrajunctional regions (Lomo & Rosenthal 1972, Jansen et al 1973, Lomo & Westgaard 1975). These experiments suggest that extrajunctional ACh receptor density is regulated by the level of electrical activity in the muscle, but that junctional receptor is controlled by a different mechanism. Recently it has been shown that localization of ACh receptor occurs very early during neuromuscular junction formation in the embryo. One spot of high ACh receptor density per muscle fiber occurs about one day after the first signs of innervation are seen (Bevan & Steinbach 1977, Burden 1977).

The distribution of ACh receptors on skeletal muscle has been extensively examined in culture. Harris et al (1971) cultured a clonal mouse nerve cell line (C1300 neuroblastoma) with a clonal rat skeletal muscle cell line (L6) and found regions of high muscle ACh sensitivity at points of neurite-muscle contact, although no chemical transmission was seen in this system (Kidokoro et al 1975). These experiments could not distinguish between a mechanism by which the nerve induces a high density of ACh receptor or one by which it associates with a preexisting region of high ACh receptor density. Kano & Shimada (1971) and Fischbach & Cohen (1973) studied innervation of chick skeletal muscle by chick spinal cord neurons in primary cultures and found localizations of muscle ACh sensitivity near nerve endings.

Fambrough et al (1974) confirmed these findings using autoradiography of rat skeletal muscle innervated by chick spinal cord neurons. These studies show that there can be an association of high ACh receptor density with nerve processes. However, these experiments did not have high enough resolution to demonstrate that the site of ACh release is located at the points of high ACh receptor density. Recently Fischbach and his collaborators (Fischbach et al 1975, Frank & Fischbach 1976, 1977, Cohen & Fischbach 1977) have devised methods to localize sites of ACh release. Two techniques have been used:

1. Neurites are stimulated with an extracellular electrode near points of contact with the muscle. Tetrodotoxin is used to prevent the spread of current along the neurite by the action potential mechanism. ACh release is detected by recording voltage changes in the muscle with an intracellular microelectrode.
2. Synaptic currents are recorded with an extracellular pipette placed against the muscle membrane. This method has a spatial resolution of 2–5 μm.

By use of these techniques in the chick culture system, sites of ACh release were found to be associated with regions of muscle membrane with high ACh sensitivity. There is, however, a circular argument in these experiments. The chick muscles used in these studies were large and had relatively low sensitivities to ACh, approximately 50–500 mV nC^{-1}. Thus it is possible that ACh release takes place at random points on the muscle membrane but can only be detected at points of high ACh receptor density where the ACh-induced currents would be large. In fact, Cohen (1976) has shown, by means of signal averaging, that ACh is also released at regions of low ACh receptor density in this chick system. Thus these experiments demonstrate that ACh release can occur at regions of both high and low ACh receptor density.

It is clear from the experiments reviewed above that ACh release is in some cases associated with high ACh receptor densities. Three possible mechanisms for the association must be considered.

1. The nerve could induce high densities of ACh receptor at points of ACh release.
2. The nerve could detect preexisting regions of high ACh receptor density and release ACh at these points. Noninnervated muscles in culture (Fischbach & Cohen 1973, Sytkowski et al 1973) and denervated mouse muscle in vivo (Ko, Anderson & Cohen 1977) develop regions of high ACh receptor density that could serve as specialized regions for innervation.
3. The nerve may release ACh at random points on the muscle. By chance, some of these ACh release sites would be at regions of high ACh receptor density.

Although none of these alternatives has been ruled out, evidence in support of the first alternative has been obtained. Anderson et al (1976, 1977, Anderson & Cohen 1977) have demonstrated in *Xenopus* cultures that spinal cord neurons can induce regions of high muscle ACh receptor density at points of nerve-muscle contact. By examining the distribution of ACh receptor on muscle cells with fluorescent α-bungarotoxin before, during, and after innervation by spinal cord neurons, they were able to show that some neurons can induce a streak of high ACh receptor density along the path of nerve-muscle cell contact. Frank & Fischbach (1976, 1977) have used electrophysiological techniques to show that nerves can induce regions of high muscle ACh sensitivity on chick muscle. They first demonstrated that the distribution of ACh sensitivity on chick muscle was stable for at least two to three days in the absence of innervation. However, upon innervation, new regions of high ACh sensitivity on the muscle at points of nerve-muscle contact were found where ACh release could be demonstrated.

A localization of AChR at the point of innervation is not, however, required, since Kidokoro et al (1975) have shown that regions of high ACh sensitivity are not associated with the majority of synapses formed on a rat skeletal muscle cell line (L6). In this rat culture system small myotubes were chosen for study and consequently they had high ACh sensitivities (1000–2000mV nC^{-1}), which presumably made it possible to detect ACh release wherever it occurred. The distribution of ACh sensitivity was determined on rat myotubes (L6) innervated by spinal cord neurons. In general, regions of high ACh sensitivity (>5X above background) were not found near nerve endings on these innervated rat myotubes. Since focal recording techniques were not used to identify sites of ACh release, it is possible that small regions of high ACh receptor density were missed in these experiments. However, a number of indirect experiments suggest the absence of ACh receptor localization at many of the sites of ACh release.

1. The mean mepp (miniature endplate potential) amplitude after a correction for muscle input resistance was low. Furthermore, the conductance change during a mepp was estimated to be very small, 4% of that found at a frog neuromuscular junction. This can be explained by the low ACh receptor density, 100 per μm^2, found on L6 myotubes.
2. The mean mepp amplitudes recorded in three different clonal cell lines of muscle innervated in culture were proportional to the average ACh receptor densities before innervation (Kidokoro & Patrick 1978).
3. The mean mepp amplitude and the conductance change did not increase with time after innervation as might be expected if the nerve made contact with the muscle and then induced ACh receptor localization (Y. Kidokoro, personal communication).
4. The mepps had a slow time course which could be mimicked by iontophoretic application of ACh onto extrajunctional regions of the muscle (Y. Kidokoro, personal communication).

It is clear from the studies described above that there is, in some culture systems, an association of a high ACh receptor density on muscle membrane at points of nerve-muscle contact. Furthermore, in the chick system it has been shown that ACh release can occur at points of high ACh receptor density, which is analogous to the situation in vivo. However, ACh can also be released at regions of low ACh receptor density, in the chick as well as the rat system. Whether or not this occurs in vivo is not known. Finally, experiments in culture demonstrate that the nerve can induce a region of high ACh receptor density on the muscle, which explains the association of high ACh receptor density with the point of nerve contact in these systems.

In innervated adult skeletal muscle the extrajunctional ACh receptor density is very low. A number of groups have looked to see whether innervation of muscle in culture results in a suppression of extrajunctional ACh receptor. Kano & Shimada (1971) reported that extrajunctional ACh sensitivity was reduced in chick muscles innervated in culture, as is the case in vivo. However, the ACh responses they obtained were very slow, seconds instead of milliseconds, and it is possible they were desensitizing the extrajunctional ACh receptors. Fischbach and his colleagues first reported that extrajunctional ACh sensitivity on innervated chick muscles was the same as that found on noninnervated muscles (Fischbach & Cohen 1973). Recently, however, they have reported higher ACh sensitivities at extrajunctional regions on innervated muscles when compared to noninnervated muscles in the same dish (Fischbach et al 1975, Cohen & Fischbach 1977). This result has not been seen in vivo but may reflect some developmental process. On the other hand, it may be an artifact created by the experimental technique. Since it is easier to see mepps or evoked responses in muscles with higher ACh sensitivities, many of the muscles with lower ACh sensitivities may in fact have been innervated but were scored as negative because the ACh-induced voltage change was too small to detect. However, Cohen & Fischbach (1977) did find an increase (1.7X) in ^{125}I-α-bungarotoxin binding on muscles near spinal cord explants in culture. Whether this effect is due to innervation and is specific to spinal cord cells remains to be determined. Two other groups found no differences in the extrajunctional ACh sensitivities of innervated, as compared to noninnervated, muscles in rat and chick cultures (Kidokoro et al 1975, Obata 1977). This conflicting data may be explained by different degrees of either spontaneous or nerve-induced muscle electrical activity in the various experiments.

Other Effects of Innervation on Muscle

Adult junctional ACh receptors are known to have many properties that are different from those of the extrajunctional ACh receptors that appear after denervation. Junctional ACh receptors are metabolically stable (Berg & Hall 1975, Chang & Huang 1975) and have a charge difference when isolated (Brockes & Hall 1975). The two ACh receptors have different channel kinetics (Neher & Sakmann 1975) and a different sensitivity to *d*-tubocurarine (Beranek & Vyskocil 1967, Brockes & Hall 1975). Kidokoro et al (1975) found no difference between the extrajunctional ACh receptors and the junctional ACh receptors with respect to their sensitivity to inactivation by *d*-tubocurarine in the rat culture system. To date no one has reported experiments designed to look for other changes in ACh receptor properties as a function of innervation in culture.

A number of investigators have looked for, and have not found, electrophysiological evidence for functional acetylcholine esterase at neuromuscular junctions formed in culture (Fischbach 1972, Kidokoro & Heinemann 1974). Shimada & Kano (1971) used histochemical methods and were not able to find evidence for acetylcholine esterase. However, Koenig (1973) was able to demonstrate histochemical evidence for acetycholine esterase, but it is not clear that this represents enzyme located at a functional synapse, since sites of ACh release were not identified.

Kidokoro et al (1975) examined the effect of innervation on the muscle action potentials in a rat skeletal muscle cell line (L6); no effect was observed on either tetrodotoxin sensitivity, which remained low, or rate of rise of the action potential after innervation. It is known from in vivo studies of rat skeletal muscle that denervation makes the muscle resistant to tetrodotoxin (Harris & Thesleff 1971).

SPECIFIC MODELS TESTED IN CELL CULTURE

Mechanism of Acetylcholine Receptor Localization

There exist two classes of models that could account for the localization of ACh receptor:

1. New ACh receptors are synthesized and preferentially inserted in the muscle membrane under the nerve ending.
2. The nerve causes an aggregation of preexisting ACh receptors at points of nerve-muscle contact; perhaps by a ligand-induced capping mechanism (for reviews see Edelman 1976, Raff 1976).

In a series of experiments, Anderson et al (1976, 1977, Anderson & Cohen 1977) demonstrated that preexisting ACh receptors aggregate to the points of nerve-muscle contact in an amphibian culture system. They labeled the ACh receptor with fluorescent labeled α-bungarotoxin, washed out the excess labeled toxin, and then added native toxin and spinal cord neurons. They showed that the nerve induced the redistribution and localization of preexisting ACh receptor (labeled prior to adding the nerve) in regions of nerve-muscle contact. These experiments do not rule out synthesis of new receptors as well, but certainly they demonstrate aggregation of preexisting receptors. One question that remains concerns specificity. Anderson et al (1976, 1977, Anderson & Cohen 1977, Axelrod et al 1976) found accumulations of ACh receptor at points of attachment of the muscle to the culture dish. This occurs in the rat culture system as well as the amphibian system. Thus, it is possible that any substance that binds tightly to the membrane will induce an aggregation of ACh receptor.

The Role of the Cholinergic System and Electrical Activity in the Localization of ACh Receptor

It has been proposed that the association of the nerve terminal with high regions of ACh receptor might require some aspect of the cholinergic system. Specifically localization of ACh receptor may involve the release of ACh by the nerve with a subsequent activation of the ACh receptor on the muscle (Thesleff 1960, Katz & Miledi 1964, Drachman 1967). This model was tested in culture utilizing clonal cell lines of nerve and muscle (Steinbach et al 1973, Steinbach & Heinemann 1973). A mouse nerve cell line (C1300, N18) that makes very low amounts of CAT, comparable to the level found in fibroblasts, was used to interact with a rat skeletal muscle cell line (L6). A specific inhibitor of the CAT enzyme was used to further reduce any ACh synthesis. Finally, the muscle ACh receptor was blocked by a high concentration of α-neurotoxin. Under these conditions, localization of ACh receptor was found at points of neurite-muscle contact. This result was confirmed and extended by experiments in an amphibian culture system in which Anderson et al (1976, 1977, Anderson & Cohen 1977) demonstrated that the nerve can induce a localization of ACh receptor on muscle in the presence of α-bungarotoxin.

The role of electrical activity in regulating ACh receptor distribution was tested in culture. Steinbach (1974b) showed that localization of ACh receptor at the point of nerve contact occurs in muscle interacted with nerve in the presence of a high concentration of K^+, which depolarizes the muscle and inactivates the action potential mechanism.

The interpretation of these experiments is that neither ACh, functional ACh receptor, nor electrical activity is necessary for ACh receptor to be associated with nerve endings.

Requirements for the Establishment of Chemical Transmission

A number of models have been proposed that suggest that electrical activity or synaptic transmission is necessary to stabilize the synapse (Hebb 1949, Stent 1973, Changeux & Danchin 1976). These ideas have been examined in cell culture. Synapses form in the presence of blockers of the action potential mechanism (Crain et al 1968, Kidokoro et al 1975, Obata 1977) and in the presence of antagonists of the ACh receptor, which block synaptic transmission (Crain & Peterson 1971, Cohen 1972, Kidokoro et al 1975, Obata 1977). Clearly these experiments in culture only apply to the initial stages of synapse formation and cannot rule out the possibility that activity is critical at later stages or in other systems.

One Neuron—One Transmitter?

Dale (1935) proposed that one neuron releases only one transmitter. This hypothesis has been tested in tissue culture and found not to be valid at early stages of synapse formation. Furshpan et al (1976) developed a microculture method in which spatially isolated neurons could be grown in contact with target cells. They studied the innervation of rat heart cells by sympathetic principal neurons in primary cultures. These experiments demonstrated that a single neuron could both inhibit and excite the heart cells. Pharmacological studies showed that the excitation was mediated by norepinephrine (NE) and the inhibition by ACh. Thus one neuron can secrete both NE and ACh. It is not yet clear whether these neurons synthesize more than one transmitter, since in the experiments described the transmitter may have been taken up from the culture medium. It is possible that in the normal course of development a neuron becomes "switched" on to one transmitter and synthesis of other transmitters is repressed.

Search for Trophic Substances

The idea that trophic factors are secreted from nerve terminals and taken up by their target tissue has enjoyed great popularity (see Guth 1968, Harris 1974). A retrograde factor secreted by the postsynaptic cell and taken up by the presynaptic neuron has also been discussed (Changeux & Danchin 1976). Such factors have been proposed to explain the many trophic effects that innervation has on muscle and the stabilization of synapses in general. One such factor, nerve growth factor, has been purified and shown to have an important role in the development of sympathetic and sensory ganglia (for review see Levi-Montalcini et al 1972). Attempts have been made to isolate factors from nerve that affect muscle function and might therefore be candidates for trophic substances. Two problems have hindered progress in this field:

1. The lack of a convenient and specific assay for such factors.
2. The demonstration that at least some of the properties of muscle reflect the degree of muscle activity and probably do not involve trophic substances.

Muscle cultures provide a convenient assay for factors that affect muscle development. A number of groups have made extracts from various parts of the nervous system and tested them for their effects on muscle development in culture. Extracts of brain, sciatic nerve, and spinal cord have been shown to increase development of muscle (Oh 1975, Festoff & Israel 1976, Hasegawa & Kuromi 1977). On the other hand, there is a report that spinal cord cells in culture produce a soluble factor that inhibits muscle development (Kagen et al 1976). Lentz (1974) has shown that extracts of whole brain can maintain acetylcholine esterase in denervated muscle in organ culture. It seems likely that no significant progress will be made until these activities are purified and chemically identified so that their mode of action can be determined at the molecular level.

SUMMARY

Cells from the nervous system can be grown in culture and exhibit many of the biophysical, biochemical, and morphological properties found in vivo. They can be induced to differentiate in culture and perform some of the interactions that occur in the nervous system, including synaptogenesis. A number of important questions relating to nervous system development and function have been approached in culture in ways that cannot be done in vivo. The use of clonal cell lines should lead to rapid progress in the analysis of the nervous system at the chemical level.

Literature Cited

Altstiel, L. D., Landsberger, F. R. 1977. Interaction of colchicine with phosphotidylcholine membranes. *Nature:* 70–72

Amano, T., Richelson, E., Nirenberg, M. 1972. Neurotransmitter synthesis by neuroblastoma clones. *Proc. Natl. Sci. USA* 69:258–63

Anderson, M. J., Cohen, M. W., Zorychta, E. 1976. Distribution of acetylcholine receptors on muscle cells cultured with and without nerve. *Proc. Can. Fed. Biol. Soc.* 19:91–100

Anderson, M. J., Cohen, M. W. 1977a. Nerve-induced and spontaneous redistribution of acetylcholine receptors on cultured muscle cells. *J. Physiol. London* In press

Anderson, M. J., Cohen, M. W., Zorychta, E. 1977. Effects of innervation on the distribution of acetylcholine receptors on cultured amphibian muscle cells. *J. Physiol. London* In press

Axelrod, D., Ravdin, P., Koppel, D. E., Schlessinger, J., Webb, W. W., Elson, E. L., Podleski, T. R. 1976. Lateral motion of fluorescently labeled acetylcholine receptors in membranes of developing muscle fibers. *Proc. Natl. Acad. Sci. USA* 73:4594–98

Axelsson, J., Thesleff, S. 1959. A study of supersensitivity in denervated mammalian skeletal muscle. *J. Physiol. London* 147:178–93

Berg, D. K., Hall, Z. 1974. Fate of α-bungarotoxin bound to acetylcholine receptors of normal and denervated muscle. *Science* 184:473–75

Berg, D. K., Hall, Z. W. 1975. Loss of α-bungarotoxin from junctional and extrajunctional acetylcholine receptors in rat diaphragm muscle in vivo and in organ culture. *J. Physiol. London* 252:771–89

Bevan, S., Miledi, R., Grampp, W. 1973. Induced transmitter release from Schwann cells and its suppression by actinomycin D. *Nature New Biol.* 241:85–86

Bevan, S., Steinbach, J. H. 1977. The distribution of alpha-bungarotoxin binding sites on mammalian skeletal muscle developing in vivo. *J. Physiol. London* 267:195–213

Beranek, R., Vyskocil, F. 1967. The action of tubocurarine and atropine on the normal and denervated rat diaphragm. *J. Physiol. London* 188:53–55

Breakefield, X. O. 1976. Neurotransmitter metabolism in murine neuroblastoma cells. *Life Sci.* 18:267–78

Breakefield, X. O., Nirenberg, M. 1974. Selection of neuroblastoma cells that synthesize certain transmitters. *Proc. Natl. Acad. Sci. USA* 71:2530–33

Brockes, J. P., Hall, Z. W. 1975. Acetylcholine receptors in normal and denervated rat diaphragm muscle. II. Comparison of junctional and extrajunctional receptors. *Biochemistry* 14:2100–6

Brownstein, M. J., Sqavedra, J. M., Axelrod, J., Zeman, G., Carpenter, D. O. 1974. Coexistence of several putative neurotransmitters in single identified neurons of aplysia. *Proc. Natl. Acad. Sci. USA* 71:4662–65

Burden, S. 1977. Development of the neuromuscular junction in chick embryo: the number, distribution, and stability of acetylcholine receptors. *Dev. Biol.* 57:317–29

Burnstock, G. 1976. Do some nerve cells release more than one transmitter? *Neuroscience* 1:239–48

Catterall, W. A. 1975. Sodium transport by the acetylcholine receptor of cultured muscle cells. *J. Biol. Chem.* 250:1776–81

Chang, C. C., Huang, M. C. 1975. Turnover of junctional and extrajunctional acetylcholine receptors of the rat diaphragm. *Nature* 253:643–44

Changeux, J-P., Danchin, A. 1976. Selective stabilisation of developing synapses as a mechanism for the specification of neuronal networks. *Nature* 264:705–12

Chou, T. C., Lee, C. Y. 1969. Effect of whole and fractionated cobra venom on sympathetic ganglionic transmission. *Eur. J. Pharmacol.* 8:326–30

Christian, C. N., Nelson, P. B., Peacock, J., Nirenberg, M. 1976. A biclonal synapse between a neuroblastoma X glioma hybrid and a clonal myogenic cell line. *Abstr. Soc. Neurosci. 6th Annu. Meet.* 1017

Cicero, T. J., Cowan, W. M., Moore, B. W., Suntzeff, V. 1970. Cellular localization of the two brain specific proteins, S-100 and 14-3-2. *Brain Res.* 18:25–34

Cohen, M. W. 1972. The development of neuromuscular connexions in the presence of d-tubocurarine. *Brain Res.* 41:457–63

Cohen, S. A. 1976. Early signs of transmitter release at neuromuscular junctions developing in culture. *Abstr. Soc. Neurosci. 6th Annu. Meet.* 1021

Cohen, S. A., Fischbach, G. D. 1973. Regulation of muscle acetylcholine sensitivity by muscle activity in cell culture. *Science* 181:76–78

Cohen, S. A., Fischbach, G. D. 1977. Clusters of acetylcholine receptors located at identified nerve-muscle synapses in vitro. *Dev. Biol.* 59:23–38

Crain, S. M., Bornstein, M. B., Peterson, E. R. 1968. Maturation of cultured embryonic CNS tissue during chronic exposure to agents which prevent bioelectric activity. *Brain Res.* 8:363–72

Crain, S. M., Peterson, E. R. 1971. Development of paired explants of fetal spinal cord and adult skeletal muscle during chronic exposure to curare and hemicholinium. *In Vitro* 6:373–89

Culp, L. A. 1976. Molecular composition and origin of substrate-attached material from normal and virus transformed cells. *J. Supramol. Struct.* 5:239–55

Culp, L. A., Buniel, J. F. 1976. Substrate-attached serum and cell proteins in adhesion of mouse fibroblasts. *J. Cell Physiol.* 88:89–106

Dale, H. H. 1935. Pharmacology and nerve endings. *Proc. R. Soc. Med.* 28:319–22

de Vitry, F., Camier, M., Czernichow, P., Benda, P., Cohen, P., Tixier-Vidal, A. 1974. Establishment of a clone of mouse hypothalamic neurosecretory cells synthesizing neurophysin and vasopressin. *Proc. Natl. Acad. Sci. USA* 71:3575–79

Devreotes, P. N., Fambrough, D. M. 1975. Acetylcholine receptor turnover in membranes of developing muscle fibers. *J. Cell Biol.* 65:335–58

Devreotes, P. N., Fambrough, D. M. 1976. Synthesis of the acetylcholine receptor by culture chick myotubes and denervated mouse extensor digitorum longus muscles. *Proc. Natl. Acad. Sci. USA* 73:161–64

Devreotes, P. N., Gardner, J. M., Fambrough, D. M. 1977. Kinetics of biosynthesis of acetylcholine receptor and subsequent incorporation into plasma membrane of cultured chick skeletal muscle. *Cell* 10:365–73

Dichter, M., Tischler, A., Greene, L. A. 1977. Nerve growth factor-induced increase in electrical excitability and acetylcholine sensitivity of a rat pheochromocytoma cell line. *Nature* 268: 501–3

Doljanski, F., Kappeller, M. 1976. Cell surface shedding—the phenomenon and its possible significance. *J. Theor. Biol.* 62:263–70

Drachman, D. B. 1967. Is acetylcholine the trophic neuromuscular substance? *Arch. Neurol. Chicago* 17:206–18

Druckrey, H., Preussmann, R., Ivankovic, S., Schmahl, D. 1967. Organotype carcinogene wirkungen bei 65 verschiedenen N-nitroso-verbindungen and BD-Ratten. *Z. Krebsforsch.* 69:103–210

Duggan, A. W., Hall, J. G., Lee, C. Y. 1976. α-Bungarotoxin, cobra neurotoxin and excitation of Renshaw cells by acetylcholine. *Brain Res.* 107:166–70

Edelman, G. M. 1976. Surface modulation in cell recognition and cell growth. *Science* 192:218–26

Eterovic, V. A., Bennett, E. L. 1974. Nicotinic cholinergic receptor in brain detected by binding of α[^{3}H] bungarotoxin. *Biochim. Biophys. Acta* 362: 346–55

Fambrough, D., Hartzell, H. C., Rash, J. E., Ritchie, A. K. 1974. Receptor properties of developing muscle. *Ann N.Y. Acad. Sci.* 228:47–62

Fambrough, D. M., Rash, J. E. 1971. Development of acetylcholine sensitivity during myogenesis. *Dev. Biol.* 26: 55–68

Fertuck, H. C., Salpeter, M. M. 1976. Quantitation of junctional and extrajunctional acetylcholine receptors by electron microscope autoradiography after ^{125}I-α-bungarotoxin binding at mouse neuromuscular junctions. *J. Cell Biol.* 69:144–58

Festoff, B. W., Israel, S. 1976. Studies of a nerve extract with trophic properties. *Abstr. Soc. Neurosci. 6th Annu. Meet.* 1040

Fischbach, G. D. 1972. Synapse formation between dissociated nerve and muscle cells in low density cell cultures. *Dev. Biol.* 28:407–29

Fischbach, G. D., Berg, D. G., Cohen, S. A., Frank, E. 1975. Enrichment of nerve-muscle synapses in spinal cord-muscle cultures and identification of relative peaks of ACh sensitivity at sites of transmitter release. *Cold Spring Harbor Symp. Quant. Biol.* 40:347–57

Fischbach, G. D., Cohen, S. A. 1973. The distribution of acetylcholine sensitivity over uninnervated and innervated muscle fibers grown in cell culture. *Dev. Biol.* 31:147–62

Fischbach, G. D., Henkart, M. P., Cohen, S. A., Breuer, A. C., Shysner, J., Neal, F. M. 1974. Studies on the development of neuromuscular junctions in cell culture. *26th Symp. Soc. Gen. Physio.* New York: Raven. pp. 259–85

Frank, E., Fischbach, G. D. 1976. The appearance of acetylcholine receptors at nerve-muscle contacts in vitro. *Abstr. Soc. Neurosci. 6th Annu. Meet.* 700

Frank, E., Fischbach, G. D. 1977. ACh receptors accumulate at newly formed nerve-muscle synapses *in vitro. Cell and Tissue Culture,* ed. J. W. Lash, M. M. Burger. New York: Raven. In press

Furmanski, P., Silverman, D. J., Lubin, M. 1971. Expression of differentiated functions in mouse neuroblastoma mediated by dibutyryl-cyclic adenosine monophosphate. *Nature* 233:413–15

Furshpan, E. J., MacLeish, P. R., O'Lague, P. H., Potter, D. D. 1976. Chemical transmission between rat sympathetic neurons and cardiac myocytes developing in microcultures: Evidence for cholinergic, adrenergic, and dual-function neurons. *Proc. Natl. Acad. Sci. USA* 73:4225–29

Garrels, J., Gibson, W. 1976. Identification and characterization of multiple forms of actin. *Cell* 9:793–805

Garrels, J., Schubert, D. 1977. Modulation of protein synthesis by NGF. *Proc. Natl. Acad. Sci. USA* Submitted

Gilman, A. G., Nirenberg, M. 1971. Regulation of adenosine 3',5'-cyclic monophosphate metabolism in cultured neuroblastoma cells. *Nature* 234:356–58

Gilman, A. G. 1972. In *Advances in Cyclic Nucleotide Research,* ed. P. Greengard, G. A. Robison, 1:389–410. New York: Raven

Greene, L. A. 1976. Binding of α-bungarotoxin to chick sympathetic ganglia: Properties of the receptor and its rate of appearance during development. *Brain Res.* 111:135–45
Greene, L. A., Sytkowski, A. J., Vogel, Z., Nirenberg, M. W. 1973. α-Bungarotoxin used as a probe for acetylcholine receptors of cultured neurons. *Nature* 243:163–66
Greene, L., Tischler, A. S. 1976. Establishment of a noradrenergic clonal line of rat adrenal pheochromocytoma cells which respond to NGF. *Proc. Natl. Acad. Sci. USA* 73:2424–28
Greengard, P. 1976. Possible role for cyclic nucleotides and phosphorylated membrane proteins in postsynaptic actions of neurotransmitters. *Nature* 260:101–8
Grimm-Jorgensen, Y., Pfeiffer, S. E., McKelvy, J. F. 1976. Metabolism of thyrotropin releasing factor in two clonal cell lines of nervous system origin. *Biochem. Biophys. Res. Comm.* 70:167–73
Grobstein, C. 1959. The differentiation of vertebrate cells. In *The Cell,* ed. J. Brachet, A. Mirsky, 1:437–96. New York: Academic
Gruenstein, E., Rich, A. 1975. Non-identity of muscle and non-muscle actins. *Biochem. Biophys. Res. Commun.* 64: 472–77
Guth, L. 1968. Trophic influence of nerve on muscle. *Physiol. Rev.* 48:645–87
Haffke, S. C., Seeds, N. 1975. Neuroblastoma: the *E coli* of neurobiology? *Life Sci.* 16:1649–58
Hanley, M. R., Catterall, G. A., Emson, P. C., Fonnum, F. 1974. Enzymatic synthesis of acetylcholine by a serotonin-containing neuron from Helix. *Nature* 251:631–33
Harris, A. J. 1974. Inductive functions of the nervous system. *Ann. Rev. Physiol.* 36:251–305
Harris, A. J., Heineman, S., Schubert, D., Tarikas, H. 1971. Trophic interaction between cloned tissue culture lines of nerve and muscle. *Nature* 231:296–301
Harris, J. B., Thesleff, S. 1971. Studies on tetrodotoxin resistant action potentials in denervated skeletal muscle. *Acta Physiol. Scand.* 83:382–88
Harrison, R. 1907. Observations on the living developing nerve fiber. *Anat. Rec.* 1:116–18
Hartzell, H. C., Fambrough, D. M. 1973. Acetylcholine receptor production and incorporation into membranes of developing muscle fibers. *Dev. Biol.* 30: 153–65
Hasegawa, S., Kuromi, H. 1977. Effects of spinal cord and other tissue extracts on resting and action potentials of organ-cultured mouse skeletal muscle. *Brain Res.* 119:133–41
Hebb, D. O. 1949. *Organization of Behavior.* New York: Wiley
Heinemann, S., Bevan, S., Kullberg, R., Lindstrom, J., Rice, J. 1977. Modulation of the acetylcholine receptor by anti-receptor antibody. *Proc. Natl. Acad. Sci. USA* 74:3090–94
Hinkley, R. E., Telser, A. G. 1974. The effects of halothane on cultured mouse neuroblastoma cells. *J. Cell. Biol.* 63:531–40
Jansen, J. K. S., Lφmo, T., Nicolaysen, Westgaard, R. H. 1973. Hyperinnervation of skeletal muscle fibres: Dependence on muscle activity. *Science* 181:559–61
Kagen, L. J., Collins, K., Roberts, R., Butt, A. 1976. Inhibition of muscle cell development in culture by cells from spinal cord due to production of low molecular weight factor. *Dev. Biol.* 48:25–34
Kano, M., Shimada, P. 1971. Innervation and acetylcholine sensitivity of skeletal muscle cells differentiated in vitro from chick embryos. *J. Cell Physiol.* 78: 233–42
Katz, B., Miledi, R. 1964. The development of acetylcholine sensitivity in nerve-free segments of skeletal muscle. *J. Physiol. London* 170:389–96
Kidokoro, Y. 1973. Development of action potentials in a clonal rat skeletal muscle cell line. *Nature New Biol.* 241:158–59
Kidokoro, Y., Heinemann, S. 1974. Synapse formation between clonal muscle cells and rat spinal cord explants. *Nature* 252:593–94
Kidokoro, Y., Heinemann, S., Schubert, D., Brandt, B. L., Klier, F. G. 1975. Synapse formation and neurotrophic effects on muscle cell lines. *Cold Spring Harbor Symp. Quant. Biol.* 40:373–88
Kidokoro, Y., Patrick, J. 1978. Correlation between miniature endplate potential amplitudes and acetylcholine receptor densities in the neuromuscular contact found in vitro. *Brain Res.* In press
Kimes, B. W., Brandt, B. L. 1976a. Characterization of two putative smooth muscle cell lines from rat thoracic aorta. *Exp. Cell Res.* 98:349–66
Kimes, B. W., Brandt, B. L. 1976b. Properties of a clonal muscle cell line from rat heart. *Exp. Cell Res.* 98:367–81

Kimes, B., Tarikas, H., Schubert, D. 1974. Neurotransmitter synthesis by two clonal cell lines: changes with culture growth and morphological differentiation. *Brain Res.* 79:291–295
Kimhi, Y., Palfrey, C., Spector, I., Barak, Y., Littauer, U. Z. 1976. Maturation of neuroblastoma cells in the presence of dimethylsulfoxide. *Proc. Natl. Acad. Sci. USA* 73:462–66
Kirkland, W. L., Burton, P. R. 1972. Cyclic adenosine monophosphate-mediated stabilization of mouse neuroblastoma cell neurite microtubules exposed to low temperature. *Nature* 240:205–7
Klee, W. A., Nirenberg, M. 1974. A neuroblastoma X glioma hybrid cell line with morphine receptors. *Proc. Natl. Acad. Sci. USA* 71:3474–77
Klier, G., Schubert, D., Heinemann, S. 1975. Ultrastructural changes accompanying the induced differentiation of clonal rat nerve and glia. *Neurobiology* 5:1–7
Ko, P. K., Anderson, M. J., Cohen, M. W. 1977. Denervated skeletal muscle fibers develop discrete patches of high acetylcholine receptor density. *Science* 196: 540–42
Koenig, J. 1973. Morphogenesis of motor end-plates *in vivo* and *in vitro*. *Brain Res.* 62:361–65
Koenig, J., Pecot-Dechavassine, M. 1971. Relations entre l'apparition des potentiels miniatures spontanés et l'ultrastructure des plaques motrices en voie de reinnervation et de neoformation chez le rat. *Brain Res.* 27:43–57
Kuno, M. 1975. Responses of spinal motor neurons to section and restoration of peripheral motor connections. *Cold Spring Harbor Symp. Quant. Biol.* 40:457–63
Lentz, T. L. 1974. Effect of brain extracts on cholinesterase activity of cultured skeletal muscle. *Exp. Neurology* 45:520–26
Levi-Montalcini, R., Angeletti, R. H., Angeletti, P. V. 1972. The nerve growth factor. *Struct. Funct. Nerv. Tissue* 5:1–38
Lϕmo, T., Rosenthal, J. 1972. Control of ACh sensitivity by muscle activity in the rat. *J. Physiol. London* 252:493–513
Lϕmo, T., Westgaard, R. H. 1975. Control of ACh sensitivity in rat muscle fibres. *Cold Spring Harbor Symp. Quant. Biol.* 40:263–74
Luduena, M. A. 1973. The growth of ganglion neurons in serum-free medium. *Dev. Biol.* 33:470–76
Lyman, G. H., Preisler, H. D., Papahadjopoulos, D. 1976. Membrane action of DMSO and other chemical inducers of Friend leukaemic cell differentiation. *Nature* 262:360–63
Matsuzawa, H., Nirenberg, M. 1975. Receptor-mediated shifts in cGMP and cAMP levels in neuroblastoma cells. *Proc. Natl. Acad. Sci. USA* 72:3472–76
McClay, D. R., Moscona, A. A. 1974. Purification of specific cell-aggregating factor from embryonic neural retina cells. *Exp. Cell Res.* 87:438–43
Merlie, J. P., Changeux, J. P., Gros, F. 1976. Acetylcholine receptor degradation measured by pulse chase labelling. *Nature* 264:74–76
Merlie, J. P., Sobel, A., Changeux, J. P., Gros, F. 1975. Synthesis of acetylcholine receptor during differentiation of cultured embryonic muscle cells. *Proc. Natl. Acad. Sci. USA* 72:4028–32
Minchin, M. C. W., Iversen, L. L. 1974. Release of ^{3}H-gamma-aminobutyric acid from glial cells in rat dorsal root ganglion. *J. Neurochem.* 21:533–41
Minna, J., Glazer, D., Nirenberg, M. 1972. Genetic dissection of neural properties using somatic cell hybrids. *Nature* 235:225–30
Monard, D., Soloman, F., Rentsch, M., Gusin, R. 1973. Glial induced morphological differentiation in neuroblastoma cells. *Proc. Natl. Acad. Sci. USA* 70: 1894–97
Morgan, J. H., Seeds, N. W. 1975. Tubulin constancy during morphological differentiation of mouse neuroblastoma cells. *J. Cell Biol.* 67:136–45
Neher, E., Sakmann, B. 1975. Voltage-dependence of drug-induced conductance in frog neuromuscular junction. *Proc. Natl. Acad. Sci. USA* 72:2140–44
Nelson, P. G. 1975. Nerve and muscle cells in culture. *Physiol. Rev.* 55:1–60
Nelson, P., Christian, C., Nirenberg, M. 1976. Synapse formation between clonal neuroblastoma X glioma hybrid cells and striated muscle cells. *Proc. Natl. Acad. Sci. USA* 73:123–27
Nurse, C. A., O'Lague, P. H. 1975. Formation of cholinergic synapses between dissociated sympathetic neurons and skeletal myotubes of the rat in cell culture. *Proc. Natl. Acad. Sci. USA* 72:1955–59
Obata, K. 1976. Synapse formation in neuron culture. *Adv. Neurol. Sci. Jpn.* 20:50–57
Obata, K. 1977. Development of neuromuscular transmission in culture with a variety of neurons and in the presence of cholinergic substances and tetrodotoxin. *Brain Res.* 119:141–53

Oh, T. H. 1975. Neurotrophic effects: Characterization of the nerve extract that stimulates muscle development in culture. *Exp. Neurol.* 46:432–38

Patrick, J., Heinemann, S. F., Lindstrom, J., Schubert, D. R., Steinbach, J. H. 1972. Appearance of acetylcholine receptors during differentiation of a myogenic cell line. *Proc. Natl. Acad. Sci. USA* 69:2762–66

Patrick, J., Lindstrom, J. 1973. Autoimmune response to acetylcholine receptor. *Science* 180:861–72

Patrick, J., Lindstrom, J., Culp, W., McMillan, J. 1973. Studies on purified acetylcholine receptor and anti-acetylcholine receptor antibody. *Proc. Natl. Acad. Sci. USA* 70:3334–38

Patrick, J., McMillan, J., Wolfson, H., O'Brien, J. O'C. 1977. Acetylcholine receptor metabolism in a non-fusing muscle cell line. *J. Biol. Chem.* 252:2143–53

Patrick, J., Stallcup, W. B. 1977a. Immunological distinction between acetylcholine receptor and the α-bungarotoxin binding component on symapthetic neurons. *Proc. Natl. Acad. Sci. USA* 74:4689–92

Patrick, J., Stallcup, B. 1977b. α-Bungarotoxin binding and cholinergic receptor function on a rat sympathetic nerve line. *J. Biol. Chem.* In press

Patterson, B., Prives, J. 1973. Appearance of acetylcholine receptor in differentiating cultures of embryonic chick breast muscle. *J. Cell. Biol.* 59:241–45

Perez, V. J., Olney, J. W., Cicero, T. J., Moore, B. W., Bahn, G. A. 1970. Wallerian degeneration in rabbit optic nerve: Cellular localization in the central nervous system of the S-100 and 14-3-2 proteins. *J. Neurochem.* 17: 511–20

Piatigorsky, J., Webster, H., Wollberg, M. 1972. Cell elongation in the cultured embryonic chick lens epithelium with and without protein synthesis. *J. Cell Biol.* 55:82–92

Prasad, K. N., Gilmer, K. N., Sahu, S. K. 1974. Demonstration of acetylcholine-sensitive adenyl cyclase in malignant neuroblastoma cells in culture. *Nature* 249:765–67

Prives, J., Silman, I., Amsterdam, A. 1976. Appearance and disappearance of acetylcholine receptor during differentiation of chick skeletal muscle in vitro. *Cell* 7:543–50

Purves, R. D. 1974. Muscarinic excitation: A microelectrophoretic study on cultured smooth muscle cells. *Br. J. Pharmacol.* 52:77–86

Raff, M. 1976. Self regulation of membrane receptors. *Nature* 259:265–66

Richelson, E. 1977. Antipsychotics block muscarinic acetylcholine receptor-mediated cyclic GMP formation in cultured mouse neuroblastoma cells. *Nature* 266:371–73

Roisen, F. J., Murphy, R. A. 1973. Neurite development in vitro. *J. Neurobiol.* 4:397–412

Schubert, D. 1974. Induced differentiation of clonal rat nerve and glial cells. *Neurobiology* 4:376–87

Schubert, D. 1976. Proteins secreted by clonal cell lines: Changes in metabolism with culture growth. *Exp. Cell Res.* 102:329–40

Schubert, D. 1977. The substrate attached material synthesized by clonal cell lines of nerve, glia and muscle. *Brain Res.* 32:337–42

Schubert, D., Carlisle, W., Look, C. 1975. Putative neurotransmitter in clonal cell lines. *Nature* 254:341–42

Schubert, D., Harris, A. J., Devine, C. E., Heinemann, S. 1974b. Characterization of a unique muscle cell line. *J. Cell Biol.* 61:398–413

Schubert, D., Harris, A. J., Heinemann, S., Kidokoro, Y., Patrick, J., Steinbach, J. H. 1973. Induced differentiation of a neuroblastoma. In *Tissue Culture of the Nervous System,* ed. G. Sato. New York: Plenum. 55 pp.

Schubert, D., Heinemann, S., Carlisle, W., Kimes, B., Patrick, J., Steinbach, J. H., Culp, W., Brandt, B. L. 1974a. Clonal cell lines from the rat central nervous system. *Nature* 249:224–27

Schubert, D., Kidokoro, Y., Heinemann, S. 1977. Cholinergic metabolism and synapse formation by a rat nerve cell line. *Proc. Natl. Acad. Sci. USA* 74:2579–83

Schubert, D., Humphreys, S., De Vitry, F., Jacob, F. 1971. Induced differentiation of a mouse neuroblastoma. *Dev. Biol.* 25:514–46

Schubert, D., Jacob, F. 1970. Bromodeoxyuridine induced differentiation of a neuroblastoma. *Proc. Natl. Acad. Sci. USA* 67:247–54

Schubert, D., LaCorbiere, M., Whitlock, C., Stallcup, W. 1977. Alteration of the nerve cell surface by nerve growth factor. *Nature.* Submitted

Schubert, D., Whitlock, C. 1977. The alteration of cellular adhesion by nerve growth factor. *Proc. Natl. Acad. Sci. USA* 74:4055–58

Seeds, N. W., Gilman, A. G., Amano, T., Nirenberg, M. W. 1970. Regulation of axon formation by clonal lines of a neural tumor. *Proc. Natl. Acad. Sci. USA* 66:160–65

Sellstrom, A., Hamberger, A. 1977. Potassium-stimulated α-aminobutyric acid release from neurons and glia. *Brain Res.* 119:189–98

Shainberg, A., Burstein, M. 1976. Decrease of acetylcholine receptor synthesis in muscle cultures by electrical stimulation. *Nature* 264:368–69

Shainberg, A., Cohen, S. A., Nelson, P. G. 1976. Induction of acetylcholine receptors in muscle cultures. *Pfluegers Arch.* 361:255–61

Shimada, Y., Fischman, D. A. 1973. Morphological and physiological evidence for the development of functional neuromuscular junctions in vitro. *Dev. Biol.* 31:200–25

Stadler, J., Franke, W. W. 1974. Characterization of the colchicine binding of membrane fractions from rat and mouse liver. *J. Cell Biol.* 60:297–303

Stallcup, W., Cohn, M. 1976. Electrical properties of a clonal cell line as determined by measurement of ion fluxes. *Exp. Cell Res.* 98:277–84

Stallcup, W. B. Generation of Na^{+} and Ca^{++} fluxes in a sympathetic nerve cell line. *J. Physiol.* Submitted

Steinbach, J. H. 1974a. Synapse formation in vitro. *Front. Neurol. Neurosci. Res.* pp. 133–41

Steinbach, J. H. 1974b. Role of muscle activity in nerve-muscle interaction in vitro. *Nature* 284:70–71

Steinbach, J. H., Harris, A. J., Patrick, J., Schubert, D., Heinemann, S. 1973. Nerve-muscle interaction in vitro. *J. Gen. Physiol.* 62:255–70

Steinbach, J. H., Heinemann, S. 1973. Nerve-muscle interaction in clonal cell culture. *Exploratory Concepts in Muscular Dystrophy II,* pp. 161–69

Steinbach, J. H., Schubert, D. 1975. Multiple modes of dibutyryl cyclic AMP induced process formation by clonal nerve glial cells. *Exp. Cell Res.* 91:449–53

Stent, G. S. 1973. A physiological mechanism for Hebb's postulate of learning. *Proc. Natl. Acad. Sci. USA* 70:997–1001

Sytkowski, A. J., Vogel, Z., Nirenberg, M. W. 1973. Development of acetylcholine receptor clusters on cultured muscle cells. *Proc. Natl. Acad. Sci. USA* 70:270–74

Thesleff, S. 1960. Supersensitivity of skeletal muscle produced by botulinum toxin. *J. Physiol. London* 151:598–607

Traber, J., Fischer, K., Buchen, C., Hambrecht, B. 1975. Muscarinic response to acetylcholine in neuroblastoma X glioma hybrid cells. *Nature* 255:558

Varon, S., Bunge, R. P. 1978. Trophic mechanisms in the peripheral nervous system. *Ann. Rev. Neurosci.* 1:327–61

Venter, J., Buonassisi, V., Bevan, S., Heinemann, S., Bevan, J. A. 1976. Hormone and neurotransmitter receptors on the initimal endothelium. In *Blood Vessels,* pp. 391–82

Vogel, Z., Sytkowski, A. J., Nirenberg, M. W. 1972. Acetylcholine receptors on muscle grown in vitro. *Proc. Natl. Acad. Sci. USA* 69:3180–84

Vogel, Z., Daniels, M. P. 1976. Ultrastructure of acetylcholine receptor clusters on cultured muscle fibers. *J. Cell Biol.* 69:501–7

West, G. J., Uki, J., Stahn, R., Herschman, H. 1977. Neurochemical properties of cell lines from N-ethyl-N-nitrosourea induced rat tumors. *Brain Res.* 130: 387–92

Witkowski, J. A., Brighton, W. D. 1972. Influence of serum on attachment of tissue cells to glass surfaces. *Exp. Cell Res.* 70:41–48

Yaffe, D. 1968. Retention of differentiation potentialities during prolonged cultivation of myogenic cells. *Proc. Natl. Acad. Sci. USA* 61:477

Yamada, K. M., Wessells, N. K. 1971. Axon elongations: effect of nerve growth factor on microtubule protein. *Exp. Cell Res.* 66:346–52

Ann. Rev. Neurosci. 1978. 1:445–71

ORGANIZATION OF NEURONAL MEMBRANES

❖11515

Karl H. Pfenninger
Department of Anatomy, Columbia University College of Physicians and Surgeons, New York, NY 10032

INTRODUCTION

Biological membranes have a twofold significance in living organisms: first as barriers, they maintain the integrity of cells and their internal compartments and second, membranes serve a mediating role between the interiors of cells or organelles and their external environments. This duality of membrane function is crucial to the cells' metabolic needs and their proper interactions as units of a tissue. This is clearly seen in the maintenance of the resting potential and the conduction of electrical signals in nerve cells. Neuronal excitability is dependent both on the conservation of an intracellular compartment that is ionically different from the extracellular space and on the precisely controlled, selective permeability of the barrier. The plasma membrane's key role in the neuron is further evidenced by the fact that short-term interaction with the environment (receptor cell) and with other cells is one of the principal neuronal functions. Neurons transmit signals to each other most frequently by localized application of a transmitter to a selected site on the target cell. This functional principle necessitates the intricate organization of the neuron, which comprises long extensions, dendrites and axons, with receptive, conducting, and transmitting portions (cf. "dynamic polarization"; Ramón y Cajal 1911). Accordingly, the neuron is enveloped by a heterogenous membrane mosaic composed of functionally different regions.

Any cell biological analysis of membranes must attempt to correlate structure, function, and chemistry. Accordingly, much of the relevant work has been carried out on isolated, more or less homogeneous membrane preparations that can be studied with a variety of techniques. The nervous system is generally not well suited to this approach because of the great variety of cell types it contains, the heterogenous character of most neuronal plasmalemmas, and the complex geometry of the neurons (and glia). However, certain types of neuronal perikarya and their plasma membranes (Hamberger & Svennerholm 1971, Norton & Poduslo 1971, Tamai et al 1971, Capps-Covey & McIlwain 1975, Henn et al 1975, Chao & Rumsby 1977,

0147-006X/78/0325-0445$01.00

Poduslo & McKhann 1977) and certain relatively homogeneous axolemmal fractions (Sheltawy & Dawson 1966, Camejo et al 1969, Holton & Easton 1971, Zambrano et al 1971, Grefrath & Reynolds 1973, Chacko et al 1974) have been successfully isolated. In addition, several workers have prepared fractions of nerve endings, with attached postsynaptic membranes, the so-called synaptosomes (for review, see Mahler et al 1975). Unfortunately all these fractions have suffered from a serious heterogeneity of membranes (due partly to contamination by intracellular elements, by membranes from non-neuronal elements, and by the presence of different neuronal types). Similarly synaptosomes have not yet lent themselves to the study of the composition and organization of any unique type of membrane not only because they are derived from many kinds of synapses, but also because each contains a combination of at least two quite different membranes—pre- and postsynaptic. For these reasons our knowledge of the structural and chemical organization of the neuronal plasmalemma is limited and to date most has been learned from the comparisons of neuronal membranes with such better-known membrane preparations as the erythrocyte plasmalemma, retinal rod outer segment membranes, myelin, and the purple membrane of *Halobacterium halobium.* In fact, almost all modern concepts in membrane biology are derived from data on these and a few other model systems.

This review does not try to offer a comprehensive overview of the field of membranes. While restricting myself to a description of the plasmalemma, I have attempted to present some novel aspects of general membrane biology that seem particularly apt to further our understanding of the neuronal plasma membrane and neuronal cell biology in general.

MODERN CONCEPTS OF MEMBRANE ORGANIZATION

As mentioned above, the most useful membranes for study are those that can be prepared in bulk as a homogeneous fraction. The best example is the human red blood cell membrane, which has been especially well studied. The results of this research have been dealt with in numerous reviews in the recent past (e.g. Singer & Nicolson 1972, Bretscher 1973, Zwaal et al 1973, Singer 1974, Steck 1974, Jamieson & Robinson 1977) and are presented here in summarized form only.

Membrane Lipid

The concept of the lipid bilayer structure of membranes, developed in the 1920s (Gorter & Grendel 1925) and later modified and extended by Davson & Danielli (1943), has undergone little change in the past fifty years (Wilkins et al 1971), except for two important recent modifications. These modifications deal with the physical state of the lipids (see below) and the chemical composition of the juxtaposed lipid monolayers. The phospholipids are now known to be distributed asymmetrically, the external leaflet containing, mainly, the choline compounds phosphatidylcholine and sphingomyelin, whereas the cytoplasmic leaflet contains most of the amino compounds phosphatidylethanolamine and phosphatidylserine (Bretscher 1972, 1973).

Membrane Protein

The most revolutionary conceptual developments have involved the arrangement of the proteins that are part of the membrane. It was originally believed that membrane proteins line the lipid bilayer on either side, a notion that seemed to be confirmed by the electron microscopic appearance of membranes as trilaminar structures. This view was an essential element in Robertson's (1959) unit membrane concept, which was later extended to all cellular membranes. However, the transport of polar solutes across the membrane was difficult to understand in such a model unless a micellar organization of the lipids (Lucy 1968, Stoeckenius & Engelman 1967) was also postulated. The first modifications of the protein-lipid-protein model had to be made when it became clear that certain membrane components, especially those that bear carbohydrates, extend beyond the limits of the osmiophilic outer membrane layer seen with the electron microscope. This finding led to the idea of the "greater membrane" (Revel & Ito 1967, Lehninger 1968), which included an external and a cytoplasmic "membrane coat" consisting of hydrophilic material. Furthermore, it was realized that the dense tramlines seen in the electron microscope represented osmium dioxide deposits near the polar ends of the lipid bilayer, rather than an osmiophilic protein entity (Riemersma 1963, Korn 1966a,b, White et al 1976; for review, see Pfenninger 1973). However, this important change in the interpretation of membrane ultrastructure did not by itself provide evidence about the actual distribution of proteins in the membrane. Greater challenge to the unit membrane model came from a number of observations of a quite different nature that were hardly compatible with the traditional protein-lipid-protein sandwich model. The wide range of protein-to-lipid ratios in different types of membrane (from about 1:4 to 3:1) virtually ruled out the possibility of an invariant lipid bilayer, always covered by protein on either side. Circular dichroism and optical rotatory dispersion measurements revealed significant amounts of α-helical protein in the membrane, which suggested that there may be hydrophobic protein regions that interact with hydrocarbon chains of the lipid (Lenard & Singer 1966, Wallach & Gordon 1968) rather than with their hydrophilic head groups. Furthermore, the freeze-fracture technique, which had meanwhile been shown to reveal intramembranous morphology (Pinto da Silva & Branton 1970, Tillack & Marchesi 1970, Wehrli et al 1970), has demonstrated the presence of globular structures within the lipid bilayer. The fact that the fracture delineates these so-called intramembranous particles indicates that they have physico-chemical properties different from those of their surroundings, i.e. the lipid film, and suggests that the interface between particles and environment is of hydrophobic nature. Finally, biochemical experiments have made it possible to identify membrane proteins that resist complete digestion with proteolytic enzymes because they are partially protected in the lipid bilayer (for review, see Zwaal et al 1973). Even more importantly, proteins have been found that can be labeled with radioactive markers from both sides of the cell membrane (see Zwaal et al 1973, Steck 1974).

This new evidence has called for a revised membrane model. The most satisfactory so far is that the membrane has a mosaiclike, bilayered lipid matrix in which proteins are partially or fully inserted, some of them floating in it like icebergs in the sea while others penetrate it completely (Singer & Nicolson 1972). Proteins must therefore be considered integral parts of a lipid-protein sheet and are excellent candidates for units of transmembrane function. Important in this model is the asymmetric distribution of both major components, lipid and protein, with carbohydrate residues facing the extracellular space, and the prevalence of hydrophobic interactions between lipid and protein. This new concept has been corroborated by a variety of studies. One integral membrane protein, erythrocyte membrane glycophorin, has been isolated and sequenced. This work has revealed a carbohydrate-bearing extracellular amino terminal, followed by a hydrophobic peptide portion, which ends intracellularly with a hydrophilic carboxyl terminal (Tomita & Marchesi 1975, Segrest 1977). Rhodopsin (e.g. Montal & Korenbrot 1976), the nicotinic acetylcholine receptor (e.g. Karlin 1975, Cohen & Changeaux 1975, Potter 1977), cytochrome b_5 from endoplasmic reticulum (Strittmatter et al 1972), and intestinal aminopeptidase (Louvard et al 1975, 1976) are other examples of well-characterized integral membrane proteins (cf. Coleman 1973). X-ray diffraction studies of gap junctions (Caspar et al 1977, Makowski et al 1977; cf. also Gilula 1978), electron diffraction studies of the purple membrane of *H. halobium* (Henderson & Unwin 1975), and immunocytochemical analyses employing antibodies directed against different portions of membrane ATPase (Kyte 1974) have further confirmed the transmembranous nature of certain protein molecules or aggregates and there is increasing evidence that the intramembranous particles revealed by freeze-fracture represent clusters of such proteins (Hong & Hubbell 1972, Grant & McConnell 1974, Segrest et al 1974, Yamanaka & Deamer 1976, Yu & Branton 1976, Fisher & Stoeckenius 1977). In the last cited study, and in many functional studies (e.g. Hong & Hubbell 1973, Racker 1973, Hazelbauer & Changeux 1974, Meissner & Fleischer 1974), reconstitution of identified membrane components to at least partially functional membranes has proved to be a particularly useful approach for studying the organization and certain intermolecular relations in membranes.

Membrane Dynamics

So far, we have only considered the static properties of membranes. An understanding of membrane dynamics adds a most important new facet. It had been noted in studies utilizing electron spin labels attached to the fatty acid portion of membrane lipids (Hubbell & McConnell 1968) that hydrocarbon tails are highly mobile within the membrane. The mobility of integral membrane proteins was later shown in a variety of experiments in which fluorescent markers had been attached to cell surface components (Frye & Edidin 1970, Edidin & Fambrough 1973; for review see Edidin 1974; Cherry 1976). The redistribution (or capping) of cell surface molecules, induced by di- or multivalent antibodies or lectins (Pernis et al 1970, Taylor et al 1971, Unanue et al 1972, Edelman 1972), has further supported the notion of membrane fluidity. These various data have been integrated in the *fluid mosaic membrane model* (Singer & Nicolson 1972), which views the lipid bilayer as a two-dimensional fluid film in which the proteins are highly mobile; while the lipid bilayer serves as a barrier between compartments, it is perforated partially or fully by proteins that may operate as channels, as receptors for hormones, transmitters, viruses or toxins, and as ligands involved in cell-to-cell interactions. As a consequence of membrane fluidity, complete freedom of movement of these functional units would tend to distribute them evenly over the cell surface according to the thermodynamic laws of diffusion. In reality, restrictive elements permit the cell to establish regional membrane specialization. A nonspecific restriction of lateral mobility can be achieved by decreasing the temperature, and a solidified state of the lipids can be reached via a broad phase transition region (Steim et al 1969, Engelman 1971; for review see Cherry 1976). (In synthetic bilayers composed of one type of phospholipid, phase transition is sharp, and the lipids change into a crystalline state). Of greater physiological significance is the fact that increasing amounts of cholesterol decrease membrane fluidity (Schreier-Muccillo et al 1973, Edidin 1974, Demel & deKruyff 1976). More specific factors influencing the lateral movement of membrane proteins are likely to be found in subplasmalemmal entities. In red blood cells, which are bounded by a membrane of uniform structure, some of the membrane components are not freely mobile until after removal of a submembranous protein, spectrin (Marchesi et al 1970, Yu & Steck 1974). However, it is also evident that the mobility of surface and intramembranous components may be independently controlled (Fowler & Branton 1977). Divalent-ligand-induced capping of lymphocyte surface antigens can be disturbed by poisoning with cytochalasin B, a drug that is believed to destroy microfilaments, and influenced by colchicine or *Vinca* alkaloids that disrupt microtubules. The latter finding has led to the hypothesis that microtubules may be involved in the anchoring of cell surface components (for review see Nicolson 1976, Edelman 1978) but, as yet, no one has demonstrated that microtubules contact, or even closely approximate the plasma membrane. Complex networks of contractile proteins (for review see Weber 1978) and an extensive maze of intracellular trabeculae (Porter 1978) have recently been identified in nonmuscle cells, and it is conceivable that these may be involved in the control of position, and in the movement, of membrane components. The occurrence of sharply localized intercellular junctions and of functionally distinct cellular regions requires the presence of some submembranous (and/or intramembranous) entities controlling the lateral distribution of membrane components and cell surface phenomena. Such controlling factors may also be involved in the transmission to the cytoplasm of signals derived from cell surface interactions and may thus be of great importance in the regulation of cellular development (Edelman et al 1974, Edelman 1978).

NEURONAL CELL POLARITY

The common view of the nerve cell (cf. Bodian 1967) is typically based on a large bipolar or multipolar vertebrate neuron (e.g. a cortical pyramidal cell, a cerebellar Purkinje cell, or a spinal cord motoneuron). According to this view one part of the cell, the "receptive portion", is formed by an elaborate dendritic tree upon which the majority of synaptic inputs terminate. More centrally, is the perikaryon, the "integrative portion," of the cell. This, in turn, is followed by the axon, which constitutes the "conducting portion" of the neuron. At the terminals of its branches, the axon forms synapses upon other cells, the axon terminal representing the "transmitting portion," where the release of neurohumor occurs (cf. dynamic polarization of the neuron: Ramón y Cajal 1911, Grundfest 1957). Although this view emphasizes the degree of regional specialization in neurons, the situation is considerably more complex in the majority of nerve cells, especially in such cases as invertebrate neurons and amacrine cells, where the same process may bear both receptive and transmitting portions.

The existence of regional membrane differences in neurons implies that there are mechanisms that establish this type of order during cellular development, and maintain it at later stages. If we assume that the neuronal plasma membrane has a comparable fluidity as the membrane of muscle cells (Edidin & Fambrough 1973), it would seem that the creation and conservation of lateral organization in the plane of the membrane are major tasks for the neuron. Furthermore, the turnover of membrane components constantly requires the insertion of new constituents in the proper places during the entire lifespan of the cells. What mechanisms are involved in this process? How are particular membrane components directed from the perikaryon to the appropriate part of the plasma membrane? Or is insertion a random process that is followed by sorting out of the elements at the membrane level? So little is known about membrane biogenesis, about the mechanism of membrane differentiation, and about the turnover of membrane constituents that for the present these questions must remain unanswered (cf. Pfenninger 1978).

GENERAL PROPERTIES OF THE NEURONAL MEMBRANE

It is legitimate to ask, to what extent data from the other membranes considered above may be applied to the membranes of such functionally different cells as neurons. Are there significant differences between those membranes and the neuronal plasmalemma? In thin sections the neuronal plasma membrane exhibits a trilaminar structure of the same 8–10 nm width as these other membranes and, similarly, freeze-fracture preparations reveal a substantial number of intramembranous particles of different sizes, most of them on the cytoplasmic leaflet. X-ray diffraction studies have confirmed the bilaminar arrangement of the lipids in neuronal membranes (Blasie et al 1972). Biochemically, neuronal plasma membranes are composed of the same lipids, mostly phospholipids and cholesterol, as other membranes (for review see Strichartz 1977), and numerous types of proteins and glycoproteins have been identified (for review see Moore 1975). Many of these components, e.g. the acetylcholine receptor (Cohen & Changeuux 1975, Karlin 1975), certain synaptic proteins (Mahler et al 1975), and the putative Na^+ channel (Benzer & Raftery 1972, Strichartz 1977), exhibit properties typical of integral membrane proteins. Perhaps the most conspicuous difference can be found with a class of glycolipids, the gangliosides, that are more prominent in the nervous system than in any other tissue (for review see Suzuki 1975). In physicochemical terms, the neuronal lipid bilayer is highly fluid as in other types of plasmalemma (the original spin-label work was actually carried out on unmyelinated nerve fibers; Hubbell &

McConnell 1968). As a result of membrane fluidity, the labeling of glycoproteins with divalent ligands leads to patching in synaptosomal plasma membranes and in other isolated plasmalemmal fragments derived from nerve tissue (Matus et al 1973, Kelly et al 1976). While a slow, retrograde movement of membrane markers has been observed on growing nerve fibers (Koda & Partlow 1976), typical patching or capping of surface components labeled with lectins has not been detected in intact neurons in tissue culture (K. H. Pfenninger, M.-F. Maylié-Pfenninger, unpublished data). In conclusion, it seems fair to say that the neuronal plasma membrane is composed and structured according to the same basic plan as other eukaryotic cell membranes.

Biochemical and Structural Correlates of Excitability

Functionally, the most distinctive feature of neuronal membranes is their excitability, and in most cases, their capacity to give rise to regenerative action potentials. Excitability is dependent upon the presence of ion channels with specific gating properties, and an ion transport system for the maintenance of membrane potential and steady state conditions, viz an ouabain-sensitive Na^+, K^+–ATPase (Hodgkin & Huxley 1952a,b, Armstrong 1975). Although these physiological facts have been known for at least two decades, the underlying biochemical, and ultrastructural correlates are far from being understood. However, with the availability of more suitable membrane preparations, with more advanced techniques for membrane protein chemistry, and by the use of channel-binding toxins, this field should now advance rapidly (cf. Strichartz 1977).

UNMYELINATED AXONS Somewhat disappointingly, plasma membrane fractions from isolated rabbit brain neurons, when analyzed by SDS gel electrophoresis (Karlsson et al 1973), exhibit a polypeptide pattern that is similar to that of glia and to a variety of nonneural cell types (e.g. Glossmann & Neville 1971). More specific data have been obtained by Grefrath & Reynolds (1973), from the olfactory nerve of the garfish. They have found that the plasmalemma of these unmyelinated axons has a surprisingly low protein content (28%; cf. Chacko et al 1974) and a relatively simple polypeptide pattern with bands ranging from 32,000 to 150,000 daltons, one of which (100,000 daltons) represents the Na^+, K^+–ATPase. The solubilization by mild detergent treatment of a tetrodotoxin-binding protein from the garfish olfactory nerve was reported at about the same time (Henderson & Wang 1972, Benzer & Raftery 1973, Narahashi 1974). The dissociation constant of this substance for tetrodotoxin (TTX) is $2.5–6 \times 10^{-9}$M, and its molecular weight seems to be of the order of several hundred thousand. Furthermore, a homogeneous, excitable membrane fraction, rich in TTX-binding sites, has been isolated from the electroplaques of the electric eel (Reed & Raftery 1976). This membrane preparation is characterized by a dissociation constant for TTX of 6×10^{-9}M and exhibits discriminating affinities for monovalent alkali metal cations. Quite recently, a method has been developed for covalent labeling with a TTX derivative (photoaffinity technique) of Na^+ channels that should greatly facilitate the isolation of TTX-binding components of the membrane (Guillory et al 1977; cf. similar developments in the analysis of K^+ channels, Hucho et al 1976). If the data summarized here can stand future critical analysis, it is likely that the Na^+ channel will emerge as at least partially composed of a cluster of proteins embedded within the lipid bilayer. Its estimated size (between 230,000 and 500,000 daltons and about 8 nm diameter; Henderson & Wang 1972, Benzer & Raftery 1973, Levinson & Ellory 1973) suggests that it should be visible as an intramembranous particle in freeze-fractured nerve membranes. This situation would not be without precedent because, in gap junctions, the intramembranous particles are known to represent the proteinaceous channels through which small molecules can readily pass between the neighboring cells (Gilula 1978). Freeze-fracture analyses of excitable membrane regions of the neuron have been started, with a view to visualization of the Na^+ channel.

Intramembranous particles in neuronal plasma membranes are rather uniformly shaped, rounded structures, ranging in diameter from ~ 5–11 nm. Their density in nonspecialized plasmalemmal regions of most neurons is estimated at about 1200–1500 μm^{-2}, which is well within the range found in most other cell types—but exceeds by more than one order of magnitude the estimated density of ion channels in these membranes. The number of Na^+ channels (measured as TTX-binding sites) has been found to be 20–30 μm^{-2} in rabbit and lobster nerve and only 3 μm^{-2} in garfish olfactory nerve (Moore et al 1967, Benzer & Raftery 1972, Colquhoun et al 1972). The density of K^+ channels has been estimated at 70 μm^{-2} in squid axons (Armstrong 1975). However, the density of Na^+, K^+–ATPase molecules appears to be considerably higher, amounting to 750 μm^{-2} in rabbit vagus nerve (Landowne & Ritchie 1970). Numerous additional channels or pumps must exist for the exchange of other ions and small molecules across the membrane. With such a diversity of proteins potentially present in the membrane, it is impossible to correlate the different functional units with intramembranous particles or specific size classes of such particles at present.

MYELINATED AXONS Regions of special excitability in neurons, such as the node of Ranvier, have also been studied by freeze-fracture (Livingston et al 1973, Rosenbluth 1976, Schnapp et al 1976, Kristol et al 1977) in the hope that a better correlation between a certain class of intramembranous particles and ion channels could be obtained. In all systems analyzed, the investigators have found an increased density of particles in the nodal, as compared to the internodal, axolemma, especially in the external membrane leaflet (cf. the increased density of particles in the external leaflet of the postsynaptic membrane discussed below). The neurogenic electric organ of the fish *Stenarchus* consists of a set of specialized axons with two main types of node of Ranvier. This system exhibits a particularly pronounced increase of particle density in the external axolemmal leaflet of certain nodes that may be excitable. It offers, perhaps, the most tempting system in which to link intramembranous particles to transmembrane ion currents (Kristol et al 1977). The particle counts reported for nodal axolemma are 1100–1200 μm^{-2} for the protoplasmic leaflet (Rosenbluth 1976, Kristol et al 1977), and somewhat higher for the external leaflet, 1200–1400 μm^{-2} (Rosenbluth 1976; Kristol et al 1977, type I nodes). Na^+ channels are much more frequent in the nodal axolemma than in the internode, perhaps by 2–3 orders of magnitude (Ritchie & Rogart 1977). However, the density estimates vary considerably, probably because of the difficulty of accurately assessing the nodal membrane, so that the density of intramembranous particles could be too low (Ritchie & Rogart 1977), or comparable (Conti et al 1976, Rosenbluth 1976), or too high (Armstrong 1975) to be correlated with the channels. The very high (12,000 μm^{-3}) density of Na^+ channels in mammalian myelinated nerve fibers reported by Ritchie & Rogart (1977) is either inaccurate or else rules out the possibility that the channel protein complexes are related to the intramembranous particles. However, it is clear that the complete set of transport and channel systems in biological membranes is highly complex, and at present intramembranous particles can only be characterized by their size and shape; thus no satisfactory understanding of structure-function relationships in the process of nerve excitation is as yet possible.

NEURONAL SURFACE PROPERTIES In the search for biochemical and structural correlates of membrane excitability, the properties of the cell surface need to be discussed. Its high carbohydrate content and negative charge have been observed by many authors (for review see Pfenninger 1973). These chemical features, although present in many cell types (e.g. Pease 1966, Rambourg & Leblond 1967, Revel & Ito 1967), are particularly striking in the initial segments of axons, at nodes of Ranvier, and synaptic clefts (Meyer 1968, 1970), all of which are areas of special excitability. Gangliosides (e.g. Suzuki 1975) are present in much greater amounts in brain grey matter than in any other tissue, and, as important plasmalemmal constituents—mainly in neurons but also in glial cells (Norton et al 1975)—contribute significantly to the cell surface content of carbohydrate and negative charge. Within neurons, gangliosides seem highly concentrated in fractions of the synaptic region and of microsomes (Wolfe 1961, Breckenridge et al 1972, Avrora et al 1973). The latter location may reflect the synthetic role of microsomal elements in the replacement of plasmalemmal components or may represent a contamination artifact from synaptosomal fragments. In view of their subcellular distribution, gangliosides appear less likely to play a specific role in the generation of action potentials. However, the negative charge, largely due to glycoproteins and glycolipids on neuronal (and glial) surfaces, may be an important factor influencing the dynamics of cations in the intercellular space (Schmitt & Samson 1969); but to date the two phenomena have not been causally linked.

SPECIALIZED REGIONS OF THE NEURONAL MEMBRANE

As derivatives of an epithelial formation, the neuroectoderm, neurons are capable of forming a number of specialized intercellular contacts: gap junctions (which may act as electrical synapses), intermediate junctions (puncta adherentia), both of which neurons share with nonneuronal cells; glioaxonal junctions; and synapses, which are the most specific neuronal structures. In the following sections, the most important known properties of those characteristic membrane regions are described.

Dendritic and Postsynaptic Membrane

As of today, nonjunctional dendritic membrane regions, including the nonsynaptic parts of dendritic spines, are indistinguishable from other parts of the neuronal plasmalemma, even when studied with such diverse methods as cytochemistry and freeze-fracture. Analysis of dendritic membrane proteins, which might be the most promising approach, has not yet been possible. Nor are there structural or chemical data available on possible differences between spiking and nonspiking dendritic membranes. As the prime receptive region of the neuron, the dendritic tree is the site of many synaptic terminations and thus contains a large amount of postsynaptic membrane.

POSTSYNAPTIC MEMBRANE The most important functional elements of the postsynaptic membrane are the receptors for the related neurotransmitters and the associated ion channels for the generation of synaptic potentials. Although extensive data are available on the ultrastructural and biochemical properties of the acetylcholine receptor in electric organs and in muscle, virtually nothing is known about the organization of receptors in neuronal membranes beyond their pharmacological characteristics. The properties of the acetylcholine receptor have been frequently reviewed (e.g. Cohen & Changeaux 1975, Karlin 1975, Potter 1977, Changeux 1978), so that a brief summary of some of its characteristics will suffice. In the electric organs of *Torpedo* and *Electrophorus* the acetylcholine receptor consists of a number of protein subunits that range in size from ~ 40,000 to 60,000 daltons and form aggregates with a molecular weight of ~ 300,000. The subunits are arranged to form a doughnut-shaped structure that is inserted into the receptive zone of the membrane. This can be seen by negative staining of isolated electrocyte membrane fragments as well as by freeze-fracture (Cartaud et al 1973, Nickel & Potter 1973, Potter & Smith 1977). It seems likely that other acetylcholine receptors, and possibly the receptors for other neurotransmitters, have similar properties, and probably resemble other receptor types, such as the insulin receptor (Cuatrecasas 1974).

POSTSYNAPTIC DENSITY The most conspicuous structural element of many postsynaptic membranes is a subjunctional density composed of a fuzzy, filamentous material (Palade & Palay 1954, Palay 1956, 1958, DeRobertis, 1958, Pfenninger 1973). Ultrastructural and cytochemical analyses have indicated that this material is composed of proteinaceous filaments that extend from the plasma membrane for about 20–50 nm into the cytoplasm (Bloom & Aghajanian 1968, Pfenninger 1971a,b, Cotman & Taylor 1972, Pfenninger 1973). Within the same types of postsynaptic membrane, freeze-fracture studies have revealed arrays of large intramembranous particles that are located on the external leaflet of the membrane (Sandri et al 1972). It has been suggested (Landis & Reese 1974, Landis et al 1974), that these clusters of particles can be seen only in the so-called "S" or "Gray type I" synapses (containing spherical vesicles and characterized by pronounced postsynaptic densities) which are believed to be excitatory in function (Gray 1959, Uchizono 1965, Bodian 1966).

The biochemical analysis of postsynaptic membranes has made great progress in recent years, especially since the development of procedures for the subfractionation of synaptosomes (DeRobertis et al 1967, Cotman et al 1971, 1974, Davis & Bloom 1973, Mahler et al 1975, Cohen et al 1977). Mild treatment with detergents initially leaves the junctional region intact, but more extensive treatment solubilizes the presynaptic membrane and its densities as well as part of the postsynaptic membrane. In this way it has been possible to obtain reasonably pure fractions of postsynaptic densities as judged by electron microscopy (Cotman et al 1974, Cohen et al 1977). At present it is impossible to assess to what extent proteins that are not part of the postsynaptic density proper, such as synaptic cleft proteins, are still included in this fraction or to what extent proteins that belong to the postsynaptic density but are more loosely bound to it have been lost during the preparation procedure. Although complex, the protein composition of postsynaptic densities, as analyzed by SDS polyacrylamide gel electrophoresis, is much simpler than that of synaptosomal plasmamembranes, which is composed of a vast number of bands (Banker et al 1972, Wannamaker & Kornguth 1973, Barondes 1974, Mahler et al 1975, Wang & Mahler 1976, Cohen et al 1977, Kelly & Cotman 1977). The postsynaptic density gel pattern consists of about 15 major proteins, with molecular weights in the range of 26,000–100,000, and probably an additional 10 minor proteins. In a particularly thorough study, Siekevitz and his collaborators (Cohen et al 1977, Blomberg et al 1977) have characterized the predominant protein as having a molecular weight of 51,000 daltons (which is in reasonable agreement with the 52,000 and 53,000 values given by others; Banker et al 1972, Kelly & Cotman 1977). This protein is not identical with another major protein of 52,000 daltons found in synaptic membrane fractions (including the plasma membrane from synaptosomes). The predominant postsynaptic density protein is very similar to neurofilament protein (Blomberg et al 1977, Yen et al 1977). However, it has been suggested, on the basis of immunoprecipitation studies, that only a fraction of this band may represent neurofilament protein, which presumably comigrates with an as yet unknown postsynaptic density component (Yen et al 1977). Other efforts to identify postsynaptic proteins have led to the suggestion that tubulin (55,000 mol wt) may be a major component of synaptic membranes or of the postsynaptic density (Feit et al 1971, Blitz & Fine 1974, Bhattacharyya & Wolff 1975, Matus et al 1975, Wang & Mahler 1976). However, more recent analyses have shown that tubulin is, at best, a minor component of the postsynaptic density (Blomberg et al 1977, Yen et al 1977). By contrast, actin and possibly other contractile proteins of the actin-myosin complex seem to represent consistent elements of the postsynaptic density (Blitz & Fine 1974, Blomberg et al 1977). Collectively these findings (as well as the suggested filamentous nature of the 51,000 dalton protein) are in good agreement with the early electron microscopical description of the fibrillar nature of postsynaptic dense material (for review see Pfenninger 1973). Some of the proteins of high molecular weight isolated from postsynaptic densities exhibit lectin-binding sites (Gurd 1977a,b, Kelly & Cotman 1977; cf. Churchill et al 1976). This is of particular importance because the lectin-binding sugar residues are located exclusively in the synaptic clefts in intact synaptosomes[1] (Matus et al 1973, Bittiger & Schnebli 1974, Cotman & Taylor 1974, Kelly et al 1976), whereas the postsynaptic density fraction contains mainly cytoplasmically oriented elements of the postsynaptic membrane. This suggests that certain glycoproteins span the postsynaptic membrane and are firmly connected with proteins of the postsynaptic density (cf Gurd 1977b). Attempts to characterize functionally the proteins isolated from postsynaptic membranes have so far failed to demonstrate enzyme activities. However, studies on putative "second messenger" mechanisms have provided increasing evidence for the existence of a cyclic nucleotide system in postsynaptic membranes (Greengard 1976). The enzymes adenylate cyclase, phosphodiesterase, and protein kinase have all been found in high concentration in synaptic membrane preparations (Cheung & Salganicoff 1967, DeRobertis et al 1967, Maeno et al 1971, Ueda et al 1975). Cytochemical studies have also localized phosphodiesterase activity to the area of the postsynaptic density (Florendo et al 1971). Furthermore, endogenous protein phosphorylation is now evident in various synaptic membrane preparations (Weller & Morgan 1976, Ueda et al 1973), and in at least one case appears to involve the acetylcholine receptor (Gordon et al 1977). By analogy with other systems (e.g. erythrocytes or bladder epithelium,) in which there is a correlation between the phosphorylation of an integral membrane protein and transmembranous ion flux (Rudolf & Greengard 1974, Walton et al 1975), it has been suggested that protein phosphorylation might play an important role both in the short-term generation of postsynaptic potentials, and in the regulation of long-term synaptic effects (Greengard 1976).[2]

In the preceding discussion, little attention has been paid to the fact that postsynaptic membrane specializations differ markedly in different types of synapse (for review see Pfenninger 1973). For example, the much thinner postsynaptic density seen in F- or Gray's type II synapses (with flattened vesicles) some, at least, of which are inhibitory in nature (Gray 1959, Uchizono 1965, Bodian 1966), should be mentioned.[3] And it should also be noted that postsynaptic densities may be absent altogether, and replaced by subsurface cisternae (Rosenbluth 1962); these are

[1]In a recent publication, Matus & Walters (1976) described binding of concanavalin A (a lectin specific for glucosyl and mannosyl residues) to postsynaptic density material, i.e. on the cytoplasmic side of the postsynaptic membrane, in type II but not in type I synapses. This finding contradicts today's widely accepted view that most, if not all, complex carbohydrate of the plasmalemma is disposed on the external surface, and it is in disagreement with earlier cytochemical studies on the synapse. The authors have failed to present (*a*) control experiments as well as (*b*) quantitative data and (*c*) have not excluded the possibility of (selective) trapping of extrinsic carbohydrate in postsynaptic densities (sucrose from isolation procedures [cf. Churchill et al 1967], soluble glycoproteins from homogenate).

[2]The reader should keep in mind that the information has been gathered from a wide variety of systems. Furthermore, even the purest postsynaptic density fraction is heterogeneous because it is derived from the heterogeneous synaptic population of the CNS.

[3]These synapses have also been called "symmetric" because of the relatively poor visibility of the postsynaptic density (Colonnier 1968). This term, however, is a misnomer because chemical synapses are never strictly symmetrical, either structurally or functionally.

flattened membrane sacks, morphologically similar to smooth endoplasmic reticulum, that are found underlying the postsynaptic membrane at a well-defined interval. In other cases the subsynaptic membrane density may be very extensive with an array of globular structures of similar texture immediately beneath it—the so-called subjunctional bodies (Milhaud & Pappas 1966, Akert et al 1967). In spite of their apparent specificity, postsynaptic densities appear, morphologically, to be closely related to the submembranous densities found in nonsynaptic junctions such as intermediate junctions (puncta and zonulae adherentia) and desmosomes. In the latter, submembranous densities are also of a proteinaceous nature, but they seem to be more resistant to proteolytic digestion than postsynaptic densities and are quite frequently stratified parallel to the junctional membrane (Pfenninger 1971b, 1973).

Considering the similarity between postsynaptic and other subjunctional densities, and in view of their variability (cf. also neuromuscular junctions; e.g. Couteaux 1958, Rosenbluth 1973, 1974, 1975, Fertuck & Salpeter 1976), it seems unlikely that the postsynaptic density is a specific synaptic structure uniquely concerned with chemical transmission. Rather, it may represent a structure that holds the apposed junctional regions together and anchors the postsynaptic membrane to the underlying cytoplasm (cf. desmosomes). This view is supported by the fact that the densities survive detergent treatment and can be isolated as more or less intact entities; such resistance indicates that the components of the postsynaptic density are closely packed and tightly held together (cf. Kelly & Cotman 1976).

CLEFT MATERIAL Another important element associated with the postsynaptic membrane is formed by substances that extend into the synaptic cleft. This "cleft material", which forms the extracellular portion of the greater membrane (also called "membrane coat" or "glycocalyx") is of considerable significance because it marks the limits of the target cell and thus the site of physical interaction of this cell with the presynaptic nerve terminal. It is at this level that cellular recognition must occur during the development of the nervous system (cf Pfenninger & Rees 1976), even though the composition of the cell surface is markedly different in the mature synapse compared to that found at the time the future pre- and postsynaptic membranes first make contact (Pfenninger & Rees 1976, Pfenninger 1978). Thus, the well-developed layer of post- (and pre-) synaptic cell surface material is a consequence of synaptogenesis rather than the substratum of neuronal recognition that leads to it (Pfenninger & Rees 1976). In mature synapses, the pre- and postsynaptic moieties of the cleft material must be responsible for synaptic adhesion and maintenance. Synaptic cleft material is fibrillar in nature (Van der Loos 1963, DeRobertis 1964, Gray 1966, Pfenninger 1971b). However, the frequently described thick strands crossing the synaptic gap (cf Kornguth 1974, Elfvin 1976) are probably the result of fixation-induced clumping of more delicate elements (Pfenninger 1971b). At neuromuscular junctions Rosenbluth (1973, 1974) has described discrete knoblike structures protruding from the sarcolemma into the junctional cleft. However, no similar elements have been detected on the postsynaptic membrane of interneuronal contacts even though, in the latter, the postsynaptic cleft component seems somewhat better developed than its presynaptic counterpart (e.g. Pfenninger 1973, Cotman & Banker 1974).

Apart from its cross-synaptic striations, the cleft material is stratified parallel to the junctional membranes and forms at least three distinctive layers. This is best seen with a cytochemical technique involving the use of bismuth iodide, uranyl- and lead staining (BIUL), but excluding osmium tetroxide (Pfenninger et al 1969, Pfenninger 1971a). The more distal cleft material, which is osmiophilic and binds colloidal iron (Pappas & Purpura 1966, Pfenninger 1973), excludes the heavy metals of the BIUL procedure, whereas both the pre- and postsynaptic proximal moieties stain well with BIUL and for this reason are thought to be rich in polar groups (Pfenninger 1971a, cf. Bloom & Aghajanian 1968). Experiments using protease digestion indicate that the majority of synaptic cleft substances contain a protein backbone (Bloom & Aghajanian 1968, Pfenninger 1971b, Cotman & Taylor 1972); and they also suggest that there are at least two protein classes distinguishable on the basis of their lability to different enzymes (Pfenninger 1971b). The presence of carbohydrate in the synaptic cleft material has been demonstrated by a number of cytochemical techniques (Rambourg & Leblond 1967, Meyer 1968, 1970), including labeling with lectins (Bittiger & Schnebli 1974, Cotman & Taylor 1974, Matus et al 1973). Biochemical analyses, although unsatisfactory for localizing precisely the carbohydrates, confirm these findings. As mentioned previously, some of the postsynaptic density proteins bear carbohydrate residues that seem to extend into the synaptic cleft (Churchill et al 1976, Gurd 1977b). The high ganglioside content of synaptosomes (Wolfe 1961, Breckenridge 1972, Avrora et al 1973) is probably due to the concentration of these glycolipids in the junctional region, with their carbohydrate moieties reaching into the synaptic cleft. However, it is not known whether the gangliosides are predominantly pre- or postsynaptic, or whether they are equally distributed in both. The functional role of individual synaptic cleft constituents is not known. The interaction and adhesion of junctional partners must involve this material; the adhesion mechanism might be polyionic in nature (Pfenninger 1971b) but this has not yet been confirmed. Undoubtedly, the cleft material permits easy (and perhaps even facilitated) diffusion of ions and neurotransmitter, and it is clear that the receptor for the neurotransmitter as well as such degradative enzymes as acetylcholine esterase must be among the components exposed on the postsynaptic membrane surface (cf. Salpeter 1969, Fertuck & Salpeter 1976).

SUMMARY Glycoproteins and glycolipids associated with the postsynaptic membrane appear to extend part of their peptide chain, and probably all of their carbohydrate, into the extracellular space to form the postsynaptic moiety of the synaptic cleft material. Also inserted into this membrane are some of the filamentous proteins that constitute the postsynaptic density and presumably ion channels and receptor molecules, with at least the former penetrating the membrane. Where present enzymes for second messenger formation (such as adenylate cyclase) must be linked to a receptor, either directly or via transmembranous intermediaries. Several of these components probably constitute the morphologically recognizable intramembranous particle clusters that are found in some types of postsynaptic membrane.

Biochemical analyses also suggest that contractile properties may be associated with the postsynaptic membrane which may account for the increased curvature or bending of postsynaptic membranes in stimulated synapses (Streit et al 1972, Pysh & Wiley 1974).

Perikaryal and Axonal Plasmalemma

Apart from certain specialized areas of the axolemma to be discussed below, the nonjunctional portions of perikaryal plasmalemma, and most of the axolemma, seem to be unexceptional. However, there are two types of highly specialized axolemma: (*a*) the presynaptic membrane of nerve terminals, and (*b*) the axolemma at nodes of Ranvier and the initial axon segment.

NODE OF RANVIER AND INITIAL SEGMENT The membranes in these two areas have similar properties. Their special features include the presence of an osmiophilic undercoating (Elfvin 1961, Peters 1966, Peters et al 1968, Palay et al 1968), an increased density of intramembranous particles (see above), a particularly dense, thick layer of extracellular carbohydrate (Meyer 1970), a cytochemically stainable network that reaches deep into the cytoplasm (Quick & Waxman 1977, Waxman & Quick 1978) and immunocytochemically demonstrable ATPase (Wood et al 1976). The relationship between these structural and cytochemical characteristics and the special excitability of these regions is not clear, but they do indicate a high degree of membrane differentiation which is probably associated with a large number of special proteins, glycoproteins, and glycolipids.

GLIO-AXONAL JUNCTION The nodal regions are bounded on either side by glio-axonal junctions, that share certain characteristics with gap and septate junctions (Livingston et al 1973, Dermietzel 1974, Rosenbluth, 1976, Schnapp et al 1976). Of particular note are the rows of large particles that seem to interconnect the axolemma with the plasma membrane of the glial loops and are represented in thin sections as discrete densities crossing the intercellular space (Elfvin 1961, Peters 1966, Hirano & Dembitzer 1967, 1969). Although the intercellular gap is appreciably wider, this appearance is reminiscent of the connecting elements seen in gap junctions. The other striking feature of these glio-axonal junctions is the subunit structure found in the matrix of the external axolemmal leaflet. This may reflect a special organization of membrane lipids or proteins or some other intrinsic membrane property of unknown significance. It is unclear to what extent these junctions act as an insulating unit for the node or paranodal region, and whether they play some role in communication between the surrounding glial sheaths and the axon. Interestingly, this type of junction has only been seen in nerve tissue, and only between neurons and myelin-forming Schwann cells or oligodendrocytes.

GROWING AXONS The membrane characteristics of the axon are quite different during its out-growth and in its mature state, and this is probably true also of dendrites. While in thin sections the axolemma of growing neurites looks no different from the mature plasma membrane and carries a complex set of carbohydrates on its surface (Pfenninger & Maylié-Pfenninger 1976, Pfenninger & Rees 1976), freeze-fracture studies have revealed that there is a decreasing gradient in the density of intramembranous particles from the perikaryon to the growing tip (Pfenninger & Bunge 1974). Undoubtedly the paucity of particles in the distal part of the axon reflects specific, though as yet unknown, membrane properties. It has been suggested that this may be related to the process of plasmalemmal expansion by a mechanism of distal insertion of preformed, particle-free membrane (Pfenninger & Reese 1976, Pfenninger 1978).

Presynaptic Membrane

The presynaptic membrane of a neuron is analagous to the apical plasma membrane of an exocrine cell. It appears to be the only site at which secretory vesicles can fuse with the plasmalemma to discharge their contents into the extracellular space by the process of exocytosis (cf Pfenninger 1977). Accordingly, current investigations of the presynaptic membrane are, to a large extent, aimed at the identification of those regions of the membrane which serve as fusion sites for synaptic vesicles.

MOLECULAR BIOLOGY OF A RELEASE SITE In contrast to the apical membrane of exocrine cells which, to date, cannot be distinguished from other plasmalemmal regions, the presynaptic membrane has several distinctive morphological and cytochemical characteristics. Presynaptic membranes (except for those in the peripheral autonomic nervous system) always exhibit cytoplasmic densities, either in the form of an array of dense projections (Gray 1963) or as dense bars (Couteaux & Pécot-Dechavassine 1970, Pfenninger 1973, 1977, Akert & Peper 1975, Wood et al 1977). These densities are particularly clear after treatment with ethanolic phosphotungstic acid and bismuth iodide (BIUL) (Bloom & Aghajanian 1966, 1968, Pfenninger et al 1969). Like the postsynaptic density, the presynaptic dense projections are composed of filamentous proteinaceous material (Bloom & Aghajanian 1968, Pfenninger 1971b). Synaptic vesicles are always closely associated with the densities, and in most synapses in the vertebrate CNS they together form a closely packed assembly that has been termed the "presynaptic vesicular grid" (Akert et al 1969). In the immediate vicinity of the presynaptic densities, where the presynaptic membrane is free of cytoplasmic specializations, freeze-fracture reveals what are known as "vesicle attachment sites" (VAS)—the loci of vesicular attachment to the presynaptic membrane and of exocytotic neurotransmitter release (Pfenninger et al 1971, 1972, Streit et al 1972, Dreyer et al 1973, Heuser et al 1974, Pfenninger & Rovainen 1974). Undoubtedly, at these sites the presynaptic membrane must have specific properties to permit vesicle-plasmalemma interaction. Selective inhibition of transmitter release (e.g. by tetanus toxin-Curtis & DeGroat 1968), suggests that there may be a specific molecule that is responsible for initiating exocytosis. From the Ca^{2+} requirement of transmitter release (Katz & Miledi 1965, Douglas 1966), one would expect this molecule, or some other molecule closely associated with it, to have Ca^{2+} –binding properties. Indeed, proteins with a relatively high affinity for Ca^{2+} have recently been isolated from synaptic membrane preparations (Abood et al 1977). One of these molecules (mol wt 16,000) has been found to have 4 binding sites for Ca^{2+} (K_m 1.5×10^{-5}M). Unfortunately it has not yet been possible to isolate presynaptic membranes: as we have seen, their disintegration during the preparation of postsynaptic densities indicates that they are much more fragile than their postjunctional counterparts. However, from a comparison of the membrane proteins that can be labeled enzymatically with ^{125}I in intact synaptosomes (i.e. proteins exposed on the outside) with those that can only be labeled after osmotic shock (i.e. proteins exposed only inside the presynaptic bag) it is evident that there are a number of interior proteins (in addition to 3 transmembranous proteins, with molecular weights of 31,000, 50,000 and 145,000) that are potentially important for the specific functional needs of the presynaptic membrane (Mahler et al 1975). Brain tissue is known to be rich in contractile proteins, and some of them have been found in nerve endings (LeBeux & Willemot 1975, Berl 1975). It is possible, therefore, that an actomyosinlike complex may be involved in the process of transmitter release (Berl et al 1973), but as yet the relationship between these proteins and the presynaptic membrane is unknown.

CLEFT MATERIAL Like its postsynaptic counterpart, the protein- and carbohydrate-rich presynaptic moiety of the cleft material must be involved in the maintenance of the synaptic contact. To what extent this moiety is different from the postsynaptic cleft components is not known at present because so far all the relevant biochemical evidence is derived from preparations of complete "junctional sandwiches". The question whether presynaptic membranes contain receptors for the transmitter which they release is currently the subject of some controversy. At the neuromuscular junction the binding of horseradish-peroxidase–labeled α-bungarotoxin results not only in the staining of the subjunctional membrane but also in the deposition of reaction product on the nerve terminal membrane (Lentz & Chester 1977, Lentz et al 1977); however, radioiodinated-toxin binding reveals the receptor only on the postsynaptic membrane (Fertuck & Salpeter 1976).

SUMMARY In summary, the presynaptic membrane, like its postjunctional counterpart, is rich in membrane components that extend into the intercellular space, as well as into the cytoplasm in those areas where presynaptic densities are seen. A greater number of intramembranous particles is found within the presynaptic membrane as compared to the nonjunctional portion of the axolemma (Pfenninger et al 1972). These particles are of different sizes (from about 5 to 14 nm), and the largest seem to be more numerous after enhanced synaptic activity (Venzin et al 1977). At the neuromuscular junction, the particles form a double row on either side of the presynaptic density (Dreyer et al 1973, Heuser et al 1974, Akert & Peper 1975), but no such correlation has been observed between the particles and the presynaptic dense projections at interneuronal synapses. Although the functional significance of the presynaptic densities is not known, comparative morphology of synapses, and especially the close association of the densities with synaptic vesicles and vesicle attachment (release) sites, strongly suggests that they play a key role in the release mechanism and/or in the metabolism of vesicle contents (cf Akert & Peper 1975, Pfenninger 1977).

The exocytotic release of neurotransmitters involves at least a temporary fusion of the vesicular and presynaptic membranes, and it is probable that after the vesicles have been reused several times, they may become incorporated into the synaptic axolemma (Pfenninger & Rovainen 1974, cf Ceccarelli & Hurlbut 1975, Holtzman 1977, Heuser 1978). In agreement with this hypothesis, Breckenridge & Morgan (1972) have reported the occurrence of glycoproteins common to both membrane types; however analyses of this type can hardly be conclusive, because it is always difficult to rule out cross-contamination between vesicular and plasmalemmal fractions. One difficulty exists if irreversible fusion of vesicle and plasma membrane occurs at some stage during the exocytotic process: the incorporation of vesicles would greatly disturb the organization of the presynaptic membrane. If this happens, special mechanisms must exist to protect the presynaptic membrane from increasing disarray as a consequence of continued transmitter release. It is possible that one of the functions of the pre- and/or postsynaptic densities, together with cleft material, is the localization and close packaging of transmitter release sites, to the area directly opposite a well-defined patch of postsynaptic membrane equipped with the correct type of receptor; in the CNS this is probably an important spatial consideration.

CONCLUDING REMARKS

One goal of this review has been to illustrate the enormous complexity of neuronal membrane organization. It should be evident from the data presented here that this complexity is of two kinds: (1) in composition and architecture especially at sites of membrane specialization; and (2) in neuronal geometry, for which the term *complex lateral membrane organization,* might be appropriate. Clearly, this second feature has severely hampered progress in the field. However, it is evident that modern views of membrane organization (based on non-neuronal cells) can be readily, and probably legitimately, applied to neurons, and certainly the concepts of a fluid lipid bilayer, of integral and transmembranous proteins, of membrane asymmetry, and of restricted lateral diffusion by submembranous elements are applicable to neurons, and collectively will certainly further our understanding of the specific properties of the neuronal membrane. But it would be unwise, to disguise the fact that our knowledge of neuronal membranes, in terms of molecular organization, is still very limited. Future progress in this area will depend largely on the use of adequate biological systems; the discovery of marker molecules or enzyme activities specific for certain membrane types; the development of new, and more sophisticated, membrane isolation procedures; and the development of more sensitive microchemical techniques.

ACKNOWLEDGMENTS

Expert editorial assistance by Ms. Linda Siegel and Ms. Christine Wade during the preparation of this manuscript is herewith gratefully acknowledged.

Literature Cited

Abood, L. G., Hong, J. S., Takeda, F., Tometsko, A. M. 1977. Preparation and characterization of calcium binding and other hydrophobic proteins from synaptic membranes. *Biochim. Biophys. Acta* 443:414–27

Akert, K., Moor, H., Pfenninger, K., Sandri, C. 1969. Contributions of new impregnation methods and freeze-etching to the problem of synaptic fine structure. In Mechanisms of Synaptic Transmission, ed. K. Akert, P. Waser *Prog. Brain Res.* 31:223–40

Akert, K., Peper, K. 1975. Ultrastructure of chemical synapses: A comparison between presynaptic membrane complexes of the motor endplate and the synaptic junction in the central nervous system. In *Golgi Centennial Symposium Proceedings,* ed. M. Santini. New York: Raven. pp. 521–27

Akert, K., Pfenninger, K., Sandri, C. 1967. Crest synapses with subjunctional bodies in the subfornical organ. *Brain Res.* 5:118–21

Armstrong, C. M. 1975. Ionic pores, gates and gating currents. *Q. Rev. Biophys.* 7:179–210

Avrora, N. F., Chenykaeva, E. Y., Obukhova, E. L. 1973. Ganglioside composition and content of rat-brain subcellular fractions. *J. Neurochem.* 20:997–1004

Banker, G., Crain, B., Cotman, C. W. 1972. Molecular weights of the polypeptide chains of synaptic plasma membranes. *Brain Res.* 42:508–13

Barondes, S. H. 1974. Synaptic macromolecules: identification and metabolism. *Ann. Rev. Biochem.* 43:147–68

Benzer, T. I., Raftery, M. A. 1972. Partial characterization of a tetrodotoxin-binding component from nerve membrane. *Proc. Natl. Acad. Sci. USA* 69:3634–37

Benzer, T. I., Raftery, M. A. 1973. Solubilization and partial characterization of the tetrodotoxin binding component from nerve axons. *Biochem. Biophys. Res. Commun.* 51:939–44

Berl, S. 1975. The actomyosin-like system in nervous tissue. In *The Nervous System. Vol. 1: The Basic Neurosciences,* ed. D. B. Tower, pp. 565–73. New York: Raven Press

Berl, S., Puszkin, S., Nicklas, W. J. 1973. Actomyosin-like protein in brain. *Science* 179:441–46

Bhattacharyya, B., Wolff, J. 1975. Membrane-bound tubulin in brain and thyroid tissue. *J. Biol. Chem.* 250:7639–46

Bittiger, H., Schnebli, H. P. 1974. Binding of concanavalin A and ricin to synaptic junctions of rat brain. *Nature* 249: 370–71

Blasie, J. K., Goldman, D. E., Chacko, G., Dewey, M. M. 1972. X-ray diffraction studies on axonal membranes from garfish olfactory nerve. *Biophysical Society Annu. Meet. Abstracts* 16:253a

Blitz, A. L., Fine, R. E. 1974. Muscle-like contractile proteins and tubulin in synaptosomes. *Proc. Natl. Acad. Sci. USA* 71:4472–76

Blomberg, F., Cohen, R. S., Siekevitz, P. 1977. The structure of postsynaptic densities isolated from dog cerebral cortex. II. Characterization and arrangement of some of the major proteins within the structure. *J. Cell Biol.* 74:204–25

Bloom, F. E., Aghajanian, G. K. 1966. Cytochemistry of synapses: a selective staining method for electron microscopy. *Science* 154:1575–77

Bloom, F. E., Aghajanian, G. K. 1968. Fine structural and cytochemical analysis of the staining of synaptic junctions with phosphotungstic acid. *J. Ultrastruct. Res.* 22:361–75

Bodian, D. 1966. Electron microscopy: Two major synaptic types on spinal motoneurons. *Science* 151:1093–94

Bodian, D. 1967. Neurons, Circuits and Neuroglia. In *The Neurosciences, A Study Program,* ed. G. C. Quarton, T. Melnechuk, F. O. Schmitt, pp. 6–24. New York: Rockefeller Univ. Press

Breckenridge, W. C., Gombos, G., Morgan, I. G. 1972. The lipid composition of adult rat brain synaptosomal plasma membranes. *Biochim. Biophys. Acta* 266:695–707

Breckenridge, W. C., Morgan, I. G. 1972. Common glycoproteins of synaptic vesicles and the synaptosomal plasma membrane. *FEBS Lett.* 22:253–256

Bretscher, M. S. 1972. Phosphatidyl-ethanolamine: Differential labelling in intact cells and cell ghosts of human erythrocytes by a membrane-impermeable reagent. *J. Mol. Biol.* 71:523–28

Bretscher, M. S. 1973. Membrane structure: some general principles. *Science* 181: 622–29

Camejo, B., Villegas, G. M., Barnola, F. V., Villegas, R. 1969. Characterization of two different membrane fractions isolated from the first stellar nerves of the

squid Dosidicuo gigas. *Biochim. Biophys. Acta* 193:247–59
Capps-Covey, P., McIlwain, D. L. 1975. Bulk isolation of large ventral spinal neurons. *J. Neurochem.* 25:517–21
Cartaud, J., Benedetti, E. L., Cohen, J. B., Meunier, J.-C., Changeux, J.-P. 1973. Presence of lattice structure in membrane fragments rich in nicotinic receptor proteins from the electric organ of Torpedo marmorata. *FEBS Lett.* 33:109–13
Caspar, D. L. D., Goodenough, D. A., Makowski, L., Phillips, W. C. 1977. Gap junction structures. I. Correlated electron microscopy and x-ray diffraction. *J. Cell Biol.* 74:605–28
Ceccarelli, B., Hurlbut, W. P. 1975. Transmitter release and the vesicle hypothesis See Akert & Peper 1975, pp. 529–45
Chacko, G. K., Goldman, D. E., Malhorta, H. C., Dewey, M. M. 1974. Isolation and characterization of plasma membrane fractions from garfish Lepisosteus osseus olfactory nerve. *J. Cell Biol.* 62:831–43
Changeux, J.-P. 1978. Molecular interactions in mature and developing synaptic contacts. In *The Neurosciences. Fourth Study Program.* ed. F. O. Schmitt, F. G. Worden. Cambridge, Mass: MIT. In press
Chao, S.-W., Rumsby, M. G. 1977. Preparation of astrocytes, neurones, and oligodendrocytes from the same rat brain. *Brain Res.* 124:347–51
Cherry, R. J., 1976. Protein and lipid mobility in biological and model membranes. In *Biological Membranes,* ed. D. Chapman, D. F. H. Wallach, Vol. 3, pp. 47–102. New York: Academic
Cheung, W. Y., Salganicoff, L. 1967. Cyclic 3',5'-nucleotide phosphodiesterase: Localization and latent activity in rat brain. *Nature* 214:90–91
Churchill, L., Cotman, C. W., Banker, G., Kelly, P., Shannon, L. 1976. Carbohydrate composition of central nervous system synapses. Analysis of isolated synaptic junctional complexes and postsynaptic densities. *Biochim. Biophys. Acta* 448:57–72
Cohen, V. B., Changeux, J.-P. 1975. The cholinergic receptor protein in its membrane environment. *Ann. Rev. Pharmacol.* 15:83–108
Cohen, R. S., Blomberg, F., Berzins, K., Siekevitz, P. 1977. The structure of postsynaptic densities isolated from dog cerebral cortex. I. Overall morphology and protein composition. *J. Cell Biol.* 74:181–203
Coleman, R. 1973. Membrane-bound enzymes and membrane ultrastructure. *Biochem. Biophys. Acta* 300:1–30
Colonnier, M. 1968. Synaptic patterns on different cell types in the different laminae of the cat visual cortex, an electron microscope study. *Brain Res.* 9:268–87
Colquhoun, D., Henderson, R., Ritchie, J. M. 1972. The binding of labelled tetrodotoxin to non-myelinated nerve fibres. *J. Physiol. London* 227:95–120
Conti, F., Hille, B., Neumcke, B., Nonner, W., Stämpfli, R. 1976. Conductance of the sodium channel in myelinated nerve fibres with modified sodium inactivation. *J. Physiol. London* 262:729–42
Cotman, C. W., Banker, G. A. 1974. The making of a synapse. In *Reviews of Neuroscience,* ed. S. Ehrenpreis, I. J. Kopin, Vol. 1, pp. 1–62. New York: Raven Press
Cotman, C. W., Banker, G., Churchill, L., Taylor, D. 1974. Isolation of postsynaptic densities from rat brain. *J. Cell Biol.* 63:441–55
Cotman, C. W., Levy, W., Banker, G., Taylor, D. 1971. An ultrastructural and chemical analysis of the effect of Triton X-100 on synaptic plasma membranes. *Biochem. Biophys. Acta* 249:406–18
Cotman, C. W., Taylor, D. 1972. Isolation and structural studies on synaptic complexes from rat brain. *J. Cell Biol.* 55:696–711
Cotman, C. W., Taylor, D. 1974. Localization and characterization of concanavalin A receptors in the synaptic cleft. *J. Cell Biol.* 62:236–42
Couteaux, R. 1958. Morphological and cytochemical observations on the postsynaptic membrane at motor end-plates and ganglionic synapses. *Exp. Cell Res. Suppl.* 5:294–322
Couteaux, R., Pécot-Dechavassine, M. 1970. Vésicules synaptiques et poches au niveau des "zones actives" de la junction neuromusculaire. *C. R. Acad. Sci Ser. D.* 271:2346–49
Cuatrecasas, P. 1974. Membrane receptors. *Ann. Rev. Biochem.* 43:169–214
Curtis, D. R., DeGroat, W. C. 1968. Tetanus toxin and spinal inhibition. *Brain Res.* 10:208–212
Davis, G., Bloom, F. E. 1973. Isolation of synaptic junctional complexes from rat brain. *Brain Res.* 62:135–53
Davson, H., Danielli, J. F. 1943. The permeability of natural membranes. Cambridge: Cambridge Univ. Press
Demel, R. A., De Kruyff, B. 1976. The func-

tion of sterols in membranes. *Biochem. Biophys. Acta* 457:109–32

Dermietzel, R. 1974. Junctions in the central nervous system of the cat. *Cell Tiss. Res.* 148:576–87

De Robertis, E. 1958. Submicroscopic morphology and function of the synapse. *Exp. Cell Res. Suppl.* 5:347–69

De Robertis, E. 1964. *Histophysiology of Synapses and Neurosecretion.* New York: Pergamon

De Robertis, E., Azcorra, J. M., Fiszer, S. 1967. Ulstrastructure and cholinergic binding capacity of junctional complexes isolated from rat brain. *Brain Res.* 5:45–46

De Robertis, E., DeLores Arnaiz, G. R., Alberici, M., Butcher, R. W., Sutherland, E. W. 1967. Subcellular distribution of adenyl cyclase and cyclic phosphodiesterase in rat brain cortex. *J. Biol. Chem.* 242:3487–93

Douglas, W. W. 1966. Calcium-dependent links in stimulus-secretion coupling in the adrenal medulla and neurohypophysis. *Mechanisms of Release of Biogenic Amines.* U. S. Von Euler, S. Rosell, B. Uvnäs. pp. 267–90. Oxford: Pergamon

Dreyer, F., Peper, K., Akert, K., Sandri, C., Moor, H. 1973. Ultrastructure of the "active zone" in the frog neuromuscular junction. *Brain Res.* 62:373–80

Edelman, G. M. 1978. Cell surface modulation and transmembrane control. In *The Neurosciences, Fourth Study Program.* ed. F. O. Schmitt, F. G. Worden. Cambridge: MIT Press. In press

Edelman, G. M., Spear, P. G., Rutishauser, U., Yahara, I. 1974. Receptor specificity and mitogenesis in lymphocyte populations. In *The Cell Surface in Development,* ed. A. A. Moscona, pp. 141–64. New York: Wiley

Edidin, M. 1974. Rotational and translational diffusion in membranes. *Ann. Rev. Biophys. Bioeng.* 3:179–201

Edidin, M., Fambrough, D. 1973. Fluidity of the surface of cultured muscle fibers. Rapid lateral diffusion of marked surface antigens. *J. Cell Biol.* 57:27–37

Elfvin, L.-G. 1961. Electron microscopic investigation of the plasma membrane and myelin sheath of autonomic nerve fibers in the cat. *J. Ultrastruct. Res.* 5:388–407

Elfvin, L.-G. 1976. The ultrastructure of neuronal contacts. *Prog. Neurobiol.* 8:45–79

Engelman, D. M. 1971. Lipid bilayer structure in the membrane of Mycoplasma laidlawii. *J. Mol. Biol.* 58:153–65

Feit, H., Dutton, G. R., Barondes, S. H., Shelanski, M. L. 1971. Microtubule protein. Identification and transport to nerve endings. *J. Cell Biol.* 51:138–48

Fertuck, H. C., Salpeter, M. M. 1976. Quantitation of junctional and extrajunctional acetylcholine receptors by electron microscope autoradioautography after ^{125}I-α-bungarotoxin binding at mouse neuromuscular junctions. *J. Cell Biol.* 69:144–58

Fisher, V. A., Stoeckenius, W. 1977. Freeze-fractured purple membrane particles: protein content. *Science* 197:72–74

Florendo, N. T., Barrnett, R. J., Greengard, R. 1971. Cyclic 3',5'-nucleotide phosphodiesterase: cytochemical localization in cerebral cortex. *Science* 173: 745–47

Fowler, V., Branton, J. 1977. Lateral mobility of human erythrocyte integral membrane proteins. *Nature* 268:23–26

Frye, L. D., Edidin, M. 1970. The rapid intermixing of cell surface antigens after formation of mouse-human heterokaryons. *J. Cell Sci.* 7:319–35

Gilula, N. B. 1978. Electrotonic junctions. In *The Neurosciences. Fourth Study Program,* ed. F. O. Schmitt, F. G. Worden. Cambridge: MIT Press. In press

Glossmann, H., Neville, D. M. 1971. Glycoproteins of cell surfaces. A comparative study of three different cell surfaces of the rat. *J. Biol. Chem.* 246:6339–46

Gordon, A. S., Davis, C. G., Diamond, I. 1977. Phosphorylation of membrane proteins at a cholinergic synapse. *Proc. Natl. Acad. Sci. USA* 74:263–67

Gorter, E., Grendel, F. 1925. On bimolecular layers of lipids on the chromocytes of the blood. *J. Exp. Med.* 41:439–43

Grant, C. W. M., McConnell, H. M. 1974. Glycoproteins in lipid bilayers. *Proc. Natl. Acad. Sci. USA* 71:4653–57

Gray, E. G. 1959. Axosomatic and axodendritic synapses of cerebral cortex: an electron microscope study. *J. Anat.* 93:420–33

Gray, E. G. 1963. Electron microscopy of presynaptic organelles of the spinal cord. *J. Anat.* 97:101–6

Gray, E. G. 1966. Problems of interpreting the fine structure of vertebrate and invertebrate synapses. *Int. Rev. Gen. Exp. Zool.* 2:139–70

Greengard, P. 1976. Possible role for cyclic nucleotides and phosphorylated membrane proteins in postsynaptic actions of neurotransmitters. *Nature* 260: 101–8

Grefrath, S. P., Reynolds, J. A. 1973. Polypeptide components of an excitable plasma membrane. *J. Biol. Chem.* 248:6091–94

Grundfest, H. 1957. Electrical inexcitability of synapses and some consequences in the central nervous system. *Physiol. Rev.* 37:337–61

Guillory, R. J., Rayner, M. D., D'Arrigo, J. S. 1977. Covalent labeling of the tetrodotoxin receptor in excitable membranes. *Science* 196:883–85

Gurd, J. W. 1977a. Synaptic plasma membrane glycoproteins: Molecular identification of lectin receptors. *Biochemistry* 16:369–74

Gurd, J. W. 1977b. Identification of lectin receptors associated with rat brain postsynaptic densities. *Brain Res.* 126: 154–59

Hamberger, A., Svennerholm, L. 1971. Composition of gangliosides and phospholipids of neuronal and glial cell enriched fractions. *J. Neurochem.* 18:1821–29

Hazelbauer, G. L., Changeux, J.-P. 1974. Reconstitution of a chemically excitable membrane. *Proc. Natl. Acad. Sci. USA* 71:1479–83

Henderson, R., Unwin, P. N. T. 1975. Three-dimensional model of purple membrane obtained by electron microscopy. *Nature* 257:28–32

Henderson, R., Wang, J. H. 1972. Solubilization of a specific tetrodotoxin-binding component from garfish olfactory nerve membrane. *Biochemistry* 11:4565–69

Henn, F. A., Hansson, H.-A., Hamberger, A. 1975. Preparation of plasma membrane from isolated neurons. *J. Cell Biol.* 53:654–61

Heuser, J. E. 1978. The structural basis of neurosecretion. In *The Neurosciences. Fourth Study Program,* ed. F. O. Schmitt, F. G. Worden. Cambridge: MIT Press. In press

Heuser, J. E., Reese, T. S., Landis, D. M. D. 1974. Functional changes in frog neuromuscular junctions studied with freeze-fracture. *J. Neurocytol.* 3:109–31

Hirano, A., Dembitzer, H. M. 1967. A structural analysis of the myelin sheath in the central nervous system. *J. Cell Biol.* 34:555–67

Hirano, A., Dembitzer, H. M. 1969. The transverse bands as a means of access to the periaxonal space of the central myelinated nerve fiber. *J. Ultrastruct. Res.* 28:141–49

Hodgkin, A. L., Huxley, A. F. 1952a. Currents carried by sodium and potassium ions through the membrane of the giant axon of Loligo. *J. Physiol. London* 116:449–72

Hodgkin, A. L., Huxley, A. F. 1952b. The components of membrane conductance in the giant axon of Loligo. *J. Physiol. London* 116:473–96

Holton, J. B., Easton, D. M. 1971. Major lipids of non-myelinated (olfactory) and myelinated (trigeminal) nerve of garfish Lepisosteus osseus. *Biochem. Biophys. Acta* 239:61–70

Holtzman, E. 1977. The origin and fate of secretory packages, especially synaptic vesicles. *Neuroscience* 2:327–55

Hong, K., Hubbell, W. L. 1972. Preparation and properties of phospholipid bilayers containing rhodopsin. *Proc. Natl. Acad. Sci. USA* 69:2617–21

Hong, K., Hubbell, W. L. 1973. Lipid requirements for rhodopsin regenerability. *Biochemistry* 12:4517–23

Hubbell, W. L., McConnell, H. M. 1968. Spin-label studies of the excitable membranes of nerve and muscle. *Proc. Natl. Acad. Sci. USA* 61:12–16

Hucho, F., Bergman, C., Dubois, J. M., Rojas, E., Kiefer, H. 1976. Selective inhibition of potassium conductance in node of Ranvier with a photoaffinity label derived from tetraethylammonium. *Nature* 260:802–4

Jamieson, G. A., Robinson, D. M., eds. 1977. *Mammalian Cell Membranes,* Vol. 1–5. London & Boston: Butterworth

Karlin, A. 1975. The acetylcholine receptor: isolation and characterization. In *The Nervous System. Vol. 1: The Basic Neurosciences,* ed. D. B. Tower, pp. 323–31. New York: Raven

Karlsson, J. O., Hamberger, A., Henn, F. A. 1973. Polypeptide composition of membranes derived from neuronal and glial cells. *Biochem. Biophys. Acta* 298: 219–29

Katz, B., Miledi, R. 1965. The effect of calcium on acetylcholine release from motor nerve terminals. *Proc. R. Soc. Ser. B* 161:496–503

Kelly, P. T., Cotman, C. W. 1976. Intermolecular disulfide bonds at central nervous system synaptic junctions. *Biochem. Biophys. Res. Commun.* 73: 858–64

Kelly, P. T., Cotman, C. W. 1977. Identification of glycoproteins and proteins at synapses in the central nervous system. *J. Biol. Chem.* 252:786–93

Kelly, P. T., Cotman, C. W., Gentry, C., Nicolson, G. L. 1976. Distribution and mobility of lectin receptors on synaptic membranes of identified neurons in the central nervous system. *J. Cell Biol.* 71:487–96

Koda, L. Y., Partlow, L. M. 1976 Membrane marker movement on sympathetic axons in tissue culture. *J. Neurobiol.* 7:157–72

Korn, E. D. 1966a. Synthesis of bis(methyl 9,10-dihydrostearate)—osmate from methyl oleate and osmium tetroxide under conditions used for fixation of biological material. *Biochem. Biophys. Acta* 116:317–24

Korn, E. D. 1966b. Modification of oleic acid during fixation of amoeba by osmium tetroxide. *Biochem. Biophys. Acta* 116: 325–35

Kornguth, S. E. 1974. The synapse: a perspective from in situ and in vitro studies. In *Reviews of Neuroscience* ed. S. Ehrenpreis, I. J. Kopin, Vol. 1, pp. 63–114. New York: Raven Press

Kornguth, S. E., Sunderland, E. 1975. Isolation and partial characterization of a tubulin-like protein from human and swine synaptosomal membranes. *Biochem. Biophys. Acta* 393:100–114

Kristol, C., Akert, K., Sandri, C., Wyss, U. R., Bennett, M. V. L., Moor, H. 1977. The Ranvier nodes in the neurogenic electric organ of the knifefish *Stenarchus:* A freeze-etching study on the distribution of membrane-associated particles. *Brain Res.* 125:197–212

Kyte, J. 1974. The reactions of sodium and potassium ion-activated adenosine triphosphatase with specific antibodies. Implications for the mechanism of active transport. *J. Biol. Chem.* 249: 3652–60

Landis, D. M. D., Reese, L. S. 1974. Differences in membrane structure between excitatory and inhibitory synapses in the cerebellar cortex. *J. Comp. Neurol.* 155:93–126

Landis, D. M. D., Reese, L. S., Raviola, E. 1974. Differences in membrane structure between excitatory and inhibitory components of the reciprocal synapse in the olfactory bulb. *J. Comp. Neurol.* 155:67–92

Landowne, D., Ritchie, J. M. 1970. The binding of tritiated ouabain to mammalism non-myelinated nerve fibres. *J. Physiol. London* 207:529–37

LeBeux, Y., Willemot, J. 1975. An ultrastructural study of microfilaments in rat brain by means of E-PTA staining and heavy meromyosin labeling. II. The synapses. *Cell Tissue Res.* 160: 37–68

Lehninger, A. L. 1968. The neuronal membrane. *Proc. Natl. Acad. Sci. USA* 60:1069–80

Lenard, J., Singer, S. J. 1966. Protein conformation in cell membrane preparations as studied by optical rotatory dispersion and circular dichroism. *Proc. Natl. Acad. Sci. USA* 56:1828–35

Lentz, T. L., Chester, J. 1977. Localization of acetylcholine receptors in central synapses. *J. Cell Biol.* 75:258–67

Lentz, T. L., Mazurkiewicz, J. E., Rosenthal, J. 1977. Cytochemical localization of acetylcholine receptors at the neuromuscular junction by means of horseradish peroxidase-labeled α-bungarotoxin. *Brain Res.* 132:423–42

Levinson, S. L., Ellory, J. C. 1973. Molecular size of the tetrodotoxin binding site estimated by irradiation inactivation. *Nature New Biol.* 245:122–23

Livingston, R. B., Pfenninger, K., Moor, H., Akert, K. 1973. Specialized paranodal and internodal glial-axonal junctions in the peripheral and central nervous system: a freeze-etching study. *Brain Res.* 58:1–24

Louvard, D., Maroux, S., Desnuelle, P. 1975. Topological studies on the hydrolases bound to the intestinal brush border membrane. II. Interactions of free and bound aminopeptidase with a specific antibody. *Biochem. Biophys. Acta* 389: 389–400

Louvard, D., Semeriva, M., Maroux, S. 1976. The brush-border intestinal amino-peptidase, a transmembrane protein as probed by macromolecular photolabelling. *J. Mol. Biol.* 106:1023–35

Lucy, J. A. 1968. Theoretical and experimental models for biological membranes. In *Biological Membranes,* ed. D. Chapman, pp. 233–88. London, New New York: Academic

Maeno, H., Johnson, E. M., Greengard, P. 1971. Subcellular distribution of adenosine 3',5'-monophosphate-dependent protein kinase in rat brain. *J. Biol. Chem.* 246:134–42

Mahler, H. R., Gurd, J. W., Wang, Y.-J. 1975. Molecular topography of the synapse. In *The Nervous System, Vol. 1: The Basic Neurosciences,* ed. D. B. Tower, pp. 455–66. New York: Raven

Makowski, L., Caspar, D. L. D., Phillips, W. C., Goodenough, D. A. 1977. Gap junction structures. II. Analysis of the X-ray diffraction data. *J. Cell Biol.* 74:629–45

Marchesi, S. L., Steers, E., Marchesi, V. T., Tillack, T. W. 1970. Physical and chemical properties of a protein isolated from red cell membranes. *Biochemistry* 9:50–57

Matus, A., DePetris, S., Raff, M. C. 1973. Mobility of concanavalin A receptors in myelin and synaptic membranes. *Nature New Biol.* 244:278–80
Matus, A. I., Walters, B. B. 1976. Type 1 and 2 synaptic junctions: differences in distribution of concanavalin A binding sites and stability of the junctional adhesion. *Brain Res.* 108:249–56
Matus, A. I., Walters, B. B., Mughal, S. 1975. Immunohistochemical demonstration of tubulin associated with microtubules and synaptic junctions in mammalian brain. *J. Neurocytol.* 4:733–44
Meissner, G., Fleischer, S. 1974. Dissociation and reconstitution of functional sarcoplasmic reticulum vesicles. *J. Biol. Chem.* 249:302–9
Meyer, W. J. 1968. Extracellular material of the brain demonstrated with phosphotungstic acid. *J. Cell Biol.* 39:155–56a
Meyer, W. J. 1970. *Distribution of phosphotungstic acid–stained carbohydrate moieties in the brain.* Dissertation, Univ. Calif., Los Angeles
Milhaud, M., Pappas, G. D. 1966. Postsynaptic bodies in the habenula and interpeduncular nuclei of the cat. *J. Cell Biol.* 30:437–41
Montal, M., Korenbrot, J. I. 1976. Rhodospin in cell membranes and the process of phototransduction. In *The Enzymes of Biological Membranes, Vol. 4: Electron Transport Systems and Receptors,* ed. A. Martonosi, pp. 365–405. New York: Plenum
Moore, B. W. 1975. Membrane proteins in the nervous system. In *The Nervous System, Vol. 1: The Basic Neurosciences,* ed. D. B. Tower, pp. 503–14. New York: Raven
Moore, J. W., Narahashi, T., Shaw, T. I. 1967. An upper limit to the number of sodium channels in nerve membrane? *J. Physiol. London* 188:99–105
Narahashi, T. 1974. Chemicals as tools in the study of excitable membranes. *Physiol. Rev.* 54:813–89
Nickel, E., Potter, L. T. 1973. Ultrastructure of isolated membranes of *Torpedo* electric tissue. *Brain Res.* 57:508–17
Nicholson, G. L. 1976. Transmembrane control of the receptors on normal and tumor cells. I. Cytoplasmic influence over cell surface components. *Biochim. Biophys. Acta* 457:57–108
Norton, W. T., Poduslo, S. E. 1971. Neuronal perikarya and astroglia of rat brain: chemical composition during myelination. *J. Lipid Res.* 12:84–90
Norton, W. T., Abe, T., Poduslo, S. E., DeVries, G. H. 1975. The lipid compostion of isolated brain cells and axons. *J. Neurosci. Res.* 1:57–75
Palade, G. E., Palay, S. L. 1954. Electron microscope observations of interneuronal and neuromuscular synapses. *Anat. Rec.* 118:335–36
Palay, S. L. 1956. Synapses in the central nervous system. *J. Biophys. Biochem. Cytol.* 2(Suppl.):193–202
Palay, S. L. 1958. The morphology of synapses in the central nervous system. *Exp. Cell Res.* 5(Suppl.):275–93
Palay, S. L., Sotelo, C., Peters, A., Orkand, P. M. 1968. The axon hillock and the initial segment. *J. Cell Biol.* 38:193–201
Pappas, G. D., Purpura, D. P. 1966. Distribution of colloidal particles in extracellular space and synaptic cleft substance of mammalism cerebral cortex. *Nature* 210:1391–92
Pease, D. C. 1966. Polysaccharides associated with the exterior surface of epithelial cells: Kidney, intestine, brain. *J. Ultrastruct. Res.* 15:555–88
Pernis, B., Forni, L., Amante, L. 1970. Immunoglobulin spots on the surface of rabbit lymphocytes. *J. Exp. Med.* 132:1001–18
Peters, A. 1966. The node of Ranvier in the central nervous system. *Quart. J. Exp. Physiol.* 51:229–36
Peters, A., Proskauer, C. C., Kaiserman-Abramof, I. R. 1968. The small pyramidal neuron of the rat cerebral cortex. The axon hillock and initial segment. *J. Cell Biol.* 39:604–19
Pfenninger, K. H. 1971a. The cytochemistry of synaptic densities. I. An analysis of the bismuth iodide impregnation method. *J. Ultrastruct. Res.* 34:103–22
Pfenninger, K. H. 1971b. The cytochemistry of synaptic densities. II. Proteinaceous components and mechanism of synaptic connectivity. *J. Ultrastruct. Res.* 35:451–75
Pfenninger, K. H. 1973. Synaptic morphology and cytochemistry. *Prog. Histochem. Cytochem.* 5:1–86
Pfenninger, K. H. 1977. Cytology of the chemical synapse: Morphological correlates of synaptic function. In *Neurotransmitter Functions,* ed. W. S. Fields, pp. 27–57. Miami, Fla: Symposia Specialists
Pfenninger, K. H. 1978. Synaptic membrane differentiation. In *The Neurosciences, Fourth Study Program.* Cambridge, Mass: MIT Press. In press
Pfenninger, K. H., Akert, K., Moor, H., Sandri, C. 1971. Freeze-fracturing of presy-

naptic membranes in the central nervous system. *Philos. Trans. R. Soc. London Ser. B.* 261:387

Pfenninger, K. H., Akert, K., Moor, H., Sandri, C. 1972. The fine structure of freeze-fractured presynaptic membranes. *J. Neurocytol.* 1:129–49

Pfenninger, K. H., Bunge, R. P. 1974. Freeze-fracturing of nerve growth cones and young fibers. A study of developing plasma membrane. *J. Cell Biol.* 63:180–96

Pfenninger, K. H., Maylié-Pfenninger, M.-F. 1976. Differential lectin receptor content on the surface of nerve growth cones of different origin. *Neurosci. Abstr.* II(1):224

Pfenninger, K. H., Rees, R. P. 1976. From the growth cone to the synapse: Properties of membranes involved in synapse formation. In *Neuronal Recognition,* ed. S. H. Barondes, pp. 131–78. New York: Plenum

Pfenninger, K. H., Rovainen, C. M. 1974. Stimulation- and calcium-dependence of vesicle attachment sites in the presynaptic membrane; a freeze-cleave study on the lamprey spinal cord. *Brain Res.* 72:1–23

Pfenninger, K. H., Sandri, C., Akert, K., Eugster, C. H. 1969. Contribution to the problem of structural organization of the presynaptic area. *Brain Res.* 12:10–18

Pinto da Silva, P., Branton, D. 1970. Membrane splitting in freeze-etching. *J. Cell Biol.* 45:598–605

Poduslo, S. E., McKhann, G. M. 1977. Maintenance of neurons isolated in bulk from rat brain: incorporation of radiolabeled substrates. *Brain Res.* 132:107–20

Porter, K. R. 1978. The Cytoskeleton. In *The Neurosciences, Fourth Study Program.* Cambridge, Mass: MIT Press. In press

Potter, L. T. 1977. Molecular properties of acetylcholine receptor-channel molecules, and an oligomeric model for their activation, inactivation, and desensitization. *Proc. Aust. Physiol. Pharmacol. Soc.* 8:55–74

Potter, L. T., Smith, D. S. 1977. Postsynaptic membranes in the electric tissue of *Narcine:* 1. Organization and innervation of electric cells. Fine structure of nicotinic receptor-channel molecules revealed by transmission microscopy. *Tissue Cell.* 9:In press

Pysh, J. J., Willey, G. G. 1974. Synaptic vesicle depletion and recovery in cat sympathetic ganglia stimulated in vitro. Evidence for transmitter secretion by exocytosis. *J. Cell Biol.* 60:365–74

Quick, D. C., Waxman, S. G. 1977. Specific staining of the axon membrane at nodes of Ranvier with ferric ion and ferrocyanide. *J. Neurol. Sci.* 31:1–11

Racker, E. 1973. A new procedure for the reconstitution of biologically active phospholipid vesicles. *Biochem. Biophys. Res. Comm.* 55:224–30

Rambourg, A., Leblond, C. P. 1967. Electron microscope observations on the carbohydrate-rich cell coat present at the surface of cells in the rat. *J. Cell Biol.* 32:27–53

Ramón y Cajal, S. 1911. *Histologie du système nerveux de l'homme et des vertébrés.* Vol. 2 (Reprinted by Consejo Superior de Investigaciones Cientificas, Instituto Ramón y Cajal, Madrid, 1955). Paris: A. Maloine

Reed, J. K., Raftery, M. A. 1976. Properties of the tetrodotoxin binding component in plasma membranes isolated from *Electrophorus electricus. Biochemistry* 15:944–53

Revel, J. P., Ito, S. 1967. The surface components of cells. In *The Specificity of Cell Surfaces,* ed. B. D. Davis, L. Warren. pp. 211–34. Englewood Cliffs, N.J.: Prentice-Hall

Riemersma, J. C. 1963. Osmium tetroxide fixation of lipids: nature of the reaction products. *J. Histochem. Cytochem.* 11:436–42

Ritchie, J. M., Rogart, R. B. 1977. The density of sodium channels in mammalian myelinated nerve fibers and the nature of the axonal membrane under the myelin sheath. *Proc. Natl. Acad. Sci. USA* 74:211–15

Robertson, J. D. 1959. The ultrastructure of cell membranes and their derivatives. *Biochem. Soc. Symp.* 16:3–43

Rosenbluth, J. 1962. Subsurface cisterns and their relationship to the neuronal plasma membrane. *J. Cell Biol.* 13: 405–22

Rosenbluth, J. 1973. Postjunctional membrane specialization at cholinergic myoneural junctions in the leech. *J. Comp. Neurol.* 151:399–406

Rosenbluth, J. 1974. Substructure of amphibian motor endplate. Evidence for a granular component projecting from the outer surface of the receptive membrane. *J. Cell Biol.* 62:755–66

Rosenbluth, J. 1975. Synaptic membrane structure in *Torpedo* electric organ. *J. Neurocytol.* 4:697–712

Rosenbluth, J. 1976. Intramembranous particle distribution at the node of Ranvier and adjacent axolemma in myelinated axons of the frog brain. *J. Neurocytol.* 5:731–45

Rudolph, S. A., Greengard, P. 1974. Regulation of protein phosphorylation and membrane permeability by β-adrenergic agents and cyclic adenosine 3',5'-monophosphate in the avian erythrocyte. *J. Biol. Chem.* 249:5684–87

Salpeter, M. M. 1969. Electron microscope radioautography as a quantitative tool in enzyme cytochemistry. II. The distribution of DFP-reactive sites at motor endplates of a vertebrate twitch muscle. *J. Cell Biol.* 42:122–34

Sandri, C., Akert, K., Livingston, R. B., Moor, H. 1972. Particle aggregation in freeze-etched postsynaptic membranes. *Brain Res.* 41:1–16

Schmitt, F. O., Samson, F. E. 1969. Brain Cell Microenvironment. Neurosci. *Res. Prog. Bull.* 7:277–417

Schnapp, B., Peracchia, C., Mugnaini, E. 1976. The paranodal axo-glial junction in the central nervous system studied with thin sections and freeze-fracture. *Neuroscience* 1:181–90

Schreier-Muccillo, S., Marsh, D., Dugas, H., Schneider, H., Smith, I. C. P. 1973. A spin probe study of the influence of cholesterol on motion and orientation of phospholipids in oriented multibilayers and vesicles. *Chem. Phys. Lipids* 10: 11–27

Segrest, J. P. 1977. The erythrocyte: topomolecular anatomy of MN-glycoprotein. In *Mammalian Cell Membranes, Vol. 3: Surface Membranes of Specific Cell Types.* ed. G. A. Jamieson, D. M. Robinson, pp. 1–26. London/Boston: Butterworths

Segrest, J. P., Gulik-Krzywicki, T., Sardet, C. 1974. Association of the membrane-penetrating polypeptide segment of the human erythrocyte MN-glycoprotein with phospholipid bilayers. I. Formation of freeze-etch intramembranous particles. *Proc. Natl. Acad. Sci. USA* 71:3294–98

Sheetz, M. P., Singer, S. J. 1977. On the mechanism of ATP-induced shape changes in human erythrocyte membranes. I. Role of the spectrin complex. *J. Cell Biol.* 73:638–46

Sheltawy, A., Dawson, R. M. C. 1966. The polyphosphoinositides and other lipids of peripheral nerves. *Biochem. J.* 100:12–18

Singer, S. J. 1974. The molecular organization of membranes. *Ann. Rev. Biochem.* 43:805–33

Singer, S. J., Nicholson, G. L. 1972. The fluid mosaic model of the structure of cell membranes. *Science* 175:720–31

Steck, T. L. 1974. The organization of proteins in the human red blood cell membrane. *J. Cell Biol.* 62:1–19

Steim, J. M., Tourtelotte, M. E., Reinert, J. C., McElhaney, R. N., Rader, R. L. 1969. Calorimetric evidence for the liquid-crystalline state of lipids in a biomembrane. *Proc. Natl. Acad. Sci. USA* 63:104–9

Stoeckenius, W., Engelman, D. M. 1967 Current models for the structure of biological membranes. *J. Cell Biol.* 42:613–46

Streit, P., Akert, K., Sandri, C., Livingston. R. B., Moor, H. 1972. Dynamic ultrastructure of presynaptic membranes at nerve terminals in the spinal cord of rats. Anesthetized and unanesthetized preparations compared. *Brain Res.* 48:11–26

Strichartz, G. R. 1977. The composition and structure of excitable nerve membrane. In *Mammalian Cell Membranes, Vol. 3: Surface Membranes of Specific Cell Types,* ed. G. A. Jamieson, D. M. Robinson, pp. 172–205. London/Boston: Butterworths

Strittmatter, P., Rogers, M. J., Spatz, L. 1972. The binding of cytochrome b_5 to liver microsomes. *J. Biol. Chem.* 247:7188–94

Suzuki, K. 1975. Sphingolipid of the nervous system. In *The Nervous System, Vol. 1: The Basic Neurosciences,* ed. D. B. Tower, pp. 483–91. New York: Raven

Tamai, Y., Matsukawa, S., Satake, M. 1971. Gangliosides in neuron. *J. Biochem. Tokyo* 69:235–38

Taylor, R. B., Duffus, P. H., Raff, M. C., DePetris, S. 1971. Redistribution and pinocytosis of lymphocyte surface immunoglobulin molecules induced by anti-immunoglobulin antibody. *Nature New Biol.* 233:225–29

Tillack, T. W., Marchesi, V. T. 1970. Demonstration of the outer surface of freeze-etched red blood cell membranes. *J. Cell Biol.* 45:649–53

Tomita, M., Marchesi, V. T. 1975. Aminoacid sequence and oligosaccharide attachment sites of human erythrocyte glycophorin. *Proc. Natl. Acad. Sci. USA* 72:2964–68

Uchizono, K. 1965. Characteristics of excitatory and inhibitory synapses in the cental nervous system of the cat. *Nature* 207:642–43

Unanue, E. R., Perkins. W. D., Karnovsky, M. J. 1972. Ligand-induced movement of lymphocyte membrane macromolecules. I. Analysis by immunofluorescence and ultrastructural radioautography. *J. Exp. Med.* 136:885–906

Ueda, T., Maeno, H., Greengard, P. 1973. Regulation of endogenous phosphorylation of specific proteins in synaptic membrane fractions from rat brain by adenosine 3',5'-monophosphate. *J. Biol. Chem.* 248:8295-8305

Ueda, T., Rudolph, S. A., Greengard, P. 1975. Solubilization of a phosphoprotein and its associated cyclic AMP-dependent protein kinase and phosphoprotein phosphatase from synaptic membrane fractions, and some kinetic evidence for their existence as a complex. *Arch. Biochem. Biophys.* 170:492–503

Van Der Loos, H. 1963. Fine structure of synapses in the cerebral cortex. *Z. Zellforsch.* 60:815–25

Venzin, M., Sandri, C., Akert, K., Wyss, U. R. 1977. Membrane associated particles of the presynaptic active zone in rat spinal cord. A morphometric analyses. *Brain Res.* 130:393–404

Wallach, D. F. H., Gordon, A. 1968. Lipid protein interactions in cellular membranes. *Fed. Proc.* 27:1263–68

Walton, K. G., DeLorenzo, R. J., Curran, P. F., Greengard, P. 1975. Regulation of protein phosphorylation and sodium transport in toad bladder. *J. Gen. Physiol.* 65:153–77

Wang, Y. J., Mahler, H. R. 1976. Topography of the synaptosomal membrane. *J. Cell Biol.* 71:639–58

Wannamaker, B. B., Kornguth, S. E. 1973. Electrophoretic patterns of proteins from isolated synapses of human and swine brain. *Biochem. Biophys. Acta* 303:333–37

Waxman, S. G., Quick, D. C. 1978. Intra-axonal ferric ion-ferrocyanide staining of nodes of Ranvier and initial segments in central myelinated fibers. *Brain Res.* 150:In press

Weber, K. 1978. Microfilaments and microtubules studies by indirect immunofluorescence microscopy in tissue culture cells. In *The Neurosciences, Fourth Study Program,* ed. F. O. Schmitt, F. G. Worden. Cambridge, Mass: MIT Press. In press

Wehrli, E., Mühlethaler, K., Moor, H. 1970. Membrane structure as seen with a double replica method for freeze fracturing. *Exp. Cell Res.* 59:336–39

Weller, M., Morgan, I. G. 1976. Localization in the synaptic junction of the cyclic-AMP stimulated intrinsic protein kinase activity of synaptosomal plasma membranes. *Biochim. Biophys. Acta* 433:223–28

White, D. L., Andrews, S. B., Faller, J. W., Barrnett, R. J. 1976. The chemical nature of osmium tetroxide fixation and staining of membranes by X-ray photoelectron spectroscopy. *Biochim. Biophys. Acta* 436:577–92

Wilkins, M. H. F., Blaurock, A. E., Engelman, D. M. 1971. Bilayer structure in membranes. *Nature New Biol.* 230: 72–76

Wolfe, L. S. 1961. The distribution of gangliosides in subcellular fractions of guinea-pig cerebral cortex. *Biochem. J.* 79:348–55

Wood, J. G., Jean, D. H., Whitaker, J. N., McLaughlin, B. J., Albers, R. W. 1976. Fine structural localization of the Na^+, K^+-activated ATPase in knifefish brain. *Neurosci. Abstr.* II(1):421

Wood, M. R., Pfenninger, K. H., Cohen, M. J. 1977. Two types of presynaptic configurations in insect central synapses: An ultrastructural analysis. *Brain Res.* 130:25–45

Yahara, I., Edelman, G. M. 1972. Restriction of mobility of lymphocyte immunoglobulin receptors by concanavalin A. *Proc. Natl. Acad. Sci. USA* 69:608–12

Yamanaka, N., Deamer, D. W. 1976. Protease digestion of membranes. Ultrastructural and biochemical effects. *Biochim. Biophys. Acta* 426:132–47

Yen, S.-H., Liem, R. K. H., Kelly, P. T., Cotman, C. W., Shelanski, M. L. 1977. Membrane linked proteins at CNS synapses. *Brain Res.* 132:172–75

Yu, J., Branton, D. 1976. Reconstitution of intramembrane particles in recombinants of erythrocyte protein band 3 and lipid: Effects of spectrin-actin association. *Proc. Natl. Acad. Sci. USA* 73:3891–95

Yu, J., Steck, T. L. 1974. Association between the major red cell membrane penetrating protein and two inner surface proteins. *Fed. Proc.* 33:1532

Zambrano, F., Cellino, M., Canessa-Fischer, M. 1971. The molecular organization of nerve membranes. IV. The lipid composition of plasma membranes from squid retinal axons. *J. Membrane Biol.* 6:280–303

Zwaal, R. F. A., Roelofsen, B., Colley, C. M. 1973. Localization of red cell membrane constituents. *Biochim. Biophys. Acta* 300:159–82

AUTHOR INDEX

I

S

SUBJECT INDEX